Communications in Computer and Information Science 1963

Rationale

The CCIS series is devoted to the publication of proceedings of computer science conferences. Its aim is to efficiently disseminate original research results in informatics in printed and electronic form. While the focus is on publication of peer-reviewed full papers presenting mature work, inclusion of reviewed short papers reporting on work in progress is welcome, too. Besides globally relevant meetings with internationally representative program committees guaranteeing a strict peer-reviewing and paper selection process, conferences run by societies or of high regional or national relevance are also considered for publication.

Topics

The topical scope of CCIS spans the entire spectrum of informatics ranging from foundational topics in the theory of computing to information and communications science and technology and a broad variety of interdisciplinary application fields.

Information for Volume Editors and Authors

Publication in CCIS is free of charge. No royalties are paid, however, we offer registered conference participants temporary free access to the online version of the conference proceedings on SpringerLink (http://link.springer.com) by means of an http referrer from the conference website and/or a number of complimentary printed copies, as specified in the official acceptance email of the event.

CCIS proceedings can be published in time for distribution at conferences or as post-proceedings, and delivered in the form of printed books and/or electronically as USBs and/or e-content licenses for accessing proceedings at SpringerLink. Furthermore, CCIS proceedings are included in the CCIS electronic book series hosted in the SpringerLink digital library at http://link.springer.com/bookseries/7899. Conferences publishing in CCIS are allowed to use Online Conference Service (OCS) for managing the whole proceedings lifecycle (from submission and reviewing to preparing for publication) free of charge.

Publication process

The language of publication is exclusively English. Authors publishing in CCIS have to sign the Springer CCIS copyright transfer form, however, they are free to use their material published in CCIS for substantially changed, more elaborate subsequent publications elsewhere. For the preparation of the camera-ready papers/files, authors have to strictly adhere to the Springer CCIS Authors' Instructions and are strongly encouraged to use the CCIS LaTeX style files or templates.

Abstracting/Indexing

CCIS is abstracted/indexed in DBLP, Google Scholar, EI-Compendex, Mathematical Reviews, SCImago, Scopus. CCIS volumes are also submitted for the inclusion in ISI Proceedings.

How to start

To start the evaluation of your proposal for inclusion in the CCIS series, please send an e-mail to ccis@springer.com.

Biao Luo · Long Cheng · Zheng-Guang Wu ·
Hongyi Li · Chaojie Li
Editors

Neural Information Processing

30th International Conference, ICONIP 2023
Changsha, China, November 20–23, 2023
Proceedings, Part IX

Springer

Editors
Biao Luo
School of Automation
Central South University
Changsha, China

Long Cheng
Institute of Automation
Chinese Academy of Sciences
Beijing, China

Zheng-Guang Wu
Institute of Cyber-Systems and Control
Zhejiang University
Hangzhou, China

Hongyi Li
School of Automation
Guangdong University of Technology
Guangzhou, China

Chaojie Li
School of Electrical Engineering
and Telecommunications
UNSW Sydney
Sydney, NSW, Australia

ISSN 1865-0929 ISSN 1865-0937 (electronic)
Communications in Computer and Information Science
ISBN 978-981-99-8137-3 ISBN 978-981-99-8138-0 (eBook)
https://doi.org/10.1007/978-981-99-8138-0

This Springer imprint is published by the registered company Springer Nature Singapore Pte Ltd.
The registered company address is: 152 Beach Road, #21-01/04 Gateway East, Singapore 189721, Singapore

Organization

Honorary Chair

Weihua Gui	Central South University, China

Advisory Chairs

Jonathan Chan	King Mongkut's University of Technology Thonburi, Thailand
Zeng-Guang Hou	Chinese Academy of Sciences, China
Nikola Kasabov	Auckland University of Technology, New Zealand
Derong Liu	Southern University of Science and Technology, China
Seiichi Ozawa	Kobe University, Japan
Kevin Wong	Murdoch University, Australia

General Chairs

Tingwen Huang	Texas A&M University at Qatar, Qatar
Chunhua Yang	Central South University, China

Program Chairs

Biao Luo	Central South University, China
Long Cheng	Chinese Academy of Sciences, China
Zheng-Guang Wu	Zhejiang University, China
Hongyi Li	Guangdong University of Technology, China
Chaojie Li	University of New South Wales, Australia

Technical Chairs

Xing He	Southwest University, China
Keke Huang	Central South University, China
Huaqing Li	Southwest University, China
Qi Zhou	Guangdong University of Technology, China

Local Arrangement Chairs

Wenfeng Hu	Central South University, China
Bei Sun	Central South University, China

Finance Chairs

Fanbiao Li	Central South University, China
Hayaru Shouno	University of Electro-Communications, Japan
Xiaojun Zhou	Central South University, China

Special Session Chairs

Hongjing Liang	University of Electronic Science and Technology, China
Paul S. Pang	Federation University, Australia
Qiankun Song	Chongqing Jiaotong University, China
Lin Xiao	Hunan Normal University, China

Tutorial Chairs

Min Liu	Hunan University, China
M. Tanveer	Indian Institute of Technology Indore, India
Guanghui Wen	Southeast University, China

Publicity Chairs

Sabri Arik	Istanbul University-Cerrahpaşa, Turkey
Sung-Bae Cho	Yonsei University, South Korea
Maryam Doborjeh	Auckland University of Technology, New Zealand
El-Sayed M. El-Alfy	King Fahd University of Petroleum and Minerals, Saudi Arabia
Ashish Ghosh	Indian Statistical Institute, India
Chuandong Li	Southwest University, China
Weng Kin Lai	Tunku Abdul Rahman University of Management & Technology, Malaysia
Chu Kiong Loo	University of Malaya, Malaysia
Qinmin Yang	Zhejiang University, China
Zhigang Zeng	Huazhong University of Science and Technology, China

Publication Chairs

Zhiwen Chen	Central South University, China
Andrew Chi-Sing Leung	City University of Hong Kong, China
Xin Wang	Southwest University, China
Xiaofeng Yuan	Central South University, China

Secretaries

Yun Feng	Hunan University, China
Bingchuan Wang	Central South University, China

Webmasters

Tianmeng Hu	Central South University, China
Xianzhe Liu	Xiangtan University, China

Program Committee

Rohit Agarwal	UiT The Arctic University of Norway, Norway
Hasin Ahmed	Gauhati University, India
Harith Al-Sahaf	Victoria University of Wellington, New Zealand
Brad Alexander	University of Adelaide, Australia
Mashaan Alshammari	Independent Researcher, Saudi Arabia
Sabri Arik	Istanbul University, Turkey
Ravneet Singh Arora	Block Inc., USA
Zeyar Aung	Khalifa University of Science and Technology, UAE
Monowar Bhuyan	Umeå University, Sweden
Jingguo Bi	Beijing University of Posts and Telecommunications, China
Xu Bin	Northwestern Polytechnical University, China
Marcin Blachnik	Silesian University of Technology, Poland
Paul Black	Federation University, Australia
Anoop C. S.	Govt. Engineering College, India
Ning Cai	Beijing University of Posts and Telecommunications, China
Siripinyo Chantamunee	Walailak University, Thailand
Hangjun Che	City University of Hong Kong, China

Wei-Wei Che	Qingdao University, China
Huabin Chen	Nanchang University, China
Jinpeng Chen	Beijing University of Posts & Telecommunications, China
Ke-Jia Chen	Nanjing University of Posts and Telecommunications, China
Lv Chen	Shandong Normal University, China
Qiuyuan Chen	Tencent Technology, China
Wei-Neng Chen	South China University of Technology, China
Yufei Chen	Tongji University, China
Long Cheng	Institute of Automation, China
Yongli Cheng	Fuzhou University, China
Sung-Bae Cho	Yonsei University, South Korea
Ruikai Cui	Australian National University, Australia
Jianhua Dai	Hunan Normal University, China
Tao Dai	Tsinghua University, China
Yuxin Ding	Harbin Institute of Technology, China
Bo Dong	Xi'an Jiaotong University, China
Shanling Dong	Zhejiang University, China
Sidong Feng	Monash University, Australia
Yuming Feng	Chongqing Three Gorges University, China
Yun Feng	Hunan University, China
Junjie Fu	Southeast University, China
Yanggeng Fu	Fuzhou University, China
Ninnart Fuengfusin	Kyushu Institute of Technology, Japan
Thippa Reddy Gadekallu	VIT University, India
Ruobin Gao	Nanyang Technological University, Singapore
Tom Gedeon	Curtin University, Australia
Kam Meng Goh	Tunku Abdul Rahman University of Management and Technology, Malaysia
Zbigniew Gomolka	University of Rzeszow, Poland
Shengrong Gong	Changshu Institute of Technology, China
Xiaodong Gu	Fudan University, China
Zhihao Gu	Shanghai Jiao Tong University, China
Changlu Guo	Budapest University of Technology and Economics, Hungary
Weixin Han	Northwestern Polytechnical University, China
Xing He	Southwest University, China
Akira Hirose	University of Tokyo, Japan
Yin Hongwei	Huzhou Normal University, China
Md Zakir Hossain	Curtin University, Australia
Zengguang Hou	Chinese Academy of Sciences, China

Lu Hu	Jiangsu University, China
Zeke Zexi Hu	University of Sydney, Australia
He Huang	Soochow University, China
Junjian Huang	Chongqing University of Education, China
Kaizhu Huang	Duke Kunshan University, China
David Iclanzan	Sapientia University, Romania
Radu Tudor Ionescu	University of Bucharest, Romania
Asim Iqbal	Cornell University, USA
Syed Islam	Edith Cowan University, Australia
Kazunori Iwata	Hiroshima City University, Japan
Junkai Ji	Shenzhen University, China
Yi Ji	Soochow University, China
Canghong Jin	Zhejiang University, China
Xiaoyang Kang	Fudan University, China
Mutsumi Kimura	Ryukoku University, Japan
Masahiro Kohjima	NTT, Japan
Damian Kordos	Rzeszow University of Technology, Poland
Marek Kraft	Poznań University of Technology, Poland
Lov Kumar	NIT Kurukshetra, India
Weng Kin Lai	Tunku Abdul Rahman University of Management & Technology, Malaysia
Xinyi Le	Shanghai Jiao Tong University, China
Bin Li	University of Science and Technology of China, China
Hongfei Li	Xinjiang University, China
Houcheng Li	Chinese Academy of Sciences, China
Huaqing Li	Southwest University, China
Jianfeng Li	Southwest University, China
Jun Li	Nanjing Normal University, China
Kan Li	Beijing Institute of Technology, China
Peifeng Li	Soochow University, China
Wenye Li	Chinese University of Hong Kong, China
Xiangyu Li	Beijing Jiaotong University, China
Yantao Li	Chongqing University, China
Yaoman Li	Chinese University of Hong Kong, China
Yinlin Li	Chinese Academy of Sciences, China
Yuan Li	Academy of Military Science, China
Yun Li	Nanjing University of Posts and Telecommunications, China
Zhidong Li	University of Technology Sydney, Australia
Zhixin Li	Guangxi Normal University, China
Zhongyi Li	Beihang University, China

Ziqiang Li	University of Tokyo, Japan
Xianghong Lin	Northwest Normal University, China
Yang Lin	University of Sydney, Australia
Huawen Liu	Zhejiang Normal University, China
Jian-Wei Liu	China University of Petroleum, China
Jun Liu	Chengdu University of Information Technology, China
Junxiu Liu	Guangxi Normal University, China
Tommy Liu	Australian National University, Australia
Wen Liu	Chinese University of Hong Kong, China
Yan Liu	Taikang Insurance Group, China
Yang Liu	Guangdong University of Technology, China
Yaozhong Liu	Australian National University, Australia
Yong Liu	Heilongjiang University, China
Yubao Liu	Sun Yat-sen University, China
Yunlong Liu	Xiamen University, China
Zhe Liu	Jiangsu University, China
Zhen Liu	Chinese Academy of Sciences, China
Zhi-Yong Liu	Chinese Academy of Sciences, China
Ma Lizhuang	Shanghai Jiao Tong University, China
Chu-Kiong Loo	University of Malaya, Malaysia
Vasco Lopes	Universidade da Beira Interior, Portugal
Hongtao Lu	Shanghai Jiao Tong University, China
Wenpeng Lu	Qilu University of Technology, China
Biao Luo	Central South University, China
Ye Luo	Tongji University, China
Jiancheng Lv	Sichuan University, China
Yuezu Lv	Beijing Institute of Technology, China
Huifang Ma	Northwest Normal University, China
Jinwen Ma	Peking University, China
Jyoti Maggu	Thapar Institute of Engineering and Technology Patiala, India
Adnan Mahmood	Macquarie University, Australia
Mufti Mahmud	University of Padova, Italy
Krishanu Maity	Indian Institute of Technology Patna, India
Srimanta Mandal	DA-IICT, India
Wang Manning	Fudan University, China
Piotr Milczarski	Lodz University of Technology, Poland
Malek Mouhoub	University of Regina, Canada
Nankun Mu	Chongqing University, China
Wenlong Ni	Jiangxi Normal University, China
Anupiya Nugaliyadde	Murdoch University, Australia

Toshiaki Omori	Kobe University, Japan
Babatunde Onasanya	University of Ibadan, Nigeria
Manisha Padala	Indian Institute of Science, India
Sarbani Palit	Indian Statistical Institute, India
Paul Pang	Federation University, Australia
Rasmita Panigrahi	Giet University, India
Kitsuchart Pasupa	King Mongkut's Institute of Technology Ladkrabang, Thailand
Dipanjyoti Paul	Ohio State University, USA
Hu Peng	Jiujiang University, China
Kebin Peng	University of Texas at San Antonio, USA
Dawid Połap	Silesian University of Technology, Poland
Zhong Qian	Soochow University, China
Sitian Qin	Harbin Institute of Technology at Weihai, China
Toshimichi Saito	Hosei University, Japan
Fumiaki Saitoh	Chiba Institute of Technology, Japan
Naoyuki Sato	Future University Hakodate, Japan
Chandni Saxena	Chinese University of Hong Kong, China
Jiaxing Shang	Chongqing University, China
Lin Shang	Nanjing University, China
Jie Shao	University of Science and Technology of China, China
Yin Sheng	Huazhong University of Science and Technology, China
Liu Sheng-Lan	Dalian University of Technology, China
Hayaru Shouno	University of Electro-Communications, Japan
Gautam Srivastava	Brandon University, Canada
Jianbo Su	Shanghai Jiao Tong University, China
Jianhua Su	Institute of Automation, China
Xiangdong Su	Inner Mongolia University, China
Daiki Suehiro	Kyushu University, Japan
Basem Suleiman	University of New South Wales, Australia
Ning Sun	Shandong Normal University, China
Shiliang Sun	East China Normal University, China
Chunyu Tan	Anhui University, China
Gouhei Tanaka	University of Tokyo, Japan
Maolin Tang	Queensland University of Technology, Australia
Shu Tian	University of Science and Technology Beijing, China
Shikui Tu	Shanghai Jiao Tong University, China
Nancy Victor	Vellore Institute of Technology, India
Petra Vidnerová	Institute of Computer Science, Czech Republic

Shanchuan Wan	University of Tokyo, Japan
Tao Wan	Beihang University, China
Ying Wan	Southeast University, China
Bangjun Wang	Soochow University, China
Hao Wang	Shanghai University, China
Huamin Wang	Southwest University, China
Hui Wang	Nanchang Institute of Technology, China
Huiwei Wang	Southwest University, China
Jianzong Wang	Ping An Technology, China
Lei Wang	National University of Defense Technology, China
Lin Wang	University of Jinan, China
Shi Lin Wang	Shanghai Jiao Tong University, China
Wei Wang	Shenzhen MSU-BIT University, China
Weiqun Wang	Chinese Academy of Sciences, China
Xiaoyu Wang	Tokyo Institute of Technology, Japan
Xin Wang	Southwest University, China
Xin Wang	Southwest University, China
Yan Wang	Chinese Academy of Sciences, China
Yan Wang	Sichuan University, China
Yonghua Wang	Guangdong University of Technology, China
Yongyu Wang	JD Logistics, China
Zhenhua Wang	Northwest A&F University, China
Zi-Peng Wang	Beijing University of Technology, China
Hongxi Wei	Inner Mongolia University, China
Guanghui Wen	Southeast University, China
Guoguang Wen	Beijing Jiaotong University, China
Ka-Chun Wong	City University of Hong Kong, China
Anna Wróblewska	Warsaw University of Technology, Poland
Fengge Wu	Institute of Software, Chinese Academy of Sciences, China
Ji Wu	Tsinghua University, China
Wei Wu	Inner Mongolia University, China
Yue Wu	Shanghai Jiao Tong University, China
Likun Xia	Capital Normal University, China
Lin Xiao	Hunan Normal University, China
Qiang Xiao	Huazhong University of Science and Technology, China
Hao Xiong	Macquarie University, Australia
Dongpo Xu	Northeast Normal University, China
Hua Xu	Tsinghua University, China
Jianhua Xu	Nanjing Normal University, China

Xinyue Xu	Hong Kong University of Science and Technology, China
Yong Xu	Beijing Institute of Technology, China
Ngo Xuan Bach	Posts and Telecommunications Institute of Technology, Vietnam
Hao Xue	University of New South Wales, Australia
Yang Xujun	Chongqing Jiaotong University, China
Haitian Yang	Chinese Academy of Sciences, China
Jie Yang	Shanghai Jiao Tong University, China
Minghao Yang	Chinese Academy of Sciences, China
Peipei Yang	Chinese Academy of Science, China
Zhiyuan Yang	City University of Hong Kong, China
Wangshu Yao	Soochow University, China
Ming Yin	Guangdong University of Technology, China
Qiang Yu	Tianjin University, China
Wenxin Yu	Southwest University of Science and Technology, China
Yun-Hao Yuan	Yangzhou University, China
Xiaodong Yue	Shanghai University, China
Paweł Zawistowski	Warsaw University of Technology, Poland
Hui Zeng	Southwest University of Science and Technology, China
Wang Zengyunwang	Hunan First Normal University, China
Daren Zha	Institute of Information Engineering, China
Zhi-Hui Zhan	South China University of Technology, China
Baojie Zhang	Chongqing Three Gorges University, China
Canlong Zhang	Guangxi Normal University, China
Guixuan Zhang	Chinese Academy of Science, China
Jianming Zhang	Changsha University of Science and Technology, China
Li Zhang	Soochow University, China
Wei Zhang	Southwest University, China
Wenbing Zhang	Yangzhou University, China
Xiang Zhang	National University of Defense Technology, China
Xiaofang Zhang	Soochow University, China
Xiaowang Zhang	Tianjin University, China
Xinglong Zhang	National University of Defense Technology, China
Dongdong Zhao	Wuhan University of Technology, China
Xiang Zhao	National University of Defense Technology, China
Xu Zhao	Shanghai Jiao Tong University, China

Liping Zheng	Hefei University of Technology, China
Yan Zheng	Kyushu University, Japan
Baojiang Zhong	Soochow University, China
Guoqiang Zhong	Ocean University of China, China
Jialing Zhou	Nanjing University of Science and Technology, China
Wenan Zhou	PCN&CAD Center, China
Xiao-Hu Zhou	Institute of Automation, China
Xinyu Zhou	Jiangxi Normal University, China
Quanxin Zhu	Nanjing Normal University, China
Yuanheng Zhu	Chinese Academy of Sciences, China
Xiaotian Zhuang	JD Logistics, China
Dongsheng Zou	Chongqing University, China

Contents – Part IX

Theory and Algorithms

Self-adaptive Inverse Soft-Q Learning for Imitation 3
Zhuo Wang, Quan Liu, and Xiongzhen Zhang

Membership Inference Attacks Against Medical Databases 15
Tianxiang Xu, Chang Liu, Kun Zhang, and Jianlin Zhang

Application of ALMM Technology to Intelligent Control System for a Fleet of Unmanned Aerial Vehicles 26
Ewa Zeslawska, Zbigniew Gomolka, and Ewa Dydek-Dyduch

Bloomfilter-Based Practical Kernelization Algorithms for Minimum Satisfiability 38
Chao Xu, Liting Dai, and Kang Liu

TPTGAN: Two-Path Transformer-Based Generative Adversarial Network Using Joint Magnitude Masking and Complex Spectral Mapping for Speech Enhancement 48
Zhaoyi Liu, Zhuohang Jiang, Wendian Luo, Zhuoyao Fan, Haoda Di, Yufan Long, and Haizhou Wang

Sample Selection Based on Uncertainty for Combating Label Noise 62
Shuohui Hao, Zhe Liu, Yuqing Song, Yi Liu, Kai Han, Victor S. Sheng, and Yan Zhu

Design of a Multimodal Short Video Classification Model 75
Xinyan Cao, He Yan, Changjin Li, Yongjian Zhao, Jinming Che, Wei Ren, Jinlong Lin, and Jian Cao

DGNN: Dependency Graph Neural Network for Multimodal Emotion Recognition in Conversation 86
Zhen Zhang, Xin Wang, Lifeng Yuan, Gongxun Miao, Mengqiu Liu, Wenhao Yun, and Guohua Wu

Global Exponential Synchronization of Quaternion-Valued Neural Networks via Quantized Control 100
Jiaqiang Huang, Junjian Huang, Jinyue Yang, and Yao Zhong

Improving SLDS Performance Using Explicit Duration Variables with Infinite Support ... 112
Mikołaj Słupiński and Piotr Lipiński

Policy Representation Opponent Shaping via Contrastive Learning ... 124
Yuming Chen and Yuanheng Zhu

FHSI-GNN: Fusion Hierarchical Structure Information Graph Neural Network for Extractive Long Documents Summarization ... 136
Zhen Zhang, Wenhao Yun, Xiyuan Jia, Qiyun Lv, Hao Ni, Xin Wang, and Guohua Wu

How to Support Sport Management with Decision Systems? Swimming Athletes Assessment Study Sase ... 150
Jakub Więckowski and Wojciech Sałabun

Differential Private (Random) Decision Tree Without Adding Noise ... 162
Ryo Nojima and Lihua Wang

Cognitive Neurosciences

Pushing the Boundaries of Chinese Painting Classification on Limited Datasets: Introducing a Novel Transformer Architecture with Enhanced Feature Extraction ... 177
Haiming Zhao, Jiejie Chen, Ping Jiang, Tianrui Wu, and Zhuzhu Zhang

Topological Dynamics of Functional Neural Network Graphs During Reinforcement Learning ... 190
Matthew Muller, Steve Kroon, and Stephan Chalup

Quantized SGD in Federated Learning: Communication, Optimization and Generalization ... 205
Shiyu Liu, Linsen Wei, Shaogao Lv, and Zenglin Xu

Many Is Better Than One: Multiple Covariation Learning for Latent Multiview Representation ... 218
Yun-Hao Yuan, Pengwei Qian, Jin Li, Jipeng Qiang, Yi Zhu, and Yun Li

Explainable Sparse Associative Self-optimizing Neural Networks for Classification ... 229
Adrian Horzyk, Jakub Kosno, Daniel Bulanda, and Janusz A. Starzyk

Efficient Attention for Domain Generalization ... 245
Zhongqiang Zhang, Ge Liu, Fuhan Cai, Duo Liu, and Xiangzhong Fang

Adaptive Accelerated Gradient Algorithm for Training Fully Complex-Valued Dendritic Neuron Model ... 258
Yuelin Wang and He Huang

Interpreting Decision Process in Offline Reinforcement Learning for Interactive Recommendation Systems ... 270
Zoya Volovikova, Petr Kuderov, and Aleksandr I. Panov

A Novel Framework for Forecasting Mental Stress Levels Based on Physiological Signals ... 287
Yifan Li, Binghua Li, Jinhong Ding, Yuan Feng, Ming Ma, Zerui Han, Yehan Xu, and Likun Xia

Correlation-Distance Graph Learning for Treatment Response Prediction from rs-fMRI ... 298
Francis Xiatian Zhang, Sisi Zheng, Hubert P. H. Shum, Haozheng Zhang, Nan Song, Mingkang Song, and Hongxiao Jia

Measuring Cognitive Load: Leveraging fNIRS and Machine Learning for Classification of Workload Levels ... 313
Mehshan Ahmed Khan, Houshyar Asadi, Thuong Hoang, Chee Peng Lim, and Saeid Nahavandi

Enhanced Motor Imagery Based Brain-Computer Interface via Vibration Stimulation and Robotic Glove for Post-Stroke Rehabilitation ... 326
Jianqiang Su, Jiaxing Wang, Weiqun Wang, Yihan Wang, and Zeng-Guang Hou

MTSAN-MI: Multiscale Temporal-Spatial Convolutional Self-attention Network for Motor Imagery Classification ... 338
Junkongshuai Wang, Yangjie Luo, Lu Wang, Lihua Zhang, and Xiaoyang Kang

How Do Native and Non-native Listeners Differ? Investigation with Dominant Frequency Bands in Auditory Evoked Potential ... 350
Yifan Zhou, Md Rakibul Hasan, Md Mahbub Hasan, Ali Zia, and Md Zakir Hossain

A Stealth Security Hardening Method Based on SSD Firmware Function Extension ... 362
Xiao Yu, Zhao Li, Xu Qiao, Yuan Tan, Yuanzhang Li, and Li Zhang

Attention-Based Deep Convolutional Network for Speech Recognition Under Multi-scene Noise Environment ... 376
Chuanwu Yang, Shuo Ye, Zhishu Lin, Qinmu Peng, Jiamiao Xu, Peipei Yuan, Yuetian Wang, and Xinge You

Discourse-Aware Causal Emotion Entailment ... 389
Dexin Kong, Nan Yu, Yun Yuan, Xin Shi, Chen Gong, and Guohong Fu

DAformer: Transformer with Domain Adversarial Adaptation for EEG-Based Emotion Recognition with Live-Oil Paintings ... 402
Zhong-Wei Jin, Jia-Wen Liu, Wei-Long Zheng, and Bao-Liang Lu

Time-Frequency Transformer: A Novel Time Frequency Joint Learning Method for Speech Emotion Recognition ... 415
Yong Wang, Cheng Lu, Yuan Zong, Hailun Lian, Yan Zhao, and Sunan Li

Asymptotic Spatiotemporal Averaging of the Power of EEG Signals for Schizophrenia Diagnostics ... 428
Włodzisław Duch, Krzysztof Tołpa, Ewa Ratajczak, Marcin Hajnowski, Łukasz Furman, and Luís A. Alexandre

Human Centred Computing

Non-contact Respiratory Flow Extraction from Infrared Images Using Balanced Data Classification ... 443
Ali Roozbehi, Mahsa Mohaghegh, and Vahid Reza Nafisi

The Construction of DNA Coding Sets by an Intelligent Optimization Algorithm: TMOL-TSO ... 455
Yongxu Yan, Wentao Wang, Zhihui Fu, and Jun Tian

Heterogeneous Graph Fusion with Adversarial Learning for Recommendation Service ... 470
Jiaxi Wang, Tong Mo, and Weiping Li

SSVEP Data Augmentation Based on Filter Band Masking and Random Phase Erasing ... 483
Yudong Pan, Ning Li, Lianjin Xiong, Yiqian Luo, and Yangsong Zhang

ONEI: Unveiling Route and Phase of Breathing from Snoring Sounds ... 494
Xinhong Li, Baoai Han, Li Xiao, Xiuping Yang, Weiping Tu, Xiong Chen, Weiyan Yi, Jie Lin, Yuhong Yang, and Yanzhen Ren

MVCAL: Multi View Clustering for Active Learning ... 506
Yi Fan, Biao Jiang, Di Chen, and Yu-Bin Yang

Extraction of One Time Point Dynamic Group Features via Tucker Decomposition of Multi-subject FMRI Data: Application to Schizophrenia 518
Yue Han, Qiu-Hua Lin, Li-Dan Kuang, Ying-Guang Hao, Wei-Xing Li, Xiao-Feng Gong, and Vince D. Calhoun

Modeling Both Collaborative and Temporal Information for Sequential Recommendation 528
Jinyue Dai, Jie Shao, Zhiyi Deng, Hongcai He, and Feiyu Chen

Multi-level Attention Network with Weather Suppression for All-Weather Action Detection in UAV Rescue Scenarios 540
Yao Liu, Binghao Li, Claude Sammut, and Lina Yao

Learning Dense UV Completion for 3D Human Mesh Recovery 558
Qingping Sun, Yanjun Wang, Zhenni Wang, and Chi-Sing Leung

Correction to: Correlation-Distance Graph Learning for Treatment Response Prediction from rs-fMRI C1
Francis Xiatian Zhang, Sisi Zheng, Hubert P. H. Shum, Haozheng Zhang, Nan Song, Mingkang Song, and Hongxiao Jia

Author Index 571

Theory and Algorithms

Self-adaptive Inverse Soft-Q Learning for Imitation

Zhuo Wang[1], Quan Liu[1,2(✉)], and Xiongzhen Zhang[1]

[1] School of Computer and Technology, Soochow University, Suzhou, Jiangsu 215006, China
{20215227102,20214227044}@stu.suda.edu.cn

[2] Provincial Key Laboratory for Computer Information Processing Technology, Soochow University, Suzhou, Jiangsu 215006, China
quanliu@suda.edu.cn

Abstract. As a powerful method for solving sequential decision problems, imitation learning (IL) aims to generate policy similar to expert behavior by imitating demonstrations. However, the quality of demonstrations directly limits the performance of the agent imitation policy. To solve this problem, self-adaptive inverse soft-Q learning for imitation (SAIQL) is proposed. SAIQL proposes a novel three-level buffer system by introducing an online excellent buffer based on the expert buffer and the normal buffer. Trajectories from interactions with superior performance are stored in the online excellent buffer. When the amount of data in the online excellent buffer and the expert buffer is equal, the former data will be cleaned and transferred to the latter, ensuring that demonstrations in the expert buffer are continuously optimized. Finally, we compare SAIQL with up-to-date IL methods in both the continuous control and the Atari tasks. The experimental results show the superiority of SAIQL. It improves the quality of expert demonstrations and the utilization of trajectories.

Keywords: Imitation Learning · Sub-Optimal Expert Demonstration · Self-Adaptability

1 Introduction

Imitation learning [2,19] (IL), an important branch of deep reinforcement learning [12], has been widely used in the fields of autonomous driving [11], healthcare [18], and simulation [13]. As the research continues, IL is becoming a hot topic for more researchers in related fields. Such methods imitate the expert behavior through demonstrations to obtain the policy with similar or even better performance than the expert.

Supported by National Natural Science Foundation of China (61772355, 61702055, 61876217, 62176175). Project Funded by the Priority Academic Program Development of Jiangsu Higher Education Institutions (PAPD).

B. Luo et al. (Eds.): ICONIP 2023, CCIS 1963, pp. 3–14, 2024.
https://doi.org/10.1007/978-981-99-8138-0_1

IL methods have a long history, with early work [17] using supervised learning to match the policy behavior with the expert. Later, Abbeel et al. [1,21] interpreted IL as an inverse reinforcement learning (IRL) process to recover expert policy. In addition, Dvijotham et al. [4,14] found formal equivalence between IRL and IL using the inverse bellman operator.

IL techniques are experiencing rapid development. However, the performance of IL methods is directly limited by the quality of expert demonstrations [20]. It may not be possible to obtain optimal expert demonstrations within the limited budget in practice. Since IL methods can not discriminate the expert demonstrations, the convergence speed and convergence results will be influenced when the demonstrations are sub-optimal. In addition, this situation will lead IL methods to learn only sub-optimal policy. At the same time, the reinforcement learning process trained by IL methods generates many trajectories, among which there is no shortage of data that outperforms the sub-optimal expert demonstrations. However, existing IL methods usually use only a normal buffer to store some online trajectories in recent times. The quality is not distinguished in this buffer [7,15]. This way does not effectively use the superior trajectory of the training process. According to the above, IL methods are limited by sub-optimal expert demonstrations and the inefficient use of online superior trajectories.

We propose the self-adaptive inverse soft-Q learning for imitation (SAIQL) method to improve imitation demonstration quality and trajectory utilization. The main contributions of this paper are as follows:

(1) We introduce an online excellent buffer to form a three-level buffer system. It ensures that the stored trajectory is satisfied with high-performance requirements and can provide better imitation data for the method.
(2) We propose a buffer self-adaptively update rule. When the amount of data in the online excellent buffer is equal to the expert buffer, the data will be transferred to the expert buffer to become new demonstrations.
(3) We validate our method and up-to-date IL baselines on the continuous control tasks and Atari tasks. The experimental results show that the SAIQL method is superior to other methods, and the effectiveness of each module is verified by ablation experiments.

2 Background

2.1 IL

Up to now, IL research can be divided into the following stages.

The first is the behavioural cloning (BC) method [16], which treats the IL problem as a supervised problem and therefore only maximizes the probability of the expert behavior under the learned policy. It ignores the sequential decision problem and has the problem of compounding error.

By introducing the environment's dynamics, IL problems can be framed naturally as IRL problems. Most of those methods use generative adversarial networks (GAN) [6]. Lots of methods inspired by this idea have been proposed, all of which explicitly or implicitly use the reward to express the environment's dynamics when learning policy [7,10,15]. Kostrikov et al. [9] and Barde et al. [3] have improved IL methods based on GAN. However, such methods learn unstably, are sensitive to hyperparameters or implementation details, and are usually difficult to use in practice.

Garg et al. [5] suggest that most of the difficulties of IL methods once resulted from representing the IL problem motivated by IRL as a min-max problem with reward and policy [1] and made improvements based on this. Then they propose the IQL [5] converts the min-max problem into a simple single-function minimization problem.

2.2 Exploration in IL

Exploration is a major component of reinforcement learning. Agents interact with the environment and observe the outcomes of their actions to improve themselves. Through introducing exploration into IL, encourage the agent to learn from sub-optimal expert demonstrations and then continuously optimize expert data [20]. The goal of such methods can be expressed as

$$\max J(\pi) = -D(\mathrm{d}_\pi, d_\mathrm{E}) + D(\mathrm{d}_\pi, d_B), \tag{1}$$

where D is the statistical distance, common ones are f-divergence, KL-divergence, JS-divergence, etc. d_π, d_E, d_B are the density distributions of the learned policy, expert policy, and method-generated data, respectively. $-D(\mathrm{d}_\pi, d_\mathrm{E})$ is the regular IL objective to find the policy to recover the expert density distribution. $D(\mathrm{d}_\pi, d_B)$ is the goal of the method of self-exploration, for which maximization facilitates access to states rarely used by the learned policy. In other words, such methods encourage more exploration than traditional IL methods. By introducing exploration into IL methods, the overall performance of IL methods can be improved by encouraging methods to explore a larger unknown space.

3 Method

3.1 Three-Level Buffer System and Self-adaptively Update

To effectively utilize online trajectories, we propose a three-level buffer system for IL methods and specify the self-adaptively update, as shown in Fig. 1.

Firstly, an online excellent buffer B_O is introduced for storing some of the online generated trajectories better than demonstrations. The expert buffer B_E is used to store the previously acquired expert demonstrations, in which the data is usually sub-optimal. The normal buffer B_N will store the latest experience trajectories in the learning process. These buffers are combined to form the three-level buffer system. When the experience trajectories with better performance than the average performance of B_E are obtained, they will be stored in B_O.

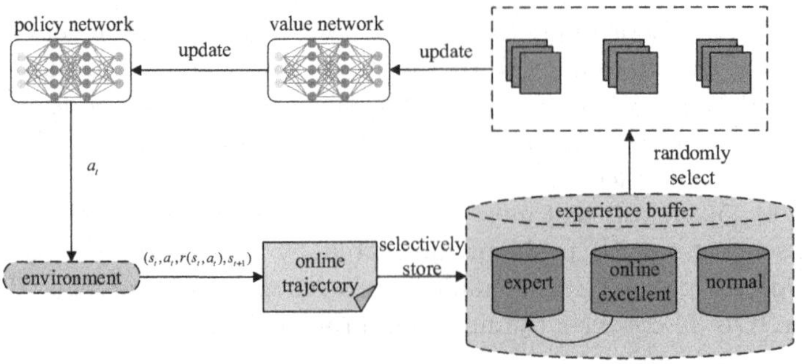

Fig. 1. Three-level buffer system and self-adaptively update diagram.

So the trajectories we store in the buffer are more than adequate for high-performance requirements.

We constructed the data interaction between B_E and B_O to selectively store the better data in a self-adaptive way. To improve buffer utilization efficiency, the contents of B_O and B_E are self-adaptively adjusted. When the amount of data in B_O is equal to B_E, the original sub-optimal demonstration in B_E is discarded, and the data in B_O is transferred to B_E as new expert demonstrations. After that, B_O is cleared to store the new trajectory data that is better than current expert demonstrations.

The quality of the experience trajectories in B_E and B_O is continuously improved through the learning process, and the density distribution of B_E is close to the oracle distribution [20].

Theorem 1. *For any deterministic policy, its generated rewards for the corresponding trajectory indicate that the policy distribution and oracle distribution are consistent.*

When the amount of data in B_O is equal to B_E, set their corresponding policies to be π_O and π_E. $\forall \psi \in B_O$, $\exists r(\psi) > \overline{r}_e$, so we have $\mathbb{E}_{\psi \sim \pi_O}(r(\psi)) > \mathbb{E}_{\psi \sim \pi_E}(r(\psi))$, where r is the reward function, $\overline{r}_e$ is the average reward value of the corresponding expert demonstration.

Then, for any π_x and π_y, if their cumulative rewards have $\mathbb{E}_{\psi \sim \pi_x}(r(\psi)) > \mathbb{E}_{\psi \sim \pi_y}(r(\psi))$, it means $D(\pi_x(a|s), \pi^*(a|s)) < D(\pi_y(a|s), \pi^*(a|s))$, where a is the action taken, s is the current state, and π^* is the optimal policy. Consequently, π_O is closer to π^* than π_E.

Therefore, sub-optimal demonstrations in B_E are replaced by better trajectories generated by B_O in the training process. The distribution obtained by B_E becomes closer and closer to the optimal state, and the buffer advantage of the replacement is gradually improved.

3.2 IQL

To properly combine the idea with the IL method, we instantiate it using the IQL. IQL is a dynamics-aware IL method that avoids the adversarial training framework and expresses reward and policy implicitly by learning a single Q function. It enables exploration-driven IL as it accomplishes IL tasks in a similar way to reinforcement learning.

In the discrete control task, the optimal value of IQL is

$$\max_{Q\in\Omega} \Gamma^*(Q) = \mathbb{E}_{\rho_E}[g(Q(s,a) - \gamma \mathbb{E}_{s'\sim P(\cdot|s,a)} V^*(s'))] - (1-\gamma)\mathbb{E}_{\rho_0}[V^*(s_0)]. \quad (2)$$

Among these, $V^*(s) = \log \sum_a \exp Q(s,a)$, $g(x)$ is a convex function. For each Q value in the discrete task, the corresponding reward value obtained is $r_{s,a} = Q(s,a) - \gamma \mathbb{E}_{s'\sim(\cdot|s,a)}[\log \sum_{a'} \exp Q(s',a')]$, which corresponds to each other.

In the continuous control task, for a policy, the IRL objective in the IQL method is

$$\Gamma(\pi, Q) = \mathbb{E}_{\rho_E}[g(Q - \gamma \mathbb{E}_{s'\sim P(\cdot|s,a)} V_\pi(s'))] - (1-\gamma)\mathbb{E}_{\rho_0}[V_\pi(s_0)]. \quad (3)$$

For each π, optimize Q by maximizing $\Gamma(\pi, Q)$. For each Q, apply the SAC update style to make π approximate π_Q.

3.3 SAIQL

Algorithm 1 SAIQL

Input: Initialize expert buffer B_E, online excellent buffer B_O, normal buffer B_N, batch size N, average reward value of B_E as $\overline{r}_e$, Q function parameter Q_θ, Optional policy π_ϕ, maximum time step $t_{\max}$

Output: Q_θ and π_ϕ

```
for t in t_max do
    Sample trajectory ψ ~ π_φ
    if r(ψ) > r̄_e then
        B_O ← ψ
    end if
    if the amount of B_O is equal to B_E then
        Empty B_E
        B_E ← B_O
        Update r̄_e
        Empty B_O
    end if
    if update then
        {(a_i, s_i, ...)}_{i=1}^{α} ~ B_E, {(a_i, s_i, ...)}_{i=1}^{β} ~ B_O, {(a_i, s_i, ...)}_{i=1}^{N/2} ~ B_N
        Train the Q function according to Eq. 2 and Eq. 3
        θ_{t+1} ← θ_t − α_Q ∇_θ[−Γ(θ)] (Q-learning uses V*, SAC uses V^{π_φ})
        φ_{t+1} ← φ_t − α_π ∇_{φ s,a}[Q(s,a) − log π_θ(a|s)](SAC)
        Update improvement policy using SAC-style actors π_φ: φ_{t+1} ← φ_t −
        ∂_π ∇_φ E_{s,a}[Q(s,a) − log π_φ(a|s)]
    end if
end for
```

Combining all the components forms our SAIQL method as shown in Algorithm 1. SAIQL maintains three main experience buffers B_E, B_O, and B_N. It unites the following two components: the Q that indirectly provides the reward value and the policy π that maximizes the reward value.

The SAIQL method uses IQL methods to implement exploration-driven IL. During the training process, we sample the trajectory of the current method. If it is better than current expert demonstrations, then it is stored in B_O. As the training continues, the amount of data in B_O becomes larger. Once the amount of data in B_O is the same as in B_E, both buffers are updated self-adaptively.

When updating the method, we randomly select trajectories from B_E, B_O, and B_N according to proportion, and optimize the method by Eq. 2 and Eq. 3. Set α and β to be the number of trajectories in B_E and B_O in the sampling process. The two together form half of the selected data. So even if the data in B_O has not yet been transferred to B_E, we can still take advantage of the superior trajectories in time.

As training continues, expert demonstrations will iterate, providing superior trajectory guidance for the method. In this way, the method performance can eventually exceed that of original expert demonstrations.

4 Experiments

4.1 Environments and Settings

The Gym platform is a guideline platform developed by OpenAI for reinforcement learning methods. Since the method considers the case of sub-optimal expert demonstrations, we select Hopper, HalfCheetah, Ant, and Walker2d tasks from the continuous control task environments. In addition, the Breakout and Space Invaders tasks are also selected for experiments in more challenging Atari tasks.

Parameters. All the experiments adopt the same parameter settings as [8], including various parameters such as learning rate and the number of expert trajectories, to ensure the fairness as well as the validity of the experiments.

Table 1. The average reward for expert demonstrations in various environments.

Task name	Reward value
Hopper-v2	3298.47
HalfCheetah-v2	5863.04
Ant-v2	3862.63
Walker-v2	3974.05
Breakout	264.18
Space Invaders	647.03

Expert Demonstrations Dataset. In all experiment environments, expert policies are trained in the same way. The trained expert policies sample the environment to obtain expert demonstrations.

The dataset contains 25 expert trajectories. Each trajectory contains state, action, reward value, data length, and other information. The trajectory is 1500 state-action pairs sequences, and each sequence contains 1000 state-action pairs. Table 1 shows the average reward of expert demonstrations in different experiment environments.

Evaluation Standards. In experiments, the method performance is measured using the historical average reward value and the error rate of the reward value. The historical average reward value $r_t = \frac{1}{t}\sum_{i=0}^{t}\bar{r}_i$, where $\bar{r}_i$ is the average of all random seed policies evaluated at moment i.

The reward value rate is the error between the learned policy and the expected reward of the expert policy. And $error = \overline{r}_\pi/\overline{r}_\mathrm{e}$, where $\overline{r}_\pi$ is the average reward value of the learning policy evaluation.

4.2 Experimental Results

Performance in Continuous Tasks. As shown in Fig. 2, SAIQL is the only method with better trajectory utilization and performance among all continuous tasks. The SAIQL method can efficiently and effectively utilize the sub-optimal demonstrations and online excellent trajectories provided by B_E and B_O to recover expert policy with fewer steps. The proposed method can recover sub-optimal expert performance with 200,000 steps, while the IQL and DAC [9] methods require 300,000 or 500,000 steps and still fail to fully recover. GAIL, on the other hand, can recover only 80% in the Hopper environment and is almost ineffective in other environments.

In addition to this, the SAIQL method is effective in exploring the environment. After recovering the sub-optimal expert demonstrations, online excellent trajectories still provide guidance. Unlike other comparison methods, the SAIQL method can continue exploring the environment using online excellent trajectories after the expert performance. It enables better performance through B_O. Expert performance can be achieved even in sparsely rewarded environments. However, methods such as IQL rarely surpass demonstrations and can not effectively exploit the online experience.

The rate of reward value between the historical maximum value and the expert value for each method in the continuous environments is shown in Table 2. SAIQL achieves the best rate in all experiment environments and outperforms experts. The worst-performing HalfCheetah-v2 environment also managed to achieve 1.15 multiple of expert demonstrations, and the most efficient performance in Ant-v2 with the rate of 1.47. The rest of the experiment methods rarely surpassed the expert performance, with the GAIL method not even recovering the sub-optimal expert, reaching only 0.11 at the lowest level.

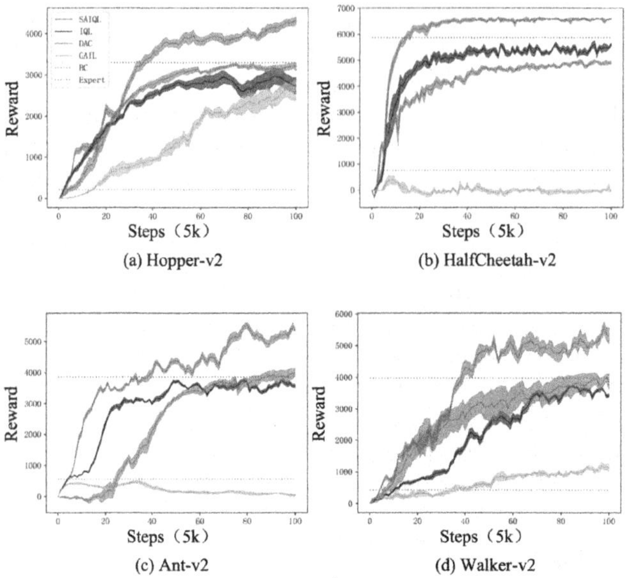

Fig. 2. Results of reward value for continuous experimental tasks.

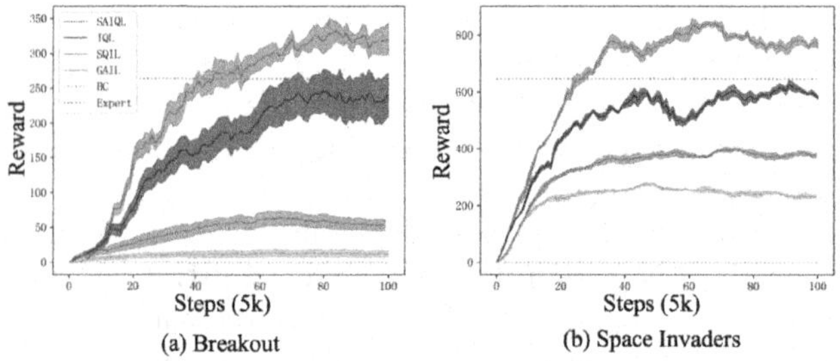

Fig. 3. Results of reward value for Atari experimental tasks.

Table 2. Rate of reward value for different methods in continuous environments.

Task	Hopper-v2	HalfCheetah-v2	Ant-v2	Walker2d-v2
SAIQL	1.31(±0.06)	1.15(±0.01)	1.47(±0.05)	1.4(±0.10)
IQL	1.02(±0.03)	0.99(±0.02)	1.01(±0.04)	0.97(±0.04)
GAIL	0.87(±0.14)	0.11(±0.10)	0.13(±0.02)	0.34(±0.05)
DAC	1.00(±0.02)	0.86(±0.01)	1.07(±0.09)	1.02(±0.08)

Performance in Atari Tasks. The performance of the SAIQL method and the comparison method in the Atari environments is shown in Fig. 3. The SAIQL method still shows excellent performance in selected Atari environments. In the Breakout environment, our method recovers the sub-optimal expert policy at about 200,000 steps. In contrast, the IQL method can not fully achieve SAIQL performance after 300,000 steps. SQIL [15] can only reach 17% at 500,000 steps. The worst is the GAIL method, which is almost ineffective in Breakout environment. In the Space Invaders environment, the SAIQL method achieves sub-optimal demonstrations after 100,000 steps. While the IQL method requires 400,000 steps to achieve, the SQIL method and the GAIL method eventually fail to achieve expert performance.

The rate of reward value for each method in Atari environments, as shown in Table 3. As can be seen from the table, the SAIQL method performs optimally in both of the two environments, with a stable rate of about 1.30. The IQL method, on the other hand, could only produce 0.89 in Breakout. SQIL and GAIL methods could not reach the demonstrations in both experimental methods, much less surpass the sub-optimal expert.

Table 3. Rate of reward value for different methods in Atari environments.

Task	Breakout	Space Invaders
SAIQL	1.30(±0.11)	1.31(±0.01)
IQL	0.89(±0.04)	0.98(±0.02)
SQIL	0.25(±.13)	0.63(±0.01)
GAIL	0.06(±0.05)	0.43(±0.01)

Ablation Studies. Furthermore, additional experiments to verify the indispensability of the proposed three-level buffer system and the self-adaptively update in the method are carried out. That is, we demonstrate the contribution of different components of the SAIQL method to the method by removing them. As shown in Fig. 4, in the continuous control task, we consider three variants of the SAIQL method without updating B_E (SAIQL-A), deleting the B_O (SAIQL-B) and using only the normal experience buffer (SAIQL-C).

As can be seen in the figure, method performance decreases without updating B_E. However, it still recovers and outperforms sub-optimal expert demonstrations. The SAIQL method uses much more than expert demonstration data, so even when the expert demonstration data is not updated, B_O can still provide excellent online experience trajectories. However, due to the lack of self-adaptively updates, the method convergence speed is reduced. In addition, when the B_O is removed, the method will perform inferiorly to the SAIQL. The method performs slightly lower in the same training steps due to the underutilization of environmental excellent trajectories. However, sub-optimal expert demonstrations also provide trajectories for the method that can eventually approach expert performance.

When both buffers are removed, the method can only learn from B_N and can not distinguish between different levels of experience. Therefore, the overall quality of demonstrations is low. This eventually leads to poor method performance, which eventually fails to achieve expert performance and can only achieve sub-optimal expert performance at 50%.

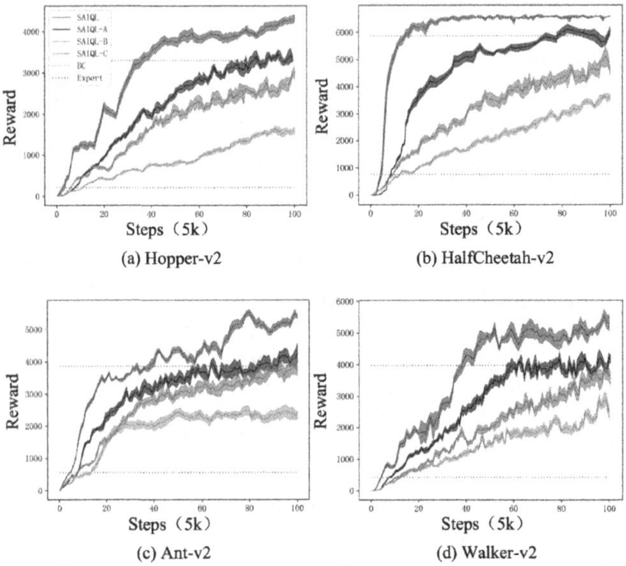

Fig. 4. Performance of different variants of the SAIQL method.

4.3 Analysis of the Effect of SAIQL Method

In this section, the SAIQL method is demonstrated to be superior to other IL methods in continuous action tasks and Atari tasks. The experimental results also show that the SAIQL method outperforms other comparison methods in all tasks in terms of convergence speed, reward value, and the rate of reward value. The proposed SAIQL method converges significantly faster than other comparison methods at 500,000 training steps. At the same time, it can achieve a higher historical average reward than other methods with the same number of iterations. In terms of final results, the SAIQL method outperforms sub-optimal expert performance and achieves higher performance.

5 Conclusion and Future Work

We propose the SAIQL method for improving IL methods in the case of sub-optimal expert demonstrations, where the methods converge to sub-optimal performance. To address this challenging problem, we encourage exploration-driven IL by proposing a three-level buffer system and self-adaptively update. Specifically, we introduce an online excellent buffer for storing experience data that perform well during the method learning process. And during the training of the method, the data in buffers are updated self-adaptively. Thus buffers can provide demonstrations of higher quality for the method. Better demonstrations for the IL method are provided by using experience trajectories that perform well in the learning process. By continuously improving the quality, the overall performance of the method is improved. And the feasibility, stability, and superiority of the proposed method are verified experimentally. In the future, we will consider applying our method in more practical situations.

References

1. Abbeel, P., Ng, A.Y.: Apprenticeship learning via inverse reinforcement learning. In: Proceedings of the Twenty-First International Conference on Machine Learning, p. 1 (2004)
2. Arora, S., Doshi, P.: A survey of inverse reinforcement learning: challenges, methods and progress. Artif. Intell. **297**, 103500 (2021)
3. Barde, P., Roy, J., Jeon, W., Pineau, J., Pal, C., Nowrouzezahrai, D.: Adversarial soft advantage fitting: imitation learning without policy optimization. In: Advances in Neural Information Processing Systems, vol. 33, pp. 12334–12344 (2020)
4. Dvijotham, K., Todorov, E.: Inverse optimal control with linearly-solvable MDPs. In: Proceedings of the 27th International Conference on Machine Learning, ICML 2010, pp. 335–342 (2010)
5. Garg, D., Chakraborty, S., Cundy, C., Song, J., Ermon, S.: IQ-learn: inverse soft-Q learning for imitation. In: Advances in Neural Information Processing Systems, vol. 34, pp. 4028–4039 (2021)
6. Goodfellow, I.J., et al.: Generative adversarial nets, pp. 2672–2680 (2014)
7. Ho, J., Ermon, S.: Generative adversarial imitation learning. In: Advances in Neural Information Processing Systems, vol. 29 (2016)
8. Imani, M., Ghoreishi, S.F.: Scalable inverse reinforcement learning through multifidelity Bayesian optimization. IEEE Trans. Neural Netw. Learn. Syst. **33**(8), 4125–4132 (2021)
9. Kostrikov, I., Agrawal, K.K., Dwibedi, D., Levine, S., Tompson, J.: Discriminator-actor-critic: addressing sample inefficiency and reward bias in adversarial imitation learning. arXiv preprint arXiv:1809.02925 (2018)
10. Kostrikov, I., Fergus, R., Tompson, J., Nachum, O.: Offline reinforcement learning with fisher divergence critic regularization. In: International Conference on Machine Learning, vol. 139, pp. 5774–5783. PMLR (2021)
11. Le Mero, L., Yi, D., Dianati, M., Mouzakitis, A.: A survey on imitation learning techniques for end-to-end autonomous vehicles. IEEE Trans. Intell. Transp. Syst. **23**, 14128–14147 (2022)

12. Liu, Q., et al.: A survey on deep reinforcement learning. Chin. J. Comput. **41**(1), 1–27 (2018)
13. Mohammed, H., Sayed, T., Bigazzi, A.: Microscopic modeling of cyclists on off-street paths: a stochastic imitation learning approach. Transportmetrica A Transp. Sci. **18**(3), 345–366 (2022)
14. Piot, B., Geist, M., Pietquin, O.: Bridging the gap between imitation learning and inverse reinforcement learning. IEEE Trans. Neural Netw. Learn. Syst. **28**(8), 1814–1826 (2016)
15. Reddy, S., Dragan, A.D., Levine, S.: SQIL: imitation learning via reinforcement learning with sparse rewards. arXiv preprint arXiv:1905.11108 (2019)
16. Ross, S., Bagnell, D.: Efficient reductions for imitation learning. In: Proceedings of the Thirteenth International Conference on Artificial Intelligence and Statistics, vol. 9, pp. 661–668. JMLR Workshop and Conference Proceedings (2010)
17. Sammut, C., Hurst, S., Kedzier, D., Michie, D.: Learning to fly. In: Machine Learning Proceedings 1992, pp. 385–393. Elsevier (1992)
18. Wang, L., et al.: Adversarial cooperative imitation learning for dynamic treatment regimes. In: Proceedings of the Web Conference 2020, pp. 1785–1795 (2020)
19. Zhang, K., Yu, Y.: Methodologies for imitation learning via inverse reinforcement learning: a review. J. Comput. Res. Develop. **56**(2), 254–261 (2019)
20. Zhu, Z., Lin, K., Dai, B., Zhou, J.: Self-adaptive imitation learning: Learning tasks with delayed rewards from sub-optimal demonstrations. In: Proceedings of the AAAI Conference on Artificial Intelligence, vol. 36, pp. 9269–9277 (2022)
21. Ziebart, B.D., Maas, A.L., Bagnell, J.A., Dey, A.K., et al.: Maximum entropy inverse reinforcement learning. In: AAAI, Chicago, IL, USA, vol. 8, pp. 1433–1438 (2008)

Membership Inference Attacks Against Medical Databases

Tianxiang Xu[1], Chang Liu[2], Kun Zhang[2], and Jianlin Zhang[2](✉)

[1] School of Information Science and Engineering, University of Jinan, Jinan, China
[2] Chinese Health Medical Information and Big Data Association-Information and Application Security Agency, Beijing, China
CHMIAASA@outlook.com

Abstract. Membership inference is a powerful attack to privacy databases especially for medical data. Existing attack model utilizes the shadow model to inference the private members in the privacy data-sets and information, which can damage the benefits of the data owners and may cause serious data leakage. However, existing defence are concentrated on the encryption methods, which ignore the inference can also cause the unacceptable loss in the real applications. In this work, we propose a novel inference attack model, which utilizes a shadow model to simulate the division system in the medical database and subsequently infer the members in the medical databases. Moreover, the established shadow inference model can classify the labels of medical data and obtain the privacy members in the medical databases. In contrast with traditional inference attacks, we apply the attack in the medical databases rather than recommendation system or machine learning classifiers. From our extensive simulation and comparison with traditional inference attacks, we can observe the proposed model can achieve the attacks in the medical data with reasonable attack accuracy and acceptable computation costs.

Keywords: Membership Inference Attack · Medical Database · Privacy Information Security · Shadow Model

1 Introduction

As one of the most prevalent and privacy information applications in existing databases, medical data systems have been utilized in tremendous hospitals or medical companies. Medical databases have become a crucial utilization in the field of healthcare through providing or collecting a centralized and organized repository of medical information. Furthermore, these databases contain a wealth of patient data, clinical records, sensitive research investigations and other medical-related information that can be accessed and utilized by trusted users [1]. The advent of digital technology and the increasing adoption of electronic health records have significantly contributed to the growth and development of medical databases. The most essential component is that the medical data system contains the privacy information about patients and other non-member information, which is vulnerable by utilizing inference attacks [2].

B. Luo et al. (Eds.): ICONIP 2023, CCIS 1963, pp. 15–25, 2024.
https://doi.org/10.1007/978-981-99-8138-0_2

Medical databases are devised to store, manage and retrieve vast amounts of healthcare information, which can only access by authorized users. Indeed, the utilization of medical databases can offer numerous advantages to healthcare stakeholders [3]. As for healthcare providers, accessing these databases can assist to dispose comprehensive patient profiles, enabling more accurate diagnoses, personalized treatment plans and improve patient outcomes [4]. Researchers can leverage medical databases to conduct large-scale studies, identify patterns and generate evidence-based insights for medical advancements and innovations.

Initially, medical data systems concentrate on the classification issue, which can enhance the classification effectiveness and accuracy when system contains tremendous sensitive data [5]. With the rethinking of security, numerous encryption methods are proposed to dispose the sensitive information leakage and have made extraordinary success in sensitive information defenses [6-9]. Moreover, the attackers also develop effective method to obtain the sensitive information including the injection, statistical learning and cross-attacks. Unlike membership inference attacks, these attacks concentrate on accessing the content of sensitive information and ignore the utilization of these sensitive information [10].

With the continuous growing dependence on machine learning models for various applications, concerns about privacy and security have become increasingly prominent. One particular area of concern is membership inference attacks, which target the confidentiality of individual records within a trained machine learning model [11]. Membership inference attacks can assist to determine whether a particular data point was part of the training data-set used to establish the model. These attacks exploit vulnerabilities in the model's outputs to infer the presence or absence of specific records, posing a significant threat to data privacy and confidentiality. Membership inference attacks rely on the observation that machine learning models often encode information about the training data in their learned representations and decision boundaries [12–14]. Following Fig. 1 shows the general framework of membership inference attack.

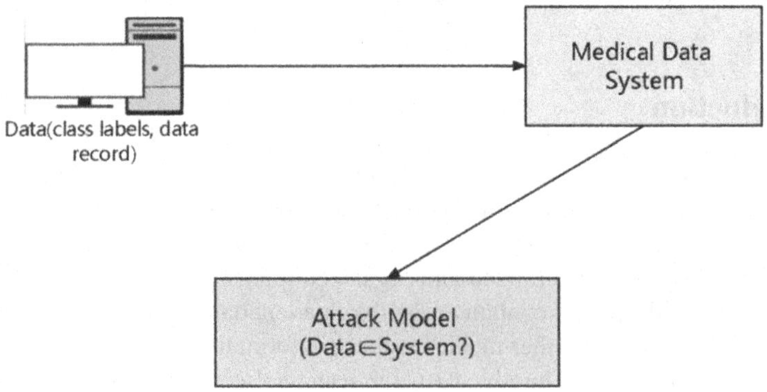

Fig. 1. General framework of inference in medical data system.

The consequences of successful membership inference attacks can be severe. They can compromise sensitive information, breach privacy rights, and undermine trust in machine learning applications. In healthcare, for example, membership inference attacks could reveal the inclusion of individuals medical records in the training data of a predictive model, violating their privacy and confidentiality [15]. Similarly, in financial or legal domains, these attacks could expose sensitive personal or financial information. Indeed, mitigating membership inference attacks poses a significant challenge [16]. Traditional privacy-preserving techniques, such as data anonymization or differential privacy, may not be sufficient to protect against these attacks since they focus on hiding individual records rather than preventing membership inference. Adversaries can exploit subtle patterns in model outputs to deduce membership information, even when the data has been anonymized [17].

To achieve above inference attacks in medical system, we propose a novel framework of membership inference attacks by utilizing the shadow model and simulated similar distributions of sensitive data. Following items describe the primary contributions of about our proposed model:

- To best of our knowledge, we are initially applied the membership inference attack in the medical databases and achieve the attack with acceptable success ratio. Subsequently, the comparison with traditional attacks indicates the proposed model can achieve reasonable effectiveness in aspect of inference accuracy. From our extensive simulation in the medical data, the developed model can classify the members and non-members features and can assist model to attack high-order and valuable information from medical databases.
- Our model utilizes the shadow classification mechanism and shadow data-set, which has the similar outputs and data distribution in the target data-sets and model, respectively. Therefore, the proposed model computation costs are greatly reduced by applying the shadow mechanism in the simulation of attack.
- We propose the model can distinguish the sensitive members in tremendous data, which can assist the attack model to access more valuable information. Indeed, a defense method against the proposed model is also provided in the experimental analysis and achieve useful prevention utilization.

2 Preliminaries

In this section, we initially illustrate the existing attack methods for medical databases and introduce the used medical data-sets in the simulation process.

Initially, we introduce several related attacks mechanism in existing medical data system. Injection attacks (IA) involve the insertion of malicious code or unauthorized commands into a database system, leveraging vulnerabilities in input validation or inadequate security measures. These attacks can lead to unauthorized access, modification, or destruction of sensitive data, compromising patient privacy and potentially jeopardizing their health and well-being [18].

Indeed, the statistical learning (SL) attacks leverage the outputs and behaviors of machine learning algorithms to extract sensitive information about individuals, even if their explicit data is not accessible. By analyzing the statistical patterns and correlations learned by the models, attackers can deduce potentially sensitive attributes or infer membership in the training data-set. This raises concerns about the disclosure of personal medical information and the potential misuse of such information [19]. The implications of statistical learning attacks on medical databases are far-reaching. Patient privacy is of paramount importance in the healthcare domain, as medical records often contain highly sensitive information, including personal identifiers, diagnoses, treatment plans, and genetic data.

Indeed, cross-attacks (CA) also known as cross-site attacks or cross-system attacks, occur when an attacker targets one system or database to gain unauthorized access to another system or database. In the context of medical databases, cross-attacks involve exploiting vulnerabilities in the interfaces or communication protocols that enable data exchange between different healthcare systems. By exploiting these vulnerabilities, attackers can traverse across interconnected systems to extract or manipulate sensitive medical data. The consequences of successful cross-attacks on medical databases can be severe. Patient privacy is a fundamental aspect of healthcare, as medical records often contain highly sensitive information, including personal identifiers, medical histories, diagnoses, treatments, and genetic data [20].

3 Attack Model

In this section, we systematically describe the proposed membership inference attack by showing each inner component in the sequence of model execution. Generally, the model contain dual separate sections including the shadow training and the inference attack.

Following Fig. 2 demonstrates the two components and inner structure of the proposed model. As the framework shown, we can significantly observe the model initially generates the similar distribution of users data and the shadow model imitates the outputs of the target medical system. Subsequently, an inference network is utilized to distinguish the members in the target data-sets and system.

3.1 Shadow Model

After showing the general structure of proposed model, we introduce the shadow model component in following items.

- Define the Shadow Model: Determine the architecture and characteristics of the shadow model, which will be used to mimic the target model. The shadow model should have a similar structure but does not need to match it exactly. Gather a data-set that consists of both public data (non-sensitive) and private data (sensitive) samples using the similar distribution with the target data-set. Ensure the simulated data-set is representative of the target model training data.
- Split the Data-set: Divide the data-set into three subsets: training, validation, and testing. The training set will be used to train the shadow model, the validation set to tune hyper-parameters, and the testing set to evaluate the performance of proposed model. Utilizing the training set to train the shadow model. The training process should be similar to the one used for training the target model, but on a smaller scale. Adjust hyper-parameters, such as learning rate and batch size, using the validation set to optimize the shadow model performance.

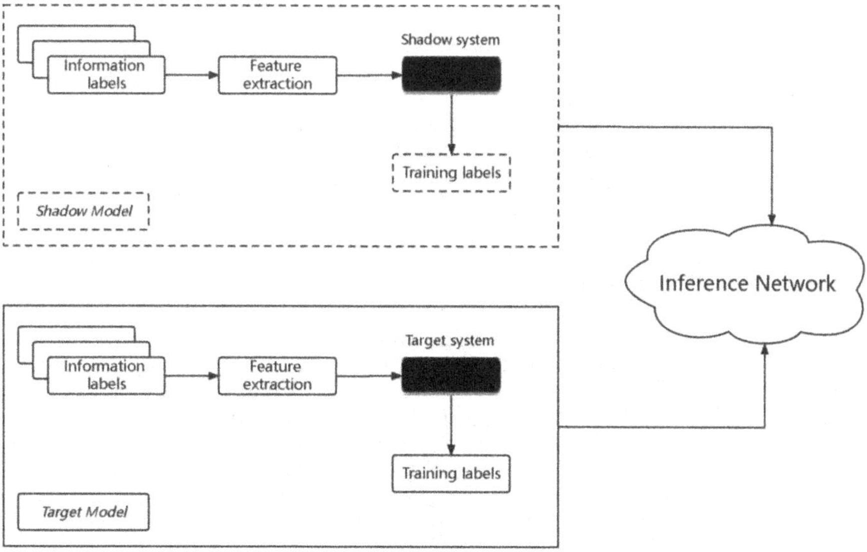

Fig. 2. Membership inference attack model execution framework illustration.

3.2 Inference Network

The inference network is response for searching the suspicious members data using the minimal loss function of shadow model and the target system outputs, which is described in following Eq. 1.

$$L(inf) = MIN[dist(Shad(I\prime), Targ(I))] \tag{1}$$

After the description of optimization module, we introduce the structure of inference network with Multi-layer Perceptron. Following Fig. 3 shows the construction of proposed inference network.

3.3 Members Identification

The final procedure of proposed model is the identification of members from the outputs of inference network, which may contain the non-members in the optimization process. Therefore, we need to eliminate the large characteristic difference results from previous inference network. Following Algorithm 1 describes the execution method of members identification with the method of maximum gradient calculation.

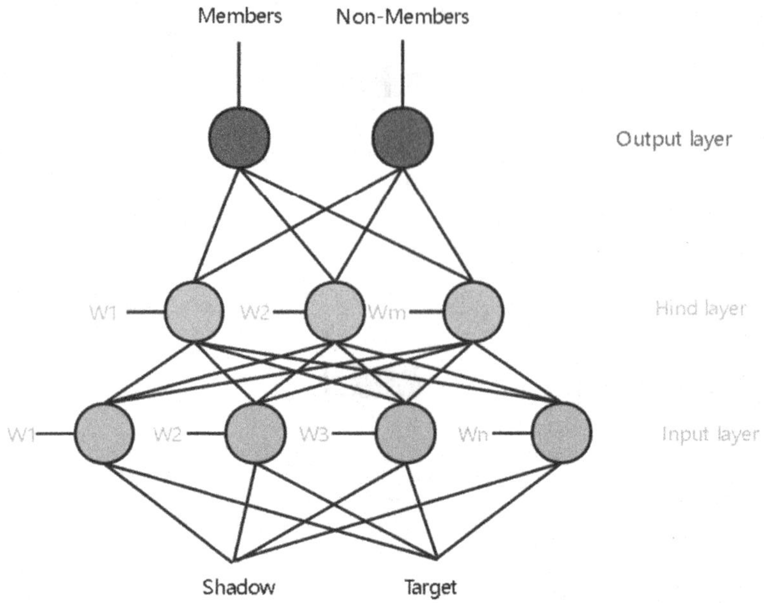

Fig. 3. Construction of inference network.

Algorithm 1: Members Data Identification
Input: Suspicious members set **Sus**, Gradient values γ_i
Output: Members information set **M**
for each member j do:
for each non-member i do:
calculate the gradient of members and non-members.
end.
select the maximum amount set as the members.
end.
Return the selections.

4 Experiments

In this section, we initially introduce the related simulation environment that are utilized in the evaluation of proposed model. Subsequently, we estimate the performance of our proposed attack with previous method to evaluate the attack accuracy and computation costs metrics.

4.1 Experimental Setups

Initially, we introduce the metrics of evaluation of proposed model. Following items contain the evaluation indicators.

A. **Attack Accuracy:** The accuracy of an attack depends on multiple factors, including the skill and capabilities of the attacker, the vulnerabilities and weaknesses present in the target system, the effectiveness of defensive measures in place, and the attack surface that can be exploited. Understanding attack accuracy provides valuable insights into the level of risk associated with specific attack vectors and guides the allocation of resources and efforts toward securing vulnerable areas.
B. **Computation Cost:** The concept of computation cost is closely linked to algorithmic efficiency and complexity. Algorithms that are designed to minimize computation cost, such as through the use of optimized data structures, parallel processing, or algorithmic shortcuts, can significantly reduce the time and resources required for computation. This optimization is particularly crucial in domains where real-time or near real-time processing is essential, such as financial transactions, healthcare diagnostics, and online recommendation systems. Computation cost is not limited to individual algorithms or computations.
C. **Members Classification Accuracy:** The accuracy of member classification is typically evaluated by comparing the model's predictions against ground truth labels or known class assignments in a data-set. It is calculated as the percentage of correctly classified instances divided by the total number of instances in the data-set. A high classification accuracy indicates that the model is effectively discerning the underlying patterns and characteristics that differentiate one class from another. The choice of classification algorithm and the features used for classification heavily influence member classification accuracy. Machine learning techniques such as logistic regression, support vector machines, decision trees, and deep neural networks are commonly employed to build classification models. These models are trained on labeled data, where the relationship between input features and target classes is learned to make accurate predictions on unseen data.

Indeed, we introduce the medical data-set, which have become invaluable resources in the field of healthcare and medical research. These datasets contain a wealth of information derived from various sources, including electronic health records (EHRs), medical imaging, genomic data, clinical trials, and more. They provide a comprehensive and structured collection of medical information that enables advancements in diagnostics, treatment strategies, disease understanding, and healthcare delivery. Medical datasets are curated repositories of patient health information, encompassing diverse attributes such as demographic details, medical histories, diagnoses, treatments, laboratory results, imaging data, and genetic profiles. These datasets serve as a foundation for conducting epidemiological studies, clinical trials, predictive modeling, and other data-driven research endeavors. They play a pivotal role in driving evidence-based medicine, improving patient outcomes, and shaping healthcare policies and practices.

Additionally, we summarize the primary used parameters and corresponding utilization of proposed model as following Table 1 demonstration.

The volume and complexity of medical data have been significantly amplified by the advent of digital healthcare systems and advancements in medical technology. Electronic health records, for instance, have transitioned from paper-based records to comprehensive digital repositories, capturing and consolidating patient information across multiple

Table 1. Summary of primary notations used in this paper.

Notation	Description
Shad, Targ	Shadow and target model of the medical data system
I	Input simulation data
γ	Gradient values of each data
L	Loss function

healthcare providers and settings. This wealth of data presents unprecedented opportunities for researchers, healthcare providers, and data scientists to extract valuable insights and patterns.

4.2 Attack Success Ratio

The experimental results for attack success ratios are shown in Fig. 4. We can observe that the performance of other three methods are extremely unacceptable and our proposed protocol gets reasonable attack success ratios compared with other three methods.

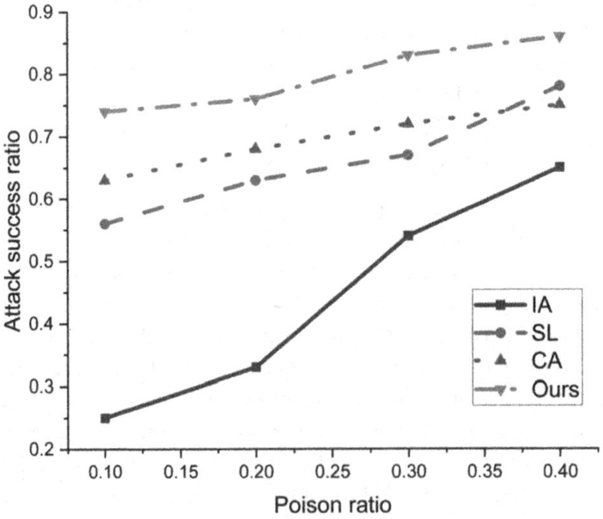

Fig. 4. Attack success ratios comparison results.

4.3 Costs Evaluations

At last, the computation cost among the proposed components and corresponding comparison protocols are shown in Table 2. It can be found that the computation costs of our proposed protocol obtains lowest costs compared with other methods, which indicates that our trained model can directly use for disposing the attack member information. This is caused that our method utilizes the shadow model, which can significantly decrease the attack computation period.

Table 2. Performance of computation costs comparison results.

Owner Data Size(KB)	Computation Cost(Second)			
	Ours	IA	SL	CA
40	2.1	2.7	2.9	2.7
60	4.3	4.6	4.6	4.4
80	6.4	7.2	8.4	7.3
100	8.2	8.6	10.2	9.9

Membership inference attacks pose a significant threat to the privacy and security of medical databases. These attacks aim to determine whether an individual's data is present in a given data-set, effectively breaching the confidentiality of sensitive patient information. Conducting an experimental analysis of membership inference attacks against medical databases is crucial for understanding the vulnerabilities and risks associated with such attacks, and for devising effective countermeasures to protect patient privacy.

4.4 Members Classification Accuracy

At last, the members classification accuracy is another essential metric for estimating the performance of proposed method. Following Fig. 5 shows the comparison of members classification accuracy comparison results with existing membership inference attack methods LOMIA [11] and MIAsec [17], which are introduced in previous introduction section.

From our extensive experiment analysis, we can highlight the significant threat posed by membership inference attacks against medical databases. The results demonstrate that attackers can potentially infer membership with varying degrees of accuracy, potentially compromising patient privacy. To mitigate such attacks, robust defense mechanisms must be employed, considering the sensitivity of medical data and the importance of maintaining patient confidentiality. Continual research and evaluation of attack methodologies and defense strategies are essential to safeguard sensitive medical information and protect individuals privacy rights in an increasingly interconnected healthcare landscape.

Additionally, the defense for inference is important for existing medical data system. We could utilize the data perturbation techniques involve modifying or perturbing the data in a way that preserves statistical properties but introduces randomness and uncertainty. Perturbation can be applied to various attributes, such as demographics, diagnoses, or laboratory results. Techniques like adding random noise, generalization, and aggregation can help protect against membership inference attacks by making it harder for an attacker to distinguish between members and non-members based on individual records. From our simulation, the data perturbation can prevent almost 60%

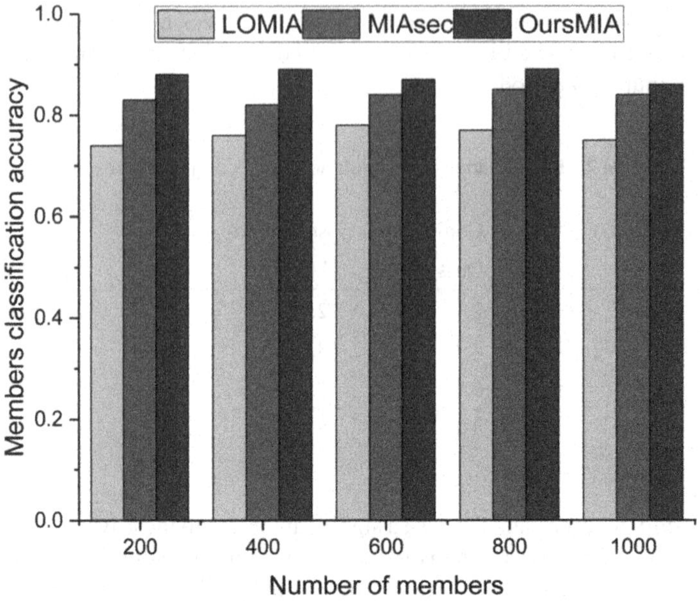

Fig. 5. Membership inference success ratio comparison results.

inferences from existing LOMIA and MIAsec attacks and the detection about our attack approximately reaches 30%.

5 Conclusion

In conclusion, membership inference attacks against medical databases pose significant threats to the privacy and security of individuals' health information. These attacks exploit vulnerabilities in machine learning models trained on sensitive medical data, allowing adversaries to determine whether an individual's data was included in the training set. The consequences of successful membership inference attacks can be severe, as they can lead to the identification of individuals, unauthorized access to their medical records, and potential misuse of their personal information. The increasing adoption of machine learning techniques in the healthcare sector has provided numerous benefits, such as improved diagnostics and personalized treatment options. However, it has also brought attention to the privacy risks associated with the use of sensitive medical data. Membership inference attacks exploit the unique patterns and statistical properties of individual data points present in the training set, enabling attackers to infer membership status with high accuracy.

References

1. Hai, P.H., Hoa, V.N.: A novel ensemble of support vector machines for improving medical data classification. Engineering Innovations **6710** (2023)

2. Wang, J., Li, X.: Secure medical data collection in the internet of medical things based on local differential privacy. Electronics **12**(2) (2023)
3. Aneta, P., Elina, V., Witold, M.: Medical data transformations in healthcare systems with the use of natural language processing algorithms. Applied Sciences **13**(2) (2023)
4. Oh, W., Nadkami, G.N.: Federated learning in health care using structured medical data. Advances in Kidney Disease and Health **30**(1), 416 (2023)
5. Yuan, Y., Ma, M., Luo, L., et al.: B-SSMD: a fine-grained secure sharing scheme of medical data based on blockchain. Security and Communication Networks (2022)
6. Xi, P., Zhang, X., Wang, L., et al.: A review of blockchain-based secure sharing of healthcare data. Applied Sciences **12**(15), 7912 (2022)
7. Ngabo, D., Wang, D., Iwendi, C., et al.: Blockchain-based security mechanism for the medical data at fog computing architecture of internet of things. Electronics **10**(17), 2110 (2021)
8. Lv, Z., Piccialli, F.: The security of medical data on internet based on differential privacy technology. ACM Transactions on Internet Technology (2020)
9. Yang, W., Wang, S., Hu, J., et al.: Securing mobile healthcare data: a smart card based cancelable finger-vein bio-cryptosystem. IEEE Access **6**, 3693936947 (2018)
10. Patil, A., Ashwini, D.V., Rashmi, P.T.R., et al.: A mobile cloud based approach for secure medical data management. International Journal of Computer Applications **119**(5) (2015)
11. Zhang, G., Liu, B., Zhu, T., et al.: Label-only membership inference attacks and defenses in semantic segmentation models. IEEE Transactions on Dependable and Secure Computing **20**(2), 14351449 (2023)
12. Hu, L., Li, J., Lin, G., et al.: Defending against membership inference attacks with high utility by GAN. IEEE Transactions on Dependable and Secure Computing **20**(3) (2023)
13. Di, W., Saiyu, Q., Yong, Q., et al.: Understanding and defending against White-box membership inference attack in deep learning. Knowledge-Based Systems **259**, 110014 (2023)
14. Kulynych, B., Yaghini, M., Cherubin, G., et al.: Disparate vulnerability to membership inference attacks. Proceedings on Privacy Enhancing Technologies **2022**(1) (2021)
15. Jia, J., Salem, A., Backes, M., et al.: MemGuard: Defending against Black-Box Membership Inference Attacks via Adversarial Examples. CoRR, abs/1909.10594 (2019)
16. Truex, S., Liu, L., Gursoy, M.E., et al.: Demystifying membership inference attacks in machine learning as a service. IEEE Transactions on Services Computing (2019)
17. Chen, W., Gaoyang, L., Haojun, H., et al.: MIASec: enabling data indistinguishability against membership inference attacks in MLaaS. IEEE Transactions on Sustainable Computing **5**(3), 365376 (2019)
18. Halfond, W.G., Viegas, J., Orso, A.: A classification of SQL-injection attacks and countermeasures. Proceedings of the IEEE International Symposium on Secure Software Engineering. IEEE **1**, 13–15 (2006)
19. Grubbs, P., Lacharité, M.S., Minaud, B., et al.: Learning to reconstruct: Statistical learning theory and encrypted database attacks. 2019 IEEE Symposium on Security and Privacy (SP). IEEE, pp. 1067–1083 (2019)
20. Fang, M., Damer, N., Boutros, F., et al.: Cross-database and cross-attack iris presentation attack detection using micro stripes analyses. Image Vis. Comput.Comput. **105**, 104057 (2021)

Application of ALMM Technology to Intelligent Control System for a Fleet of Unmanned Aerial Vehicles

Ewa Zeslawska[1(✉)], Zbigniew Gomolka[1(✉)], and Ewa Dydek-Dyduch[2]

[1] College of Natural Sciences, University of Rzeszow, Pigonia St. 1, 35-959 Rzeszow, Poland
{ezeslawska,zgomolka}@ur.edu.pl
[2] Computer Science and Biomedical Engineering, AGH University of Science and Technology, 30-010 Cracow, Poland
edd@agh.edu.pl

Abstract. The article is related to an intelligent information system for managing a fleet of Unmanned Aerial Vehicles (UAVs) while taking into account various dynamically changing constraints. Variability of the time intervals that are available for flights over certain areas located close to the airports is one of the essential constraints. The system must be developed very flexibly to easily accommodate changing both flight destinations and performance conditions. The authors propose application of Algebraic Logic Meta Modelling (ALMM) technology to design and implement the models and algorithms used in this system. The article presents part of the research carried out by the authors during the design of the aforementioned system. An Algebraic Logic (AL) model of UAV flights optimization scheduling problem is given. The execution of overflights in the circumpolar zone with the criterion of minimizing the total completion time of all tasks C_{max} is described. Then a hybrid algorithm for solving this problem and the results of the experiments carried out are presented. The component nature of the proposed approach allows easy transposition of the models and algorithms in case of more complex, additional assumption and restrictions referring to manage flights in real conditions.

Keywords: Dynamic task scheduling and optimization · Unmanned Aerial Vehicle · algebraic logical model · Intelligent decision technology

1 Introduction

The main objective of the proposed research is to develop modern air traffic synchronization and scheduling technology, taking into account the optimization of logistics tasks for unmanned flying systems. Existing air traffic monitoring and control systems (e.g., European ATM Master Plan, Pegasus) do not provide capabilities for fully autonomous systems for intelligent synchronization and

B. Luo et al. (Eds.): ICONIP 2023, CCIS 1963, pp. 26–37, 2024.
https://doi.org/10.1007/978-981-99-8138-0_3

scheduling of air traffic [1,2,17]. Smooth management of air traffic taking into account the criterion of safety, reduction of CO_2 emissions, elimination of noise pollution around airports while minimizing transportation costs is a complex problem with which research teams of the most renowned research centers and institutions around the world are struggling. A system that generates sequences of air traffic control decisions that will depend on dynamically changing process state coordinates would allow to overcome this problem increasing the improvement of flight safety and environmental protection while reducing transportation costs.

Classical discrete optimization algorithms do not provide algorithmic solutions in situations of dynamically changing process structure or emerging disturbances of various types that affect the process. In a typical aircraft logistics task optimization process, the process is affected by problems related to downtime (e.g., aircraft failures or delays in flight tasks), continuous control of fleet aircraft, their availability, optimum task planning, optimal use of the available potential, and available airspace capacity. In current air task planning systems, this is still a manual process that is supported by additional technical means [5–7,14,16]. The proposed technology makes it possible to synchronize and schedule traffic taking into account the time constraints therein and dynamically occurring disturbances, which are included in the algebraic-logical model describing the process. It has been shown in the research carried out so far that the use of mathematical modeling of real production processes allows the construction of intelligent systems that ensure that optimal or acceptable solutions are obtained in a given system [8,12].

The market lacks systems using synchronization and scheduling of air traffic with consideration of logistics task optimization for unmanned flying systems, which allows dynamic task scheduling under conditions of changing system resources [3,11,13,15]. The proposed technology refers to domestic and foreign discrete production processes (DPP) optimization systems, which, due to their versatility (Ms Dynamics, IBM Smart Business, Optima) and targeting a different audience, do not allow for application in the areas and tasks that are the purpose of this research. The key innovative element of the proposed approach is the use of a formalized algebraic-logical description of the optimized process. In the tasks of scheduling and synchronizing flight tasks, this allows the models to account for both changes in process structure and external disturbances while maintaining an automatic search for acceptable and/or optimal scheduling solutions in such situations. This represents a key technological advantage over classical algorithms, which require the redefinition of new process parameters, which in turn involves each time a complete modification of the software and its implementation under new operating conditions.

2 ALMM Technology to a Fleet of UAV

The proposed technology is related to the formalization and analysis of the problem of DPP based on the algebraic–logic paradigm of the meta-model of multistage decision-making processes. ALMM provides a unified model for creating a formal record of information and constraints associated with deterministic problems, the solution of which can be represented as a sequence of decisions. An algebraic logical meta-model is called a discrete process P which can be described by six components $P = (U, S, s_0, f, S_N, S_G)$, where: $f : U \times S \rightarrow S$ is a partial function (defined only for certain pairs), $(u, s) \in U \times S$ called a transition function, where: the set U is called the set of control decisions or control signals; the set $S = X \times T$ is called the set of generalized states, $T \subset R^+$ – a subset of non-negative real numbers representing moments in time, the set X is called the set of proper states. The sets U, X, T are non-empty.

The transition function f is defined by two functions f_x and f_t, where: $f_x : U \times X \times T \rightarrow X$ determines the next proper state, $f_t : U \times X \times T \rightarrow T$ determines the next moment of time and satisfies the following condition $\Delta t = f_t(u, x, t) - t > 0$ and takes a finite value. The generalized initial state $s_0 = (x_0, t_0)$, $s_0 \in S$, the set of nonadmissible states $S_N \subset S$, and the non-empty set of generalized admissible states, i.e., the states in which the process should find itself as a result of the action of the proper controls $S_G \subset S$ [9].

The main purpose of the ALMM system is not only to solve discrete optomechanical problems, but also to help the user choose the appropriate method and algorithm to solve them. The papers [8,10,12] describe the idea of the ALMM system and its formal description. The paper [4] presents the concept of using the Protégé environment to extract knowledge about the properties and characteristics of discrete problems described by ALMM technology. The authors, building on previous work on solvers based on the component structure using ALMM technology, invited the use of a description of the properties of individual problems to create a corresponding knowledge base. The paper [9] presents an implementation of a knowledge base supporting an intelligent system for solving optimization problems, in particular discrete manufacturing process optimization problems called intelligent Algebraic-Logical Meta-Model Solver. The paper [8] proposes a paradigm for modelling multistage decision-making processes using a machine learning method for solving NP-hard discrete optimization problems, in particular, planning and scheduling. The proposed learning method uses a purpose-built local multi-criteria optimization problem, which is solved using scalarization.

2.1 Description of the Problem at Hand

Classical discrete optimization algorithms do not provide algorithmic solutions in situations of dynamically changing process structure or emerging disturbances of various types that affect the process. The proposed optimization approach may have its applications in the fast-growing aerospace industry and in industries such as: task scheduling for a set of parallel machines, mining of mined fields, optimization of logistics tasks for unmanned aerial vehicles, synchronization and scheduling of air traffic with consideration of optimization of logistics tasks for unmanned aerial systems, in the process in which synchronization of discrete production processes affects the reduction of production costs, while maintaining the continuity of operational processes, in planning the selection of materials and raw materials in the process for complex production processes with consideration of set optimization criteria. The essence of the theory work is modeling combined with simulations for the problem of synchronization and scheduling of air traffic, taking into account the optimization of logistics tasks for unmanned aircraft systems (UAS). The feasibility of the model seems to be achievable assuming that the scheduling of unmanned aircraft traffic and air traffic interacts with it in the form of a disturbance.

In the transport problem under consideration, tasks are performed by a fleet of 3 to 10 UAVs. We assume that there is a defined take-off point, where UAVs start the process of waiting for task assignment (start air service), defined are the air destinations to which drones transport hypothetical cargo. Assigned tasks have predicted execution times, acceptable deadlines for their execution (depending on the complexity of the mathematical model, they could also have a defined priority). The task is performed continuously cannot be interrupted. The number of tasks to be performed changes dynamically, they arrive in real time. Figure 1 shows a schematic diagram of a system controlling a discrete transportation process for flexibly acting on a sequence of control decisions during changes occurring in the system's resources or in the structure of the state vector of the logistics process under consideration.

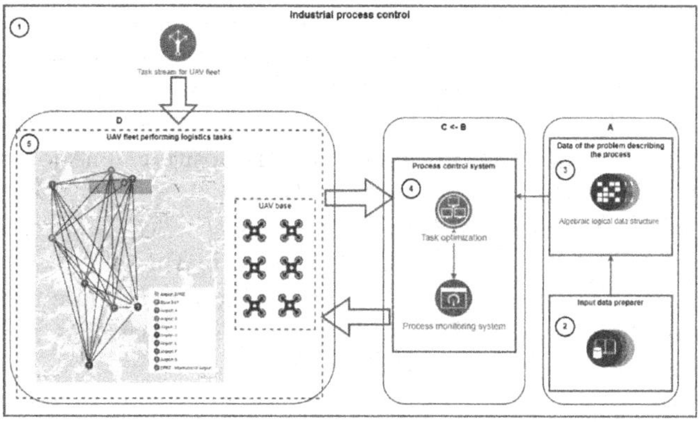

Fig. 1. Diagram of the system controlling the logistics process

Main block 1 responsible for the control of the transport process. Block 2 responsible for extracting the properties of the input data, which are necessary for the formation of the algebraic-logical description in block 3. Block 3 is responsible for transfer the properties of the input data provided by block 2 into the process control system 4. The format of the transmitted data takes into account the algebraic-logical structure describing the process. In addition, all other modules that make up block 3 are compatible in terms of exchanging information about the properties of the process during the execution of the tasks entered into block 1. The process monitoring system for allows dynamically determine the quantities: U, S, s_0, f, S_N, S_G. These quantities are necessary for on-line determination of subsequent decisions controlling the process. Block No. 5 is a supervised transport process, which in the present case is a fleet of UAVs. Individual UAV receive decisions controlling the operation of their autopilots, which determine their mode of operation, flight parameters, etc. What is an advantage of the proposed solution over other systems is that the set of decisions controlling the process changes dynamically due to the adopted algebraic–logical formula for describing its contents. Changes in the simulation models are made automatically and immediately with the emerging information about the change of constraints.

3 AL Model of the Considered Transport Process with Limitations on Available UAV Operating Time Periods

A finite set of tasks is given, represented by natural numbers $J = \{1, 2, \ldots, n\}$ and function $p = J \rightarrow R^{+}$, which determines processing times. Tasks are to be performed using identical UAVs in parallel, and exclusion of work is not allowed. UAV operation is possible only in designated available time periods $\left[B^1, F^1\right], \left[B^2, F^2\right], \ldots, \left[B^{ILs}, F^{ILs}\right]$, B denotes given slot beginning time, F denotes given slot end tine and ILs is the number of available time periods also refered to as slots. It is necessary to find the assignment of tasks to the UAV and the sequence of task execution, so that all tasks are completed before the end of the last period, i.e. the time of F^{ILs}. For notation convenience we introduce a flight number 0 with processing time of $p(0) = 0$ to represent no job assigned to a UAV. For all other jobs the processing time is positive. The job set plus the zero job will be denoted as J^{*}, hence $J^{*} = J \cup \{0\}$.

Let's define the elements that define the process $P = (U, S, s_0, f, S_N, S_G)$ and its interpretation. The actual state x is determined by the set of completed tasks and the states of individual UAVs. The decision is to determine the next tasks to be completed by individual UAVs. Thus, the decision values are the names (indexes) of the selected flight tasks. Formally:

A set of proper states $X = X^0 \times X^1 \times X^2 \times \cdots \times X^m$, the proper state is described by the vector $x = \left(x^0, x^1, x^2, \ldots x^m\right)$, where: $x^0 \subset 2^J$ – set of completed tasks, $x^k = \left(\beta^k, \tau^k\right)$, $k = 1, 2, \ldots, m$ – k-th UAV status,$\beta^k J^{*} \setminus x^0$, $k = 1, 2, \ldots, m$ – the number of the flight task that is currently being performed by k-th UAVs,$\tau^k \subset \mathbb{R}$, $k = 1, 2, \ldots, m$ – remaining time to complete the job (if $\tau^k \geq 0$); time $\tau^k \leq 0$ describes the state of the process when the task for the UAV has been completed and the UAV can receive another task, then $\beta^k = 0$. The initial state $s_0 = (x_0, t_0)$ is a state in which $t_0 = 0$ and $x_0 = (\varnothing, (0,0), (0,0), \ldots, (0,0))$. This means that no UAV has been assigned a task to perform and no flight task has been completed.

Set of goal states $S_G = \left\{(x,t) : x^0 = J \wedge t \leq F^{ILs}\right\}$ – describes a process that is in the goal state: all tasks are completed, and the time is earlier than or equal to the end time of the last F^{ILs} availability period. The set of nonadmissible states S_N represents two situations. First, it includes states (x, t) for which the time t does not belong to any availability period $[B^r, F^r]$, $r = 1, 2, \ldots, ILs$ while there is a UAV in which there is an uncompleted task, or such states in which not all tasks have been completed while the time is later than the end time of the last F^{ILs} slot (the last available period). For the formal notation, let us define a set of tasks processed by the UAV in state s (currently processed task) $J_M(s)$.

$$J_M(s) = \bigcup_{k=1}^{m} \beta^k \setminus \{0\} \tag{1}$$

Set of not admissible state:

$$\begin{aligned} S_N = \{ (x,t) : & \left(J_M(x,t) \neq \varnothing \wedge F^r < t < B^{r+1}, 1 \leq r \leq Ils - 1 \right) \\ & \cup \left(\exists_{j \in J} \notin x^0 \wedge t > F^{ILs} \right) \} \end{aligned} \tag{2}$$

Set of decisions $U \subset J^* \times J^* \cdots \times J^*$, $u = \left(u^1, u^2, \ldots, u^k\right)$ with the coordinate u^k denote the task number assigned to the k–th UAV as a result of the decision or no task assignment denoted as $u^k = 0$. Set of possible decisions $U_p(x,t) = \varnothing$ if $F^r < t < B^{r+1}$, $1 \leq rILs - 1 \cup t > F^{ILs}$, that is, when time t does not belong to any of the avail-able periods of time. So let's further assume that t belongs to one of the available slots. To define a possible decision set $U_p(s)$ let us introduce the notion of a decision state of a UAV and the notion of admissible set of decisions in a state s. We say that a state $s = (x,t)$ is a decision state for the $k-th$ given if for the state occurs $\tau^k \leq 0$, that is if a job processed by the $k-th$ UAV is completed while no new flight is assigned yet. Then, for a correctly developed model, $\beta^k = 0$. The decision state set for UAV k will be denoted as $S_d(k)$. Based on the definition of the set of nonadmissible states S_N, we can define the set of decisions admissible in state s denoted as $U_d(s) = U_d^1(s) \times U_d^2(s) \times \cdots \times U_d^n(s)$ as those that do not directly carry out the process from state s to a set of nonadmissible states.

$$U_d(s) = \{u \in U_p(s) : f(u,s) \notin S_N\} \tag{3}$$

Sets of possible decisions $U_p(s) = U_p^1(s) \times U_p^2(s) \times \cdots \times U_p^m(s)$, are defined by following properties:

- if $s \in S_d(k)$ then sets $U_p^k(s) \subset \left\{ \left(J \setminus x^0 \setminus J_M(s) \right) \cup \{0\} \right\}$ for $k = 1, 2, \cdots, m$ (only jobs that have not been completed yet and are not currently processed can be assigned for processing),
- if a given state $s = (x,t)$ is not a decision state for the $k-th$ UAV, that is $\tau^k > 0$, then set of possible decisions is $U_p^k(s) = \left\{\beta^k\right\}$,
- $u^1 = u^k \Leftrightarrow u^1 = 0 \ \wedge u^k = 0$ for k, $k = 1, 2, \ldots, m$, $k \neq 1$ (the same job cannot be assigned simultaneously to two different UAV; this property will be satisfied as long as $U_p(s)$ and the transition function are defined properly),
- We distinguish a special decision $u^* = (0, 0, \ldots, 0)$, which belongs to the set $U_p(s)$ only in exceptional states (x,t), namely when t "does not belong" to the last slot, the state is not the goal state (target state), it is the decision state for any UAV, and any other decision from the set U_p would be an nonadmissible

decision due to violation of the time constraint of the availability period. Correctly defining the transition function for the u^* decision allows further trajectories to be generated in subsequent permissible time periods without violating this constraint.

$$u^* \in U_p(s) \Leftrightarrow \bigvee_{1\le k\le m} (s \in S_d(k) \wedge s \notin S_G) \wedge \left(\forall_{u\neq u^*} (u \in U_p(s) \Rightarrow u \notin U_d(s)) \right) \quad (4)$$

Let us determine the transition function $f(u_i, x_i, t_i) = (x_{i+1}, t_{i+1})$. We need to bear in mind that the transition function is defined as a pair of $f = (f_x, f_t)$, $f_x(u_i, x_i, t_i) = x_{i+1}$, $f_t(u_i, x_i, t_i) = t_{i+1}$. Let $u_i = (u_i^1, \ldots, u_i^m)$, $x = (x_i^0, x_i^1, \ldots, x_i^m)$, $f_t = (u_i, x_i, t_i) = t_{i+1} = t_i + \Delta t_i$. We will define separately the transition function for $u \neq u^*$ and $u = u^*$. Let $u \neq u^*$ then $\Delta t_i = min^+ \{p(u_i^1), p(u_i^3), \ldots, p(u_i^m), \tau^1, \tau^2, \ldots, \tau^m\}$ where min^+ denotes minimization only for elements of positive values. The procedure for determining the next state for the accepted decision is related to the definition of the f_x function. For the process contemplated here, a transition function can be expressed as a vector function of the $f_x = (f_x^0, f_x^1, \ldots, f_x^m)^T$. Algorithms for functions: $f_x^0(u_i, s_i) = x_{i+1}^0$ and $f_x^k(u_i, s_i) = x_{i+1}^k$ for $k = 1, 2, \ldots, m$ differ.

In order to be able to provide an algorithm for the f_0^x function, let us define a subset of flight (job indices) completed in state $s_{i+1} = (x_{i+1}, t_{i+1})$ as a result of a decision u_i taken in a state x_i denoted as $J_{comp}(u_i, x_i, t_i)$. Such a set is a sum of two subsets: a subset of task started earlier (being processed) $J_{comp-x}(u_i, x_i, t_i)$, with completion time of Δt_i, and a subset $J_{comp-u}(u_i, x_i, t_i)$, containing jobs assigned under decision u_i, with processing time of Δt_i.

$$J_{cmp-x}(u_i, x_i, t_i) = \{\beta_i^k : \tau_i^k = \Delta t_i,\ where\ x_i^k = (\beta_i^k, \tau_i^k),\ 1 \le k \le m\} \quad (5)$$

$$J_{cmp-u}(u_i, x_i, t_i) = \{u_i^k : p(u_i^k) = \Delta t_i,\ 1 \le k \le m\} \quad (6)$$

$$J_{cmp}(u_i, x_i, t_i) = J_{cmp-x}(u_i, x_i, t_i) \cup J_{cmp-u}(u_i, x_i, t_i) \quad (7)$$

$$x_{i+1}^0 = x_i^i \cup J_{cmp}(u_i, x_i, t_i) \quad (8)$$

Algorithm for function $f_x^k(u_i, s_i) = x_{i+1}^k = (\beta_{i+1}^k, \tau_{i+1}^k)$ for $k = 1, 2, \ldots, m$ will be:

- if state s_i is not a decision state for the $k - th$ UAV i.e. $s_i \notin S_d(k)$, then $U_p^k(s_i) = \{\beta_i^k\}$, $\beta_{i+1}^k = \beta_i^k$ and $\tau_{i+1}^k = \tau_i^k - \Delta t$,
- if state s_i is a decision state for the $k - th$ UAV, that is $s_i \in S_d(k)$, then: $\beta_{i+1}^k = u_i^k$, $\tau_{i+1}^k = p(u_i^k)$.

The same in an abbreviated form:

$$x_{i+1}^{k} = \begin{Bmatrix} \left(\beta_i^k, \tau_i^k - \Delta t\right) \ for \ s_i \notin S_d\left(k\right) \\ \left(u_i^k, p\left(u_i^k\right)\right) \ for \ s_i \in S_d\left(k\right) \end{Bmatrix} \tag{9}$$

Let $u = u^*$ then, for the $r - th$ slot time $r \neq ILs$, $\Delta t = B^{r+1}$, $f_t\left(x_i, t_i\right) = t_{i+1} = B^{r+1}$, $f_x\left(x_i, t_i\right) = \left(x_{i+1}^0, x_{i+1}^1, \ldots, x_{i+1}^m\right)$, where $x_{i+1}^0 = x_i^0$, $x_{i+1}^k = (0, 0)$ for $k = 1, 2, \ldots, m$. For the last slot $r = ILs$, $u^* \notin U_p\left(s\right)$ (if for any state from this slot U_d is an empty set, then the trajectory is not admissible)

4 Results

The task of controlling the process of executing dynamically incoming aerial tasks by a fleet of drones that move concurrently according to the distribution of nodal points - individual aerial destinations - is defined. The aviation destinations for individual tasks can dynamically appear in the system, according to the properties of the process state vector. The drone fleet consisted of nine unmanned aerial vehicles (UAV1–UAV9). UAVs take off and land at a dedicated airport, nodal points are known. The tasks to be executed arrive in real time and are respectively defined as $J_1, J_2, \ldots, J_n$ tasks, which have a structure of one-stage and multistage target nodes. The execution of the assigned individual tasks for the UAV is carried out respectively at specific time intervals depending on the currently prevailing conditions and external interference, as well as taking into account conditions related to possible non-safe/collision situations of the UAV. Disturbances are due to the continuous traffic taking place at the airport and weather conditions. Examples of trajectory classes from the set of incoming flight tasks are shown in Table 1, while Fig. 2 shows the exemplary schedule of assigned tasks to the UAVs.

Table 1. Selected classes of flight trajectories for a set of incoming tasks.

No. of task	Trajectory class	The task to be accomplished (L – airport take-off/landing, A – F - nodal points)
J_1	one-stage	L–A –L
J_2	multistage	L – E – D – F – L
J_3	one-stage	L – C – L
J_4	multistage	L – D – B –A –L
J_5	multistage	L– A – B – C – L
J_6	multistage	L – E – F – E – L
J_7	multistage	L – F – E – C – D – L
J_8	multistage	L – E – C – A – L

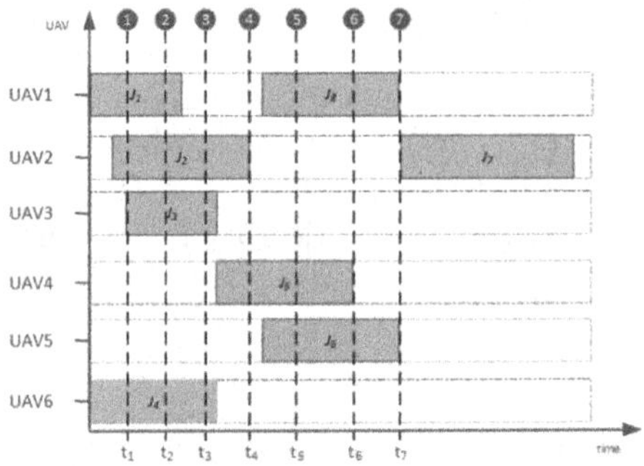

Fig. 2. Gantt chart for scheduling aviation tasks

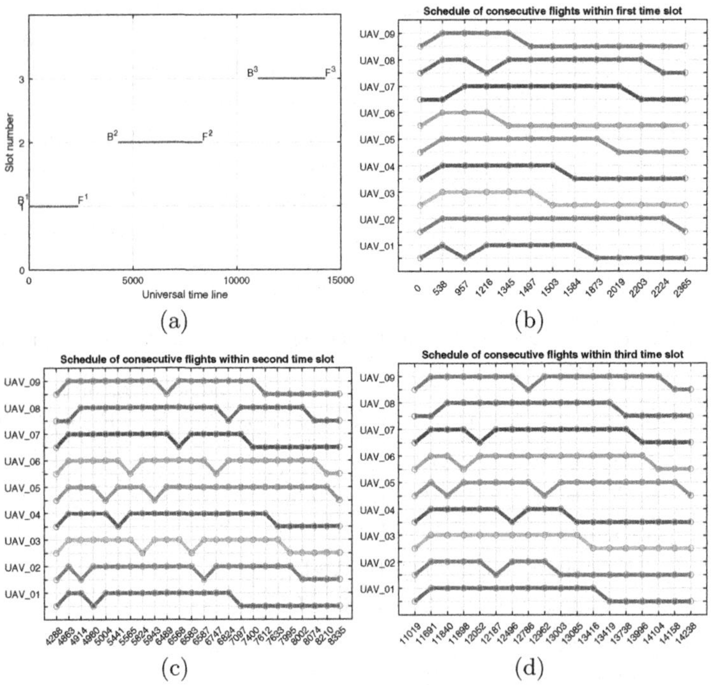

Fig. 3. A set of process trajectories generated in three time slot windows for performing set of flights tasks by a fleet of 9 drones

For the problem under consideration, a number of tests were conducted for different classes of trajectories, number of drones and tasks. These are only those trajectories that allow a process to reach the goal state. Unacceptable trajectories also referred to as non-admissible trajectories, which bring the process to forbidden states, for example, as a result of over-timing the task to be executed, have been omitted. However, they can provide additional information to search for acceptable solutions to protect the process from entering a forbidden states. Figure 3 show the results of simulation studies showing the solution trajectory for 52 tasks performed by a fleet drones, respectively, taking into account time constraints due to the availability of time slot windows. At this stage, the optimization criterion adopted is the shortest execution time for all tasks. As can be observed, the proposed system dynamically adapts to the form of the structure of the process state vector, allowing the search for optimal solutions. According to the assumptions of the ALMM model, all flight tasks have a processing time less than the completion time of the last available slot. During the real-time flights performed, all data were transmitted and collected, which were used in the process control system block, which is responsible for entering the properties of the data provided. Figure 4 shows the obtained trajectories of flights at individual moments of time (t_1, t_2, t_3).

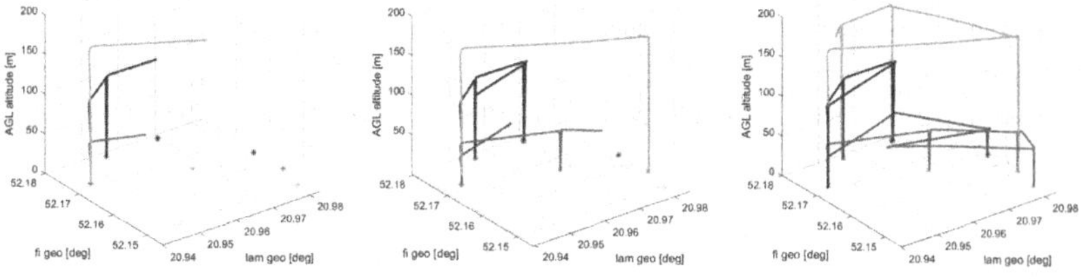

(a) (b) (c)

Fig. 4. Task completion status for t_1 (a), t_2 (b), t_3 (c) respectively

The proposed algebraic-logic technology makes it possible to optimize logistics tasks taking into account the time constraints in them and dynamically occurring disturbances of a certain type, which are included in the model. As part of the work carried out, it was shown that the use of mathematical modeling of real production processes using algebraic–logic technology allows the construction of intelligent algorithms that ensure the obtaining of solutions that are optimal or acceptable in a given system.

5 Conclusion

The proposed discrete optimization method for a fleet of UAVs has the unique ability of a system controlling a discrete industrial process to flexibly influence a sequence of control decisions during changes occurring in the system's resources or in the state vector structure of a given process. The task of the technology is to provide solutions (exact or approximate) for NP-hard discrete optimization problems, using the terminology of algebraic-logic description. The main idea of the proposed solution is to generate one or more task execution optimization trajectories (including partial trajectories, if necessary). The generation of trajectories is controlled by various general methods and/or specific optimization algorithms or algorithms for finding acceptable solutions. As a solution, the system provides a decision sequence (which determines the trajectory going from the initial state s_0 to the state belonging to S_G) or indicates that no solution has been found. The proposed approach provides functionality for problem modeling, problem optimization, process simulation for a given decision sequence and presentation of results (including visualization). A model of the discrete process of scheduling logistics tasks for unmanned aircraft reflecting the model of the real system was developed and described using ALMM technology. The proposed technology allows for intelligent scheduling of tasks carried out by a fleet of UAVs, and enables optimization of logistics tasks taking into account time constraints in them and dynamically occurring disturbances of a certain type, which were taken into account in the model.

The next work related to the Intelligent ALMM Technology will be related to the development of a set of flight tasks for real geographic destinations located in near airport zones for which it will be possible to introduce disruptions, e.g. time limitations, slots made available, changing number of aircraft on operational standby. Carry out a series of flights in real conditions using the existing drone fleet. Evaluate the effectiveness of intelligent optimization using the designed ALMM model of the decision-making process

References

1. Bolanowski, M., Paszkiewicz, A., Rumak, B.: Coarse traffic classification for high-bandwidth connections in a computer network using deep learning techniques. In: Barolli, L., Yim, K., Enokido, T. (eds.) CISIS 2021. LNNS, vol. 278, pp. 131–141. Springer, Cham (2021). https://doi.org/10.1007/978-3-030-79725-6_13
2. Cwiklak, J., Krasuski, K., Ambroziak, R., Wach, J.: Selected aspects of air navigation security (In Polish). Lotnicza Akademia Wojskowa (2023)
3. Djenouri, Y., Belhadi, A., Srivastava, G., Lin, J.C.W.: Hybrid graph convolution neural network and branch-and-bound optimization for traffic flow forecasting. Futur. Gener. Comput. Syst. **139**, 100–108 (2023). https://doi.org/10.1016/j.future.2022.09.018
4. Dudek-Dyduch, E., Gomolka, Z., Twarog, B., Zeslawska, E.: The concept of the ALMM solver knowledge base retrieval using Protégé environment. In: Zamojski, W., Mazurkiewicz, J., Sugier, J., Walkowiak, T., Kacprzyk, J. (eds.) DepCoS-RELCOMEX 2019. AISC, vol. 987, pp. 177–185. Springer, Cham (2020). https://doi.org/10.1007/978-3-030-19501-4_17
5. Ekici, A.: A large neighborhood search algorithm and lower bounds for the variable-sized bin packing problem with conflicts. Eur. J. Oper. Res. **308**(3), 1007–1020 (2023). https://doi.org/10.1016/j.ejor.2022.12.042

6. Fleszar, K.: A MILP model and two heuristics for the bin packing problem with conflicts and item fragmentation. Eur. J. Oper. Res. **303**(1), 37–53 (2022). https://doi.org/10.1016/j.ejor.2022.02.014
7. Fleszar, K.: A new MILP model and fast heuristics for the variable-sized bin packing problem with time windows. Comput. Ind. Eng. **175**, 108849 (2023). https://doi.org/10.1016/j.cie.2022.108849
8. Gomolka, Z., Dudek-Dyduch, E., Zeslawska, E.: Generalization of ALMM based learning method for planning and scheduling. Appl. Sci. **12**(24), 12766 (2022). https://doi.org/10.3390/app122412766
9. Gomolka, Z., Twarog, B., Zeslawska, E., Dudek-Dyduch, E.: Knowledge base component of intelligent ALMM system based on the ontology approach. Exp. Syst. Appl. **199**, 116975 (2022)
10. Grobler-Debska, K., Kucharska, E., Baranowski, J.: Formal scheduling method for zero-defect manufacturing. Int. J. Adv. Manuf. Technol., 4139–4159 (2022). https://doi.org/10.21203/rs.3.rs-315238/v1
11. Haouari, M., Mhiri, M.: Lower and upper bounding procedures for the bin packing problem with concave loading cost. Eur. J. Oper. Res. (2023). https://doi.org/10.1016/j.ejor.2023.06.028
12. Korzonek, S., Dudek-Dyduch, E.: Component library of problem models for ALMM solver. J. Inf. Telecommun. **1**(3), 224–240 (2017). https://doi.org/10.1080/24751839.2017.1347418
13. Martinovic, J., Strasdat, N., Valério de Carvalho, J., Furini, F.: A combinatorial flow-based formulation for temporal bin packing problems. Eur. J. Oper. Res. **307**(2), 554–574 (2023)
14. McLain, T.W., Beard, R.W.: Coordination variables, coordination functions, and cooperative timing missions. J. Guid. Control. Dyn. **28**(1), 150–161 (2005)
15. Missaoui, A., Ruiz, R.: A parameter-less iterated greedy method for the hybrid flowshop scheduling problem with setup times and due date windows. Eur. J. Oper. Res. **303**(1), 99–113 (2022). https://doi.org/10.1016/j.ejor.2022.02.019
16. Moshref-Javadi, M., Winkenbach, M.: Applications and research avenues for drone-based models in logistics: a classification and review. Exp. Syst. Appl. **177**, 114854 (2021)
17. SESAR: eATM Portal. https://www.atmmasterplan.eu/

Bloomfilter-Based Practical Kernelization Algorithms for Minimum Satisfiability

Chao Xu([✉]), Liting Dai, and Kang Liu

School of Computer and Communication Engineering, Changsha University of Science and Technology, Changsha 410003, Hunan, People's Republic of China
xuchaofay@163.com

Abstract. Minimum Satisfiability problem (briefly, given a CNF formula, find an assignment satisfying the minimum number of clauses) has raised much attention recently. In the theoretical point of view, Minimum Satisfiability problem is fixed-parameterized, by transforming into Vertex Cover. However, such kind of transformation would be time-consuming, which takes $O(m^2 \cdot n)$ times to transform into Vertex Cover. We first present a $O(m^2)$ filtering algorithm to transform MinSAT into Vertex cover with low false positive rate, by utilizing Bloom Filter structure. And then, instead of transformation to Vertex Cover, we present a practical kernelization rule directly on the original formula which takes time of $O(L \cdot d(F))$, with a kernel size of $k^2 + k$.

Keywords: MinSAT · Bloomfilter · Kernelization

1 Introduction

Over a decade, optimization versions of Satisfiability problem have been an interesting challenging task in theoretical computer science. A most famous and extensively studied optimization version of Satisfiability problem is the Maximum Satisfiability problem(MaxSAT), whose parameterized version is defined as follows.

> MaxSAT: Given a CNF formula F and an integer k (the *parameter*), is there an assignment to the variables in F that satisfies at least k clauses in F?

Garey *et al.* [1] first proved it is NP-hard, and Jianer *et al.* [2], according to Mahajan and Raman's preprocessing rules (also called reduction rules) [3], presented a kernel of size $O(k^2)$ literals, and the latest upper bound of exact and parameterized algorithm is $O^*(1.325^k)$ [4][1].

[1] Use the notation $O^*(c^k)$ to denote the bound $c^k n^{O(1)}$, where c is a positive number and n is the instance size.

This paper was supported by Supported by the National Natural Science Foundation of China under Grants (62002032, 62302060, 62372066) and Natural Science Foundation of Hunan Province of China under grant 2022JJ30620.

B. Luo et al. (Eds.): ICONIP 2023, CCIS 1963, pp. 38–47, 2024.
https://doi.org/10.1007/978-981-99-8138-0_4

However, it is well-known that the parameter k for the Maximum Satisfiability problem is quite large, such that k is always not less than half of the number of clauses [3]. Consequently, such parameterized version may not be efficient enough as the exact algorithms in terms of the number of clauses.

Here, we investigate another optimization version of Satisfiability problem, with smaller parameter (as the number of the satisfied clauses): the Minimum Satisfiability problem. This problem is to find an assignment satisfying the minimum number of clauses. In 1994, Kohli *et al.* [5] firstly proved that this problem is NP-hard and gave an approximation algorithm with performance guarantee equal to the maximum number of literals in a clause. Later, Marathe *et al.* [6] gave a 2-approximation algorithm for this problem, and Arif *et al.* [7] gave a $2(1 - 1/2^k)$-approximation algorithm for the Minimum k-Satisfiability problem (i.e. at most k literals in each clause).

Since 2002, researchers have started to fill the gap between the Minimum Satisfiability problem and its applications, including the Answer Set Programming [7] and the conditional minisum approval voting [8]. They have presented a few practical heuristic algorithms for the Minimum Satisfiability problem [9–13]. Moreover, Li *et al.* [13] presented a branch-and-bound algorithm for a general version of the Minimum Satisfiability problem–weighted partial MinSAT problem, and gave a famous weighted partial MinSAT solver–MinSatz. An instance of the Weighted Partial MinSAT problem is a CNF formula ϕ where some clauses are hard and the remaining ones are soft, every clause have a weight. The goal is to find an assignment which satisfies all hard clauses and minimizes the sum of weights of the satisfied soft clauses.

In 1999, Mahajan and Raman [3] in the conclusion section, which introduces some open problems, mentioned that the Minimum Satisfiability problem is fixed-parameterized, i.e., it is solvable in time of $O^*(f(k))$ for a function f that only depends on the number k of satisfied clauses. Bliznets et. al. [14] show that partial MinSAT is W[1]-hard when parameterized by $h + k$, where h is the number of hard clauses and k is the number of satisfied soft clauses.

Formally, the (parameterized) Minimum Satisfiability problem (in short, MinSAT) is defined as follows.

> MinSAT: Given a CNF formula F and an integer k (the *parameter*), is there an assignment to the variables in F that satisfies at most k clauses in F?

One motivation of this paper is on how to find a practical kernelization algorithm for the MinSAT problem. Since up to our acknowledgement, the kernelization algorithm by transforming MinSAT to Vertex Cover problem cost too much time, and it is not applicable for practical solver. Note that most of the preprocessing rules, including resolution principles [15], for the Satisfiability problem and the Maximum Satisfiability problem, are not directly applicable for the MinSAT problem. This paper obtains a kernel of size $k^2 + k$ clauses, by presenting a preprocessing rule for the MinSAT problem.

2 Preliminary

A (Boolean) *variable* x can be assigned value either 1 (TRUE) or 0 (FALSE). A variable x has two corresponding literals: the *positive literal* x and the *negative literal* $\bar{x}$, which is called the *literals* of x. A *clause* C is a disjunction of a set of literals, which can be regarded as a set of the literals. Hence, we may write $C_1 = zC_2$ to indicate that the clause C_1 consists of the literal z plus all literals in the clause C_2, and use C_1C_2 to denote the clause that consists of all literals in either C_1 or C_2, or both. Without loss of generality, we assume that a literal can appear in a clause at most once. A clause C is *satisfied* by an assignment if under the assignment, at least one literal in C gets a value 1. A (CNF Boolean) *formula* F is a conjunction of clauses C_1, ..., C_m, which can be regarded as a collection of the clauses. The formula F is *satisfied* by an assignment to the variables in the formula if all clauses in F are satisfied by the assignment. Throughout this paper, denote by n the number of variables and by m the number of clauses in a formula.

A literal z is an (i, j)*-literal* in a formula F if z appears i times and $\bar{z}$ appears j times in the formula F. A variable x is an (i, j)*-variable* if the positive literal x is an (i, j)-literal. Therefore, a variable x has degree h, also called an h*-variable*, if x is an (i, j)-variable such that $h = i + j$. A variable is an h^+*-variable* if its degree is at least h.

The *size* of a clause C is the number of literals in C. A clause is an h*-clause* if its size is h, and an h^+*-clause* if its size is at least h. A clause is *unit* if its size is 1 and is *non-unit* if its size is larger than 1. The *size* of a CNF formula F is equal to the sum of the sizes of the clauses in F.

An instance (F, k) of the MINSAT problem asks whether there is an assignment to the variables in a given CNF formula F that satisfies at most k clauses in F.

Hashing refers to mapping multiple elements to the corresponding location of the storage cell through a Hash function. The query of elements can be quickly realized by hashing, so it is widely used in various fields of computers. The Bloom Filter was proposed by Bloom in 1970 [19] and consisted of a binary vector and $r \geq 2$ hash mapping functions. Element e through r mapping functions $h_1(e), \cdots, h_r(e)$ is mapped to a binary vector H(e)=$\sum_{k=1}^{r} 2^{h_k(e)}$. Nevertheless, in this paper, we make a small change to Bloom Filter: a hash function corresponds to a binary vector. Alternatively, we cut a full binary vector into 2 segments, and each hash function corresponds to a segment of the binary vector.

3 Kernelization

A *parameterized problem* is a subset $L \subseteq \Sigma^* \times N$ over a finite alphabet Σ. Given a parameterized problem L, a *kernelization* of L is a polynomial-time algorithm that maps an instance (x, k) to an instance (x', k') (the *kernel*) such that: (1) $(x, k) \in L$ if and only if $(x', k') \in L$, (2) k' is bounded by a polynomial $h(k)$ of k, and (3) $|x'|$ is bounded by another polynomial $g(k)$ of k. It is famous to see that a decidable parameterized problem L is fixed-parameter tractable if and only if it has a *kernel*. By strengthening Condition (2) with $k' \leq k$, we obtain a *proper kernel*.

3.1 Previous Kernelization

We now show that any instance of the MinSAT problem can be solved by reducing the problem to the vertex cover problem.

Definition 1 *([6]). Let I be an instance of MinSAT consisting of the clause set C_I, and variable set X_I. The auxiliary graph $G_I(V_I, E_I)$ corresponding to I is constructed as follows. The node set V_I is in one-to-one correspondence with the clause set C_I. For any two nodes v_i and v_j in V_I, the edge (v_i, v_j) is in E_I if and only if the corresponding clauses c_i and c_j are such that there is a variable $x \in X_I$ that appears in uncomplemented form in c_i and in complemented form in c_j or vice versa*

Proposition 1 *([16]). There is an algorithm of running time $O(kn + k^3)$ that, given an instance (G, k) of the Vertex Cover problem where $|G| = n$, constructs another instance (G_1,k_1) of Vertex Cover problem with $k_1 \leq k$ and $|G_1| \leq 2k_1$.*

Lemma 1 *([6]). Let I be an instance of MinSAT with clause set C_t and let G_t be the corresponding auxiliary graph.*
(1) Given any truth assignment for which the number of satisfied clauses of the MinSAT instance I is equal to k, we can find a vertex cover of size k for G_t.
(2) Given any vertex cover C' of size k for G_t, we can find a truth assignment that satisfies at most k clauses of the MinSAT instance I.

It is noted that for the auxiliary graph $G_I(V_I, E_I)$, V_I, corresponding to a set of m clauses, contains m vertices, instead of n vertices; and the edge set E_I might contain at many as $O(m^2)$ edges.

Now we show that to our best acknowledge, this transformation would need time $O(m^2 n)$, using KMP algorithm [17] (Fig. 1).

Theorem 1. *The kernelization algorithm with kernel of size $2k$ runs in time of $O(max\{m^2 n, km + k^3\})$, by transforming MinSAT problem into Vertex Cover problem.*

Proof. By constructing the auxiliary graph, for each two clauses C_i and C_j, check whether there is any complementing pair of literals, if there is, then there is an edge in the auxiliary graph. According to KMP algorithm, it checks for complementing pair of literals in time of $O(n)$. Thus, it needs $O(m^2 n)$ running time to construct the auxiliary graph.

Then since there are m vertices in the auxiliary graph, according to Proposition 1, it needs $O(km + k^3)$ running time to get a kernel.

It is noted that for each non-trivial instance, $k \geq m/2$, so that the kernelization algorithm runs in time of $O(max\{m^2 n, km + k^3\})$.

Algorithm Trans-MinSAT-VC(F, k)
INPUT: a parameterized instance of CNF formula F with m clauses, n variables and a parameter k
OUTPUT: a parameterized instance of Vertex Cover with auxiliary graph $G_I(V_I, E_I)$, which contains vertex set V_I with m vertices, Edge set E_I, and a parameter k

1. **for each** clause C_i
2. **for each other** clause C_j
3. comparing C_i with C_j using KMP algorithm.
4. **if** there is some literal $x_i \in C_i$ and $\bar{x}_i \in C_j$;
5. **then** let edge $E_I[i, j] == 1$ in auxiliary graph $G_I(V_I, E_I)$;

Fig. 1. The algorithm MINSAT problem reduced to the VERTEX COVER problem

Remark that since the number m of clauses might be as many as $c \cdot n^2$, the transformation algorithm **Trans-MinSAT-VC** would be run in time as many as $O(n^5)$. In this way, since the number n of variables in the benchmark of MinSAT/MaxSAT problem[2] would be as many as one million, this transformation algorithm would not be practical to reduce redundant clauses, although the size of it's kernel is $2k$.

3.2 Our Bloomfilter-Based Kernelization with Auxiliary Graph Transformation

In this part, we try to apply Bloom filter technique to accelerate the kernelization algorithm of size $2k$, by utilizing four hash arrays for each clause. For each clause $C_i = x_{i1}\bar{x}_{i2}\cdots x_{ij}$ in F, two arrays, with distinct hash functions H_1 and H_2, represent the literals in the clauses; and the other two hash arrays, with distinct hash functions H_1 and H_2, represent all the complement literals in the clauses.

Denote $H_1(C_i)$ (resp. $H_2(C_i)$) be an array such that each literal $l \in x_{i1}\bar{x}_{i2}\cdots x_{ij}$ as a number is hashed by hash function $H_1(l)$ (resp. $H_2(l)$) into the array. Also denote $H_1(\bar{C}_i)$ (resp. $H_2(\bar{C}_i)$) be an array such that each complement literal in C_i, that is $l' = \bar{x}_{i1}x_{i2}\cdots\bar{x}_{ij}$ in $\bar{C}_i$ as a number, is hashed by hash function $H_1(l')$ (resp. $H_2(l')$) into the array.

Reminded that when $C_i = \bar{x}_5x_{10}x_{18}\bar{x}_{21}x_{47}$, $H_1(\bar{C}_i)$ is to hash$x_5\bar{x}_{10}\bar{x}_{18}x_{21}\bar{x}_{47}$ into a new array, different from the Hash array $H_1(C_i)$. According to Fig. 2, for the literal x_{26}, $H_1(x_{26}) = 1$ means that the literal x_{26} belongs to clause C_i and presents a false positive. According to Fig. 3, for the literal x_{26}, $H_1(x_{26}) = 1$ but $H_2(x_{26}) = 0$, it can be judged that literal x_{26} is not in the clause C_i. Note that the existence of the element is determined only if two hash values are 1.

[2] https://maxsat-evaluations.github.io/.

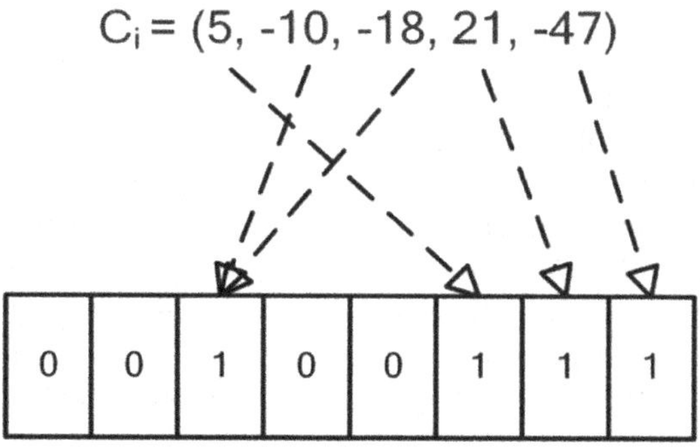

Fig. 2. A Hash array $H_1(C_i)$, where $C_i = \bar{x}_5 x_{10} x_{18} \bar{x}_{21} x_{47}$.

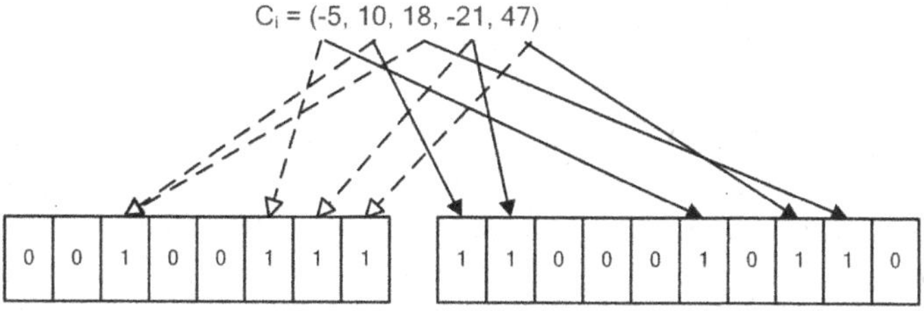

Fig. 3. Two Hash arrays $H_1(C_i)$ and $H_2(C_i)$, where $C_i = \bar{x}_5 x_{10} x_{18} \bar{x}_{21} x_{47}$. (The dashed line represents the hash array $H_1(C_i)$, and the solid line represents the hash array $H_2(C_i)$)

Corollary 1 *([18,19]). By utilizing Bloom filter, two clauses with complemented literals would be found in time $O(1)$ by CVDA-Bloom algorithm in the rate up to 90%.*

Theorem 2. *By utilizing Bloom Filter structure, for most of the cases, the Transformation algorithm Trans-MinSAT-VC(F, k) would be run in time of $O(m^2)$, with low false positive rate.*

Proof. *For each clause $C_i = x_{i1}\bar{x}_{i2}\cdots x_{ij}$, we set two hash arrays $H_1(C_i), H_2(C_i)$, with two distinct hash functions, to represent all its literals $x_{i1}\bar{x}_{i2}\cdots x_{ij}$; and set another two hash arrays $H_1(\bar{C}_i), H_2(\bar{C}_i)$ to represent all its complemented literals $\bar{x}_{i1}x_{i2}\cdots\bar{x}_{ij}$ in $\bar{C}_i$. Therefore, by properly set hash functions and the size of hash arrays, according to the experiment [18], up to 90% of two clauses without complemented literals could be filtered in time of $O(1)$, instead of utilizing KMP algorithm in time of $O(n)$. By constructing the auxiliary graph, for each two vertices, which represent corresponding two clauses, we might check in time $O(1)$ whether one clause contains some literal x_i, and the other contains its complemented literal. Therefore, the Transformation algorithm Trans-MinSAT-VC(F, k) would be run in time of $O(m^2)$.*

3.3 Practical Kernelization

Since the transformation by auxiliary graph costs too much time, and for most of the practical MaxSAT solver, for efficiency, it is preferred directly preprocessing on the formula instead of auxiliary graphs.

Here, our kernelization algorithm runs a *preprocessing rule* exhaustively. For each *preprocessing rule*, we show that it transforms, in polynomial time, an instance (F, k) of the MINSAT problem into another instance (F', k') with $k \geq k'$ such that (F, k) is a Yes-instance if and only if (F', k') is a Yes-instance.

R-Rule 1. If there is a clause $C = (x_1 x_2 \cdots x_l)$, such that there are at least $k+1$ clauses containing at least one literal from $\bar{x}_1, \cdots, \bar{x}_l$, then reduces clause C in the given formula, so that $(F \wedge C, k) \rightarrow (F, k-1)$.

Lemma 2. *R-Rule 1 transforms the instance $(F \wedge C, k)$ of the* MINSAT *problem into an instance $(F, k-1)$ such that $(F \wedge C, k)$ is a Yes-instance if and only if $(F, k-1)$ is a Yes-instance.*

Proof. If $(F, k-1)$ is a Yes-instance, then $(F \wedge C, k)$ is a Yes-instance. If $(F, k-1)$ is a Yes-instance, then at least one clause is not satisfied, so that at least one literal from $\bar{x}_1, \cdots, \bar{x}_l$ is not satisfied. Therefore, $(F \wedge C, k)$ is a Yes-instance.

Suppose that $(F \wedge C, k)$ is a Yes-instance but $(F, k-1)$ is a No-instance (i.e. there is an assignment σ_1 that satisfies at most k clauses in $F \wedge C$, but assignment σ_1 satisfies at least k clauses in F). Hence, assignment σ_1 satisfies k clauses in F and $\sigma_1(C) = 0$. However, when $C = 0$, there is at least $k+1$ clauses containing at least one literal from $\bar{x}_1$ to $\bar{x}_l$, denoted that $C = (x_1 \cdots x_l)$, so that assignment σ_1 satisfies at least the $k+1$ clauses in $F \wedge C$, which contradicts the fact that assignment σ_1 satisfies at most k clauses in $F \wedge C$. Therefore, $(F \wedge C, k)$ is a Yes-instance, then $(F, k-1)$ is a Yes-instance.

Let $d(l)$ be the number of occurrence of literal l. Let $d(F)$ be the maximum number of occurrence of each literal in formula F. Now we show that R-Rule 1 runs in time $O(L \cdot d(F))$.

Lemma 3. *R-Rule 1 runs in time $O(L \cdot d(F))$, using an auxiliary space A of size $2n + m$.*

Proof. Firstly, construct the position set P_{x_i} of clauses containing each literal x_i in linear time O(L), similar to the method using in Maxsatz solver. For example, for the i-th clause $C_i = (x_1 \cdots x_l)$, fill the number i into the lists $P_{x_i}, \cdots, P_{x_l}$.

Then, for each clause $C_i = (x_1 \cdots x_l)$, let $Cl[i][m]$ be an auxiliary array to count the number of clauses containing literal from set $\{\bar{x}_1, \cdots, \bar{x}_l\}$, and $Cl[i][0]$ is the number of clauses containing literal from set $\{\bar{x}_1, \cdots, \bar{x}_l\}$. Firstly, $Cl[i][0...m] = \{0\}$. For each literal x_j in the clause $(x_1 \cdots x_l)$, we know the position of each clause j' containing $\bar{x}_j$, with $j' \in P_{\bar{x}_i}$. Thus, For each clause j' containing $\bar{x}_j$, if auxiliary array $Cl[i][j'] == 0$, set 1 to auxiliary array $Cl[i][j']$ and add 1 to $Cl[i][0]$. If $Cl[i][0] \geq k$, then clause C_i is redundant, R-Rule 1 should be applied and delete clause C_i. As a whole, for each literal in the given *formula F, add at most $d(F)$ times. As a result, the running time of R-Rule 1 is $O(L \cdot d(F))$.*

Now we show the problem kernel in this section. Once R-Rule 1 is applicable on a formula F, we apply the rule, which either decreases the parameter value k. A formula F is *irreducible* if R-Rule 1 is applicable on F. It is obvious that each of R-Rules 1 takes polynomial time, and that these rules can be applied at most polynomial times. Thus, with a polynomial-time preprocessing, we can always reduce a given instance into an irreducible instance. Therefore, without loss of generality, we can assume that the formula F in our discussion is always irreducible.

Next, we show the irreducible instance admits a kernel of size $k^2 + k$.

Theorem 3 (problem kernel). *Given a formula F and a positive integer k, then in polynomial time(R-Rule 1), we can compute a formula G consisting of m' clauses and a positive $k' \leq k$ with $m' \leq k'^2 + k'$, such that F has an assignment satisfying at most k clauses if and only if G has an assignment satisfying at most k' clauses.*

Proof Since R-Rule 1 is safe, in polynomial time, we can compute a irreducible formula G consisting of m' clauses and a positive $k' \leq k$, such that F has an assignment satisfying at most k clauses if and only if G has an assignment satisfying at most k' clauses.

Now we show $m' \leq k'^2 + k'$. Let G be an irreducible Yes-instance for MinSAT problem, such that there is an assignment A satisfying at most k' clauses in G. Since G is an irreducible Yes-instance of MinSAT problem, no pure literal, so that for each variable x in G, there is a satisfied literal (i.e. literal x if $A(x) = 1$, or $\bar{x}$ if $A(x) = 0$) of variable x contained in the k' clauses. For each clause $C = (x_1 \cdots x_l)$ satisfied by A, w.l.o.g., suppose there are i satisfied literals $A(x_1) = \cdots = A(x_i) = 1$ and $A(x_{i+1}) = \cdots = A(x_l) = 0$, $1 \leq i \leq l$, then all clauses containing any literal from x_1 to x_i and from $\bar{x}_{i+1}$ to $\bar{x}_l$, are contained in the k' clauses satisfied by A. Since R-Rule 1 is not applicable in G, so that there is at most k' clauses containing at least one literal from $\bar{x}_1$ to $\bar{x}_i$. Hence, for each clause $C = (x_1 x_2 \cdots x_l)$ satisfied by A with $A(x_1) = \cdots = A(x_i) = 1$ and $A(x_{i+1}) = \cdots = A(x_l) = 0$, at most k' clauses containing a variable from x_1 to x_i not satisfied by A. Note that there are k' clauses in G satisfied by A, and for each variable in G, there is a satisfied literal by A of this variable in the satisfied clauses. Therefore, at most k'^2 clauses in G not satisfied by A. Besides the satisfied clauses, the number m' of clauses is at most $k'^2 + k'$.

Therefore, in polynomial-time, we simplify the problem instance, and obtain the problem kernel of size $k^2 + k$ clauses, i.e. after preprocessing rules, either there is at least $k+1$ clauses satisfied and returns No-instance, or there is a quick solution satisfied at most k clauses and returns Yes-instance, or the number of clauses left are not more than $k^2 + k$

4 Conclusion and Discussion

In this paper, for the MinSAT problem, firstly we directly utilizing Bloom Filter to accelerate the transformation algorithm from $O(m^2n)$ to $O(m^2)$, with low false positive rate. Secondly, by introducing a preprocessing rules a faster kernelization algorithm of running time of $O(L \cdot d(F))$ is presented, with a larger kernel size $O(k^2)$ clauses. Although the kernel size is larger, but its running time turns to be more acceptable than the previous one, which transforms MinSAT into Vertex Cover.

It is noted that, by using auxiliary graph, it could transform MinSAT problem into Vertex Cover problem, and there is a kernel of size $2k$ vertices for Vertex Cover problem. However, these $2k$ vertices equal to $2n$ variables. And in terms of the number of clauses, the kernel is still $O(k^2)$ clauses. By the way, by constructing the auxiliary graph, for each variable x_i, there might be m/p_i positive literals x_i, and m/n_i negative literals $\bar{x}_i$, we might check $(m/p_i) \cdot (m/n_i) \cdot n$ times to construct the auxiliary graph. Thus, it need at least $O(m^2 \cdot n)$ running time to construct the auxiliary graph. As a result, the kernelization algorithm with a kernel of k^2 clauses by using auxiliary graph is not practical in Minimum Satisfiability, since there are always tens of thousands of clauses (the number m too large) in most of the benchmarks in MinSAT Competition. Meanwhile, in our kernelization algorithm, although with a larger kernel of size $k^2 + k$, it is practically usable, since it only needs time of $O(L \cdot d(F))$.

References

1. Garey, M., Johnson, D.: Computers and Intractability: A Guide to the Theory of NP-Completeness. W.H. Freeman and Company, New York (1979)
2. Chen, J., Kanj, I.: Improved exact algorithms for Max-SAT. Discret. Appl. Math. **142**(1–3), 17–27 (2004)
3. Mahajan, M., Raman, V.: Parameterizing above guaranteed values: MaxSat and MaxCut. J. Algorithms **31**(2), 335–354 (1999)
4. Chen, J., Xu, C., Wang, J.: Dealing with 4-variables by resolution: an improved MaxSAT algorithm. Theoret. Comput. Sci. **670**, 33–44 (2017)
5. Kohli, R., Krishnamurti, R., Mirchandani, P.: The minimum satisfiability problem. SIAM J. Discret. Math. **7**(2), 275–283 (1994)
6. Marathe, M.V., Ravi, S.S.: On approximation algorithms for the minimum satisfiability problem. Inf. Process. Lett. **58**(1), 23–29 (1996)
7. Arif, U., Benkoczi, R., Gaur, D.R., et al.: A primal-dual approximation algorithm for Minsat. Discret. Appl. Math. **319**, 372–381 (2022)
8. Markakis, E., Papasotiropoulos, G.: Computational aspects of conditional minisum approval voting in elections with interdependent issues. In: Proceedings of the 22nd International Joint Conference on Artificial Intelligence, pp. 304–310 (2020)
9. Ansotegui, C., Li, C.M., Manyá, F., et al.: A SAT-based approach to MinSAT. In: CCIA, pp. 185–189 (2012)
10. Li, C.M., Xiao, F., Manyá, F.: A resolution calculus for MinSAT. Logic J. IGPL **29**(1), 28–44 (2021)

11. Heras, F., Morgado, A., Planes, J., et al.: Iterative SAT solving for minimum satisfiability. In: 2012 IEEE 24th International Conference on Tools with Artificial Intelligence, Athens Greece, pp. 922–927. IEEE(2012)
12. Li, C.M., Manyà, F., Quan, Z., Zhu, Z.: Exact MinSAT solving. In: Strichman, O., Szeider, S. (eds.) SAT 2010. LNCS, vol. 6175, pp. 363–368. Springer, Heidelberg (2010). https://doi.org/10.1007/978-3-642-14186-7_33
13. Li, C.M., Zhu, Z., Manya, F., et al.: Minimum satisfiability and its applications. In: Proceedings of the Twenty-Second International Joint Conference on Artificial Intelligence, pp. 605–610. AAAI Press (2011)
14. Bliznets, I., Sagunov, D., Simonov, K.: Fine-grained complexity of partial minimum satisfiability. In: Proceedings of the Thirty-First International Joint Conference on Artificial Intelligence, pp. 1774–1780 (2022)
15. Davis, M., Putnam, H.: A computing procedure for quantification theory. J. ACM **7**(3), 201–215 (1960)
16. Chen, J., Kanj, I., Jia, W.: Vertex cover: further observations and further improvements. J. Algorithms **41**(2), 280–301 (2001)
17. Knuth, D.E., Morris, J.H., Pratt, V.R.: Fast pattern matching in strings. SIAM J. Comput. **6**(2), 323–350 (1997)
18. Wang, Z., Liu, K., Xu, C.: A bloom filter-based algorithm for fast detection of common variables. In: Proceedings of the 1st International Conference on the Frontiers of Robotics and Software Engineering (FRSE) (2023, accepted)
19. Bloom, B.H.: Space/time tradeoffs in hash coding with allowable errors. Commun. ACM **13**(7), 422–426 (1970)

TPTGAN: Two-Path Transformer-Based Generative Adversarial Network Using Joint Magnitude Masking and Complex Spectral Mapping for Speech Enhancement

Zhaoyi Liu[1], Zhuohang Jiang[2], Wendian Luo[2], Zhuoyao Fan[2], Haoda Di[2], Yufan Long[1], and Haizhou Wang[1](✉)

[1] School of Cyber Science and Engineering, Sichuan University, Chengdu 610207, China
{iamv,longyufan}@stu.scu.edu.cn, whzh.nc@scu.edu.cn
[2] College of Computer Science, Sichuan University, Chengdu 610207, China
{jzh,luowendian,fanzhuoyao,dihaoda}@stu.scu.edu.cn

Abstract. In recent studies, conformer is extensively employed in speech enhancement. Nevertheless, it continues to confront the challenge of excessive suppression, especially in human-to-machine communication, attributed to the unintended loss of target speech during noise filtering. While these methods may yield higher Perceptual Evaluation of Voice Quality (PESQ) scores, they often exhibit limited effectiveness in improving the signal-to-noise ratio of speech which is proved vital in automatic speech recognition. In this paper, we propose a two-path transformer-based metric generative adversarial network (TPTGAN) for speech enhancement in the time-frequency domain. The generator consists of an encoder, a two-stage transformer module, a magnitude mask decoder and a complex spectrum decoder. Encoder and two-path transformers characterize the magnitude and complex spectra of the inputs and model both sub-band and full-band information of the time-frequency spectrogram. The estimation of magnitude and complex spectrum is decoupled in the decoder, and then the enhanced speech is reconstructed in conjunction with the phase information. Through the implementation of intelligent training strategies and structural adjustments, we have successfully showcased the remarkable efficacy of the transformer model in speech enhancement tasks. The experimental results on the Voice Bank+DEMAND dataset illustrate that TPTGAN shows superior performance compared to existing state-of-the-art methods, with SSNR of 11.63 and PESQ of 3.35, which alleviates the problem of excessive suppression, while the complexity of the model (1.03M) is significantly reduced.

Keywords: Speech enhancement · Deep learning · Two-path transformer · Generative adversarial networks

B. Luo et al. (Eds.): ICONIP 2023, CCIS 1963, pp. 48–61, 2024.
https://doi.org/10.1007/978-981-99-8138-0_5

1 Introduction

Speech enhancement plays a crucial role in various front-end tasks associated with speech processing, including automatic speech recognition (ASR), telecommunications systems, and hearing aid devices [1]. In recent years, deep learning has emerged as a powerful tool and has been extensively applied in speech enhancement research [2,3]. Popular deep learning architectures, such as convolutional neural networks (CNNs) and recurrent neural networks (RNNs), have been employed to address this challenge.

Currently, research on speech enhancement primarily focuses on two methods: time-domain-based and time-frequency-domain-based approaches. In the time-domain-based method, the original speech waveform is directly enhanced without undergoing any transformations. This approach aims to estimate a clean waveform from the noisy raw data [4,5]. Transformer models have shown remarkable effectiveness in speech enhancement within this paradigm [6]. However, the time-domain approach has limitations since it lacks a direct frequency representation of speech. Consequently, it may fail to capture subtle speech phonetics accurately. Additionally, complex deep computation frameworks are often required due to the large number of inputs associated with the original waveform [7,8]. On the other hand, the time-frequency-domain-based method utilizes the Short-Time Fourier Transform (STFT) to represent speech as a spectral magnitude, and the enhanced speech is reconstructed using Inverse Short-Time Fourier Transform (ISTFT) [9,10]. The importance of phase information in improving speech quality has been recognized [11]. However, most traditional time-frequency-domain-based models tend to focus solely on spectral magnitude and overlook the phase information. In response to this issue, some methods are proposed to enhance the complex spectrogram, which includes both real and imaginary parts, thereby implicitly enhancing both magnitude and phase [12]. Nevertheless, the compensatory effects between magnitude and phase often result in inaccurate magnitude estimates [13].

Speech enhancement models serve various applications and goals, each with its specific requirements. For instance, in interpersonal communication, the focus lies on the quality and comprehensibility of speech. In speech recognition preprocessing, the objective is to enhance accuracy. Therefore, training speech enhancement models with specific targets can improved performance. In order to address this challenge, MetricGAN [14] utilizes a generative adversarial network to learn target criteria through discriminators, thus evaluating the performance of the generator. On this basis, conformer architectures have been employed in both the time-domain [15] and time-frequency domain [16] speech enhancement networks, demonstrating promising results. However, the strong learning capacity of conformer models can lead to significant over-suppression, wherein the target speech signal is excessively attenuated along with the noisy components. This phenomenon poses a significant challenge, particularly in pre-processing for speech recognition tasks, where the loss of target speech can be detrimental.

In this context, existing speech enhancement methods still face the following challenges:

- Difficult to strike a balance between model complexity and denoising effect for end devices with limited resources.
- No targeted training for specific tasks and requirements such as human-machine communication or interpersonal communication.
- Over-learning leads to significant over-suppression, such that the target speech signal is over-attenuated together with the noise component.

1.1 Contributions

Taking into account the aforementioned challenges and drawing inspiration from prior research, we propose a novel approach utilizing a generative adversarial network based on two-path transformer. In summary, this paper makes the following contributions:

1. We conducted a comprehensive investigation into the structures of the encoder and decoder that are compatible with the two-path transformer with the objective to thoroughly explore the performance potential of the transformer in the context of speech enhancement. Simultaneously, we significantly mitigate the model's complexity, rendering it more efficient and viable for real-world applications.
2. We concentrated on details of the training strategy for the generative adversarial network, incorporating the information from the raw noisy speech into the discriminator, which effectively improves performance for targeted tasks.
3. While prioritizing high speech quality, we have further enhanced the model's capability to preserve target speech phonetics, mitigating the problem of excessive suppression to some extent. Additionally, we have made our code[1] publicly available, intending to facilitate future research endeavors.

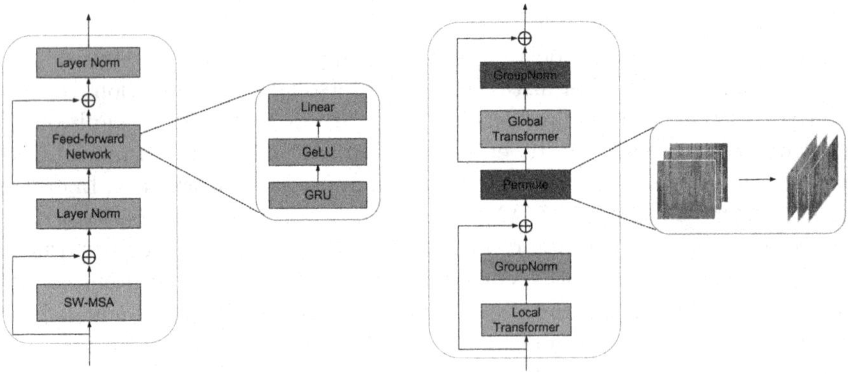

(a) Improved transformer. (b) Two-path transformer block.

Fig. 1. The overall architecture of the transformer block.

[1] https://github.com/TAEV1/TPTGAN.

2 Improved Transformer

As shown in Fig. 1(a), we improve the self-attention in [17]. It is expressed by the following equation [18]:

$$Q_i = IW_i^Q, K_i = IW_i^K, V_i = IW_i^V \tag{1}$$

$$head_i = Attention\left(Q_i, K_i, V_i\right) = softmax\left(\frac{Q_i K_i^T}{\sqrt{d}}\right) V_i \tag{2}$$

$$MultiHead = Concat(head_1, ..., head_h)W^O \tag{3}$$

$$Mid = LayerNorm(I + MultiHead) \tag{4}$$

where $I \in \mathbb{R}^{l \times d}$ is the input sequence of length l and dimension d, $i = 1, 2, ..., h$; $Q_i, K_i, V_i \in \mathbb{R}^{l \times d/h}$ are the mapped queries, keys, and values, respectively; $W_i^Q, W_i^K, W_i^V \in \mathbb{R}^{d \times d/h}$ and $W^O \in \mathbb{R}^{d \times d}$ are linear transformation matrices.

The output of the multi-head attention block is then processed through a feed-forward network with GeLU [19] to obtain the final output of the improved transformer encoder, to which residual connections and layer normalization [20] are also added. The procedure is defined as follows:

$$FFN(Mid) = GeLU(GRU(Mid))W_1 + b_1 \tag{5}$$

$$Output = LayerNorm(Mid + FFN) \tag{6}$$

where $FFN(\cdot)$ represents the output of the position-wise feed-forward network, and $W_1 \in \mathbb{R}^{d_{ff} \times d}$, $b_1 \in R^d$ and $d_{ff} = 4 \times d$.

2.1 Two-Path Transformer Block

We propose a two-path transformer block based on the above improved transformer. As shown in Fig. 1(b), there are local transformers and global transformers that extract local and global context information, respectively. Local transformers are first applied to a single block to process local information in parallel, and then global transformers are used to fuse information from local transformers to learn global dependencies. In addition, each transformer is then grouped normalized and the residual connection is utilized.

3 Proposed Network

In this section, we propose a magnitude, frequency-domain two-path transformer network for the speech enhancement task. As shown in Fig. 2, our proposed generator consists of an encoder, a two-path transformer module, a magnitude masking module, and a complex spectrogram decoder.

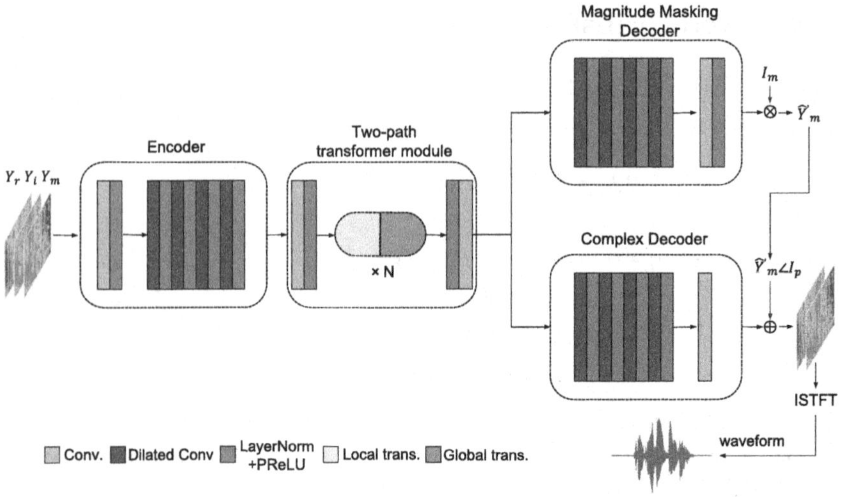

Fig. 2. The overall architecture of generator.

3.1 Encoder

For the distorted speech waveform $y \in \mathbb{R}^{L\times 1}$, the STFT operation first converts the waveform to a complex spectrogram $Y_o \in \mathbb{R}^{T\times F\times 2}$, where T and F represent the time and frequency dimensions, respectively. Then the compressed spectrogram Y is obtained by power-law compression [16]:

$$Y = |Y_o|^c e^{jY_p} = Y_m e^{jY_p} = Y_r + jY_i \tag{7}$$

where Y_m, Y_p, Y_r, and Y_i represent the magnitude, phase, real and imaginary parts of the compressed spectrogram, respectively. The c is the compression index, which ranges from 0 to 1. Here we set c = 0.3 as in reference [21]. The real and imaginary parts Y_r and Y_i are then concatenated with the magnitude Y_m as inputs to the model. The shape of input feature is $Y \in \mathbb{R}^{B\times 3\times T\times F}$, where B represents the batch size. The encoder consists of a convolution layer and a DenseNet block [22]. The 1×1 convolution layer increases the channel of the input feature from 3 to C. The DenseNet block contains four convolution blocks with dense residual connections, each with an expansion factor set to $\{1, 2, 4, 8\}$. Both the convolution layer and the DenseNet block are followed by a LayerNorm and a PReLU activation [23].

3.2 Two-Path Transformer Module

The module consists of an input convolutional layer, N stacked two-path transformer (TPT) blocks, and an output convolution layer. Prior to TPT, we used a 1×1 convolution layer to halve the channel of the output feature of encoder to $C_0 = C/2$ to reduce complexity. Each TPT consists of an intra-transformer and an inter-transformer where the sub-band information is modeled within the intra-transformer and the full-band information is modeled by inter-transformer. Finally, a 1×1 convolution layer is used to restore the channel to C. Each convolution layer is followed by a PReLU activation.

3.3 Decoder

Magnitude Masking decoder uses the output of two-path transformer module (TPTM) to extract the magnitude mask. It is designed to predict a mask that is multiplied by the input magnitude Y_m to predict $\hat{X}'_m$. The masking decoder consists of a convolution layer and a DenseNet, which is similar in structure to the encoder, but the layer structure is distributed symmetrically. DenseNet is used to decode information from the output of TPTM, and then compress the channel to 1 by a 1×1 convolution layer to obtain a prediction mask. The complex spectrum decoder directly predicts the real and imaginary parts of the spectrogram. Its structure is roughly the same as that of the Magnitude Masking Decoder, containing a DenseNet block and a convolution layer. But the difference is that there is no activation function after the convolution layer in Complex Decoder. Firstly, mask $\hat{X}'_m$ is combined with noise phase Y_p to obtain a complex spectrum with magnitude enhancement. Then we sum the elements with the output of the complex decoder $(\hat{X}'_r, \hat{X}'_i)$ to obtain the final complex spectrum:

$$\hat{X}_r = \hat{X}'_m \cos(Y_p) + \hat{X}'_r \quad \hat{X}_i = \hat{X}'_m \sin(Y_p) + \hat{X}'_i \tag{8}$$

Then reverse the power-law compression on the complex spectrum $(\hat{X}_r, \hat{X}_i)$, and ISTFT are applied to obtain the time domain signal $\hat{X}$.

3.4 Discriminator

Here, we use a discriminator architecture similar to CMGAN [16], as well as PESQ scores as labels. The discriminator consists of four convolutional blocks. Each block starts with a convolutional layer, followed by LayerNorm and PReLU activation. After the convolutional blocks, global average pooling is followed by two feed-forward layers and a learnable Sigmoid function [24]. The discriminator is then trained to estimate the maximum normalized PESQ score (=1) by taking the two inputs as clean magnitudes.

3.5 Loss Function

As inspired by Sherif et al. [21], we use a linear combination of complex loss $Loss_{RI}$ and magnitude loss $Loss_{Mag}$ in the Time-Frequency (TF) domain:

$$Loss_{TF} = \alpha Loss_{Mag} + (1 - \alpha) Loss_{RI} \tag{9}$$

$$Loss_{Mag} = \mathbb{E}_{X_m, \hat{X}_m} \left[\left\| X_m - \hat{X}_m \right\|^2 \right] \tag{10}$$

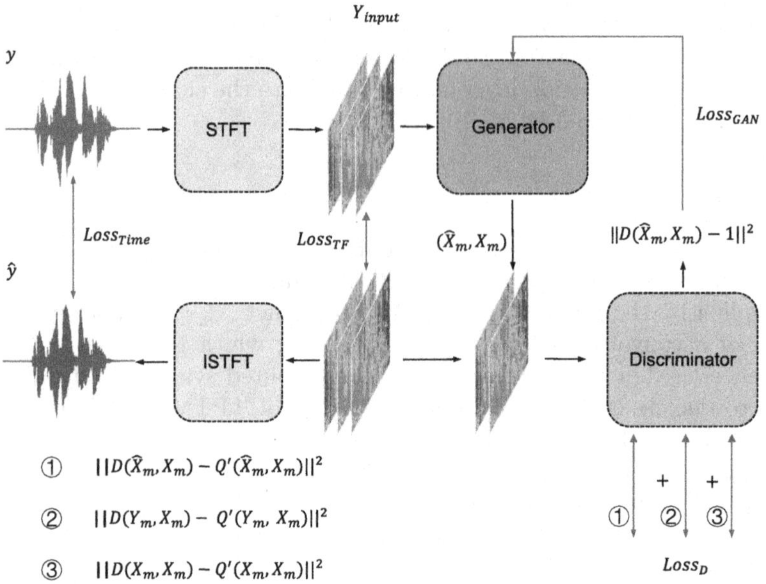

Fig. 3. Loss of adversarial generator and discriminator.

$$Loss_{RI} = \mathbb{E}_{X_r,\hat{X}_r}\left[\left\|X_r - \hat{X}_r\right\|^2\right] + \mathbb{E}_{X_i,\hat{X}_i}\left[\left\|X_i - \hat{X}_i\right\|^2\right] \tag{11}$$

where α is the selected weight. In addition, we also utilize $Loss_{time}$ to improve the restored speech quality:

$$Loss_{Time} = \mathbb{E}_{x,\hat{x}}\left[\|x - \hat{x}\|_1\right] \tag{12}$$

where $\hat{x}$ is the enhanced waveform and x is the clean target waveform. Moreover, similar to least-square GANs [25], the adversarial training is following a min-min optimization task over the discriminator loss L_D and the corresponding generator loss L_{GAN} expressed as follows:

$$Loss_{GAN} = \mathbb{E}_{X_m,\hat{X}_m}[\|D(X_m - \hat{X}_m) - 1\|^2] \tag{13}$$

where D refers to the discriminator. Therefore, the final loss of generator is as follows:

$$Loss = \beta_1 Loss_{TF} + \beta_2 Loss_{Time} + \beta_3 Loss_{GAN} \tag{14}$$

where β_1, β_2 and β_3 are the weights of the corresponding loss. Moreover, including noisy speech when training the discriminator is turned out to stabilize the learning process [24]. Thus, we adopt the same strategy for training the discriminator as well:

$$\begin{aligned} Loss_D = \mathbb{E}_{x,y}[&(D(y,y) - Q'(y,y))^2 \\ &+ (D(G(x),y) - Q'((G(x),y))^2 \\ &+ (D(x,y) - Q'(x,y))^2] \end{aligned} \tag{15}$$

where G stands for generator and x, y represent noisy and clean speech, respectively. The $Q^{'}$ refers to the normalized PESQ score which we normalize to the range from zero to one. The entire loss calculation process can be described in Fig. 3.

4 Experiments

4.1 Datasets

We evaluate our proposed model on a standard speech dataset [2], selecting it from the speech library corpus [26], which includes 11,572 utterances for the training set of 28 speakers (14 females and 14 males) and 824 utterances for the test set. Noisy speech is generated by 10 types of noise (8 from the DEMAND dataset [27] and 2 artificially generated) with signal-to-noise ratios of 15 dB, 10 dB, 5 dB, and 0 dB for training, and 5 invisible noises at SNRs of 17.5 dB, 12.5 dB, 7.5 dB, and 2.5 dB for testing. In our experiment, all utterances are resampled to 16 kHz.

4.2 Experimental Setup

The window length and frame shift of STFT and ISTFT are 25 ms and 6.25 ms, respectively. The number of two-path transformer block N, batch size B, and channel C is set to 4, 4, and 64, respectively. The weight factor in the loss function is set to $\alpha = 0.9$, $\beta_1 = 1$, $\beta_2 = 0.2$ and $\beta_3 = 0.01$. In the training stage, we train the model to 100 epochs. We use Adam as the optimizer. Dynamic strategies [28] are used for both generator and discriminator to adjust the learning rate during the training stage and the discriminator always maintains a learning rate twice as high as the generator.

$$\begin{aligned} &lr_G = k_1 \cdot d_{model}^{-0.5} \cdot n \cdot warmups^{-1.5}, \quad n < warmup \\ &lr_G = k_2 \cdot 0.98^{\lfloor \frac{epoch}{2} \rfloor}, \quad n > warmup \\ &lr_D = 2 \cdot lr_G \end{aligned} \tag{16}$$

where n is the number of steps, d_{model} represents the input size of the transformer, and k_1, k_2 are adjustable scalars. In this paper, k_1, k_2, d_{model}, and $warmup$ are set to 0.2, $4e^{-4}$, 64, 4000, respectively.

4.3 Evaluation Metrics

We evaluate the proposed speech enhancement model by several objective indicators: Perceptual Evaluation of Voice Quality (PESQ) [29], with scores ranging from −0.5 to 4.5. We also employ subjective mean opinion scores (MOSs) [30], including CSIG for signal distortion, CBAK for noise distortion evaluation, and COVL for overall quality assessment. And all the MOS range from 1 to 5. A segmented signal-to-noise ratio (SSNR) [31] is also used, with values ranging from −10 to 35.

Table 1. Comparison with other methods on the VoiceBank+Demand dataset. Feature Type denotes the type of the model input, where W, C, M and P are short for Waveform, Complex, Magnitude and Phase, separately.

Method	Year	Feature Type	Para. (Million)	SSNR	WB-PESQ	CSIG	CBAK	COVL
Noisy	N/A	N/A	N/A	1.68	1.97	3.34	2.44	2.63
SEGAN [7]	2017	W	97.47	7.73	2.16	3.48	2.94	2.8
Wave-U-Net [33]	2018	W	10.0	9.97	2.40	3.52	3.24	2.96
MetricGAN [14]	2019	M	N/A	N/A	2.86	3.99	3.18	3.42
PHASEN [3]	2020	M+P	N/A	10.8	2.99	4.21	3.55	3.62
TSTNN [6]	2021	W	0.92	9.7	2.96	4.1	<u>3.77</u>	3.52
MetricGAN+ [24]	2021	M	N/A	N/A	3.15	4.14	3.16	3.64
DB-AIAT [32]	2022	C+M	2.81	<u>10.79</u>	3.31	**4.61**	3.75	3.96
DPT-FSNet [18]	2022	C	<u>0.91</u>	N/A	<u>3.33</u>	4.58	3.72	<u>4.00</u>
DeepFilterNet2 [34]	2022	C	2.31	N/A	3.08	4.30	3.40	3.70
THLNet [35]	2023	C	**0.58**	N/A	3.04	4.27	3.59	3.66
Ours	2023	C+M	1.03	**11.63**	**3.35**	<u>4.59</u>	**3.83**	**4.02**

4.4 Results Analysis

Table 1 presents a comparison of the results obtained by our proposed model on the same dataset with those of other existing state-of-the-art (SOTA) methods. The data reveal that our model surpasses the majority of existing methods. Remarkably, despite containing only 1.03 million parameters, 1.7 times fewer than DB-AIAT [32], our model exhibits a significant improvement in SSNR (11.63), PESQ (3.35) and even achieves state-of-the-art performance. This enhancement partially mitigates the issue of excessive suppression. The Figs. 4, 5, 6 and 7 show an example of a test speech, where Clean, Noisy and Enhanced represent the original pure target speech, the noised speech to be enhanced and the speech enhanced by our model, respectively. It can be seen that the noise is effectively eliminated, while the signal of the target voice is greatly preserved.

4.5 Denoising Ablation Study

TPT Blocks. In our initial experiment, we remove the discriminator and train the generator separately in order to investigate the impact of the number of

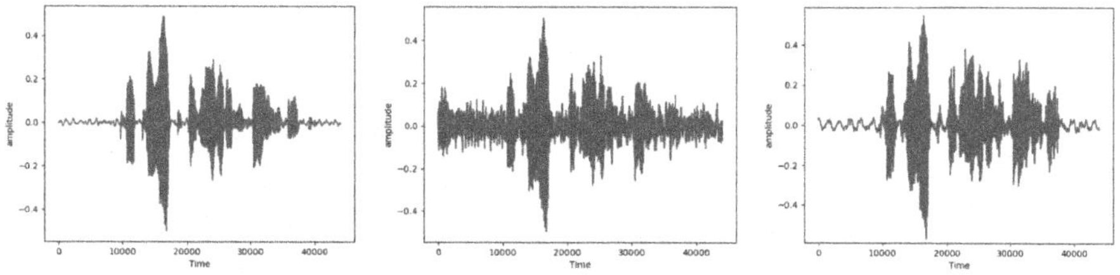

(a) Clean (b) Noisy (c) Enhanced

Fig. 4. Waveform of clean, noisy and enhanced speech.

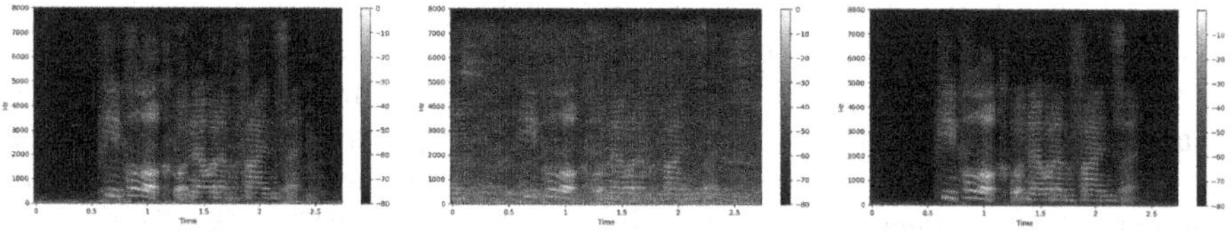

(a) Clean (b) Noisy (c) Enhanced

Fig. 5. Linear spectrogram of clean, noisy and enhanced speech.

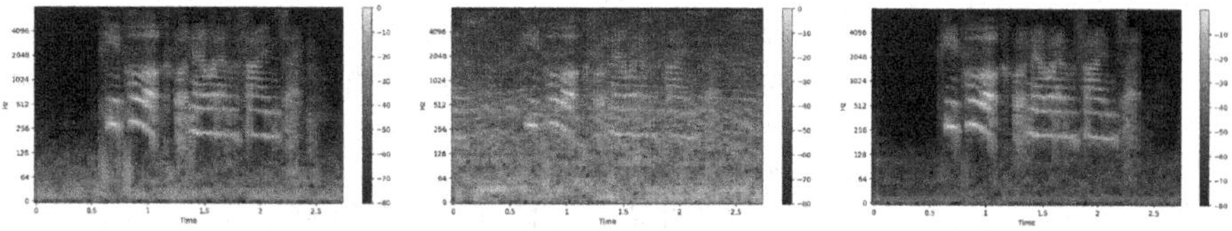

(a) Clean (b) Noisy (c) Enhanced

Fig. 6. Logarithmic spectrogram of clean, noisy and enhanced speech.

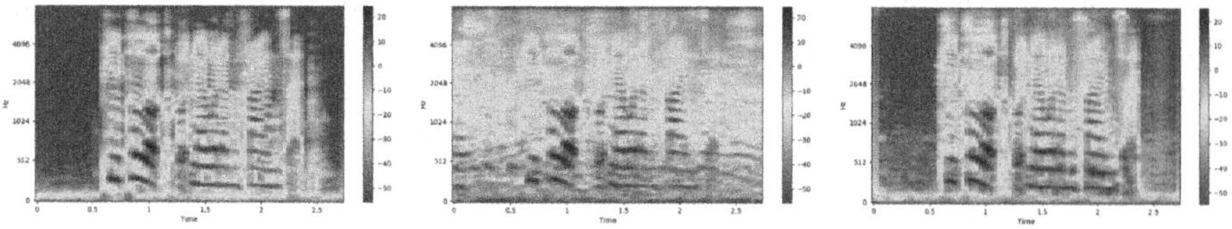

(a) Clean (b) Noisy (c) Enhanced

Fig. 7. Mel spectrogram of clean, noisy and enhanced speech.

two-path transformer blocks on the performance of our proposed model. Table 2 demonstrates that even with a small number of blocks, our method already achieves significant results, comparable to other SOTA methods like MetricGAN+ [24]. To strike a balance between performance and complexity, we gradually increased the number of blocks until the model's performance reached a highly satisfactory level, which occurred after incorporating four blocks.

Table 2. The influence of TPT blocks on objective scores.

Blocks	SSNR	WB-PESQ	CSIG	CBAK	COVL
2	11.65	3.21	4.54	**3.79**	3.90
3	11.63	3.15	4.53	3.78	3.86
4	**11.66**	**3.23**	**4.58**	**3.79**	**3.92**

Denoising Study. As shown in Table 3, we conduct an ablation study to verify our design choices. Firstly we remove the discriminator and train the generator separately (No. Disc). The two-path transformer blocks was set to 4. The results show that the SSNR score of the model reaches the highest without the GAN network architecture, but the other target scores are mediocre. Secondly, we replace the TPT blocks using the conformer module proposed by Cao et al. [16]. The number of blocks is still 4 and the model architecture remains unchanged. This result shows a decrease in all target scores. Finally, TPTGAN denotes that we build the network using TPT and utilize GAN network for training. The comparison results show that the PESQ score of the model increase by 0.12 under the premise of maintaining a high SSNR score, while other target scores are effectively improved. At the same time, we study the impact of different discriminator loss calculation strategies. TPTGAN-1 denotes that the discriminator loss does not contain the raw noise information, that is, its loss function is:

$$\begin{aligned} Loss_D = \mathbb{E}_{x,y}[(D(y,y) - Q'(y,y))^2 \\ + (D(G(x),y) - Q'(G(x),y))^2] \end{aligned} \tag{17}$$

where G stands for generator and x, y represent noisy and clean speech, respectively, $Q^{'}$ refers to the normalized PESQ score which we normalize to the range from 0 to 1. The comparison of results indicates that incorporating the raw noise information into the discriminator leads to an improvement of 0.21 in SSNR and 0.03 in PESQ scores, while other target scores also exhibit significant improvement.

Table 3. Results of the denoising study.

Method	SSNR	WB-PESQ	CSIG	CBAK	COVL
Noisy	1.68	1.97	3.34	2.44	2.63
No. Disc	**11.66**	3.23	4.58	3.79	3.92
Conformer-Based	11.55	3.13	4.49	3.74	3.83
TPTGAN-1	11.42	3.32	**4.60**	3.82	3.98
TPTGAN	11.63	**3.35**	4.59	**3.83**	**4.02**

5 Conclusions

In this paper, we propose a two-path transformer-based generative adversarial network for extracting both local and global information to enhance speech in the time-frequency domain. Further, we investigate the extraction and modeling capabilities of two-path transformers for integrating magnitude spectrum and complex spectrum fusion information. Our experimental results on the Voice Bank+DEMAND dataset demonstrate that our proposed method alleviates the problem of excessive suppression (SSNR of 11.63, PESQ of 3.35), achieving superior or competitive performance against other SOTA methods, while the number of parameters are further reduced (1.03 M).

Acknowledgements. This work is supported by the National Natural Science Foundation of China (NSFC) under grant No. 61802271. In addition, this work is also partially supported by the National Key Research and Development Program of China under grant No. 2022YFC3303101 and Key Research and Development Program of Science and Technology Department of Sichuan Province under grant No. 2023YFG0145.

References

1. Zheng, C., Peng, X., Zhang, Y., Srinivasan, S., Lu, Y.: Interactive speech and noise modeling for speech enhancement. In: Proceedings of the AAAI Conference on Artificial Intelligence, vol. 35, no. 16, pp. 14549–14557 (2021)
2. Valentini-Botinhao, C., Wang, X., Takaki, S., Yamagishi, J.: Investigating RNN-based speech enhancement methods for noise-robust text-to-speech. In: Proceedings of ISCA Workshop on Speech Synthesis Workshop, pp. 146–152 (2016)
3. Yin, D., Luo, C., Xiong, Z., Zeng, W.: PHASEN: a phase-and-harmonics-aware speech enhancement network. In: Proceedings of the AAAI Conference on Artificial Intelligence, vol. 34, no. 05, pp. 9458–9465 (2020)
4. Rethage, D., Pons, J., Serra, X.: A Wavenet for speech denoising. In: Proceedings of International Conference on Acoustics, Speech and Signal Processing, pp. 5069–5073 (2018)
5. Pandey, A., Wang, D.: TCNN: temporal convolutional neural network for real-time speech enhancement in the time domain. In: Proceedings of International Conference on Acoustics, Speech and Signal Processing, pp. 6875–6879 (2019)
6. Wang, K., He, B., Zhu, W.-P.: TSTNN: two-stage transformer based neural network for speech enhancement in the time domain. In: Proceedings of International Conference on Acoustics, Speech and Signal Processing, pp. 7098–7102 (2021)
7. Pascual, S., Bonafonte, A., Serra, J.: SEGAN: speech enhancement generative adversarial network. In: Proceedings of Interspeech, pp. 3642–3646 (2017)
8. Defossez, A., Synnaeve, G., Adi, Y.: Real time speech enhancement in the waveform domain. In: Proceedings of Interspeech, pp. 3291–3295 (2020)
9. Fu, S.-W., Hu, T., Tsao, Y., Lu, X.: Complex spectrogram enhancement by convolutional neural network with multi-metrics learning. In: Proceedings of International Workshop on Machine Learning for Signal Processing, pp. 1–6 (2017)
10. Soni, M.H., Shah, N., Patil, H.A.: Time-frequency masking-based speech enhancement using generative adversarial network. In: Proceedings of International Conference on Acoustics, Speech and Signal Processing, pp. 5039–5043 (2018)

11. Takahashi, N., Agrawal, P., Goswami, N., Mitsufuji, Y.: PhaseNet: discretized phase modeling with deep neural networks for audio source separation. In: Proceedings of INTERSPEECH, pp. 2713–2717 (2018)
12. Tan, K., Wang, D.: Learning complex spectral mapping with gated convolutional recurrent networks for monaural speech enhancement. In: IEEE/ACM Transactions on Audio, Speech, and Language Processing, vol. 28, pp. 380–390 (2020)
13. Wang, Z.-Q., Wichern, G., Le Roux, J.: On the compensation between magnitude and phase in speech separation. IEEE Sig. Process. Lett. **28**, 2018–2022 (2021)
14. Fu, S.-W., Liao, C.-F., Tsao, Y., Lin, S.D.: MetricGAN: generative adversarial networks based black-box metric scores optimization for speech enhancement. In: Proceedings of International Conference on Machine Learning, pp. 2031–2041 (2019)
15. Kim, E., Seo, H.: SE-conformer: time-domain speech enhancement using conformer. In: Proceedings of Interspeech, pp. 2736–2740 (2021)
16. Cao, R., Abdulatif, S., Yang, B.: CMGAN: conformer-based metric GAN for speech enhancement. In: Proceedings of INTERSPEECH, pp. 936–940 (2022)
17. Vaswani, A., et al.: Attention is all you need. In: Proceedings of International Conference on Neural Information Processing Systems, pp. 6000–6010 (2017)
18. Dang, F., Chen, H., Zhang, P.: DPT-FSNet: dual-path transformer based full-band and sub-band fusion network for speech enhancement. In: Proceedings of International Conference on Acoustics, Speech and Signal Processing, pp. 6857–6861 (2022)
19. Hendrycks, D., Gimpel, K.: Bridging nonlinearities and stochastic regularizers with gaussian error linear units. arXiv preprint arXiv:1606.08415 (2016)
20. Ba, J.L., Kiros, J.R., Hinton, G.E.: Layer normalization. arXiv preprint arXiv:1607.06450 (2016)
21. Braun, S., Tashev, I.: A consolidated view of loss functions for supervised deep learning-based speech enhancement. In: Proceedings of International Conference on Telecommunications and Signal Processing, pp. 72–76 (2021)
22. Pandey, A., Wang, D.: Densely connected neural network with dilated convolutions for real-time speech enhancement in the time domain. In: Proceedings of International Conference on Acoustics, Speech and Signal Processing, pp. 6629–6633 (2020)
23. He, K., Zhang, X., Ren, S., Sun, J.: Delving deep into rectifiers: surpassing human-level performance on ImageNet classification. In: Proceedings of International Conference on Computer Vision, pp. 1026–1034 (2015)
24. Fu, S.-W., et al.: MetricGAN+: an improved version of MetricGAN for speech enhancement. In: Proceedings of INTERSPEECH, pp. 201–205 (2021)
25. Mao, X., et al.: Least squares generative adversarial networks. In: Proceedings of International Conference on Computer Vision, pp. 2813–2821 (2017)
26. Veaux, C., Yamagishi, J., King, S.: The voice bank corpus: design, collection and data analysis of a large regional accent speech database. In: Proceedings of International Conference Oriental COCOSDA held jointly with Conference on Asian Spoken Language Research and Evaluation, pp. 1–4 (2013)
27. Joachim, T., Ito, N., Vincent, E.: The diverse environments multi-channel acoustic noise database: a database of multichannel environmental noise recordings. J. Acoust. Soc. Am. **133**(5), 3591–3591 (2013)
28. Chen, J., Mao, Q., Liu, D.: Dual-path transformer network: direct context-aware modeling for end-to-end monaural speech separation. In: Proceedings of INTERSPEECH, pp. 2642–2646 (2020)
29. Loizou, P.C.: Speech Enhancement: Theory and Practice, 2nd edn. CRC Press, Boca Raton (2013)

30. Hu, Y., Loizou, P.C.: Evaluation of objective quality measures for speech enhancement. IEEE Trans. Audio Speech Lang. Process. **16**(1), 229–238 (2008)
31. Hansen, J.H.L., Pellom, B.L.: An effective quality evaluation protocol for speech enhancement algorithms. In: Proceedings of International Conference on Spoken Language Processing (1998)
32. Yu, G., Li, A., Zheng, C., Guo, Y., Wang, Y., Wang, H.: Dual-branch attention-in-attention transformer for single-channel speech enhancement. In Proceedings of International Conference on Acoustics, Speech and Signal Processing, pp. 7847–7851 (2022)
33. Macartney, C., Weyde, T.: Improved speech enhancement with the wave-U-Net. arXiv preprint arXiv:1811.11307 (2018)
34. Schröter, H., Escalante-B., A.N., Rosenkranz, T., Maier, A.: DeepFilterNet2: towards real-time speech enhancement on embedded devices for full-band audio. arXiv preprint arXiv:2205.05474 (2022)
35. Dang, F., Hu, Q., Zhang, P.: THLNet: two-stage heterogeneous lightweight network for monaural speech enhancement. arXiv preprint arXiv:2301.07939 (2023)

Sample Selection Based on Uncertainty for Combating Label Noise

Shuohui Hao[1], Zhe Liu[1](✉), Yuqing Song[1], Yi Liu[1], Kai Han[1], Victor S. Sheng[2], and Yan Zhu[3]

[1] The School of Computer Science and Communication Engineering, Jiangsu University, Zhenjiang, Jiangsu, China
1000004088@ujs.edu.cn

[2] The Department of Computer Science, Texas Tech University, Lubbock, TX, USA

[3] [3]The Department of Imaging, Affiliated Hospital of Jiangsu University, Zhenjiang, China

Abstract. Automatic segmentation of medical images plays a crucial role in scientific research and healthcare. Obtaining large-scale training datasets with high-quality manual annotations poses challenges in many clinical applications. Utilizing noisy datasets has become increasingly important, but label noise significantly affects the performance of deep learning models. Sample selection is an effective method for handling label noise. In this study, we propose a medical image segmentation framework based on entropy estimation uncertainty for sample selection to address datasets with noisy labels. Specifically, after sample selection, parallel training of two networks and cross-model information exchange are employed for collaborative optimization learning. Based on the exchanged information, sample selection is performed using entropy estimation uncertainty, following a carefully designed schedule for gradual label filtering and correction of noisy labels. The framework is flexible in terms of the precise deep neural network (DNN) models used. Method analysis and empirical evaluation demonstrate that our approach exhibits superior performance on open datasets with noisy annotations. The sample selection method outperforms small loss criterion approaches, and the segmentation results surpass those of traditional fully supervised models. Our framework provides a valuable solution for effectively handling noisy label datasets in medical image segmentation tasks.

Keywords: Deep Learning · Noise Label · Sample Selection · Label Correction · Computer Vision

1 Introduction

Deep learning has had a significant impact on various branches of medicine, particularly medical imaging, and its influence will continue to grow [2]. It is considered a crucial tool in medical research and practice, shaping the future of healthcare. Semantic segmentation, a technique that extracts meaningful regions

B. Luo et al. (Eds.): ICONIP 2023, CCIS 1963, pp. 62–74, 2024.
https://doi.org/10.1007/978-981-99-8138-0_6

from images, enables crucial applications like anatomical research, disease diagnosis, treatment planning, and prognosis monitoring [5]. As medical imaging data grows exponentially, automated segmentation algorithms are necessary to aid doctors in accurate and timely diagnoses [17].

The applicability of deep learning methods in clinical practice is limited due to their heavy reliance on training data, particularly annotated labels. Annotating medical images is a time-consuming, labor-intensive, and expensive process. Depending on the complexity of the region and local anatomical structures to be segmented, annotating a single image may take minutes to hours. Furthermore, label noise is inevitable in the practical application of deep learning models. It has been reported that the proportion of noisy labels in real-world datasets ranges from 8.0% to 38.5% [14]. This noise may result from annotation errors and discrepancies among annotators. Therefore, the lack of large-scale and high-quality annotated datasets has been identified as a major constraint in applying supervised deep learning to medical imaging tasks. Achieving robust generalization performance in the presence of label noise poses a key challenge.

Currently, there are several solutions available for learning with noisy labels, such as robust loss functions and sample selection methods. Robust loss functions that are resilient to label noise have been proposed in [1,18]. The results demonstrate that these methods enhance the noise robustness of deep neural networks. However, they tend to perform well only in simple scenarios. Additionally, modifying the loss function can lead to increased convergence time. Sample selection methods are currently the most effective approach for addressing noisy labels [22]. Prominent examples include MentorNet [7], Co-teaching [4], JoCoR [19], O2U-net [6] and CJC-Net [22]. However, this approach is susceptible to the accumulation of error due to incorrect sample selection. Additionally, directly removing samples with noise can disrupt the distribution of the dataset.

In this study, we propose a medical image segmentation framework based on entropy estimation uncertainty for sample selection to handle datasets with noisy labels. We introduce a cross-model self-correction method for efficient network learning. Specifically, cross-model cooperative optimization learning is achieved through parallel training of two networks and information exchange across models. Based on the exchanged information, sample selection is performed based on entropy estimation uncertainty, following a carefully designed schedule, to gradually perform self-label filtering and correction of low-cost noisy labels in cascading local and global steps. The framework is flexible in terms of the precise deep neural network (DNN) models to be used. The main contributions of this paper can be summarized in three aspects:

1) We propose a medical image segmentation framework adapted to sample selection, including cross-model self-correction of original labels and the segmentation of medical images.

2) We propose a sample selection method based on entropy estimation uncertainty, addressing the inherent limitations of discarding all unselected examples in sample selection.

3) We extensively evaluate our approach on widely accepted publicly available

datasets of data and labels, demonstrating clear improvements in performance compared to state-of-the-art methods.

2 Related Work

The memory properties of deep neural networks (DNNs) have been explored theoretically and empirically to identify clean examples from noisy training data [21]. When noise labels occur, DNN weights deviate from their initial values as overfitting happens, while they remain close to the initial weights in the early stages of training. The memory effect has also been observed in empirical studies [15], as DNNs tend to initially learn simple and general patterns and gradually overfit to all patterns containing noise. Therefore, sample selection methods are commonly employed to design robust training approaches to address the issue of noisy labels. Recent methods often involve multiple training rounds or utilize multiple DNNs collaborating with each other.

2.1 Multi-round Learning

Without maintaining additional DNNs, multiround learning iteratively improves the selected set of clean examples by repeating training rounds. ITLM [13] alternates between selecting examples with true labels at the current moment and retraining DNNs using them, iteratively minimizing the pruned loss. In each training round, only a small fraction of low-loss samples obtained in the current round is used to retrain the DNNs in the next round. O2U-Net [6] repeats the entire training process with cyclic learning rates until sufficient loss statistics for each example are collected. Next, only the DNN is retrained on clean data, where examples with erroneous labels are detected and removed based on the statistical data. TSS-net [12] combines sample selection and semi-supervised learning. In the first stage, noise samples are separated from clean samples through cyclic training. In the second stage, noisy samples are used as unlabeled data, and clean samples are used as labeled data for semi-supervised learning. The selected clean set is primarily refined through iterative rounds of multiround learning, progressively expanding and purifying it. However, as the number of training rounds increases, the computational cost of training grows linearly.

2.2 Multi-network Learning

The sample selection process is guided by the mentor network in the case of collaborative learning or the peer network in the case of co-training. MentorNet [7] pretrains a teacher network to select clean instances and guides the training of a student network. Co-teaching [4] selects clean instances from the network to provide parameter updates for their peer networks. Co-teaching+ [20] and JoCoR [19] also emphasize the selection of low-loss samples, where each DNN selects a certain number of low-loss samples and feeds them back to the peer DNN for further training, explicitly assisting the collaborative training process of the two networks. JoCoR [19] introduces a regularization term into the same loss function, reducing the discrepancy between the two base classifiers and selecting reliable data. Co-learning [16] combines supervised learning and self-supervised learning in a cooperative manner, imposing similarity constraints on the shared feature encoder to maximize consistency. AsyCo [11] achieves new prediction divergence by using different training strategies, where one model is trained using multi-class learning and the other model is trained using multi-label learning. However, this approach is affected by cumulative errors from the selection of erroneous noise. Moreover, directly removing noisy samples disrupts the distribution of the dataset [1], leading to partial exploration of the training data. These approaches avoid the risk of overcorrection by simply excluding unreliable samples. Although these methods have made significant progress, they overlook the potential supervisory signal from training samples with high loss.

Our proposed sample selection method based on entropy estimation uncertainty can leverage all noisy labels, ensuring the learning of the distribution of the original dataset. Furthermore, our network framework also corrects the noisy labels, improving the segmentation results of the network for medical images.

3 Method

Our proposed framework is a deep learning framework that enables accurate image segmentation on imperfect training datasets. We introduce a sample selection method and a cross-model self-label correction mechanism to effectively utilize low-cost noisy training data labels. The precise segmentation network we employ features a classical encoder-decoder structure with multiple streams to extract image features from different modalities. The network architecture is depicted in Fig. 1, and the entire process is divided into three steps. The first step is uncertainty estimation for sample selection, the second step is local label filtering, and the third step is label correction.

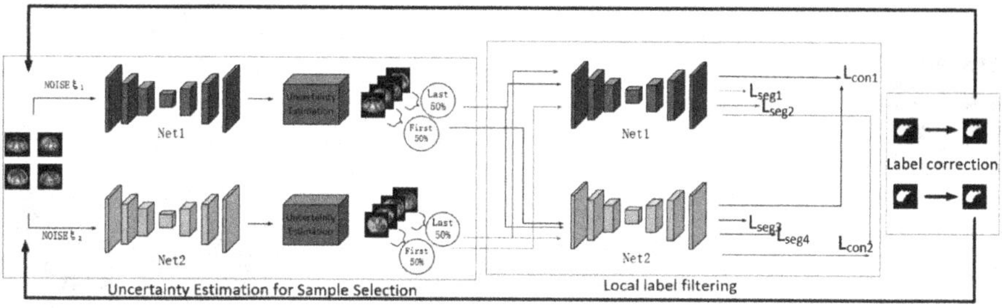

Fig. 1. Framework for Medical Image Segmentation based on Uncertainty Estimation using Entropy

3.1 Uncertainty Estimation for Sample Selection

If noise is present in the labeled input, the predicted targets from the network model may lack reliability. To address this, we have devised an uncertainty-aware approach that enables the network model to progressively learn from more dependable targets. When provided with a batch of training images, the network model not only generates target predictions but also estimates the uncertainty associated with each target. By employing a consistency loss, we optimize the peer models to prioritize confident targets based on the estimated uncertainty.

For each input pixel, we execute T random forward passes using two models-one with random data augmentation and the other with input Gaussian noise. Consequently, for every pixel in the input, we obtain a set of softmax probability vectors:$\{P_t\}_{t=1}^{T}$. To approximate uncertainty, we adopt predictive entropy as a metric, as it possesses a fixed range. In a formal sense, predictive entropy can be described as follows:

$$\mu = \frac{1}{T}\sum_{t} P_t^c \tag{1}$$

$$u = -\sum_{c} \mu log\mu \tag{2}$$

where P_t^c is the probability of the c-th class in the t-th time prediction.

As shown in Fig. 2, a batch of input images undergo random data augmentation and are fed into two peer networks with the addition of Gaussian noise. The entropy of each image is calculated using Eq. 1 and Eq. 2, and the results from each network are sorted based on the entropy values to approximate the ranking of uncertainty. In this step, the two networks operate independently without influencing each other.

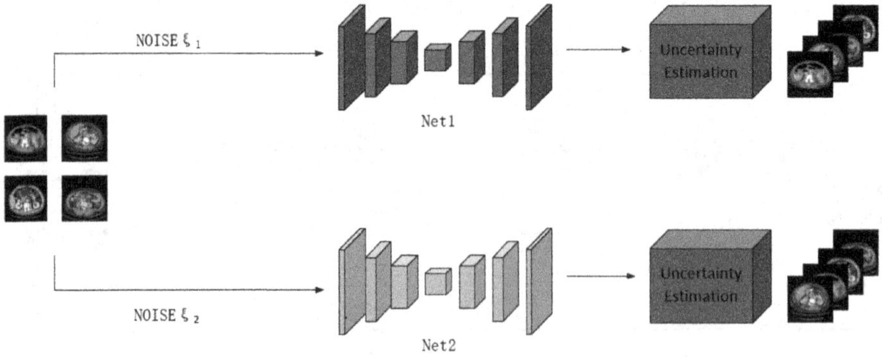

Fig. 2. Module for Uncertainty Estimation Using Entropy Estimation

3.2 Local Label Filtering

The second step is local label filtering. In each iteration, the corresponding model selects training samples with higher uncertainty (entropy) defined as a percentage and identifies them as suspicious samples with low-quality labels. For these samples, the final loss function is computed as the weighted sum of segmentation loss and consistency loss. In this work, the commonly used combination of Dice loss and cross-entropy loss is chosen as the segmentation loss (Eq. 3). The consistency loss is introduced as consistency regularization into the network.

Sample selection is based on uncertainty rankings. The consistency loss is computed between the network output of suspicious noisy samples and the segmentation results from the peer network. It is implemented as mean squared error (MSE) loss (Eq. 4). Samples with smaller expected segmentation loss are considered more reliable and only have the segmentation loss computed. The network parameters are updated based on their respective losses.

$$L_{seg}\left(y, y^{'}\right) = L_{Dice}\left(y, y^{'}\right) + \alpha \cdot L_{CE}\left(y, y^{'}\right) \tag{3}$$

$$L_{cor}\left(y^{'}, y^{''}\right) = \frac{1}{2N}\sum_{i=1}^{N} \parallel y_i^{'} - y_i^{''} \parallel^2 \tag{4}$$

where $y^{'}$ is the prediction of the network, y is the reference,$y^{''}$ is the result of peer-to-peer network segmentation, N refers to the total number of pixels in the image, α is a constant to balance the different losses (we set $\alpha = 1$ in all experiments unless otherwise specified).

Taking Net1 as an example, the total loss of the network (Eq. 5)

$$L_1 = L_{seg1} + (1 - \lambda)L_{seg2} + \lambda L_{con2} \tag{5}$$

where λ is the minimum value between q divided by the square of k and 1, where q represents the current epoch and k is the warm-up epoch.

3.3 Label Correction

The third step is label correction. After each epoch, the dice coefficient (DSC) is calculated for the entire training set and sorted. Samples with the lowest DSC, defined as a percentage (25% in our experiments), have their labels updated under certain criteria. In this work, labels are updated if the training epoch is smaller than the defined warm epoch; otherwise, labels are updated every 10 epochs.

The underlying principles of the designed filtering and correction steps are related to the network's memory patterns often observed in natural image analysis. Deep neural networks tend to learn simple patterns before memorizing all samples, and real data examples are easier to fit than examples with noisy labels. Here, a similar memory pattern of the network for medical image segmentation is discovered. By training neural networks with traditional supervised learning methods using noisy labels, a positive correlation is observed between the network's output DSC calculated with noisy labels and the noisy label-to-high-quality label DSC, particularly during the initial training period. In other words, within the considered training epochs, the network fails to memorize samples containing large label noise but can learn patterns from samples with low label noise. On the other hand, when the network converges, all samples are well memorized. Based on this observation, a schedule for updating noisy labels is designed. When the training epoch is smaller than the defined warm epoch, suspicious noisy labels are updated every 10 epochs. The first criterion is based on the observed network memory behavior mentioned above, and studies show that model performance is not sensitive to the value of the warm epoch within a large range. The implementation of the second criterion is because the network's performance becomes relatively stable after the initial training epoch, requiring less frequent label updates. Therefore, in our experiments, labels are updated every 10 training epochs.

4 Experiments and Results

4.1 Dataset

The dataset consists of abdominal MR images from the CHAOS challenge, specifically images [8,9]. These images were acquired using a 1.5 T Philips MRI system. The image matrix size is 256 × 256, with in-plane resolution ranging from 1.36 to 1.89 mm. The slice thickness varies between 5.5 mm and 9 mm.

Each 3D image comprises 26 to 50 slices. We utilized the T1-DUAL images. For each patient, there is an in-phase image and an out-of-phase image, which can be treated as multi-modal inputs. The challenge provides a total of 40 patient data, with 20 cases (647 T1-DUAL images, 647 samples) having high-quality annotations and 20 cases (653 T1-DUAL images, 653 samples) without annotations. Annotations of the images are performed by radiology experts and experienced medical imaging scientists. In our experimental setup, only the labels from one randomly selected case (30 labeled image samples) were used, while pseudo-labels for the remaining data (29 cases, including 9 randomly selected cases from the provided labeled cases and 20 unlabeled cases) were generated by the network trained on the labeled samples. The remaining 10 labeled cases were used for model testing.

4.2 Implementation Details

U-Net [3] was used as the network segmentation model in this experiment. During the encoding process, the input image passes through 5 downsampling blocks sequentially, with 4 max-pooling operations in between to extract multi-level image features [10]. For multi-modal inputs, multiple downsampling streams are used to extract features from different modalities. The extracted features are then combined in a multi-level fusion manner. During the decoding process, the extracted image features go through four upsampling blocks, with a bilinear upsampling operation after each upsampling block, to generate segmentation outputs of the same size as the input. The downsampling and upsampling blocks have similar components: two convolutional layers with a 3×3 kernel, batch normalization, and ReLU activation. The low-level features from the encoder are introduced into the decoder through skip connections and feature concatenation. Finally, two features corresponding to the background and target segmentation maps are generated through 1×1 convolutions, and segmentation probability maps are generated through softmax activation. Unless otherwise stated, we set the image size to 256×256, the warmup epoch to 10, the batch size to 4, and maintain a constant learning rate of 0.001 throughout all experiments. All experiments of our proposed method were performed on a workstation with Intel Core i9 9900X @ 3.5 GHz, 128 GB RTX, NVIDIA 2080 Ti GPU, Ubuntu 18.04.1, Python 3.8.

4.3 Evaluation Metrics

Different metrics can be used to characterize the segmentation results. For our experiments, we choose the commonly utilized Dice score (Dice similarity coefficient, DSC), relative area/volume difference (RAVD), average symmetric surface distance (ASSD). Higher DSC values and lower RAVD, and ASSD indicate more accurate segmentation results.

$$DSC = \frac{2TP}{2TP + FP + FN} \tag{6}$$

$$RAVD = \frac{FP - FN}{TP + FN} \tag{7}$$

$$ASSD = \frac{\sum_{a \in S(y')} min_{b \in S(y)} \mid a - b \mid + \sum_{b \in S(y)} min_{a \in S(y')} \mid b - a \mid}{\mid S(y) \mid + \mid S(y') \mid} \tag{8}$$

where TP, FP, and FN refer to true positive predictions, false positive predictions, and false negative predictions, respectively. $S(y')$ and $S(y)$ indicate the boundary points on the predicted segmentations and reference segmentations.

4.4 Ablation Study

The CHAOS dataset, which is used for liver segmentation, was introduced to investigate the effectiveness of our method in terms of NLL (Normalized Liver Lesion). The network was trained using annotated images of varying sizes and quality. As expected, a larger number of training samples resulted in better segmentation results. Compared to the network trained using all 10 labeled cases (S02_331_0_F in Table 1, consisting of 331 image samples, DSC: 88.3%,), the segmentation performance of the network trained with only 1 labeled case (S01_30_0_F in Table 1, consisting of 30 image samples) was significantly lower (DSC: 71.2%). Therefore, when training deep learning models in a fully supervised manner, a large dataset is required to achieve satisfactory results.

To expand the training dataset, the model trained using the minimum annotated training data (1 case with 30 image samples) generated low-quality noisy labels. Then, the network was trained using the expanded dataset. In this work, a pixel-wise counting method was employed to estimate the number of low-quality or noisy labels in an image-based manner. Experiments were conducted on different numbers of labeled training image samples, and the results consistently confirmed the effectiveness of our method (see Table 1 for experimental results). When the number of unlabeled samples was the same, our sample selection method outperformed the small-loss criterion-based sample selection method (comparison between S07_30_954_S and S08_30_954_E in Table 1). Furthermore, as the number of unlabeled training samples increased, the performance of our method continued to improve (comparison between S05_30_301_E and S08_30_954_E in Table 1). This characteristic is particularly important for clinical applications as collecting unlabeled data is easier. The visualization of the corrected labels and model outputs is shown in the figure. Overall, the corrected labels are closer to high-quality annotations (Fig. 3), and the segmentation results generated by our method are more accurate compared to the corresponding baseline network (Fig. 4).

Table 1. Segmentation results of networks under different settings

Settings	Train HQA	Train LQA	Small loss	Entropy	DSC (%)	RAVD (%)	ASSD (mm)
S01_30_0_F	30	0	NO	NO	71.2	40.1	16.4
S02_331_0_F	331	0	NO	NO	88.3	10.8	5.16
S03_30_301_F	30	301	NO	NO	78.4	19.2	11.9
S04_30_301_S	30	301	YES	NO	79.9	18.5	10.8
S05_30_301_E	30	301	NO	YES	80.3	17.2	9.66
S06_30_954_F	30	954	NO	NO	80.2	19.1	8.24
S07_30_954_S	30	954	YES	NO	84.7	10.4	5.79
S08_30_954_E	30	954	NO	YES	87.2	10.1	5.41

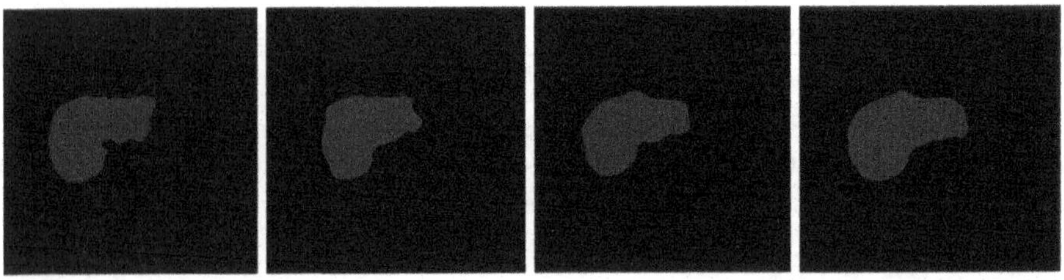

Fig. 3. Example results of training data label correction. The grey regions in the four images from left to right correspond to the high-quality label, the low-quality label utilized to train the model, and the self-corrected labels of the two networks, respectively.

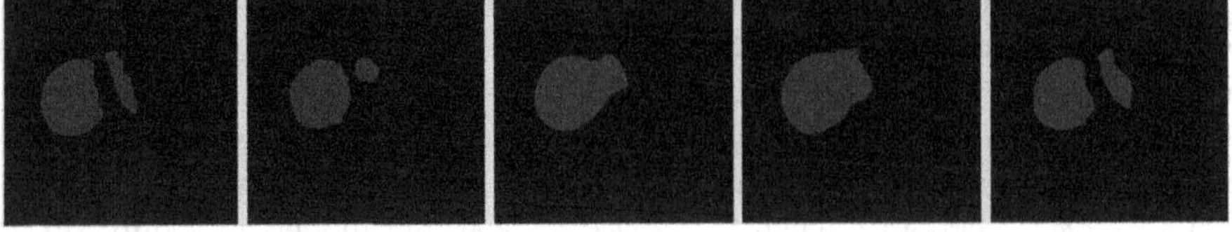

Fig. 4. Example segmentation results. The first is the corresponding high-quality labels. The second to the last are the results achieved under settings S01, S02, S07 and S08 in Table 1.

Note: HQA and LQA indicate high-quality and low-quality annotations. LQAs are generated by the model trained using the data provided with HQAs. Context is included in the notation of the experimental setting. For the setting S03_30_301_F, 30 refers to 30 training samples with HQAs, 301 means 301 training samples with LQAs, F indicates training with the conventional fully-supervised learning approach.

Table 2 presents a comparison between different approaches: using Unet [3] alone, using Unet with small-loss criterion, using Unet with entropy-based sample selection, and using the network framework of Fig. 1 for image segmentation. This comparison validates the effectiveness of our proposed sample selection method, which involves ranking samples based on entropy estimation. The results clearly demonstrate that utilizing entropy estimation for sample selection outperforms the small-loss criterion. When compared to using Unet alone or incorporating the small-loss criterion, our approach achieves superior segmentation performance. This indicates that considering uncertainty through entropy estimation provides a more effective and reliable approach for selecting informative samples during training. The improvement in performance can be attributed to the fact that the entropy-based sample selection method effectively identifies samples with low-quality or noisy labels, leading to a more accurate and robust training process. By prioritizing samples based on their uncertainty, the network focuses on challenging examples that are crucial for learning and avoids overfitting to noisy or ambiguous samples.

Table 2. Comparison of Sample Selection Methods

Method	Train HQA	Train LQA	DSC (%)	RAVD (%)	ASSD (mm)
Unet [3]	30	954	80.2	19.8	8.24
Unet [3]+Small loss	30	954	84.8	13.6	6.88
Unet [3]+Entropy	30	954	86.7	10.7	5.98
Ours	30	954	87.2	10.1	5.41

Table 3. Selection of Parameter T

Parameter T	Each poach time(s)	DSC (%)
2	84	82.6
4	157	87.7
8	300	87.4

The proposed medical image segmentation framework based on entropy estimation for sample selection was evaluated with different values of parameter T in an ablation study. The training dataset consisted of 30 high-quality labeled image samples and 954 low-quality labeled image samples. The test dataset included 10 cases with a total of 305 image samples.Table 3 presents the results of the study. Considering both the training time and the DSC, a trade-off needs to be made. Based on the results, selecting T = 4 appears to be a reasonable choice as it provides a good balance between training efficiency and segmentation performance.

With a higher DSC compared to T = 2, the increase in training time is still manageable. Therefore, T = 4 was selected as the experimental parameter for subsequent evaluations and analysis. This choice allows for an acceptable compromise between computational efficiency and segmentation accuracy, making it a practical parameter for the proposed framework.

4.5 Comparison with the State-of-the-Art Methods

Table 4 provides a comparison between our method and state-of-the-art approaches on the Chaos dataset, which includes 30 high-quality labels and 954 low-quality labels. The reported accuracy represents the average performance over the last 10 epochs. It can be observed that our method outperforms the existing four state-of-the-art methods in terms of Dice Similarity Coefficient (DSC), and achieving lower values in relative area/volume difference and average symmetric surface distance, compared to the four existing state-of-the-art methods. Based on these three evaluation metrics, it can be concluded that our method performs better in terms of segmentation results on the noisy dataset. Figure 5 sequentially presents medical images and their corresponding high-quality labels, segmentation results of four state-of-the-art methods, and the segmentation results of our method. Through the quantitative

Table 4. Comparison with the state-of-the-art methods

Method	DSC (%)	RAVD (%)	ASSD (mm)
Co-teaching [4]	81.4	16.8	6.47
MentorNet [7]	82.2	17.4	6.44
Co-learning [16]	83.6	13.9	6.74
TSS-Net [12]	85.8	10.6	5.93
Ours	87.2	10.1	5.41

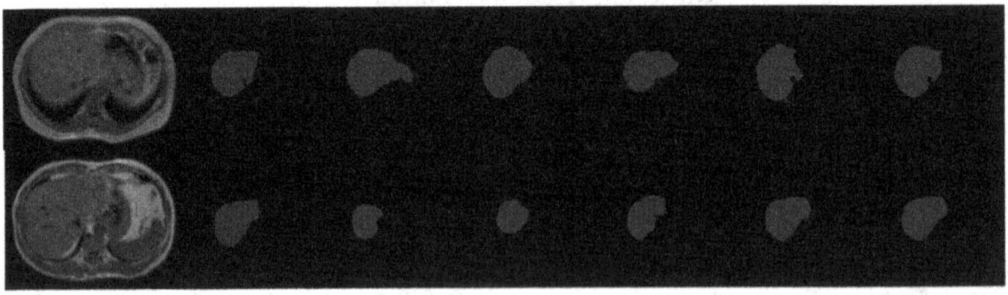

Fig. 5. The segmentation results compared to state-of-the-art methods. From top to bottom and left to right, the sequence consists of medical images and their corresponding high-quality labels, segmentation results of co-teaching [4], segmentation results of MentorNet [7], segmentation results of Co-learning [16], segmentation results of TSS-Net [12], and segmentation results of our method.

analysis in Table 4 and the segmentation results shown in Fig. 5, it is evident that our method surpasses previous approaches, demonstrating its robustness in handling noisy labels.

5 Conclusion

In this paper, we introduced an open-source framework for medical image segmentation based on uncertainty-guided sample selection using entropy estimation. Our goal is not to establish a more complex model for fully supervised learning but rather to create a framework that can function effectively even with limited labeled data, thereby reducing the reliance on time-consuming and expensive manual annotations when applying AI to medical imaging. Extensive experiments were conducted using publicly available datasets, and the results demonstrate that under the same training conditions, our framework outperforms traditional fully supervised models, and our sample selection method achieves higher accuracy than the small-loss criterion.

In conclusion, the sample selection method and segmentation framework proposed in this study can enhance the robustness and generalization capability of deep learning models, ultimately contributing to improving the medical image diagnostic workflow.

Acknowledgment. This work was supported by the National Natural Science Foundation of China (62276116, 61976106); Six talent peaks project in Jiangsu Province (DZXX-122); Jiangsu Province Graduate Research Innovation Program (KYCX23_3677).

References

1. Arazo, E., Ortego, D., Albert, P., O'Connor, N., McGuinness, K.: Unsupervised label noise modeling and loss correction. In: International Conference on Machine Learning, pp. 312–321. PMLR (2019)
2. Ching, T., Himmelstein, D.S., Beaulieu-Jones, B.K., Kalinin, A.A., Do, B.T., Way, G.P., Ferrero, E., Agapow, P.M., Zietz, M., Hoffman, M.M., et al.: Opportunities and obstacles for deep learning in biology and medicine. J. R. Soc. Interface **15**(141), 20170387 (2018)
3. Falk, T., et al.: U-net: deep learning for cell counting, detection, and morphometry. Nat. Methods **16**(1), 67–70 (2019)
4. Han, B., et al.: Co-teaching: robust training of deep neural networks with extremely noisy labels. In: Advances in Neural Information Processing Systems **31** (2018)
5. Hollon, T.C., et al.: Near real-time intraoperative brain tumor diagnosis using stimulated Raman histology and deep neural networks. Nature Med. **26**(1), 52–58 (2020)
6. Huang, J., Qu, L., Jia, R., Zhao, B.: O2u-net: a simple noisy label detection approach for deep neural networks. In: Proceedings of the IEEE/CVF International Conference on Computer Vision, pp. 3326–3334 (2019)

7. Jiang, L., Zhou, Z., Leung, T., Li, L.J., Fei-Fei, L.: Mentornet: learning data-driven curriculum for very deep neural networks on corrupted labels. In: International Conference on Machine Learning, pp. 2304–2313. PMLR (2018)
8. Kavur, A.E., Gezer, N.S., Barış, M., Aslan, S., Conze, P.H., Groza, V., Pham, D.D., Chatterjee, S., Ernst, P., Özkan, S., et al.: Chaos challenge-combined (ct-mr) healthy abdominal organ segmentation. Med. Image Anal. **69**, 101950 (2021)
9. Kavur, A.E., et al.: Comparison of semi-automatic and deep learning-based automatic methods for liver segmentation in living liver transplant donors. Diagn. Interv. Radiol. **26**(1), 11 (2020)
10. Li, C., Sun, H., Liu, Z., Wang, M., Zheng, H., Wang, S.: Learning cross-modal deep representations for multi-modal MR image segmentation. In: Shen, D., Liu, T., Peters, T.M., Staib, L.H., Essert, C., Zhou, S., Yap, P.-T., Khan, A. (eds.) MICCAI 2019. LNCS, vol. 11765, pp. 57–65. Springer, Cham (2019). https://doi.org/10.1007/978-3-030-32245-8_7
11. Liu, F., Chen, Y., Wang, C., Tain, Y., Carneiro, G.: Asymmetric co-teaching with multi-view consensus for noisy label learning. arXiv preprint arXiv:2301.01143 (2023)
12. Lyu, X., Wang, J., Zeng, T., Li, X., Chen, J., Wang, X., Xu, Z.: Tss-net: two-stage with sample selection and semi-supervised net for deep learning with noisy labels. In: Third International Conference on Intelligent Computing and Human-Computer Interaction (ICHCI 2022), vol. 12509, pp. 575–584. SPIE (2023)
13. Shen, Y., Sanghavi, S.: Learning with bad training data via iterative trimmed loss minimization. In: International Conference on Machine Learning, pp. 5739–5748. PMLR (2019)
14. Song, H., Kim, M., Lee, J.G.: Selfie: refurbishing unclean samples for robust deep learning. In: International Conference on Machine Learning, pp. 5907–5915. PMLR (2019)
15. Song, H., Kim, M., Park, D., Lee, J.G.: How does early stopping help generalization against label noise? arXiv preprint arXiv:1911.08059 (2019)
16. Tan, C., Xia, J., Wu, L., Li, S.Z.: Co-learning: learning from noisy labels with self-supervision. In: Proceedings of the 29th ACM International Conference on Multimedia, pp. 1405–1413 (2021)
17. Tang, H., et al.: Clinically applicable deep learning framework for organs at risk delineation in CT images. Nature Mach. Intell. **1**(10), 480–491 (2019)
18. Wang, Y., Ma, X., Chen, Z., Luo, Y., Yi, J., Bailey, J.: Symmetric cross entropy for robust learning with noisy labels. In: Proceedings of the IEEE/CVF International Conference on Computer Vision, pp. 322–330 (2019)
19. Wei, H., Feng, L., Chen, X., An, B.: Combating noisy labels by agreement: A joint training method with co-regularization. In: Proceedings of the IEEE/CVF Conference on Computer Vision and Pattern Recognition, pp. 13726–13735 (2020)
20. Yu, X., Han, B., Yao, J., Niu, G., Tsang, I., Sugiyama, M.: How does disagreement help generalization against label corruption? In: International Conference on Machine Learning, pp. 7164–7173. PMLR (2019)
21. Zhang, C., Bengio, S., Hardt, M., Mozer, M.C., Singer, Y.: Identity crisis: memorization and generalization under extreme overparameterization. arXiv preprint arXiv:1902.04698 (2019)
22. Zhang, Q., et al.: Cjc-net: a cyclical training method with joint loss and co-teaching strategy net for deep learning under noisy labels. Inf. Sci. **579**, 186–198 (2021)

Design of a Multimodal Short Video Classification Model

Xinyan Cao, He Yan, Changjin Li, Yongjian Zhao, Jinming Che, Wei Ren, Jinlong Lin, and Jian Cao(✉)

Peking University, Beijing, China
{caoxinyan,caoxinyan,yongjianzhao,chejinming, renw}@stu.pku.edu.cn, yanhe.ian@pku.edu.cn, {linjl, caojian}@ss.pku.edu.cn

Abstract. With the development of mobile Internet, a large amount of short video data is generated on the Internet. The urgent problem of short video classification is how to better fuse the information of different multimodal information. This paper proposes a short video multimodal fusion (SV-MF) scheme based on deep learning combined with pre-trained models to complete the classification task of short video. The main innovations of the SV-MF scheme are as follows: (1) We find that text modalities contain higher-order information and tend to perform better than audio and visual modalities, and with the use of pre-trained language models, text modalities have been further improved in multimodal video classification. (2) Due to the strong semantic representation ability of text. The SVMF scheme proposes a local fusion method based on Transformer for low-order visual and audio modal information to alleviate the information deviation caused by multi-mode fusion. (3) The SV-MF scheme proposes a post processing strategy based on keywords to further improve the classification accuracy of the model. Experimental results based on a multimodal short video classification dataset derived from social networks show that the performance of the SV-MF scheme is better than the previous video fusion scheme.

Keywords: Multimodal Classification · Deep Learning · Attention Mechanism

1 Introduction

With the development of mobile communication technology, short video has become a more extensive information carrier [1]. The existing short video classification models cannot complete the short video classification task well. After AlexNet [2] won the 2012 ImageNet Challenge, deep learning began to attract more attention. In the field of video, deep learning has not surpassed traditional methods, which was not widely used in the video domain until the two-stream network [3] was proposed. The two-stream network slightly outperforms traditional methods on the UCF-101 [4] and HMDB-51 [5] datasets, but it is more important contribution is to provide researchers with a new idea. However,

Yan, H. and Cao, X. contributed equally to this paper and should be considered as co-first authors.

B. Luo et al. (Eds.): ICONIP 2023, CCIS 1963, pp. 75–85, 2024.
https://doi.org/10.1007/978-981-99-8138-0_7

the method has the disadvantage that it takes a lot of time to extract the dense optical flow. Attention Cluster [6] believes that timing information is not important for classification. Attention Cluster uses an attention mechanism to cluster the feature sequences of images, optical flow, and audio. The representations of the three modalities are concatenated and then input into the classifier. With the successful application of large-scale pre-training techniques in the field of natural language processing, researchers have also begun to explore the application of pre-trained models in joint representation of vision and language [7].

Although the above methods have achieved satisfactory results in multimodal video classification, we have found that the above methods still have the following shortcomings: (1) There is still no general mature scheme for multimodal short video classification. (2) There is no high-quality short video dataset for multimodal short video classification. (3) How to effectively combine different modal information in short videos, and how to design model parameters and objective functions have not been well resolved.

The rest of this paper is organized as follows: In Sect. 2, the SV-MF scheme is proposed. Then, the experiment result is proposed in Sect. 3. Finally, Sect. 4 provides conclusions.

2 The Design of SV-MF Scheme

This section first introduces the overall framework of the SVMF scheme, secondly introduces the feature extraction methods of different modalities, then introduces the design of local fusion and multi-modal fusion, and finally presents a model independent post-processing method.

2.1 Overall Framework

The overall framework of SV-MF scheme is shown in Fig. 1, including the following parts: (1) The first part preprocesses the original video, including obtaining cover, extracting frame, extracting audio information, and obtaining title information. (2) The second part is to extract the features of different modes through different feature extraction methods from three kinds of pre-processed modal data. (3) The third part is the local fusion of multi-frame image features and audio features. (4) The fourth part is the integration of multiple modes, and then connected with multi-layer perceptron to get the final prediction results.

As shown in Fig. 1, the multi-frame images are input to ResNeXt [8] to extract visual feature vector. The audio information in the video was input to the pre-trained VGGish [9] network to extract the 128-dimensional audio feature vector. The title text of the video is first serialized according to the thesaurus provided by the pre-training model, and then the serialized data is input into BERT [10] to obtain the feature vector. The extracted visual and audio features are sequences composed of multiple features vectors. Then local fusion of these two feature modes is performed to obtain their joint representation. Finally, the joint representation, video cover feature and text feature vector are fused.

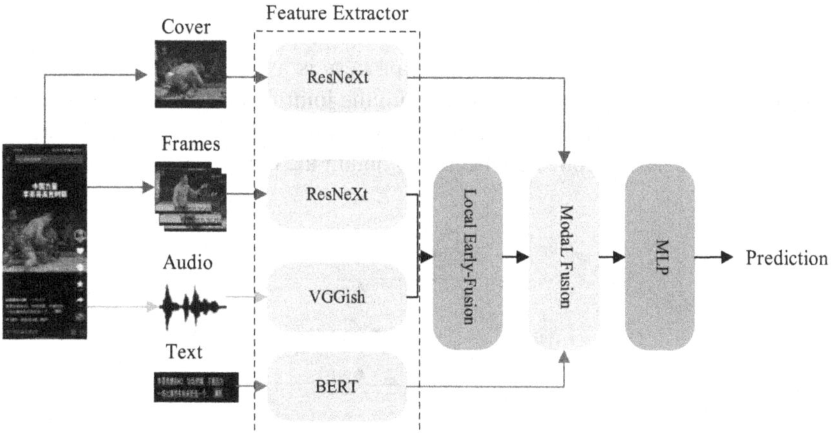

Fig. 1. The overall framework of SV-MF scheme.

2.2 Feature Extraction

Visual feature: This article preserves only the first 10 s of each video. A typical video frame rate is 30 frames per second, so 10 s equals 300 frames. This paper chooses the ResNeXt network model combining ResNet [11] and Inception to extract the visual feature.

Audio feature: For audio features, we use standard pipeline for processing. The processing includes framing, audio resampling, spectrogram, calculating Mel spectrum and logarithmic operations. Then VGGish is used to extract audio features.

Text feature: We use Chinese-BERT-wwm which is pre-trained on the Chinese corpus and uses the whole word mask strategy. The original BERT uses a character-level masking strategy for Chinese in the Masked Language Model task, while the whole word mask is to do Chinese word segmentation first, and then mask the Chinese word, which is a better pre-training strategy.

2.3 Local Fusion

Compared to text, visual and audio contain lower-level information that is often more complex and harder to use. Therefore, the SV-MF scheme proposes a local forward fusion method for multi-modal feature fusion as shown in Fig. 2, which firstly fuses low-order video features and audio features to obtain joint representation, and then fuses them with higher-order semantic text features to obtain output. The local fusion steps are as follows: first, the inputs of different dimensions are mapped to the same dimensions through the position-wise feed-forward module. After that, the Transformer with parameter sharing is used to extract features, then the visual and audio features are spliced together as a joint representation of the two kinds of information.

2.4 Multi-Modal Fusion

As shown in Fig. 3, the multi-modal fusion process is as follows: First, cross-modal Transformer is used to fuse the text feature with the joint feature. Then, it is fused with the cover feature vector and the text feature vector through the self-attention mechanism. Finally, multi-layer perceptron is connected to obtain the output.

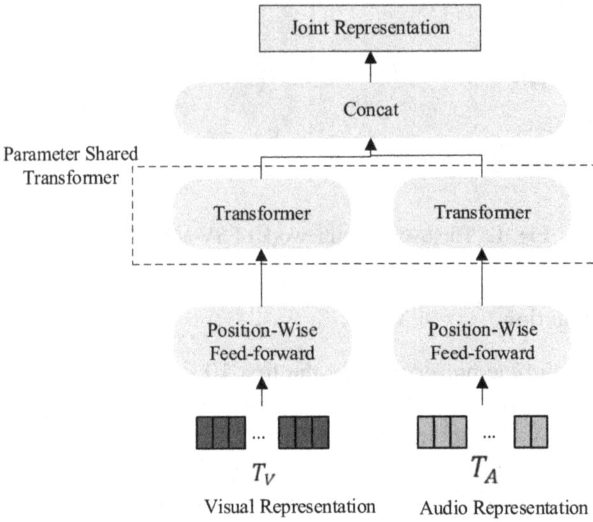

Fig. 2. Local fusion module.

The feature vectors of different modalities are almost misaligned for the following reasons: (1) different modal length sequences are different. (2) The same index in different modal sequences does not necessarily represent information at the same time. (3) The feature extraction methods of different modalities are different, and the dimensions of modal feature vectors are different. To solve the above problems, the calculation process of multi-modal fusion is as follows. The two modalities are denoted as α and β, and the modal representation vectors are denoted as X_α and X_β. T_α and d_α represent the sequence length and characteristic dimension of modal α respectively. T_β and d_β represent the sequence length and characteristic dimension of modal β respectively. So, there are $X_\alpha \in \mathbb{R}^{T_\alpha \times d_\alpha}$ and $X_\beta \in \mathbb{R}^{T_\beta \times d_\beta}$.

The queries in the attention mechanism are denoted as $Q_\alpha = X_\alpha W_{Q_\alpha}$, the keys are denoted as $K_\beta = X_\beta W_{K_\beta}$, the values are denoted as $V_\beta = X_\beta W_{V_\beta}$, where $W_{Q_\alpha} \in \mathbb{R}^{d_\alpha \times d_k}$, $W_{K_\beta} \in \mathbb{R}^{d_\beta \times d_k}$ and $W_{V_\beta} \in \mathbb{R}^{d_\beta \times d_v}$ is the learnable matrix parameter. The attention weight of queries of modal α to keys of modal β is calculated as follows formula (1).

$$att_{\alpha \to \beta} = \text{softmax}\left(\frac{Q_\alpha K_\beta^\top}{\sqrt{d_k}}\right) = \text{softmax}\left(\frac{X_\alpha W_{Q_\alpha} W_{K_\beta}^\top X_\beta^\top}{\sqrt{d_k}}\right) \tag{1}$$

After obtaining the cross-modal attention weight, $CM_{\beta\rightarrow\alpha}$ is obtained by formula (2), which represents the information extracted by modality α from modality β.

$$CM_{\beta\rightarrow\alpha}(X_\alpha, X_\beta) = att_{\alpha\rightarrow\beta}V_\beta = att_{\alpha\rightarrow\beta}X_\beta W_{V_\beta} \tag{2}$$

$CM_{\beta\rightarrow\alpha}$ and Q_α have the same sequence length but represent information from modal β because it weights V_β.

In fact, Eq. (1) obtains an attention matrix, where the (i, j) element represents the attention weight from the information at the i th moment in modal α to the j th moment in modal β. This method of obtaining a set of attention weights for both sequences is also known as the single-headed attention mechanism.

The model based solely on attention does not consider the location information. If the order of each token in the input sequence is adjusted or disrupted, the output calculated by attention weight will not change. To solve this problem, Transformer [12] for machine translation manually adds an initialized position code vector for different positions, but it is not a trainable parameter. In BERT, the position code vector is initialized randomly, and then the model learns the position code vector by itself. This section uses fixed position encoding vector and initializes it before training. For an input sequence X with length T and dimension d, the initialization strategy of position coding is shown in the following formula (3) and (4).

$$PE\left[i, 2j\right] = \sin\left(\frac{i}{10000^{\frac{2j}{d}}}\right) \tag{3}$$

$$\mathrm{PE}\left[i, 2j+1\right] = \cos\left(\frac{i}{10000^{\frac{2j}{d}}}\right) \tag{4}$$

The i is the i th position in the sequence, $i = 1, \ldots, T \circ j$ is the j th, $j = 0, \ldots, \frac{d}{2}$. The position coding vector presents sine or cosine changes in the characteristic dimension, and the element value of the vector is between $[-1, 1]$.

Analogous to Transformer, this paper combines cross-modal attention modules, residual connection, position coding and layer normalization into a cross-modal Transformer module that allows one modal to receive information from other modalities during training. As shown in Fig. 4, when using a cross-modal Transformer, multiple layers are stacked together to extract higher-order combined features. As shown in Eqs. (5), (6) and (7), in the multi-layer cross-modal Transformer, the forward calculation is performed sequentially from the first layer to the D layer.

$$Z^{[0]}_{\alpha\rightarrow\beta} = Z^{[0]}_\beta \tag{5}$$

$$\begin{aligned}\hat{Z}^{[i]}_{\alpha\rightarrow\beta} = \mathrm{CM}^{[i],\mathrm{multi-head}}_{\alpha\rightarrow\beta}\left(\mathrm{LNorm}\left(Z^{[0]}_\alpha\right)\right) \\ +\mathrm{LNorm}\left(Z^{[i-1]}_{\alpha\rightarrow\beta}\right)\end{aligned} \tag{6}$$

$$Z^{[i]}_{\alpha\rightarrow\beta} = f_{\theta^{[i]}_{\alpha\rightarrow\beta}}\left(\mathrm{LNorm}\left(\hat{Z}^{[i]}_{\alpha\rightarrow\beta}\right)\right) + \mathrm{LNorm}\left(\hat{Z}^{[i]}_{\alpha\rightarrow\beta}\right) \tag{7}$$

where $i = 1, \ldots, D$ represents the i-th layer, $Z^{[0]}_{\alpha\rightarrow\beta}$ represents the input of the first layer, $Z^{[i]}_{\alpha\rightarrow\beta}$ represents the output of the i-th layer, f_θ represents the parameters of the

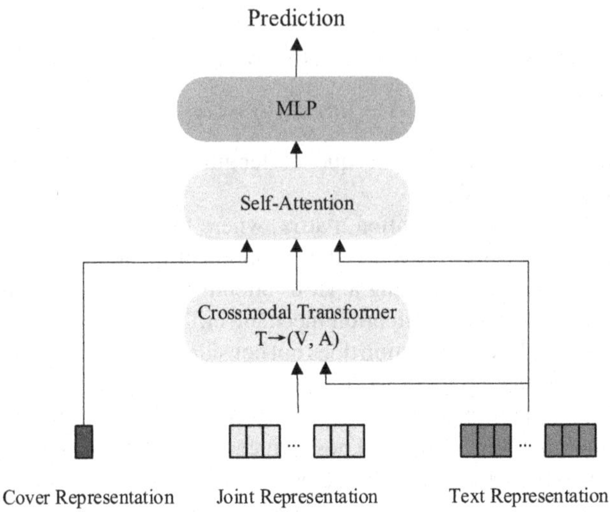

Fig. 3. Multi-modal fusion module.

feed-forward sublayer in the module, $CM_{\alpha \to \beta}^{[i],\text{multi-head}}$ represents the cross – modal information extracted by multi – head attention at layer i, and LNorm represents layer normalization.

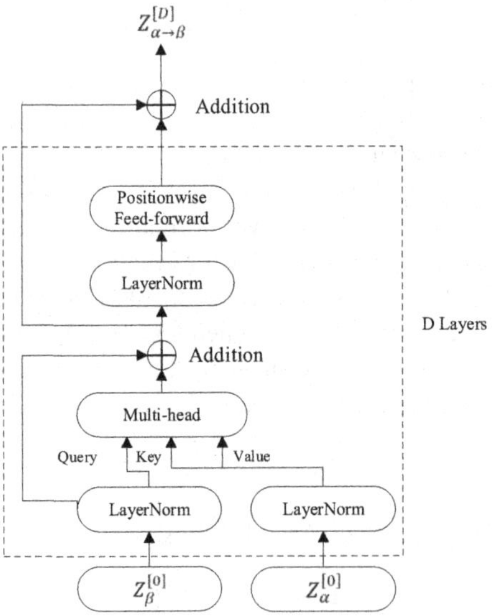

Fig. 4. The cross-modal transformer module.

The self-attention mechanism is used to combine the information from different modalities through the cross-modal attention module of text to audio and visual joint representation. The self-attention mechanism itself does not include location encoding. Since the position information has been added to the model in the shallow module, in the final fusion stage, there is no order between the information of different modalities, so the self-attention mechanism is used here for fusion.

2.5 Loss Function

This paper is a short video multi-classification problem, so the cross-entropy function is used as the loss function of the model. For a single sample, the predicted probability for each class is denoted by the vector ŷ, the true label is recorded as a vector y after one-hot encoding. Equation (8) is the loss function of a single sample.

$$l(y) = -\sum_{i=1}^{K} y_i \log \hat{y}_i \tag{8}$$

Softmax indicates the possibility of predicting each category. Equation (9) shows the calculation process of Softmax, z_i represents the i th element in the output vector of the fully connected layer, K represents the number of categories of the classification result. S_i represents the predicted probability of the i th class of the output.

$$S_i = \frac{e^{z_i}}{\sum_{k=1}^{K} e^{z_k}} \tag{9}$$

2.6 Post-Processing

The post-processing steps are as follows: Firstly, keywords of distinct categories are collected in numerous ways and a thesaurus is established. For a sample, after obtaining the scores of distinct categories through the model, it is judged whether it hits various categories of thesaurus. If the keywords in the thesaurus are not hit, the category with the highest score is directly taken as the prediction result; if the keyword in the thesaurus is hit, and the score of the category is higher than a certain threshold, the category is directly used as the classification result; if the keyword is hit, but the score is not higher than the specified threshold, the category with the highest score is still taken as the result. If there are multiple categories that meet the hit keywords at the same time, and the score meets the threshold requirements, the category with the highest score is taken as the prediction result.

Figure 5 shows the post-processing correction process. If the category with the highest score is directly taken as the prediction result, the model will classify the video as "Game". However, in the post-processing process, the video hits the keyword "Inuyasha" in the "anime" thesaurus, and the model's score for the "anime" category is higher than a certain threshold. So post-processing corrects the result to a "Quadratic Element". Checking the annotations of the video, post-processing gives the correct class, but the class with the highest score is not the correct class.

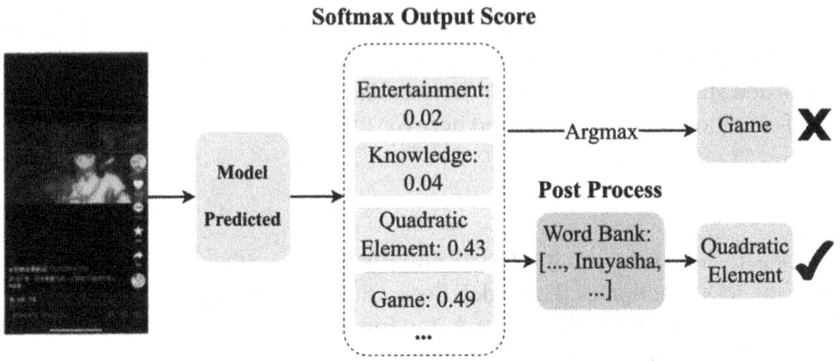

Fig. 5. The post-processing correction process.

3 Experiment

This section presents a multimodal short video classification dataset. The multi-modal method in SV-MF scheme is carried out on this dataset, which proves the effectiveness of the multi-modal method in SV-MF scheme for short video classification.

3.1 Dataset Introduction

The data set used in this paper includes 38,271 short videos with an average duration of about 58 s. The data set includes eight categories of entertainment, knowledge, anime, games, food, sports, fashion, and music. We use search engines to collect videos, and manually check the collected video tags, and use popular categories of short video platforms to label videos. There is only one label per video in the experiment. Map each category to a numerical value, using 0–8 to represent its corresponding category.

3.2 Implementation Details

The experiment of this paper is implemented with a single RTX3090 GPU (graphics processing units). The evaluation metric in this paper adopts the multi-class accuracy rate. Due to the small number of categories in the dataset, only the Top-1 accuracy rate is used. Formula 10 is the calculation method: the number of correctly predicted samples N_{right} divided by the total number of samples in the test set N_{all}.

$$accuracy = \frac{N_{right}}{N_{all}} \tag{10}$$

Xavier [13] initialization is to initialize the parameters to a uniform distribution with mean 0 and variance σ. However, Xavier initialization does not apply to ReLU and Sigmoid activation functions. For the network sub-module that is not pre-trained, the fully connected layer in it uses the Xavier initialization method, and the convolutional network uses the Kaiming [14] initialization method.

This paper uses the Mini Batch gradient descent algorithm to update the neural network parameters. To speed up the convergence of the network, the Adam optimization algorithm [15] was chosen, and the Momentum hyper parameter was set to 0.9. At the same time, the periodic decay strategy is used for model training, and the change of the learning rate during the training process is shown in Fig. 6.

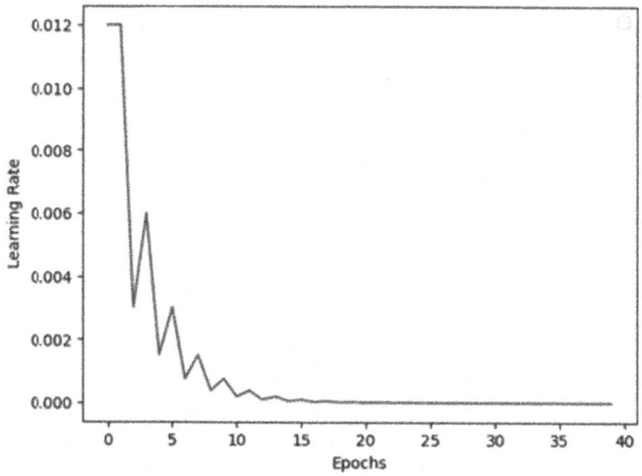

Fig. 6. The change of the learning rate.

3.3 Comparative Analysis of Experimental Results

This section conducts experiments on the (SV-MF) scheme proposed in this paper.

Table 1. Multimodal Model Experiment

Fusion method	Modal fusion method	Accuracy
Late-Fusion	Average fusion	0.692
	Weighted fusion	0.723
Early-Fusion	NetVLAD + Concat	0.726
	Attention clustering + Concat	0.730
SNHNet	SNHNet	0.757
	SNHNet-v0 (without Siamese Transformer)	0.743
	SNHNet-v1 (without Cross-modal Transformer)	0.749

Comparisons uses a text-to-audio and visual cross-modal Transformer and the local fusion structure. To reflect the difference, in this section, one Transformer used in local fusion is converted to two different Transformers for experiments, the model is recorded as SNHNet-v0, and the model that removes the cross-modal. Transformer is recorded as SNHNet-v1. After removing this module, the three-way input is only spliced in the modal fusion part, and then input to the subsequent self-attention module. The experimental results are shown in Table 1.

According to the experimental results, the multi-modal fusion effect is better than the single-modal model. SNHNet outperforms the two fusion methods of front-end and back-end, and only uses the cross-modal Transformer to achieve superior results. Using different Transformers in local fusion has a greater impact on the model than removing the cross-modal Transformer. Both methods jointly strengthen the visual and audio feature representations, further improving model accuracy.

3.4 Experimental Results of Different Frame Extraction Strategies

This section uses SNHNet to compare the effect of different frame rates on the model by adjusting the frame density of the video during pre-processing. The model defaults to extracting 30 frames from a 10-s video, which means 3 FPS. According to the experimental results shown in Table 2, increasing the FPS from 3 to 6 can achieve better results.

Table 2. Experimental results of different frame density

FPS	1	3	6	9	12
Accuracy	0.726	0.757	0.759	0.752	0.747

3.5 Post Processing

Table 3 shows the experimental results of post-processing. Although the post-processing method only uses a simple form of knowledge such as keywords, it has achieved reliable results and brought a stable improvement. For the different models experimented in this section, the Accuracy is improved by 2 to 4 percentage points.

Table 3. The experimental results of post-processing

Model	SNHNet	Weighted fusion	Attention clustering + Concat
Accuracy (before)	0.757	0.723	0.730
Accuracy (after)	0.778	0.753	0.752

4 Conclusion

This paper aims at the shortcomings of previous studies, an SV-MF scheme based on deep learning combined with pre-trained models is proposed to complete the classification task of short video. The SV-MF scheme newly designs a hybrid fusion model for multimodal short video classification, and proposed a local fusion method based on Transformer, so that low-level visual and audio information can be locally fused which improves the effect of short video classification. We made a real dataset derived from Chinese social platform for short video classification to conduct experiments. The SV-MF scheme is compared with the single modal classification model and other classification models with different fusion methods, which verifies that the SV-MF scheme is better than previous short video classification strategies.

References

1. Shutsko, A.: User-generated short video content in social media. a case study of TikTok. International Conference on Human-Computer Interaction. Springer, pp. 108–125 (2020)
2. Krizhevsky, A., Sutskever, I., Hinton, G.E.: Imagenet classification with deep convolutional neural networks. Advances in Neural Information Processing Systems, 25 (2012)
3. Simonyan, K., Zisserman, A.: Two-stream convolutional networks for action recognition in videos. Advances in Neural Information Processing Systems, 1 (2014)
4. Soomro K, Zamir A R, Shah M. UCF101: A Dataset of 101 Human Actions Classes From Videos in The Wild[J]. Computer Science, 2012
5. Kuehne, H., Jhuang, H., Stiefelhagen, R., et al.: HMDB: A Large Video Database for Human Motion Recognition. Springer, Berlin Heidelberg (2013)
6. Long, X., Gan, C., De Melo, G., et al.: Attention clusters: Purely attention based local feature integration for video classification. Proceedings of the IEEE Conference on Computer Vision and Pattern Recognition, pp. 7834–7843 (2018)
7. Li, L.H., Yatskar, M., Yin, D., et al.: Visualbert: A Simple and Performant Baseline for Vision and Language. arXiv preprint arXiv:1908.03557 (2019)
8. Devlin, J., Chang, M.W., Lee, K., et al.: Bert: Pre-Training of Deep Bidirectional Transformers for Language Understanding. arXiv preprint arXiv:1810.04805 (2018)
9. Xie, S., Girshick, R., Dollár, P., et al.: Aggregated residual transformations for deep neural networks. Proceedings of the IEEE Conference on Computer Vision and Pattern Recognition, pp. 1492–1500 (2017)
10. Hershey, S., Chaudhuri, S., Ellis, D.P.W., et al.: CNN architectures for large-scale audio classification. 2017 IEEE International Conference on Acoustics, Speech and Signal Processing (ICASSP). IEEE, pp. 131–135 (2017)
11. Vaswani, A., Shazeer, N., Parmar, N., et al.: Attention is all you need. Advances in Neural Information Processing Systems **30** (2017)
12. He, K., Zhang, X., Ren, S., et al.: Deep residual learning for image recognition. Proceedings of the IEEE Conference on Computer Vision and Pattern Recognition, pp. 770–778 (2016)
13. Glorot, X., Bengio, Y.: Understanding the difficulty of training deep feedforward neural networks. JMLR Workshop and Conference Proceedings, pp. 249–256 (2010)
14. He, K., Zhang, X., Ren, S., et al.: Delving deep into rectifiers: Surpassing human-level performance on imagenet classification. Proceedings of the IEEE International Conference on Computer Vision, pp. 1026–1034 (2015)
15. Kingma, D.P., Ba, J.: Adam: A Method for Stochastic Optimization. arXiv preprint arXiv: 1412.6980 (2014)

DGNN: Dependency Graph Neural Network for Multimodal Emotion Recognition in Conversation

Zhen Zhang[1], Xin Wang[1], Lifeng Yuan[1(✉)], Gongxun Miao[1], Mengqiu Liu[1], Wenhao Yun[1], and Guohua Wu[1,2]

[1] School of Cyberspace Security, Hangzhou Dianzi University, Hangzhou 310018, China
{zhangzhen,wangxluo,yuanlifeng,miaogx,liumengqiu,yunwenhao,wugh}@hdu.edu.cn
[2] Data Security Governance Zhejiang Engineering Research Center, Hangzhou Dianzi University, Hangzhou 310018, China

Abstract. For emotion recognition in conversation (ERC), the modeling of conversational dependency plays a crucial role. Existing methods often directly connect multimodal information and then build a graph neural network based on a fixed number of past and future utterances. The former leads to the lack of interaction between modalities, and the latter is less consistent with the logic of the conversation. Therefore, in order to better build conversational dependency, we propose a Dependency Graph Neural Network (DGNN) for ERC. First, we present a cross-modal fusion transformer for modeling dependency between different modalities of the same utterance. Then, we design a directed graph neural network model based on the adaptive window for modeling dependency between different utterances. The results of the extensive experiments on two benchmark datasets demonstrate the superiority of the proposed model.

Keywords: Emotion recognition in conversation · Cross-modal fusion transformer · Adaptive window

1 Introduction

Emotion recognition in conversation (ERC) is a new task in natural language processing (NLP) that aims to determine the emotion label of each utterance in a conversation by analyzing multimodal information, including textual content, facial expressions, and audio signals. With the tremendous growth of conversational data on social media platforms and other sources in recent years, ERC has huge potential value in many fields, including medical diagnosis [1], opinion mining [10], fake news detection [3], and dialogue generation [7]. Therefore, ERC has attracted attention of many researchers.

The factors that influence the emotion of utterance are mainly historical conversation. As shown in Fig. 1, if speaker A does not know the historical conversational information, it is difficult to determine the emotion label of his 4-th

B. Luo et al. (Eds.): ICONIP 2023, CCIS 1963, pp. 86–99, 2024.
https://doi.org/10.1007/978-981-99-8138-0_8

utterance. Furthermore, textual information alone is not sufficient to convey enough information, audio and visual information can also express emotions to some extent. Therefore, in ERC research, many researchers have focused on modeling conversation dependency [4] with multimodal data to achieve better emotion recognition results.

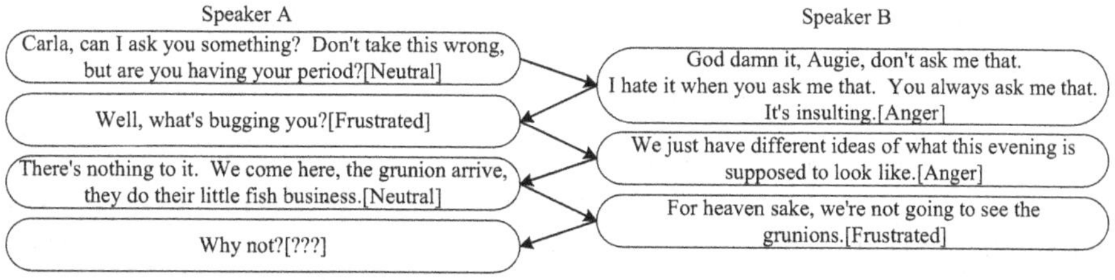

Fig. 1. An example of a conversation from the IEMOCAP dataset.

In order to model contextual information, many Recurrent Neural Networks (RNN)-based methods have been proposed in recent years. RNN-based methods mainly model dependency based on the temporal sequence of the conversation, such as CMN proposed by Hazarika et al. [11], and DialogueRNN proposed by Majumder et al. [20]. However, these methods tend to utilize information from the nearest utterances, with less influence from farther utterances. Moreover, when dealing with multimodal feature inputs, these methods often simply concatenate data without considering their interactions. Compared with RNN, Graph Neural Network (GNN) can view utterances as nodes in a graph, and can model contextual information by updating node states. So GNN often performs better in ERC task. For example, Ghosal et al. [9] proposed DialogueGCN, Hu et al. [14] proposed MMGCN. However, these methods often capture past and future utterances in the fixed window of the current utterance to construct a graph. The results of emotion recognition show that future utterances can affect the current utterance. But from the logicality of conversation, the emotion of the current utterance can only be inferred from past utterances. This is because the conversation is a process of one-way and irreversible information transmission.

Based on the limitations of existing methods, we propose a **D**ependency **G**raph **N**eural **N**etwork (DGNN), which includes four stages. It fully utilizes the intra-utterance (i.e., different modality information of the same utterance) and inter-utterance dependency. Specifically, first, we use three unimodal feature extractors to obtain three modality features separately. Then we input them into the Cross-Modal Fusion Transformer (CMFT) to obtain the fused feature representation. We use the audio and visual modalities as a complement to the textual modality. Next, according to the actual conversation structure, we construct an Adaptive Window Graph Neural Network (AWGNN) to model dependency between utterances. In addition, when aggregating information, we add the relational position encoding to the model. Finally, we use a Multi-Layer Perceptron (MLP) to obtain a predictive emotion label. Experimental results on two benchmark datasets demonstrate the superiority of the proposed model.

The contributions of this paper are as follows: (1) We propose DGNN, a novel Dependency Graph Neural Network, to better model conversational dependency for ERC; (2) We propose a cross-modal fusion transformer to fuse features from different modalities, which can enhance complementarity between modalities; (3) We propose an adaptive window graph neural network to construct dependency between utterances in a more logical way.

2 Related Work

Emotion recognition in conversation is a research hotspot in NLP due to its potential applications in extensive areas, and many researchers have proposed different methods. Based on the network models used in these methods, they can mainly be divided into two categories.

RNN-Based Methods. DialogueRNN [20] models speaker and sequence information by using three different GRUs. COSMIC [8] is structurally similar to DialogueRNN, but utilizes new common knowledge information to improve performance. DialogueCRN [13] fully understands conversational context from a cognitive perspective. BiERU [18] simplifies the three-step task into two steps, by merging the conversation information into utterance representations while obtaining them. These methods directly perform modal fusion, which may lack modality interaction. Therefore, this paper proposes a modal-fusion strategy to better utilize different modality information.

GNN-Based Methods. DialogueGCN [9] constructs relational graph attention network to capture long-range context dependencies in conversation. MMDFN [12] designs a graph-based dynamic fusion module to fuse multimodal features based on MMGCN [14]. Ra-GAN [15] incorporates relational position encoding into the RGCN [22] structure to capture speaker dependency and the sequential information. DAG-ERC [23] utilizes directed acyclic graphs to better model the intrinsic structure within a conversation. GraphCFC [17] alleviates the heterogeneity gap problem in multimodal fusion by utilizing multiple subspace extractors and pairwise cross-modal complementation strategy. These methods construct GNN based on a fixed number of past and future utterances. However, after considering the logical coherence and the complex diversity of actual conversations, this paper proposes a novel GNN construction strategy.

3 Methodology

The framework of our proposed model is shown in Fig. 2. We divide the process of emotion recognition into four steps: feature extraction, cross-modal dependency modeling, utterance dependency modeling, and emotion classifier.

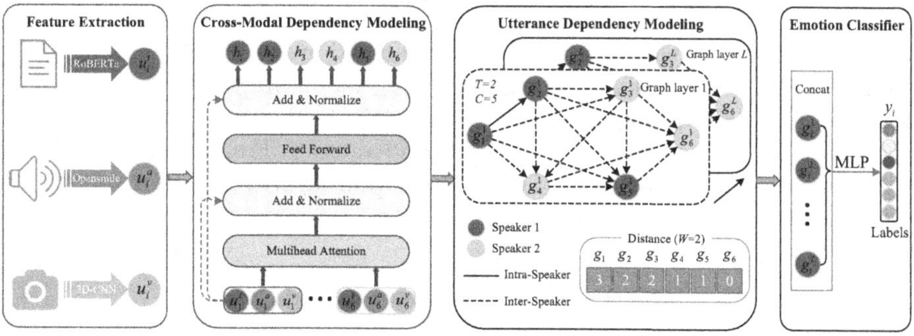

Fig. 2. The framework of DGNN.

3.1 Problem Definition

Give a conversation $U = \{u_1, u_2, \ldots, u_N\}$, where $u_i = \{u_i^t, u_i^a, u_i^v\}$, N is the number of utterances in the conversation. u_i^t, u_i^a, u_i^v represent textual (t), audio (a), and visual (v) modality feature of utterance u_i. S is the number of speakers, $P = \{p_1, p_2, \ldots, p_S\} (S \geq 2)$, the utterance u_i is spoken by speaker p_s, where $p_s = \varphi(u_i)$, φ represents the mapping relationship between utterance and speaker. The goal of multimodal emotion recognition in conversation is to predict the emotion label y_i of each utterance u_i, such as happy, angry, sad, etc.

3.2 Feature Extraction

Textual Feature Extraction. We use the pre-trained language model RoBERTa [19] to extract textual features. RoBERTa is an enhanced version of BRET [5] with larger number of model parameters, and more training data. By encoding Textual using RoBERTa, we obtain a 1024 dimensions feature vector.

Audio Feature Extraction. We use OpenSMILE [6] to extract audio features, which is a modular and flexible feature extractor. The IS10 paraling config file is used to extract 1582 features for each audio segment. And then we use the convolutional neural network to reduce the dimensionality to 1024 dimensions.

Visual Feature Extraction. We use the deep 3D-CNN model to extract visual features, which helps to capture emotional changes and extract more expressive feature vectors. We also extract 1024 dimensions visual feature vector.

3.3 Cross-Modal Dependency Modeling

We use audio and visual modalities as a complement to textual modality. We utilize the Cross-Modal Fusion Transformer (CMFT) to build the dependencies between different modalities. As shown in Eq. (1), we use audio feature u_i^a as query Q of the multi-head attention layer, and use textual feature u_i^t as key K and value V. After calculating multi-head attention, we obtain textual-audio fusion feature T_a. Similarly, as shown in Eq. (2), we obtain textual-visual fusion feature T_v by using visual feature u_i^v as query Q of other multi-head attention layer and using textual feature u_i^t as key K and value V. Finally, we add T_a and T_v together to obtain output T_{av} of the attention layer. And the structure of multi-head attention layer is shown in Fig. 3:

$$T_{a_m} = softmax\left(\frac{(u_i^a)_m * (u_i^t)_m}{\sqrt{d_k}}\right)(u_i^t)_m \tag{1}$$

$$T_{v_m} = softmax\left(\frac{(u_i^v)_m * (u_i^t)_m}{\sqrt{d_k}}\right)(u_i^t)_m \tag{2}$$

where T_{a_m} and T_{v_m} are the m-th attention head of T_a and T_v, d_k denotes the dimension of K. The output of the multi-head attention layer can be calculated as follows:

$$T_{av} = W^{ta}(||_{m=1}^{M} T_{a_m}) + W^{tv}(||_{m=1}^{M} T_{v_m}) \tag{3}$$

where $||$ denotes concatenation operation. M is the number of attention heads $(1 \leq m \leq M)$, W^{ta} and W^{tv} are two learnable weight matrices.

Referring to the structure of Transformer encoder, We calculate residual of T_{av} and u_i^t, then input the result into normalization layer to obtain the sub-layer output $\tilde{T_{av}}$. Finally, we input $\tilde{T_{av}}$ to feedforward layer and then calculate residual with u_i^t. The result is input to normalization layer to get the final output h_i of CMFT. The residual calculated twice with u_i^t further enhances the representation of textual modality information.

$$\tilde{T_{av}} = LayerNorm(u_i^t + T_{av}) \tag{4}$$

$$h_i = LayerNorm(u_i^t + FeedForward(\tilde{T_{av}})) \tag{5}$$

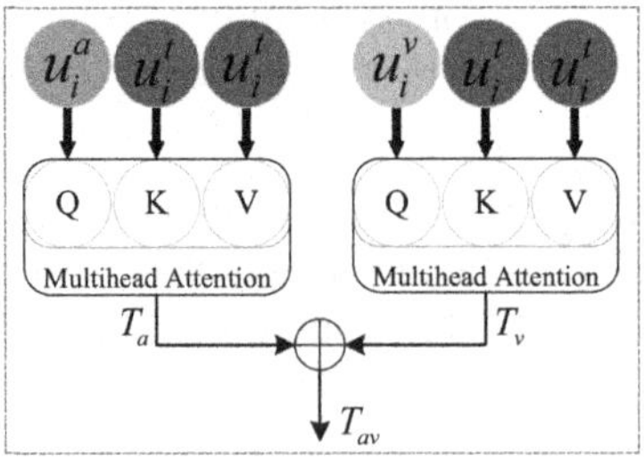

Fig. 3. The framework of multi-head attention.

3.4 Utterance Dependency Modeling

After obtaining the multimodal fused features, we construct a new Adaptive Windows Graph Neural Network (AWGNN) based on the actual conversation structure, which is more adaptable to various complex conversation situations.

Graph Architecture. We construct a directed graph $\mathcal{G} = (\mathcal{V}, \mathcal{E}, \mathcal{R})$ for a conversation consisting of N utterances. The nodes in the graph are the utterances in conversation, i.e., $\mathcal{V} = \{u_1, u_2, \ldots, u_N\}$. $\forall j < i, (v_j, v_i, r_{ji}) \in \mathcal{E}$ represents the propagation of information from u_j to u_i, where $r_{ji} \in \mathcal{R}$ is a relation type.

Nodes. We initialize the output h_i of CMFT as feature vector g_i for each node v_i. Through the stacked graphical layers, we use the aggregation algorithm to obtain the output feature g_i^l, where l is the number of layers in the stack.

Edges. We focus only on the influence of past utterances on current utterance. To tackle complex and variable conversation, we propose a novel adaptive window algorithm instead of using a fixed-sized window. This algorithm has three cases. *Case1:* we first select a candidate past utterances window C for the current utterance. If the number of consecutive utterances O by one speaker in C past utterances is greater than the threshold window T, then the actual window size $W = min(O, C)$. *Case2:* if O less than equal T in C past utterances, but the number of utterances by one speaker is at least twice as much as any other speakers, or at most half any other speakers, the actual window size $W = T$. *Case3:* if neither of the above two cases applies, the actual window size $W = T$. When calculating the distance D_{ij} between utterances, we use absolute position for the first two cases and relational position [15] for the third case. Finally, when the distance is less than equal W, a directed edge is constructed. We show three examples of graph construction in Fig. 4, corresponding to the three cases.

Edge Types. There are two types of edges, indicating whether the connected utterances are spoken by the same speaker or different speaker, i.e., $\mathcal{R} = \{1, 0\}$.

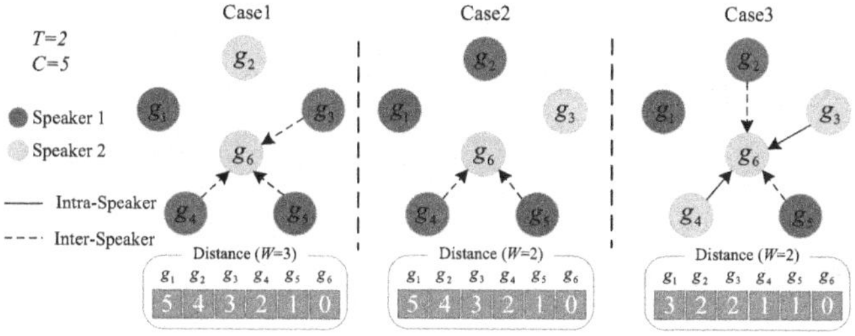

Fig. 4. Three examples of graph construction, the table shows the distances for each utterance to the current utterance u_6.

Aggregation Information. For each node i in each layer of graph, we aggregate information from its neighbors at the same layer and combine with the information from the node i at the previous layer. The edge weights α_{ij}^l is computed by the attention mechanism, and inspired by Ishiwatari et al. [15], we add position encoding into the attention mechanism. For the utterances u_i in layer l, the edge weights can be calculated as follows:

$$D_{ij} = E_d\left(\sigma(u_i - u_j)\right) \tag{6}$$

$$\alpha_{ij}^l = Softmax_{j \in N_i}\left(LRL\left(X^T\left[W_a^l g_i^{l-1} \parallel W_b^l\left(g_j^l + D_{ij}\right)\right]\right)\right) \tag{7}$$

where D_{ij} represents the distance representation between utterances u_i and u_j, which is encoded by the trainable embedding matrix E_d. σ represents the mapping function between the position of utterances and conversation. N_i is the set of all Neighborhoods of node i. LRL denotes the *LeakyRelu* activation function. X is a parameterized weight vector and T represents transpose. W_a^l and W_b^l are trainable parameters. $||$ denotes concatenation operation.

When aggregating information, we not only gather information based on edge weights, but also introduce the edge relationship types to distinguish the influence of different speakers:

$$G_i^l = \sum_{j \in \mathcal{N}_i} \alpha_{ij} W_{r_{ij}}^l g_j^l \tag{8}$$

where $W_{r_{ij}}^l \in \{W_1^l, W_0^l\}$, is the trainable parameters for different edge types.

After obtaining the aggregated information, we combine G_i^l with the representation g_i^{l-1} of u_i in the previous layer using a gated recurrent unit (GRU) to update the representation of u_i in the current layer:

$$F_i^l = GRU_f^l(g_i^{l-1}, G_i^l) \tag{9}$$

According to the method in DAG-ERC [23], we know that using only Eq. (9), the contextual information is not fully utilized. Therefore, in order to interact the utterance u_i sufficiently with its neighborhoods, we reverse g_i^{l-1} and G_i^l to simulate the historical information flow as shown in Eq. (10):

$$B_i^l = GRU_b^l(G_i^l, g_i^{l-1}) \tag{10}$$

Finally, the representation of u_i at layer l is obtained by adding F_i^l and B_i^l:

$$g_i^l = F_i^l + B_i^l \tag{11}$$

3.5 Emotion Classifier

After concatenating the representations of the utterance u_i at all layers, we input it into a multi-layer perceptron (MLP) to obtain the predicted emotion label:

$$Z_i = ReLU\big(W_g g_i + b_g\big) \tag{12}$$
$$P_i = Softmax(W_z Z_i + b_z) \tag{13}$$
$$y_i = \arg\max_m(P_i[m]) \tag{14}$$

where W_g and W_z are weight parameters, b_g and b_z are bias parameters. We use cross-entropy along with L2 regularization as the loss function for training:

$$\mathcal{L} = -\frac{1}{\sum_{l=1}^{L} n(l)} \sum_{i=1}^{L} \sum_{j=1}^{N_i} logP_{i,j}(\hat{y}_{i,j}) + \mu||\Theta_{re}||_2 \tag{15}$$

where $\mathcal{L}$ refers to the number of conversations in the training set, N_i is the number of utterances in the i-th conversation, $P_{i,j}$ denotes the probability distribution of the predicted emotion label of utterance j in utterance i, $\hat{y}_{i,j}$ denotes the ground truth label corresponding to utterance j, μ is the weight parameter of $L2$ regularization, and Θ_{re} is all trainable parameters.

4 Experiments Setting

4.1 Datasets

We evaluate our proposed model DGNN on two benchmark datasets, IEMOCAP [2] and MELD [21]. The statistic of them are shown in Table 1. IEMOCAP is a multimodal dataset that includes video, audio and textual transcription of conversations from 10 actors. The dataset contains 151 conversations and 7433 utterances. There are six emotion labels: *happy, angry, sad, neutral, excited, and frustrated.* MELD is extracted from Friends TV series, which is also a multimodal dataset that includes audio, video and textual. The dataset contains 1433 conversations, 13708 utterances and 304 different speakers. It includes seven emotion labels: *neutral, surprise, sadness, joy, angry, fear, and disgust.*

Table 1. The statistics of IEMOCAP and MELD datasets.

Dataset	# Conversations			# Uterrances		
	Train	Val	Test	Train	Val	Test
IEMOCAP	120		31	5810		1623
MELD	1038	114	280	9989	1109	2610

4.2 Baseline Methods

CMN [11] directly connect three modalities of information as input to the model and use two GRUs to model the conversation of two speakers.
DialogueRNN [20] uses three GRUs to track the state and sequence information of each speaker in conversation, and performs emotion classification based on this information.
DialogueCRN [13] designs a multi-round inference module to extract and integrate emotional clues and fully understand context information from a cognitive perspective.
DialogueGCN [9] constructs the conversation as a graph structure and uses the dependencies between speakers and within speakers to improve context understanding. We extend its input to be multimodal.
MMGCN [14] utilizes the dependencies between multiple modalities and speakers information. It constructs a fully connected graph in each modality and connects nodes corresponding to the same utterance in different modalities.
DAG-ERC [23] combines the advantages of RNN and GNN, constructs the conversation as a directed acyclic graph. We expand its input to be multimodal for better comparison with our proposed model.
MMDFN [12] designs a graph-based dynamic fusion module to fuse multimodal context information and aggregate intra-modal and inter-modal context information in a specific semantic space at each layer.
COGMEN [16] models local and global information based on GNN and uses Transformer for modeling speaker relations.
GraphCFC [17] propose a cross modal feature complementarity module, which can effectively model contextual and interactive information.

4.3 Implementation Details

We implement the proposed DGNN model in the PyTorch framework, and all experiments are run on GeForce RTX 3090. Using AdamW as the optimizer with an initial learning rate of 1e-4, and the L2 regularization parameter is 1e-5. The number of attention heads for CMFT is 8, and the number of AWGNN layers is 4. For IEMOCAP dataset, the learning rate is 5e-4, the dropout is 0.2, and the epoch is 30. For MELD dataset, the learning rate is 1e-5, the dropout rate is 0.2, and the epoch is 70. The reported results of our implemented models are all based on the average score of 5 random runs on the test set.

5 Results and Analysis

5.1 Overall Performance

On the IEMOCAP and MELD dataset, we compare our proposed DGNN with the baseline methods in Sect. 4.2, and all the experimental results are shown in Table 2. We use the weighted average F1 score (wa-F1) and accuracy score (Acc) as evaluation metric, and record the F1 score for each emotion in detail.

We can see that the proposed model DGNN achieves the best Acc and wa-F1 on two datasets, demonstrating its superiority. And the DGNN model has higher F1 scores in most emotions than the other models when each emotion is observed separately. On the IEMOCAP dataset, sad and excited have a higher F1 score, while happy has a lower score. On the MELD dataset, neutral has a higher score, while sad and anger have a lower scores. By looking at these two datasets, we find that they suffer from class imbalance. This imbalance can lead to the model being biased towards the majority classes, struggling to correctly classify the minority class.

There is a significant improvement on the IEMOCAP dataset, but not on MELD. After comparing the differences between the two datasets, we find that the MELD dataset has more speakers and shorter conversation than the IEMOCAP dataset. And the MELD dataset is intercepted from a TV series, some of conversations is incoherent. So DGNN may not capture the contextual information well on the MELD dataset.

In addition, we note that the performance of RNN-based methods is usually worse than that of GNN-based methods. This suggests that GNN-based methods can better capture the dependencies between utterances.

Table 2. Overall performance of all methods on two datasets. All results are in the multimodal setting. IEMOCAP has 6 labels, and MELD has 5 labels (where *Fear* and *Disgust* are not reported due to their unsatisfactory results). Best performances are highlighted in bold, and some of the missing experimental data are indicated by "-".

Methods	IEMOCAP								MELD						
	Happy	*Sad*	*Neutral*	*Angry*	*Excited*	*Frustrated*	Acc	wa-F1	*Neutral*	*Surprise*	*Sadness*	*Joy*	*Anger*	Acc	wa-F1
CMN	30.38	62.41	52.39	59.83	60.25	60.69	56.56	56.13	–	–	–	–	–	–	–
DialogueRNN	33.18	78.80	59.21	65.28	71.86	58.91	63.40	62.75	76.79	47.69	20.41	50.92	45.52	60.31	57.66
DialogueCRN	**53.23**	83.37	62.96	66.09	75.40	66.07	67.16	67.21	77.01	50.10	26.63	52.77	45.15	61.11	58.67
DialogueGCN	51.57	80.48	57.69	53.95	72.81	57.33	63.22	62.89	75.97	46.05	19.60	51.20	40.83	58.62	56.36
MMGCN	40.78	78.65	62.76	68.28	74.24	61.92	65.93	65.71	76.33	48.15	26.74	53.02	46.09	60.42	58.31
DAG-ERC	51.15	81.99	67.70	68.64	67.97	68.32	68.56	68.69	77.18	58.27	37.36	60.04	49.24	64.02	63.61
MM-DFN	42.22	78.98	66.42	69.77	75.56	66.33	68.21	68.18	**77.76**	50.69	22.93	54.78	47.82	62.49	59.46
COGMEN	51.9	81.7	68.6	66.0	75.3	58.2	68.2	67.6	76.82	55.41	35.78	58.98	48.64	63.21	62.51
GraphCFC	43.08	84.99	64.70	**71.35**	78.86	63.70	69.13	68.91	76.98	49.36	26.89	51.88	47.59	61.42	58.86
DGNN	47.15	**85.09**	**69.13**	63.70	**79.03**	**68.93**	**70.90**	**70.51**	77.74	**58.44**	**39.59**	**60.62**	**49.83**	**64.48**	**64.05**

5.2 Ablation Study

To better study the impact of each module in DGNN, we conducted experiments by replacing CMFT with directly connecting three modalities and replacing AWGNN with a fixed-window GNN. The experimental results show in Table 3. When CMFT or AWGNN is replaced, resulting in a significant performance decrease on both datasets. When both CMFT and AWGNN are replaced, the performance drop the most. In conclusion, both of these modules contribute to improving the performance of DGNN. In addition, we observe that the performance of IEMOCAP dataset drop more than MELD dataset, which may be due to DGNN is not very suitable for MELD. This further validates the conclusion in Sect. 5.1.

Table 3. Performance (wa-F1) of ablation study on two datasets.

Method	IEMOCAP	MELD
DGNN	**70.51**	**64.05**
- CMFT	68.86 (↓1.84)	63.01 (↓1.04)
- AWGNN	69.15 (↓1.37)	63.57 (↓0.48)
- CMFT and AWGNN	68.38 (↓2.14)	62.62 (↓1.43)

5.3 Various Modality Settings

The performance of DGNN on IEMOCAP and MELD dataset with different modality settings is shown in Table 4. For the unimodal data, the textual modality performs significantly better than the other two modalities. The performance of multimodal data is significantly better than that of unimodal data. The performance is improved when audio and visual modalities are fused with textual modality, and the performance is slightly better when textual and audio are fused. The performance the fusion of the three modalities is naturally the best.

Table 4. Performance (wa-F1) of different modality settings.

Modality setting	IEMOCAP	MELD
Textual	68.70	62.77
Audio	47.35	42.11
Video	32.68	33.93
T+A	69.59	63.62
T+V	68.87	63.33
T+V+A	**70.51**	**64.05**

5.4 Effect of Candidate Window Size and Threshold Window Size

The performance of DGNN on IEMOCAP dataset with different combinations of threshold window size (T) and candidate window size (C) are shown in Fig. 5. We set T to $\{2, 3, \ldots, 10\}$ and C to $\{10, 12, \ldots, 26\}$, and obtain a total of 81 results. As T and C increase, the overall trend of performance is to first increase and then decrease. When T is 6 and C is 18, we obtain the best Acc and wa-f1. It can be inferred that too many or too few utterances can both reduce DGNN comprehension. $T = 6$ and $C = 18$ are the moderate parameter that can be applied to majority of conversations on IEMOCAP dataset. Some experiments conducted on MELD dataset also demonstrate this.

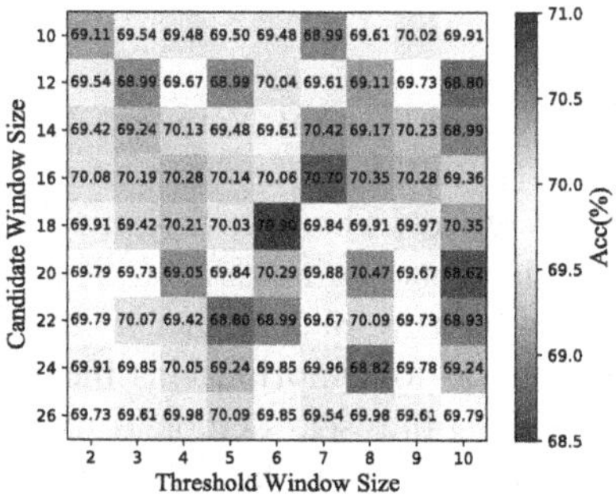

(a) Performance change of Acc.

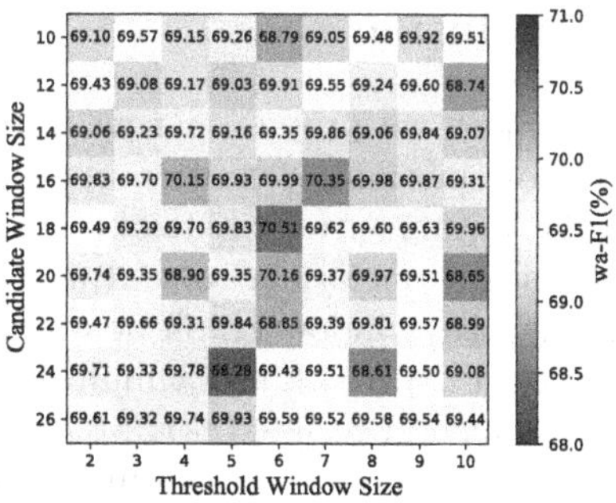

(b) Performance change of wa-F1.

Fig. 5. Effect of candidate window size and threshold window size on IEMOCAP dataset.

5.5 Case Study

As shown in Fig. 6a, common phrases such as "yeah", "Okay" and "I know" are often identified as *Neutral*, but their true emotion in conversation may be *Sad*. Therefore, sometimes it's difficult to correctly recognize emotion in the single-textual modality model. However, with the combination of audio and visual modalities, these utterances may be correctly identified as *Sad*.

As shown in Fig. 6b, we want to determine the emotion of utterance 7. However, if we use other model to judge based on three past and future utterances, it will be recognized as *Neutral*. But after communicating sufficiently with past utterances based on AWGNN, utterance 7 can be correctly identified as *Sad*.

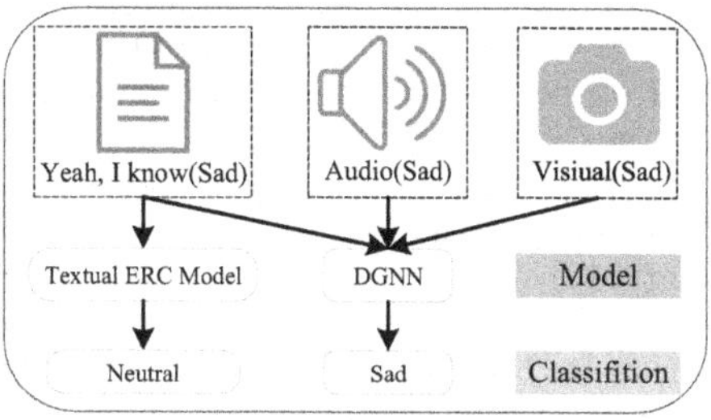

(a) An case of multimodal.

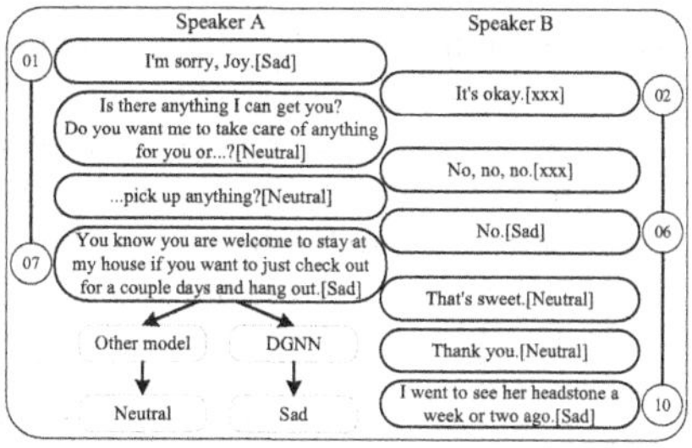

(b) An case of AWGNN.

Fig. 6. Two cases of ERC on IEMOCAP dataset. In (b), unknown emotions are indicated by "xxx".

6 Conclusion

In this paper, we propose a new Dependency Graph Neural Network, named DGNN, for emotion recognition in conversation (ERC). Specifically, we first use unimodal feature extractors to obtain modality features. Then, we design a cross-modal fusion transformer to capture dependencies between modalities, followed by an adaptive window graph neural network to capture dependencies between utterances. Finally, we use a multi-layer perceptron for emotion classification. The experimental results on IEMOCAP and MELD datasets demonstrate that our proposed model outperforms other methods, and can effectively capture intra-modal information and inter-utterance information. In future work, we intend to further improve the model so that it can be better applied to emotion recognition in multi-party short conversation.

Acknowledgements. This research was supported by "Pioneer" and "Leading Goose" R&D Program of Zhejiang (Grant No. 2023C03203, 2023C03180, 2022C03174).

References

1. Bhavya, S., Nayak, D.S., Dmello, R.C., Nayak, A., Bangera, S.S.: Machine learning applied to speech emotion analysis for depression recognition. In: 2023 International Conference for Advancement in Technology (ICONAT), pp. 1–5 (2023)
2. Busso, C., et al.: IEMOCAP: interactive emotional dyadic motion capture database. Lang. Resour. Eval. **42**(4), 335–359 (2008)
3. Cevallos, M., De Biase, M., Vocaturo, E., Zumpano, E.: Fake news detection on COVID 19 tweets via supervised learning approach. In: 2022 IEEE International Conference on Bioinformatics and Biomedicine (BIBM), pp. 2765–2772 (2022)
4. Deng, J., Ren, F.: A survey of textual emotion recognition and its challenges. IEEE Trans. Affect. Comput. **14**(1), 49–67 (2021)
5. Devlin, J., Chang, M.W., Lee, K., Toutanova, K.: BERT: pre-training of deep bidirectional transformers for language understanding. In: Proceedings of the 2019 Conference of the North American Chapter of the Association for Computational Linguistics: Human Language Technologies, pp. 4171–4186 (2019)
6. Eyben, F., Wöllmer, M., Schuller, B.: OpenSMILE: the Munich versatile and fast open-source audio feature extractor. In: Proceedings of the 18th ACM International Conference on Multimedia, pp. 1459–1462 (2010)
7. Gao, P., Han, D., Zhou, R., Zhang, X., Wang, Z.: CAB: empathetic dialogue generation with cognition, affection and behavior. In: Database Systems for Advanced Applications: 28th International Conference, pp. 597–606 (2023)
8. Ghosal, D., Majumder, N., Gelbukh, A., Mihalcea, R., Poria, S.: COSMIC: COmmonSense knowledge for emotion identification in conversations. In: Findings of the Association for Computational Linguistics: EMNLP 2020, pp. 2470–2481 (2020)
9. Ghosal, D., Majumder, N., Poria, S., Chhaya, N., Gelbukh, A.: DialogueGCN: A graph convolutional neural network for emotion recognition in conversation. In: Proceedings of the 2019 Conference on Empirical Methods in Natural Language Processing, pp. 154–164 (2019)
10. Ghosal, S., Jain, A.: HateCircle and unsupervised hate speech detection incorporating emotion and contextual semantic. ACM Trans. Asian Low-Resour. Lang. Inf. Process. **22**(4), 2375–4699 (2022)

11. Hazarika, D., Poria, S., Zadeh, A., Cambria, E., Morency, L.P., Zimmermann, R.: Conversational memory network for emotion recognition in dyadic dialogue videos. In: Proceedings of the 2018 conference of the Association for Computational Linguistics. vol. 2018, pp. 2122–2132 (2018)
12. Hu, D., Hou, X., Wei, L., Jiang, L., Mo, Y.: MM-DFN: multimodal dynamic fusion network for emotion recognition in conversations. In: ICASSP 2022–2022 IEEE International Conference on Acoustics, Speech and Signal Processing (ICASSP), pp. 7037–7041 (2022)
13. Hu, D., Wei, L., Huai, X.: DialogueCRN: contextual reasoning networks for emotion recognition in conversations. In: Proceedings of the 59th Annual Meeting of the Association for Computational Linguistics, pp. 2470–2481 (2021)
14. Hu, J., Liu, Y., Zhao, J., Jin, Q.: MMGCN: multimodal fusion via deep graph convolution network for emotion recognition in conversation. In: Proceedings of the 59th Annual Meeting of the Association for Computational Linguistics, pp. 5666–5675 (2021)
15. Ishiwatari, T., Yasuda, Y., Miyazaki, T., Goto, J.: Relation-aware graph attention networks with relational position encodings for emotion recognition in conversations. In: Proceedings of the 2020 Conference on Empirical Methods in Natural Language Processing (EMNLP), pp. 7360–7370 (2020)
16. Joshi, A., Bhat, A., Jain, A., Singh, A., Modi, A.: COGMEN: COntextualized GNN based multimodal emotion recognitioN. In: Proceedings of the 2022 Conference of the North American Chapter of the Association for Computational Linguistics: Human Language Technologies, pp. 4148–4164 (2022)
17. Li, J., Wang, X., Lv, G., Zeng, Z.: GraphCFC: A directed graph based cross-modal feature complementation approach for multimodal conversational emotion recognition. IEEE Transactions on Multimedia (2023)
18. Li, W., Shao, W., Ji, S., Cambria, E.: BiERU: bidirectional emotional recurrent unit for conversational sentiment analysis. Neurocomputing **467**(7), 73–82 (2022)
19. Liu, Y., et al.: RoBERTa: A robustly optimized BERT pretraining approach. arXiv preprint arXiv:1907.11692 (2019)
20. Majumder, N., Poria, S., Hazarika, D., Mihalcea, R., Gelbukh, A., Cambria, E.: DialogueRNN: an attentive RNN for emotion detection in conversations. In: Proceedings of the AAAI Conference on Artificial Intelligence. vol. 33, pp. 6818–6825 (2019)
21. Poria, S., Hazarika, D., Majumder, N., Naik, G., Cambria, E., Mihalcea, R.: MELD: a multimodal multi-party dataset for emotion recognition in conversations. In: Proceedings of the 57th Annual Meeting of the Association for Computational Linguistics, pp. 527–536 (2019)
22. Schlichtkrull, M., Kipf, T.N., Bloem, P., Van Den Berg, R., Titov, I., Welling, M.: Modeling relational data with graph convolutional networks. In: The Semantic Web: 15th International Conference, pp. 593–607 (2018)
23. Shen, W., Wu, S., Yang, Y., Quan, X.: Directed acyclic graph network for conversational emotion recognition. In: Proceedings of the 59th Annual Meeting of the Association for Computational Linguistics (ACL), pp. 1551–1560 (2021)

Global Exponential Synchronization of Quaternion-Valued Neural Networks via Quantized Control

Jiaqiang Huang[1], Junjian Huang[1,2(✉)], Jinyue Yang[1], and Yao Zhong[1]

[1] College of Electronic and Information Engineering, Southwest University, Chongqing 400715, China
[2] College of Electronic and Information Engineering, Southwest University, Chongqing 400715, China
hmomu@sina.com

Abstract. In this paper, quantized controllers are designed to implement global exponential synchronization of quaternion-valued neural networks. Firstly, based on Hamilton's principle, the quaternion-valued neural networks are decomposed into four equivalent real-valued neural networks. Then, utilizing the principles of Lyapunov stability and matrix inequality theory, the drive-response synchronization method is utilized to obtain the result on exponential synchronization of quaternion-valued neural networks. Finally, the effectiveness of the proposed method is verified by numerical simulation examples.

Keywords: Exponential Synchronization · Quantized control · Quaternion-valued neural networks · Lyapunov function

1 Introduction

Neural networks have been widely investigated over the last few decades for applications in domains such as pattern recognition [1], picture encryption [3], and information processing [2]. Quaternion-valued neural networks have received attention as a key technique to improve image processing capabilities [4]. In the image domain, as an extension of complex numbers, quaternions have one real and three imaginary parts [5]; Unlike real-valued neural networks, quaternion-valued neural networks can be used to process four-dimensional data and are also applicable to one-dimensional data or two-dimensional data [6]. For example, Liu uses quaternions to represent R(red), G(green), and B(blue) in image pixels. Wei combines color image RGB color channels and luminance with quaternions. The imaginary part's color channels and luminance are used as coefficients to obtain a complete quaternion [8]. Therefore, it is necessary to study the quaternion-valued neural networks corresponding to the quaternion domain in this paper.

This work is supported by the Key Project of Science and Technology Research Program of Chongqing Education Commission of China (No. KJZD-K202300205).

B. Luo et al. (Eds.): ICONIP 2023, CCIS 1963, pp. 100–111, 2024.
https://doi.org/10.1007/978-981-99-8138-0_9

So far, the dynamical properties of neural networks have been extensively studied, such as stability, dissipativity, nilpotency, and synchronization. Combining the research experience of real-valued neural networks to study quaternion-valued neural networks will make the study of the above properties more in-depth. As a result, researchers have focused on the correlated dynamical features of quaternion neural networks [7]. The synchronization phenomena have been extensively explored in several domains, including information science [9], chaos-based communication networks [11], and chaotic encryption [10]. In practice, researchers desire chaotic systems to achieve not only synchronization but also rapid synchronization [13]. Because it has a higher convergence rate than asymptotic synchronization, exponential synchronization is of interest in many fields [14].

Synchronization is an essential phenomenon in neural networks, which refers to the phenomenon that two or more systems make the final dynamic behaviors converge to the same state by interacting with each other. Due to its successful applications in image processing and information science, synchronization and its applications have been a hot topic of extensive and intensive research in academia and industry in the past decades. Many kinds of control strategies have been proposed to effectively facilitate the synchronization of complex networks, such as feedback control [15], quantized control [16], and event-triggered control [17]. It is worth mentioning that the capacity and bandwidth of the communication channel usually limit signal transmission. In real applications, the control signal can be quantized before transmission to improve communication efficiency [18].

Based on the above discussion, this paper focuses on the exponential synchronization problem of quaternion-valued neural networks based on quantized control. Using quaternion algebra and Hamilton's rules, the quaternion-valued neural networks are transformed into four real-valued networks. Suitable Lyapunov functions and controllers are constructed, and sufficient conditions for exponential synchronization are derived. The main contributions of this paper are as follows.

(1) An exponential synchronization criterion for high-dimensional neural networks is derived.

(2) The quantized control controller is designed by using an appropriate lemma, thus saving the control cost.

The remainder of the paper is structured as follows. The quaternion-valued neural networks model is introduced in the second section, along with the relevant lemmas and assumptions. The third section designs a synchronization controller based on quantized control and theoretically derives the exponential synchronization of the quaternion-valued neural networks. The fourth section uses numerical simulations to validate the controller's efficacy and the suggested theory's accuracy. Finally, in the fifth section, the conclusion is presented.

Notations: $\mathbb{R}^n$ and $\mathbb{Q}^n$ denote the n-tuple vectors in the real field and quaternion skew field, respectively. $\mathbb{Q}^{n\times n}$ is an $n \times n$ matrix whose entries are quaternion numbers. The quaternion-valued function is denoted by $z(t) =$

$z^R(t) + iz^I(t) + jz^J(t) + kz^K(t)$ where i, j, k are standard imaginary units in Q which satisfy the Hamilton rules: $ij = -ji = k, jk = -kj = i, ki = -ik = j, i^2 = j^2 = k^2 = ijk = -1$ and $z(t) \in \mathbb{Q}$.

2 Preliminaries

We consider a class of quaternion-valued neural networks with the following form:

$$\dot{z}(t) = -Az(t) + Bf(z(t)) + I, \tag{1}$$

where $z(t) = (z_1(t), ..., z_n(t))^\top \in \mathbb{Q}$ represents the state of the neuron, $f(z(\cdot)) = (f_1(z_1(\cdot)), ..., f_n(z_n(\cdot)))^\top : \mathbb{Q}^n \to \mathbb{Q}^n$ represents the activation function. $A = diag(A_1, A_2, ..., A_n) \in \mathbb{R}^{n\times n}$ stands for the neuron self-inhibition, $B \in \mathbb{Q}^{n\times n}$ is the connection weight matrix and $I = (I_1, ..., I_n)^\top \in \mathbb{Q}$ is the outer input.

Assumption 1. For computational convenience, let the inputs, outputs, and state variables of a quaternion-valued neural network $z(t)$, $f(z(t))$, I be decomposed into the following form:

$$\begin{gathered} z(t) = z^R(t) + iz^I(t) + jz^J(t) + kz^K(t), \\ f(z) = f^R(z^R) + if^I(z^I) + jf^J(z^J) + kf^K(z^K), \\ I = I^R + iI^I + jI^J + kI^K. \end{gathered}$$

According to the definition of Assumption 1 and Hamilton's rule, The drive system (1) can be transformed into the following four real-valued neural networks.

$$\begin{aligned} \dot{z}^R(t) =& -Az^R(t) + B^Rf^R(z^R(t)) - B^If^I(z^I(t)) - B^Jf^J(z^J(t)) - B^Kf^K(z^K(t)) + I^R(t), \\ \dot{z}^I(t) =& -Az^I(t) + B^Rf^I(z^I(t)) + B^If^R(z^R(t)) + B^Jf^K(z^K(t)) - B^Kf^J(z^J(t)) + I^I(t), \\ \dot{z}^J(t) =& -Az^J(t) + B^Rf^J(z^J(t)) + B^Jf^R(z^R(t)) + B^Kf^I(z^I(t)) - B^If^K(z^K(t)) + I^J(t), \\ \dot{z}^K(t) =& -Az^K(t) + B^Rf^K(z^K(t)) + B^If^J(z^J(t)) + B^Kf^R(z^R(t)) - B^Jf^I(z^I(t)) + I^K(t), \end{aligned} \tag{2}$$

The initial value of the system (1) are $z(t) = \phi(v)$, $\phi(v)$ is bounded and continuous on $v \in [-\tau, 0]$.

In this paper, using the drive-response approach, the response system is defined as shown below:

$$\dot{s}(t) = -As(t) + Bf\left(s(t)\right) + I + u(t), \tag{3}$$

where $s(t) = (s_1(t), ..., s_n(t))^\top \in \mathbb{Q}$ represents the state of the neurons, $f(s(\cdot)) = (f_1(s_1(\cdot)), ..., f_n(s_n(\cdot)))^\top : \mathbb{Q}^n \to \mathbb{Q}^n$ represents the activation function. $A = diag(A_1, A_2, ..., A_n) \in \mathbb{R}^{n\times n}$ stands for the neuron self-inhibition, $B \in \mathbb{Q}^{n\times n}$

is the connection weight matrix and $I = (I_1, ..., I_n)^\mathsf{T} \in \mathbb{Q}$ is the outer input. $u_p(t) = u_p^R(t) + iu_p^I(t) + ju_p^J(t) + ku_p^K(t) \in \mathbb{Q}^n$ is the controller.

Similarly, the response system (3) is transformed into the following four real-valued neural networks according to Assumption 1.

$$
\begin{aligned}
\dot{s}^R(t) &= -As^R(t) + B^R f^R(s^R(t)) - B^I f^I(s^I(t)) - B^J f^J(s^J(t)) - B^K f^K(s^K(t)) + I^R(t) + u^R(t),\\
\dot{s}^I(t) &= -As^I(t) + B^R f^I(s^I(t)) + B^I f^R(s^R(t)) + B^J f^K(s^K(t)) - B^K f^J(s^J(t)) + I^I(t) + u^I(t),\\
\dot{s}^J(t) &= -As^J(t) + B^R f^J(s^J(t)) + B^J f^R(s^R(t)) + B^K f^I(s^I(t)) - B^I f^K(s^K(t)) + I^J(t) + u^J(t),\\
\dot{s}^K(t) &= -As^K(t) + B^R f^K(s^K(t)) + B^I f^J(s^J(t)) + B^K f^R(s^R(t)) - B^J f^I(s^I(t)) + I^K(t) + u^K(t),
\end{aligned} \tag{4}
$$

where the initial value of the system (1) are $s(t) = \psi(v)$, $\psi(v)$ is bounded and continuous on $v \in [-\tau, 0]$.

Assumption 2. [24] For $z_q(t), s_q(t) \in \mathbb{Q}^n$, the activation function satisfies the Lipschitz condition, if the constants $M_q^R > 0, M_q^I > 0, M_q^J > 0, M_q^K > 0$ are satisfied, we have.

$$
\begin{aligned}
|f(s_q^R(x)) - f(z_q^R(y))| &\le M_q^R |s_q^R(x) - z_q^R(y)|,\\
|f(s_q^I(x)) - f(z_q^I(y))| &\le M_q^I |s_q^I(x) - z_q^I(y)|,\\
|f(s_q^J(x)) - f(z_q^J(y))| &\le M_q^J |s_q^J(x) - z_q^J(y)|,\\
|f(s_q^K(x)) - f(z_q^K(y))| &\le M_q^K |s_q^K(x) - z_q^K(y)|,
\end{aligned}
$$

where $\forall t \in \mathbb{R}$, $q = 1, 2, ..., n$ and $z_q^R(t)$, $z_q^I(t)$, $z_q^J(t)$, $z_q^K(t)$, $s_q^R(t)$, $s_q^I(t)$, $s_q^J(t)$, $s_q^K(t) \in \mathbb{R}$.

Define the synchronization error system as $e(t) = z(t) - s(t)$, where $f(e(t)) = f(z(t)) - f(s(t)), e(v) = \phi(v) - \psi(v)$ is the initial value of the error system (5). The synchronization error dynamics system is shown below:

$$
\begin{aligned}
\dot{e}^R(t) &= -Ae^R(t) + B^R(f^R(s^R(t)) - f^R(z^R(t))) - B^I(f^I(s^I(t)) - f^I(z^I(t)))\\
&\quad -B^J(f^J(s^J(t)) - f^J(z^J(t))) - B^K(f^K(s^K(t)) - f^K(z^K(t))) + u^R(t)\\
\dot{e}^I(t) &= -Ae^I(t) + B^R(f^I(s^I(t)) - f^I(z^I(t))) + B^I(f^R(s^R(t)) - f^R(z^R(t)))\\
&\quad +B^J(f^K(s^K(t)) - f^K(z^K(t))) - B^K(f^J(s^J(t)) - f^J(z^J(t))) + u^I(t),\\
\dot{e}^J(t) &= -Ae^J(t) + B^R(f^J(s^J(t)) - f^J(z^J(t))) + B^J(f^R(s^R(t)) - f^R(z^R(t)))\\
&\quad +B^K(f^I(s^I(t)) - f^I(z^I(t))) - B^I(f^K(s^K(t)) - f^K(z^K(t))) + u^J(t),\\
\dot{e}^K(t) &= -Ae_p^K(t) + B^R(f^K(s^K(t)) - f^K(z^K(t))) + B^I(f^J(s^J(t)) - f^J(z^J(t)))\\
&\quad + B^K(f^R(s^R(t)) - f^R(z^R(t))) - B^J(f^I(s^I(t)) - f^I(z^I(t))) + u^K(t),
\end{aligned} \tag{5}
$$

Definition 1. [20] If there exist constants $\alpha > 0, \beta > 0$ such that the errors system satisfy the following inequality:

$$
||e(t)|| \le \alpha \sup_{-\tau \le s \le 0} ||e(s)|| e^{(-\beta t)}
$$

then the driving system (1) and the response system (3) are said to be globally exponential synchronization.

Lemma 1. *[22] Let $\tau \leq 0$ be a constant and $V(t)$ be a non-negative continuous function defined for $[-\tau, \infty]$ which satisfies $\dot{V}(t) \leq \sup_{t-\tau \leq s \leq s} -pV(t)$ for $t \leq 0$, where p are constants. If $p > 0$, then $V(t) \leq \sup_{-\tau \leq s \leq 0} e^{(-pt)} fort > 0$, where p is a unique positive root of the equation.*

Lemma 2. *[21] Suppose Ω_1, Ω_2, Ω_3 are real matrices of arbitrary dimension and $\omega_3 > 0$, then for $x, y \in \mathbb{R}$ we have*

$$2x^{\mathsf{T}}\Omega_1^{\mathsf{T}}\Omega_2^{\mathsf{T}}y \leq x^{\mathsf{T}}\Omega_1^{\mathsf{T}}\Omega_3 x + y^{\mathsf{T}}\Omega_2^{\mathsf{T}}\Omega_3^{-1}\Omega_2 y \tag{6}$$

3 Main Results

In this section, a new quantizer controller is designed. This controller enables the exponential synchronization of the drive-response system. The controller is shown as follows:

$$u_p^R(t) = -\zeta_p^R q(e_p^R(t)), u_p^I(t) = -\zeta_p^I q(e_p^I(t)), u_p^J(t) = -\zeta_p^J q(e_p^J(t)), u_p^K(t) = -\zeta_p^K q(e_p^K(t)), \tag{7}$$

where $\zeta_p^R > 0, \zeta_p^I > 0, \zeta_p^J > 0, \zeta_p^K > 0$ and $\zeta_p^R, \zeta_p^I, \zeta_p^J, \zeta_p^K \in \mathbb{R}$. For $\forall \tau \in \mathbb{R}$, the quantizer $q(\tau)$ is constructed as follows [12]:

$$\mathrm{q}(\tau) = \begin{cases} \omega_i, & if\ \frac{1}{1+\delta}\omega_i < \tau \leq \frac{1}{1-\delta}\omega_i, \\ 0, & if\ \tau = 0, \\ -q(-\tau), & if\ \tau < 0, \end{cases}$$

where $\omega = \{\pm\omega_i : \omega_i = \rho^i\omega_0, 0 < \rho < 1, i = 0, \pm 1, \pm 2, \ldots\} \cup \{0\}$ with a sufficiently large constant $\omega_0 > 0$ and $\delta = \frac{1-\rho}{1+\rho}$. Based on the analysis and discussion in [19], if $\Delta \in [-\delta, \delta]$ exists in Eq.(9), then $q(\tau) = (1 + \Delta)\tau, \forall\tau \in \mathbb{R}$. If $\Lambda(t) = diag(\Lambda_1(t), \ldots\Lambda_n(t))$, and $\Lambda_i(t) \in [-\delta, \delta]$, $i = 1, 2 \ldots n$, then

$$q(e_p^\varsigma(t)) = (I_n + \Lambda(t))e_p^\varsigma(t), (\varsigma = R, I, J, K). \tag{8}$$

using the controller (7) and Eq.(8), then

$$-q(e_p^\varsigma(t)) \leq \ -(1-\delta)e_p^\varsigma(t), (\varsigma = R, I, J, K)$$

Theorem 1. *Under Assumption 1 and Assumption 2, as well as the following inequality holds, the synchronization error system (5) will achieve exponential synchronization*

$$\zeta^\varsigma \geq \frac{\varphi}{1-\delta}(\varsigma = R, I, J, K), \tag{9}$$

where $\varphi = \Big(-A + 2M^\varsigma + \frac{1}{2}\Big(B^R(B^R)^{\mathsf{T}} + B^I[(B^I)]^{\mathsf{T}} + B^J(B^J)^{\mathsf{T}} + B^K(B^K)^{\mathsf{T}}\Big)$, *$\delta$ is the quantization function parameter, $\zeta^R > 0$, $\zeta^I > 0$, $\zeta^J > 0$ and $\zeta^K > 0$.*

Proof. Consider the following Lyapunov function:

$$V(t) = V_1(t) + V_2(t) + V_3(t) + V_4(t), \tag{10}$$

Where

$$V_1(t) = \frac{1}{2}(e^R(t))^\top e^R(t), \qquad V_2(t) = \frac{1}{2}(e^I(t))^\top e^I(t),$$
$$V_3(t) = \frac{1}{2}(e^J(t))^\top e^J(t), \qquad V_4(t) = \frac{1}{2}(e^K(t))^\top e^K(t), \qquad .$$

Differentiating V(t) along the solution of (5), one obtains

$$\dot{V}(t) =(e^R(t))^\top \dot{e}^R(t) + (e^I(t))^\top \dot{e}^I(t) + (e^J(t))^\top \dot{e}^J(t) + (e^K(t))^\top \dot{e}^K(t). \tag{11}$$

Derivative for Lyapunov function V (t) under the quantized controller (7),

$$\dot{V}_1(t) = (e^R(t))^\top \Big(- Ae^R(t) + B^R(f^R(s^R(t)) - f^R(z^R(t))) - B^I(f^I(s^I(t))$$
$$-f^I(z^I(t))) - B^J(f^J(s^J(t)) - f^J(z^J(t))) - B^K(f^K(s^K(t)) - f^K(z^K(t))) + u^R(t)\Big),$$

The following inequality can be obtained from Lemma 2

$$[e^\alpha(t)]^\top B^\beta f^\gamma(e^{(\gamma)}(t)) \le \frac{1}{2}[e^\alpha(t)]^\top B^\beta (B^\beta)^\top e^\alpha(t) + \frac{1}{2}(f^\gamma(e^{(\gamma)}(t)))^\top f^\gamma(e^{(\gamma)}(t)), \tag{12}$$

where $\alpha, \beta, \gamma = R, I, J, K$.
By Assumption 2, there exists a diagonal matrix $\widetilde{M}$ such that the following equation holds:

$$[f^\varsigma(e^{(\varsigma)}(t))]^\top f^\varsigma(e^{(\varsigma)}(t)) \le [e^\varsigma(t)]^\top \widetilde{M}^\varsigma e^{(\varsigma)}(t), \tag{13}$$

where $\widetilde{M} = diag\big((M_1^\varsigma(t))^2, ..., (M_1^\varsigma(t))^2\big), \varsigma = R, I, J, K$.

Combining the inequalities (12), (13) with the equation (8), the following inequalities can be obtained

$$\le(e^R(t))^\top \Big(- Ae^R(t) + \frac{1}{2}B^R(B^R)^\top e^R(t) + \frac{1}{2}M^R e^R(t) + \frac{1}{2}B^I(B^I)^\top e^R(t) + \frac{1}{2}M^I e^R(t)$$
$$+\frac{1}{2}B^J(B^J)^\top e^R(t) + \frac{1}{2}M^J e^R(t) + \frac{1}{2}B^K(B^K)^\top e^R(t) + \frac{1}{2}M^K e^R(t) - (1-\delta)\zeta^R e^R(t)\Big). \tag{14}$$

With similar techniques,

$$\dot{V}_2 \le (e^I(t))^\top \Big(- Ae^I(t) + \frac{1}{2}B^R(B^R)^\top e^I(t) + \frac{1}{2}M^R e^I(t) + \frac{1}{2}B^I(B^I)^\top e^I(t) + \frac{1}{2}M^I e^I(t)$$
$$+ \frac{1}{2}B^J(B^J)^\top e^I(t) + \frac{1}{2}M^J e^I(t) + \frac{1}{2}B^K(B^K)^\top e^I(t) + \frac{1}{2}M^K e^I(t) - (1-\delta)\zeta^I e^I(t)\Big). \tag{15}$$

Similarly,

$$\dot{V}_3(t) \leq (e^J(t))^\mathsf{T}\Big(- Ae^J(t) + \frac{1}{2}B^R(B^R)^\mathsf{T}e^J(t) + \frac{1}{2}M^R e^J(t) + \frac{1}{2}B^I(B^I)^\mathsf{T}e^J(t) + \frac{1}{2}M^I e^J(t)$$
$$+ \frac{1}{2}B^J(B^J)^\mathsf{T}e^J(t) + \frac{1}{2}M^J e^J(t) + \frac{1}{2}B^K(B^K)^\mathsf{T}e^J(t) + \frac{1}{2}M^K e^J(t) - (1-\delta)\varsigma^J e^J(t)\Big). \quad (16)$$

Similarly,

$$\dot{V}_4(t) \leq (e^K(t))^\mathsf{T}\Big(- Ae^K(t) + \frac{1}{2}B^R(B^R)^\mathsf{T}e^K(t) + \frac{1}{2}M^R e^K(t) + \frac{1}{2}B^I(B^I)^\mathsf{T}e^K(t) + \frac{1}{2}M^I e^K(t)$$
$$+ \frac{1}{2}B^J(B^J)^\mathsf{T}e^K(t) + \frac{1}{2}M^J e^K(t) + \frac{1}{2}B^K(B^K)^\mathsf{T}e^K(t) + \frac{1}{2}M^K e^K(t) - (1-\delta)\varsigma^K e^K(t)\Big). \quad (17)$$

From equations (14) - (17), we have:

$$\dot{V}(t) \leq [e^R(t)]^\mathsf{T}\Big(- A - \varsigma^R(1-\delta) + 2M^R + \frac{1}{2}\Big(B^R(B^R)^\mathsf{T} + B^I[(B^I)]^\mathsf{T} + B^J(B^J)^\mathsf{T} + B^K(B^K)^\mathsf{T}\Big)\Big)e^R(t)$$
$$+ [e^I(t)]^\mathsf{T}\Big(- A - \varsigma^I(1-\delta) + 2M^I + \frac{1}{2}\Big(B^R(B^R)^\mathsf{T} + B^I[(B^I)]^\mathsf{T} + B^J(B^J)^\mathsf{T} + B^K(B^K)^\mathsf{T}\Big)\Big)e^R(t)$$
$$+ [e^J(t)]^\mathsf{T}\Big(- A - \varsigma^J(1-\delta) + 2M^J + \frac{1}{2}\Big(B^R(B^R)^\mathsf{T} + B^I[(B^I)]^\mathsf{T} + B^J(B^J)^\mathsf{T} + B^K(B^K)^\mathsf{T}\Big)\Big)e^R(t)$$
$$+ [e^K(t)]^\mathsf{T}\Big(- A - \varsigma^K(1-\delta) + 2M^K + \frac{1}{2}\Big(B^R(B^R)^\mathsf{T} + B^I[(B^I)]^\mathsf{T} + B^J(B^J)^\mathsf{T} + B^K(B^K)^\mathsf{T}\Big)\Big)e^R(t),$$
$$(18)$$

The above inequality can be rewritten as

$$\dot{V}(t) \leq -\theta V(t),$$

where $\theta = -\Big(- A - \zeta\varsigma(1-\delta) + 2M^\varsigma + \frac{1}{2}\Big(B^R(B^R)^\mathsf{T} + B^I[(B^I)]^\mathsf{T} + B^J(B^J)^\mathsf{T} + B^K(B^K)^\mathsf{T}\Big)$.

Combining Definition 1 and Lemma 1 yields

$$||e(t)|| \leq \sup_{-\tau\leq s\leq 0} ||e(s)||e^{(-\theta t)},$$

where θ is the only solution of the equation; thus, the drive system (1) and the response system (3) is globally exponential synchronization under the controller (7). This completes the proof.

Remark 1. In this paper, the relevant parameters in the drive system (1) and response system (3), as well as the controller gain and inequality parameters, are quaternions. Therefore, the correlation properties of quaternions should be fully considered in the simulation.

Remark 2. The synchronization criterion is derived based on Lyapunov stability theory and inequality (9) in the case of satisfying Assumptions 1 and 2. It is worth noting that the sufficient conditions for the exponential synchronization of systems (1) and (3) are independent of the delay parameters but depend on the inequality parameters of the system and the controller gain.

4 Numerical Simulations

In this section, numerical simulations are performed based on the above theoretical results, and the experimental results obtained verify the designed controller's effectiveness and the proposed theory's correctness. In the simulations, the quantizer density is taken as $\rho = 0.6$.

Example 1. Consider the following quaternion-valued neural network as the drive system [23].

$$\dot{z}_i(t) = -A_{ij}z_i(t) + B_{ij}f(z_i(t)) + I_i, \tag{19}$$

To achieve synchronization, the following response system is designed:

$$\dot{s}_i(t) = -A_{ij}s_i(t) + B_{ij}f(s_i(t)) + I_i + u_i(t), \tag{20}$$

where U is equivalent to the controller of Eq.(8). The activation functions are

$$f(z(t)) = \begin{pmatrix} f_1^R(z_1^R) + if_1^I(z_1^I) + jf_1^J(z_1^J) + kf_1^K(z_1^K) \\ f_2^R(z_2^R) + if_2^I(z_2^I) + jf_2^J(z_2^J) + kf_2^K(z_2^K) \end{pmatrix},$$

and $f_p(z_p^\varsigma) = tanh(z_p^\varsigma)$, $p = 1, 2$, $\varsigma = R, I, J, K$. In addition, the parameters of the system (19) are

$$A = \begin{pmatrix} 1.2 & 0 \\ 0 & 1.1 \end{pmatrix},$$

$$I = \begin{pmatrix} 1.2 - 1.2i + 1.3j + 1.4k \\ 1.4 + 1.3i + 1.4j - 1.3k \end{pmatrix},$$

$$B = \begin{pmatrix} 2 + 1.2i + 1.5j + 1.3k & -1 - 4.5i - 2.3j - 3.2k \\ 1 + 1.7i + 1.4j + 1.6k & 0.8 + 0.9i + 0.6j + 0.5k \end{pmatrix}.$$

The initial states of systems (19) and (20) are shown below. $z_1(0) = 0.5 + 3i + 2.5j + 3.5k$, $z_2(0) = -1.5 - 0.1i - 2.5j - 3.5k$, $s_1(0) = -1.3 - 2i - 2j - 2.5k$, $s_2(0) = 0.8 - 3.5i + 2j + 3k$. Figure 1–4 simulate the state trajectories of the drive system (1) and the response system (3) in the uncontrolled state. From the figures, it can be concluded that the drive-response system is not synchronized.

According to Assumption 2 $M_q^R > 0$, $M_q^I > 0$, $M_q^J > 0$ and $M_q^K > 0$, setting $M_q^R = M_q^I = M_q^J = M_q^K = 1$ and $q = 1, 2$. The controller parameters shown in equation (7) are taken as $\zeta_i^R = 52.10, \zeta_i^I = 50.16, \zeta_i^J = 50.16, \zeta_i^K = 57.36$ and $i = 1, 2$. The quantization density of the controller $0 < \rho < 1$, and then $\delta = \frac{1-\rho}{1+\rho} = 0.25$. Calculated by inequalities (9), one obtains $\theta_1^\varsigma > 12.93$, $\theta_2^\varsigma > 52.99$, $\varsigma = R, I, J, K$. Then choosing $\theta_1^\varsigma > 13$, $\theta_2^\varsigma > 53$. In the above inequality,

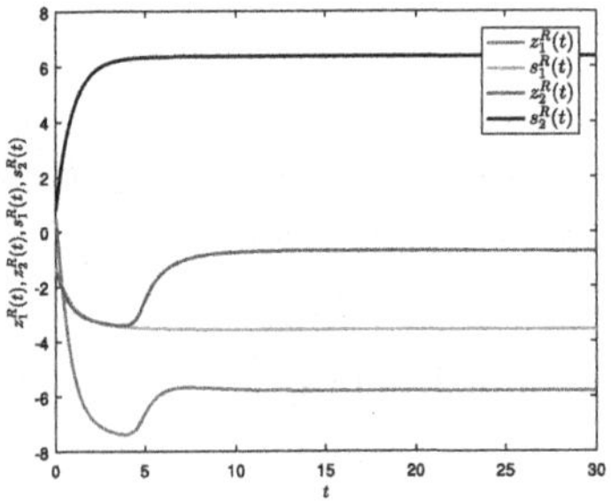

Fig. 1. State trajectories of the uncontrolled drive system (19) and response system (20). $(z_1^R, s_1^R, z_2^R, s_2^R)$

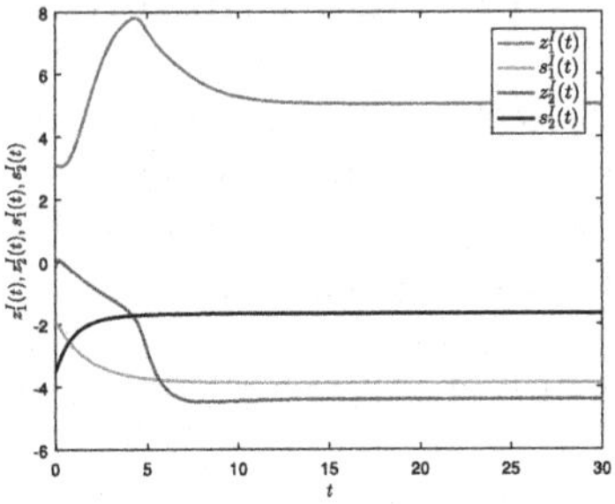

Fig. 2. State trajectories of the uncontrolled drive system (19) and response system (20). $(z_1^K, s_1^K, z_2^K, s_2^K)$

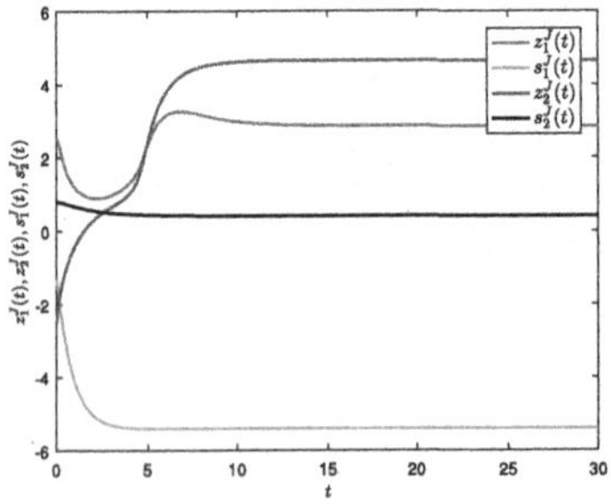

Fig. 3. State trajectories of the uncontrolled drive system (19) and response system (20). $(z_1^J, s_1^J, z_2^J, s_2^J)$

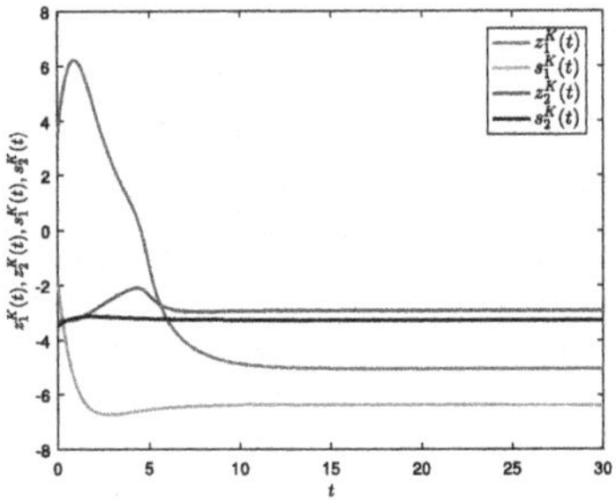

Fig. 4. State trajectories of the uncontrolled drive system (19) and response system (20). $(z_1^K, s_1^K, z_2^K, s_2^K)$

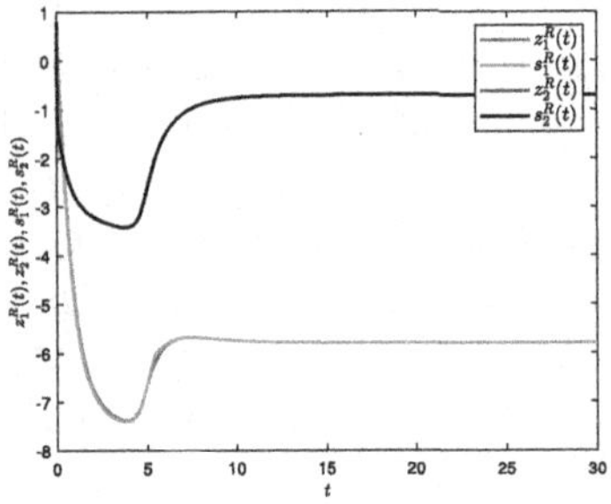

Fig. 5. State trajectories of the controlled drive system (19) and response system (20). $(z_1^R, s_1^R, z_2^R, s_2^R)$

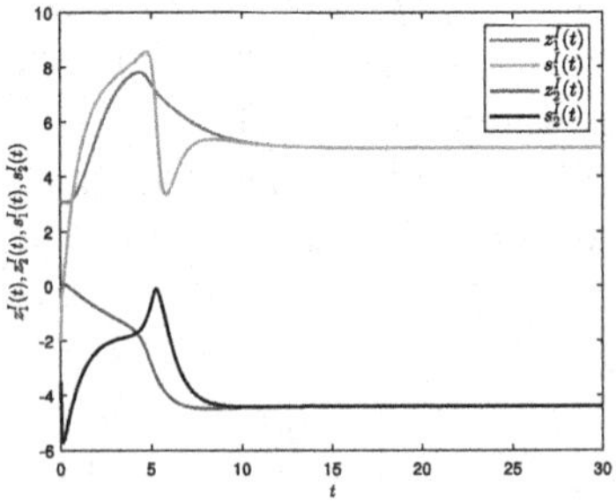

Fig. 6. State trajectories of the controlled drive system (19) and response system (20). $(z_1^I, s_1^I, z_2^I, s_2^I)$

any figure that satisfies the inequality is allowed. The state trajectories of the driven system (1) and the response system (3) after being controlled are shown in Fig. 5–8. It can be seen that the drive system is synchronized with the response system. The trajectory of the synchronization error system is shown in Fig. 9.

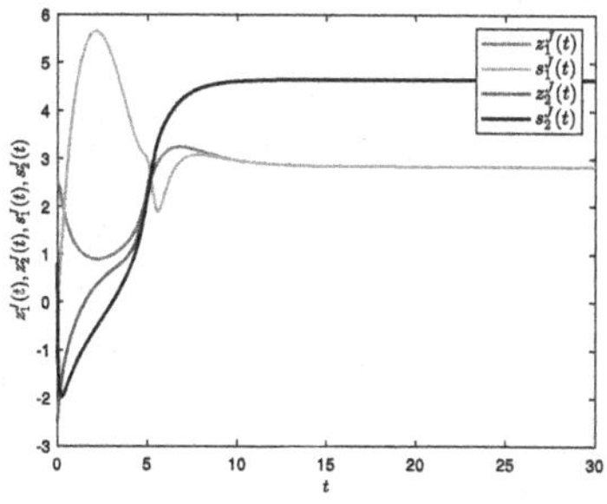

Fig. 7. State trajectories of the controlled drive system (19) and response system (20). $(z_1^J, s_1^J, z_2^J, s_2^J)$

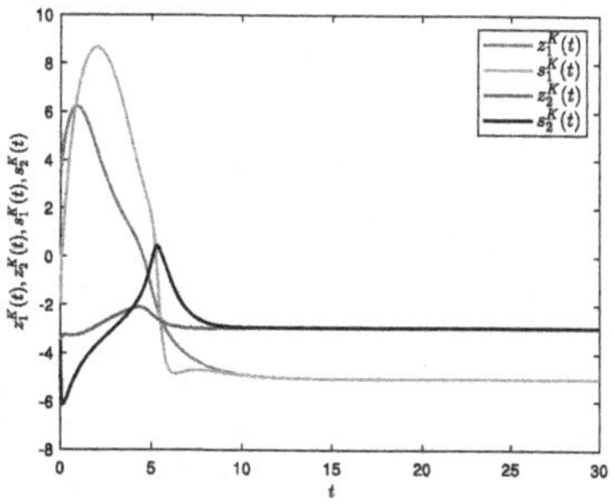

Fig. 8. State trajectories of the controlled drive system (19) and response system (20). $(z_1^K, s_1^K, z_2^K, s_2^K))$

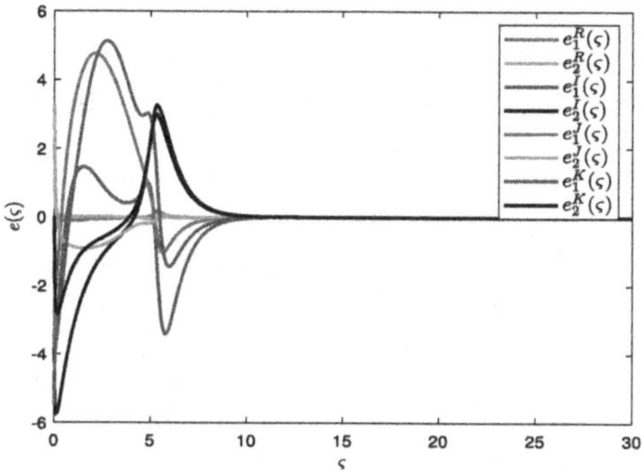

Fig. 9. synchronization error system trajectory

5 Conclusions

In this paper, the exponential synchronization problem of quaternion-valued neural networks is studied using a quantized control approach. Hamilton's law is used to transform the quaternion-valued neural network into an equivalent real-valued neural network system, a controller based on the quantization strategy is designed, and the discriminant criterion for achieving global exponential synchronization of the driving and response systems is given. Finally, the validity of the proposed design scheme is verified by numerical simulation, which enriches the research related to the analysis of the dynamical behavior of quaternion-valued neural networks. Future work will consider fixed-time synchronization of quaternion-valued neural networks with time-varying delays.

References

1. Hoppensteadt, F.C., Izhikevich, E.M.: Pattern recognition via synchronization in phase-locked loop neural networks. IEEE Trans. Neural Netw. **11**(3), 734–738 (2000)
2. Luo, F.-L., Unbehauen, R.: Applied neural networks for signal processing. Cambridge University Press (1998)
3. Yang, W., Huang, J., He, X., Wen, S., Huang, T.: Finite-time synchronization of neural networks with proportional delays for RGB-D image protection. IEEE Transactions on Neural Networks and Learning Systems, pp. 1–12 (2022)
4. Liu, L., Lei, M., Bao, H.: Event-triggered quantized quasisynchronization of uncertain quaternion-valued chaotic neural networks with time-varying delay for image encryption. IEEE Trans. Cybern. **53**(5), 3325–3336 (2022)
5. Wei, W., Wang, Y.: Color image retrieval based on quaternion and deep features. IEEE Access **7**, 126:430–126:438 (2019)
6. Chen, X., Song, Q., Li, Z.: Design and analysis of quaternion-valued neural networks for associative memories. IEEE Trans. Syst. Man Cybern. Syst. **48**(12), 2305–2314 (2018)
7. Liu, Y., Zhang, D., Lou, J., Lu, J., Cao, J.: Stability analysis of quaternion-valued neural networks: decomposition and direct approaches. IEEE Trans. Neural Netw. Learn. Syst. **29**(9), 4201–4211 (2017)
8. Wei, W., Wang, Y.: Color image retrieval based on quaternion and deep features. IEEE Access **7**, 126:430–126:438 (2019)
9. Kanter, I., Kinzel, W., Kanter, E.: Secure exchange of information by synchronization of neural networks. Europhys. Lett. **57**(1), 141 (2002)
10. Lakshmanan, S., Prakash, M., Lim, C.P., Rakkiyappan, R., Balasubramaniam, P., Nahavandi, S.: Synchronization of an inertial neural network with time-varying delays and its application to secure communication. IEEE Trans. Neural Netw. Learn. Syst. **29**(1), 195–207 (2016)
11. Sarkar, A.: Multilayer neural network synchronized secured session key based encryption in wireless communication. IAES Int. J. Artif. Intell. **8**(1), 44 (2019)
12. Yang, W., Huang, J., Wang, X.: Fixed-time synchronization of neural networks with parameter uncertainties via quantized intermittent control. Neural Process. Lett. **54**(3), 2303–2318 (2022)
13. Wei, R., Cao, J.: Global exponential synchronization of quaternion-valued memristive neural networks with time delays. Nonlinear Anal. Model. Control **25**(1), 36–56 (2020)
14. Yuan, M., Wang, W., Wang, Z., Luo, X., Kurths, J.: Exponential synchronization of delayed memristor-based uncertain complex-valued neural networks for image protection. IEEE Trans. Neural Netw. Learn. Syst. **32**(1), 151–165 (2020)
15. Lv, X., Li, X., Cao, J., Duan, P.: Exponential synchronization of neural networks via feedback control in complex environment. Complexity **2018**, 1–13 (2018)

16. Feng, Y., Xiong, X., Tang, R., Yang, X.: Exponential synchronization of inertial neural networks with mixed delays via quantized pinning control. Neurocomputing **310**, 165–171 (2018)
17. Li, X., Zhang, W., Fang, J.-A., Li, H.: Event-triggered exponential synchronization for complex-valued memristive neural networks with time-varying delays. IEEE Trans. Neural Netw. Learn. Syst. **31**(10), 4104–4116 (2019)
18. Zhang, W., Li, C., Yang, S., Yang, X.: Synchronization criteria for neural networks with proportional delays via quantized control. Nonlinear Dyn. **94**, 541–551 (2018)
19. Xu, C., Yang, X., Lu, J., Feng, J., Alsaadi, F.E., Hayat, T.: Finite-time synchronization of networks via quantized intermittent pinning control. IEEE Trans. Cybern. **48**(10), 3021–3027 (2017)
20. Cheng, Y., Shi, Y.: Exponential synchronization of quaternion-valued memristor-based neural networks with time-varying delays. In: International Journal of Adaptive Control and Signal Processing (2023)
21. Hu, J., Wang, J.: Global stability of complex-valued recurrent neural networks with time-delays. IEEE Trans. Neural Netw. Learn. Syst. **23**(6), 853–865 (2012)
22. Cheng, C.-J., Liao, T.-L., Hwang, C.-C.: Exponential synchronization of a class of chaotic neural networks. Chaos, Solitons Fractals **24**(1), 197–206 (2005)
23. Deng, H., Bao, H.: Fixed-time synchronization of quaternion-valued neural networks. Phys. A **527**, 121351 (2019)
24. Yang, W., Huang, J., Wen, S., He, X.: Fixed-time synchronization of neural networks with time delay via quantized intermittent control. Asian J. Control **25**(3), 1823–1833 (2023)

Improving SLDS Performance Using Explicit Duration Variables with Infinite Support

Mikołaj Słupiński (✉) and Piotr Lipiński

Computational Intelligence Research Group, Institute of Computer Science, University of Wrocław, Wrocław, Poland
{mikolaj.slupinski,piotr.lipinski}@cs.uni.wroc.pl

Abstract. Switching Linear Dynamical Systems (SLDS) are probabilistic graphical models used both for self-supervised segmentation and dimensionality reduction. Despite their modeling capabilities, SLDS are particularly hard to train. They oftentimes over-segment the timeseries or completely ignore some of the states, reducing the usefulness of the acquired segmentation.

To improve the segmentation in Switching Linear Dynamical Systems, we introduce explicit-duration variables with infinite support. We extend the Beam Sampling algorithm to perform the efficient inference allowing for a duration distribution with infinite support. We conduct experiments on three benchmarks (two already prevalent in the state-space model literature and one demonstrating behavior in a sparse setting) that test the correctness and efficiency of our solution.

Keywords: Bayesian Inference · Switching Linear Dynamical Systems · Explicit Duration

1 Introduction

Spatiotemporal data segmentation plays a crucial role in modern data analysis. Besides simple clustering methods that treat the data as stationary, ignoring the variability of their characteristics over time, some of the popular approaches consider state-space models that treat the data as coming from different sub-models related to different possible states of the phenomenon under study and try to describe not only such sub-models but also the manner in which they change when the states change. A commonly known example is Hidden Markov Models (HMMs) [17] which use discrete hidden states, or more and more popular, Switching Linear Dynamical Systems (SLDS) [9,11] which use both discrete hidden states (a general state of the phenomena) and continuous hidden states (a detailed position of the phenomena).

Despite the fact that SLDS are powerful models that allow self-supervised segmentation, they are particularly difficult to train in a self-supervised manner.

B. Luo et al. (Eds.): ICONIP 2023, CCIS 1963, pp. 112–123, 2024.
https://doi.org/10.1007/978-981-99-8138-0_10

Their main problem is not using all of the latent discrete states [1,5] or performing noisy segmentation [1,2]. This leads to underfitted models, which do not express the whole complexity of the phenomenon. Additionally, even with full utilization of discrete states models tend to produce noisy segmentations [1,2].

To deal with the first issue, previous work concerning Switching Dynamical Systems (SDS) used modeling explicit durations [1,2]. However, as far as we are concerned, no previous work regarding SDS utilized explicit duration variables with infinite duration support.

As we show by series of experiments, using explicit duration variables leads to smoother time-series segmentation combined with more diverse state utilization. Moreover, modeling duration times with Poisson distribution combined with the removal of self-transition gives results comparable with previously used categorical duration times. Compared to the countdown approach [1] used previously in explicit duration modeling in SDS, it leads to more interpretable duration variables.

The main contributions of this paper are the application of the infinite support duration model in parametric Explicit Duration Switching Linear Dynamical Systems (EDSLDS) models, an algorithm for inference, and an evaluation on a number of benchmark datasets for time-series segmentation.

2 Background

Our work builds on discoveries in the domains of switching state-space models and explicit duration modeling.

Switching State-Space Models. There is a very large selection of models using both state-space modeling and discrete Markovian states; however, one may argue that Switching Linear Dynamical Systems are the most basic building block for more sophisticated constructions.

The Switching Linear Dynamical System [9] is the generative model, defined as follows. At each time $t = 1, 2, \ldots, T$ there is a discrete latent state $s_t \in \{1, 2, \ldots, K\}$ following Markovian dynamics,

$$s_{t+1} \mid s_t, \{\boldsymbol{\pi}_k\}_{k=1}^{K} \sim \boldsymbol{\pi}_{s_t}, \tag{1}$$

where $\{\boldsymbol{\pi}_k\}_{k=1}^{K}$ is the Markov transition matrix and $\boldsymbol{\pi}_k \in [0,1]^K$ is its k-th row. In addition, a continuous latent state $\mathbf{x}_t \in \mathbb{R}^M$ follows conditionally linear dynamics, where the discrete state s_t determines the linear dynamical system used at time t:

$$\mathbf{x}_{t+1} = \mathbf{A}_{s_{t+1}} \mathbf{x}_t + \mathbf{v}_t, \quad \mathbf{v}_t \overset{\text{iid}}{\sim} \mathcal{N}\left(0, \mathbf{Q}_{s_{t+1}}\right), \tag{2}$$

for matrices $\mathbf{A}_k, \mathbf{Q}_k \in \mathbb{R}^{M \times M}$ (where $\mathbf{A}_k$ is linear transformation matrix between subsequent continuous states, and $\mathbf{Q}_k$ is their covariance) for $k = 1, 2, \ldots, K$. Finally, at each time t a linear Gaussian observation $\mathbf{y}_t \in \mathbb{R}^N$ is generated from the corresponding latent continuous state,

$$\mathbf{y}_t = \mathbf{C}_{s_t} \mathbf{x}_t + \mathbf{w}_t, \quad \mathbf{w}_t \overset{\text{iid}}{\sim} \mathcal{N}\left(0, \mathbf{S}_{s_t}\right), \tag{3}$$

for $\mathbf{C}_k \in \mathbb{R}^{N\times M}, \mathbf{S}_k \in \mathbb{R}^{N\times N}$, where $\mathbf{C}_k$ is an emission matrix an $\mathbf{S}_k$ is covariance of the measurement noise. The system parameters comprise the discrete Markov transition matrix and the library of linear dynamical system matrices, which we write as

$$\theta = \{(\boldsymbol{\pi}_k, \mathbf{A}_k, \mathbf{Q}_k, \mathbf{C}_k, \mathbf{S}_k)\}_{k=1}^{K} . \tag{4}$$

To learn an SLDS using Bayesian inference [6], we place conjugate Dirichlet (Dir) priors on each row of the transition matrix and conjugate matrix normal inverse Wishart (MNIW) priors on the linear dynamical system parameters, writing

$$\begin{gathered} \boldsymbol{\pi}_k \Big| \alpha \overset{\text{iid}}{\sim} \text{Dir}(\alpha), \quad \mathbf{A}_k, \mathbf{Q}_k \Big| \lambda \overset{\text{iid}}{\sim} \text{MNIW}(\lambda), \\ \mathbf{C}_k, \mathbf{S}_k \mid \eta \overset{\text{iid}}{\sim} \text{MNIW}(\eta), \end{gathered} \tag{5}$$

where α, λ, and η denote hyperparameters.

SLDS serve as the backbone for more sophisticated models, which have been investigated in recent years. In [10] the authors proposed novel methods for learning Recurrent Switching Linear Dynamical Systems (RSLDS). They used information on the position in the latent hyperplane to influence the probability of future hidden states, thus incorporating nonstationarity into the state distribution (namely $p(s_t|s_{t-1}, x_{t-1}) \neq p(s_t|s_{t-1})$)).

A similar approach was applied by [5]. However, the authors replaced linear transformations with neural networks.

Models incorporating duration variables into both Switching Linear Dynamical Systems [2,3,14] or Switching Nonlinear Dynamical Systems [1] used finite support for duration variable.

Explicit Duration Modeling. One may observe that using classic Markovian transitions, the duration time of a single state always follows the geometric distribution, and that may not be true for real-life data.

Explicit Duration Switching Dynamical Systems are a family of models introducing additional random variables to model the switch duration distribution explicitly. Explicit duration variables have been applied to both HMMs [4,17] and Switching Dynamical Systems (SDS) with continuous states [2,3,14]. Several methods have been proposed in the literature for modeling the duration of the switch, for example, using decreasing or increasing count and duration indicator variables [2,3].

In most of the cases of SDS with an explicit state duration variable, the duration variable is modeled using a categorical variable with support $\{1, 2, \ldots, D_{max}\}$ [1–3].

3 Problem Statement

There are two main issues when trying to fit an SLDS in a fully self-supervised setting. First, the model does not utilize all possible states, for instance, in transportation mode detection "walking" and "cycling" may collapse to a single state. Secondly, SLDS does noisy segmentation, i.e. there is a single occurrence of wrong label. For instance, the model outputs ten minutes car ride, one second bicycle ride, and ten minutes car again.

We identify the assumption of geometric duration distribution as one of the causes of the problem. In the classical hidden Markov setting, when modeling long sequences, we often get the same sequence of states, so that the transition matrix is nearly diagonal. A simple example is the use of sensors on smartphones and fitness trackers to solve the problem of identifying means of transportation. Since they probe at high frequencies (>1 Hz), there is a large correlation between subsequent measurements. Mode transitions occur at much lower frequencies, so the transition matrix tends to be diagonal. This illustrates why it may be beneficial to model duration times explicitly.

As we mentioned earlier, explicit duration modeling was previously applied in a setting of SDS models, using categorical distribution. However, this approach has several drawbacks.

First of all, we must have some prior knowledge about the state duration, because we have to define the support $\{1, \ldots, D_{max}\}$. It may happen that we do not have any prior knowledge about segment durations or our assumptions about D_{max} were simply wrong. For this reason, we would prefer a distribution with infinite support.

To mitigate the problem with too small D_{max} and allow for larger model flexibility, one may allow for adding loops, by which we mean setting the transition probability $P(s_t = s|s_{t-1} = s) > 0$. The main drawback of this solution is that the explicit duration variable is no more "explicit". So we prefer to remove the loops (namely set $P(s_t = s|s_{t-1} = s) = 0$), to model actual state durations. As we will show in the experiments part, EDSLDS formulated with categorical duration times struggle to learn dynamics when loops are removed.

Finally, we have to store KD_{max} parameters, which may be troublesome for larger numbers of states.

For these exact reasons, we propose to model the explicit duration using the Poisson distribution without self-transition. We chose Poisson specifically because it is unimodal, has infinite support, and belongs to the exponential family, meaning that inference is simple.

In their work [1,5], the authors observed that SDS models fitted using variational inference are especially prone to posterior collapse, where they completely ignore latent states. Recent developments in the domain of variational inference [16] suggest that it may be the issue of nonidentifiability of these models, and SLDS without some additional constraints are not identifiable (for instance, we can rotate the latent space). For this reason, we prefer MCMC algorithms in inference and training of SLDS models.

To the best of our knowledge, none of the previous work investigated infinite support duration modeling in the context of SLDS. We are aware of these in the context of HMM (both in parametric [4] and in nonparametric modeling [8]). One can argue that SLDS can be viewed as a HMM whose emissions are state-space models. However, it is not straightforward how the idea implemented in one setting will work in practice in the other. In the case of HMM, the discrete states are responsible for the segmentation of point-wise measurements, whereas in SLDS we are rather segmenting the whole relationships between subsequent measurements.

Model Formulation. Let's consider the graphical model defined as (similar to [1] but we use explicit duration variables instead of count-down variables)

$$\begin{aligned} p(\mathbf{y}_{1:T}, \mathbf{x}_{1:T}, s_{1:T}, d_{1:T}) &= p(\mathbf{y}_1 \mid \mathbf{x}_1)p(\mathbf{x}_1 \mid s_1)p(d_1 \mid s_1)p(s_1) \\ &\cdot [\prod_{t=2}^{T} p(\mathbf{y}_t \mid \mathbf{x}_t)p(\mathbf{x}_t \mid \mathbf{x}_{t-1}, s_t) \cdot p(s_t \mid s_{t-1}, d_t)p(d_t \mid s_t, d_{t-1})]. \end{aligned} \tag{6}$$

Analogously to regular SLDS, the probabilities follow distributions defined below

$$\begin{gathered} p(s_1) = \mathrm{Cat}(s_1; \boldsymbol{\pi_0}) \quad p(d_1|s_1) = Dur_{s_1}(d_t - 1) \\ p(\mathbf{x}_1 \mid s_1) = \mathcal{N}(\mathbf{x}_1; \boldsymbol{\mu}_{s_1}, \boldsymbol{\Sigma}_{s_1}) \quad p(\mathbf{x}_t \mid \mathbf{x}_{t-1}, s_t) = \mathcal{N}(\mathbf{x}_t; \boldsymbol{A}_{s_t}\mathbf{x}_{t-1}, \boldsymbol{Q}_{s_t}) \\ p(\mathbf{y}_t \mid \mathbf{x}_t, s_t) = \mathcal{N}(\mathbf{y}_t; \boldsymbol{C}_{s_t}\mathbf{x}_t, \boldsymbol{S}_{s_t}), \end{gathered} \tag{7}$$

where π_0 is a vector of initial state probabilities, $\boldsymbol{\mu}_{s_1}^{init} := \boldsymbol{A}_{s_1}\boldsymbol{\mu}_{s_1}$ and $\boldsymbol{\Sigma}_{s_1}^{init} := \boldsymbol{A}_{s_1}\boldsymbol{\Sigma}_{s_1}\boldsymbol{A}_{s_1}^T + \boldsymbol{Q}_{s_1}$.

The transition of duration d_t and state s_t variables is defined as

$$p(d_t \mid s_t, d_{t-1}) = \begin{cases} \delta_{(d_{t-1}-1)} & \text{if} \quad d_{t-1} > 1, \\ Dur_{s_t}(d_t - 1) & \text{if} \quad d_{t-1} = 1 \end{cases} \tag{8}$$

where Dur_{s_t} is duration distribution in state s_t, $\delta_{(x)}$ is Dirac's delta, and

$$p(s_t \mid s_{t-1}, d_{t-1}) = \begin{cases} \delta_{s_{t-1}} & \text{if} \quad d_{t-1} > 1, \\ \boldsymbol{\pi}(s_{t-1}) & \text{if} \quad d_{t-1} = 1 \end{cases} \tag{9}$$

where $\boldsymbol{\pi}(i)$ is discrete transition matrix.

This model formulation allows us to efficiently represent state durations. Contrary to [1], we chose the decreasing count variable instead of the rising counter because in our case it allowed easier model formulation. For a comparison of these two modeling approaches, please refer to [2].

We test two duration models:

Categorical: there is a finite support of duration time, namely $\{1, \ldots, D_{max}\}$. In this case $Dur_{s_t} = \mathrm{Cat}\left(s_t; \boldsymbol{\pi}_{s_t}^{Dur}\right)$, and the prior is set to $\boldsymbol{\pi}_k^{Dur}|\alpha_{dur}^{Cat} \overset{\text{iid}}{\sim} \mathrm{Dir}(\alpha_{dur}^{Cat})$.

Poisson: $Dur_{s_t}(d_t - 1) = Poisson(\mu_{s_t}, d_t)$, with Gamma prior

$$\mu_k \overset{\text{iid}}{\sim} \mathrm{Gamma}(\alpha_{dur}^{Poi}, \beta_{dur}^{Poi}).$$

We chose Poisson because of the reasons explained in the previous section. However, at this point, we have to note that the inference algorithm would also work for any other duration distribution with infinite support. Especially if we are interested in multimodal explicit duration distributions, we can use a mixture of Poisson distributions.

Inference and Learning. In most of the proposed inference methods, the forward-backward algorithm was used to obtain the distribution of discrete latent factors [1,5,18] or at least the forward probability was computed [10]. We compute this probability dynamically, which requires $O(K^2T)$ time, where K is the number of states, and T is the length of the time series.

Let us define $z_t := (s_t, d_t)$, then our model becomes SLDS with an infinite number of states. In the case of the Poisson duration model, we cannot directly compute the forward probability, because z_t has infinite support.

One simple solution is to discard the states z_t with a small enough probability; however, as we will show later, this results in poor performance.

To overcome this problem, we use a trick known from the inference of infinite HMM models [15], also used for Hidden Semi-Markov Models [4]. The whole algorithm was generalized to work for SLDS-based models, called beam sampling.

As the literature extensively covers the Gibbs sampling technique for estimating the parameters of SLDS models and its various variations [6,10,12], we will focus on explaining the specific sampling scheme employed to approximate the forward probability.

First, for each t we introduce an auxiliary variable u_t with conditional distribution $u_t \sim$ Uniform $(0, p(z_t|z_{t-1}))$. Namely, let's sample

$$p\left(u_t \mid z_t, z_{t-1}\right) = \frac{\mathbb{I}\left(0 < u_t < p\left(z_t \mid z_{t-1}\right)\right)}{p\left(z_t \mid z_{t-1}\right)}. \tag{10}$$

It allows us to limit each sum in the forward pass to a finite number of terms while still allowing all possible durations. This mechanism is the application of a mechanism called "slice sampling" [13].

We sample the sequence $\mathbf{x}_{1:T}$ and treat it as pseudo-observations. We can get an approximation of the forward probability:

$$\begin{aligned} \hat{\alpha}_t\left(z_t\right) &= p\left(z_t, \mathbf{y}_{1:t}, \mathbf{x}_{1:t}, u_{1:t}\right) = \sum_{z_{t-1}} p\left(z_t, z_{t-1}, \mathbf{x}_{1:t}, \mathbf{y}_{1:t}, u_{1:t}\right) \\ &\propto \sum_{z_{t-1}} p\left(u_t \mid z_t, z_{t-1}\right) \cdot p\left(z_t, z_{t-1}, \mathbf{x}_{1:t}, \mathbf{y}_{1:t}, u_{1:t-1}\right) \\ &= \sum_{z_{t-1}} \mathbb{I}\left(0 < u_t < p\left(z_t \mid z_{t-1}\right)\right) \cdot p\left(\mathbf{y}_t, \mathbf{x}_t \mid z_t\right) \hat{\alpha}_{t-1}\left(z_{t-1}\right). \end{aligned} \tag{11}$$

The states are sampled backward using

$$\begin{aligned} & p\left(z_{t-1} \mid z_t, \mathbf{x}_{1:t}, \mathbf{y}_{1:T}, u_{1:T}\right) \propto p\left(z_t, z_{t-1}, \mathbf{x}_{1:t}, \mathbf{y}_{1:T}, u_{1:T}\right) \\ & \propto p\left(u_t \mid z_t, z_{t-1}\right) p\left(z_t \mid z_{t-1}\right) \hat{\alpha}_{t-1}\left(z_{t-1}\right) \\ & \propto \mathbb{I}\left(0 < u_t < p\left(z_t \mid z_{t-1}\right)\right) \hat{\alpha}_{t-1}\left(z_{t-1}\right). \end{aligned} \tag{12}$$

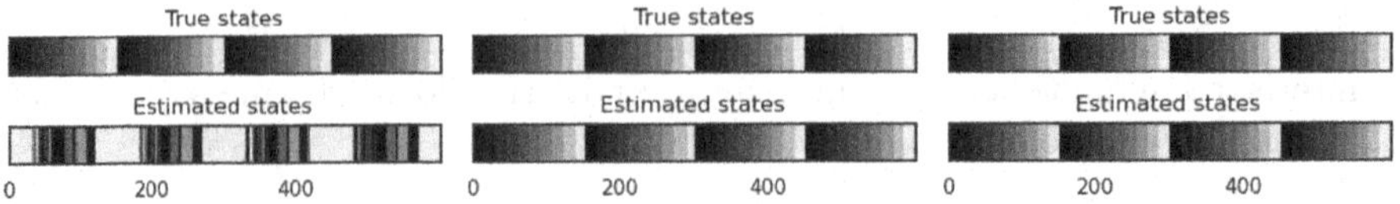

Fig. 1. Comparison of "Spikes" data segmentation (subplots b and c are indistinguishable at that scale). a) SLDS segmentation using Laplace EM algorithm, b) Poisson EDSLDS Non-loopy segmentation, c) Categorical Loopy EDSLDS segmentation

4 Experiments

In this section, we evaluate our method. Since it is difficult to evaluate segmentation without labels, we use two synthetic datasets, where we know the ground truth (i.e. "Spikes" and "Bouncing Ball"), for quantitative evaluation. We also qualitatively evaluate the segmentation of a Lorenz attractor.

We decided on the above three datasets for several reasons. They are all simple enough to gain insight into what the model is learning. Although the data set "Spikes" is simple, it is challenging enough for SLDS. The "Bouncing Ball" is a basic benchmark that often appears as a toy problem; however, it allows us to show how the modeling scheme works in an edge case where we have only two states. The Lorenz Attractor is an example more related to the real-life world and serves as a nice visual representation of both dimensionality reduction and segmentation capabilities. Because inference requires costly sequential sampling, we limited ourselves to only three data sets, which seems to be standard in the domain of switching state-space models [10,12].

In each case, we fit the model to the data and then sample the underlying states. Since the model is unidentifiable, the state labels have no meaning, so we post-process them by selecting the permutation that maximized the accuracy.

It is a common practice to have equal emission for all the states, i.e. $\forall_k C_k = C$ (this setup is used, for instance, by [5,6,10]). We also decided to do that in our experiments for faster convergence.

In the experiments below, by Categorical EDSLDS we denote Explicit Duration Switching Linear Dynamical System with categorical duration time, by Poisson EDSLDS with duration time modeled by Poisson distribution. As a baseline, we use SLDS implemented by us using Gibbs to infer parameters and implementation by S. Linderman fitted with black-box variational inference (BBVI) and Laplace EM algorithm[1]. By "Truncated" we refer to models fitted without Beam Sampling, but truncating the states with low probability. By "Loopy", we mean models that allow for self-transitions.

[1] We used implementation of RSLDS and SLDS provided by Linderman et al. https://github.com/lindermanlab/ssm.

Table 1. (a) The results of "Spikes" data segmentation after 200 steps. (b) The best results of "Spikes" data segmentation coming from the whole history.

(a)

	Weighted F1	Accuracy
SLDS (Laplace EM)	0.18	0.24
SLDS (BBVI)	0.18	0.24
SLDS (MCMC)	0.33	0.46
SLDS (MCMC) without init	0.30	0.44
Poisson EDSLDS	0.94	0.94
Poisson EDSLDS Loopy	0.83	0.86
Poisson EDSDLDS Truncated	0.28	0.42
Categorical EDSLDS	0.14	0.30
Categorical EDSLDS Loopy	**0.96**	**0.96**

(b)

	Weighted F1	Accuracy
SLDS (Ours)	0.37	0.48
SLDS (Ours) without init	0.46	0.49
Poisson EDSLDS	0.94	0.94
Poisson EDSLDS Loopy	0.84	0.87
Poisson EDSLDS Truncated	0.30	0.44
Categorical EDSLDS	0.31	0.33
Categorical EDSLDS Loopy	**0.97**	**0.97**

The prototype code was implemented in Python3 using NumPy and SciPy. All the computations were performed on a PC with AMD Ryzen Threadripper 3970X 32-Core Processor.

Spikes. *Motivation:* We chose this benchmark because it is straightforward to segment by humans but is challenging for SLDS independently of the fitting algorithm. It comes from the fact that the signal is relatively high-dimensional and sparse.

The data was created given the pattern: $X_d(t) = \mathbb{I}_d(t)|\sin(\frac{t\pi}{N})|$, where $\mathbb{I}_d(t) := \mathbb{I}((d-1)N < (t \mod DN) \leq dN)$, $X \in \mathbb{R}^D$, $N \in \mathbb{N}_+$, where D is the number of dimensions and the number of hidden states, N is the length of a single spike, $t \in \{1, \ldots, T\}$ is the current time point. In this case $s_t = d$ if $\mathbb{I}_d(t) = 1$. They all have a constant duration time, so the duration time is not geometrically distributed.

We noticed that in the case of this dataset, Poisson EDSLDS benefits from a preinitialization, i.e. for $n_{init} = 100$ iterations we approximate the forward probability by the HMM with emission probability equal to $p(\mathbf{y}_t \mid s_t) = p(\mathbf{y}_t|\mathbf{x}_t) \cdot \mathcal{N}\left(\mathbf{x}_t; \boldsymbol{A}_{s_t}\boldsymbol{\mu}_{s_t}, \boldsymbol{A}_{s_t}\boldsymbol{\Sigma}_{s_t}\boldsymbol{A}_{s_t}^T + \boldsymbol{Q}_{s_t}\right)$. However, it seems that it does not bring any improvement to plain SLDS.

In our setup, we used $D = 10$, $N = 15$, and 2400 data points.

Table 1(a) presents the results of segmentation after 200 sampling steps (or 2000 variational inference steps). The best score was achieved by the model with loops and categorical duration distribution.

However, the Poisson duration model also managed to perform this task quite well, both with and without loops. It may be worth noting here that the Poisson duration model may be a nice alternative to categorical duration modeling if we do not know what might be actual D_{max}. The trained model can also serve as an initialization for the categorical duration.

Interestingly, Poisson EDSLDS without loops fails to learn the true duration distribution, while Categorical EDSLDS seems to benefit from incorporating loops.

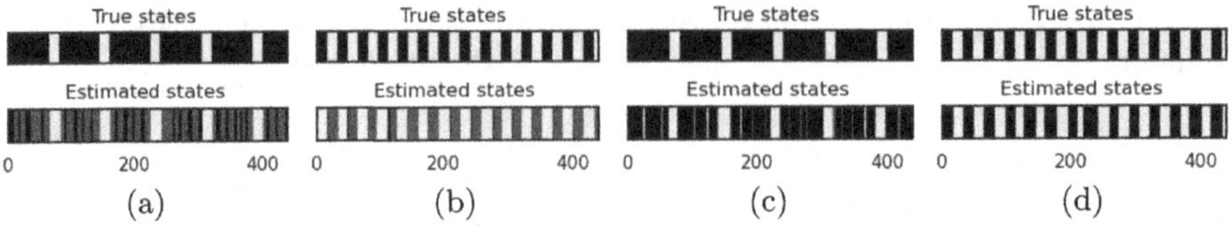

Fig. 2. Visualization of the segmentation of "ball" movement when both of the segments are relatively short. As we can see, the model behaves relatively correctly even after merging the states. (a) Uneven bouncing three state segmentation, (b) Even bouncing three state segmentation, (c) Uneven bouncing two state segmentation, (d) Even bouncing two state segmentation

The Poisson EDSLDS and Categorical EDSLDS with loops were able to use all of the possible states. With well-initialized SLDS all the states were used, but three of them were used less than eighteen times. SLDS trained with BBVI and Laplace EM collapsed to eight states. The loopy variant of Poisson collapsed to nine states. The most prevalent case of collapse occurs in the case of a truncated Poisson, which collapses to five states, and Categorical EDSLDS with no loops, collapsing to three states. Because we noticed that some models would score higher accuracy if sampling was finished earlier, we also present their maximum accuracy scores. These results are presented in Table 1(b). The qualitative comparison can be seen in Fig. 1.

Bouncing Ball. *Motivation:* We have to keep in mind here that two hidden states are a special case for the nonloopy models. In this situation, the transition matrix becomes $\begin{pmatrix} 0 & 1 \\ 1 & 0 \end{pmatrix}$. Because of that, those models suffer from poor performance while using only two states. Similarly to [7] and [5], we simulate the movement of a bouncing ball. The underlying system switches between two operating modes. We experimented with the number of states and the "speed" of motion.

In our setup "ball" is a point moving from point $[1, 2]^T$ to point $[16, 32]^T$ and back to point $[1, 2]^T$ following a linear dynamic.

In the first case, we tried 64 steps of "going up" and 16 steps of "going down". The whole time series had 7920 time points. The inference with three states took six hours and 150 steps). The results of this experiment are presented in Table 2(a).

If we still want to use the non-loopy Poisson model (because we want Poisson parameters to be interpretable), we can use three states to initialize the model. In our experiments, two of the state modes have almost identical continuous state linear transformation matrices (the Frobenius norm of their difference is equal to 0.01). By treating those two states as one, we get an accuracy equal to **0.98** (see Fig. 2(a)). If we try to use it as initialization for the two-state model, the accuracy tends to drop to 0.87 (see Fig. 2(c)). This is probably the effect of the variance of the Poisson distribution which is equal to its mean. That lets us think that using distribution giving more control in the future will be a better idea.

Table 2. Results of segmentation of (a) uneven (b) even bouncing ball.

(a)

	Weighted F1	Accuracy
Poisson EDSLDS Loopy	**0.97**	**0.97**
SLDS (Ours)	0.72	0.80
Categorical EDSLDS	0.92	0.92
Categorical EDSLDS Loopy	**0.97**	**0.97**
Poisson EDSLDS	0.77	0.79

(b)

	Weighted F1	Accuracy
Poisson EDSLDS Loopy	0.91	0.90
SLDS (Ours)	0.82	0.82
Categorical EDSLDS	0.73	0.73
Categorical EDSLDS Loopy	**0.97**	**0.97**
Poisson EDSLDS	0.77	0.77

In the second case, we tried 16 steps of "going up" and 16 steps of "going down". The whole time series had 3120 time points. This time the reduction of accuracy after merging was not as prevalent. By treating two states as one in the case of Poisson EDSLDS we get an accuracy equal to **0.98**, but after treating the learned model as the initialization for the two-state model it drops to 0.97 (which is the same as for the Categorical EDSLDS model). The segmentation results are presented in Table 2(b).

Lorenz Attractor. *Motivation:* Lorenz Attractor is one of the most popular data sets used to benchmark nonlinear dynamical systems, appearing in many works on state space modeling and switching regime detection [1,10,12,18].

Its nonlinear dynamics are given by:

$$\frac{\mathrm{d}x}{\mathrm{d}t} = \sigma(y - x), \quad \frac{\mathrm{d}y}{\mathrm{d}t} = x(\rho - z) - y, \quad \frac{\mathrm{d}z}{\mathrm{d}t} = xy - \beta z. \tag{13}$$

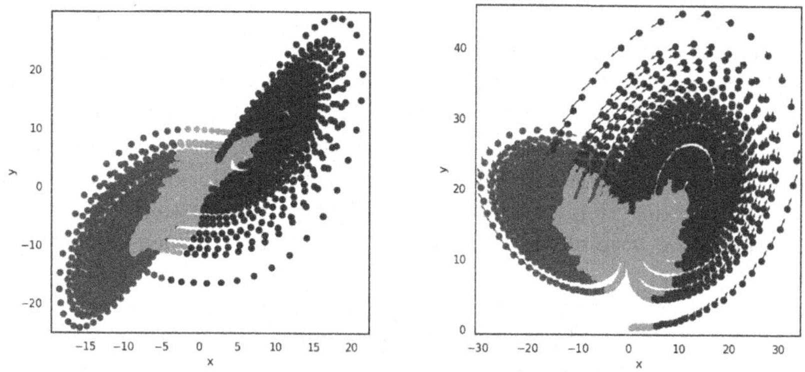

(a) Segmented emission of Lorenz Attractor

(b) The learned latent 2D space

Fig. 3. The visualization of learned Lorenz Attractor model. The dimensionality was reduced from three to two dimensions. The arrows in the latent continuous space visualization show the direction of linear transformations. The color used represents a latent discrete state.

We trained models on 5000 timepoints sampled for $\sigma = 10$, $\rho = 28$, $\beta = 2.667$. For the reasons described above, we decided to use three hidden states.

The results of the segmentation using Poisson EDSLDS are presented in Fig. 3(a). One can see that the model distinguished between the left and right "wings" and a special state for switching between planes. Figure 3(b) presents the two-dimensional latent space. It is similar to the projection of the Lorenz attractor on the xz plane.

Compared to the results presented by [10] our segmentation, even though demands one more state, is much smoother, in the sense that it does not do single-state switches.

The segmentations obtained by the other models are presented in the supplement.

Experiments Summary. The "Spikes" dataset was proved to be challenging, resulting in "state collapse" for models without explicit duration modeling.

Due to the construction of the transition matrix in the case of models not allowing loops, we evaluated them on the special case with just two hidden states. Despite the fact that there were two states with linear dynamics (i.e. going up and down), Poisson EDSLDS did not collapse into two states, which further shows its "state collapse" avoiding properties.

The last experiment shows that explicit duration in SLDS can smoothen state segmentation. Furthermore, it allows visually comparing with other variations of dynamical systems, since the Lorenz attractor is a popular benchmark [10].

5 Conclusion and Future Work

There are many examples of spatio-temporal data that show persistent and recurring patterns across the regime. Modeling both states and their explicit durations helps us better understand the process.

As shown, adding duration variables allows for smoother state segmentation and allows the model to use all possible states. Using distributions with infinite support, such as Poisson, allows us to efficiently model duration time without as much memory usage as categorical distribution. The proposed solution may be combined with recurrent connections to add duration modeling similar to RSLDS.

A natural next step would be applying the newly developed models to real-world problems.

Although we argued that the MCMC methods seem to give better performance than variational inference, the latter is significantly faster than the former. Given this, it may be beneficial to develop a variational inference algorithm for EDSLDS models with infinite support of state duration.

In this work, we only analyzed the model performance for the Poisson distribution of the duration. In the future, one may try to model it given, for instance, Negative Binomial distribution, to have control over the variance.

Another interesting direction is to apply the model in the (semi-)supervised setting. Thanks to the choice of marginal distribution, it is straightforward to condition on explicit label information.

Acknowledgement. This work was supported by the Polish National Science Centre (NCN) under grant OPUS-18 no. 2019/35/B/ST6/04379.

References

1. Ansari, A.F., et al.: Deep explicit duration switching models for time series. In: NeurIPS, pp. 29949–29961 (2021)
2. Chiappa, S.: Explicit-Duration Markov switching models. Found. Trends ® Mach. Learn. **7**(6), 803–886 (2014)
3. Chiappa, S., Peters, J.: Movement extraction by detecting dynamics switches and repetitions. In: NeurIPS, vol. 23. Curran Associates, Inc. (2010)
4. Dewar, M., Wiggins, C., Wood, F.: Inference in hidden Markov models with explicit state duration distributions. IEEE Signal Process. Lett. **19**(4), 235–238 (2012)
5. Dong, Z., Seybold, B., Murphy, K., Bui, H.: Collapsed amortized variational inference for switching nonlinear dynamical systems. In: ICML, pp. 2638–2647 (2020)
6. Fox, E., Sudderth, E.B., Jordan, M.I., Willsky, A.S.: Bayesian nonparametric inference of switching dynamic linear models. IEEE Trans. Signal Process. **59**(4), 1569–1585 (2011)
7. Johnson, M.J., Duvenaud, D.K., Wiltschko, A., Adams, R.P., Datta, S.R.: Composing graphical models with neural networks for structured representations and fast inference. In: NIPS (2016)
8. Johnson, M.J., Willsky, A.S.: Bayesian nonparametric hidden semi-Markov models. J. Mach. Learn. Res. **14**(1), 673–701 (2013)
9. Kim, C.J.: Dynamic linear models with Markov-switching. J. Econ. **60**(1), 1–22 (1994)
10. Linderman, S., Johnson, M., Miller, A., Adams, R., Blei, D., Paninski, L.: Bayesian learning and inference in recurrent switching linear dynamical systems. In: AISTATS, pp. 914–922 (2017)
11. Murphy, K.: Switching Kalman Filters (1998)
12. Nassar, J., Linderman, S.W., Bugallo, M.F., Park, I.M.: Tree-structured recurrent switching linear dynamical systems for multi-scale modeling. In: ICLR (2019)
13. Neal, R.M.: Slice sampling. Ann. Stat. **31**(3), 705–767 (2003)
14. Oh, S., Rehg, J., Balch, T., Dellaert, F.: Learning and inferring motion patterns using parametric segmental switching linear dynamic systems. Int. J. Comput. Vis. 77 (2008)
15. Van Gael, J., Saatci, Y., Teh, Y.W., Ghahramani, Z.: Beam sampling for the infinite hidden Markov model. In: ICML, pp. 1088–1095 (2008)
16. Wang, Y., Blei, D., Cunningham, J.P.: Posterior collapse and latent variable non-identifiability. In: NeurIPS, pp. 5443–5455 (2021)
17. Yu, S.Z.: Hidden Semi-Markov Models: Theory, Algorithms and Applications. Computer Science Reviews and Trends. Elsevier, Amsterdam; Boston (2016)
18. Zoltowski, D., Pillow, J., Linderman, S.: A general recurrent state space framework for modeling neural dynamics during decision-making. In: ICML, pp. 11680–11691 (2020)

Policy Representation Opponent Shaping via Contrastive Learning

Yuming Chen[1,3](✉) and Yuanheng Zhu[1,2]

[1] State Key Laboratory of Multimodal Artificial Intelligence Systems, Institute of Automation, Chinese Academy of Sciences, Beijing 100190, China
chenyuming@ucass.edu.cn

[2] School of Artificial Intelligence, University of Chinese Academy of Sciences, Beijing 100049, China
yuanheng.zhu@ia.ac.cn

[3] School of Computer Science, University of Birmingham, Birmingham, UK

Abstract. To acquire results with higher social welfare in social dilemmas, agents need to maintain cooperation. Independent agents manage to navigate social dilemmas via opponent shaping. However, opponent shaping needs extra information of opponent. It is not always accessible in mixed tasks if agents are decentralized. To address this, We present PROS, which runs in a fully-independent setting and needs no extra information. PROS shapes the opponent with an extended policy that takes the opponent's dynamics as additional input. Instead of receiving policy from the opponent, we discriminate the policy representation via contrastive learning. In terms of experiments, PROS reaches the optimal Nash equilibrium in iterated prisoners' dilemma (IPD) and shows the same ability to maintain cooperation in Coin Game, a highly-dimensional version of IPD. The source code is available on https://github.com/RandSF/Policy-Representation-Opponent-Shaping.

Keywords: multi-agent systems · reinforcement learning · contrastive learning

1 Introduction

Besides fully-cooperative tasks [9] and zero-sum tasks [29], mixed cooperative-competitive tasks [3,16] are gathering importance in multi-agent reinforcement learning (MARL), including self-driving cars [10], multi-robot control [12], etc. One major meaningful type of mixed task is the social dilemma, where there is a trade-off between the benefits for specific individual agents and higher social welfare, the average reward in MARL, for the system. To maintain high welfare, self-interested agents are expected to cooperate in social dilemmas. MARL agents are able to perform good policy in many mixed tasks with centralized

Y. Chen—The majority of this work YC carried out during the internship at the Institute of Automation of the Chinese Academy of Sciences.

B. Luo et al. (Eds.): ICONIP 2023, CCIS 1963, pp. 124–135, 2024.
https://doi.org/10.1007/978-981-99-8138-0_11

training decentralized execution (CTDE) paradigm [16,26,28]. Such algorithms require a centralized aggregator who can obtain all information about the system. Learning of the aggregator becomes hard when the number of agents in the system goes larger. Is it able to learn cooperation spontaneously for decentralized independent agents?

Specifically, independent agents can address such social dilemmas via opponent shaping [5,13,27]. Opponent shaping takes the opponent's policy as a function of its own policy and updates its own policy by shaping the learning process of the opponent. In iterated prisoners' dilemma [1], where two agents play the prisoners' dilemma for infinite times, learning with opponent learning awareness (LOLA) [5] is the first autonomous learning agent to discover *tit-for-tat* (TFT) [1], i.e. performing cooperation first and copying the action opponent performed last time. Model-free opponent shaping (M-FOS) [17] learns a meta policy to defeat another algorithm via meta-game and succeeds to shape LOLA in IPD.

However, LOLA requires white-box access to the opponent's policy and assumes the opponent is a naive learner. Another weakness of LOLA is the low sample efficiency. M-FOS fails to learn a meta policy in large-scale environments, and alternatively, it learns a conditional vector to replace the meta policy from the global information of the system. Such information is not always accessible to independent agents.

In this paper, we introduce policy representation opponent shaping (PROS) to resolve the above problems. Without obtaining the opponent's action, PROS represents the opponent's dynamic change via contrastive learning to implement opponent shaping. Our contributions are two-fold as followed:

- We propose a framework that formulates the learning process in multi-agent environments as a meta-game. Agents can implement opponent shaping by learning an opponent-dynamics conditional policy in the game.
- We derive PROS, which uses a contrastive learning algorithm to learn the opponent-dynamics representation under a fully-independent setting, i.e. agents are trained decentralized and have no access to information about the opponent. With a proper learning objective, PROS learns faster than other algorithms with latent representation.

We show the feasibility to shape the opponent when the opponent's policy is unknown. Then we explain why the dynamics of the opponent can be discriminated with no extra information, and present a contrastive learning module to achieve such discrimination. In terms of experiments, We evaluate PROS in IPD, iterated matching pennies (IMP), and coin game. The result shows that PROS leads to equilibria with higher social welfare compared with baselines.

2 Relative Work

Opponent Modeling. To navigate the social dilemma as an independent agent, a number of works focus on opponent modeling, which reconstructs some aspect of the policy of the opponent. One straightforward idea is modeling the opponent's policy and using the predicted action [18]. The family of learning and influencing latent intent(LILI) [10,20,25] learns a latent policy of the other agents by high-level representation in order to influence the long-term behaviour of partner [10]. Comparing opponent shaping and PROS, such work pays more attention to cooperative environments and requires agents to learn how to exploit or co-adapt with the other agents.

Opponent Shaping: By influencing the future actions of the opponent, opponent shaping exploits the opponent whose policy is known. The family of LOLA updates the policy with the awareness that its own policy influences the opponent's learning process. Most of the work [13,23,27] focus on the convergence or consistency in self-play and assume white-box access to the opponent's model. To relax this assumption, M-FOS [17] learns in a meta-game, where each meta-step is an episode of the underlying[1] game. Agents take inner policy as a meta-action from meta-policy. Directly sampling inner policy from meta-policy is intractable in high-dimensional games and it has to replace meta-policy with cross-episode conditional vectors without theoretical guarantee. PROS uses latent representation instead of meta-policy, thus it learns in the same way in games on any scale.

Contrastive Learning. Technically, PROS uses contrastive learning to learn the representation of policy. There is a large body of research on applying contrastive learning to reinforcement learning, like pixel representation learning [11,15] and the exploration [7,21]. Contrastive learning is also used in meta-learning [14] and policy representation [6]., which are fields relative to our work. However, PROS is under a fully decentralized training process, that is, PROS uses no information from the opponent.

3 Background

3.1 Learning Process as an Repeated Game

In this section, we describe the learning process of agents and formulate the process to be an repeated game.

We consider several players learn to optimise their own individual objectives in a game $\mathcal{G} = \langle \mathcal{I}, \mathcal{S}, \mathcal{A}, \Omega, \mathcal{O}, \mathcal{P}, r, \gamma \rangle$, where $\mathcal{I} = \{1, 2, \ldots, n\}$ is the set of n players. $\mathcal{S}$ denotes the state space. $\mathcal{A} = \times_{i \in \mathcal{I}} \mathcal{A}^i$ is the action space. Observation space $\Omega = \times_{i \in \mathcal{I}} \Omega^i$ and observation function $\mathcal{O} : \mathcal{S} \times \mathcal{A} \times \Omega \mapsto [0, 1]$ decide the observation of players. $r = \times_{i \in \mathcal{I}} r^i$ represents the reward function. $\mathcal{P}$ is the transition function of the state.

[1] In this paper, we use 'underlying' or 'inner' to refer to the game agents actually playing, and 'meta' or 'outer' to refer to the meta-game we introduce here.

$\mathcal{G}$ can be regarded as a partially observable Markov decision process (POMDP). An agent learns in the POMDP to obtain an optimal policy π^i : $\mathcal{O}^i \mapsto \mathcal{A}^i$ that maximises the objective, for simplicity, we set it to be accumulated discounted rewards, i.e. discounted return $R^i(\pi^i, \pi^{-i}) := \sum_{t=0}^{T} \gamma^t r^i(s_t, a_t)$, where actions are sampled from policy $\pi^i(o_t^i)$, $\pi^{-i}(o_t^{-i})$ respectively. R^i is a mapping from joint policy space to the reals. An RL agent updates its parameterized policy π_ϕ^i based on the discounted return, which can be formalized to be

$$\pi_{k+1}^i = \pi_k^i + \alpha^i \nabla_\phi R^i(\pi^i | \pi_k^{-i})|_{\pi_k^i} \tag{1}$$

where α^i is the learning rate and ϕ is the policy parameters of agent i. We will omit parameter ϕ at the notation of policy if the policy π_k^i refers to those in a certain epoch k. It should be noted that policies π in Eq. (1) are not always regarded as a function of ϕ.

The return R^i implicitly depends on the game $\mathcal{G}$ besides policies. Thus the underlying game can be regarded as a function mapping joint policy to returns. We can formulate the learning process from initial policies to Nash equilibrium as a repeated game, which is the *meta-game*. In each round, agents perform a policy, get returns and update their policies.At the round k of the meta-game, agents interact in the underlying game $\mathcal{G}$ with policy π_k^i, π_k^{-i} and get discounted inner return $R_k^i = R^i(\pi_k^i, \pi_k^{-i})$, $R_k^{-i} = R^{-i}(\pi_k^{-i}, \pi_k^i)$.

A naive learner's learning process can be seen as learning to maxising R^i, regardless of the opponent, that is, the opponent policy π^{-i} is *not* the function of ϕ. For a naive learner, Eq. (1) becomes

$$\pi_{k+1}^i = \pi_k^i + \alpha^i \nabla_\phi R^i(\pi^i(\phi) | \pi_k^{-i})|_{\pi_k^i}$$

where $\pi^i(\phi)$ means π^i is parameterized by ϕ.

LOLA-like learners [5,13,23,27] have awareness of the opponent's policy and have white-box access to it. That is, LOLA-like learners know how the opponent updates, i.e., the opponent policy π^{-i} is accessible and $\pi_{k+1}^{-i} - \pi_k^{-i}$ (not π^{-i}) is determined by ϕ. Thus Eq. (1) becomes

$$\begin{aligned}\pi_{k+1}^i &= \pi_k^i + \alpha^i \nabla_\phi R^i(\pi^i(\phi), \pi^{-i})|_{(\pi_k^i, \pi_{k+1}^{-i})} \\ &= \pi_k^i + \alpha^i \nabla_\phi \left[\nabla_1 R^i(\pi^i(\phi), \pi^{-i}) + \left(\Delta\pi_k^{-i}(\phi)\right)^\top \nabla_2 R^i(\pi^i(\phi), \pi^{-i})\right]_{(\pi_k^i, \pi_k^{-i})}\end{aligned}$$

according to the Taylor expansion, where ∇_1 refers to the gradient w.r.t. the first variant and ∇_2 refers to the gradient w.r.t. the second variant. $\Delta\pi_k^{-i}$ is $\pi_{k+1}^{-i} - \pi_k^{-i}$.

One improvement of LOLA is M-FOS, which learns the optimal meta-policy via PPO [22] or Genetic Algorithm. An M-FOS learner maintains a meta-policy $\tilde{\pi}$ and directly samples policy π from it. M-FOS learner learns the meta-policy by optimising the discounted accumulation of returns

$$\tilde{\pi}_{k+1}^i = \tilde{\pi}_k^i + \alpha^i \nabla_\pi \sum_{k=0}^{K} \eta^k R^i(\pi_k^i, \pi_k^{-i}) \tag{2}$$

where η is the meta-discount factor set and K is the estimation of rounds to reach the equilibrium. Both of them are hyperparameters.

Since policy π^i is directly sampled before each interaction round, original M-FOS can only work in tabular environments where state and action spaces are not large. In large-scale environments, M-FOS uses a conditional vector c to avoid directly generating policy π^i or accessing opponent policy π^{-i}. In round k, conditional vector c_k is encoded from the trajectory of the last round

$$c_k = \mathcal{E}(\tau_{k-1}, c_{k-1}), \qquad \tau_{k-1} := \{(o_t^i, o_t^{-i}, a_t^i, a_t^{-i}, r_t^i, r_t^{-i})\}$$

where $\mathcal{E}$ is the vector encoder. Action is sampled from policy π_k^i, i.e. $a_t^i \sim \pi_k^i(o_t^i, c_k^i)$.

Both of the vector encoder $\mathcal{E}$ and policy π_ϕ are updated according to Eq. (2) every T rounds.

4 Method

M-FOS uses different implementations in tabular environments and highly dimensional environments since π^{-i} is infinitely dimensional in large-scale environments. To overcome such weakness, PROS learns an opponent-dynamics conditional policy $\pi_\phi^i(o_t^i, z_t)$ to implement opponent shaping. The opponent dynamic z is inferred from local trajectories $\tau^i = \{(o_t^i, a_t^i, r_t^i)\}$ via contrastive learning. Such a policy-representing module is agnostic to RL algorithms, thus it can work as a plug-in in RL algorithms. The network structure is illustrated in Fig. 1.

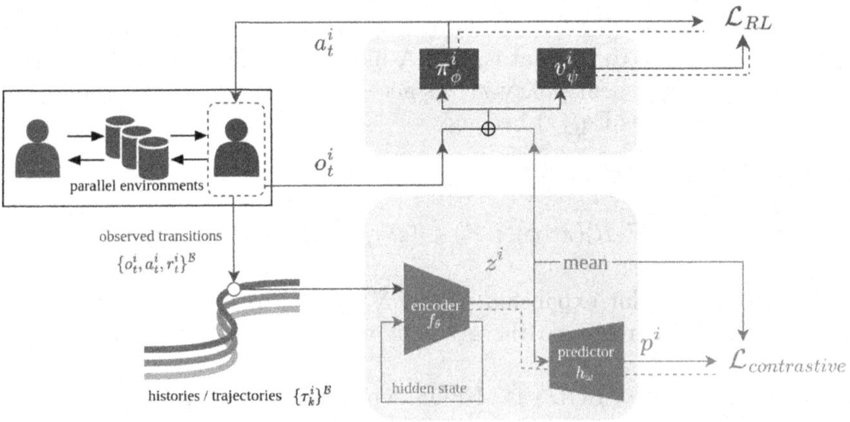

Fig. 1. Training of policy representation encoder. The blue area is the RL module (we use PPO in this paper), and the green area is the Dynamics encode module. The learning of opponent-dynamics encoder and policy are decoupled in terms of gradients, which is denoted as the dotted line. (Color figure online)

4.1 Learning via Opponent-Dynamics Conditional Policy

PROS learns a latent z from local trajectory to have *meta-learning awareness* instead of learning the meta-game directly. The policy makes dicisions based on observation o^i and latent z.

Different latent z refer to different opponent's policies. However, we do not require the invertibility of z. There is no guarantee that π^{-i} can be reconstructed from z. It is a significant difference between PROS and works involving opponent modeling.

Similar to many algorithms in meta-learning, PROS optimises the accumulation of discounted inner returns of multiple episodes to estimate both the current policy's performance and the current learning step's performance.

$$\sum_{k=0}^{K} \gamma^k R^i(\pi^i_{j+k} | \pi^{-i}_{j+k}) \tag{3}$$

where K is the memory capacity of the agent and γ plays the same role as η in M-FOS.

Algorithm 1: PROS-PPO

Initialize encoder parameters θ, predictor parameters ω, actor parameters ϕ, and critic parameters ψ;
Initialize policy representation vector z;
while *true* **do**
    for $k = 0$ **to** K **do**
        reset buffer $\mathcal{D}$;
        reset parallel environment;
        for $t = 0$ **to** T **do**
            sample action $a^i_t \sim \pi_\phi(o^i_t, z)$;
            apply the action to get new observation o^i_{t+1} and reward r^i_t;
            encode current dynamics $z \leftarrow f_\theta(o^i_t, a^i_t, r^i_t, z)$;
            update parameters θ and ω with $\mathcal{L}_{contrastive}(z)$;
            `// the loss is computed according to Equation (5)`
            store data (o^i_t, a^i_t, r^i_t) to buffer $\mathcal{D}$;
        end
    end
    update parameters ϕ and ψ with data in $\mathcal{D}$ according to PPO;
    `// the returns of adjoint episodes accumulate according to Equation (3)`
end

4.2 Policy Representation via Contrastive Learning

In terms of inferring the opponent's dynamics in a fully-independent setting, the policy of opponent is unknown and agent needs to learn the opponent's policy from its local trajectory, i.e. the observation, action, and reward sequences $\{(o_t^i, a_t^i, r_t^i)\}_{t=0}^T$, where T is the length of the trajectory. To solve this problem, we use contrastive learning to learn the dynamic change caused by the opponent's policy. Rather than a CPC [19] training style used in CURL [11] to obtain the ability to predict the future of dynamics, we expect that the encoder is able to distinguish the dynamics itself, thus we train it in the way of policy representation [6].

Dynamics Discrimination. In an independent setting, agent does not know in advance any prior knowledge about the opponent. That is, agent has no assumption on the opponent's algorithm Γ^{-i}, and no access to the grounded-true trajectory node $(o_t^{-i}, a_t^{-i}, r_t^{-i})$ of the opponent.

If the action of the opponent a^{-i} is unknown, it is impossible to reconstruct the transition probability $P(s'|a^i, a^{-i}, s)$ of inner environments. However, discriminating the dynamics of the inner game, instead of directly reconstructing policy of the opponent, is theoretically feasible. With a fixed opponent policy, the mixed dynamics

$$P_{\pi^{-i}}(s'|a^i, s) := \sum_{a^{-i}} P(s'|a^i, a^{-i}, s)\pi^{-i}(a^{-i}|o^{-i}) \tag{4}$$

is invariant. And we assume that the opponent's policy satisfies the below property.

Property: There are not two different policies π_1^{-i}, π_2^{-i} satisfying the condition that for all state-action tuple (s', a^i, s), $P_{\pi_1^{-i}}(s'|a^i, s) = P_{\pi_2^{-i}}(s'|a^i, s)$.

We say two policies are different if there is some action a^{-i} such that $\pi_1^{-i}(a^{-i}|o^{-i}) \neq \pi_2^{-i}(a^{-i}|o^{-i})$. If for all state s, the observation o^i is different, we can discriminate the mixed dynamics with trajectories $\{\tau_k^i|\tau_k^i = (o_t^i, a_t^i, r_t^i)_{t=1}^{T_k}\}_{k=1}^K$. Such discrimination is the policy representation z. It can be captured by an auto-regressive encoder which takes history node (o_t^i, a_t^i, r_t^i) and the last output z_{t-1} as input.

Negative Sample. Another problem is contrastive learning requires negative samples. In terms of policy representaion [6], negative sample is a trajectory generated with a different opponent's policy, which is not always available in MARL as most environments are symmetric games and the homogeneous agents play the same policy repeatedly at the stable fixed point of the game.

We use SimSiam [4], a contrastive learning algorithm without negative samples, as the policy encoder. Following the notation in [4], we use f_θ to denote the encoder and h_ω the predictor. They are parameterized by θ and ϕ respectively and train as Fig. 1 shows.

Data Augmentation. Data augmentation remarkably improves the performance of the model. It generates new samples from one sample without causing significant semantic alterations. In policy representation, the semantics is the mixed dynamics, thus trajectories from the same policy can be regarded as augmented data. To further relax the assumption on the opponent, we do not require the opponent to perform the sample policy in one episode. Instead, PROS learns the representation at every time step. The feasibility is shown by Hjelm et al. [8]. The contrastive loss is

$$\mathcal{L} = \frac{1}{|\mathcal{B}|} \sum_{b=0}^{|\mathcal{B}|} \frac{h_\omega(f_\theta(x_b)) \cdot \perp(\frac{1}{|\mathcal{B}|}\sum_{d=0}^{|\mathcal{B}|} f_\theta(x_d)^\top)}{\|h_\omega(f_\theta(x_b))\| \cdot \perp\|\frac{1}{|\mathcal{B}|}\sum_{d=0}^{|\mathcal{B}|} f_\theta(x_d)\|} \tag{5}$$

where $\mathcal{B}$ refers to the batch of data, which are sampled from synchronous parallel environments at each step, $x_b = (o_t^i, a_t^i, r_t^i)_b$ is the data from a parallel environment, and $\perp$ is the stop gradient operator. The training process is shown as Algorithm 1.

5 Experiment

In this section, we perform experiments to evaluate the ability of PROS to maintain cooperation in mixed cooperative-competitive environments with different scales.

5.1 Experiment Setting

We compare PROS with four baseline algorithms:

- **PPO** [22]. We use iPPO rather than MAPPO [26] to optimise the policy of a self-interested agent without awareness of opponent.
- **M-FOS** [17]. M-FOS represents agents with opponent learning awareness without white-box access to the opponent's policy. It uses PPO as the RL backbone. To evaluate its scalability, We use M-FOS with conditional vector in all experiments.
- **LILI** [25]. LILI learns the latent policy of the other agents, which is similar to PROS but focuses on cooperative environments. It uses SAC as the RL backbone.

There are some works relative to PROS but we do not use them as baselines for some reason. LOLA-DiCE [5] requires approximating the updating of opponent's policy, which involves additional interactions with the imagined policy of opponent between episodes playing with the true opponent. This leads to LOLA-DiCE needing a simulator of the game. FURTHER [10], which covers techniques used in opponent shaping and latent representation, is proposed from the aspect of Active Markov Game, unlike all other works.

All the actor and critic networks are MLPs since all the following environments are fully observable. Networks of different algorithms have the same width and depth.

Table 1. Payoff matrix of two iterated games

	C	D
C	(-1,-1)	(-3, 0)
D	(0, -3)	(-2,-2)

(a) Prisoners' dilemma

	H	T
H	(+1,-1)	(-1,+1)
T	(-1,+1)	(+1,-1)

(b) Matching pennies

5.2 Iterated Matrix Games

Following prior work [5] on opponent shaping, we first study one-step-memory iterated prisoners' dilemma (IPD) and iterated matching pennies (IMP). The properties of the game are well studied [1].

In iterated matrix games, players play the same normal-form game repeatedly and maximise the discounted cumulative reward. The unique Nash equilibrium in IMP is uniformly selecting action. However, the folk theorem shows that there are infinite Nash equilibria in IPD, which are different in welfare. Table 1 shows the reward at each step.

5.3 Coin Game

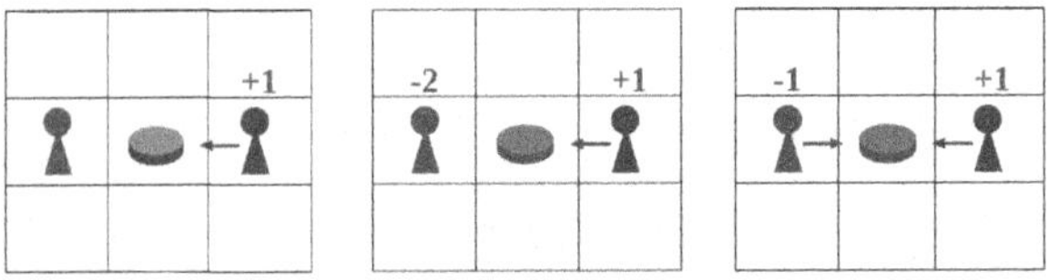

Fig. 2. Three situations of collecting a coin. (Color figure online)

We investigate the scalability and efficiency of PROS in coin game, first proposed by Lerer and Peysakhovich [12]. The coin game is a Markovian version of prisoners' dilemma, where two agents, colored red and blue respectively, are tasked with collecting coins in a 3×3 grid. Coins are either red or blue and spawn with the other color after being picked up. One Agent gains a +1 reward once it picks up a coin of any color. When agents pick up a coin of different color, the other agent gains a –2 reward. In coin game, a fully cooperative policy is always picking up coins of the same color and a selfish policy is greedily picking up all coins (Fig. 2).

5.4 Results

Iterated Prisoners' Dilemma. In IPD, each algorithm plays with each other. We visualize the mean of both agents' rewards at each training step to measure the ability to learn to cooperate in an environment mainly guided by competing. Specifically, different mean rewards refer to different equilibria.

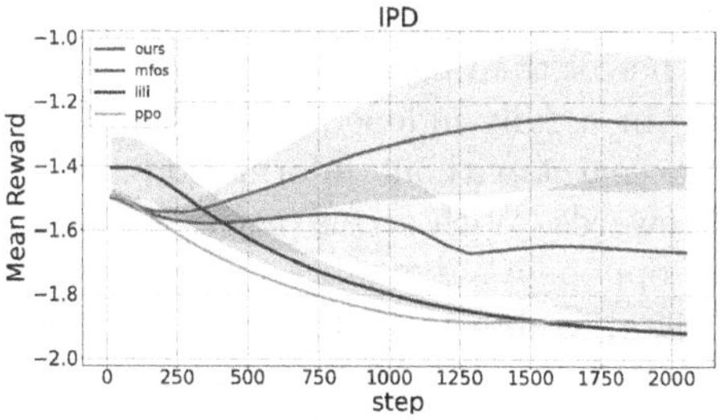

Fig. 3. The mean reward of both agents during training

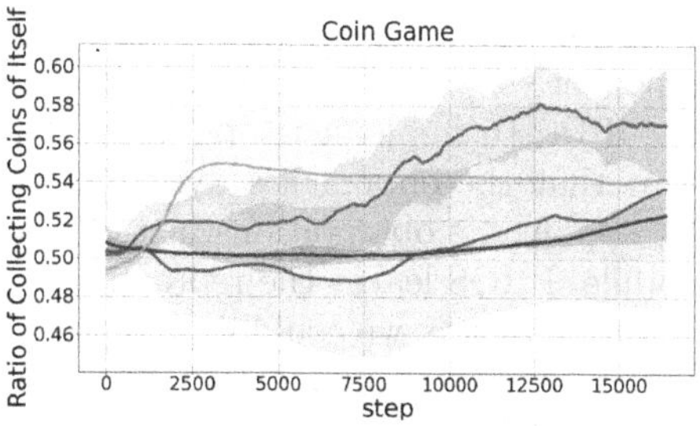

Fig. 4. The empirical percentage of picking up coins of the same colours.

Figure 3 shows that only PROS achieves long-term cooperation. M-FOS with conditional vector is able to learn to cooperate but underperforms the original version that is available in such a tabular environment (the mean reward of the original version is –1.01 as shown in [17]).

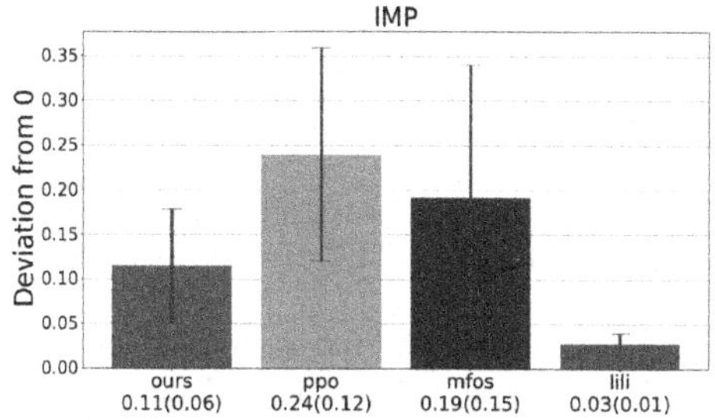

Fig. 5. Reward deviation from zero. Error bars show the SEM. The values of mean deviation is shown at the bottom of figure, with value of SEM in the brace.

Iterated Matching Pennies. With the same learning rate, PROS has a minimum reward deviation from zero compared with two other on-policy algorithms, PPO and M-FOS. Such deviation characterizes the distance to the Nash equilibrium. LILI has the best result as it is off-policy with higher sample efficiency than others. Figure 5 shows that PROS has the best result and the lowest standard error about the mean (SEM) among on-policy algorithms.

Coin Game. The ratio of coins with the same color in all collected coins can be seen as the probability of cooperation during training. Figure 4 shows that the cooperation probability of PROS, M-FOS, and LILI all increase over training, but PROS learns faster with a lower standard deviation. PPO seems to outperform LILI and M-FOS but the performance of PPO highly depends on the initial state, making it unstable. The ratios of all algorithms do not reach 1. The same problem has been reported in other works [5,27]. It is possibly caused by a large amount of redundancy in the neural network parameters [5].

6 Conclusion and Future Work

We have proposed PROS that applied opponent shaping [5] in cooperative-competitive mixed tasks with fully-independent settings. Although PROS does not require assumptions or extra information about opponents such as other opponent shaping algorithms, it cooperates better than baselines in mixed tasks like IPD. To implement opponent shaping in a fully-independent setting, we show that the opponent's dynamics can be learnt without information about the opponent's observations, actions, and rewards. With a contrastive learning module, PROS learns the representation of opponent's dynamics.

In the future, we could investigate the learning ability of PROS as decentralized agents in continuous decision tasks such as MPE [16], Multi-Agent MuJoCo [24] and Robotic Controlling [2], or use the representation to enhance the ability of agents trained in a CTDE paradigm.

Acknowledgements. This work was supported in part by the Strategic Priority Research Program of Chinese Academy of Sciences under Grant XDA27030400; in part by the National Natural Science Foundation of China under Grant 62293541; in part by the International Partnership Program of Chinese Academy of Sciences under Grant 104GJHZ2022013GC; and in part by the Youth Innovation Promotion Association CAS.

References

1. Axelrod, R., Hamilton, W.D.: The evolution of cooperation. Science **211**(4489), 1390–1396 (1981)
2. Chai, J., Chen, W., Zhu, Y., Yao, Z.X., Zhao, D.: A hierarchical deep reinforcement learning framework for 6-DOF UCAV air-to-air combat. IEEE Trans. Syst. Man Cybern. Syst. **53**, 5417–5429 (2023)
3. Chai, J., et al.: UNMAS: multiagent reinforcement learning for unshaped cooperative scenarios. IEEE Trans. Neural Netw. Learn. Syst. **34**, 2093–2104 (2021)
4. Chen, X., He, K.: Exploring simple siamese representation learning. In: Proceedings of the IEEE/CVF Conference on Computer Vision and Pattern Recognition, pp. 15750–15758 (2021)
5. Foerster, J., Chen, R.Y., Al-Shedivat, M., Whiteson, S., Abbeel, P., Mordatch, I.: Learning with opponent-learning awareness. In: Proceedings of the 17th International Conference on Autonomous Agents and MultiAgent Systems (2018)
6. Grover, A., Al-Shedivat, M., Gupta, J., Burda, Y., Edwards, H.: Learning policy representations in multiagent systems. In: International Conference on Machine Learning, pp. 1802–1811. PMLR (2018)
7. Guo, Z., et al.: Byol-explore: exploration by bootstrapped prediction. Adv. Neural. Inf. Process. Syst. **35**, 31855–31870 (2022)
8. Hjelm, R.D., et al.: Learning deep representations by mutual information estimation and maximization. In: International Conference on Learning Representations (2019)
9. Hu, G., Zhu, Y., Zhao, D., Zhao, M., Hao, J.: Event-triggered communication network with limited-bandwidth constraint for multi-agent reinforcement learning. IEEE Trans. Neural Netw. Learn. Syst. **34**(8), 3966–3978 (2023)
10. Kim, D.K., et al.: Influencing long-term behavior in multiagent reinforcement learning. Adv. Neural. Inf. Process. Syst. **35**, 18808–18821 (2022)
11. Laskin, M., Srinivas, A., Abbeel, P.: Curl: contrastive unsupervised representations for reinforcement learning. In: International Conference on Machine Learning, pp. 5639–5650. PMLR (2020)

12. Lerer, A., Peysakhovich, A.: Maintaining cooperation in complex social dilemmas using deep reinforcement learning. arXiv preprint arXiv:1707.01068 (2017)
13. Letcher, A., Foerster, J., Balduzzi, D., Rocktäschel, T., Whiteson, S.: Stable opponent shaping in differentiable games. In: International Conference on Learning Representations (2019)
14. Li, L., Yang, R., Luo, D.: FOCAL: efficient fully-offline meta-reinforcement learning via distance metric learning and behavior regularization. In: International Conference on Learning Representations (2021)
15. Liu, M., Li, L., Hao, S., Zhu, Y., Zhao, D.: Soft contrastive learning with Q-irrelevance abstraction for reinforcement learning. IEEE Trans. Cogn. Dev. Syst. **15**, 1463–1473 (2022)
16. Lowe, R., Wu, Y.I., Tamar, A., Harb, J., Pieter Abbeel, O., Mordatch, I.: Multi-agent actor-critic for mixed cooperative-competitive environments. Adv. Neural Inf. Process. Syst. **30** (2017)
17. Lu, C., Willi, T., De Witt, C.A.S., Foerster, J.: Model-free opponent shaping. In: International Conference on Machine Learning, pp. 14398–14411. PMLR (2022)
18. Mealing, R., Shapiro, J.L.: Opponent modeling by expectation-maximization and sequence prediction in simplified poker. IEEE Trans. Comput. Intell. AI Games **9**(1), 11–24 (2015)
19. Oord, A.V.D., Li, Y., Vinyals, O.: Representation learning with contrastive predictive coding. arXiv preprint arXiv:1807.03748 (2018)
20. Parekh, S., Losey, D.P.: Learning latent representations to co-adapt to humans. In: Autonomous Robots, pp. 1–26 (2023)
21. Richemond, P.H., et al.: Byol works even without batch statistics. arXiv preprint arXiv:2010.10241 (2020)
22. Schulman, J., Wolski, F., Dhariwal, P., Radford, A., Klimov, O.: Proximal policy optimization algorithms. arXiv preprint arXiv:1707.06347 (2017)
23. Willi, T., Letcher, A.H., Treutlein, J., Foerster, J.: Cola: consistent learning with opponent-learning awareness. In: International Conference on Machine Learning, pp. 23804–23831. PMLR (2022)
24. de Witt, C.S., Peng, B., Kamienny, P., Torr, P.H.S., Böhmer, W., Whiteson, S.: Deep multi-agent reinforcement learning for decentralized continuous cooperative control. CoRR (2020)
25. Xie, A., Losey, D., Tolsma, R., Finn, C., Sadigh, D.: Learning latent representations to influence multi-agent interaction. In: Conference on Robot Learning. PMLR (2021)
26. Yu, C., et al.: The surprising effectiveness of PPO in cooperative multi-agent games. Adv. Neural. Inf. Process. Syst. **35**, 24611–24624 (2022)
27. Zhao, S., Lu, C., Grosse, R.B., Foerster, J.: Proximal learning with opponent-learning awareness. Adv. Neural Inf. Process. Syst. **35**, 1–13 (2022)
28. Zhu, Y., Li, W., Zhao, M., Hao, J., Zhao, D.: Empirical policy optimization for n-player Markov games. IEEE Trans. Cybern. **53**, 6443–6455 (2022)
29. Zhu, Y., Zhao, D.: Online minimax Q network learning for two-player zero-sum Markov games. IEEE Trans. Neural Netw. Learn. Syst. **33**(3), 1228–1241 (2020)

FHSI-GNN: Fusion Hierarchical Structure Information Graph Neural Network for Extractive Long Documents Summarization

Zhen Zhang[1], Wenhao Yun[1], Xiyuan Jia[1(✉)], Qiyun Lv[1], Hao Ni[1], Xin Wang[1], and Guohua Wu[1,2]

[1] School of Cyberspace Security, Hangzhou Dianzi University, Hangzhou 310018, China
{zhangzhen,yunwenhao,jiaxiyuan,lvqiyun,haoni,wangxluo,wugh}@hdu.edu.cn
[2] Data Security Governance Zhejiang Engineering Research Center, Hangzhou Dianzi University, Hangzhou 310018, China

Abstract. Extractive text summarization aims to select salient sentences from documents. However, most existing extractive methods struggle to capture inter-sentence relations in long documents. In addition, the hierarchical structure information of the document is ignored. For example, some scientific documents have fixed chapters, and sentences in the same chapter have the same theme. To solve these problems, this paper proposes a Fusion Hierarchical Structure Information Graph Neural Network for Extractive Long Documents Summarization. The model constructs a section node containing sentence nodes and global information according to the document structure. It integrates the hierarchical structure information of the text and uses position information to identify sentences. The section node acts as an intermediary node for information interaction between sentences, which better enriches the relationships between sentences and has higher computational efficiency. Our model has achieved excellent results on two datasets, PubMed and arXiv. Further analysis shows that the hierarchical structure information of documents helps the model select salient content better.

Keywords: Hierarchy structure information · Extractive summarization · Graph neural network

1 Introduction

Text summarization is a hot research task in NLP that aims to extract important and concise information from lengthy text documents. Current text summarization methods can be classified into two categories: extractive and abstractive. Most abstractive summarization models generate a word-by-word summary based on understanding of the content in the original text. In contrast, extractive summarization methods focus on selecting important sentences. Abstractive

B. Luo et al. (Eds.): ICONIP 2023, CCIS 1963, pp. 136–149, 2024.
https://doi.org/10.1007/978-981-99-8138-0_12

models may produce summaries that are not fluent or grammatically accurate, while extractive summarization has a significant advantage in this regard. As a result, the extractive method still attracts much attention.

Most of the existing approaches have proved their effectiveness in shorter texts, but the desired performance is often not achieved when these methods are directly applied in long-form documents. Due to the limitation of input length of the text encoder, some longer text documents must be truncated or processed segment by segment, which leads to the loss of some salient information. And as the length of text increases, it has some unique structures. For example, scientific papers. The document is divided into different chapters, and the sentences in the same chapter have a closer relationship than the sentences in different chapters [7]. We may pay more attention to the parts titled "Methodology" and "Conclusion" rather than "Background". We can identify them using their position in the hierarchy. Sefid and Giles [20] incorporated section information into sentence vectors. Due to the limitations of the input length of the BERT model, long documents are processed segment-by-segment, so the model cannot capture the inter-sentence relationships very well.

In addition, modeling long-range of inter-sentence relationships is still a challenge [8]. In recent years, Graph Neural Networks (GNN) are commonly used to construct cross-sentence relationships in summarization tasks. Xu et al. [26] considered the discourse relationship and constructed the Rhetorical Structure Theory (RST) graph, but they relied on external tools to construct the graph structure, so the error propagation problem caused by it cannot be ignored. Doan et al. [6] constructed two types of graph structures of the documents, one is a HomoGNN with only sentence-level nodes, and the other is a HeterGNN with sentence-level and word-level nodes [23]. But they both overlook the importance of global information, and have low the calculation efficiency.

In view of the above shortcomings, we propose a **F**usion **H**ierarchical **S**tructure **I**nformation **G**raph **N**eural **N**etwork (FHSI-GNN) model for extractive summarization of long documents in this study. Firstly, we encoded the entire document with a pre-trained Longformer [1] model to obtain the representations of each sentence in the document. Longformer adopts the sparse attention mechanism, we select [CLS] token at the beginning of the sentence as the sentence representation, and put it as the input to the next layer. Moreover, some word-level tokens are selected randomly as global attention to better capture the long-range dependency of the document. Secondly, we constructed a document graph consisting of section nodes and sentence nodes according to the hierarchical structure of the document, such as the section nodes are generated by all the sentences in the section through average pooling. In addition, we incorporated the hierarchical structure information of documents into our model and updated node representation using GNN. Finally, the section and sentence nodes are fused and passed as inputs to a Feed-Forward Neural Network (FFNN) to obtain the sentence label, which is used to determine whether it is a summary sentence.

To summarize, our contributions are there-folds: (1) We propose a novel extractive summarization model for long scientific documents based on GNN, which incorporates hierarchical structure information into the process of graph propagation, and the section nodes with global information are used as an intermediary to interact with the sentence nodes to model the inter-sentence relationships. (2) We use Longformer as our document encoder and provide a novel global attention selection method as the sparse attention mechanism, which is able to generate effective sentence representations for longer documents. (3) Experimental results show that our method achieves strong competitive performance compared to the baseline models, which demonstrates the effectiveness of our method. Furthermore, we conduct ablation studies and result analysis to understand how to work of our model and where the performance gain comes from.

2 Related Work

Neural Extractive Summarization. Neural networks have achieved remarkable results in extractive summarization [27]. In recent years, pre-trained language models have provided significant gains in the performance of extractive text summarization. For example, Liu et al. [14] and Iandola et al. [12] use pre-trained models to summarize short news texts. Grail et al. [10] conducted suammarization of long documents. In this paper, we regard the summarization task as a binary-labeling problem of sentences, and encode the text based on Longformer model, which we believe has better semantic capture ability.

Graph-Based Summarization. Most of the early works on graph-based extractive text summarization were unsupervised methods, which constructed document graphs by calculating similarity scores between sentences, such as LexRank [9]. It is a current trend to construct different types of semantic units as graph nodes in the graph structure [11,13,23], such as words, sentences, entities, topics, etc. Phan et al. [16] proposed a new heterogeneous graph neural network that includes word nodes, sentence nodes, and passage nodes. Cui et al. [5] proposed a method that takes the latent topic of a document as an extra node. In this study, by aggregating sentence nodes, we build a section node based on the document hierarchical structure, which is able to better learn inter-sentence relations.

Long Document Summarization. In recent years, there has been increasing attention to summarization of long documents, especially for scientific publications. Rohde et al. [17] proposed a Transformer-based hierarchical attention model. Cui and Hu [4] adopted a sliding selector network with dynamic memory to extract summary sentences segment by segment through a sliding window. Miculicich and Han [15], Cho et al. [2] indicate the significance of paragraph structure information in the task of summarization. In contrast, our model uses the Longformer model as a text encoder with a novel sparse attention mechanism to effectively address the long-range dependency issue in documents [8]. Additionally, we incorporate the hierarchical structure information of long texts into sentence representations, such as the positional information of sentences within the hierarchical structure.

3 Model

Figure 1 shows the overall architecture of the model. Given an arbitrary document consisting of N sections $Doc = \{Sec_1, Sec_2, ..., Sec_N\}$, each section is composed of several sentences. For example, $Sec_1 = \{S_1, S_2, ..., S_i\}, Sec_N = \{S_j, ..., S_n\}$, where n is the number of all sentences in the document, and the subscript S indicates the position of the sentence in the document. The aim is to predict a sequence of binary labels $\{y_1, y_2, ..., y_n\}$, in which $y_i \epsilon \{0, 1\}$ represents whether the i-th sentence should be included in the summary sentences.

Our model mainly includes four parts, 1) Document Encoder: This part is responsible for generating sentence representations. 2) Section Encoder: It generates section representations based on the hierarchical structure of the document. 3) Graph Attention Layer: It updates node representations using Graph Attention Network. 4) Prediction Layer: It outputs labels for the sentences. Each part is described in the following subsections.

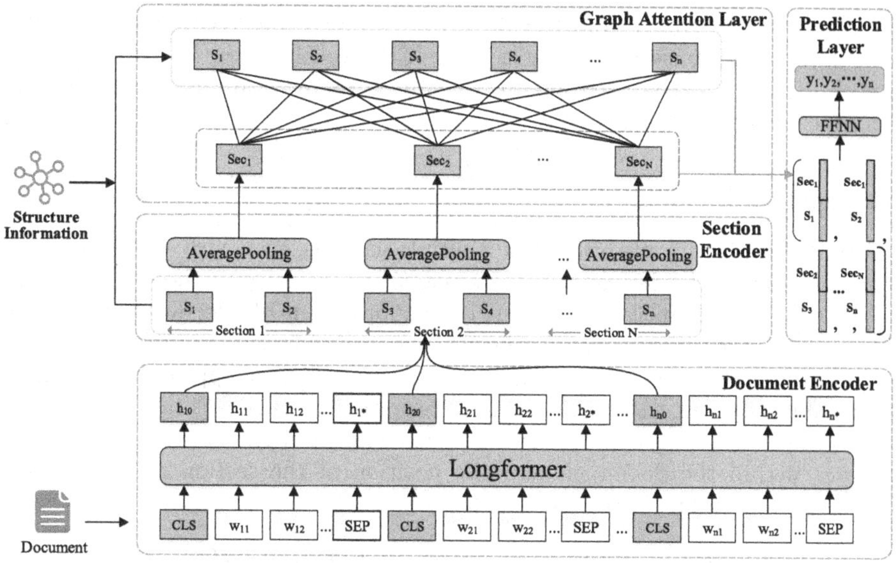

Fig. 1. Architecture of our model(FHSI-GNN). In the Longformer model, in addition to using [CLS] as global attention, we also randomly select some common word tokens as global attention. The blue square in the figure represents the sentence representations output by the document encoder. The orange square represents the section representations of the document. The green square represents the predicted sentence labels. The brackets on the far right represent the splicing of sentences and sections for dimension expansion. (Color figure online)

3.1 Document Encoder

Some Transformer-based models can only handle shorter documents because the complexity of the self-attention mechanism is proportional to the length of the input sequence. For example, in BERT [14], the maximum input sequence length is 512 tokens. We use Longformer [1] as the document encoder, which uses the sparse attention mechanism combining global attention and local attention, as shown in Fig. 2a. This sparse attention mechanism greatly reduces computational complexity and is able to handle longer documents.

In order to obtain the vector representations of the sentence, a [CLS] and [SEP] tokens are inserted at the beginning and the end of each sentence respectively. It been proven effective in previous studies. Therefore, we use [CLS] token as global attention to capture more semantic information and generate better sentence representations. Besides, we randomly select some tokens as global attention, as shown in Fig. 2b. Then, we put all the tokens into the Longformer Layer to learn the hidden states of the sentences.

$$\{h_{10}, h_{11}, ..., h_{n0}, ..., h_{n*}\} = Longformer\left(w_{10}, w_{11}, ..., w_{n0}, ..., w_{n*}\right) \quad (1)$$

where w_{ij} represents the $j\text{-}th$ word of the $i\text{-}th$ sentence. w_{i0} represents the [CLS] token at the beginning of each sentence, and w_{i*} represents the [SEP] token at the end of each sentence. h_{ij} indicates the hidden states of the corresponding token. After Longformer encoding, we select the [CLS] token $\hat{H}_S = \{h_{10}, h_{20}, ..., h_{n0}\}$ of each sentence as the corresponding sentence representations. $\hat{H}_S$ is used as the input of the next layer.

3.2 Section Encoder

Long scientific documents often have multiple sections, each with a different level of importance. To better capture the semantic information of these documents, we incorporate the hierarchical structure information of the documents after obtaining the representations $\hat{H}_S$ of each sentence, such as the position information of sentences within the hierarchical structure. In our approach, we utilize a learnable position encoding technique inspired by BERT [14]. Our method involves incorporating two types of position information: the position of each sentence within the document and the position of the section to which each sentence belongs. This is achieved as follows:

$$H_S = \{S_0, S_1, ..., S_n\} = W_1 e_S + W_2 e_{Sec} + \hat{H}_S \quad (2)$$

where e_S represents the position embedding of the sentence in the document. e_{Sec} represents the position embedding of the section where the sentence is located. S_i is the $i\text{-}th$ sentence representation, n is the number of sentences in the document. W_* is a learnable parameter.

After getting the representation of each sentence, we calculate the embedded representation of each section through AveragePooling, as follows:

$$Sec_j = AveragePooling\left(S_i, S_{i+1}, ..., S_*\right) \quad (3)$$

where S_i represents the first sentence of the j-th section. S_* represents the last sentence of the j-th section. Sec_j represents the *j-th* section representation. The set of section representation of the document is $H_{Sec} = \{Sec_1, Sec_2, ..., Sec_N\}$.

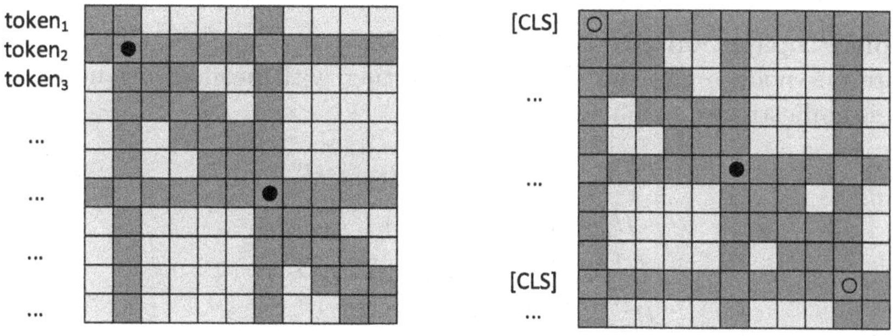

(a) (b)

Fig. 2. Sparse attention mechanism. A solid circle represents randomly selecting several tokens as global attention, and a hollow circle represents selecting the [CLS] token at the beginning of a sentence as global attention.

3.3 Graph Attention Layer

Graph Building. Let $G = \{V, E\}$ denote an arbitrary graph, where V and E represents the set of node and edge, respectively. We have defined an undirected graph consisting of two types of nodes: sentence nodes V_S and section nodes V_{Sec}. Each section node is connected to every sentence node through an edge. $V_S = \{S_1, S_2, ..., S_n\}$ represents a set of n sentence nodes, $V_{Sec} = \{Sec_1, Sec_2, ..., Sec_N\}$ represents a set of N section nodes. In comparison to the approach of connecting all pairs of sentences, our method shows higher computational efficiency, avoids unnecessary connections between unrelated nodes, and better captures semantic information.

Graph Attention Network. We encode our document graph with graph attention network(GAT). GAT updates node representations by aggregating information from their neighboring nodes, which can be denoted as:

$$z_{ij} = Leakly\mathrm{ReLu}\left(W_a \left[W_q h_i; W_k h_j\right]\right) \tag{4}$$

$$\alpha_{ij} = soft\max\left(z_{ij}\right) = \frac{\exp\left(z_{ij}\right)}{\sum\limits_{l \epsilon N_i} \exp\left(z_{il}\right)} \tag{5}$$

$$\mu_i = \sigma\left(\sum_{j \epsilon N_i} \alpha_{ij} W_v h_j\right) \tag{6}$$

$$h_i^{'} = \|_{m=q}^{M} \sigma\left(\sum_{j \epsilon N_i} \alpha_{ij}^m W_v^m h_j^m\right) \tag{7}$$

$$h_i^{''} = h_i + h_i^{'} \tag{8}$$

Where h_i denotes the representation of the i-th node, and N_i stand for its neighbor nodes. W_a, W_q, W_k and W_v are model trainable parameters. σ represents an activation function. $\|*$ represents multi-heads concatenation. $h_i^{'}$ represents the updated node representation. $h_i^{''}$ is a residual connection.

Graph Propagation. We employ GAT (Eqs. 4–8) and FFNN to iteratively update our node representations by interacting with neighboring nodes. The process is illustrated as follows:

$$U_{Sec2S}^{1} = GAT\left(H_S^0, H_{Sec}^0, H_{Sec}^0\right) \tag{9}$$

$$H_S^1 = FFNN\left(H_S^0 + U_{Sec2S}^1\right) \tag{10}$$

$$U_{S2Sec}^{1} = GAT\left(H_{Sec}^0, H_S^1, H_S^1\right) \tag{11}$$

$$H_{Sec}^1 = FFNN\left(H_{Sec}^0 + U_{S2Sec}^1\right) \tag{12}$$

where $H_S^0 = H_S$ and $H_{Sec}^0 = H_{Sec}$ represents the initialization of node vectors. We first update the representations of the sentence nodes, followed by using the updated sentence nodes to update the section nodes, completing one iteration. U_{Sec2S}^1 and U_{S2Sec}^1 respectively represent the first iteration update processes from section to sentence and from sentence to section. H_S^1 and H_{Sec}^1 represent the updated sets of sentence and section node representations, respectively.

3.4 Prediction Layer

After t rounds of iteration, we obtain the new sentence representation H_S^t and section representation H_{Sec}^t. For the i-th sentence belonging to the j-th section, we input S_i^t and Sec_j^t into an MLP classifier to calculate the summary label for the sentence. which can be denoted as:

$$\tilde{y}_i = f\left(LN\left(S_i^t \| Sec_j^t\right)\right) \tag{13}$$

where $\tilde{y}_i$ represents the predicted probability for the i-th sentence. f is a FFNN with three hidden layers and $\|$ denotes the concatenation operation. $LN(\ )$ represents Layer Normalization. We perform concatenation of the sentence representations and section representations, exploring multiple modes, and provide a detailed analysis in the experimental section.

The training objective of the model is to minimize the binary cross-entropy loss between the predicted sentence labels and the ground true labels. The calculation formula for the loss is as follows:

$$Loss = -\sum y_i \log\left(\tilde{y}_i\right) + \left(1 - y_i\right)\log\left(1 - \tilde{y}_i\right) \tag{14}$$

During the inference process, after obtaining the predicted probabilities for all the sentences, we rank the sentences and select the top-k sentences as the final summary. The value of k is a hyperparameter that is set based on the average length of the reference summaries.

4 Experiment

4.1 Datasets

Our model is primarily designed for hierarchical long documents, and thus we did not conduct experiments on widely studied short documents. Instead, we focused on evaluating our model on large-scale scientific paper datasets from arXiv and PubMed [3]. Table 1 provides data statistics for these relevant datasets.

Table 1. The statistics of experiential datasets.

Datasets	#Doc			#Avg Tokens	
	Train	Val	Test	Doc.	Sum
arXiv	203,037	6,436	6,440	4,938	220
PubMed	119,224	6,633	6,658	3,016	203

4.2 Models for Comparison

We evaluate our approach using recent benchmark models in this research area.

Traditional Methods. LexRank [9] is a graph-based method for extractive text summarization, It determines the importance of sentences mainly by calculating their similarity and extracts the most salient ones. In contrast, LSA [21] identifies important sentences by analyzing semantic similarity and matrix factorization. Another approach to extractive summarization is SumBasic [22], which is statistics-based and calculates the probability and importance of each word in the text to generate effective summaries.

Abstractive Summarization Models. Pointer Generator Network (PGN) [19] is a seq2seq framework that employs attention, coverage and copy mechanisms. On the other hand, Discourse-Aware [3] framework utilizes a hierarchical encoder and a decoder to capture the discourse structure of long documents. This results in improved performance, particularly for long documents with complex discourse structure.

Extractive Summarization Models. ExtSum-LG [25] is a transformer-based model that addresses the problem of redundancy in summaries. Another model, Match-Sum [28] is BERT-based and selects appropriate sentences for summaries through text matching. HiStruct [18] encodes section titles and sentence positions, and then integrates the obtained embeddings into the sentences. SSN-DM [4] and Topic-GraphSum [5] are implemented based on graphs.

4.3 Experiment Setup

The document encoder uses the "longformer-base" version. We randomly select 0.2 times the number of document sentences as random global attention tokens, and the sliding window size is 1024. The section and sentence representation dimension size is set to 768. The number of iterations of GAT is 2, and the number of attention heads is set to 8.

Our model is trained for 100,000 steps on two NVIDIA GPUs (RTX 3090) with 24GB of memory. The optimizer is NOAM. The batch size changes according to the length of the document. Due to limited resources, Longformer was not fine-tuned during training. We select seven sentences as the final summary.

4.4 Main Results

Table 2 shows the results of different models on two datasets. We use ROUGE F1 scores as the evaluation metric, where unigram overlap (R-1) and bigram overlap (R-2) are used to evaluate informativeness, and longest common subsequence (R-L) is used to evaluate fluency. We divide the results into four parts for presentation. The first part mainly consists of some traditional extractive summarization methods and Oracle. The second part shows the results of the abstractive summarization model. The third part shows the extractive summarization model. The first three models in this part do not use pre-trained models, the middle two are the results of using pre-trained models, and the last two are the results of using graph-based models. The last part shows our model.

Table 2. ROUGE F1 results on two datasets. We collected results from previous studies, where * denotes results from Cohan et al. [3], and + denotes results from Xiao and Carenini [24]. The remaining results are from their respective papers.

Model	PubMed			arXiv		
	R-1	R-2	R-L	R-1	R-2	R-L
SumBasic*	37.15	11.36	33.43	29.47	6.95	26.30
LexRank*	39.19	13.89	34.59	33.85	10.73	28.99
LSA*	33.89	9.93	29.70	29.91	7.42	25.67
Oracle+	55.05	27.48	38.66	53.88	23.05	34.90
Seq2Seq-Attention*	31.55	8.52	27.38	29.30	6.00	25.56
PGN*	35.86	10.22	29.69	32.06	9.04	25.16
Discourse-Aware*	38.93	15.37	35.21	35.80	11.05	31.80
SummaRuNNer+	43.89	18.78	30.36	42.91	16.65	28.53
Xiao annd Carenini+	44.85	19.70	31.43	43.62	17.36	29.14
ExtSum-LG [25]	45.39	20.37	40.99	44.01	17.79	39.09
Match-Sum [28]	41.21	14.91	36.75	40.59	12.98	32.64
Histruct [18]	45.76	19.64	41.34	45.22	17.67	40.16
Topic-GraphSum [5]	45.95	20.81	33.97	44.03	18.52	32.41
SSN-DM [4]	46.73	21.00	34.10	45.03	**19.03**	32.58
FHSI-GNN	**46.93**	**21.11**	**42.46**	**45.55**	18.53	**40.66**

From the evaluation results of the model in Table 2, it can be seen that our model shows strong competitiveness in this research field compared with some state-of-the-art models. For the arXiv dataset, our model's R-2 results are only lower than SSN-DM, but the R-L results are much better than SSN-DM. The phenomenon is believed to be caused by the sliding window mechanism employed by SSN-DM, which increases the accuracy of the summary but compromises its fluency. Compared with HiStruct [18], which also considers the hierarchical structure information of long scientific documents, our model also has an overall improvement.

This shows the effectiveness of our method. According to the hierarchical structure of the document, an intermediary node with global information can be established, which can better establish the relationship between sentences with lower computational complexity and solve the long-range dependency problem well. It is worth noting that we do not need to use some external semantic information or resources to improve the performance of the model, such as the latent topic used in Topic-GraphSum and the memory slots used in SSN-DM. Doing so will lead to a more complex model structure, higher computational cost, and more importantly, when wrong semantic information is incorporated, it will jeopardize the performance of the summarization model.

4.5 Ablation Study

We compared the complete model with three ablated variants. **1) w/o Random Global Attention** refers to the document encoder where only the [CLS] token is used as global attention **2) w/o Structure Information** means that the proposed model does not consider the hierarchical structure information of the document. **3)w/o Section Node** means to remove the section encoder and not use the section node in the graph attention layer. Additionally, we removed all methods as Baseline.

Table 3. Ablation Study on two datasets.

Model	PubMed			arXiv		
	R-1	R-2	R-L	R-1	R-2	R-L
Baseline	41.36	14.87	36.81	37.96	10.77	33.41
w/o Random Global Attention	46.78	20.95	42.33	45.42	18.31	40.43
w/o Structure Information	46.17	20.21	41.75	44.82	17.34	39.78
w/o Section Node	46.05	19.86	41.59	44.87	17.29	39 82
FHSI-GNN	**46.93**	**21.11**	**42.46**	**45.55**	**18.53**	**40.66**

Table 3 shows the results of different variants on two datasets. From the table, we can observe that our full model outperforms all variants, which shows that the proposed method is effective. When the Section Node is removed, the model's performance drops more seriously, which shows that our method can better establish the relationship between sentences.

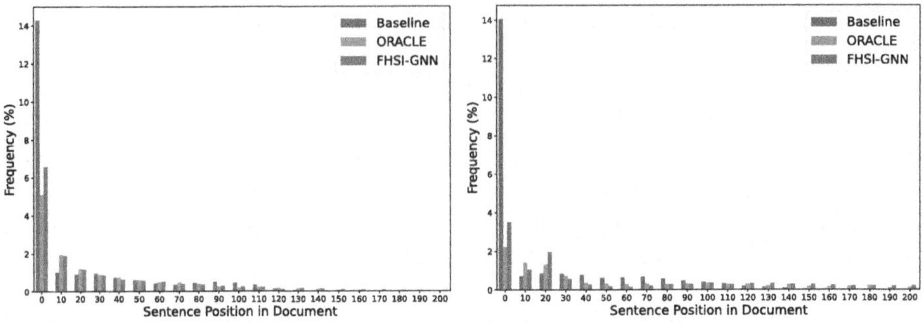

(a) PubMed (b) arXiv

Fig. 3. Proportions of the extracted sentences at each linear position. Due to space constraints, the sentences are displayed at intervals of 10.

To further analyze the model's output, we show the sentence position distribution diagram of the extracted summary in Fig. 3. The figure shows the distribution of summary sentences extracted by Baseline, ORACLE, and our model. It can be seen intuitively that our model's output is close to the ORACLE summary and that salient sentences are located at the end.

In the Prediction Layer, we discuss the impact of the four splicing methods of the updated sentence representations and section representations on our results. As shown in the left figure of Fig. 4, from left to right, Mode A represents only sentence representation. Mode B represents adding sentence representation and section representation. Mode C splicing sentence representation and section representation to expand the dimension. Mode D splicing the first three modes.

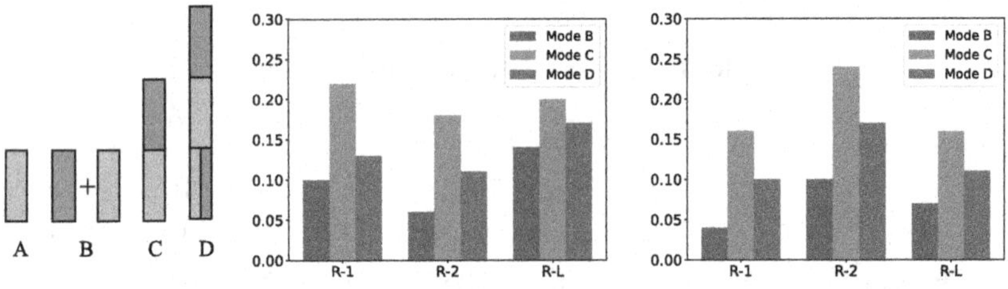

PubMed arXiv

Fig. 4. Comparison results of different splicing methods on two datasets. The blue squares in the figure on the left represent sentence representations, and the orange squares represent section representations. The results of the two graphs on the right refer to the degree of improvement in Mode A.

The ROUGE results of Mode A on PubMed and arXiv are 46.71/20.93/42.26 and 45.39/18.29/40.50. We take the results of this mode as the baseline of this part of the study and verify the improvement effect of other modes by comparing with it. Figure 4 show that different splicing methods also produce different effects. We believe that one of the main reasons is that the updated section nodes have richer semantic information, which helps the classifier to better extract summary sentences. However, when too much information is spliced, it will lead to a decrease in the result, which may be due to the addition of some interference information that affects the model's judgment. We ended up taking the third one Mode C as the splicing method in our model.

5 Conclusions

In this paper, we explore the impact of document hierarchy on the extractive summarization task. We propose an extractive long documents summarization model that is fused with hierarchy structure information graph neural network. The model builds section nodes according to document hierarchy structure and uses position information to identify sentences. It enriches the relationships between sentences through graph neural network. The results on two datasets show that our model has strong competitive results. We do not rely on any extra external semantic information, which makes it more robust to long documents. In the future, we intend to shift the focus of our work to abstractive text summarization and explore the impact of our hierarchical structure on abstractive long text summarization.

Acknowledgements. This research was supported by "Pioneer" and "Leading Goose" R&D Program of Zhejiang (Grant No. 2023C03203, 2023C03180, 2022C03174).

References

1. Beltagy, I., Peters, M.E., Cohan, A.: Longformer: the long-document transformer. arXiv preprint arXiv:2004.05150 (2020)
2. Cho, S., Song, K., Wang, X., Liu, F., Yu, D.: Toward unifying text segmentation and long document summarization. In: Proceedings of the 2022 Conference on Empirical Methods in Natural Language Processing, pp. 106–118 (Dec 2022)
3. Cohan, A., Dernoncourt, F., Kim, D.S., Bui, T., Kim, S., Chang, W., Goharian, N.: A discourse-aware attention model for abstractive summarization of long documents. arXiv preprint arXiv:1804.05685 (2018)
4. Cui, P., Hu, L.: Sliding selector network with dynamic memory for extractive summarization of long documents. In: Proceedings of the 2021 Conference of the North American Chapter of the Association for Computational Linguistics: Human Language Technologies, pp. 5881–5891 (2021)
5. Cui, P., Hu, L., Liu, Y.: Enhancing extractive text summarization with topic-aware graph neural networks. arXiv preprint arXiv:2010.06253 (2020)
6. Doan, X.D., Le Nguyen, M., Bui, K.H.N.: Multi graph neural network for extractive long document summarization. In: Proceedings of the 29th International Conference on Computational Linguistics, pp. 5870–5875 (2022)
7. Dong, Y., Mircea, A., Cheung, J.C.: Discourse-aware unsupervised summarization of long scientific documents. arXiv preprint arXiv:2005.00513 (2020)

8. Dong, Z., Tang, T., Li, L., Zhao, W.X.: A survey on long text modeling with transformers. arXiv preprint arXiv:2302.14502 (2023)
9. Erkan, G., Radev, D.R.: Lexrank: graph-based lexical centrality as salience in text summarization. J. Artif. Intell. Res. **22**, 457–479 (2004)
10. Grail, Q., Perez, J., Gaussier, E.: Globalizing bert-based transformer architectures for long document summarization. In: Proceedings of the 16th Conference of the European Chapter of the Association for Computational Linguistics: Main Volume, pp. 1792–1810 (2021)
11. Huang, Y.J., Kurohashi, S.: Extractive summarization considering discourse and coreference relations based on heterogeneous graph. In: Proceedings of the 16th Conference of the European Chapter of the Association for Computational Linguistics: Main Volume, pp. 3046–3052 (2021)
12. Iandola, F.N., Shaw, A.E., Krishna, R., Keutzer, K.W.: Squeezebert: what can computer vision teach NLP about efficient neural networks? arXiv preprint arXiv:2006.11316 (2020)
13. Jing, B., You, Z., Yang, T., Fan, W., Tong, H.: Multiplex graph neural network for extractive text summarization. arXiv preprint arXiv:2108.12870 (2021)
14. Liu, Y., Lapata, M.: Text summarization with pretrained encoders. arXiv preprint arXiv:1908.08345 (2019)
15. Miculicich, L., Han, B.: Document summarization with text segmentation. arXiv preprint arXiv:2301.08817 (2023)
16. Phan, T.A., Nguyen, N.D.N., Bui, K.H.N.: Hetergraphlongsum: heterogeneous graph neural network with passage aggregation for extractive long document summarization. In: Proceedings of the 29th International Conference on Computational Linguistics, pp. 6248–6258 (2022)
17. Rohde, T., Wu, X., Liu, Y.: Hierarchical learning for generation with long source sequences. arXiv preprint arXiv:2104.07545 (2021)
18. Ruan, Q., Ostendorff, M., Rehm, G.: HiStruct+: improving extractive text summarization with hierarchical structure information. In: Findings of the Association for Computational Linguistics: ACL 2022, pp. 1292–1308 (May 2022)
19. See, A., Liu, P.J., Manning, C.D.: Get to the point: summarization with pointer-generator networks. arXiv preprint arXiv:1704.04368 (2017)
20. Sefid, A., Giles, C.L.: Scibertsum: extractive summarization for scientific documents. In: Document Analysis Systems: 15th IAPR International Workshop, DAS 2022, La Rochelle, 22–25 May 2022, Proceedings, pp. 688–701 (2022)
21. Steinberger, J., et al.: Using latent semantic analysis in text summarization and summary evaluation. Proc. ISIM **4**(93–100), 8 (2004)
22. Vanderwende, L., Suzuki, H., Brockett, C., Nenkova, A.: Beyond sumbasic: task-focused summarization with sentence simplification and lexical expansion. Inf. Process. Manag. **43**(6), 1606–1618 (2007)
23. Wang, D., Liu, P., Zheng, Y., Qiu, X., Huang, X.: Heterogeneous graph neural networks for extractive document summarization. In: Proceedings of the 58th Annual Meeting of the Association for Computational Linguistics, pp. 6209–6219 (2020)
24. Xiao, W., Carenini, G.: Extractive summarization of long documents by combining global and local context. arXiv preprint arXiv:1909.08089 (2019)
25. Xiao, W., Carenini, G.: Systematically exploring redundancy reduction in summarizing long documents. arXiv preprint arXiv:2012.00052 (2020)
26. Xu, J., Gan, Z., Cheng, Y., Liu, J.: Discourse-aware neural extractive text summarization. arXiv preprint arXiv:1910.14142 (2019)

27. Yadav, A.K., Singh, A., Dhiman, M., Kaundal, R., Verma, A., Yadav, D.: Extractive text summarization using deep learning approach. Int. J. Inf. Technol. **14**(5), 2407–2415 (2022)
28. Zhong, M., Liu, P., Chen, Y., Wang, D., Qiu, X., Huang, X.: Extractive summarization as text matching. In: Proceedings of the 58th Annual Meeting of the Association for Computational Linguistics, pp. 6197–6208 (2020)

How to Support Sport Management with Decision Systems? Swimming Athletes Assessment Study Sase

Jakub Więckowski(✉) and Wojciech Sałabun

West Pomeranian University of Technology in Szczecin, ul. Żołnierska 49, 71-210 Szczecin, Poland
{jakub-wieckowski,wojeciech.salabun}@zut.edu.pl

Abstract. Information systems in sports play an increasingly important role due to the opportunities and benefits they present to sports clubs. The purpose of these systems is to assist in decision-making processes concerning marketing, the selection of training parameters, recovery methods, or the selection of team members, among others. It increases interest in decision support systems in this area, allowing clubs to gain an advantage over their competitors. In this paper, we propose a research approach for creating models to evaluate the performance of swimming athletes based on their physical parameters and the sport level they represent. Due to the uncertainties involved, data presented in the form of Triangular Fuzzy Numbers (TFN) were used in the Fuzzy Technique for Order Preference by Similarity to an Ideal Solution (TOPSIS) method to obtain a ranking of the athletes. A sensitivity analysis for the exclusion of subsequent criteria was also carried out. The results obtained were compared regarding selecting the different significance of the criteria weights. An additional six Fuzzy Multi-Criteria Decision Analysis (MCDA) methods were used for a comprehensive analysis, and the results showed that the proposed averaged ranking is a reasonable solution. The proposed approach can be used to evaluate players from different sports so that sports clubs can recruit athletes with high-performance potential.

Keywords: Sport management · Decision-making · Swimming · Support systems · Fuzzy MCDA · weighting methods · Athletes performance

1 Introduction

Sports management is an important aspect of leading sports clubs, where many factors influence the club's overall performance [8]. The selection of training parameters, recovery methods, marketing activities, the formation of a coaching team, or the selection of athletes for a team is among the many elements that comprise the entirety of team management [27]. By making sensible and rational choices, it is possible to increase the potential achieved by the team [5]. However,

B. Luo et al. (Eds.): ICONIP 2023, CCIS 1963, pp. 150–161, 2024.
https://doi.org/10.1007/978-981-99-8138-0_13

the number of elements that make up the bottom line is considerable. It makes using dedicated information systems to support team decision-making increasingly common [11]. It is crucial, as it allows the use of the latest technology in the sport and can significantly improve teams' performance using such solutions. Using these systems, exploring potential opportunities, making different marketing or team formation choices, studying the impact of different parameters in the training process, and much more is possible [13]. All these factors make it worthwhile to use modern solutions in the sports management process.

Multi-Criteria Decision Analysis (MCDA) solutions are popular in problems involving multiple criteria influencing the evaluation [9]. They allow evaluating the decision variants under consideration based on an identified set of criteria to identify the most preferred solutions. In turn, it helps minimize unsuitable choices and maximize profits against losses [23]. Due to the differences arising from how the chosen multi-criteria methods work, it is important to remember that the proposed solutions may vary considerably [19]. Therefore, examining different techniques within a given decision-making problem is worthwhile to identify proposed outcomes based on different research approaches. As research from the literature has shown, MCDA methods are readily used in sports [7], sport management [2], sports events [12], player evaluation [16], or training factors [4]. Additionally, information systems are used in sports areas such as football [6], e-sports [22], swimming [25], or basketball [1]. It demonstrates that the combination of sports and dedicated decision support systems is a popular topic and is worth developing as it can significantly contribute to the quality of sports clubs. Moreover, since uncertainty often occurs in decision problems, fuzzy extensions are applied to model the data [10,14].

In this paper, we propose an assessment model to examine swimmers' performance regarding their attractiveness to sports clubs. The evaluation is performed based on the Fuzzy TOPSIS method with equal criteria weights. The sensitivity analysis with excluding subsequent criteria, is applied to indicate if the results differ. Moreover, three objective weighting techniques, namely Shannon entropy, standard deviation, and variance methods, are used to determine if varied criteria importance would influence the determined rankings. To compare the obtained results, six selected Fuzzy MCDA methods are applied, namely Additive Ratio Assessment (ARAS), Complex Proportional Assessment (COPRAS), Evaluation based on Distance from Average Solution (EDAS), Multi-Attributive Border Approximation area Comparison (MABAC), Multi-Objective Optimization Method by Ratio Analysis (MOORA), and Operational Competitiveness Ratings (OCRA). The purpose of the study is to indicate if different criteria importance could significantly impact the proposed solutions. In practical terms, different criteria weights reflect different preferences of sports clubs, making it important to determine the approach that can be used when searching for athletes with great potential and adjusting the search process with team requirements. The main contributions of the study are:

- comparison of different criteria weighting methods in practical problem of swimmers performance assessment

- verification of sensitivity analysis approach of removing subsequent criteria in the problem to examine the robustness of the results in the fuzzy decision problem

The rest of the paper is organized as follows. Section 2 presents the preliminaries regarding the Triangular Fuzzy Numbers and Fuzzy TOPSIS method. Section 3 describes the study case with criteria identified to assess swimmers' performance and decision matrix. In Sect. 4, the results obtained from the performed experiments are presented and described. Finally, Sect. 5 includes the conclusions drawn from the research with further directions of work.

2 Preliminaries

2.1 Triangular Fuzzy Numbers

To represent uncertainty in decision problems, Fuzzy Set Theory and its extensions can be used. It is an important element of modeling uncertain data, where measures are inaccurate and data are not exactly known or varied. Lofti Zadeh introduced the main assumptions of the Fuzzy Set Theory [28]. Some of the core elements are described below:

The Fuzzy Set and the Membership Function - the characteristic function μ_A of a crisp set $A \subseteq X$ assigns a value of either 0 or 1 to each member of X, as well as the crisp sets only allow a full membership ($\mu_A(x) = 1$) or no membership at all ($\mu_A(x) = 0$). This function can be generalized to a function $\mu_{\tilde{A}}$ so that the value assigned to the element of the universal set X falls within a specified range, i.e. $\mu_{\tilde{A}} : X \rightarrow [0, 1]$. The assigned value indicates the degree of membership of the element in the set A. The function $\mu_{\tilde{A}}$ is called a membership function and the set $\tilde{A} = (x, \mu_{\tilde{A}}(x))$, where $x \in X$, defined by $\mu_{\tilde{A}}(x)$ for each $x \in X$ is called a fuzzy set.

The Triangular Fuzzy Number (TFN) - a fuzzy set $\tilde{A}$, defined on the universal set of real numbers $\Re$, is told to be a triangular fuzzy number $\tilde{A}(a, m, b)$ if its membership function has the following form (1):

$$\mu_{\tilde{A}}(x, a, m, b) = \begin{cases} 0 & x \leq a \\ \frac{x-a}{m-a} & a \leq x \leq m \\ 1 & x = m \\ \frac{b-x}{b-m} & m \leq x \leq b \\ 0 & x \geq b \end{cases} \tag{1}$$

and the following characteristics (2), (3):

$$x_1, x_2 \in [a, b] \wedge x_2 > x_1 \Rightarrow \mu_{\tilde{A}}(x_2) > \mu_{\tilde{A}}(x_1) \tag{2}$$

$$x_1, x_2 \in [b, c] \wedge x_2 > x_1 \Rightarrow \mu_{\tilde{A}}(x_2) > \mu_{\tilde{A}}(x_1) \tag{3}$$

The Support of a TFN - the support of a TFN $\tilde{A}$ is defined as a crisp subset of the $\tilde{A}$ set in which all elements have a non-zero membership value in the $\tilde{A}$ set (4):

$$S(\tilde{A}) = x : \mu_{\tilde{A}}(x) > 0 = [a, b] \tag{4}$$

The Core of a TFN - the core of a TFN $\tilde{A}$ is a singleton (one-element fuzzy set) with the membership value equal to 1 (5):

$$C(\tilde{A}) = x : \mu_{\tilde{A}}(x) = 1 = m \tag{5}$$

The Fuzzy Rule - the single fuzzy rule can be based on the Modus Ponens tautology. The reasoning process uses the $IF - THEN, OR$ and AND logical connectives.

2.2 Fuzzy Technique for Order Preference by Similarity to an Ideal Solution Method

The Technique for Order Preference by Similarity to an Ideal Solution (TOPSIS) is one of the most popular MCDA methods and can be used to assess considered decision variants [3]. It is based on the distances to an ideal solution: Positive Ideal Solution (PIS) and Negative Ideal Solution (NIS). Extending the standard TOPSIS method with fuzzy logic allows for greater flexibility in this technique and applying it to problems with uncertain data [17]. One of the proposed solutions is based on Triangular Fuzzy Numbers (TFN). To introduce the main steps of the Fuzzy TOPSIS method, subsequent steps of operation are described as follows:

Step 1. Determination of the triangular fuzzy decision matrix, which contains m alternatives and n criteria ($i = 1, 2, \ldots, m$ and $j = 1, 2, \ldots, n$), where x_{ij} is represented as Triangular Fuzzy Number (x^L, x^M, x^U).

Step 2. Calculation of the normalized fuzzy decision matrix $\tilde{X}$, where for profit criteria the values are calculated as (6):

$$\tilde{x}_{ij} = \left(\frac{x_{ij}^L}{x_j^*}, \frac{x_{ij}^M}{x_j^*}, \frac{x_{ij}^U}{x_j^*} \right); \quad x_j^* = \max_i \left\{ x_{ij}^U \right\} \tag{6}$$

and for cost criteria, the values are with the formula (7):

$$\tilde{x}_{ij} = \left(\frac{x_j^-}{x_{ij}^U}, \frac{x_j^-}{x_{ij}^M}, \frac{x_j^-}{x_{ij}^L} \right); \quad x_j^- = \min_i \left\{ x_{ij}^L \right\} \tag{7}$$

Step 3. Calculation of the weighted normalized fuzzy decision matrix $\widetilde{\tilde{X}}$, where $\widetilde{\tilde{x}}_{ij} = \tilde{x}_{ij} \times \widetilde{w}_j, \quad i = 0, 1, \ldots, m, j = 1, 2, \ldots, n.$

Step 4. Determination of the Fuzzy Positive Ideal Solution (FPIS) (8) and Fuzzy Negative Ideal Solution (FNIS) (9):

$$A^* = \left(\widetilde{\widetilde{x}}_1^*, \widetilde{\widetilde{x}}_2^*, \cdots, \widetilde{\widetilde{x}}_n^*\right); \quad \widetilde{\widetilde{x}}_j^* = \max_i \left\{\widetilde{\widetilde{x}}_{ij}\right\} \tag{8}$$

$$A^- = \left(\widetilde{\widetilde{x}}_1^-, \widetilde{\widetilde{x}}_2^-, \cdots, \widetilde{\widetilde{x}}_n^-\right); \quad \widetilde{\widetilde{x}}_j^- = \min_i \left\{\widetilde{\widetilde{x}}_{ij}\right\} \tag{9}$$

Step 5. Calculation of the distance from each alternative to the FPIS and FNIS as follows (10):

$$D_i^* = \sum_{j=1}^{n} d\left(\widetilde{\widetilde{x}}_{ij}, \widetilde{\widetilde{x}}_j^*\right) \quad D_i^- = \sum_{j=1}^{n} d\left(\widetilde{\widetilde{x}}_{ij}, \widetilde{\widetilde{x}}_j^-\right) \tag{10}$$

Step 6. Determination of the Closeness Coefficient CC_i for each alternative (11):

$$CC_i = \frac{D_i^-}{D_i^- + D_i^*} \tag{11}$$

3 Study Case

Sports management raises many aspects that must be controlled to achieve the best possible results. Due to the development of information systems and the increasing demand to support the training process in many sports, such systems are readily used. Their main principle is to analyze the available data on the condition of the athlete, club, or team in order to identify possible directions for training, development, or areas that need to be strengthened to achieve better results. One area of sports management is athlete selection, which aims to identify the best and most promising athlete who can achieve significant results in the future. It is an important area, as selecting the most promising athletes increases their potential and allows clubs or teams with such players to perform better.

This problem also arises in swimming, where many athletes start their swimming training young and then continue their development over the years. For sports clubs managing their group of athletes, it is important to strengthen their team and thus search for potentially the most promising athletes among swimmers in all age categories. To enable the identification of the best swimmers from among those in training, it is necessary to take into account a number of parameters that are fundamental to swimming training, as well as those that determine an athlete's potential and current sporting level.

Due to the existence of multiple criteria defining the quality of swimming athletes' performances, Multi-Criteria Decision Analysis (MCDA) can be used to analyze their performance and determine an assessment of their potential, which allows assessments to be made based on a defined set of criteria. In Table 1, 11 criteria are presented to identify the motor and physical parameters of swimming athletes, through which it is possible to assess their sporting level. The criteria set that was taken into account for the performed assessment was presented in [20]. Parameters such as height (C_1) [15], foot length (C_4) [26], swimming technique (C_6) [21], or maximum heart rate (C_8) [18] were taken into account, among others. These and the other criteria significantly influence swimmers' performance, allowing them to be used to evaluate athletes. Table 1 presents a type of preferred criterion value direction, and the significance of that criterion is expressed numerically.

Table 1. Set of criteria selected for swimmer performance assessment

C_i	C_1	C_2	C_3	C_4	C_5	C_6	C_7	C_8	C_9	C_{10}	C_{11}
Type	Cost	Profit	Cost	Profit	Profit	Profit	Profit	Profit	Cost	Cost	Profit
Weight	0.09	0.09	0.09	0.09	0.09	0.09	0.09	0.09	0.09	0.09	0.09

Based on the defined set of criteria, a decision matrix for the problem of evaluating swimmers' results is presented in Table 2. The data for the 5 athletes were represented using Triangular Fuzzy Numbers (TFN), which allow for the representation of uncertain data in the problem. Due to the measurement uncertainties that arise and the fluctuation of values over time, modeling them with fuzzy numbers allows potential differences in the input data to be considered. In the case where the values for the players during the study were constant, they were represented using singletons, where the values for TFN(a, m, b) were equal.

The study included five swimming athletes whose ages were between $16 - 18$ years. On the other hand, as shown by the decision matrix data, the values determining their physical parameters varied considerably between swimmers, especially the attributes concerning weight (C_1), swimming technique (C_7), or fat index (C_9). Then, the defined fuzzy decision matrix was used to analyze the athletes' performance evaluation using fuzzy MCDA methods. To examine the relevance of each criterion in the problem, equal weights were initially used for all factors, and then the selected methods for objective determination of criterion weights were used. The main part of the experiments was based on the Fuzzy TOPSIS method, which was used to determine the ranking of the swimmers. The selected methods for determining criteria weights were then examined to determine the influence of parameter relevance on the results obtained. The results were compared with six methods from the MCDA group. The study also used sensitivity analysis in conjunction with the Fuzzy TOPSIS method on the impact of excluding subsequent criteria from the problem and examining the resulting rankings. The description of the methods and references to the papers in which the used Multi-Criteria Decision Analysis methods are available in the PyFDM package, which was used as a calculation tool in the research [24].

Table 2. Fuzzy decision matrix determined for swimmer performance assessment

C_i	A_1	A_2	A_3	A_4	A_5
C_1	[78.5, 79.7, 81.2]	[90.5, 92.9, 94.1]	[86.5, 87.4, 88.3]	[72.1, 73.5, 74.6]	[80.2, 81.2, 83.4]
hline C_2	[186, 186, 186]	[194, 194, 194]	[188, 188, 188]	[186, 186, 186]	[184, 184, 184]
C_3	[18, 18, 18]	[18, 18, 18]	[17, 17, 17]	[18, 18, 18]	[16, 16, 16]
C_4	[29.0, 29.0, 29.0]	[31.5, 31.5, 31.5]	[29.5, 29.5, 29.5]	[29.0, 29.0, 29.0]	[29.0, 29.0, 29.0]
C_5	[1.01, 1.02, 1.03]	[1.00, 1.03, 1.04]	[1.00, 1.02, 1.03]	[1.00, 1.01, 1.04]	[0.99, 1.00, 1.02]
C_6	[6, 8, 9]	[6, 7, 8]	[8, 9, 10]	[8, 9, 10]	[7, 7, 9]
C_7	[8, 9, 10]	[5, 6, 7]	[6, 7, 8]	[6, 7, 8]	[7, 8, 8]
C_8	[205, 210, 215]	[210, 215, 220]	[205, 210, 215]	[200, 205, 210]	[195, 200, 205]
C_9	[5.5, 6.3, 7.0]	[8.2, 9.6, 11.4]	[6.9, 7.8, 8.5]	[5.9, 6.6, 7.4]	[11.4, 12.4, 13.2]
C_{10}	[0.062, 0.084, 0.102]	[0.118, 0.129, 0.147]	[0.097, 0.104, 0.111]	[0.085, 0.091, 0.103]	[0.139, 0.181, 0.214]
C_{11}	[717, 717, 717]	[697, 697, 697]	[787, 787, 787]	[757, 757, 757]	[661, 661, 661]

4 Results

Table 3 shows the preference values with the ranking order of the athletes obtained by using the Fuzzy TOPSIS method with equal-value criteria weights, shown in Table 1. It can be seen that the differences between the determined preference values are subtle, especially between the ranked first A_1 (0.81) and the second-ranked A_4 (0.80). A similar difference can be seen in fourth and fifth place, where athletes A_2 and A_5 were separated by an equal difference of 0.01.

Table 3. Results obtained from the Fuzzy TOPSIS method with equal weights of criteria

Fuzzy TOPSIS	A_1	A_2	A_3	A_4	A_5
Preferences	0.081	0.073	0.078	0.080	0.072
Ranking	1	4	3	2	5

Figure 1 shows a visualization of the rankings obtained based on the sensitivity analysis approach, excluding individual criteria in the problem. The results for the Fuzzy TOPSIS method show that for equal relevance of criteria in the problem, the exclusion of individual criteria impacts the swimmers' proposed ranking order. Notably, changes are apparent for swimmers ranked 1st and 2nd and 4th and 5th, i.e., those between whom the difference in preference score was slight. It can be seen that athlete A_4, was better rated by the model when the flexibility parameter C_7 was ignored, where athlete A_4 has the lowest value among the compared swimmers. Additionally, athlete A_5 would be ranked higher than athlete A_2 if the criteria foot length (C_4), maximum heart rate (C_8), fat index (C_9), and fat-muscle ratio (C_{10}), respectively, were omitted from the model. In contrast, the rating of athlete A_3 did not change regardless of the exclusion of individual criteria in the problem.

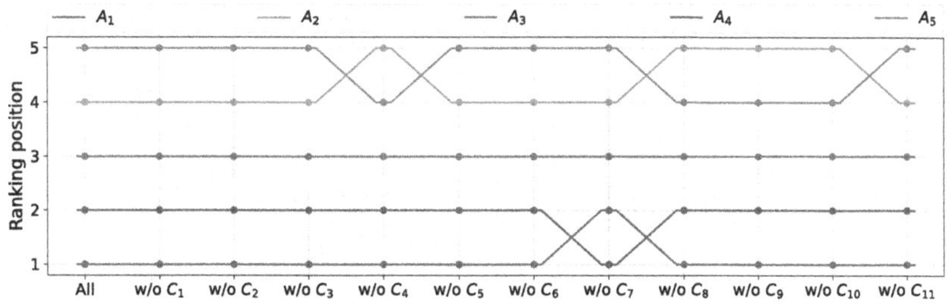

Fig. 1. Flow of rankings for the Fuzzy TOPSIS method (equal weights) regarding excluding subsequent criteria

Table 4 shows the rankings calculated for the selected six Fuzzy MCDA methods using equal weights for criteria. What is worth noting is that differences in the performance of the Fuzzy MCDA methods may cause discrepancies in the proposed rankings. As can be seen in Table 4, five of the six methods propose identical rankings of swimmers. In contrast, the Fuzzy MABAC method ranked the players A_1 and A_3 in a different order, with the former ranked 3rd and the latter ranked 1st. In contrast, the resulting proposals for the order of athletes are highly consistent and coherent.

Table 4. Rankings obtained from selected Fuzzy MCDA methods with equal criteria weights

Method	A_1	A_2	A_3	A_4	A_5
Fuzzy ARAS	1	4	3	2	5
Fuzzy COPRAS	1	4	3	2	5
Fuzzy EDAS	1	4	3	2	5
Fuzzy MABAC	3	4	1	2	5
Fuzzy MOORA	1	4	3	2	5
Fuzzy OCRA	1	4	3	2	5

To determine the effect of changing the relevance of the criteria in the problem on the proposed order of classification of the examined athletes, the study used 3 selected methods to determine criteria weights objectively. The Shannon entropy, standard deviation, and variance weighting methods were used to compare with the results obtained for equal criteria weights. It reflects how a sports club prioritizes individual athletes' physical parameters differently. Therefore, they can adjust athlete searches and evaluations according to their preferences and the specifics of club management. Figure 2 shows a visualization of the resulting rankings for the Fuzzy TOPSIS method using the selected weighting methods. It can be seen that the proposed results are highly consistent, with a discrepancy regarding the ranking position of player A_1 and A_3 for equal criteria weights.

A similar experiment was carried out on the remaining six Fuzzy MCDA methods to indicate how the other player evaluation techniques would respond

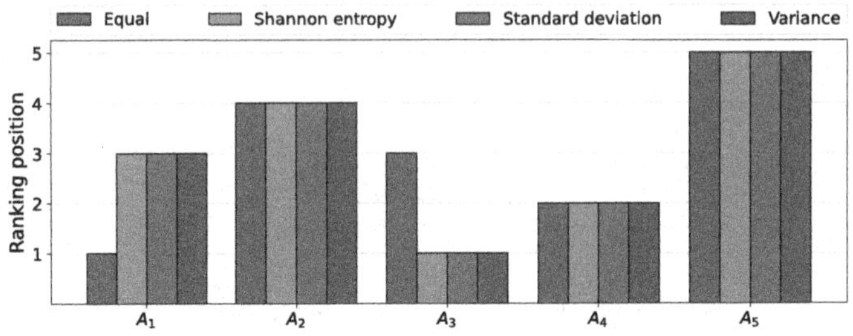

Fig. 2. Rankings obtained from selected objective weighting methods for the Fuzzy TOPSIS method

to a change in the significance of the parameters in the problem. Figure 3 shows the designated swimmers' positions for the Fuzzy ARAS, Fuzzy COPRAS, Fuzzy EDAS, Fuzzy MABAC, Fuzzy MOORA, and Fuzzy OCRA methods. It can be seen that the 5 methods have discrepancies consistent with those noted in the Fuzzy TOPSIS method. In contrast, the Fuzzy MABAC method classified the A_1 and A_2 athletes differently when the significance of the criteria was determined by the Shannon entropy method.

The average rankings determined for each of the examined Fuzzy MCDA methods and weights are presented in Table 5 to determine a compromise ranking for the examined methods and weights. The rankings indicated by the tested methods for determining the criteria weights were averaged. It can be seen that, despite the discrepancies in the individual ratings of the alternatives for the selected methods for determining parameter significance, the average ranking for each of the examined methods is the same. It makes it possible to conclude with a high degree of certainty from the tested criteria relevance that the ranking of swimmers $A_3 > A_4 > A_1 > A_2 > A_5$ is a reasonable solution.

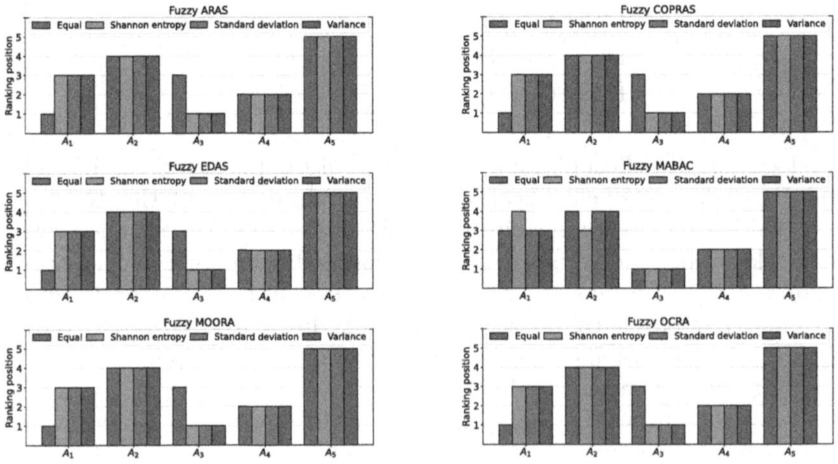

Fig. 3. Comparison of rankings obtained with six selected Fuzzy MCDA methods for different weighting techniques

Table 5. Mean rankings obtained from selected Fuzzy MCDA methods and different weighting techniques

Method	A_1	A_2	A_3	A_4	A_5
Fuzzy TOPSIS	3	4	1	2	5
Fuzzy ARAS	3	4	1	2	5
Fuzzy EDAS	3	4	1	2	5
Fuzzy COPRAS	3	4	1	2	5
Fuzzy MABAC	3	4	1	2	5
Fuzzy MOORA	3	4	1	2	5
Fuzzy OCRA	3	4	1	2	5

On the other hand, it is worth mentioning that the study used objective methods to determine the criterion weights, which base their operation on a statistical analysis of the data in the decision matrix. Due to the different preferences of clubs and sports teams and their different operational and training goals, it would be worthwhile to include a definition of the relevance of the parameters in the problem based on the needs of these clubs in order to identify the most preferred athletes for them.

5 Conclusions

Managing sports and sports clubs is a demanding task that many teams face. Various factors have to be taken into account, from coaching to regeneration, the coaching squad, and the group of players being trained. Dedicated information systems can help throughout the process. In this article, we focused on evaluating the performance of swimming athletes based on a set of physical parameters and the current sport level of the athletes, which were used in a Multi-Criteria Decision Analysis. Using the selected Fuzzy MCDA methods, it was possible to indicate the proposed classification ranking of the tested athletes. In addition, possible changes in the ranking when excluding individual criteria in the problem were indicated. An examination of the selected methods for the objective determination of criteria weights made it possible to conclude that the relevance of individual parameters influences the proposed results. Despite the differences appearing in the individual ratings obtained from the selected Fuzzy MCDA methods, the averaged positional ranking indicated an unambiguous classification that can be considered the most rational choice with the criteria relevance taken into account.

For future directions, it is worth considering applying subjective weighting methods to determine the criteria relevance for better customization of results for the specific needs of the swimming clubs. Moreover, it would be meaningful to compare the proposed approach with other Machine Learning (ML) methods.

The set of criteria considered in the problem could also be adjusted for the specific needs of the sports club. It is also worth comparing if different normalization techniques and distance measures can also cause changes in proposed rankings.

References

1. Ballı, S., Korukoğlu, S.: Development of a fuzzy decision support framework for complex multi-attribute decision problems: a case study for the selection of Skilful basketball players. Expert. Syst. **31**(1), 56–69 (2014)
2. Barajas, A., Castro-Limeres, O., Gasparetto, T.: Application of MCDA to evaluate financial fair play and financial stability in European football clubs. J. Sports Econ. Manage. **7**(3), 143–164 (2017)
3. Behzadian, M., Otaghsara, S.K., Yazdani, M., Ignatius, J.: A state-of the-art survey of TOPSIS applications. Expert Syst. Appl. **39**(17), 13051–13069 (2012)
4. Dadelo, S., Turskis, Z., Zavadskas, E.K., Dadeliene, R.: Multi-criteria assessment and ranking system of sport team formation based on objective-measured values of criteria set. Expert Syst. Appl. **41**(14), 6106–6113 (2014)
5. Dixon, M.A., Bruening, J.E.: Perspectives on work-family conflict in sport: an integrated approach. Sport Manage. Rev. **8**(3), 227–253 (2005)
6. Gökgöz, F., Yalçın, E.: A comparative multi criteria decision analysis of football teams: evidence on FIFA world cup. Team Perform. Manage. Int. J. **27**(3/4), 177–191 (2021)
7. Gunduz, M., Tehemar, S.R.: Assessment of delay factors in construction of sport facilities through multi criteria decision making. Product. Plan. Control **31**(15), 1291–1302 (2020)
8. Hoye, R., Smith, A.C., Nicholson, M., Stewart, B.: Sport Management: Principles and Applications. Routledge, London (2015)
9. Ishizaka, A., Nemery, P.: Multi-criteria Decision Analysis: Methods and Software. John Wiley & Sons, Hoboken (2013)
10. Jiang, X.P.: Algorithms for multiple attribute group decision making with intuitionistic 2-tuple linguistic information and its application. Int. J. Knowl. Based Intell. Eng. Syst. **26**(1), 37–45 (2022)
11. Kondratenko, Y.P.: Robotics, automation and information systems: future perspectives and correlation with culture, sport and life science. In: Gil-Lafuente, A.M., Zopounidis, C. (eds.) Decision Making and Knowledge Decision Support Systems. LNEMS, vol. 675, pp. 43–55. Springer, Cham (2015). https://doi.org/10.1007/978-3-319-03907-7_6
12. Lee, S., Juravich, M.: Multi criteria decision-making: ticket sales outsourcing in an NCAA division I athletic department. Case Stud. Sport Manage. **6**(1), 31–38 (2017)
13. Liebermann, D.G., Katz, L., Hughes, M.D., Bartlett, R.M., McClements, J., Franks, I.M.: Advances in the application of information technology to sport performance. J. Sports Sci. **20**(10), 755–769 (2002)
14. Liu, Q.: TOPSIS model for evaluating the corporate environmental performance under intuitionistic fuzzy environment. Int. J. Knowl. Based Intell. Eng. Syst. **26**(2), 149–157 (2022)
15. Mallett, A., Bellinger, P., Derave, W., Osborne, M., Minahan, C.: The age, height, and body mass of Olympic swimmers: a 50-year review and update. Int. J. Sports Sci. Coach. **16**(1), 210–223 (2021)

16. Mu, E.: Who really won the FIFA 2014 golden ball award?: What sports can learn from multi-criteria decision analysis. Int. J. Sport Manage. Mark. **16**(3–6), 239–258 (2016)
17. Nădăban, S., Dzitac, S., Dzitac, I.: Fuzzy TOPSIS: a general view. Proc. Comput. Sci. **91**, 823–831 (2016)
18. Olstad, B.H., Bjørlykke, V., Olstad, D.S.: Maximal heart rate for swimmers. Sports **7**(11), 235 (2019)
19. Sałabun, W., Watróbski, J., Shekhovtsov, A.: Are MCDA methods benchmarkable? A comparative study of topsis, vikor, copras, and promethee ii methods. Symmetry **12**(9), 1549 (2020)
20. Sałabun, W., Więckowski, J., Watróbski, J.: Swimmer assessment model (SWAM): expert system supporting sport potential measurement. IEEE Access **10**, 5051–5068 (2022)
21. Strzala, M., Tyka, A.: Physical endurance, somatic indices and swimming technique parameters as determinants of front crawl swimming speed at short distances in young swimmers. Medicina Sportiva **13**(2), 99–107 (2009)
22. Urbaniak, K., Watróbski, J., Sałabun, W.: Identification of players ranking in e-sport. Appl. Sci. **10**(19), 6768 (2020)
23. Vadde, S., Zeid, A., Kamarthi, S.V.: Pricing decisions in a multi-criteria setting for product recovery facilities. Omega **39**(2), 186–193 (2011)
24. Więckowski, J., Kizielewicz, B., Sałabun, W.: pyFDM: a python library for uncertainty decision analysis methods. SoftwareX **20**, 101271 (2022)
25. Więckowski, J., Kołodziejczyk, J.: Swimming progression evaluation by assessment model based on the comet method. Proc. Comput. Sci. **176**, 3514–3523 (2020)
26. Więckowski, J., Watróbski, J., Baczkiewicz, A., Kizielewicz, B., Shekhovtsov, A., Sałabun, W.: Multi-criteria assessment of swimmers' predispositions to compete in swimming styles. In: 2021 International Conference on Decision Aid Sciences and Application (DASA), pp. 127–131. IEEE (2021)
27. Woratschek, H., Horbel, C., Popp, B.: The sport value framework-a new fundamental logic for analyses in sport management. Eur. Sport Manag. Q. **14**(1), 6–24 (2014)
28. Zadeh, L.A.: Fuzzy logic. Computer **21**(4), 83–93 (1988)

Differential Private (Random) Decision Tree Without Adding Noise

Ryo Nojima[1,2] and Lihua Wang[2(✉)]

[1] College of Information Science and Engineering, Ritsumeikan University, Shiga, Japan
ryo-no@fc.ritsumei.ac.jp

[2] National Institute of Information and Communications Technology, Tokyo 184-8795, Japan
lh-wang@nict.go.jp

https://www.ritsumei.ac.jp/ise/teacher/detail/?id=249,
https://www.ritsumei.ac.jp/ise/,
https://sfl.nict.go.jp/en/people/lihua-wang.html,
https://sfl.nict.go.jp/en/

Abstract. The decision tree is a typical algorithm in machine learning and has multiple expanded variations. However, regarding privacy, few in the variations reached practical level due to many challenges on balancing privacy preservation and performance. In this paper, we propose a method of applying privacy preservation to the (random) decision tree, which is a variation of the expanded decision tree proposed by Fan et al. in 2003, to achieve the following goals:

- Model training with data belonging to multiple organizations and concealing these data among organizations.
- No leakage of training data from trained models.

Keywords: Differential privacy · k-Anonymity · Homomorphic encryption · Sampling

1 Introduction

In recent years, the development in data analysis technologies and computing power has resulted in a vigorous utilization of various types of data held by organizations. However, in most cases, such data contain personal information. Therefore, privacy preservation is a primary concern. Typical privacy preservation approaches include anonymity, statistics and encryption, etc. A big problem in privacy preservation is the trade-off among accuracy, security and processing speed. Generally, it is highly difficult to preserve the accuracy and security while achieving sufficient processing speed.

Motivations. In this paper, we propose a method of applying privacy preservation to the (random) decision tree, which is a typical algorithm in machine learning with multiple expanded variations. However, regarding privacy, few of these variations have been put into practice. Usually the following problems need to be considered in privacy preservation methods:

B. Luo et al. (Eds.): ICONIP 2023, CCIS 1963, pp. 162–174, 2024.
https://doi.org/10.1007/978-981-99-8138-0_14

- What amount of personal data will leak from the trained model?
- Is it possible to conceal these data among organizations?

To overcome these problems, we propose a privacy preservation method based on the random decision tree, which is a variation of the expanded decision tree proposed by Fan et al. [6].

Contributions. The proposed method has two advantages:

(1) Leakage of personal data from a trained model can be prevented to a certain level.
(2) The model is trained with data concealed from other organizations.

Homomorphic encryption and commitment have been used to achieve advantage (2). Regarding advantage (1), many related works have been reported. In one of them, differential privacy was realized by adding Laplace noise. In this study, instead of adding Laplace noise, we propose a method of enhancing the security of a random decision tree by removing the leaf containing fewer data than those contained in other leaves of the decision tree. This method meets the k-anonymization property of the decision tree. Moreover, after performing sampling, differential privacy can be achieved. The basic concept is to use the method reported in [8]. Intuitively, the model in which k-anonymity is performed after sampling achieves differential privacy, as in [8]. In this study, the process of removing the leaf containing fewer data than those contained in other leaves of the decision tree is essentially the same as the k-anonymization process.

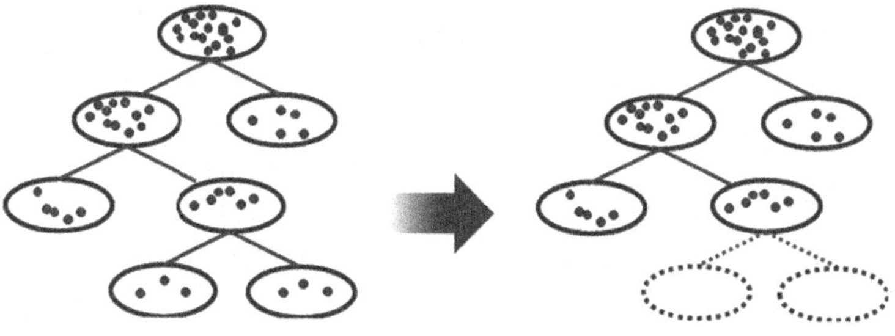

Fig. 1. k-anonymization in a decision tree: Removing the leaves containing less than k amount of data

In [2], a method that preserves the privacy of distributed extremely randomized trees was proposed. The aim was to conceal data from other organizations. Despite the efficiency of this method, if there are two colluding adversaries, security is not preserved. In contrast, there is not such weakness in our method.

The paper is organized as follows. In Sect. 2, the random decision tree proposed by Fan et al. [6] is introduced, and the importance and necessity of "removing the leaf containing fewer data than those contained in other leaves" is examined. The differential privacy and sampling are also defined in this section. In

---

Algorithm 1. Train in the Random Decision Tree [6]

Input: Training data $D = \{(\mathbf{x}_1, t_1), \ldots, (\mathbf{x}_n, t_n)\}$, the set of features $X = \{F_1, \ldots, F_m\}$, number of random decision trees to be generated N
Output: Random decision trees $T_1, \ldots, T_N$
1: **for** $i \in \{1, \ldots, N\}$ **do**
2:    $T_i = \text{BuildTreeStructure}(\text{root}, \mathbf{x})$
3: **end for**
4: **for** $i \in \{1, \ldots, N\}$ **do**
5:    $T_i = \text{UpdateStatics}(T_i, D)$
6: **end for**
7: **return** $T_1, \ldots, T_N$
8:
9: $\text{BuildTreeStructure}(\text{node}, X)$
10: **if** $(X \neq \emptyset)$ **then**
11:    Set the node as a leaf
12: **else**
13:    $F \leftarrow X$
14:    /* Set F as the feature of the node. Also, assume the number of values $\{d_i\}$ F has is c. Sub-node$_i$ is generated for each d_i */
15:    **for** $i \in \{1, \ldots, c\}$ **do**
16:       $T = \text{BuildTreeStructure}(\text{node}_i, X - F)$
17:    **end for**
18: **end if**
19:
20: $\text{UpdateStatistics}(T, D)$
21: Set $n[\ell, t] = 0$ for all leaves ℓ and labels t.
22: **for** $(\mathbf{x}, t) \in D$ **do**
23:    Find the leaf ℓ corresponding to $\mathbf{x}$, and set $n[\ell, t] = n[\ell, t] + 1$.
24: **end for**

Sect. 3, the proposed approach is described and the security achieved is demonstrated. The results showing the amount of efficiency lost due to privacy preservation are shown in Sect. 4. Finally, Sect. 5 concludes the paper.

2   Preliminaries

2.1   Basic Approach

The random decision tree was proposed by Fan et al. [6]. Intriguingly, in their approach, the tree is randomly generated without depending on the data. Furthermore, the sufficient performance can be achieved by appropriately selecting the parameters.

The shape of a normal (not random) decision tree depends on the data, which may cause private information leakage from the tree. However, the random decision tree avoids this leakage due to its totally random generation. Therefore, its

Algorithm 2. Classify in the Random Decision Tree [6]

Input: $\{T_1, \ldots, T_N\}, \mathbf{x}$

return $\sum_{i=1}^{N} n_i[\ell, t] / \left(\sum_t \sum_{i=1}^{N} n_i[\ell, t] \right)$, where ℓ denotes the leaf corresponding to $\mathbf{x}$.

Algorithm 3. Train in the proposal

Input: Training data $D = \{(\mathbf{x}_1, t_1), \ldots, (\mathbf{x}_n, t_n)\}$, the set of features $X = \{F_1, \ldots, F_m\}$, number of random decision trees to be generated N

Output: Random decision trees $T_1, \ldots, T_N$

for $i \in \{1, \ldots, N\}$ **do**
  $T_i =$ `BuildTreeStructure`(root, X)
end for
for $i \in \{1, \ldots, N\}$ **do**
  $D_i \leftarrow \emptyset$
  Regarding each $(\mathbf{x}, t) \in D$, $D_i \leftarrow D_i \cup (\mathbf{x}, t)$ with probability β
  $T_i =$ `UpdateStatics-`k`-anon`(T_i, D_i)
end for
return $T_1, \ldots, T_N$

`UpdateStatistics-`k`-anon`(T_i, D_i)
Set $n_i[\ell, t] = 0$ for all leaves ℓ and labels t.
for $(\mathbf{x}, t) \in D$ **do**
  Find the leaf ℓ corresponding to $\mathbf{x}$, and set $n_i[\ell, t] = n_i[\ell, t] + 1$.
end for
for All pairs of leaf and label (ℓ, t) **do**
  if $n_i[\ell, t] < k$ **then**
    $n_i[\ell, t] \leftarrow 0$ /*Removing leaves with fewer data*/
  end if
end for

performance is expected to match the performance achieved by other proposed security methods.

The random decision tree is described in Algorithms 1 and 2. Algorithm 1 shows that the generated tree does not depend on dataset D, except for n created by `UpdateStatistics`. Here, n denotes the 2D array, which represents the number of feature vectors reaching each leaf. However, privacy preservation is necessary since n depends on D. Considering each parameter, we assume that the depth of the tree is

$$\frac{\text{dimension of feature vectors}}{2}$$

and the number of trees is 10.

A Potentially Problematic Case: Despite the availability of the random decision tree described above, a problem may arise if the tree is too deep in terms of the dataset size.

For example, for a random decision tree with a pair $(\mathbf{x}, t)$ of feature vector and label as the training data, if $\sum_t n_i[\ell, t] = 1$ at leaf ℓ where $\mathbf{x}$ reaches, then the $(\mathbf{x}, t)$ pair is leaked from the model. Let $n[\ell, t]$ denote the amount of data reached at leaf ℓ of label t, the $n_i[\ell, t]$ here is $n[\ell, t]$ for tree T_i. Generally, it can be stated that the trained model contains statistical data instead of personal data. However, there is a risk of leakage of a massive amount of personal data from such a model, considering parameters such as the tree depth.

The differential privacy method was proposed by Jagannathan et al. [7] to overcome this problem. In their method, Laplace noise was added. Namely, let

$$n_i[\ell, t] + \text{Laplace Noise},$$

for all trees T_i, all leafs ℓ and all labels t.

However, even for this method, if $n_i[\ell, t]$ is small for a certain T_i, leaf ℓ, and label t, it can be regarded as merely a method of processing personal data. A general way to handle such a case is **removing the rare data**, i.e., "removing the leaf containing fewer data" [1]. In other words, as shown in Fig. 1, a threshold k is set, and the following operation is performed for all (i, ℓ, t):

$$n_i[\ell, t] \leftarrow \begin{cases} n_i[\ell, t], & \text{when } n_i[\ell, t] \geq k; \\ 0, & \text{otherwise.} \end{cases}$$

This study shows that the above processing and the sampling achieve **differential privacy** without adding noise.

2.2 Security-Related Definitions

This study adopts *differential privacy* to evaluate the security efficiency of the model.

Definition 1. ((ϵ, δ) -Differential Privacy, (ϵ, δ) -DP [4]) A randomized algorithm A satisfies (ϵ, δ)-DP, if any pair of neighboring datasets D and D', and for any $O \subseteq \mathsf{Range}(A)$:

$$\Pr[A(D) \in O] \leq e^{\epsilon} \cdot \Pr[A(D') \in O] + \delta.$$

When studying differential privacy, it is assumed that the attacker knows all the elements in D. However, such an assumption may not be realistic. This is taken into consideration, and the following definition is given in [8]:

Definition 2. (DP under sampling, $(\beta, \epsilon, \delta)$ -DPS) An algorithm A satisfies $(\beta, \epsilon, \delta)$-DPS if and only if the algorithm A^{β} satisfies (ϵ, δ)-DP, where A^{β} denotes the algorithm used to initially sample each tuple in the input dataset with probability β; then A is applied to the the sampled dataset.

In other words, the definition focuses on the output of A by inputting D', which is the result obtained after sampling dataset D. Hence, the attacker may know D, but not D'.

3 Proposed Method

3.1 Differentially Private Random Decision Tree Under Sampling

The proposed method is shown as Algorithm 3. Its differences from existing methods are the following:

- Training using D_i, which is the result obtained after sampling dataset D of each tree T_i with probability β.
- For a threshold k, if there exists a tree i, a leaf ℓ, and a label t, satisfying $n_i[\ell, t] < k$, then let $n_i[\ell, t]$ equal to 0.

3.2 Security: Strongly-Safe k-Anonymization Meets Differential Privacy

In [8], it was shown that performing k-anonymization after sampling achieves differential privacy. We initially introduce the necessary items for the security evaluation presented below since this study is based on the result of [8].

Definition 3. (Strongly-safe k -anonymization algorithm [8]**)** $\mathcal{T}$ denotes a set, and function g has $\mathcal{D} \rightarrow \mathcal{T}$. When g does not depend on the input $D \subseteq \mathcal{D}$, g is constant. The strongly-safe k-anonymization algorithm A with input $D \subseteq \mathcal{D}$ is defined as follows:

- Calculate $Y_1 = \{g(\mathbf{x})|\mathbf{x} \in D\}$.
- $Y_2 = \{(y, |\{\mathbf{x} \in D | g(\mathbf{x}) = y\}|)|y \in Y_1\}$.
- For each element in $Y_2 = \{(y, c)\}$, if $c < k$, then the element is set as $(y, 0)$, and the result is set as Y_3.

Assume that $f(j; n, \beta)$ denotes the probability mass function, namely, the probability to succeed for j times when trying n times, where the probability of success for one trial is β. Furthermore, the cumulative distribution function is expressed as follows:

$$F(j; n, \beta) = \sum_{i=0}^{j} f(i; n, \beta).$$

Theorem 1. (Theorem 5 in *[8]***)** *Any strongly-safe k-anonymization algorithm satisfies $(\beta, \epsilon, \delta)$-DPS for any $0 < \beta < 1$, $\epsilon \geq -\ln(1-\beta)$, and*

$$\delta = d(k, \beta, \epsilon) = \max_{n: n \geq \lceil \frac{k}{\gamma} - 1 \rceil} \sum_{j > \gamma n}^{n} f(j; n, \beta),$$

where $\gamma = \frac{e^{\epsilon} - 1 + \beta}{e^{\epsilon}}$.

$(\mathbf{x}, t) \in D$ is applied to random decision tree T_i and set the reached leaf as ℓ. If g_i is defined as

$$g_i((\mathbf{x}, t)) = (\ell, t),$$

Table 1. Value of $N\delta$ for $k = 5, 10, 20$ and $N = 10$

	$k = 5$			$k = 10$			$k = 20$		
$N\epsilon \setminus \beta$	0.01	0.1	0.4	0.01	0.1	0.4	0.01	0.1	0.4
1.0	0.001	-	-	$4.66 * 10^{-7}$	0.355	-	$1.20 * 10^{-13}$	0.045	-
2.0	$5.52 * 10^{-5}$	0.352	-	$1.08 * 10^{-9}$	0.034	-	$7.00 * 10^{-19}$	0.000	-
3.0	$7.68 * 10^{-6}$	0.127	-	$2.72 * 10^{-11}$	0.005	-	$4.75 * 10^{-22}$	$7.82 * 10^{-6}$	0.426
4.0	$1.86 * 10^{-6}$	0.043	-	$1.68 * 10^{-12}$	0.001	0.583	$1.92 * 10^{-24}$	$2.37 * 10^{-7}$	0.134
5.0	$7.47 * 10^{-7}$	0.028	0.994	$2.85 * 10^{-13}$	0.000	0.348	$3.54 * 10^{-26}$	$1.61 * 10^{-8}$	0.051
6.0	$2.417 * 10^{-7}$	0.009	0.963	$3.19 * 10^{-14}$	$3.93 * 10^{-5}$	0.191	$7.71 * 10^{-28}$	$1.03 * 10^{-9}$	0.016
7.0	$1.22 * 10^{-7}$	0.009	0.963	$8.51 * 10^{-15}$	$2.05 * 10^{-5}$	0.175	$5.75 * 10^{-29}$	$1.48 * 10^{-10}$	0.008
8.0	$1.22 * 10^{-7}$	0.004	0.498	$4.07 * 10^{-15}$	$4.53 * 10^{-6}$	0.093	$6.27 * 10^{-30}$	$1.56 * 10^{-11}$	0.003
9.0	$5.47 * 10^{-8}$	0.002	0.410	$7.58 * 10^{-16}$	$1.87 * 10^{-6}$	0.078	$5.06 * 10^{-31}$	$2.82 * 10^{-12}$	0.001

then g_i is apparently constant since g_i does not depend on D. Therefore, the $n_i[\ell, t]$, which is generated using g_i can be regarded as the strongly-safe k-anonymization, and Theorem 1 can be applied.

The above Theorem 1 can be applied in its original form when there is one T_i, i.e., the number of trees $N = 1$. The following Theorem 2 can be applied when $N > 1$.

Theorem 2. (Theorem 3.16 in *[5])* *Assume A_i be an (ϵ_i, δ_i)-DP algorithm for $1 \leq i \leq N$. Then*

$$A(D) := (A_1(D), A_2(D), \ldots, A_N(D))$$

satisfies $(\sum_{i=1}^{N} \epsilon_i, \sum_{i=1}^{N} \delta_i)$-DP.

In Algorithm 3, each T_i is selected randomly, and sampling is performed for each tree. Hence, the following conclusion can be derived.

Corollary 1. *The proposed algorithm satisfies $(\beta, N\epsilon, N\delta)$-DPS, for any $0 < \beta < 1$, $\epsilon \geq -\ln(1-\beta)$ and*

$$\delta = d(k, \beta, \epsilon) = \max_{n: n \geq \lceil \frac{k}{\gamma} - 1 \rceil} \sum_{j > \gamma n}^{n} f(j; n, \beta), \quad (1)$$

where $\gamma = \frac{e^{\epsilon} - 1 + \beta}{e^{\epsilon}}$.

Table 1 shows the relationship, derived from Eq.(1), between β and $N\epsilon$ in determining the value of $N\delta$ when k and N are fixed. The cells in the table represent the value of $N\delta$. For k and N, we choose $(k, N) = (5, 10)$, $(k, N) = (10, 10)$ and $(k, N) = (20, 10)$, shown in (a), (b) and (c) of Table 1, respectively.

3.3 Other Cases

Regarding the Case where D is the Sampling Result: As described above, we assume that the attacker knows that all elements in D are strongly safe.

Therefore, we consider the case, where D is the result obtained after sampling $\mathcal{D}$ with probability α. The point for this case is how to select $D_1, ..., D_N$ from D. When selecting D_i by sampling from $\mathcal{D}$ with probability β, for a certain $(\mathbf{x}, t) \in \mathcal{D}$, the probability of being not selected all the time is $(1-\beta)^N$. Hence, the probability of being selected at least once is $1-(1-\beta)^N$. We set $1-(1-\beta)^N$ as α, and D as the set which is the sampling result of $\mathcal{D}$ with probability α. For $(\mathbf{x}, t) \in D$, the probability of $(\mathbf{x}, t) \in D_i$ is the following.

$$\begin{aligned} &\Pr[(\mathbf{x}, t) \in D_i \mid (\mathbf{x}, t) \in D] \\ &= \frac{\Pr[(\mathbf{x}, t) \in D_i \wedge (\mathbf{x}, t) \in D]}{\Pr[(\mathbf{x}, t) \in D]} \\ &= \frac{\beta}{\alpha}. \end{aligned}$$

As a result, a simple method of selecting $D_1, ..., D_N$ from D is as following:

- For a given α, calculate β which satisfies

$$\alpha = 1 - (1-\beta)^N.$$

- For $1 \leq i \leq N$, set $D_i \leftarrow \emptyset$.
- For all $(\mathbf{x}, t) \in D$ and all i, set

$$D_i \leftarrow D_i \cup \{(\mathbf{x}, t)\}.$$

  with probability β/α.

The generated $D_1, ..., D_N$ are used to form random decision trees $T_1, ..., T_N$.

Anonymous Collaborative Learning: The learning method, where the data are concealed [10] assumes M participants, $P_1, \ldots, P_M$. Each P_i generates trees randomly until an acceptable result is obtained. Then, this result is used. However, theoretically, assuming a semi-honest attacker, the process can fall into a loop and become endless. In such a case, we present a simpler method that uses commitment and threshold additive homomorphic encryption:

- Set $(\mathsf{Com}, \mathsf{Dcom})$ denoting commitment; and
- $(\mathsf{Enc}, \mathsf{Dec})$ denoting the (f-out-of-M)-threshold additive homomorphic encryption such that both the additive homomophic computation[1]

$$\mathsf{Enc}(X) * \mathsf{Enc}(Y) = \mathsf{Enc}(X+Y)$$

  and the (f-out-of-M)-threshold decryption[2] are possible.

We describe our method Step-by-Step below:

[1] We use the Pedersen commitment and the expanded ElGamal encryption (plaintext m is encoded as g^m).

[2] (f-out-of-M)-threshold decryption means any f out of M participants cooperate can decrypt the ciphertexts, but any participants less than f cannot.

Step 1. Each participant P_j selects a random number r_j and computes

$$\mathsf{Com}(r_j).$$

For $1 \leq j \leq M$, when all $\mathsf{Com}(r_j)$ are obtained, r_j is decommitted with Dcom and

$$r = \sum r_j$$

is computed.

Step 2. P_j activates `BuildTreeStructure` with r, i.e., let $\mathbf{x} = r$ in line-2 of Algorithm 1 and run

$$T_i = \texttt{BuildTreeStructure}(\text{root}, \mathbf{x})$$

to obtain Tree's structure.

Step 3. P_j uses `UpdateStatistics` to generate $n_{j,i}[\ell, t]$. It then uses an (f-out-of-M)-threshold additive homomorphic encryption to compute

$$\mathsf{Enc}(n_{j,i}[\ell, t]),$$

then broadcast

$$\{(j, \ell, t, i, \mathsf{Enc}(n_{j,i}[\ell, t]))\}.$$

Step 4. P_j runs an additive homomorphic computation to obtain the amount of data belonging to leaf ℓ of tree T_i

$$(\ell, t, i, \mathsf{Enc}(n_i[\ell, t])).$$

Here,

$$n_i[\ell, t] = \sum_j n_{j,i}[\ell, t].$$

Step 5. Any f out of M participants, e.g., $\{P_{j_1}, ..., P_{j_f}\} \subseteq \{P_1, P_2, ..., P_M\}$, cooperate to decrypt

$$\mathsf{Enc}(n_i[\ell, t])$$

to obtain threshold values $n_i[\ell, t]$ for tree i.

Remark 1. Details on how to combine the above cases with the method described in Sect. 3.1 will be described in the full version of this study. Since any participant (even a decryptor participant) might be a semi-honest attacker, it is necessary to devise a method to prevent an attacker from obtaining $n_i[\ell, t]$ in Step 5, the information about the amount of data at leaf node ℓ. Therefore, when checking if $n_i[\ell, t] \geq k$ as in the method described in Sect. 3.1, we will use a secure comparison method to resolve the issue.

Table 2. Evaluation of the Random Decision Tree Performance

Depth of Tree	3	4	5
Accuracy	0.973	0.984	0.981

4 Efficiency Verification of the Proposed Method

The efficiency of the proposed method was verified using the Nursery [9] and Adult datasets [3].

Nursery dataset: The Nursery dataset consists of 8 attributes and 12,960 records. Besides, 3 labels were used as in [7]. The training results obtained using the original decision tree (ID3), the random decision tree, and the proposed method are presented below. Python 3 was used in the calculations.

- The accuracy obtained using ID3 was 98.19% [7].
- The accuracy obtained using the original random decision tree for different tree depths is shown in Table 2.
- The accuracy of the proposed method was examined by setting of $\delta = 0.4$. The depth of the tree was set in the $4-8$ range, and the number of trees was set in the $8-11$ range. In addition, k, which controls the anonymity, was set in the $0-10$ range.
  - As shown in Table 3, for a tree depth equal to 4, the accuracy obtained is not good compared to the corresponding accuracy shown in Table 2.
  - As shown in Table 3, for a tree depth equal to 6, the accuracy obtained is similar compared to the corresponding accuracy shown in Table 2, even in the case where the model was anonymized.
  - As shown in Table 3, for a tree depth equal to 8, the accuracy obtained is significantly affected by the anonymized parameter k.
- As shown in Table 3, although the number of usable hyperparameters is small, the performance of the proposed method is similar to the performance of the non-anonymized random decision tree.

Adult dataset: The Adult dataset consists of 14 attributes, 48,842 records, and 2 labels. The training results obtained using ID3 and the proposed method are presented below:

- The accuracy obtained using ID3 was 83%.
- The accuracy of the proposed method was examined by setting $\delta = 0.4$. The depth of the tree was set in the $6-8$ range, and the number of trees was set in the $8-11$ range. In addition, k, which controls the anonymity, was set in the $0-10$ range. As shown in Table 4, the achieved accuracy for is $k = 5$ is 81.3%.

In summary, the accuracy achieved by the proposed method is slightly inferior to the accuracy achieved by ID3 (see Fig. 2).

Table 3. Accuracy of the Proposed Method Using the Nursery Dataset

	Depth = 4		Depth = 6		Depth = 8	
k	#Trees	Accuracy	#Trees	Accuracy	#Trees	Accuracy
0	8	0.891	8	0.959	8	0.983
5	8	0.891	8	0.957	8	0.647
10	8	0.878	8	0.957	8	0.337
0	9	0.809	9	0.967	9	0.983
5	9	0.809	9	0.966	9	0.689
10	9	0.901	9	0.956	9	0.334
0	10	0.866	10	0.968	10	0.983
5	10	0.866	10	0.966	10	0.706
10	10	0.903	10	0.961	10	0.335
0	11	0.912	11	0.971	11	0.986
5	11	0.912	11	0.970	11	0.750
10	11	0.899	11	0.958	11	0.340

Table 4. Accuracy of the Proposed method Using the Adult Dataset

	Depth = 6		Depth = 7		Depth = 8	
k	#Trees	Accuracy	#Trees	Accuracy	#Trees	Accuracy
0	8	0.773	8	0.781	8	0.806
5	8	0.773	8	0.781	8	0.807
10	8	0.780	8	0.794	8	0.801
0	9	0.772	9	0.789	9	0.813
5	9	0.772	9	0.789	9	0.813
10	9	0.786	9	0.804	9	0.812
0	10	0.783	10	0.795	10	0.791
5	10	0.783	10	0.795	10	0.791
10	10	0.785	10	0.786	10	0.803
0	11	0.780	11	0.786	11	0.807
5	11	0.780	11	0.786	11	0.807
10	11	0.773	11	0.793	11	0.797

For example, by applying the proposed method to the Nursery dataset and setting the depth equal to 6, the accuracy achieved for $k = 5$ and $k = 10$ exceeds 95%, which is very close to that obtained using ID3. Similarly, by applying the proposed method to the Adult dataset and setting the depth equal to 8, the accuracy achieved for $k = 5$ and $k = 10$ exceeds 80%, which is also very close to that obtained using the original ID3.

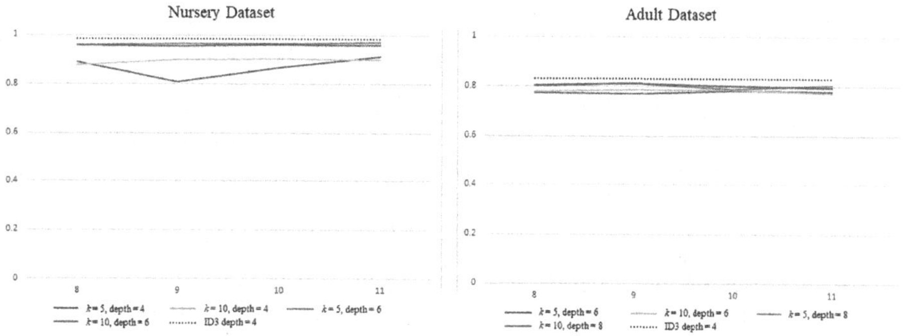

Fig. 2. Accuracy Comparison of Our Proposal with k-anonymity to the Original ID3

5 Concluding Remarks

In this study, we proposed a privacy preservation method based on the random decision tree proposed in [6]. The proposed algorithm applied sampling and k-anonymity to the original random decision tree and achieved differential privacy. The advantage of the proposed method is its ability to achieve differential privacy without adding Laplace noise. However, we believe that apart from the random decision tree, there exist other algorithms can achieve differential privacy. Thus, the future work is to discover more realistic methods.

Acknowledgments. This work was supported in part by JST CREST Grant Number JPMJCR21M1, and JSPS KAKENHI Grant Number JP20K11826, Japan.

References

1. Mobile kukan toukei (guidelines). https://www.intage.co.jp/english/service/platform/mobile-kukan-toukei/ (https://www.docomo.ne.jp/english/binary/pdf/service/world/inroaming/inroaming_service/Mobile_Kukan_Toukei_Guidelines.pdf)
2. Aminifar, A., Rabbi, F., Pun, K.I., Lamo, Y.: Privacy preserving distributed extremely randomized trees. In: Hung, C., Hong, J., Bechini, A., Song, E. (eds.) SAC '21: The 36th ACM/SIGAPP Symposium on Applied Computing, Virtual Event, Republic of Korea, March 22–26, 2021, pp. 1102–1105. ACM (2021). https://doi.org/10.1145/3412841.3442110
3. Becker, B., Kohavi, R.: Adult. UCI Machine Learning Repository (1996)
4. Dwork, C., Kenthapadi, K., McSherry, F., Mironov, I., Naor, M.: Our data, ourselves: privacy via distributed noise generation. In: Vaudenay, S. (ed.) EUROCRYPT 2006. LNCS, vol. 4004, pp. 486–503. Springer, Heidelberg (2006). https://doi.org/10.1007/11761679_29
5. Dwork, C., Roth, A.: The algorithmic foundations of differential privacy. Found. Trends Theor. Comput. Sci. **9**(3–4), 211–407 (2014). https://doi.org/10.1561/0400000042

6. Fan, W., Wang, H., Yu, P.S., Ma, S.: Is random model better? on its accuracy and efficiency. In: Proceedings of the 3rd IEEE International Conference on Data Mining (ICDM 2003), 19–22 December 2003, Melbourne, Florida, USA, pp. 51–58. IEEE Computer Society (2003). https://doi.org/10.1109/ICDM.2003.1250902
7. Jagannathan, G., Pillaipakkamnatt, K., Wright, R.N.: A practical differentially private random decision tree classifier. Trans. Data Priv. **5**(1), 273–295 (2012). www.tdp.cat/issues11/abs.a082a11.php
8. Li, N., Qardaji, W.H., Su, D.: On sampling, anonymization, and differential privacy or, k-anonymization meets differential privacy. In: Youm, H.Y., Won, Y. (eds.) 7th ACM Symposium on Information, Compuer and Communications Security, ASIACCS '12, Seoul, Korea, May 2–4, 2012, pp. 32–33. ACM (2012). https://doi.org/10.1145/2414456.2414474
9. Rajkovic, V.: Nursery. UCI Mach. Learn. Reposit. (1997). https://doi.org/10.24432/C5P88W
10. Vaidya, J., Shafiq, B., Fan, W., Mehmood, D., Lorenzi, D.: A random decision tree framework for privacy-preserving data mining. IEEE Trans. Dependable Secur. Comput. **11**(5), 399–411 (2014). https://doi.org/10.1109/TDSC.2013.43

Cognitive Neurosciences

Pushing the Boundaries of Chinese Painting Classification on Limited Datasets: Introducing a Novel Transformer Architecture with Enhanced Feature Extraction

Haiming Zhao[1], Jiejie Chen[1(✉)], Ping Jiang[2], Tianrui Wu[1], and Zhuzhu Zhang[1]

[1] School of Computer and Information Engineering, Hubei Normal University, Huangshi 435000, China
chenjiejie118@126.com
[2] School of Computer, Hubei PolyTechnic University, Huangshi 435000, China
jiangping@hbpu.edu.cn

Abstract. This study ventures into the relatively unexplored field of Chinese painting classification. The paper addresses challenges such as limited datasets, feature map redundancy, and inherent limitations of the Transformer model by curating a high-resolution dataset of Chinese paintings from various dynasties. To enhance efficiency, we introduce a strategy for selecting channels that capture unique features of these paintings and propose a novel architecture. This architecture uses local self-attention with sliding windows and two residual connections for spatial token mixing and channel feature transformation. Experiments show our architecture's superior accuracy in classifying small datasets. The research advances the field of Chinese painting classification, demonstrating deep learning models' potential in this area. The dataset and code can be accessed at: https://github.com/qwerty0814/gangan.

Keywords: Chinese Painting Classification · High-Resolution Dataset · Channel Selection Strategy · Local Self-Attention

1 Introduction

The study of art, particularly painting classification, has evolved with technology's advent, using deep learning for art appreciation and recommendation. However, research on Chinese painting classification is limited despite its cultural significance and unique aesthetic values. Traditional methods struggle with large-scale data, necessitating a universally applicable classification method. Chinese paintings incorporate distinct techniques and subjects, leading to diverse compositions. However, building a large-scale annotated dataset is challenging due to the art form's complexity, limited quantity, and high artistic value. Small

B. Luo et al. (Eds.): ICONIP 2023, CCIS 1963, pp. 177–189, 2024.
https://doi.org/10.1007/978-981-99-8138-0_15

datasets cannot cover all Chinese painting styles and elements, and multi-scale features like lines and brush strokes challenge models [1]. Moreover, imbalanced category distribution in these datasets can affect performance.

Improving Chinese painting research requires a comprehensive, large-scale dataset that includes different styles, elements, and high-resolution images to capture multi-scale features. This will enhance deep learning model performance in tasks like Chinese painting image classification [2]. Redundancy in feature mapping, caused by diverse compositions and imbalanced distribution, decreases model performance and wastes computational resources, and needs addressing.

The Transformer model performs well on large-scale datasets but may not be accurate with small datasets due to overfitting or underfitting risks. To address this, we created a dataset of Chinese paintings from 11 dynasties, considered common themes and elements, and constructed a dataset with over 900 bird's-eye view images. We reduced redundancy in feature mapping by extracting features using a subset of channels and developed a new network structure to improve classification accuracy on small datasets. Experiments showed the effectiveness of our diverse Chinese painting dataset, efficient feature extraction methods, and optimized network.

Our contributions can be summarized as follows:

- We have successfully constructed two comprehensive and diverse Chinese painting datasets. The first dataset includes 4,747 paintings from 11 dynasties while the second focuses on bird paintings with over 900 images. These datasets are currently the most comprehensive available for Chinese painting analysis.
- We've developed an effective feature extraction method that optimizes channel combination to minimize redundancy, thereby enhancing model efficiency and accuracy.
- Additionally, we've designed an innovative architecture that accurately captures painting details, improving classification and recognition accuracy even on smaller datasets.
- We conducted extensive experiments on Chinese painting and bird painting datasets, validating the effectiveness and low redundancy computation of our channel reordering and grouping, as well as the expanded local self-attention

2 Related Work

2.1 Efficient Feature Map Compression

Efficient feature map compression has gained attention for reducing computational complexity and improving model performance. Methods like channel pruning [3] and sparsity-constrained optimization [4] have been proposed to reduce feature map redundancy. Channel-wise attention mechanisms dynamically adjust channel importance to compress feature maps. Advanced techniques integrate compression methods, such as spatial and channel-wise correlations

[5] and channel pruning with knowledge distillation [6]. Some studies combine multiple techniques, like channel pruning, quantization, and tensor decomposition. Joint optimization strategies optimize channel selection and quantization. Our study builds on these achievements, proposing a tailored method for Chinese painting analysis that combines channel selection and compression techniques to improve efficiency and accuracy.

2.2 Boosting Transformer Performance on Small Datasets

Enhancing transformer performance on small datasets has been a critical research focus, with approaches including data augmentation, architecture modification [7], and transfer learning techniques. Efforts have been made to design more efficient transformer architectures suited for small datasets. Lightweight transformer models with fewer layers and attention heads have been proposed [8], while novel architectures combining local and global attention mechanisms have been developed to capture both short and long-range dependencies. Our study aims to build upon these achievements by proposing a novel method tailored for Chinese painting analysis, leveraging the latest advancements in transformer models and small dataset adaptation, focusing on effectively combining data augmentation and architecture modifications for improved efficiency and accuracy.

3 Method

3.1 FCTNet Block as a Basic Operator

As depicted in Fig. 1(a), the FCTNet Block uses a design akin to the Swin Transformer, enhancing accuracy by separating spatial token mixing (STM) and channel feature transformation into two residual connections. These consist of a normalization layer and a double residual link structure, differing from the traditional ResNet or ConvNeXt block design. To minimize redundancy and optimize data use, we first execute channel reordering and grouping. This increases data capacity and selects a subset of channels for subsequent processes. The following STM module has fewer parameters as it only processes a subset of channels. Coupled with increased training data from channel reordering, this allows better utilization of small datasets, fully exploiting transformers' potential. Additionally, an extended local self-attention mechanism approximates convolutional operations, capturing local and global features effectively. Overall, the FCTNet Block's design enhances data usage in small datasets, improving accuracy while efficiently capturing both local and global features in artworks.

3.2 Channel Reordering and Grouping

Below, we will demonstrate the method of channel reordering and grouping. As shown in Fig. 1(b), there is a high similarity between feature maps from different channels. This is achieved by introducing a partial ratio parameter (r) which

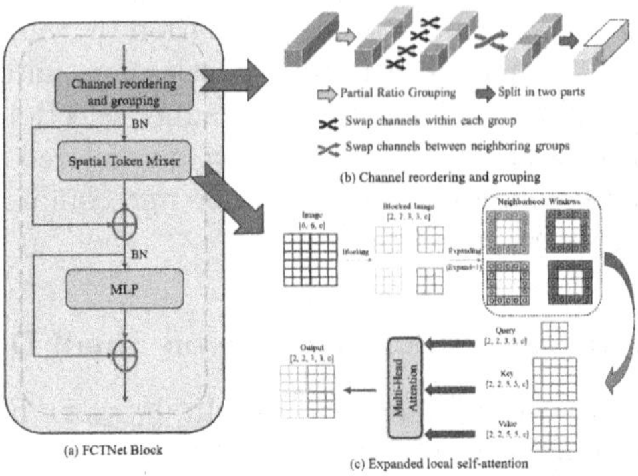

Fig. 1. The FCTNet block utilizes two residual connections to separate the spatial token mixing (STM) and channel feature transformation parts. It starts with channel reordering and grouping, followed by the spatial token mixing module

divides the total number of channels (C) into groups, each containing C/r channels. Within each group, we perform a channel reversal - the first channel becomes the last and so forth. We also conduct channel swapping between adjacent groups, thus rearranging the channels within feature maps to diversify features.

Our experimental results demonstrate that this reordering method outperforms other random reordering methods. The processed channels are divided by the module, with one group passed to subsequent STM and MLP sections. To maintain feature integrity, this passed group is then concatenated with the remaining groups. The channel reordering and grouping method also reduces computational redundancy as it only processes a portion of the input channels, leaving the rest unchanged. Therefore, the FLOPs for the local self-attention can be computed using Formula 1.

$$\begin{aligned}&3\times H\times W\times C\times C+h\times H\times W\times N\times N\times C\\&+h\times H\times W\times N\times N\times C+h\times H\times W\times N\times N\end{aligned}\tag{1}$$

To calculate the total FLOPs of the local self-attention after processing with channel reordering and grouping modules, as shown in Eq. 2

$$\begin{aligned}&3\times H\times W\times\frac{C}{r}\times\frac{C}{r}+h\times H\times W\times N\times N\times\frac{C}{r}\\&+h\times H\times W\times N\times N\times\frac{C}{r}+h\times H\times W\times N\times N\end{aligned}\tag{2}$$

Assuming the size of the input feature map is H, W, the size of the local window is $N\times N$, and the number of heads is h.

Our study revealed that as the value of rr increases, the number of channels in the transmission group diminishes, effectively minimizing redundant computations and implicitly boosting the volume of training data. However, we found that a larger r doesn't necessarily yield better outcomes. Experiments were conducted to validate this finding. For varying data capacities, we discovered that harnessing the implicit data increase through larger r values can maximize local self-attention potential. Yet, when r is overly large, it results in fewer parameters within the local self-attention. This insufficient parameter count might struggle to encapsulate complex data information. Therefore, choosing an appropriate r value requires a balance between reducing computational redundancy and preserving sufficient parameters for the local self-attention. This insight carries substantial implications for model performance enhancement and should be contemplated during neural network architecture design.

3.3 Expanded Local Self-attention

As shown in Fig. 1(c), it is challenging for our expanded self-attention to adopt sliding window methods like convolutional operations. This is because when the window slides with a stride of 's', the total number of sliding steps required to traverse the entire image can be calculated using Formula 3.

$$((W - K)/S + 1) \times ((H - K)/S + 1) \tag{3}$$

The described process traverses the entire image without freeing memory, potentially causing Out-Of-Memory (OOM) issues. As inferred from Eq. 3, increasing the stride of the sliding window significantly reduces memory usage [9]. To achieve this, we adopt a Swin Transformer-like strategy, partitioning the image into multiple blocks and performing local self-attention computations on each block. To prevent information loss from only considering intra-block information, we devise a method to gather extra-block data. Specifically, we expand pixels around the original image's blocks, collecting information beyond the local scope, effectively simulating convolution's sliding window operation.

We first segment the image into non-overlapping blocks and compute queries (q) from them. Each query corresponds to a specific position in the input feature map, or a pixel block. Keys (k) and values (v) are extracted from the extended-pixel input feature map. Each key relates to a pixel block and its surrounding extended pixels, while each value represents the information content in the key [9]. Our expanded self-attention mechanism improves hardware utilization, enhancing the speed-accuracy trade-off. Despite using local attention, the receptive field for each pixel can become quite large, depending on the number of extended pixels. For instance, with a block size of 7 and 3 extended pixels, the receptive field reaches 13×13. The receptive field (f) calculation formula is presented in Formula 4.

$$f = N + 2e \tag{4}$$

The variable e represents the number of expanded pixels. Larger receptive field helps to handle larger images and achieve better results.

4 FCTNet as a General Backbone

Taking our channel permutation, grouping, and expanded self-attention as primary architectural components, we propose a novel family of neural networks named FCTNet, specifically tailored for Chinese painting data. The overall architecture is depicted in Fig. 2. The network comprises four hierarchical stages, each prefaced by either an embedding layer (a regular Conv 4 × 4, stride 4) or a merging layer (a regular Conv 2 × 2, stride 2), employed for spatial down-sampling and channel number expansion. Each stage consists of a stack of FCTNet blocks. More blocks are placed in the last two stages due to their lower memory usage and higher FLOPS.

To cater to different sizes of Chinese painting datasets under varying computational budgets, we provide small and base variants of FCTNet, termed FCTNet-s and FasterNet respectively. While their architectures are similar, their depth and number of heads differ. For FCTNet-s, the block numbers in four stages are (2, 2, 6, 2), while for FCTNet they are (2, 2, 18, 2). Correspondingly, the head numbers are (3, 6, 12, 24) as the cost of heads increases with resolution. Our attention receptive field is 13 × 13 ($N = 7, e = 3$).

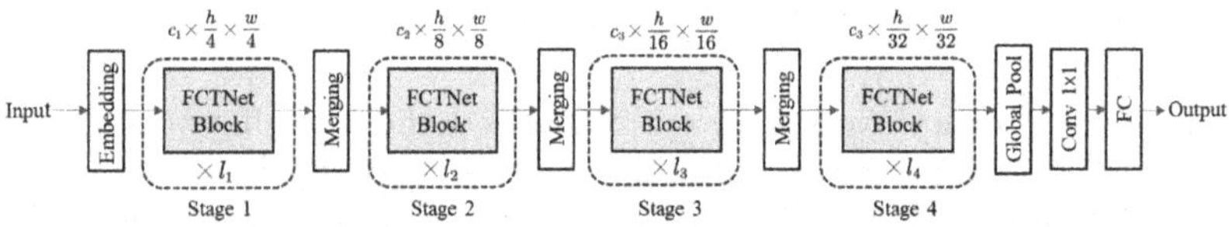

Fig. 2. The overall architecture of our FCTNet consists of four hierarchical stages, each preceded by an embedding or merging layer. Each stage comprises a stack of FCTNet blocks, with the last three layers dedicated to feature classification

5 Experimental Results

5.1 Datasets

We carried out painting classification experiments on our custom-built Chinese and bird painting datasets, supported by the School of Fine Arts at Hubei Normal University and sourced from the Chinese Treasure Museum. The datasets were split into training, validation, and testing sets in an 8:1:1 ratio. While constructing these datasets, we accounted for two key factors: firstly, to represent various eras, we gathered paintings from different dynasties, spanning from the Five Dynasties period to the modern era, encompassing a total of 11 dynasties. Secondly, we included classic paintings from each dynasty, collected by painters such as those behind "Along the River During the Qingming Festival" from the Song Dynasty, Xu Beihong's "Horse" from the modern era, and Qi Baishi's "Shrimp". Consequently, our Chinese painting dataset comprises 4,747 works, covering diverse themes like flower and bird paintings, figure paintings, landscape paintings, animal paintings, and various techniques like meticulous and freehand brushwork. Our bird painting dataset contains 934 images from six dynasties, with the number of paintings from each dynasty displayed in Fig. 3. These datasets provide a rich and varied sample set for our experiments, allowing us to accurately assess the model's performance in the Chinese painting classification task.

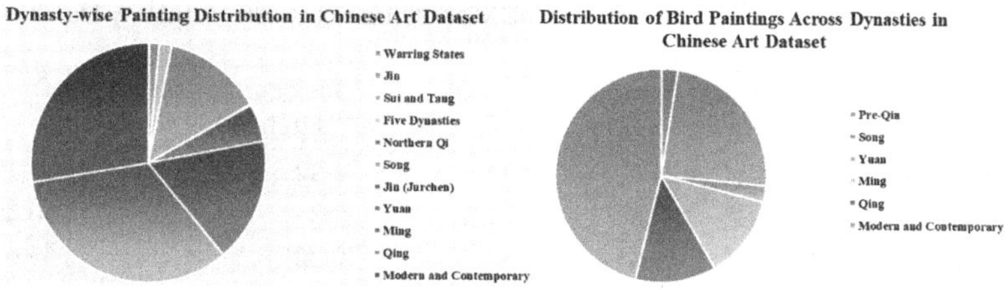

Fig. 3. The composition and number of categories in Chinese painting and bird painting datasets are divided based on dynasties

5.2 Implementation Details

To implement training, we used the same hyperparameters for all architectures and model sizes for simplicity, although fine-tuning could potentially improve our performance. All models were trained for 300 epochs on high-resolution images from the Chinese painting training dataset and 150 epochs on the bird painting training dataset. For data augmentation, we employed minimal augmentation techniques (random cropping and horizontal flipping) along with RandomErasing, without using grayscale or other color-changing data augmentations. Our experiments demonstrated that color transformations made it difficult for the model to converge and reduced accuracy for painting classification tasks. We utilized the AdamW optimizer for training with a learning rate of 4e−4. The learning rate schedule included a 20-epoch linear warm-up followed by cosine decay. The batch size was set to 24, and the weight decay was 0.05. Layer Scale was applied with an initial value of 1e−6.

5.3 Results

Chinese Painting Dataset. Table 1 provides a comprehensive comparison of various models' performance in terms of Top-1 accuracy and computational efficiency on the Chinese painting dataset. FCTNet stands out with a Top-1 accuracy of 68.9%, making it the best-performing model. Compared to other models such as ViT-B/16 and swan-s, FCTNet demonstrates superior accuracy while maintaining relatively lower parameter count (24.8M) and FLOPs (4.53G). It is worth noting that FCTNet outperforms EfficientNetv2-M, which has higher parameter count (55M) and FLOPs (5.4G), further emphasizing FCTNet's ability to achieve a good balance between model complexity and performance. Additionally, FCTNet exhibits high throughput on GPUs with a value of 2.04 (Batch/T), indicating its efficient processing capability Table 1.

Bird Painting Dataset. Table 2 showcases the performance of another set of models on the bird painting dataset, where FCTNet-S emerges as the best-performing model with a Top-1 accuracy of 85.7 % and an f1 score of 0.864. Compared to models like ResNet-34 and MobileViT-XS, FCTNet-S not only achieves

Table 1. Performance Comparison of Models on Chinese Painting Dataset

Model	Image Size	Params(M)	FLOPs (G)	Throughput on GPU (Batch/T)	Top-1 Acc
PVTv2-B2 [10]	224^2	25.4	3.90	2.22	39.1%
ViT-B/16 [11]	224^2	86.6	17.6	2.3	50.5%
Swin-S [12]	224^2	49.6	8.77	2.22	66.6%
MobileViT-S [13]	224^2	5.6	1.65	1.92	63.1%
EdgeNeXt-S [14]	224^2	5.6	1.26	1.91	62.9%
ConvNeXt-T [15]	224^2	28.6	4.47	1.99	65.7%
EfficientNetv2-M [16]	224^2	55	5.4	2.34	68.6%
Mobilenetv2 × 1.4 [17]	224^2	6.1	0.58	1.94	68.8%
FasterNet-T2 [18]	224^2	15.0	1.90	2.04	67.9%
FCTNet	224^2	24.8	4.53	2.04	**68.9%**

higher accuracy and f1 score but also maintains fewer parameters (15.97M) and lower FLOPs (2.36G). This result highlights the effectiveness of FCTNet-S in balancing performance and computational efficiency. Furthermore, FCTNet-S outperforms EfficientNetv2-S and Mobilenetv2, further showcasing its advantages in the Chinese painting image classification task. FCTNet-S exhibits a throughput of 1.97 (Batch/T) on GPUs, demonstrating its high processing efficiency.

Table 2. Performance Comparison of Models on Bird Painting Dataset

Model	Image Size	Params(M)	FLOPs (G)	Throughput on GPU (Batch/T)	Top-1 Acc	F1-score
PVTv2-B1	224^2	13.1	2.1	1.86	79.8%	0.799
ViT-B/16	224^2	86.6	17.6	2.3	77.6%	0.783
Swin-T	224^2	28.3	4.51	2.04	82.7%	0.831
ResNet-34	224^2	21.78	3.68	1.9	85.7%	0.859
MobileViT-XS	224^2	2.3	1.03	1.93	85.1%	0.861
EdgeNeXt-XS	224^2	2.3	0.54	1.83	80.6%	0.810
ConvNeXt-T	224^2	28.6	4.47	1.99	62.2%	0.607
EfficientNetv2-S	224^2	24	2.9	2.19	81.6%	0.821
Mobilenetv2	224^2	3.5	0.3	1.90	85.7%	0.862
FasterNet-T1	224^2	7.06	0.85	1.93	78.6%	0.796
FCTNet-S	224^2	15.97	2.36	1.97	**85.7%**	**0.864**

In conclusion, experimental results demonstrate outstanding performance of FCTNet and FCTNet-S in terms of Top-1 accuracy and f1 score while maintaining reasonable computational complexity. This exceptional performance validates the effectiveness of our proposed models in the Chinese painting image classification task and highlights their ability to strike a balance between performance and efficiency.

5.4 Ablation Experiment

Effect of Data Augmentation on Performance. To investigate the impact of data augmentation on performance, we conducted experiments using two different network architectures, ResNet-34 and Swin-T, employing a variety of data augmentation techniques. The augmentation techniques utilized in this study included baseline operations such as random scaling, cropping, and horizontal flipping, as well as additional techniques such as random erasing, color jitter, grayscale, and RandAugment. ACC and loss curves were plotted for each augmentation technique. As shown in Fig. 4 The experimental results revealed significant effects of color transformations, such as color jitter and grayscale, on the classification task for Chinese paintings. Interestingly, these color-based augmentations led to a decrease in model accuracy, and in some cases, RandAugment and grayscale even hindered model convergence. On the other hand, the application of random erasing proved to be beneficial, resulting in reduced accuracy and loss.

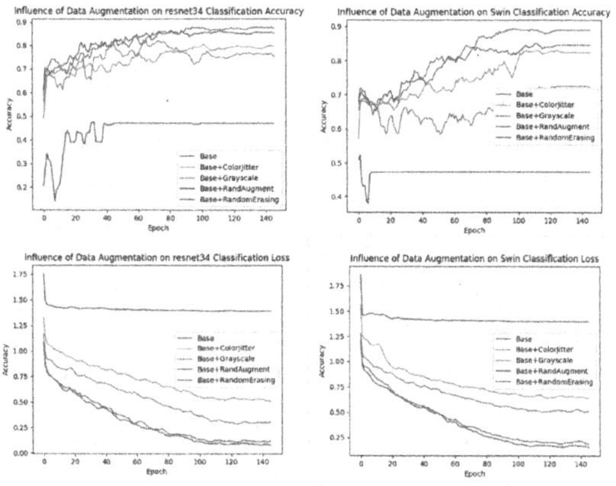

Fig. 4. Impact of Data Augmentation Techniques on Performance of ResNet-34 and Swin-T Models for Chinese Painting Classification

Effect of Channel Reordering on Performance. Table 3 data reveals the impact of varying channel reordering methods and block designs on performance. We set the baseline model's (without channel reordering) performance as our reference, then compared two residual connection methods. The dual residual connection method yielded a 62.6% accuracy on the baseline model, while the single residual connection method attained a 58.9% accuracy. We further experimented with several random permutation methods to gauge their performance impacts. With dual residual connections, randomly permuting all channels (Random 1) decreased accuracy by 5.4 points to 57.2%. Channel rolling (Random 2) increased accuracy by 2.2 points to 64.8%. The best performance came from channel group reversal and swapping (Random 3), improving accuracy by 6.3 points to 68.9%.

Finally, rolling a small segment of channels (Random 4) enhanced accuracy by 4.2 points to 66.8%. In conclusion, the experimental results from Table 3 indicate that channel reordering methods and block designs significantly influence performance. Notably, under the dual residual connection scenario, channel group reversal and swapping (Random 3) demonstrated the highest performance improvement.

Table 3. Performance Impact of Different Channel Reordering Methods on Chinese Painting Classification

Method	Dual Residual Connections	Baseline Δ	Singular Residual Connection	Baseline Δ
Baseline	62.6	0.0	58.9	0.0
Random 1	57.2	-5.4	60.5	1.6
Random 2	64.8	2.2	63.4	4.5
Random 3	68.9	**6.3**	64.8	**5.9**
Random 4	66.8	4.2	64.3	5.4

Effect of Block Designs on Performance. Our ablative experiments systematically evaluated the influence of Block Designs on performance. As Table 4 illustrates, specific ratios like r = 4 enhanced accuracy and throughput, whereas for r = 1, local attention reduced to standard local attention. Increasing r improved accuracy, implying that training with more data is beneficial. However, excessive r values lessened the effectiveness of local attention in capturing spatial features. Regarding normalization layers, despite similar parameters and FLOPs, Layer Normalization (LN) didn't converge during training. When comparing ReLU and GELU activation functions, no significant accuracy differences emerged, though ReLU performed slightly lower. We also tested micro-designs by comparing positional encoding and masking and removing padding masks. Including positional encoding and masking didn't significantly affect accuracy, but without padding mask, accuracy notably decreased. Lastly, the Multi-Layer Perceptron's (MLP) role in the local attention module was examined. Removing MLP considerably decreased both the number of parameters and FLOPs, as well as accuracy. By increasing the MLP ratio, we maintained stable accuracy while augmenting parameters and FLOPs, indicating a balance can be struck between computational complexity and performance.

Table 4. Performance Impact of Block Designs on Chinese Painting Classification

Ablation	Variant	Param.(M)	FLOPs (G)	Throughput on GPU(Batch/T)	Acc
Partial ratio	$c/r = 1$	36.6	9.16	2.17	61.8
	$c/r = 1/2$	28.8	6.07	2.12	65.2
	$c/r = 1/4$	24.8	4.53	2.11	68.9
	$c/r = 1/8$	22.9	3.76	2.06	67.2
Normalization	BN	24.8	4.53	2.11	68.9
	LN	24.8	4.51	2.12	failed
Activation	ReLU	24.8	4.53	2.10	68.1
	GELU	24.8	4.53	2.11	68.9
Micro Design	no pos+mask	24.8	4.53	2.13	69.1
	pos+mask	24.8	4.53	2.11	68.9
	no padding mask	24.8	4.53	2.10	61.8
	no mlp	9.1	1.74	2.02	56.7
	mlp ratio = 2	24.8	4.53	2.11	68.9
	mlp ratio = 4	40.6	7.32	2.19	67.7

6 Conclusion

This study significantly advances Chinese painting classification through the creation of comprehensive datasets, a refined feature extraction method, and the design of an innovative transformer architecture. We have assembled two extensive Chinese painting datasets, offering invaluable resources for research and improving understanding of Chinese art styles. Our feature extraction technique mitigates feature map redundancy, thereby enhancing model efficiency and precision-a crucial advancement in this field. Our novel transformer architecture effectively captures intricate painting details, increasing classification accuracy even with smaller datasets. This highlights the potential of Transformer models in Chinese painting classification. Through rigorous experiments, we've confirmed our methods' effectiveness and our channel reordering and grouping's low redundancy computation. Looking forward, we plan to enlarge our datasets and refine our methodologies, exploring more applications of our approach in art appreciation and recommendation to contribute further to understanding and appreciating Chinese paintings.

Acknowledgments. This work is supported by the Natural Science Foundation of China under Grant 61976085 and 62273136. The post-graduate innovation research project of Hubei Normal University 20220551 and 2023Z102.

References

1. Jiang, W., et al.: MtffNet: a multi-task feature fusion framework for Chinese painting classification. Cogn. Comput. **13**, 1287–1296 (2021)
2. Xue, A.: End-to-end Chinese landscape painting creation using generative adversarial networks. In: Proceedings of the IEEE/CVF Winter Conference on Applications of Computer Vision, pp. 3863–3871 (2021)
3. Li, Y., Adamczewski, K., Li, W., Gu, S., Timofte, R., Van Gool, L.: Revisiting random channel pruning for neural network compression. In: Proceedings of the IEEE/CVF Conference on Computer Vision and Pattern Recognition, pp. 191–201 (2022)
4. Liu, J., et al.: Discrimination-aware network pruning for deep model compression. IEEE Trans. Pattern Anal. Mach. Intell. **44**(8), 4035–4051 (2021)
5. Li, B., Ren, H., Jiang, X., Miao, F., Feng, F., Jin, L.: SCEP-a new image dimensional emotion recognition model based on spatial and channel-wise attention mechanisms. IEEE Access **9**, 25278–25290 (2021)
6. Yoon, D., Park, J., Cho, D.: Lightweight alpha matting network using distillation-based channel pruning. In: Proceedings of the Asian Conference on Computer Vision, pp. 1368–1384 (2022)
7. Wu, Z., Liu, Z., Lin, J., Lin, Y., Han, S.: Efficient transformer for mobile applicatoins. In: International Conference on Learning Representitive (ICLR) (2020)
8. Shi, W., Xu, J., Gao, P.: Ssformer: a lightweight transformer for semantic segmentation. In: IEEE 24th International Workshop on Multimedia Signal Processing (MMSP), pp. 1–5. IEEE (2022)
9. Vaswani, A., Ramachandran, P., Srinivas, A., Parmar, N., Hechtman, B.A., Shlens, J.: Scaling local self-attention for parameter efficient visual backbones. In: 2021 IEEE/CVF Conference on Computer Vision and Pattern Recognition (CVPR), pp. 12889–12899 (2021). www.api.semanticscholar.org/CorpusID:232320340
10. Wang, W., et al.: PVT v2: improved baselines with pyramid vision transformer. Comput. Vis. Media **8**(3), 415–424 (2022)
11. Dosovitskiy, A., et al.: An image is worth 16 ×16 words: transformers for image recognition at scale. arXiv preprint arXiv:2010.11929 (2020)
12. Liu, Z., et al.: Swin transformer: hierarchical vision transformer using shifted windows. In: 2021 IEEE/CVF International Conference on Computer Vision (ICCV), pp. 9992–10002 (2021). www.api.semanticscholar.org/CorpusID:232352874
13. Mehta, S., Rastegari, M.: MobileViT: light-weight, general-purpose, and mobile-friendly vision transformer. arXiv preprint arXiv:2110.02178, (2021)
14. Maaz, M., et al.: EdgeNeXt: efficiently amalgamated CNN-transformer architecture for mobile vision applications. In: Karlinsky, L., Michaeli, T., Nishino, K. (eds.) Computer Vision – ECCV 2022 Workshops. ECCV 2022. Lecture Notes in Computer Science, vol. 13807, pp. 3–20. Springer, Cham (2022). https://doi.org/10.1007/978-3-031-25082-8_1
15. Liu, Z., Mao, H., Wu, C.-Y., Feichtenhofer, C., Darrell, T., Xie, S.: A convnet for the 2020s. In: Proceedings of the IEEE/CVF Conference on Computer Vision and Pattern Recognition, pp. 11976–11986 (2022)
16. Tan, M., Le, Q.: EfficientNetV2: smaller models and faster training. In: International Conference on Machine Learning, pp. 10096–10106. PMLR (2021)

17. Sandler, M., Howard, A., Zhu, M., Zhmoginov, A., Chen, L.-C.: MobileNetV2: inverted residuals and linear bottlenecks. In: Proceedings of the IEEE Conference on Computer Vision and Pattern Recognition, pp. 4510–4520 (2018)
18. Chen, J., et al.: Run, don't walk: chasing higher flops for faster neural networks. In: Proceedings of the IEEE/CVF Conference on Computer Vision and Pattern Recognition, pp. 12021–12031 (2023)

Topological Dynamics of Functional Neural Network Graphs During Reinforcement Learning

Matthew Muller[1(✉)], Steve Kroon[2], and Stephan Chalup[1]

[1] The University of Newcastle, Callaghan 2308, Australia
matthew.muller@uon.edu.au, stephan.chalup@newcastle.edu.au
[2] Computer Science Division, Stellenbosch University, Stellenbosch, South Africa
kroon@sun.ac.za

Abstract. This study investigates the topological structures of neural network activation graphs, with a focus on detecting higher-order Betti numbers during reinforcement learning. The paper presents visualisations of the neurotopological dynamics of reinforcement learning agents both during and after training, which are useful for different dynamics analyses which we explore in this work. Two applications are considered: frame-by-frame analysis of agent neurotopology and tracking per-neuron presence in cavity boundaries over training steps. The experimental analysis suggests that higher-order Betti numbers found in a neural network's functional graph can be associated with learning more complex behaviours.

1 Introduction

In recent years, there has been growing interest in understanding the functioning of neural networks [18]. One way to study this, which we will use in this paper, is by analysing the topology of the manifolds associated with the dynamical systems of neural activations during control tasks. Betti numbers are topological invariants that provide information about the number of connected components, holes, and higher-dimensional voids of a topological space or manifold [9,17]. Persistent homology offers solutions to determine the Betti numbers, denoted by β_n, in applications where manifolds are approximated by point clouds. This is done by converting the set of sample points into a complex, computing topological features, and tracking their evolution over increasing scales. Persistent homology captures the birth, life, and death of topological features in a dataset, and the lifespan of these topological features indicates the significance of the structure [2]. This work uses the flag complex and the Vietoris-Rips complex [14,29].

A cocycle is a mathematical concept used to express obstructions, such as the inability to integrate a differential equation on a closed manifold. In algebraic

This research was partially supported by the Australian Government through the ARC's Discovery Projects funding scheme (project DP210103304). The first author was supported by a PhD scholarship from the University of Newcastle.

B. Luo et al. (Eds.): ICONIP 2023, CCIS 1963, pp. 190–204, 2024.
https://doi.org/10.1007/978-981-99-8138-0_16

topology, a cocycle is a closed cochain, meaning it vanishes on boundary chains. This provides a means to determine which discrete elements form a boundary, permitting analysis of these elements.

Reinforcement Learning (RL) considers how agents should act in an environment to maximise a reward signal. An RL agent learns by interacting with its environment, receiving feedback in the form of rewards or penalties, and adjusting its policy to improve future performance. For key algorithms, methods, and theoretical underpinnings of RL, see [26].

Existing work has attempted to incorporate neural network topology by calculating β_0 [1] and β_1 [30] numbers using the Vietoris-Rips complex. Such methods are limited by the computational complexity of calculating Betti numbers, which makes it challenging to extend these methods to β_2 and β_3 [15]. This is unfortunate since these higher-dimensional Betti numbers have been shown to be important in the study of biological neural networks, where they have been shown to be characteristic of functional structure [24].

To address this gap in investigative tools, we have developed a novel approach to exploring the topological structures of (feed-forward) neural network activation graphs with support for higher-order Betti numbers, specifically β_2 and β_3, to improve understanding of neural network functionality. Our method can be applied at any point during a model's life-cycle and only requires the neural activations from a single sample. Our approach differs from previous methods with respect to how the network's neural activations are processed, how the complexes are constructed, and how the resulting topology is collected and presented. We illustrate our method on the half-cheetah, humanoid, ant, and reacher tasks from the MuJoCo library [27]. Furthermore, we show that neural networks gradually develop topological structure in a top-to-bottom fashion (i.e. starting near the output layer and progressing towards the input layer. Finally, we investigate the relationship between higher-dimensional activity and learning. Our finding that higher-order Betti numbers regularly develop is of particular significance, because it suggests that the neural network's activation graph naturally develops a higher-dimensional learning structure for sufficiently complex tasks, similar to what is observed in biological neural networks [24]. Our approach is able to compute higher-order Betti numbers by constructing an overall sparser graph with Algorithm 1, reducing the required computation.

The proposed approach aims to contribute to the field of explainable AI, and specifically neural network interpretability, through the detection of higher-order Betti numbers and the exploration of neurotopological dynamics during reinforcement learning tasks. By analysing the topology of neural networks while they are learning, rather than only after they have been trained, our method permits a more dynamic understanding of how the network processes information and incorporates knowledge. These findings can be especially beneficial in applications where interpretability is crucial, such as in safety-critical systems, or investigating adversarial attacks [8].

2 Related Work

Approaches to incorporating topology into neural network research generally fall into one of two categories: determining the topological properties of data external to the model (e.g. [5,19,20]), or calculating the topology of the model itself (e.g. [6–8,15]).

Recent exploration into the potential value of topological features in neural networks has yielded promising results. For example, research has shown that topological features can identify adversarial samples, which are inputs specifically crafted to fool a neural network [8,12]. Additionally, topological features have been used to predict testing error, which can be used to improve the generalisation of neural networks [7]. Topological features have also been used to determine appropriate timing for early stopping to prevent overfitting [6]. Topological augmentation of machine learning has demonstrable benefits and untapped potential, but it faces a similar black-box problem to neural networks, where the dynamics are not well understood.

The construction of flag complexes from neural network weights has seen recent activity in [30], where a flag complex was constructed on a static neural network using the transitive closure of the subgraph of the neural network with positive weights, e.g. $v_1 \rightarrow v_2 \rightarrow v_3$ would generate a connection between vertices v_1 and v_3. They found that the resulting β_1 of these flag complexes was correlated with excessive neurons for a given task and task difficulty. Our approach differs in that our graph construction algorithm inserts intra-layer transient connections based on a similarity metric, and we consider connections regardless of sign due to our focus on activation strength. The sparser resulting graph in our approach also allows us to compute the higher-order Betti numbers β_2 and β_3.

Despite the lack of association between topological features and specific neural network function within current literature, topology holds significant promise for reliable model interpretation due to its ability to effectively and robustly process high-dimensional and noisy data. The use of topology in neural network interpretation, particularly through the detection of higher-order Betti numbers, may provide insight into decision-making processes and assist with identifying contributing factors for failure cases.

3 Approach

This work proposes a graph construction method facilitating topological analysis of neural networks, and presents novel visualisations of neurotopological dynamics of reinforcement learning agents during and after training. While our findings inspire various hypotheses on the dynamics of the presented neurotopology, a more comprehensive evaluation is needed to draw sound empirical conclusions. The experiments are intended to provide preliminary evidence of relationships between the derived neurotopology and the learning performance of the agent.

Functional graphs derived from correlation coefficients based on neural network activations [8] have been employed to extract basic topological features, such as β_0 and β_1. However, the computational limitations of persistent homology become more pronounced when one attempts to scale these analyses to higher dimensions, and the high ratio of edges to nodes in complexes derived from correlation coefficients exacerbates this poor scaling [10, 25, 28].

While topological analysis has strategies for subsampling, such as greedy permutation [4], directly applying such methods to neural networks is a coarse approach since they function by removing a neuron (along with all its connections) at a time. Neural network pruning [16] differs in that the pruning targets are edges rather than neurons. Our proposed method resembles an edge pruning technique, where the result is a binary adjacency matrix. We do this by filtering out edges with weighted activations below an adjustable activation threshold t. After this filtering process, if two neurons (m_1, m_2) within the same layer have incoming weighted activations (p_{n,m_1}, p_{n,m_2}) from the same source neuron n, a connection is added between those neurons if the difference $|p_{n,m_1} - p_{n,m_2}|$ is less than the difference threshold d. This approach is motivated by a characteristic of correlation coefficient activity graphs [6], where connections are added between highly correlated nodes in the same layer.

Research on the formation of cliques and cavities in simulated neural tissue reveals that correlated activity is preferentially concentrated in directed simplices, and that neurons become bound into cliques and cavities due to correlated activity in response to stimuli [24]. These observations lend support to our approach of considering both neural activations and virtual connections, which assist in identifying important topological features of the neural network.

In light of these insights, we propose Algorithm 1 for deriving a functional graph from multi-layer perceptron activations, enabling fast topological analysis of actively learning neural networks. This algorithm reduces topological computation time by generating sparser graphs than previous approaches, while also integrating aspects inspired by biological neural networks. By incorporating these elements, our method offers a promising avenue for deriving the topology of learned neural network knowledge.

3.1 Functional Graph Extraction

Our proposed method for functional graph extraction, as defined in Algorithm 1, can be employed at any stage of a model's life-cycle. By adjusting filter thresholds, users can balance analysis speed against topological accuracy. In this section, we elaborate on the function and motivation behind Algorithm 1.

Our method uses both the positive and the negative activations from the forward pass of the neural network. While determining the differences between activations for comparison with the difference threshold d, we preserve the original signs of the activations. However, we verify only whether the activation magnitudes exceed the activation threshold t. In the scenarios discussed in this paper, we set an activation threshold value of $t = 0.6$ and a minimum difference threshold value of $d = 0.2$.

A significant advantage of our method is its ability to analyse neural activations from a single sample, enabling the assessment of the neurotopologi-

---

Algorithm 1. Functional Graph Construction from Neural Network

---

Input: MLP consisting of layers $L_1, \ldots, L_K$, with weight of edge from n to m indicated by $w_{n,m}$; activation threshold t; minimum distance threshold d; input x considered as layer L_0.

Output: functional activation graph corresponding to forward pass of x through the MLP

1: Initialize M as empty set
2: **for** k in range(K) **do**
3:    **for** each node n in L_k **do**
4:       $a_n \leftarrow$ activation of n (available from forward pass)
5:       **for** (n, m) in outgoing_edges(n) **do**
6:          $p_{n,m} \leftarrow a_n w_{n,m}$
7:          **if** $|p_{n,m}| \geq t$ **then**
8:             M.add (n, m)
9:          **end if**
10:       **end for**
11:       **for** pairs $(n, m_1), (n, m_2)$ of edges **in** outgoing_edges(n) **do**
12:          **if** (n, m_1) **exists in** M **and** (n, m_2) **exists in** M **then**
13:             **if** $|a_n(w_{n,m_1} - w_{n,m_2})| < d$ **then**
14:                M.add(m_1, m_2)
15:             **end if**
16:          **end if**
17:       **end for**
18:    **end for**
19: **end for**
20: return M

---

cal impacts of individual samples. This is particularly advantageous in cases involving limited data, adversarial attacks, or the evaluation of unlabelled data. Although our graph extraction approach could be applied somewhat more generally, we focus on applying our method to reinforcement learning tasks with ReLU multi-layer perceptron (MLP) networks.

3.2   Methodology

We developed two scenarios for investigating the neurotopological behavior of reinforcement learning agents in order to assess the utility and verify the functionality of our functional graph extraction method. These scenarios aim to illustrate how our method can be applied during or after training, and to substantiate that the derived neurotopological structures are representative of a learned internal structure and not spurious. Of particular note is the emergence of higher-order Betti numbers in artificial neural networks (ANNs), for which the neurotopological dynamics have not been previously explored.

Scenario 1: Frame-by-frame Analysis of Agent Neurotopology Traditionally, when calculating the topology of data in the context of neural networks, the calculations are performed on discrete samples that are otherwise disconnected. An underexplored avenue of research involves sequential data, such as the steps comprising an episode of an agent interacting with an environment. Calculating neurotopology for sequential data lends itself to synchronising the neurotopological data with footage of the agent and allowing human observers to associate neurotopological activity with agent activity. This provides insight into the interplay between neurotopological dynamics and environmental states.

In this case, we aim to calculate the neurotopology of an RL agent for every step of the episode. We take a pre-trained RL agent, trained for five million steps, and have it complete an episode during which the neural activations are stored after each step. For each step, the neurotopology is computed by constructing the functional graph from the stored neural activations using Algorithm 1, and subsequently processing the resulting adjacency matrix with Flagser and Ripser.

The goal of this scenario is to demonstrate that ANNs exhibit neurotopological dynamics that can be paired with agent activity to provide insight into the neurotopological response to environmental stimuli.

Scenario 2: Tracking Betti Numbers and Per-Neuron Cavity Contribution During Training This scenario aims to demonstrate the following key aspects:

1 The cocycle contribution of individual neurons can be tracked over time to determine the order in which neurons become functional.
2 The Betti numbers calculated from training samples throughout training serve as a consistent indicator of learning progress and quality.
3 The derived cocycle graphs can be employed to determine the suitability of a network architecture for a given task.

A common challenge in neural network training is architecture selection, to ensure the model will learn effectively. Since a common method to choose architecture involves expensive architectural search [11], methods to improve performance estimation of this search are critical. For architecture selection, typically only the model loss over time will be used to determine architectural aptitude for a given task; however, this metric does not necessarily give a complete picture [13]. We show that our method can visualise the distribution of functionally active neurons in the network by attributing those neurons to cavities found in the functional neurotopology. This provides a means to approximate the aptitude of a model to a given task by inspecting whether cavities are concentrated towards the output or well-distributed throughout the model.

To track cavity distribution, we require a way to calculate cavities and associate them with a neuron. To this end, Ripser provides a means to calculate and return the cocycles of a Vietoris-Rips complex. In our case, a cocycle consists of the neurons that form the boundary of a cavity in the Vietoris-Rips complex, with one cocycle existing for each cavity found in the complex. For these visualisations, we restrict ourselves to the dimension one cavities of the graph to keep computational requirements tractable. There are often instances where a single neuron contributes to multiple cocycles; in such cases we consider the longest cocycle only. Therefore, for each neuron in the functional graph for a forward pass, we find the maximum length of all cocycles containing that neuron. This results in a dictionary where each key corresponds to a unique neuron, and the value represents the longest cocycle it belongs to.

Let $H = (H_{sn})$ be a matrix representing the history of maximum neuron cocycles, with $s \in S$ denoting a training step index, and $n \in N$ the index of a neuron from the neural network. We obtain this by defining C as the set of cocycles of neurons, where C_s denotes all the cocycles at step s, and $C_{s,n}$ the set of cocycles at step s containing neuron n. Then, for each step $s \in S$, we compute H_{sn} as the maximum length of the cocycles in $C_{s,n}$:

$$\forall(s \in S, n \in N) : H_{s,n} = \max\{|c| : c \in C_{s,n}\} \ . \tag{1}$$

In Eq. 1, the number of cocycles in $C_{s,n}$ for a given step s can be arbitrarily large for improved quality; however, this computation is not negligible and can be adequately approximated by calculating the cocycle for the first sample at each gradient step. Training an agent involves two alternating phases. The first phase is an action phase, where the agent makes actions based on the environment state following a fixed policy and state transitions are collected and stored in a buffer. The second phase is a training phase where experience is randomly sampled from the buffer and used to update the policy.

In our approach, we treat the matrix H as an image, with each element of the matrix representing a pixel. This yields Figs. 1a & 1b. In these figures, the darkness of a pixel indicates the size of the largest cavity a neuron forms part of: darker pixels are parts of larger cavities. The neurons appear in the image in canonical order, from bottom-to-top.

For the calculation of higher-dimensional Betti numbers over the course of training, we used Flagser. While Ripser could be employed for this purpose in principle, Flagser was significantly faster and more memory-efficient due to the stricter criteria for the flag complex clique cells, resulting in fewer cells overall. Consequently, all graphs showing β_n histories are calculated using Flagser. The β_n history graphs (Figs. 3a–3d & 4a –4d) are calculated using the same functional graphs used for calculating the cocycles.

3.3 Experimental Design and Data

Our topological analysis uses Flagser [21] and Ripser [28]. While Flagser is faster, it yields only Betti numbers, while Ripser returns the cavities' cocycles, enabling us to track associated neurons. We include cocycle information to determine the general order of functional connection development in neural networks.

For experimental reproducibility, we use the Stable-Baselines3 library [23], a collection of reinforcement learning algorithms implemented in PyTorch [22]. This library is highly configurable; Table 1 lists the configurations we used. We selected the Soft Actor-Critic (SAC) model and acquired activation weights for

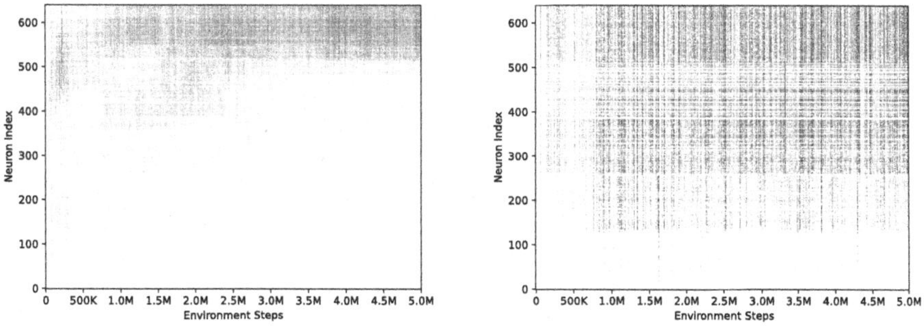

(a) 32×20 agent (b) 128×5 agent

Fig. 1. β_1 cocycle histories for the half-cheetah task, where each agent was trained for five million steps. The 32×20 agent (left) was unable to solve the environment and subsequently consolidated the functional graph towards the output layer. The 128×5 agent (right) solved the environment and, as a result, utilises a broader span of neurons in the functional graph

Table 1. Stable-Baselines3 Configuration

Parameter	Value
Model	SAC
Policy	MlpPolicy
Gamma	0.99
Buffer Size	1000000
Action Noise	normal (0, 0.1)
Learning Starts	10000
Batch Size	64
Learning Rate	0.001
Train Freq	1000
Gradient Steps	100
Net Arch	32×20, 128×5 & 32×5
Activation	ReLU

our method exclusively from the actor to analyse the model's interactions with the environment.

All the models in our demonstrations are MLPs using ReLU activation layers. These models are defined in the form {neurons per layer}×{number of layers}.

We employed MuJoCo [27] environments with OpenAI Gym [3] for training. Our trials included half-cheetah, humanoid, ant, and reacher environments. MuJoCo tasks offer realistic physics simulations and, through OpenAI Gym, provide a standardised interface for agent control and environment observation.

We chose these environments to demonstrate the variety of neurotopological structures emerging from task complexity.

We trained the agent using the soft actor-critic (SAC) method. We used 10,000 initial samples for the buffer and then performed 100 gradient steps with 64 random samples each gradient step for every 1,000 environment steps.

The evaluation environment used an AMD Ryzen 9 7950X CPU and 64GB 6200MHz DDR5 RAM running Ubuntu 20.04 under WSL2 in Windows 11. A GPU was not used for training due to the latency of transferring data to the GPU exceeding the processing time of the CPU. @

The following process was followed during the evaluation:

1. Collect neural activation data for each training step
2. Generate the adjacency matrices from the functional graphs then calculate and store all of the Betti numbers and cocycles using Ripser
3. Generate the graphs used in scenario 2
4. Run the trained agent in evaluation mode on a new episode while collecting neural activation data
5. Calculate the Betti numbers using Flagser
6. Generate the dashboards for each frame for scenario 1

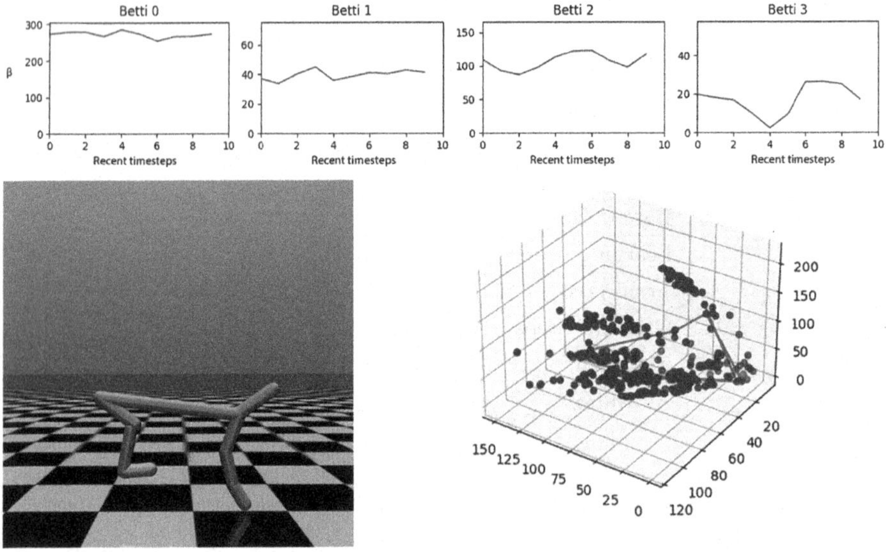

Fig. 2. Sample frame from a neurotopological dashboard of a 128 × 5 reinforcement learning agent trained for five million steps on the Half-Cheetah task. This allows for direct observation of how agent activity correlates with neurotopological changes. The bottom right plot is in a 3-dimensional latent space obtained for visualization using an autoencoder. Sample video available at https://youtu.be/wIRzgeiO6jA

4 Evaluation

This section discusses the results produced by the experiments outlined above. We first examine the results generated for the first scenario, which focused on how the Betti numbers change in relation to environmental activity during a single episode. We then discuss the results generated for the second scenario, which involved calculating both the Betti numbers of the flag complex and the maximum cocycle length for each neuron at regular training intervals.

4.1 Synchronised Episode Playback and Neurotopology

This scenario examined neurotopological data obtained from a trained agent as it interacts with the environment it was trained for. The purpose of this scenario is to illustrate how the proposed method can be used to analyse the moment-to-moment neurotopological behavior of a trained agent actively engaging with an environment.

We then extract the functional graph from these activations using Algorithm 1. For the example shown in Fig. 2, we configured Flagser to calculate up to β_3 and repeated this process for every frame in the episode. We then smoothed the Betti numbers using a moving average with a window size of four to reduce signal noise. Finally, we generated a plot of β_n for each dimension, which showed the Betti number for the current and the previous 10 frames. These plots are visible at the top of Fig. 2. Below these plots, we show the corresponding frame showing the agent state and a graph depicting a dimensionality reduction of the observation space over the episode to three dimensions. The blue dots represent environment states and the red line indicates the history of the past 10 frames, making it clear how the agent is transitioning through the environment states. The dimensionality reduction was computed using an autoencoder.

The graphs in the dashboard depicted in Fig. 2—sourced from a video of an agent acting in the half-cheetah environment—depicts synchronised graphs. In the graphs along the top of the video, representing the counts of Betti numbers $\beta_{0..3}$, a correlation is observed between the cheetah's stride and a periodic β_3 curve. While correlations for β_1 and β_2 plots exist, they are less pronounced and consistent. The β_0 counts correspond to the number of disconnected graph components and decrease as other Betti numbers increase. As per the visualisation, the stride of the cheetah appears to align with β_3 counts due to stride orchestration being retained in the activation patterns of this dimensionality. Figure 3a supports this, showing the most substantial growth in β_3 while the agent learns to run. This experiment suggests that the presence of higher-order Betti numbers may be an indicator of an agent learning complex tasks.

4.2 Development of Neurotopology During Training

The data used for this scenario are the cocycle information for the cavities of dimension 1, and the Betti numbers of the flag complex, over the entire training process. We show results for three RL agents, namely the 128×5, 32×20, and

32 × 5 models, all trained for five million steps on each of the MuJoCo environments mentioned earlier. The cocycle graphs for all tested environments are very similar. Hence we only include the cocycles for the half-cheetah environment as a representative example (Fig. 1).

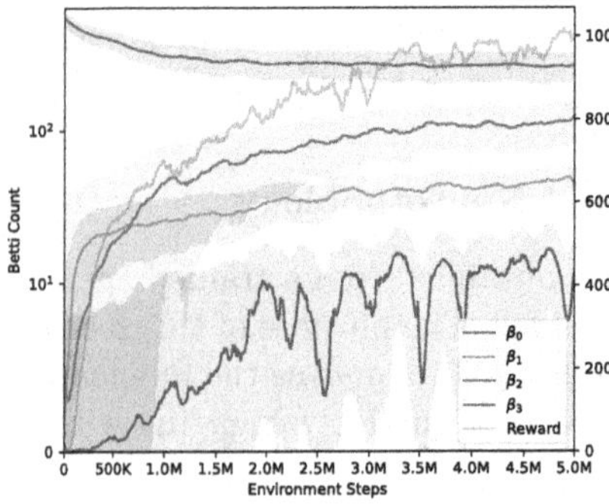

(a) 128 × 5 agent, half-cheetah

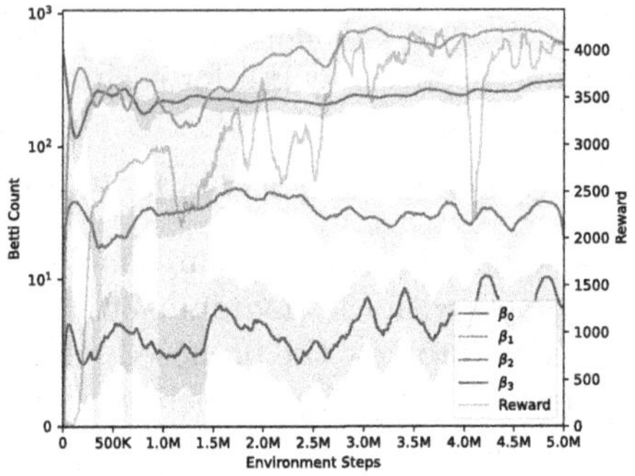

(b) 32 × 20 agent, half-cheetah

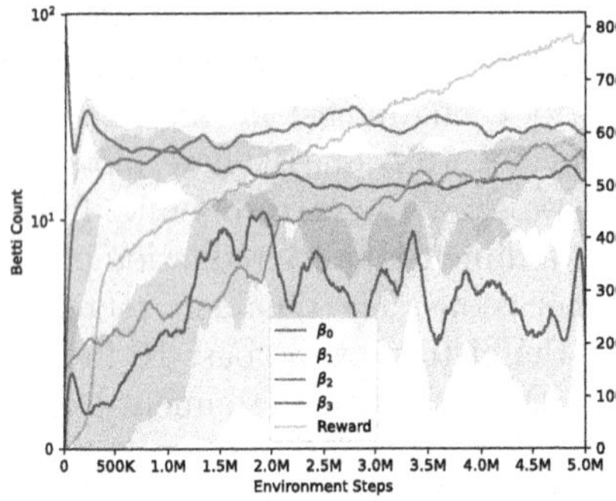

(c) 32 × 5 agent, half-cheetah

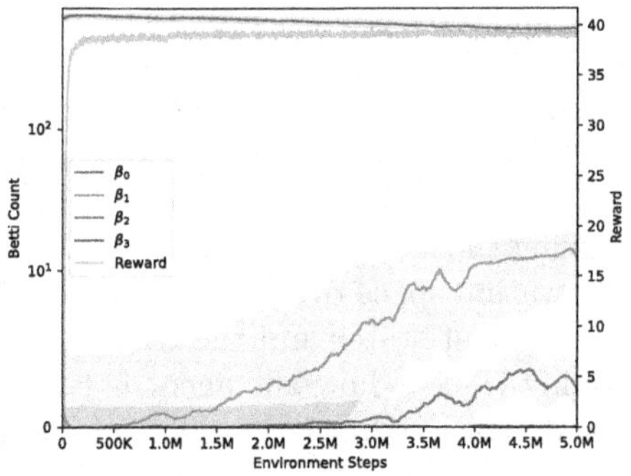

(d) 128 × 5 agent, reacher

Fig. 3. Betti number history graphs of RL agents trained for five million steps. Note how the neurotopology of the agents that solved the half-cheetah task (128 × 5 & 32 × 5) differs from that of the agent that failed the task (32 × 20). We also highlight that the Betti numbers are not superfluous: contrast the half-cheetah agents and the 128 × 5 reacher agent. The reacher environment is easily solved and has correspondingly simple neurotopology

During training for each model, we collect the neural activations and derive the associated adjacency matrices using Algorithm 1. We then take the graph M from Algorithm 1 and produce the corresponding upper-triangular adjacency matrix to calculate topological features using Flagser and cocycle graphs using Ripser. Finally, we calculate the maximum cocycle lengths for each neuron during each training step. The resulting cocycle graphs in Figs. 1a & 1b illustrate how cavities develop in the neural networks over time. Note that while these figures

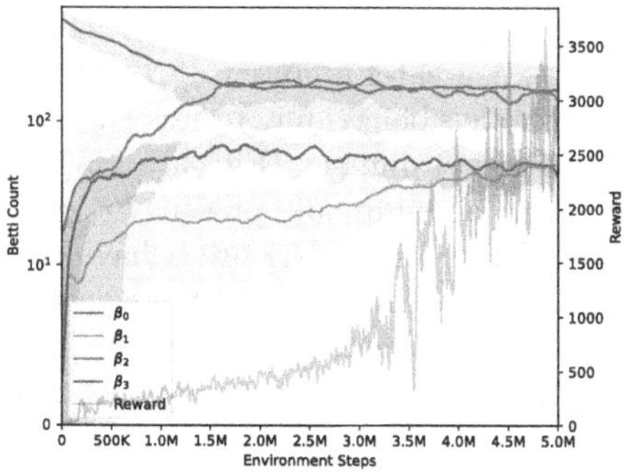

(a) 128 × 5 agent, humanoid

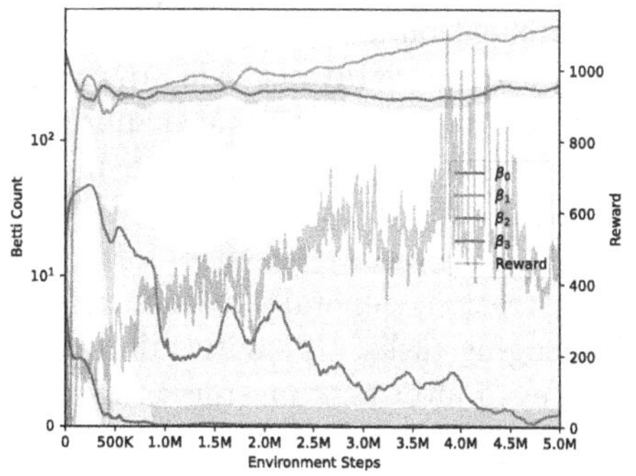

(b) 32 × 20 agent, humanoid

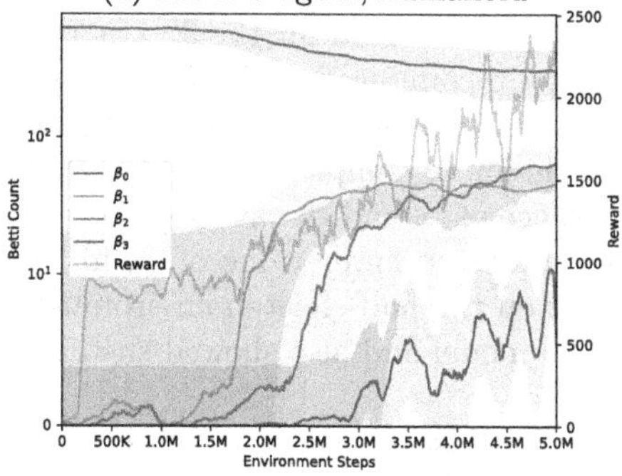

(c) 128 × 5 agent, ant

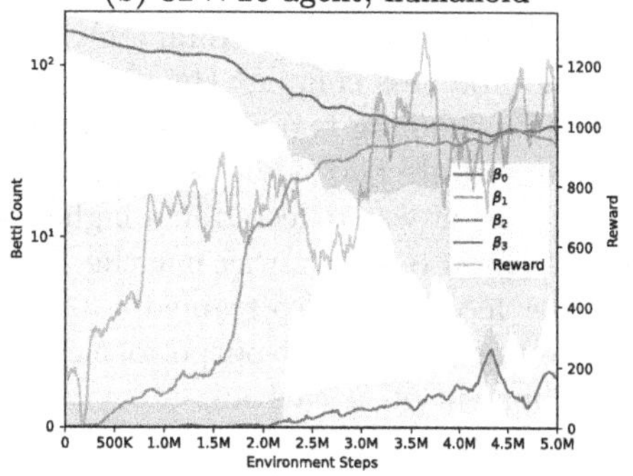

(d) 32 × 5 agent, ant

Fig. 4. Betti number history graphs of RL agents trained for five million steps. For the humanoid task, observe how the Betti number profiles differ between a successful (128×5) and an unsuccessful (32×20) agent. The ant task shows how the Betti number profiles differ when one agent (128 × 5) learns more sophisticated behaviours than the other (32 × 5)

show the same number of neurons, the number of neurons per layer are different, with the 32 × 20 and 128 × 5 models having 32 and 128 neurons per layer, respectively.

When examining the cocycle graphs, we can see broadly that the concentration of cavities is distributed unevenly throughout the layers, with a considerably higher concentration nearer the output layer. We can also observe a gradual overall development of neurotopological cavities. From the cocycle graphs, we make three main observations: neural networks develop from output-to-input, networks with excessive depth can induce topological degeneration, and that a large proportion of the cocycles form early in the training process.

The Betti number history graphs, as shown in Figs. 3a–3d & 4a–4d, provide a more dynamic interpretation compared to the cocycle graphs. These graphs depict substantial variation in all recorded Betti numbers, demonstrating distinctive characteristics pertaining to the difficulty of the environment and agent comprehension. It is evident that β_1 cavities, representing the simplest structural form, will invariably occur in any capable agent. A greater quantity of β_2 cavities typically indicates an agent acquiring more complex behaviour. In less complex environments, such as observed in the Reacher environment (Fig. 3d), cavities corresponding to β_2 and, subsequently, β_3, might be minimally present. β_3 cavities, forming after β_2 cavities form, appear to signify more intricate behavioural orchestration, generally manifesting during the development of strategies for challenging tasks. β_3 cavities do not emerge for the Reacher environment, however, even after task mastery; this is likely due to the relative simplicity of that environment.

We also trained two agents, a 128×5 (Fig. 4b) and a 32×5 (Fig. 4d) agent, on the ant environment to see how the complexity of learned behaviours interacts with the generated Betti numbers. The behaviour observed for the 32×5 agent was notably less complex than that of its 128×5 counterpart: the 32×5 agent relied on shuffling two of its legs, whereas the 128×5 agent developed a rowing action with its two side legs. While similar, the more complex behaviour of the 128×5 agent was reflected in a higher occurrence of β_2 and β_3 cavities. To verify that neuron quantity was not the sole reason for the lower value of β_3 for the smaller model, we also trained a 32×5 agent on the half-cheetah environment, where the 32×5 agent outperformed a 32×20 model and also showed sustained development of β_3 cavities.

An important takeaway from the Betti number history graphs is that complex environments appear to induce the development of higher-order Betti numbers. In cases where the higher-order Betti numbers are lower than for other agents with similar numbers of neurons, such as the case with the 32×20 agent seen in Figs. 3b & 4b, we can expect that the agent's capabilities will be limited.

5 Conclusion

This work proposed a novel approach to extracting a functional activation graph from neural network activations and producing a sparse graph for faster computation of higher-order Betti numbers. We presented scenarios that analysed reinforcement learning agents both during and after training, revealing complex neurotopological dynamics in the learning process. We provide evidence that reinforcement learning agents do not necessarily distribute internal functional structures across layers evenly, and that task complexity appears to influence the development of higher-order Betti numbers in a neural network's functional graph. Our findings also suggest that the trajectory of neurotopological dynamics during training could be indicative of an agent's potential performance. Open questions that could be investigated in future work include identifying alternative graph extraction methods that preserve the topology of the activation graph; determining the relationship between task complexity and Betti numbers; establishing how different activation functions and training optimisations (batch norm, dropout, weight decay, etc.) affect the development of Betti numbers; and determining which persistent homology methods offer the best performance. We encourage research furthering our proposed method—in particular, improved scalability may enable processing of large models and real-time neurotopological analysis.

References

1. Birdal, T., Lou, A., Guibas, L.J., Simsekli, U.: Intrinsic dimension, persistent homology and generalization in neural networks. Adv. Neural. Inf. Process. Syst. **34**, 6776–6789 (2021)
2. Botnan, M.B., Hirsch, C.: On the consistency and asymptotic normality of multiparameter persistent betti numbers. J. Appl. Comput. Topol. 1–38 (2022). https://doi.org/10.1007/s41468-022-00110-9
3. Brockman, G., et al.: Openai gym. CoRR **abs/1606.01540** (2016)
4. Cavanna, N.J., Jahanseir, M., Sheehy, D.R.: A geometric perspective on sparse filtrations. arXiv:1506.03797 (2015)
5. Chen, C., Ni, X., Bai, Q., Wang, Y.: A topological regularizer for classifiers via persistent homology. In: Chaudhuri, K., Sugiyama, M. (eds.) Proceedings of the 22nd International Conference on Artificial Intelligence and Statistics. Proceedings of Machine Learning Research, vol. 89, pp. 2573–2582. PMLR (4 2019)
6. Corneanu, C., Madadi, M., Escalera, S., Martinez, A.: Explainable early stopping for action unit recognition. In: 2020 15th IEEE International Conference on Automatic Face and Gesture Recognition (FG 2020), pp. 693–699 (2020). https://doi.org/10.1109/FG47880.2020.00080
7. Corneanu, C.A., Escalera, S., Martinez, A.M.: Computing the testing error without a testing set. In: 2020 IEEE/CVF Conference on Computer Vision and Pattern Recognition (CVPR), pp. 2674–2682 (2020). https://doi.org/10.1109/CVPR42600.2020.00275
8. Corneanu, C.A., Madadi, M., Escalera, S., Martinez, A.M.: What does it mean to learn in deep networks? and, how does one detect adversarial attacks? In: 2019 IEEE/CVF Conference on Computer Vision and Pattern Recognition (CVPR), pp. 4752–4761 (2019). https://doi.org/10.1109/CVPR.2019.00489
9. Edelsbrunner, H., Harer, J.: Computational Topology: An Introduction. American Mathematical Society, Applied Mathematics (2010)
10. Edelsbrunner, H., Parsa, S.: On the computational complexity of betti numbers: reductions from matrix rank. In: Proceedings of the Twenty-Fifth Annual ACM-SIAM Symposium on Discrete Algorithms, pp. 152–160. SIAM (2014)
11. Elsken, T., Metzen, J.H., Hutter, F.: Neural architecture search: a survey. J. Mach. Learn. Res. **20**(1), 1997–2017 (2019)
12. Gebhart, T., Schrater, P.: Adversary detection in neural networks via persistent homology. arXiv:1711.10056 (2017)
13. Geirhos, R., et al.: Shortcut learning in deep neural networks. Nature Mach. Intell. **2**(11), 665–673 (2020). https://doi.org/10.1038/s42256-020-00257-z
14. Gromov, M.: Hyperbolic groups. In: Gersten, S.M. (ed.) Essays in Group Theory, pp. 75–263. Springer New York, New York, NY (1987). https://doi.org/10.1007/978-1-4613-9586-7_3
15. Gutiérrez-Fandiño, A., Fernández, D.P., Armengol-Estapé, J., Villegas, M.: Persistent homology captures the generalization of neural networks without A validation set. arXiv:2106.00012 (2021)

16. Han, S., Pool, J., Tran, J., Dally, W.: Learning both weights and connections for efficient neural network. In: Advances in Neural Information Processing Systems 28 (2015)
17. Hatcher, A.: Algebraic Topology. Cambridge University Press, Algebraic Topology (2002)
18. Hensel, F., Moor, M., Rieck, B.: A survey of topological machine learning methods. Front. Artif. Intell. **4** (2021). https://doi.org/10.3389/frai.2021.681108
19. Hofer, C., Graf, F., Niethammer, M., Kwitt, R.: Topologically densified distributions. In: III, H.D., Singh, A. (eds.) Proceedings of the 37th International Conference on Machine Learning. Proceedings of Machine Learning Research, vol. 119, pp. 4304–4313. PMLR (7 2020)
20. Hofer, C., Kwitt, R., Niethammer, M., Uhl, A.: Deep learning with topological signatures. In: Proceedings of the 31st International Conference on Neural Information Processing Systems, pp. 1633–1643. NIPS'17, Curran Associates Inc., Red Hook, NY, USA (2017)
21. Lütgehetmann, D., Govc, D., Smith, J.P., Levi, R.: Computing persistent homology of directed flag complexes. Algorithms **13**(1) (2020). https://doi.org/10.3390/a13010019
22. Paszke, A., et al.: Pytorch: An imperative style, high-performance deep learning library. In: Wallach, H.M., Larochelle, H., Beygelzimer, A., d'Alché-Buc, F., Fox, E.B., Garnett, R. (eds.) Advances in Neural Information Processing Systems 32: Annual Conference on Neural Information Processing Systems 2019, pp. 8024–8035 (2019)
23. Raffin, A., Hill, A., Gleave, A., Kanervisto, A., Ernestus, M., Dormann, N.: Stable-baselines3: reliable reinforcement learning implementations. J. Mach. Learn. Res. **22**(268), 1–8 (2021)
24. Reimann, M.W., et al.: Cliques of neurons bound into cavities provide a missing link between structure and function. Frontiers in Computational Neuroscience **11** (2017). https://doi.org/10.3389/fncom.2017.00048
25. Rote, G., Vegter, G.: Computational topology: an introduction. In: Effective Computational Geometry for Curves and Surfaces, pp. 277–312. Springer Heidelberg (2006). https://doi.org/10.1007/978-3-540-33259-6_7
26. Sutton, R.S., Barto, A.G.: Reinforcement Learning: An Introduction. A Bradford Book, Cambridge, MA, USA (2018). https://doi.org/10.5555/3312046
27. Todorov, E., Erez, T., Tassa, Y.: Mujoco: A physics engine for model-based control. In: 2012 IEEE/RSJ International Conference on Intelligent Robots and Systems, pp. 5026–5033. IEEE (2012). https://doi.org/10.1109/IROS.2012.6386109
28. Tralie, C., Saul, N., Bar-On, R.: Ripser.py: A lean persistent homology library for python. J. Open Source Softw. **3**(29), 925 (9 2018). https://doi.org/10.21105/joss.00925
29. Vietoris, L.: Über den höheren zusammenhang kompakter räume und eine klasse von zusammenhangstreuen abbildungen. Math. Ann. **97**(1), 454–472 (1927)
30. Watanabe, S., Yamana, H.: Topological measurement of deep neural networks using persistent homology. Ann. Math. Artif. Intell. **90**(1), 75–92 (2022)

Quantized SGD in Federated Learning: Communication, Optimization and Generalization

Shiyu Liu[1], Linsen Wei[2], Shaogao Lv[3(✉)], and Zenglin Xu[4,5]

[1] University of Electronic Science and Technology of China, Sichuan, China
[2] Northwestern Polytechnical University, Xi'an, China
[3] Nanjing Audit University, Nanjing, China
lvsg716@nau.edu.cn
[4] Harbin Institute of Technology, Shenzhen, China
xuzenglin@hit.edu.cn
[5] Peng Cheng Lab, Shenzhen, China

Abstract. Federated Learning (FL) has gained significant attention due to its impressive scalability properties and its ability to preserve data privacy. To improve the communication efficiency of FL, quantization techniques have emerged as commonly used approaches. However, the introduction of randomized quantization in FL can introduce additional variance, impacting the accuracy of the models. Furthermore, few studies in the existing literature have explored the impact of quantization on the generalization ability of FL algorithms. In this paper, we focus on examining the interplay among key factors in the widely used distributed Stochastic Gradient Descent (SGD) algorithm with quantization. Specifically, we investigate the relationship between quantization level, optimization error, and generalization performance. For convex objectives, our main results reveal several trade-offs between communication efficiency, optimization error, and generalization performance. In the case of non-convex objectives, our theoretical findings indicate that the quantization level has a more significant impact on the generalization ability compared to convex cases. Moreover, our derived generalization bounds for non-convex objectives suggest that early stopping may be necessary to ensure a certain level of generalization accuracy, even when the step size in SGD is very small. Finally, we conduct several numerical experiments utilizing logistic models and deep neural networks to validate and support our theoretical findings.

Shaogao's work is partially supported by the National Natural Science Foundation of China (No.12371291), and Young and Middle-aged Academic Leaders in Jiangsu QingLan Project (2022). This work also received partial support from the National Key Research and Development Program of China (No. 2018AAA0100204), a key program of fundamental research from Shenzhen Science and Technology Innovation Commission (No. JCYJ20200109113403826), the Major Key Project of PCL (No. 2022ZD0115301), and an Open Research Project of Zhejiang Lab (NO.2022RC0AB04).

B. Luo et al. (Eds.): ICONIP 2023, CCIS 1963, pp. 205–217, 2024.
https://doi.org/10.1007/978-981-99-8138-0_17

Keywords: Federated learning · SGD · Quantization techniques · Uniform stability · Generalization

1 Introduction

Distributed algorithms have attracted great attention recently in the context of machine learning and statistics, thanks to their scalable computation, convenient storage, and safe privacy protection. Methods of this type, proposed in various settings, are mainly based on divide and conquer strategies [4,14], distributed SGD [1], split and dual methods [5], Quasi-Newton methods [21] and among others. As new variants of distributed algorithms, federated learning leads to a great interest in recent years [13], which pays more attention to data security and communication efficiency. In federated learning, training machine learning models are deployed over remote devices that often have limited communication power. Hence, communication cost between nodes is a major concern for large-scale applications and sometimes communication is much slower than computation for training complex models with huge numbers of parameters (e.g. training deep neural network), as reported in [1,15].

To improve communication efficiency in federated learning, several effective strategies to tackle this issue have been proposed recently, such as increasing the computation to communication ratio [6,17], reducing the size of the messages [1,15], and designing distributed accelerated algorithms [5,16]. In particular, a class of current approaches towards reducing the size of the messages is to communicate compressed gradient vectors or learned parameters by utilizing a quantization operator [1]. The goal of the quantization schemes is to efficiently compute either a low precision or a sparse unbiased estimate of the d-dimensional vectors. For instance, one of the notable quantization schemes is 1-bitSGD [20], which reduces each component of the gradient to just its sign, that is, only sending a constant number of bits per iteration. To preserve statistical properties of the original, such as unbiasedness and bounded variance, stochastic quantization schemes have attracted great attention recently, and particularly it can attain up to $O(\sqrt{d})$ compression [20], where d is the length of sending parameters.

With regards to algorithmic specification, much attention has been devoted to quantized stochastic gradient descent (QSGD) methods [6]. Under the hub-and-spoke architecture of parallel SGD, a central master takes responsibility for updating a solution while all the workers compute the gradient vectors in a parallel manner before broadcasting them to the master with limited communication. QSGD is fairly general and guarantees algorithmic convergence under regularity assumptions [1]. We observe that most of the aforementioned works in the literature have primarily considered numerical convergence of various distributed algorithms under convex and non-convex settings respectively, whereas their follow-up experiments were conducted in terms of test error or generalization error. Essentially, the former regards the data as fixed points, while the latter requires us to consider the uncertainty of the data that are assumed to be drawn from an underlying distribution. So one can say that most of the previous work in the literature has an obvious gap between theoretical results and numerical analysis. Moreover, note that quantization techniques are a class of randomized and lossy compression of the original, some additional loss can be induced, such as generalization performance and training error of any trained model. Understanding and balancing these trade-offs, both theoretically and empirically, are very desirable but full of considerable challenges in federated learning systems. To the best of our knowledge, there is still a lack of a comprehensive quantitative description of the interactive effects of these factors.

Given these above observations and insights, this paper focuses on understanding the trade-offs among communication cost of QSGD, optimization error, and its generalization performance. This naturally involves the choice of tools that are used to characterize generalization performance. Existing theoretical and numerical analysis indicates that SGD has an implicit regularization even if the model size is quite large [19], as a result, an algorithmic-based generalization bound is more suitable than capacity-based generalization bounds, such as VC dimension, Rademacher complexity and covering numbers. Like previous works [7,12], our analysis also employs the so-called stability notation [2] to derive sharp generalization bounds of the QSGD under standard assumptions. The main contributions of this paper can be summarized as follows:

1. We obtain tight generational bounds of QSGD using the notation of stability under convex and non-convex settings, respectively. These bounds show that the QSGD estimator can trade off the number of bits communicated per iteration with the variance added to the generalization bounds. For convex objectives, there is also a trade-off between stability and optimization.
2. Compared to the convex cases, the generalization bounds established in the non-convex cases are more sensitive to the quantization level. In addition, the non-convex results show that early stopping may be required to guarantee certain generalization accuracy, even if the step size in SGD is very small.
3. Experiments show that the generalization error generated by QSGD will become worse if the variance of the quantized variables increases. Fortunately, QSGD can reduce communication costs by enough to offset the overhead of any additional iterations to convergence.

Notations. For a vector $\mathbf{x} \in \mathbb{R}^d$, we use $\|\mathbf{x}\|_2$ to denote its Euclidean norm and $\|\mathbf{x}\|_0$ is the number of the non-zero components of $\mathbf{x}$. Given two sequences a_n and b_n, we denote $a_n = O(b_n)$ if $|a_n| \leq c_1 |b_n|$ for some absolute positive constant c_1. We write the simplified notation $[K] := \{1, 2, ..., K\}$.

2 Preliminaries

2.1 Expected Risk and Empirical Risk Minimization

The current paper only considers the following setting of supervised learning, although our analysis is also applicable to unsupervised problems with slight notional revisions. Let P be an unknown joint distribution over examples from some space $\mathcal{Z}$. Given all available observations $S = (\mathbf{z}_1, \mathbf{z}_2, ..., \mathbf{z}_N)$ of N examples drawn i.i.d. from P. The basic goal for learning is to find a model $\mathbf{w}^\star$ which minimizes $R(\mathbf{w}) = \mathbb{E}_{\mathbf{z}\sim P}[\ell(\mathbf{w};\mathbf{z})]$, where $\ell(\cdot)$ is a generalized loss function and $\ell(\mathbf{w};\mathbf{z})$ contains a standard loss defined in the context of supervised learning and the model described by a parameter $\mathbf{w}$. In machine learning, we often call $R(\cdot)$ population risk.

Since the distribution is unknown in advance, the objective $R(\mathbf{w})$ can not be computed directly. Instead, we introduce a sample-based empirical risk that seeks a proxy of $\mathbf{w}^\star$,

$$R_S(\mathbf{w}) = \frac{1}{N}\sum\nolimits_{i=1}^{N} \ell(\mathbf{w};\mathbf{z}_i). \tag{2.1}$$

Searching for a feasible solution of (2.1) often resorts to several efficient numerical optimization algorithms. Among these algorithms, SGD has been shown to be scalable, robust, and perform well across many different domains. Precisely, once we have access to stochastic gradient $\tilde{g}$ such that $\mathbb{E}[\tilde{g}(\mathbf{w})] = \nabla R(\mathbf{w})$, a standard instance of SGD iterates the following procedure to update parameters $\mathbf{w}_{t+1} = \mathbf{w}_t - \eta_t\tilde{g}(\mathbf{w}_t)$, where $\eta_t > 0$ is a step size at iteration t.

Under the standard distributed setting, we separate the total data S into equally-size K blocks, denoted by $S_k = \{\mathbf{z}_{ki}\}_{i\in[n]}$ for all $k \in [K]$. Hence each local sample size $n := N/K$. In this way, we can rewrite (2.1) as

$$R_S(\mathbf{w}) = \frac{1}{K}\sum\nolimits_{k=1}^{K} R_{S_k}(\mathbf{w}). \tag{2.2}$$

2.2 Quantization Techniques

In a communication-efficient distributed system, quantization schemes are a class of simple lossy compression of the original vector and have been successfully applied to many distributed learning and federated learning problems. Quantization operators aim at generating much sparser vectors by random sparsification techniques, which randomly mask the input vectors and only preserve a few coordinates with statistical guarantees. Our analysis depends on a general notion of quantization operators with nice statistical properties, including unbiasedness and bounded variance.

Definition 1. A class of very common quantization operators is defined as the random operator $Q : \mathbb{R}^d \to \mathbb{R}^d$ with two parameters (ω, s), satisfying

$$\mathbb{E}_\xi[Q(\mathbf{x})] = \mathbf{x}, \quad \mathbb{E}_Q[\|Q(\mathbf{x})\|_2^2] \le (1+\omega)\|\mathbf{x}\|_2^2. \tag{2.3a}$$

$$\mathbb{E}_\xi[\|Q(\mathbf{x})\|_0] = s. \tag{2.3b}$$

In other words, we require that the quantization operator is unbiased and bounded by variance. The scalar parameter s represents the level of the sparsity. In this sparse case, it suffices to encode a nonzero coordinate of $Q(\mathbf{x})$ requires $O(\log(s))$ bits. In addition, (2.3a) implies that $\mathbb{E}_Q[\|Q(\mathbf{x}) - \mathbf{x}\|_2^2] \le \omega\|\mathbf{x}\|_2^2$. It is worth noting that, the sparsity parameter s is often inversely proportional to the variance ω of Q, as indicated in the following examples. The operator Q in Definition 1 is called ω-quantization in the literature. We now give two common-used examples of ω-quantization operators that also achieve (2.3a) and (2.3b).

Example 1. *(Random sparsification)* The random sparsification operator $Q(\mathbf{x}) = {}^{d}\!/_{r} \cdot \xi \otimes \mathbf{x}$ whenever the random variable is chosen uniformly at random, $\xi \sim_{u.a.r.} \{\mathbf{y} \in \{0,1\}^d : \|\mathbf{y}\|_0 = r\}$. This definition of Q implies that $s = r$ and $\omega = {}^{d}\!/_{r} - 1$.

Example 2. *(Random dithering)* The random dithering operator over $\mathbb{R}^d$ is $Q(\mathbf{x}) = sign(\mathbf{x}) \cdot \|\mathbf{x}\|_2 \cdot \frac{1}{r}\left\lfloor r\frac{|\mathbf{x}|}{\|\mathbf{x}\|_2} + \xi \right\rfloor$, for random variable $\xi \sim_{u.a.r.} [0,1]^d$. Here $sign(\mathbf{x})$ and $|\mathbf{x}|$ are both component-wise maps from $\mathbb{R}^d \to \mathbb{R}^d$. In this case, we can check that $s = O\big(r(r+\sqrt{d})\big)$ and $\omega = O\big(\sqrt{d}/r\big)$. For proof see [1].

The traditional SGD can be easily implemented in a parallel way across the workers, each of which has access to only its own private dataset. At iteration t, the master broadcasts the current version of the global parameter to the mobile users. Each worker k then computes its local gradient vector based on its local dataset, and sparsify these gradient vectors using quantization techniques before sending them to the master. When the master takes the average of all the quantized gradient vectors, one iteration by SGD is realized.

It is worth emphasizing that, the ideas of our current research apply to a wide range of distributed learning algorithms. This paper focuses on the well-known SGD to highlight our theoretical findings: several trade-offs among communication, optimization, and generalization under a federated learning context.

3 Stability and Generalization Analysis

3.1 Theoretical Background and Technicalities

Various forms of algorithmic stability have been introduced to characterize generalization error [2,18]. To highlight the purpose of this paper, we are only interested in the uniform stability notion originally introduced by [2].

The notion of generalization can be formalized as the requirement that the empirical risk of a solution is concentrated around its expected risk. Precisely, the generalization error of a model $\mathbf{w}$ is the difference

$$R(\mathbf{w}) - R_S(\mathbf{w}). \tag{3.1}$$

When $\mathbf{w} = A(S)$ is chosen as a function of the data by a potentially randomized algorithm A, consider the expected generalization error

$$\mathbb{E}_A\big[R(\mathbf{w}) - R_S(\mathbf{w})\big]. \tag{3.2}$$

Recall that, stability is the notion that a small change in the training dataset S does not change much the solution. To this end, we define a set S' by replacing the i-th data point in S by $\mathbf{z}'_i$, given as $S' = \{\mathbf{z}_1, ..., \mathbf{z}_{i-1}, \mathbf{z}'_i, \mathbf{z}_i, ..., \mathbf{z}_N\}$.

Definition 2. A random algorithm A is β_N-uniformly stable with respect to a loss function ℓ, if for all data sets $S, S' \in \mathcal{Z}^N$, it satisfies $\sup_{\mathbf{z}}\mathbb{E}_A\big[\ell(A(S);\mathbf{z}) - \ell(A(S');\mathbf{z})\big] \le \beta_N$.

A randomized learning algorithm with uniform stability yields the following generalization error, which was proved in Theorem 2 of [22].

Lemma 1. *A uniform stable randomized algorithm* $(A(S), \beta_N)$ *with a bounded loss function* $0 \leq \ell(A(S), \mathbf{z}) \leq M$*, satisfies following generalization bound with probability at least* $1-\delta$*, over the random draw of* $S, \mathbf{z}$ *with* $\delta \in (0,1)$*,*

$$\mathbb{E}_A\big[R[A(S)] - R_S[A(S)]\big] \leq 2\beta_N + (4N\beta_N + M)\sqrt{\log(1/\delta)/2N}.$$

Note that, the generalization bound is meaningful only if the bound converges to 0 as $N \to \infty$. This occurs when β_N decays faster than $O(1/\sqrt{N})$. To provide upper bounds of uniform stability, some convexity and smoothness assumptions with respect to the loss ℓ are required. Throughout the paper, we focus on two types of loss functions: The first type of loss function $\ell(\cdot, \mathbf{z})$ is convex and α-smooth for every $\mathbf{z}$; The second type of loss function $\ell(\cdot, \mathbf{z})$ is non-convex and α-smooth for every $\mathbf{z}$. We also make use of the L-Lipschitz condition.

Definition 3. We say that f is L-Lipschitz if for all points $\mathbf{w}, \mathbf{w}'$, we have $|f(\mathbf{w}) - f(\mathbf{w}')| \leq L\|\mathbf{w} - \mathbf{w}'\|_2$. Additionally, a function f is α-smooth when satisfying $\|\nabla f(\mathbf{w}) - \nabla f(\mathbf{w}')\|_2 \leq \alpha\|\mathbf{w} - \mathbf{w}'\|_2$.

3.2 Convex Case

In this subsection, we prove stability results for QSGD under the convex setting. Using the derived stability bound, we can quantify the generalization gap between the empirical risk and the expected one in high probability.

Theorem 1. *Suppose that the loss function* $\ell(\cdot; \mathbf{z})$ *is* α*-smooth, convex and* L*-Lipschitz for all* $\mathbf{z}$*. If QSGD runs with step size* $\eta_t(1+\omega)\alpha \leq 2$ *for* T *steps. Then, QSGD satisfies uniform stability with* $\beta_N \leq 2L^2/N \times \sum_{t=1}^{T} \eta_t$*.*

Remark 1. [3] has shown that this stability upper bound can be achieved by a linear loss function, that is, this upper bound can not be further improved under a certain framework. We notice that the result in Theorem 1 is consistent with the one established in [7] without a quantization operation ($\omega = 0$).

In fact, from our proof for Theorem 1, if the constraint of $\eta_t(1+\omega)\alpha \leq 2$ does not hold, some additional error for the stability bound will be caused accordingly. In this situation, we can regard the quantity $\max\{\eta_t(1+\omega)\alpha - 2, 0\}$ as an error measurement associated with the parameter ω. Plugging the bound of β_N in Theorem 1 into Lemma 1 yields the generalization bound stated as follows.

Corollary 1. *Under the same conditions as Theorem 1 and the loss function is bounded,* $0 \leq \ell(A(S), \mathbf{z}) \leq M$ *for all* $\mathbf{z}$*. Then the following expected generalization bound of the QSGD algorithm holds with probability at least* $1-\delta$ *for any* $\delta \in (0,1)$*,*

$$\mathbb{E}_A\big[R[\mathbf{w}_T]\big] \leq \mathbb{E}_A\big[R_S[\mathbf{w}_T]\big] + O\Big(\frac{1}{N}\sum_{t=1}^{T}\eta_t + \big(\sum_{t=1}^{T}\eta_t + M\big)\sqrt{\frac{\log(1/\delta)}{N}}\Big).$$

The result in Theorem 1 implies that the stability bound decreases inversely with the size of the training set, while it increases as the step size increases. Specially, if the step size is chosen as some constant, we have to carry out early stopping ($T = o(\sqrt{N})$) to satisfy uniform stability, guaranteed when $\beta_N = o(\frac{1}{\sqrt{N}})$. In other words, our results in Corollary 1 show that for carefully designed T and η_t, QSGD updates satisfy uniform stability and prediction accuracy.

We would like to point out that, a larger parameter ω often means better communication efficiency per iteration. However, the key condition $\eta_t(1+\omega)\alpha \leq 2$ in Theorem 1 requires us to choose a smaller η_t, which in turn slows down the process of training models. For clarity, we introduce the following notation to quantify the optimization error of an algorithm.

Let $\widehat{\mathbf{w}}$ be the empirical risk minimizer of $R_S(\mathbf{w})$ over Ω. Define the expected optimization error of a model $\widetilde{\mathbf{w}}$ by $\mathcal{E}_{opt}(\widetilde{\mathbf{w}}, \ell, P, N) := \mathbb{E}_{S\sim P^N}\left[R_S[\widetilde{\mathbf{w}}] - R_S[\widehat{\mathbf{w}}]\right]$. We next define the class of all convex smooth loss functions as follows, $\mathcal{L}_c = \{\ell : \mathcal{Z}\times\Omega \to \mathbb{R} | \, \ell \, \text{is convex}, \alpha\text{-smooth}, |\Omega| = W\}$. Here $|\Omega| = W$ refers to a functional complexity that generates the learning parameter $\mathbf{w}$, appearing in minimax lower bounds for the convex smooth loss class [11]. Let $\mathcal{P}$ be the class of all possible distributions, and we are interested in a distribution-independently optimization error defined by $\mathcal{E}^{\widetilde{\mathbf{w}}}_{opt}(T, N, \mathcal{L}_c) = \sup_{\ell\in\mathcal{L}_c, P\in\mathcal{P}} \mathcal{E}_{opt}(\widetilde{\mathbf{w}}, \ell, P, N)$.

Theorem 2. *Under the loss class $\mathcal{L}_c$ with the cardinality W. For the QSGD algorithm, we consider, then there exists a universal constant $C_2 > 0$, a sample size N_0, and an iteration number $T_0 \geq 1$, such that for $T \geq T_0$, its convergence rate is lower bounded as follows, $\mathcal{E}^{\widetilde{\mathbf{w}}}_{opt}(T, N_0, \mathcal{L}_c) \geq {W^4\alpha^2}/{2L^2C_2\sum_{t=1}^{T}\eta_t}$.*

The proof for Theorem 2 is given in Appendix, which easily follows from some existing trade-off results in terms of stability and convergence, established recently by [3]. Indeed, the result derived in Theorem 3 holds for any given algorithm, not limited to SGD.

Remark 2. We also remark that even though the deceases of η_t of t guarantees small stability bound seen from Theorem 1, Theorem 2 indicates that such a chosen η_t will deteriorate the optimization error, which may result in an under-fitting of the trained model, that is, $R_S(\mathbf{w}_T)$ may be large given steps T. Consequently, given iteration times, we conclude from Corollary 1 that the QSGD algorithm has a trade-off among communication efficiency, generalization and optimization. These theoretical findings are further verified below.

3.3 Non-convex Case

In this subsection, we present our main results on parallel quantized SGD under the non-convex setting. Again, we assume that the conditions in the definition 1 hold. We also assume that ℓ is L-Lipschitz and α-smooth.

Theorem 3. *Suppose that the loss function $\ell(\cdot;\mathbf{z})$ is α-smooth and L-Lipschitz for all $\mathbf{z}$. If QSGD runs with step size $\eta_t \leq \frac{r}{t}$ for all t and any $r > 0$. Then, the QSGD algorithm satisfies uniform stability with*

$$\beta_N \leq {L^2}/{\alpha N} \times T^{\alpha r\sqrt{1+\omega}}. \tag{3.3}$$

Theorem 3 suggests that the upper bound of β_N will increase sharply as the variance parameter ω of quantization increases, provided that η_t decays with the order $O(1/t)$. To some extent, this implies that the quantization technique brings a negative effect to the stability bound, in turn affecting the generalization performance of the QSGD algorithm.

Corollary 2. *Under the same conditions as Theorem 3 and the loss function is bounded,* $0 \leq \ell(A(S), \mathbf{z}) \leq M$ *for all* $\mathbf{z}$. *Then the following expected generalization bound of the QSGD algorithm holds with probability at least* $1 - \delta$ *for any* $\delta \in (0, 1)$,

$$\mathbb{E}_A\big[R[\mathbf{w}_T]\big] \leq \mathbb{E}_A\big[R_S[\mathbf{w}_T]\big] + O\big(L^2/\alpha \times T^{\alpha r\sqrt{1+\omega}}\sqrt{\log(1/\delta)/N}\big).$$

Remark 3. We would like to remark that, the upper bound of generalization error in Corollary 2 will not hold for large T given any non-trivial step size η_t. This means that one may need to apply early stopping when running QSGD in the non-convex case, even if the step size η_t has a linear-order decay of t. Unlike the result in 2 in the convex case, providing a tight lower bound of optimization error in the non-convex case is quite challenging in the nonparametric minimax sense, as an interesting future work.

4 Numerical Experiments

In this section, we empirically validate our theoretical findings. We begin by describing the experimental setup and proceed to investigate the impact of different quantization levels on the generalization performance of QSGD in both convex and non-convex scenarios. Specifically, we analyze the trade-off between communication efficiency and generalization error in QSGD and demonstrate that QSGD can achieve a comparable training accuracy to Vanilla SGD when an appropriate quantization level is employed [24].

Datasets and Models. In our experiments, we utilize logistic regression as an example for convex problems. For the non-convex experiments, we employ CNN [10] architectures with two hidden layers and ResNet-18 [8]. The logistic regression model is trained on the MNIST [23] dataset, while the network models are trained on the FashionMNIST [23] and CIFAR-10 [9] datasets, which are popular benchmark image classification datasets. No additional preprocessing methods are applied to the datasets.

Implementation. We utilize the Quantized SGD algorithm to train image classifiers using logistic regression, CNN, and ResNet-18 models. We apply random sparsification and random dithering quantization operations to the models. For random sparsification, we consider sparsification rates of $k = r/d \in \{10\%, 20\%, 50\%, 70\%\}$ denoted as RANDk. For random dithering quantization, we use quantization levels of $s \in \{2, 4, 8\}$ denoted as QSGD$s$bits. To simulate both cross-silo and cross-device scenarios of federated learning, we set the number of workers as 20 and 100, with 5 and 10 active workers per round, respectively. The local batch size is fixed at 64. The initial learning rate is set to 0.1 and is multiplied by 0.1 when the model has completed 2/5 and 4/5 of the total number of iterations. To ensure the interpretability of the experiments, we avoid using other techniques such as momentum, weight decay, and data augmentation. We conducted each experiment 5 times with different initializations and random seeds and reported the average results.

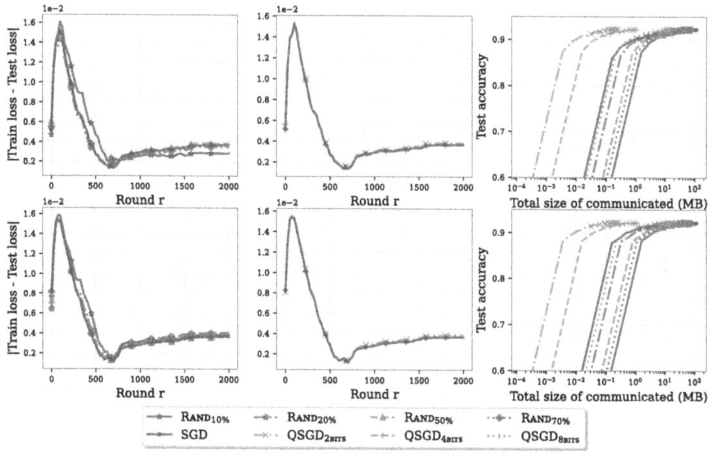

Fig. 1. Results on Logistic regression for MNIST dataset. Top: cross-silo scenarios. Bottom: cross-device scenarios

Verifying the Theory on Convex Case. We calculate the difference between the testing loss and the training loss on different quantization operations separately, as shown in Fig. 1. The right column of Fig. 1 shows the trend of achieved test accuracy of the aggregated model as the total size of communicated increases. From the left two columns of Fig. 1, for any quantization level, we see that the generalization error tends to stabilize as the epoch increases. The loss differences are generally similar: SGD $\approx$ RAND_k $\approx$ $\text{QSGD}_{s\text{bits}}$.

Note that the smaller k or s generates a much sparser vector that will be sent to the master. This also implies less communication cost. More precisely, given the epoch, the generalization error is consistent when k or s becomes smaller. This observation is almost consistent with our derived result in Theorem 2: if the constraint of $\eta_t(1+\omega)\alpha \leq 2$ holds, no additional error for the stability bound will be caused accordingly.

Verifying the Theory on Non-convex Case. With a similar numerical analysis to the above experiment, we conduct another two experiments for non-convex problems, to evaluate generalization error and training accuracy of QSGD concerning different quantization levels settings. We take 2-layers convolutional neural networks (using the ReLU activation function) as an example for non-convex problems on the FashinMNIST dataset, and the Resnet18 model on the CIFAR-10 dataset.

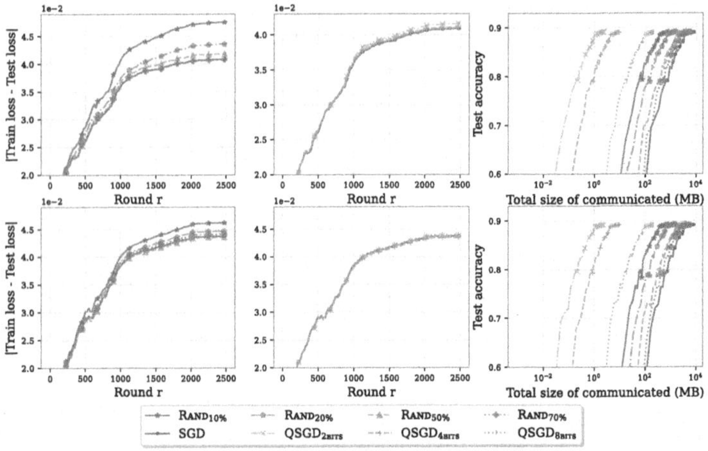

Fig. 2. Results on CNN for FashionMNIST dataset. Top: cross-silo scenarios. Bottom: cross-device scenarios

The plots in Figs. 2 and 3 demonstrate generalization error/test accuracy over epoch/total size of communicated, respectively. Two observations are obtained from those figures: (1) as the training rounds increase, the generalization gap on different quantization techniques grows; (2) the generalization can be sorted as follows: SGD $\approx$ $\text{QSGD}_{\text{8bits}}$ $\approx$ $\text{QSGD}_{\text{4bits}}$ $\approx$ $\text{RAND}_{70\%}$ $<$ $\text{QSGD}_{\text{2bits}}$ $<$ $\text{RAND}_{50\%}$ $<$ $\text{RAND}_{20\%}$ $<$ $\text{RAND}_{10\%}$. We conjecture that the generalization gaps are amplified with quantization parameters k and s, according to our theory suggesting that the generalization gaps increases with the variance ω of quantization (see Theorem 2).

In comparison to non-convex cases, we observe that the generalization error in the convex cases will increase sharply and then decrease gradually with the increase of iteration. Theoretically, explaining this phenomenon is a matter of optimization, which is beyond the scope of this work. One may observe the loss difference of the ResNet-18 model is larger than that of other models. Since the vanilla SGD converges faster, the corresponding optimization error is smaller than the quantized SGD. Therefore, the vanilla SGD may be less stable in the initial phase of training for complex large models. Compared to the convex cases, the generalization bounds established in the non-convex cases are more sensitive to the quantization level. In addition, the non-convex results show that early stopping may be required to guarantee certain generalization accuracy, even if the step size in SGD is very small. This shows that there exist several trade-offs among communication efficiency, optimization error, and generalization performance.

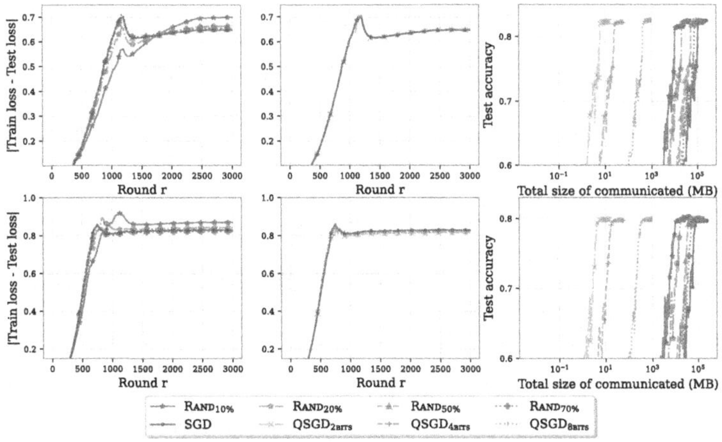

Fig. 3. Results on ResNet-18 for CIFAR-10 dataset. Top: cross-silo scenarios. Bottom: cross-device scenarios

As the communications budget increases, neural network models are more sensitive to the quantization level and we should choose a large value of k or s to reduce generalization error introduced by quantization operation. Although the quantization operation in the SGD will deteriorate generalization performance, the test accuracy is improved greatly at any certain communication cost. The quantized SGD update saves much time at commutation cost in some cases, and thus the number of iterations has increased a lot given time. This means that there exists a trade-off between communication cost and training accuracy.

5 Conclusions

This paper addresses the key aspects of communication efficiency, generalization performance, and optimization error in the context of Quantized SGD, a commonly used approach in federated learning. We have conducted rigorous convergence and stability analyses of QSGD, considering both convex and non-convex scenarios. Notably, we have established an explicit trade-off between training accuracy and stability bounds. To validate our theoretical findings, we have performed experiments using real-world datasets. Importantly, the ideas presented in this paper have broad applicability and can be extended to various distributed algorithms that involve quantization, such as federated average and federated proximal algorithms, among others.

References

1. Alistarh, D., Grubic, D., Li, J., Tomioka, R., Vojnovic, M.: QSGD: communication-efficient SGD via gradient quantization and encoding. In: Advances in Neural Information Processing Systems, vol. 30 (2017)

2. Bousquet, O., Elisseeff, A.: Stability and generalization. J. Mach. Learn. Res. **2**, 499–526 (2002)
3. Chen, Y., Jin, C., Yu, B.: Stability and convergence trade-off of iterative optimization algorithms. arXiv preprint arXiv:1804.01619 (2018)
4. Chu, C., et al.: Map-reduce for machine learning on multicore. In: Advances in Neural Information Processing Systems, vol. 19, p. 281 (2007)
5. Duchi, J.C., Agarwal, A., Wainwright, M.J.: Dual averaging for distributed optimization: convergence analysis and network scaling. IEEE Trans. Autom. Control **57**(3), 592–606 (2011)
6. Goyal, P., et al.: Accurate, large minibatch SGD: training ImageNet in 1 hour. arXiv preprint arXiv:1706.02677 (2017)
7. Hardt, M., Recht, B., Singer, Y.: Train faster, generalize better: stability of stochastic gradient descent. In: International Conference on Machine Learning, pp. 1225–1234. PMLR (2016)
8. He, K., Zhang, X., Ren, S., Sun, J.: Identity mappings in deep residual networks. In: Leibe, B., Matas, J., Sebe, N., Welling, M. (eds.) ECCV 2016. LNCS, vol. 9908, pp. 630–645. Springer, Cham (2016). https://doi.org/10.1007/978-3-319-46493-0_38
9. Krizhevsky, A., Hinton, G., et al.: Learning multiple layers of features from tiny images (2009)
10. Krizhevsky, A., Sutskever, I., Hinton, G.E.: ImageNet classification with deep convolutional neural networks. Commun. ACM **60**(6), 84–90 (2017)
11. Le Cam, L.M.: Asymptotic Methods in Statistical Theory (1986)
12. Lei, Y., Ying, Y.: Fine-grained analysis of stability and generalization for stochastic gradient descent. In: International Conference on Machine Learning, pp. 5809–5819. PMLR (2020)
13. Li, T., Sahu, A.K., Talwalkar, A., Smith, V.: Federated learning: challenges, methods, and future directions. IEEE Signal Process. Mag. **37**(3), 50–60 (2020)
14. Lin, S.B., Guo, X., Zhou, D.X.: Distributed learning with regularized least squares. J. Mach. Learn. Res. **18**(1), 3202–3232 (2017)
15. Lin, Y., Han, S., Mao, H., Wang, Y., Dally, W.J.: Deep gradient compression: reducing the communication bandwidth for distributed training. arXiv preprint arXiv:1712.01887 (2017)
16. Mahajan, D., Agrawal, N., Keerthi, S.S., Sellamanickam, S., Bottou, L.: An efficient distributed learning algorithm based on effective local functional approximations. J. Mach. Learn. Res. **19**(1), 2942–2978 (2018)
17. Mann, G., McDonald, R., Mohri, M., Silberman, N., Walker, D.D.: Efficient large-scale distributed training of conditional maximum entropy models. In: Proceedings of the 22nd International Conference on Neural Information Processing Systems, pp. 1231–1239 (2009)
18. Mukherjee, S., Niyogi, P., Poggio, T., Rifkin, R.: Learning theory: stability is sufficient for generalization and necessary and sufficient for consistency of empirical risk minimization. Adv. Comput. Math. **25**(1), 161–193 (2006)
19. Nakkiran, P., Kaplun, G., Bansal, Y., Yang, T., Barak, B., Sutskever, I.: Deep double descent: where bigger models and more data hurt. arXiv preprint arXiv:1912.02292 (2019)
20. Seide, F., Fu, H., Droppo, J., Li, G., Yu, D.: 1-bit stochastic gradient descent and its application to data-parallel distributed training of speech DNNs. In: Fifteenth Annual Conference of the International Speech Communication Association. Citeseer (2014)

21. Shamir, O., Srebro, N., Zhang, T.: Communication-efficient distributed optimization using an approximate newton-type method. In: International Conference on Machine Learning, pp. 1000–1008. PMLR (2014)
22. Verma, S., Zhang, Z.L.: Stability and generalization of graph convolutional neural networks. In: Proceedings of the 25th ACM SIGKDD International Conference on Knowledge Discovery & Data Mining, pp. 1539–1548 (2019)
23. Xiao, H., Rasul, K., Vollgraf, R.: Fashion-MNIST: a novel image dataset for benchmarking machine learning algorithms. arXiv preprint arXiv:1708.07747 (2017)
24. Zeng, D., Liang, S., Hu, X., Wang, H., Xu, Z.: FedLab: a flexible federated learning framework. J. Mach. Learn. Res. **24**, 100–1 (2023)

Many Is Better Than One: Multiple Covariation Learning for Latent Multiview Representation

Yun-Hao Yuan[1,2], Pengwei Qian[1], Jin Li[1], Jipeng Qiang[1], Yi Zhu[1], and Yun Li[1](✉)

[1] School of Information Engineering, Yangzhou University, Yangzhou, China
{yhyuan,liyun}@yzu.edu.cn
[2] School of Computer Science, Fudan University, Shanghai, China

Abstract. Canonical correlation analysis is a typical multiview representation learning technique, which utilizes within-set and between-set covariance matrices to analyze the correlation between two multidimensional datasets. However, it is quite difficult for the covariance matrix to measure the nonlinear relationship between features because of its linear structure. In this paper, we propose a multiple covariation projection (MCP) method to learn latent two-view representation, which has the ability to model the complicated feature relationship. The proposed MCP first constructs multiple general covariance matrices for modeling diverse feature relations, and then integrates them together via a linear ensemble strategy. At last, an efficient two-stage algorithm is designed for solutions. In addition, we further present a multiview MCP for dealing with the case of multiple (more than two) views. Experimental results on benchmark datasets show the effectiveness of our proposed MCP method in multiview classification and clustering tasks.

Keywords: Canonical correlation analysis · Multiview representation learning · Nonlinear relationship · Feature learning

1 Introduction

In reality, the same objects are usually represented by multiple kinds of high-dimensional features from different perspectives. For example, handwritten digits can be depicted by morphological feature and pixel information. Web pages are described by text information, image feature, and video feature. Such data with different features are referred to as multiview data in the literature. Because multiview data naturally contain the complementary and consistent information of different views, analyzing multiview data is very meaningful in many real applications. But, how to explore and exploit such information in an effective manner is still a challenging problem.

A feasible solution to the above-mentioned problem is multiview representation learning (MRL), which aims to learn an effective latent representation from multiview data for downstream tasks such as multiview classification and multiview clustering. Over the past decades, numerous MRL techniques have been

B. Luo et al. (Eds.): ICONIP 2023, CCIS 1963, pp. 218–228, 2024.
https://doi.org/10.1007/978-981-99-8138-0_18

developed based on different ideas, such as canonical correlation analysis (CCA) methods [6,19], partial least squares (PLS) methods [10,16], and deep neural network (DNN) based methods [1,7,13]. As one of the most representative MRL methods, CCA seeks a pair of linear transformations via view correlation maximization, thus obtaining highly correlated multiview representations. Due to the effectiveness of CCA, thus far its many improvements have been put forward. In general, these improvement methods fall into two categories, i.e., linear and nonlinear CCA methods.

Linear CCA methods aim to model the linear relationship between two-view high-dimensional data. For instance, discriminative CCA [9] was proposed by using the class information of training data, which minimizes the interclass correlation and at the same time, maximizes the intraclass correlation of the data. Multiset CCA (MCCA) [8] is a multiview (more than two views) extension of CCA, which maximizes the sum of the correlations of any pair of views, subject to some given quadratic constraints. Another multiview extension can be found in [18]. To reveal the nonlinear relationship hidden in two-view data, various nonlinear CCA extensions have been proposed over the past 20 years. Kernel CCA (KCCA) [2], an early nonlinear CCA technique, aims on modeling the nonlinear relationship between two-view samples through nonlinearly mapping the original inputs to higher-dimensional feature spaces. There are heretofore many KCCA improvements; see, for example, [3,12].

Another nonlinear learning paradigm is DNN-based, which has drawn great attention due to flexible representation learning. Based on DNNs, deep CCA (DCCA) [1] was proposed. DCCA utilizes two DNNs to transform two-view data and then maximizes the linear correlations between their outputs. In addition, some other DNN-based CCA methods [7,13] have also been proposed based on different considerations. Recently, a new nonlinear CCA framework [17] was presented based on covariation matrices, where a nonlinear mapping is used to project features rather than samples.

In this paper, we propose a multiple covariation projection (MCP) method (see Fig. 1) for learning latent representation, which has the ability to model the complicated relationship between different features. First, MCP builds multiple general covariance matrices (GCMs) via different kernel mappings for modeling diverse feature relations. Then, it integrates these GCMs together via using a simple linear ensemble strategy. Finally, an efficient two-stage algorithm is designed for solving MCP. In addition, we further propose a multiview MCP method for dealing with the case of multiple (more than two) views. Many experimental results on several datasets demonstrate the effectiveness of our proposed MCP method in multiview classification and clustering.

2 Related Work

2.1 General Covariance Matrix

Given two data matrices $\{X_i \in R^{d_i \times n}\}_{i=1}^2$, assume $f_t^i \in R^n$ is the feature vector associated with the t-th row of X_i, where d_i and n are separately the dimension

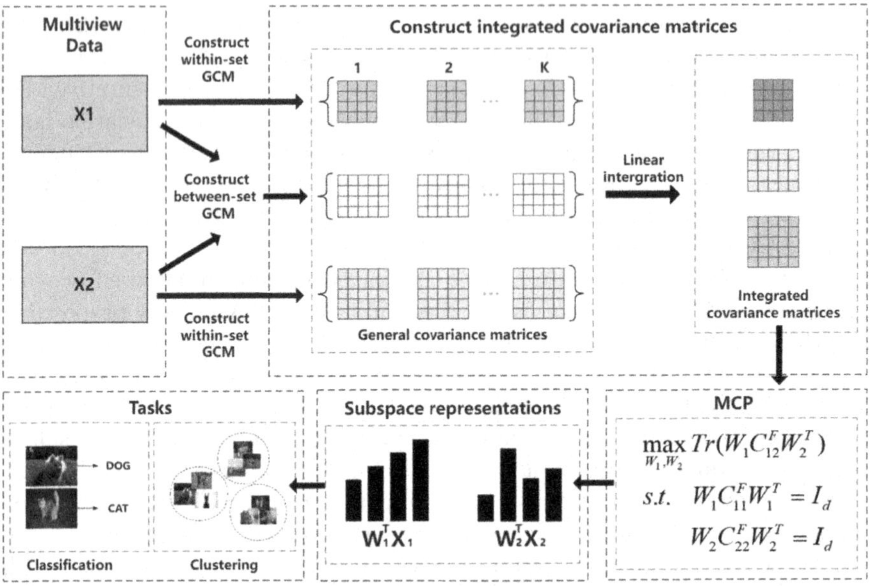

Fig. 1. The flowchart of our proposed MCP method. $X_i \in R^{d_i \times n}(i = 1, 2)$ denotes two different views. Firstly, we construct K distinct within-set and between-set general covariance matrices, respectively. Then, they are linearly integrated to form within-set and between-set integrated covariance matrices (ICMs). Finally, the MCP model is built based on these ICMs. After getting projection matrices W_1^T and W_2^T, subspace representations $W_1^T X_1$ and $W_2^T X_2$ are computed for downstream learning tasks such as classification and clustering.

and number of training samples and $t = 1, 2, \cdots, d_i$. X_i can be written as $X_i = [f_1^i, f_2^i, \cdots, f_{d_i}^i]^T$. Let $\psi(\cdot)$ be a nonlinear function which maps a feature vector from the input space R^n into a new space R^N, that is,

$$\psi(f_t^i) = \left[\psi_1(f_t^i), \psi_2(f_t^i), \cdots, \psi_N(f_t^i)\right]^T, \tag{1}$$

where N is the dimension of the new space. With (1), the general covariance matrix (GCM) [15] of views i and j is defined as

$$C_{ij}^{\psi} = \left[C_{ij}^{\psi}(z, t)\right] = \left[\psi(f_z^i)^T \psi(f_t^j)\right] \in R^{d_i \times d_j}, \tag{2}$$

where $C_{ij}^{\psi}(z, t)$ is the (z, t)-th element of GCM C_{ij}^{ψ}, $i, j = 1, 2$, $z = 1, 2, \cdots, d_i$ and $t = 1, 2, \cdots, d_j$.

2.2 CCA

CCA is a well-known two-view data analysis method. It aims to find a pair of projection vectors $\alpha \in R^{d_1}$ and $\beta \in R^{d_2}$ to maximize the correlation between canonical projections $\alpha^T X_1$ and $\beta^T X_2$. More formally, CCA can be described as the following optimization problem:

$$\begin{aligned} \max_{\alpha,\beta} \quad & \alpha^T S_{12}\beta \\ s.t. \quad & \alpha^T S_{11}\alpha = \beta^T S_{22}\beta = 1, \end{aligned} \tag{3}$$

where S_{12} is the between-set covariance matrix of X_1 and X_2, S_{11} and S_{22} are the within-set covariance matrix of X_1 and X_2, respectively. The optimization problem (3) can be solved by the following generalized eigenvalue problem:

$$\begin{bmatrix} 0 & S_{12} \\ S_{21} & 0 \end{bmatrix} \begin{bmatrix} \alpha \\ \beta \end{bmatrix} = \lambda \begin{bmatrix} S_{11} & 0 \\ 0 & S_{22} \end{bmatrix} \begin{bmatrix} \alpha \\ \beta \end{bmatrix}, \tag{4}$$

where $S_{21} = S_{12}^T$ and λ represents the generalized eigenvalue.

3 Multiple Covariation Projection (MCP)

3.1 Motivation

Traditional CCA methods utilize covariance matrices in nature to measure the relationship between two views. It is difficult for them to model the nonlinear relationship between different features rather than samples, since the covariance measure is of linear structure. On the other hand, our previous work [15] has built a general covariance matrix to model the nonlinear feature relatedness. But, it only utilizes a single GCM form for learning the latent features of the data. In fact, the general covariance matrix essentially represents the similarity between different features. In general, different GCMs may have different abilities of depicting the similarity of features. Motivated by the aforementioned considerations, we try to construct a multiple covariation learning approach in the canonical correlation framework.

3.2 Formulation

Let $\phi_c(\cdot)$ be a nonlinear mapping which projects feature vector f_t^i in the original space R^n into a new space R^{N_c}, where N_c is the dimension of the new space, $c = 1, 2, \cdots, K$, and K is the number of nonlinear mappings. According to (2), the within-set and between-set GCMs of the i-th and j-th views based on $\phi_c(\cdot)$ can be written as

$$C_{ij}^{\phi_c} = \left[C_{ij}^{\phi_c}(z,t)\right] = \left[\phi_c(f_z^i)^T \phi_c(f_t^j)\right] \in R^{d_i \times d_j}, \; c = 1, 2, \cdots, K, \tag{5}$$

where $i, j = 1, 2$, $z = 1, 2, \cdots, d_i$ and $t = 1, 2, \cdots, d_j$. To compute $C_{ij}^{\phi_c}$ in real applications, nonlinear function $\phi_c(\cdot)$ is chosen as a kernel mapping in this paper. Therefore, (5) can be transformed as

$$C_{ij}^{\phi_c} = \left[ker_c(f_z^i, f_t^j)\right] \in R^{d_i \times d_j}, \; c = 1, 2, \cdots, K, \tag{6}$$

where

$$ker_c(f_z^i, f_t^j) = \phi_c(f_z^i)^T \phi_c(f_t^j) \tag{7}$$

with $ker_c(\cdot,\cdot)$ as some kernel function. Clearly, different $C_{ij}^{\phi_c}$ will model different nonlinear associations between features of views. In other words, each $C_{ij}^{\phi_c}$ has the ability of capturing a kind of specific feature relation. As a result, it is quite necessary to integrate them together for generating a more comprehensive representation. To achieve this, we choose different types of kernel functions or ones with different parameters to first produce K individual GCMs, and then linearly combine these GCMs.

Concretely, let $\omega_1, \omega_2, \cdots, \omega_K$ be a set of nonnegative weight coefficients corresponding to K different general covariance matrices, we can define the integrated covariance matrix (ICM) as:

$$C_{ij}^F = \omega_1 C_{ij}^{\phi_1} + \omega_2 C_{ij}^{\phi_2} + \cdots + \omega_K C_{ij}^{\phi_K}, \ i,j = 1,2. \tag{8}$$

Let $W_1 \in R^{d_1 \times d}$ and $W_2 \in R^{d_2 \times d}$ be a pair of linear projection transformations, $d \leq \min(d_1, d_2)$. Based on the ICMs defined in (8), the model of our proposed multiple covariation projection (MCP) approach can be formulated as the following optimization problem:

$$\begin{aligned} &\max_{W_1, W_2, \{\omega_i\}_{i=1}^K} \ Tr(W_1^T C_{12}^F W_2) \\ &s.t. \ \ W_1^T C_{11}^F W_1 = W_2^T C_{22}^F W_2 = I_d, \\ &\qquad \sum_{i=1}^K \omega_i^2 = 1, \ \omega_i \geq 0, \ i = 1, 2, \cdots, K, \end{aligned} \tag{9}$$

where C_{12}^F, C_{11}^F, and C_{22}^F are defined in (8), $Tr(\cdot)$ denotes the trace of a matrix, and $I_d \in R^{d \times d}$ is the identity matrix.

3.3 Optimization

For optimization problem (9), we design an efficient two-stage procedure to seek its optimal solutions. In the first stage, we need to find the nonnegative weight coefficients. Inspired by [5], we make use of the following method to compute weight coefficients $\omega_1, \omega_2, \cdots, \omega_K$. To be specific, we first define the merged GCMs as:

$$C_i^{\phi_i} = \begin{bmatrix} C_{11}^{\phi_i} & C_{12}^{\phi_i} \\ C_{21}^{\phi_i} & C_{22}^{\phi_i} \end{bmatrix} \in R^{(d_1+d_2)\times(d_1+d_2)}, \ i = 1, 2, \cdots, K. \tag{10}$$

Then, we transform all the merged matrices in (10) into column vectors, i.e.,

$$c_i = vec(C_i^{\phi_i}) \in R^{(d_1+d_2)^2}, \ i = 1, 2, \cdots, K, \tag{11}$$

where $vec(\cdot)$ is an operator that transforms a matrix into a column vector.

Based on (11), we can obtain a matrix $Q = [c_1, c_2, \cdots, c_K]^T$ via arranging these K column vectors. Now, we decompose Q by a matrix approximation strategy. That is, a latent linear subspace can be found by minimizing the following loss

$$L(P, Z) = \|Q - PZ\|_F^2, \ s.t. \ P^T P = I_r, \tag{12}$$

where $\|\cdot\|_F$ represents the Frobenius norm of a matrix, $P \in R^{K \times r}$ is a linear transformation and $Z \in R^{r \times (d_1+d_2)^2}$, $r \leq \min(K, (d_1 + d_2)^2)$. Let $Z = P^T Q$. The problem of minimizing (12) can be equivalently reformulated as the following optimization problem

$$\max_P \|P^T Q\|_F^2, \ s.t. \ P^T P = I_r. \tag{13}$$

The optimization problem (13) can be solved by the standard eigenvalue decomposition of $\Sigma_Q = QQ^T$.

Theorem 1. *Suppose $\{C_i^{\phi_i}\}_{i=1}^K$ are a set of K distinct merged general covariance matrices defined in (10) and Q is a matrix containing all vectorized $C_i^{\phi_i}$. Then, all the entries of $\Sigma_Q = QQ^T$ are nonnegative.*

Proof. It is clear that the (i, j)-th entry of $\Sigma_Q = QQ^T$ is $c_i^T c_j$ for any $i, j \in \{1, 2, \cdots, K\}$. Notice that

$$c_i^T c_j = Tr(C_i^{\phi_i} C_j^{\phi_j}). \tag{14}$$

It is easy to show that the matrix $C_i^{\phi_i}$ is a symmetric and positive semidefinite matrix according to (5) and (10). Hence,

$$C_i^{\phi_i} = M_i M_i^T \tag{15}$$

must exist, where $M_i \in R^{(d_1+d_2)\times(d_1+d_2)}$, $i = 1, 2, \cdots, K$. It follows immediately from (15) that

$$\begin{aligned} Tr(C_i^{\phi_i} C_j^{\phi_j}) &= Tr(M_i M_i^T M_j M_j^T) \\ &= Tr(M_j^T M_i M_i^T M_j) \\ &= Tr\left((M_i^T M_j)^T (M_i^T M_j)\right) \\ &= Tr(M^T M) \end{aligned} \tag{16}$$

where $M = M_i^T M_j$. Since

$$Tr(M^T M) = \|M\|_F^2 \geq 0, \tag{17}$$

it follows that

$$c_i^T c_j = Tr(C_i^{\phi_i} C_j^{\phi_j}) \geq 0, \ i, j = 1, 2, \cdots, K. \tag{18}$$

It means that all the elements of Σ_Q are nonnegative. Hence, Theorem 1 is true.

According to Theorem 1, $\Sigma_Q \in R^{K\times K}$ is bound to be a nonnegative matrix. Together with the well-known Perron-Frobenius theorem, it is straightforward to show that the top eigenvalue of Σ_Q is real and nonnegative, with an associated nonnegative eigenvector. Thus, weight coefficients $\omega_1, \omega_2, \cdots, \omega_K$ can be chosen as the elements of the first column vector of P (corresponding to the top eigenvector of Σ_Q). In the second stage, substituting the obtained weight coefficients into optimization problem (9), we are able to get W_1 and W_2 by selecting the top eigenvectors of the following generalized eigenvalue problem

$$\begin{bmatrix} 0 & C_{12}^F \\ C_{21}^F & 0 \end{bmatrix} \begin{bmatrix} w_1 \\ w_2 \end{bmatrix} = \mu \begin{bmatrix} C_{11}^F & 0 \\ 0 & C_{22}^F \end{bmatrix} \begin{bmatrix} w_1 \\ w_2 \end{bmatrix}, \tag{19}$$

where μ is the generalized eigenvalue corresponding to the eigenvector $[w_1^T \; w_2^T]^T$. Once W_1 and W_2 are obtained, the latent multiview representation can be extracted and fused through the following manner [11]:

$$\mathcal{F} = \begin{bmatrix} W_1 \\ W_2 \end{bmatrix}^T \begin{bmatrix} X_1 \\ X_2 \end{bmatrix} = \sum_{i=1}^{2} W_i^T X_i. \tag{20}$$

4 Extension of MCP

MCP can handle two views well, but not deal with more than two views directly. In this section, we generalize it in the scenario of more than two views. To be specific, assume that m data matrices are given as $X_1 \in R^{d_1\times n}$, $X_2 \in R^{d_2\times n}$, $\cdots$, $X_m \in R^{d_m\times n}$, where d_i denotes the dimension of training samples in view i, $i = 1, 2, \cdots, m$. The extended version of MCP, which is referred to as multiview MCP, optimizes the following problem:

$$\begin{aligned} &\max_{\{W_i\}_{i=1}^m, \{\omega_t\}_{t=1}^K} \sum_{i=1}^{m}\sum_{j=1}^{m} Tr(W_i^T C_{ij}^F W_j) \\ &s.t. \;\; W_i^T C_{ii}^F W_i = I_d, \; i = 1, 2, \cdots, m, \\ &\qquad \sum_{t=1}^{K} \omega_t^2 = 1, \; \omega_t \geq 0, \; t = 1, 2, \cdots, K, \end{aligned} \tag{21}$$

where C_{ij}^F is defined in (8), $W_i \in R^{d_i\times d}$ denotes the projection matrix, $d \leq \min(d_1, d_2, \cdots, d_m)$, and $i, j = 1, 2, \cdots, m$.

For solutions to optimization problem (21), we follow the process in Sect. 3.3 to compute them. It should be pointed out that the calculation of W_1, W_2, $\cdots$, W_m uses an iterative procedure in [17]. More details about this procedure can be found in [17].

5 Experiments

In order to verify the effectiveness of our proposed MCP method, we implement multiview classification and clustering experiments using three popular datasets, including a document dataset [4], a handwritten digit dataset[1], and an image dataset[2]. In addition, we compare our MCP method with related methods, including CCA [6], DCCA [1], deep canonically correlated autoencoders (DCCAE) [13], and $L_{2,1}$-CCA [14].

5.1 Data Preparation

The following three well-known datasets are utilized in our experiments:

1) BBC-Sport: It is a document dataset including 544 documents from 5 distinct categories. This dataset has two different views with respective dimensions as 3183 and 3203.
2) UCI Digits (UD for short): It contains 2000 samples of 10 handwritten digits (0–9). There are 200 samples per class. Six kinds of features of handwritten digits are extracted from a collection of Dutch utility maps. That is, 76 Fourier coefficients of the character shapes (fou), 216 profile correlations (fac), 64 Karhunen-Love coefficients (kar), 240 pixel averages in 2×3 windows (pix), 47 Zernike moments (zer), and 6 morphological features (mor).
3) Caltech 101-10: It is a subset of the Caltech 101 dataset[2] and contains 850 images of 10 classes. Each class has 85 images with size of 48×48 pixels. We make use of grayscale images (Gra), Gabor feature (Gab), and LBP features to form three views.

5.2 Experimental Setting

We take advantage of RBF kernel functions with different parameters to construct different general covariance matrices. The RBF kernel parameters are selected from the set $\{10^{-6}, 10^{-5}, \cdots, 10^{4}\}$. In addition, since the excessive number of kernel functions is usually computationally expensive, we set it from 3 to 6 in our experiments.

In classification experiments, we conduct 10 independent runs for each method on the BBC-Sport and UD datasets and k-nearest neighbors (KNN) classifier with $k = 1$ is applied for the performance validation. On each dataset, we randomly select 70% samples per class to generate the training set and the rest are used to form the testing set. We directly carry out the classification test on the BBC-Sport dataset. On the UD dataset, we select any two feature sets as two views, thus having 15 pairs of views in total. The test is performed on each pair of views for each method.

In clustering experiments, we set the number of clusters to the number of classes and make use of the K-means algorithm for data clustering. We use the Caltech 101-10 dataset for this clustering test. There are a total of three pairs of different feature combinations, i.e., Gra-Gab, Gra-LBP, and Gab-LBP. On each feature pair, the accuracy (ACC) and normalized mutual information (NMI) are adopted for the clustering performance test.

1 https://archive.ics.uci.edu/dataset/72/multiple+features.
2 https://data.caltech.edu/records/mzrjq-6wc02.

Table 1. Classification accuracy of CCA, DCCA, DCCAE, $L_{2,1}$-CCA, and our MCP on the BBC-Sport and UD datasets.

Dataset	CCA	DCCA	DCCAE	$L_{2,1}$-CCA	MCP
BBC-Sport	93.78 ± 1.85	54.55 ± 1.73	82.97 ± 2.27	91.88 ± 1.05	94.81 ± 1.57
UD: fac-fou	86.33 ± 0.87	85.39 ± 0.41	84.63 ± 0.62	88.56 ± 1.09	96.90 ± 0.57
UD: fac-kar	94.91 ± 0.67	87.17 ± 0.76	88.81 ± 0.65	96.38 ± 0.46	96.58 ± 0.87
UD: fac-pix	88.78 ± 0.81	84.24 ± 0.80	88.58 ± 0.72	95.15 ± 0.49	97.08 ± 0.45
UD: fac-zer	87.95 ± 1.56	82.87 ± 0.91	82.38 ± 0.60	83.30 ± 0.49	96.12 ± 0.84
UD: fac-mor	79.43 ± 1.76	49.88 ± 1.44	54.31 ± 1.68	75.51 ± 1.06	85.80 ± 1.47
UD: fou-kar	89.05 ± 1.18	87.85 ± 0.64	86.12 ± 0.50	91.43 ± 0.65	93.30 ± 1.05
UD: fou-pix	96.76 ± 0.73	86.54 ± 1.12	82.39 ± 0.99	85.03 ± 1.22	97.35 ± 0.50
UD: fou-zer	74.78 ± 1.40	73.13 ± 0.80	75.60 ± 1.01	81.61 ± 0.82	82.30 ± 1.30
UD: fou-mor	76.86 ± 1.50	56.03 ± 0.58	46.06 ± 1.39	72.43 ± 2.61	73.60 ± 2.09
UD: kar-pix	96.01 ± 0.82	88.45 ± 0.61	86.46 ± 0.52	96.42 ± 0.44	96.78 ± 0.60
UD: kar-zer	91.80 ± 0.91	84.27 ± 0.98	84.96 ± 1.10	85.08 ± 1.07	93.70 ± 0.68
UD: kar-mor	82.42 ± 2.49	55.69 ± 0.94	51.18 ± 0.79	71.68 ± 5.14	76.00 ± 1.91
UD: pix-zer	80.90 ± 2.94	82.46 ± 1.17	82.33 ± 1.41	82.80 ± 1.02	96.37 ± 0.49
UD: pix-mor	88.38 ± 1.42	54.14 ± 0.86	53.87 ± 1.68	74.38 ± 1.58	83.87 ± 1.15
UD: zer-mor	60.10 ± 3.10	36.13 ± 1.07	40.64 ± 1.29	65.63 ± 2.18	68.30 ± 2.04

Table 2. Clustering results of CCA, DCCA, DCCAE, $L_{2,1}$-CCA, and our MCP on the Caltech 101-10 dataset.

Feature Pair	Metric	CCA	DCCA	DCCAE	$L_{2,1}$-CCA	MCP
Gra-Gab	ACC	0.5352	0.3988	0.3929	0.4376	0.6235
	NMI	0.4858	0.2399	0.2683	0.4271	0.5315
Gra-LBP	ACC	0.4964	0.3929	0.3388	0.4917	0.5882
	NMI	0.4011	0.2643	0.2055	0.4791	0.4675
Gab-LBP	ACC	0.5188	0.3176	0.3670	0.4705	0.5647
	NMI	0.4732	0.1820	0.2289	0.4463	0.4781

5.3 Result

Classification. Table 1 shows the classification accuracy values of CCA, DCCA, DCCAE, $L_{2,1}$-CCA, and our MCP method under two views on the BBC-Sport and UD datasets. As we can see from Table 1, our proposed MCP method achieves better classification accuracy than the other four methods on most cases. Both CCA and $L_{2,1}$-CCA methods perform comparably to each other on the whole. The DCCA and DCCAE methods overall perform worse than other methods, particularly when the dimension of two views is unbalanced. These classification results reveal that the learned feature representation by our MCP method is discriminative for classification purpose.

Clustering. Table 2 lists the clustering results of CCA, DCCA, DCCAE, $L_{2,1}$-CCA, and our MCP method on the Caltech 101-10 dataset. From Table 2, we can clearly see that our proposed MCP outperforms other methods under ACC and NMI metrics on most cases. MCP achieves the best ACCs and NMIs on feature pairs Gra-Gab and Gab-LBP, and the best ACC on feature pair Gra-LBP among all the methods. Only on the Gra-LBP case, our MCP under NMI metric performs worse than the $L_{2,1}$-CCA method, but comparably with it. The CCA and $L_{2,1}$-CCA methods overall achieve comparable clustering results and perform better than the DCCA and DCCAE methods. In summary, these experimental results show that the proposed MCP method is also effective for clustering purpose.

6 Conclusion

In this paper, we put forward an MCP approach for learning latent two-view representation, which has the ability to model the complicated relationship between different features. To solve the optimization problem of MCP, we design an efficient two-stage algorithm. In addition, we also present an extension of MCP to deal with more than two views. Many experimental results on three popular datasets show that our proposed MCP method is promising.

Acknowledgement. This work is supported by the National Natural Science Foundation of China under grants 62176126 and 62076217, and the China Postdoctoral Science Foundation under grant 2020M670995.

References

1. Andrew, G., Arora, R., Bilmes, J.A., Livescu, K.: Deep canonical correlation analysis. In: International Conference on Machine Learning (ICML), pp. 1247–1255. JMLR.org (2013)
2. Bach, F.R., Jordan, M.I.: Kernel independent component analysis. J. Mach. Learn. Res. **3**, 1–48 (2002)
3. Chen, L., Wang, K., Li, M., Wu, M., Pedrycz, W., Hirota, K.: K-means clustering-based kernel canonical correlation analysis for multimodal emotion recognition in human-robot interaction. IEEE Trans. Industr. Electron. **70**(1), 1016–1024 (2023)
4. Chen, M., Huang, L., Wang, C.D., Huang, D.: Multi-view clustering in latent embedding space. In: AAAI Conference on Artificial Intelligence (AAAI), pp. 3513–3520. AAAI Press (2020)
5. Gu, Y., Wang, C., You, D., Zhang, Y., Wang, S., Zhang, Y.: Representative multiple kernel learning for classification in hyperspectral imagery. IEEE Trans. Geosci. Remote Sens. **50**(7), 2852–2865 (2012)
6. Hotelling, H.: Relations between two sets of variates. Biometrika **28**(3/4), 321–377 (1936)
7. Karami, M., Schuurmans, D.: Deep probabilistic canonical correlation analysis. In: AAAI Conference on Artificial Intelligence (AAAI), pp. 8055–8063. AAAI Press (2021)
8. Kettenring, J.R.: Canonical analysis of several sets of variables. Biometrika **58**(3), 433–451 (1971)
9. Kim, T., Kittler, J., Cipolla, R.: Discriminative learning and recognition of image set classes using canonical correlations. IEEE Trans. Pattern Anal. Mach. Intell. **29**(6), 1005–1018 (2007)

10. Sharma, A., Jacobs, D.W.: Bypassing synthesis: pls for face recognition with pose, low-resolution and sketch. In: IEEE Conference on Computer Vision and Pattern Recognition (CVPR), pp. 593–600. IEEE (2011)
11. Sun, Q., Zeng, S., Liu, Y., Heng, P., Xia, D.: A new method of feature fusion and its application in image recognition. Pattern Recogn. **38**(12), 2437–2448 (2005)
12. Uurtio, V., Bhadra, S., Rousu, J.: Large-scale sparse kernel canonical correlation analysis. In: International Conference on Machine Learning (ICML), pp. 6383–6391. PMLR (2019)
13. Wang, W., Arora, R., Livescu, K., Bilmes, J.A.: On deep multi-view representation learning. In: International Conference on Machine Learning (ICML), pp. 1083–1092. JMLR.org (2015)
14. Xu, M., Zhu, Z., Zhang, X., Zhao, Y., Li, X.: Canonical correlation analysis with $L_{2,1}$-norm for multiview data representation. IEEE Trans. Cybern. **50**(11), 4772–4782 (2020)
15. Yuan, Y., Li, J., Li, Y., Gou, J., Qiang, J.: Learning unsupervised and supervised representations via general covariance. IEEE Signal Process. Lett. **28**, 145–149 (2021)
16. Yuan, Y., et al.: OPLS-SR: a novel face super-resolution learning method using orthonormalized coherent features. Inf. Sci. **561**, 52–69 (2021)
17. Yuan, Y., et al.: Learning canonical f-correlation projection for compact multi-view representation. In: IEEE/CVF Conference on Computer Vision and Pattern Recognition (CVPR), pp. 19238–19247. IEEE (2022)
18. Yuan, Y., Sun, Q., Zhou, Q., Xia, D.: A novel multiset integrated canonical correlation analysis framework and its application in feature fusion. Pattern Recogn. **44**(5), 1031–1040 (2011)
19. Zhang, L., Wang, L., Bai, Z., Li, R.: A self-consistent-field iteration for orthogonal canonical correlation analysis. IEEE Trans. Pattern Anal. Mach. Intell. **44**(2), 890–904 (2022)

Explainable Sparse Associative Self-optimizing Neural Networks for Classification

Adrian Horzyk[1(✉)], Jakub Kosno[1], Daniel Bulanda[1], and Janusz A. Starzyk[2,3]

[1] AGH University of Krakow, al. A. Mickiewicza 30, 30-059 Krakow, Poland
horzyk@agh.edu.pl, daniel@bulanda.net

[2] University of Information Technology and Management in Rzeszow, Rzeszow, Poland

[3] Ohio University, Athens, OH 45701, USA
starzykj@gmail.com

Abstract. Supervised models often suffer from a multitude of possible combinations of hyperparameters, rigid nonadaptive architectures, underfitting, overfitting, the curse of dimensionality, etc. They consume many resources and slow down the optimization process. As real-world objects are related and similar, we should adapt not only network parameters but also network structure to represent patterns and relationships better. When the network reproduces the most essential and frequent data patterns and relationships and aggregate similarities, it becomes not only efficient but also explainable and trustworthy. This paper presents a new approach to detect and represent similarities of numerical training examples to self-adapt a network structure and its parameters. Such a network will facilitate the classification by identifying hyperspace regions associated with the defined classes in a training dataset. Our approach demonstrates its ability to automatically reduce input data dimension by removing features that produce distortions and do not support the classification process. The presented adaptation algorithm uses only a few optional hyperparameters and produces a sparse associative neural network structure that fits contextually any given dataset by detecting data similarities and constructing hypercuboids in data space. The explanation of these associative adaptive techniques is followed by the comparisons of the classification results against other state-of-the-art models and methods.

Keywords: Explainable self-adaptive models · Sparse network structure self-development · Associative fuzzy representation and processing · Automatic reduction of dimensionality · Underfitting and overfitting

---

Research project partly supported by program "Excellence initiative – research university" for the AGH University of Krakow, grant IDUB 1570, and Adrian Horzyk was also partially supported by the Ministry of Education and Science (Agreement Nr 2022/WK/1).

B. Luo et al. (Eds.): ICONIP 2023, CCIS 1963, pp. 229–244, 2024.
https://doi.org/10.1007/978-981-99-8138-0_19

1 Introduction

Relationships give meaning to data, define complex objects, behaviors, and features, and play an important role in modeling the world and constituting knowledge. During training, modern computational models try to learn relationships using predefined network structures, functions, and properties. Often used predefined rigid network structures cannot fit complex training data and their relationships adequately without additional effort and many experiments, usually causing underfitting or overfitting [1]. In the real world, not everything is related or necessary to make decisions. In contrast, fully-connected layers process data in the context of all inputs of the previous layer(s), which exposes results to various inferences, decreasing the final accuracy, producing bigger losses, slowing down computations, and making models hard to explain and trust. Explainability is very desirable in human-sensitive sectors, including healthcare, employment, financial, and legal sectors, which increasingly oppose automated decision-making without an explanation of the logic involved [2]. The trend of the current model design strives for adequate explanations that can be quantitatively and qualitatively assessed by numerous metrics [4]. The attention mechanisms have shown that the limitation or decreasing the impact of some inputs can improve the network adaptation and final results [5,6]. Convolutional Neural Networks (CNN or ConvNets) successfully focus on input subareas [1]; however, their structures are still rigid, homogeneous, and not self-adapted to training data, so they are prone to underfitting and overfitting.

Sparse architectures have become more popular in recent years. In [15], the authors evaluate three state-of-the-art techniques for inducing sparsity in deep neural networks. They discuss the benefits and challenges of sparse neural networks. Paper [16] presents a benchmark for evaluating the performance of sparse neural networks. It highlights the limitations of existing benchmarks and the challenges faced by current state-of-the-art sparse networks. The authors of paper [17] propose a sparse evolutionary training procedure for adaptive neural networks. They introduce sparse bipartite layers that replace fully-connected layers based on data distributions. In [18], an algorithm inspired by the workings of a human brain is presented for training highly sparse neural networks. It compares the proposed algorithm with state-of-the-art sparse training algorithms. Sparse neural architectures also occur naturally in nature.

Brains adapt to represent data and understand their relationships, focusing on explaining things and their dependencies. These adaptations primarily involve altering neural connections and neuron parameters [9]. In contrast, artificial neural networks (ANNs) adapt by adjusting the model parameters without modifying the structure of connections to suit the specificity of training data.

Geniuses have richer connection structures in the brain regions related to their abilities and skills compared to others. They excel in discovering and relating information to produce valuable conclusions. Our brains develop new dendritic spines daily, enriching our neuronal connection structures [7,8]. ANNs require algorithms to develop their structures for different training data and automatically detect and represent essential relationships hidden in the data to prevent underfitting, overfitting, and the use of a huge number of combinations of hyperparameters that are difficult to optimize.

This paper introduces sparse associative networks ASONNv2 that automatically adapt not only the parameters but also develop the structure to given training data based on the detected relationships, making the constructed models efficient and satisfying the requirements of eXplainable AI (XAI) [3]. We improved the algorithms of ASONNv1 presented in [10, 11] and followed the ideas and approaches presented in [9]. During the introduced adaptation process, some sub-hyperspace frequent patterns are found and assigned to the defined classes, producing a sparsely connected network structure. This structure is adjusted to the detected relationships and groups of similar training examples in different sub-spaces. Such adaptation facilitates the interpretation and comprehensibility of results, enabling the drawing of conclusions, creation of rules, and trust in results.

The ASONNv2 networks can be used mainly for classification; therefore, we present comparisons to other best-performing classification methods and models. However, the main purpose of this paper is to show how adaptive sparse structures can be created and associate features, objects, and classes based on the detected relationships. The presented construction and adaptation algorithm does not require the optimization of many hyperparameters but we can only optionally check a few combinations of them. These models do not require regularization as they represent aggregated similar patterns as hypercuboids in subspaces, automatically reducing dimensionality and not being prone to overfitting. They automatically use those attributes that positively impact the classification results, solving the curse of the dimensionality problem.

2 Detecting and Representing Relationships and Patterns

The fundamental problem in knowledge representation for predictive AI models is the adequate representation of frequent dependencies and data patterns. If these patterns are under-represented or related incorrectly, models are either underfitting or overfitting (trying to memorize training data). We can use methods of frequent pattern detection known from the field of data mining [12], but modern AI models use neurons to learn representations of similar and frequent input patterns, resembling pattern-mining processes. Determining the number of neurons in subsequent layers to avoid underfitting and overfitting, requires many experiments and regularization techniques [1]. The backpropagation-based training also tries to represent data relationships, but we cannot control them and their adequacy can only be assessed through the results obtained.

In our research, we used associative structures (AGDS [10]) to construct a sparsely connected network structure representing similarities, frequencies, and other relationships between training examples. These structures help identify combinations of ranges that effectively represent the defined classes. In ASONNv2, we refined and enriched the algorithms of ASONNv1 presented in [11] and used a specific stop condition and the softmax function for class neurons.

During adaptation, we gradually enlarge so-called hypercuboids (defining aggregated combinations of feature ranges) in the input space and relate them to the defined classes. This process continues until the minimum discrimination between training examples of different classes is maintained, ensuring correct classification. Interfering attributes are detected, counted, and limited or removed from hypercuboids allowing them to exist within lower-dimensional sub-spaces of the original input data hyperspace. The hypercuboids maximize the number of features and the number of represented training examples of these classes. Conversely, they minimize the number of features from other classes and prevent the full representation of training examples of those classes.

2.1 Initial Associative Structure Construction

We start from building an **Associative Graph Data Structure (AGDS)**, described in [10], which represents the transformed training data defined by input features (for attributes $a \in \{1, ..., A\}$) and class labels (Fig. 1). This structure counts and aggregates representations of all duplicates of values of each attribute a separately and converts all feature values (v_n^a) and class labels into graph nodes: **value nodes** (VN_n^a) representing aggregated input features for all attributes separately, **object nodes** (ON_i) representing training examples (aggregated if duplicated), and **class nodes** (CN_c) representing class labels. In this structure, VNs are connected in order for each attribute, and their weights are calculated after (1)

$$w_{VN_n^a, VN_{n+1}^a} = \left(1 - \frac{|v_n^a - v_{n+1}^a|}{R^a}\right)^s \tag{1}$$

where $R^a = Max^a - Min^a$ is the range of values of attribute a, v_n^a and v_{n+1}^a are the neighbor values of attribute a represented by directly connected nodes VN_n^a and VN_{n+1}^a, and $s \in \{1, 2, 3\}$ (empirically determined as best performing) that is a hyperparameter strengthening the influence of similar values. This structure can also be used for categorical (symbolic) data. In this case, no connections are created between value nodes representing aggregated features of such attributes.

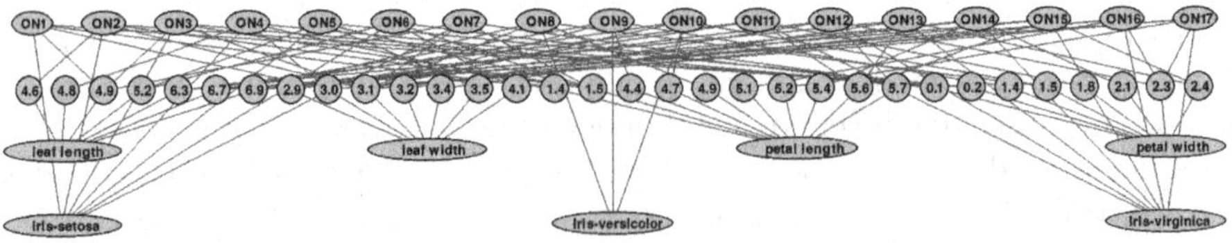

Fig. 1. The AGDS constructed for 17 training examples from the Iris dataset after [10]. The first row contains object neurons (ON_i) representing aggregated training examples. The second row has value neurons (VN_n^a) representing aggregated duplicates of attribute values. The third row is populated with nodes representing attribute names. The last row shows nodes (CN_c) that represent classes.

The connections between VNs and ONs are calculated according to formula (2), where $\leftrightarrow$ symbolizes the existence of the direct connection between them:

$$w_{VN_n^a,ON_i} = \frac{\frac{\|\{ON_j:VN_n^a \leftrightarrow ON_j \leftrightarrow CN_c \leftrightarrow ON_i\}\|}{\|\{ON_l:VN_n^a \leftrightarrow ON_l\}\|}}{\sum_{VN_m^a:VN_m^a \leftrightarrow ON_i} \frac{\|\{ON_j:VN_m^a \leftrightarrow ON_j \leftrightarrow CN_c \leftrightarrow ON_i\}\|}{\|\{ON_l:VN_m^a \leftrightarrow ON_l\}\|}} \quad (2)$$

where a, c, i, j, l, n, m are indices specifying instances of the nodes. Formula $\|\{ON_j : VN_n^a \leftrightarrow ON_j \leftrightarrow CN_c \leftrightarrow ON_i\}\|$ stands for the number of all object neurons (ON_j) which are directly connected to value neuron VN_n^a and class neuron CN_c, which has the direct connection to object neuron ON_i. It means that we are interested in counting up all object neurons of the same class as object neuron ON_i, limiting our search to those object neurons that are also connected to value neuron VN_n^a. Formula $\|\{ON_l : VN_n^a \leftrightarrow ON_l\}\|$ stands for the number of all object neurons (ON_l) that are connected to value neuron VN_n^a. All the other weights of sparse connections of this structure are set to 1.

2.2 Fuzzy Hypercuboids Aggregating Similarities

The main idea behind the construction of the introduced fuzzy hypercuboids is the aggregation of representations of similar training examples of the same classes defined in the training dataset.

Definition 1. Hypercuboid is an area of the input data subspace defined by the combination of the value subranges of all or selected attributes defining training examples.

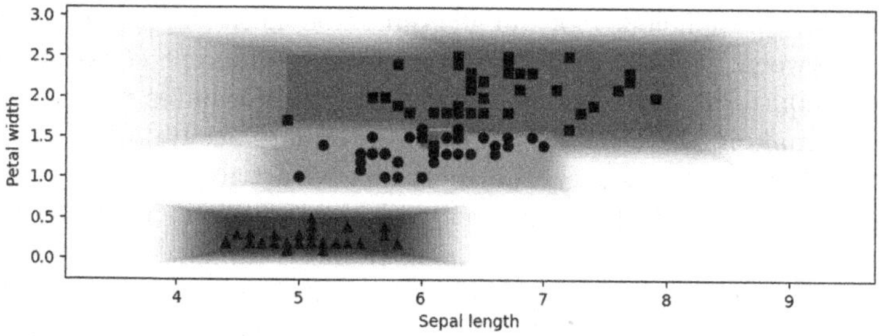

Fig. 2. View of the possible hypercuboids with fuzzy boundaries created for two selected attributes of the sample training examples of three classes, where each hypercuboid represents training examples of one class defined in the training dataset.

Each point within a hypercuboid (Fig. 2) is classified as assigned to the same class as the training example that constituted this hypercuboid. The hypercuboids representing training examples of different classes do not overlap, while the hypercuboids of the same class may overlap. The spaces between hypercuboids are covered by the fuzzy boundaries which gradually diminish in influence as we move away from the hypercuboids' vicinity. Thanks to this approach, the constructed classifiers can generalize not only within hypercuboids but also outside their boundaries, albeit with a decreasing probability of accurate classification and reflecting less certainty.

To represent hypercuboids with fuzzy boundaries, we define **fuzzy ranges** of the attribute values represented by **fuzzy range neurons** (RN_r^a). Various combinations of these ranges define hypercuboids that will be represented by **combination neurons** (KN_k). Each combination neuron is connected to a **class neuron** (CN_c) that can be connected to many such combination neurons. Hypercuboids develop gradually in an adaptation process driven by associations formed between value nodes and object nodes of the AGDS structure constructed for training data.

2.3 Adaptation of Range and Combination Neurons

First, we swap all AGDS nodes into neurons in the constructed ASONNv2 network. Next, we start developing fuzzy range neurons RN_r^a and combination neurons KN_k for similar training examples. Each object neuron (ON_i) calculates its correlations to all other object neurons (ON_j) assigned to classes other than the class to which ON_i is connected. It allows choosing **the most correlated object neuron** (ON_q) to the training examples of other classes. All connected VN_n^a to this neuron are connected to new RN_r^a neurons, which all are connected to a new KN_k neuron. The initial ranges of the new combination are defined by single values. Each RN_r^a is linked to the a attribute, which values it represents. This process repeats every time we start the construction of a new hypercuboid in the data space after finishing the construction of the previous one.

To enhance the efficiency of our classification network, we identify shared attributes between training examples belonging to different classes. This helps us eliminate contradictory examples that share all attributes but differ in-class assignments, making them impossible to discriminate accurately.

We prioritize examples with the most shared attributes as starting points for expanding hypercuboids during network construction. This choice tackles the most challenging discrimination cases. If the training data are stored in a table, the process of determining these shared attribute values would typically require quadratic computational complexity. Nonetheless, training examples are represented by the AGDS structure, which aggregates the same values and represents them by the same VNs linked to ONs representing the training examples. Therefore, we can traverse along the edges in this graph structure to calculate these numbers with linear computational complexity. This underscores the merit of using the AGDS structure instead of a tabular structure for storing training examples.

After the creation of a new combination neuron (KN_k) connected to new fuzzy range neurons (RN_r^a), we start expanding the hypercuboid in the input data space. The primary objective is to include a maximum number of training examples of the same class while excluding those from other classes and minimizing the representation of their respective features. This adaptation process is crucial for developing a small number of large hypercuboids encompassing all non-contradictory training examples of the same class. The AGDS structure is also helpful during the exploration of potential hypercuboid expansions in the data space, as it enables rapid conflict detection with examples of other classes. Fuzzy range neurons (RN_r^a) linked to the combination neuron (KN_k) can expand their ranges below the current minimum and above the current maximum values. In each hypercuboid expansion step, we try to assess which attribute expansion is the most beneficial based on unrepresented features of the training examples from the same class and those from different classes.

Definition 2. *Seeds* are all features of training examples from the same class as the one associated with the constructed hypercuboid, which are included in the expanded ranges defining this hypercuboid.

Definition 3. *Weeds* are all features of training examples from different classes than the one associated with the constructed hypercuboid, which are included in the expanded ranges defining this hypercuboid.

To achieve good generalization properties, we need to maximally expand ranges (represented by RN_r^a) by including as many *Seeds* and as few *Weeds* as possible. At every expanding step, some training examples of the same class (associated with KN_k) already have some features represented by the range neurons (RN_r^a). Consequently, adding the subsequent features of these training examples is more beneficial than adding the first features of newly or partially represented training examples from the same class. Ultimately, the objective is to add all features of as many training examples of the same class as possible to the ranges defining the constructed combination neuron (KN_k). On the other hand, we avoid including the subsequent features (*Weeds*) of training examples from different classes within the ranges of the constructed combination neuron because they would weaken its representation and decrease its discriminating properties. Therefore, the presented algorithm rewards the inclusion of the subsequent features (*Seeds*) from the already represented training examples in the ranges of the combination neuron while penalizing the inclusion of the subsequent features (*Weeds*) from training examples of other classes.

The *Seeds* of RN_r^a (3) is the sum of the squared numbers of features of object neurons of the same class as the constructed combination within this range (RN_r^a).

$$Seeds_{RN_r^a} = \sum_{\{ON_i : ON_i \leftrightarrow CN_c \leftrightarrow KN_k \leftrightarrow RN_r^a\}} \left\| \{VN_n^a : RN_r^a \leftrightarrow VN_n^a \leftrightarrow ON_i\} \right\|^2 \tag{3}$$

where the symbol $\leftrightarrow$ stands for a direct connection between the nodes.

The *Weeds* of RN_r^a (4) is the sum of the squared numbers of features of object neurons of different classes than the class of the constructed combination

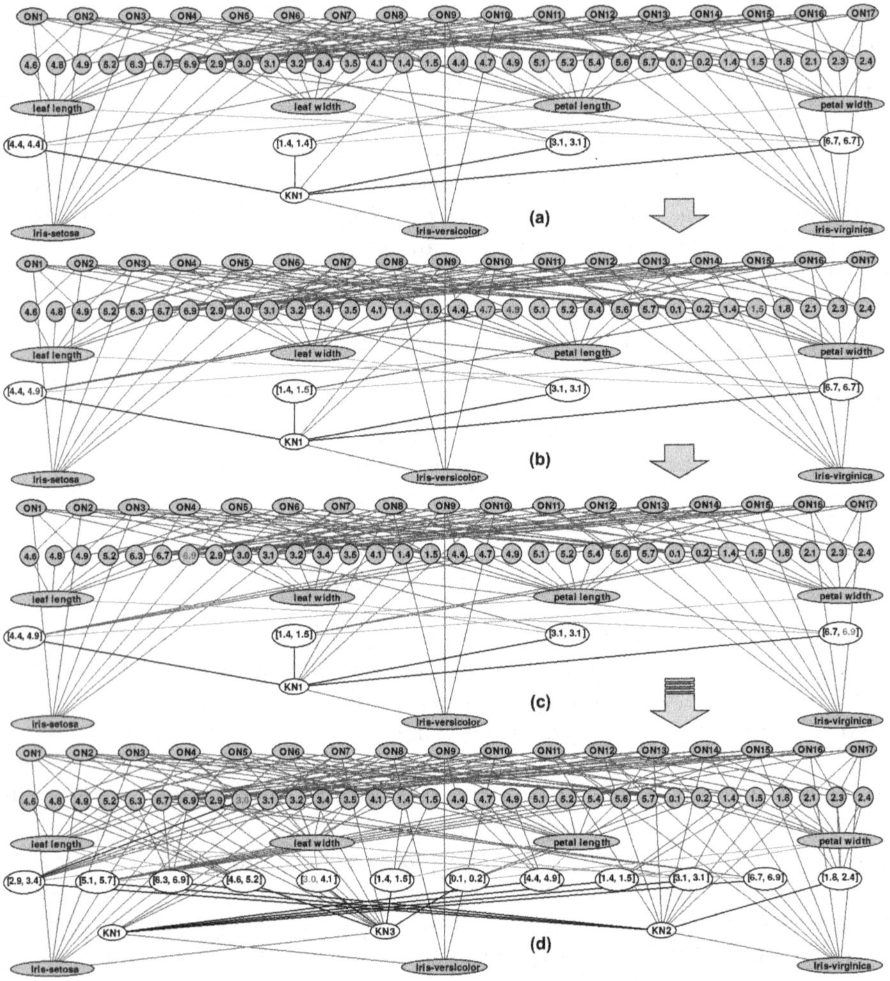

Fig. 3. Development of ASONNv2 for 17 training examples from the Iris dataset: a) Initial development stage in which combination neuron $KN1$ was initialized based on object neuron $ON9$, b) the expansion of the ranges by the neighbor values which add Seeds, c) the expansion of the ranges by the neighbor value which adds Seeds but also includes some Weeds, d) the final result of the expansion.

included in this range (RN_r^a).

$$Weeds_{RN_r^a} = \sum_{\{\widetilde{ON}_i : \widetilde{ON}_i \leftrightarrow \widetilde{CN}_e \not\leftrightarrow KN_k \leftrightarrow RN_r^a\}} \left\| \left\{ VN_n^a : RN_r^a \leftrightarrow VN_n^a \leftrightarrow \widetilde{ON}_i \right\} \right\|^2 \quad (4)$$

where the symbol $\not\leftrightarrow$ indicates that there is no direct connection between the nodes.

If an object neuron (ON_i) represents a training example assigned to the class represented by the class neuron (CN_c), then there exists a connection between them ($ON_i \leftrightarrow CN_c$). Similarly, if a combination neuron (KN_k) represents training examples of the class represented by the class neuron (CN_c), then we create a connection between them ($KN \leftrightarrow CN_c$). On the other hand, if the combination neuron KN_k is not assigned to represent training examples of class $\widetilde{CN_e}$, then there is no connection to this class neuron, which is denoted as $KN_k \not\leftrightarrow \widetilde{CN_e}$.

Example. After constructing the AGDS structure for the dataset and calculating all connections to $\widetilde{ON_i}$ representing training examples of $\widetilde{CN_e}$, we can find the object neuron ON_q that has the highest number of connections to object neurons representing training examples of classes other than its assigned class ($ON_q \leftrightarrow CN_c$). The defining values of object neuron ON_q represented by value neurons (VN_n^a) are connected to new range neurons (RN_r^a) for each attribute a creating a new combination neuron (KN_k). Initially, this combination neuron represents a point in the hyperspace. In the subsequent steps, it will expand the combined ranges to represent a hypercuboid assigned to the class to which the chosen object neuron (ON_q) is connected. Figure 3a shows an example of four new range neurons (initially representing values 4.4, 1.4, 3.1, and 6.7) and a new combination neuron ($KN1$) connected to the $Iris-versicolor$ class neuron.

2.4 Expansion of the Range Neurons Enlarging Hypercuboids

Once the range neurons (RN_r^a) are initialized by single values, they can be expanded (Fig. 3b-d). When working with numerical features, only the next bigger or smaller value can be used to expand the range neuron, while for categorical features, any value can be used (expanding a subset instead of a range). If such bigger or smaller values for any attribute define only objects from the class assigned to the combination neuron (KN_k), extensions can always be made. If a bigger or smaller value defines at least one object neuron of a different class, we must examine whether an extension is feasible. This examination is based on coefficients that describe the potential benefit of expanding the range towards smaller (5) or bigger (6) values within the existing ranges. Our goal is to find the most promising VN_n^a for one of the attributes that will expand the constructed hypercuboid in such a way that the sum of $Seeds_{RN_r^a}$ will grow the most and the sum of $Weeds_{RN_r^a}$ will grow the least (Fig. 3c-d) or at all (Fig. 3b) for all attributes defining this hypercuboid. The expansion process is controlled by all VN_n^a on the borders of all ranges. Therefore, we define coefficients (5) and (6) for both expansion directions for each RN_r^a connected to the KN^k to find the most beneficial one:

$$dir^-_{RN_r^a} = \frac{1}{\gamma_a} \sum_{VN_n^a < RN_{min}^a} dir^-_{VN_n^a} \tag{5}$$

$$dir^+_{RN^a} = \frac{1}{\gamma_a} \sum_{VN_n^a > RN_{max}^a} dir^+_{VN_n^a} \tag{6}$$

where $dir^{-}_{RN^a_r}$ denotes the expansion to the smaller value, $dir^{+}_{RN^a_r}$ denotes the expansion to the bigger value, and γ_a is the cost of the attribute a (an optional hyperparameter). The expansion direction for the a attribute can be considered if its value is positive ($dir^{-}_{RN^a_r} > 0$ or $dir^{+}_{RN^a_r} > 0$). For example, we can automatically optimize the use of various attributes, giving preference to the least expensive ones while avoiding the most expensive ones to collect. However, if certain attributes are indispensable for discrimination purposes, they can still be used. Therefore, setting their cost greater than one ($\gamma_a > 1.0$) does not exclude them from the adaptation process but gives preference to those that are cheaper. When $\gamma_a = 1.0$ for all a, then all are used with the same priority. To calculate (5) and (6), we need the following:

$$dir^{-}_{VN^a_n} = \delta^{RN^a_{min}}_{VN^a_{val}} * \left(\sum_{ON^{same}} \delta^{repr}_{ON^{same}} * \delta^{cont}_{ON^{same}} - \sum_{ON^{diff}} \delta^{cont}_{ON^{diff}} \right) \quad (7)$$

$$dir^{+}_{VN^a_n} = \delta^{RN^a_{max}}_{VN^a_{val}} * \left(\sum_{ON^{same}} \delta^{repr}_{ON^{same}} * \delta^{cont}_{ON^{same}} - \sum_{ON^{diff}} \delta^{cont}_{ON^{diff}} \right) \quad (8)$$

where

$$\delta^{RN^a_{min}}_{VN^a_{val}} = \left(1 - \frac{VN^a_{val} - RN^a_{min}}{RN^a_{range}} \right)^2 \quad (9)$$

$$\delta^{RN^a_{max}}_{VN^a_{val}} = \left(1 - \frac{RN^a_{max} - VN^a_{val}}{RN^a_{range}} \right)^2 \quad (10)$$

$$RN^a_{range} = RN^a_{max} - RN^a_{min} \quad (11)$$

$$RN^a_{max} = \max_{VN^a_{val}} \left\{ VN^a_{val} : \exists_{ON_i \leftrightarrow CN_c \leftrightarrow KN_k \leftrightarrow RN^a_r} ON_i \leftrightarrow VN^a_{val} \right\} \quad (12)$$

$$RN^a_{min} = \min_{VN^a_{val}} \left\{ VN^a_{val} : \exists_{ON_i \leftrightarrow CN_c \leftrightarrow KN_k \leftrightarrow RN^a_r} ON_i \leftrightarrow VN^a_{val} \right\} \quad (13)$$

$$\delta^{repr}_{ON^{same}} = \left(\frac{1}{1 + q^{repr}_{ON^{same}}} \right)^2 \quad (14)$$

$$q^{repr}_{ON^{same}} = \sum_{KN_k} \| \{ ON_i : ON_i \leftrightarrow KN_k \} \| \quad (15)$$

$$\delta^{cont}_{ON^{same}} = \left(\frac{1 + q^{VN^a}_{ON^{same}}}{\| RN^a_r : RN^a_r \leftrightarrow KN_k \|} \right)^2 \quad (16)$$

$$\delta^{cont}_{ON^{diff}} = \left(\frac{1 + q^{VN^a}_{ON^{diff}}}{\| RN^a_r : RN^a_r \leftrightarrow KN_k \|} \right)^2 \quad (17)$$

$$q^{VN^a}_{ON^{same}} = \sum_{RN^a_r \leftrightarrow KN_k} \| \{ VN^a_n : RN^a_r \leftrightarrow VN^a_n \leftrightarrow ON_i \leftrightarrow KN_k \} \| \quad (18)$$

$$q^{VN^a}_{ON^{diff}} = \sum_{RN^a_r \leftrightarrow KN_k} \| \{ VN^a_n : RN^a_r \leftrightarrow VN^a_n \leftrightarrow ON_i \not\leftrightarrow KN_k \} \| \quad (19)$$

In each hypercuboid expansion step, only one expansion direction (to a smaller or larger value) of one attribute is chosen. The direction is chosen based on the biggest value of coefficients (5) and (6) calculated for each attribute. We must also consider the minimum number of discriminating features to ensure the discrimination of represented training examples from training examples of other classes. To achieve this, we define minimal discrimination (20), which is calculated by subtracting the maximum number of the same attribute values used to define training examples of other classes from the total number of attributes used in the combination:

$$\begin{aligned} discrim_{min}^{KN_k} &= \|\{RN_r^a : RN_r^a \leftrightarrow KN_k\}\| \\ - \max_{ON_i \nleftrightarrow KN_k} &\|\{VN_n^a : ON_i \leftrightarrow VN_n^a \leftrightarrow ON_q\}\| \end{aligned} \tag{20}$$

This approach allows us to control the expansion of the developed hypercuboids and prevent them from exceeding the user-preferred minimum discrimination level ($discrim_{min}^{user} \geq 1$ to ensure discrimination by at least one differing feature). This hyperparameter is compared to (20), where $discrim_{min}^{KN_k} \geq discrim_{min}^{user}$ to ensure discrimination by the defined number of differing features, where usually $discrim_{min}^{user} \in \{1, 2, 3, 4\}$, which is one of the optional hyperparameters.

The expansion steps are performed until the following condition is satisfied

$$\left(dir_{RN_r^a}^{-} > 0 \vee dir_{RN^a}^{+} > 0\right) \wedge \left(discrim_{min}^{KN_k} \geq discrim_{min}^{user} \geq 1\right) = true. \tag{21}$$

The process of searching for new hypercuboids continues until all non-contradictory training examples are represented by at least one hypercuboid or a given number of hypercuboids (e.g., 1, 2, or 3), which is one of the optional hyperparameters. Therefore, if the dataset does not contain contradictory training examples, the ASONNv2 always achieves 100% accuracy for training examples (never underfits). If any of the constructed hypercuboids represent only one or very few training examples, then such hypercuboids probably represent outliers and can be removed, not interfering evaluation of test examples.

Final hypercuboids can be created for different sub-hyperspaces of the initial input data hyperspace due to the implemented optimization during the adaptation process of increasing *Seeds* and decreasing *Weeds*. On this basis, some less discriminative attributes can be automatically removed, and the final classifier can be constructed based on the most discriminative attributes (dimensions) of the input data. Hence, the presented optimization algorithm searches for the most significant combinations of attribute values that represent similarities between training examples of each class. This makes ASONNv2 models explainable, easily interpretable, and trustful because we can go back from the largest output along the edges and find the combination and the ranges that stand behind the classification. Automatic selection of the most representative attributes and their combinations improves generalization properties, solves the curse of dimensionality problem, improves the efficiency of the model, and suggests which attributes are enough and most essential for correct classification.

2.5 Sparse Associative Self-optimizing Neural Network Classifier

The sparse Associative Self-optimizing Neural Network (ASONNv2) classifier is constructed based on the created combination of neurons assigned to different classes. The output classification neurons of this network assigned to the labels of all defined classes in the dataset are connected to combination neurons representing hypercuboids aggregating and representing training examples of these classes (Fig. 4). The output neurons calculate the maxima of the connected combination neurons and use a softmax function to sharpen the final predictions and determine the probabilities of the predicted classes.

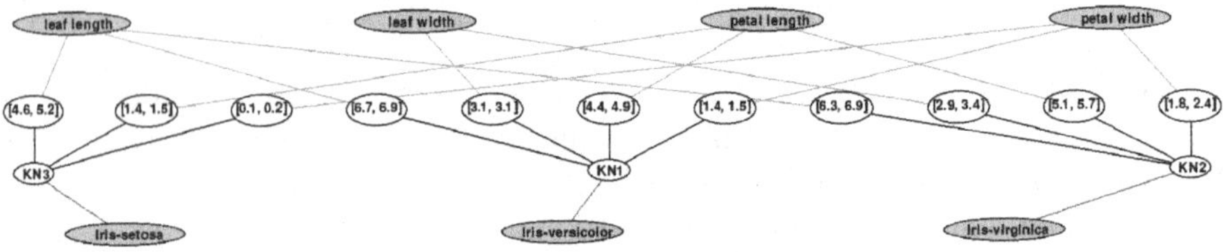

Fig. 4. ASONNv2 constructed for the set of 17 training examples of the Iris dataset.

In Fig. 4, each output neuron is connected to a single combination neuron due to the small size of the demonstrative dataset. However, in larger datasets, there would be multiple combination neurons connected to output class neurons (CN_c) representing all the defined classes in the training dataset.

Finally, we can connect fuzzy range neurons RN_r^a directly to the network inputs (without the use of value neurons VN_n^a) and fuzzify the boarders of the hypercuboids using the Gaussian-cut-hat function as defined in formula (22):

$$GCH_{RN^a}(x) = \begin{cases} 1 & if \quad RN_{Min}^a \le x \le RN_{Max}^a \\ e^{\alpha_{GCH}} & otherwise \end{cases} \tag{22}$$

where (23) defines the fuzzy coefficient of the range borders (RN_{Min}^a and RN_{Max}^a):

$$\alpha_{GCH} = \frac{1 - \left(\frac{2x - RN_{Max}^a - RN_{Min}^a}{RN_{Max}^a - RN_{Min}^a} \right)^2}{2} \tag{23}$$

The weights of connections between range neurons and combination neurons are calculated using (24) to strengthen stimuli coming from the range neurons:

$$w_{RN^a \leftrightarrow KN_k} = \frac{\varphi_{RN_r^a}}{\sum_{\{RN_r^b : RN_r^b \leftrightarrow KN_k\}} \varphi_{RN_r^b}} \tag{24}$$

where $\varphi_{RN_r^a}$ stands for the coefficient used for all fuzzy range neurons to represent unused $Seeds$ and remaining $Weeds$ in the combination.

$$\varphi_{RN_k} = \left(1 - \frac{Weeds_{RN_k}}{\varphi_{ON^{diff}}^{discr}} \right) \frac{\varphi_{ON^{same}}^{repr} + Seeds_{RN_k}}{\varphi_{ON^{same}}^{repr} + AllSeeds_{KN}} \tag{25}$$

Algorithm 1. Expansion of combination neurons

Hyperparameter: *min_discrimination* - the number of all features defining objects (training examples) subtracting the maximum number of values that the combination and each object have in common (20).

Start from building AGDS nodes and connections (Fig. 1)
asonn_not_represented := a list of ONs not connected to KNs (at this point this list contains all ONs)
while len(*asonn_not_represented*) ¿ 0 **do**
  for *object_neuron* in *asonn_not_represented* **do**
    calculate correlation with ONs from other classes - count how many ONs from other classes share 1, 2, 3, etc. values with *object_neuron*
  end for
  combination_seed := choose an ON with the biggest correlation with other classes (e.g., if an ON has the biggest number of shared values with other classes equal to S, choose an ON sharing S values with most objects from other classes. If it is not enough to choose, consider how many (S-1)-value, (S-2)-value, etc. common parts they have) and create KNs and RNs based on it (Fig. 3.a)
  has_combination_been_extended := false
  while not *has_combination_been_extended* **do**
    potential_extensions := a list of VNs with the next bigger and next smaller values for each feature
    for *value_neuron* in *potential_extensions* **do**
      if *value_neuron* is not connected to ONs of different classes than *combination_seed* **then**
        extend *combination_seed* with *value_neuron* (Fig. 3.b)
        has_combination_been_extended := true
      end if
    end for
    potential_extensions := a list of VNs with the next bigger and next smaller values for each feature
    for *value_neuron* in *potential_extensions* **do**
      For each VN, calculate (9), (10), (14), (16), and (17).
      For each VN, calculate $dir^{+}_{VN^a_n}$ (7) if this value would extend the range towards bigger values or $dir^{-}_{VN^a_n}$ (8) otherwise.
      Calculate coefficients (5) and (6).
    end for
    best_extension := choose VN from *potential_extensions* with the biggest $extension_{c}oefficient$
    if extension coefficient of *best_extension* ¿ 0 **and** discrimination after adding *best_extension* ¿= *max_discimination* **then**
      extend *combination_seed* with *value_neuron* (Fig. 3.c)
      has_combination_been_extended := true
    end if
  end while
  asonnn_not_represented := a list of unconnected ONs to KNs
end while
Remove AGDS nodes and connections (transforming Fig. 3.d to Fig. 4).

where coefficients $\varphi^{repr}_{ON^{same}}$ and $\varphi^{discr}_{ON^{diff}}$ define the numbers of represented object neurons by the combination neuron multiplied by the number of the attributes used for the representation of these combinations:

$$\varphi^{repr}_{ON_i} = \|\{ON_i : CN_c \leftrightarrow ON_i \leftrightarrow KN_k\}\| \cdot \|\{RN^a_r : RN^a_r \leftrightarrow KN_k\}\|^2 \quad (26)$$

$$\varphi^{discr}_{ON_i} = \|\{ON_i : CN_c \not\leftrightarrow ON_i \leftrightarrow KN_k\}\| \cdot \|\{RN^a_r : RN^a_r \leftrightarrow KN_k\}\|^2 \quad (27)$$

Finally, the range neurons of the ASONNv2 can be stimulated by any combination of input values, calculating outputs using (22), which subsequently stimulate combination neurons, from which class neurons take the maximum values that are used in the output softmax layer to achieve the classification results.

The algorithm used to build ASONNv2 is quite sophisticated, especially its part devoted to the expansion of combinations. To make this process clearer a pseudocode of the construction algorithm is presented in Algorithm 1. The Python implementation with detailed descriptions of this algorithm can be found in the Github repository [14].

3 Experiments, Results, and Comparisons

The performance of the proposed network, constructed and adapted by the described algorithm, was tested on typical machine learning classification tasks. During the experiments, we used 52 representative datasets from the PMLB dataset [13], which have between 32 and 3200 instances and contain between 3 and 34 binary, categorical, and numerical features and define between 2 and 10 classes. We have removed datasets from comparisons for which some of the compared models failed to produce results. The train and test sets were chosen in a ratio of 70:30.

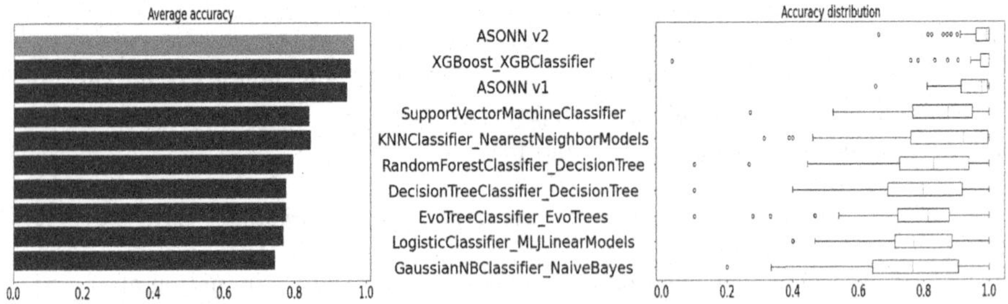

Fig. 5. Comparisons of the average performances and accuracy distribution of all tested models on 52 classification tasks.

Figure 5 compares the average accuracies achieved by the various models and presents a box plot of classification accuracy distributions of all tested models on the 52 datasets. The ASONNv2 outperformed other tested models and classification methods, averaging nearly 100% accuracy on test sets. All methods were tested on a variety of difficult datasets; hence, obvious outliers (small circles) can be observed for all methods. The best accuracy distribution was achieved for the proposed ASONNv2 classifier, which had quite a small number of outliers together with very high scores achieved. It supports the hypothesis that the networks using sparse connections, which represent essential relationships, are not prone to overfitting or underfitting, working stable and robustly.

We conducted statistical tests to see if there were statistically significant differences between the average accuracies of ASONNv1, ASONNv2, and other modern and best-performing models. ASONNv2 enriches ASONNv1 with an improved expansion algorithm for combination neurons, using the additional stop condition in which the coefficient has to be bigger than 0 according to (21) (producing smaller but better-fitted hypercuboids) and the softmax function applied to the class neuron outputs while performing classification. For each pair (ASONNv1 or ASSONv2 vs. other models), we computed an F-test to see if the two accuracy vectors (each of length 52) have equal variances. The p-values obtained were used to determine the t-test variant. The t-test showed that there is no statistically significant difference between ASONNv1, ASONNv2, and XGBoost mean accuracy (p values ≥ 0.40797). For all other pairs, the t-test showed that there were statistically significant differences between the means (p values $\leq 4.09518 \times 10^{-5}$).

4 Conclusions

This paper presented a construction algorithm of sparse ASONNv2 classification networks whose structures were contextually adapted to the detected training data relationships of training datasets. This network construction process is based on hypercuboids representing combinations of fuzzy ranges of different subsets of attribute values. This model can operate on raw numerical data without normalization or standardization. It can reduce the data space dimension, automatically reducing less discriminating attributes for each class. It can also use the cost of all attributes in the development process to use only those that are the cheapest in practical uses. This structure does not require the initialization of weights, which are calculated based on the number of occurrences and similarities. The network-building process is fully automatic, deterministic, and uses three hyperparameters (s used in (1), γ_a used in (5) and (6), and $discrim_{min}^{user}$ limiting (20)), but all of them are small natural numbers, by default equal to one. The adaptation process cannot get stuck in local minima or saddle points because there is no cost function or gradient descent training algorithm used. Due to the use of only a few contextually created connections based on the detected relationships, this network does not underfit and rarely overfits if the training dataset is sufficiently representative, which has been proved in the experiments and the presented comparisons. The comparisons to other state-of-the-art models showed that the proposed sparse associative network achieves similar accuracy to the best models (e.g., XGBoost), so it can be used as an alternative to the other models operating on vectorized data. Finally, the constructed ASONNv2 models are easily interpretable, explainable, and trustful.

References

1. Goodfellow, I., Bengio, Y., Courville, A.: Deep Learning, MIT Press (2016)
2. Ali, S., et al.: Explainable artificial intelligence (XAI): what we know and what is left to attain trustworthy artificial intelligence. Inf. Fusion **99**, 101805 (2023). Elsevier
3. Dwivedi, R., et al: Explainable AI (XAI): core ideas, techniques, and solutions. ACM Comput. Surv. **55**(9), 1–33 (2023). ACM New York, NY
4. Hedström, A., et al.: Quantus: an explainable AI toolkit for responsible evaluation of neural network explanations and beyond. J. Mach. Learn. Res. **24**(34), 1–11 (2023)
5. Vaswani, A., et al.: Attention is all you need. In: Advances in Neural Information Processing Systems, vol. 30 (2017). https://arxiv.org/abs/1706.03762
6. Subutai, A., Scheinkman, L.: How can we be so dense? The benefits of using highly sparse representations. arXiv preprint arXiv:1903.11257 (2019)
7. Runge, K., Cardoso, C., de Chevigny, A.: Dendritic Spine plasticity: function and mechanisms. Front. Synaptic Neurosci. **12**, 36 (2020). https://doi.org/10.3389/fnsyn.2020.00036. PMID: 32982715; PMCID: PMC7484486
8. Pchitskaya, E., Bezprozvanny, I.: Dendritic Spines shape analysis-classification or clusterization? Perspective. Front. Synaptic Neurosci. **12**, 31 (2020). https://doi.org/10.3389/fnsyn.2020.00031. PMID: 33117142; PMCID: PMC7561369
9. Kasabov, N.K.: Time-Space, Spiking Neural Networks and Brain-Inspired Artificial Intelligence. SSBN, vol. 7. Springer, Heidelberg (2019). https://doi.org/10.1007/978-3-662-57715-8
10. Horzyk, A: Associative graph data structures with an efficient access via AVB+trees. In: 11th International Conference on Human System Interaction (HSI), IEEE Xplore, pp. 169–175 (2018)
11. Horzyk, A.: Artificial Associative Systems and Associative Artificial Intelligence. Academic Publishing House EXIT, Warsaw (2013)
12. Linoff, G.S., Berry, M.A.: Data Mining Techniques: For Marketing, Sales, and Customer Relationship Management, 3rd Edition (2011)
13. Olson, R.S., La Cava, W., Orzechowski, P., Urbanowicz, R., Moore, J.H.: PMLB: a large benchmark suite for machine learning evaluation and comparison. BioData Min. **10**(1), 36 (2017)
14. https://github.com/jakubkosno/ASONNv2
15. Gale, T., Elsen, E., Hooker, S.: The State of Sparsity in Deep Neural Networks (2019). arXiv:1902.09574
16. Liu, S., et al.: Sparsity May Cry: Let Us Fail (Current) Sparse Neural Networks Together! (2023). arXiv preprint arXiv:2303.02141
17. Mocanu, D.C., Mocanu, E., Stone, P., et al.: Scalable training of artificial neural networks with adaptive sparse connectivity inspired by network science. Nat. Commun. **9**, 2383 (2018). https://doi.org/10.1038/s41467-018-04316-3
18. Atashgahi, Z., Pieterse, J., Liu, S., et al.: A brain-inspired algorithm for training highly sparse neural networks. Mach. Learn. **111**, 4411–4452 (2022)

Efficient Attention for Domain Generalization

Zhongqiang Zhang(✉), Ge Liu, Fuhan Cai, Duo Liu, and Xiangzhong Fang

Department of Electronic Engineering, Shanghai Jiao Tong University, Shanghai, China
zhangzhongqiang@sjtu.edu.cn

Abstract. Deep neural networks suffer severe performance degradation when encountering domain shift. Previous methods mainly focus on feature manipulation in source domains to learn transferable features to unseen domains. We propose a new perspective based on the attention mechanism, which enables the model to learn the most transferable features on source domains and dynamically focus on the most discriminative features on unseen domains. To achieve this goal, we introduce a domain-specific attention module that facilitates the identification of most transferable features in each domain. Different from channel attention, spatial information is also encoded in our module to capture global structure information of samples, which is vital for generalization performance. To minimize the parameter overhead, we also introduce a knowledge distillation formulation to train a lightweight model that has the same attention capabilities as original model. So, we align the attention weights of the student model with a specific attention weights of the teacher model that corresponding to the domain of input. The results show that the distilled model performs better than its teacher and achieves the state-of-the-art performance on several public datasets, i.e. PAC, OfficeHome and VLCS. This indicates the effectiveness and superiority of our proposed approach in terms of transfer learning and domain generalization tasks.

Keywords: Domain generalization · Attention · Knowledge Distillation

1 Introduction

In recent times, deep neural networks (DNNs) have found extensive applications in computation vision tasks. Nevertheless, DNNs suffer severe performance drop when training and testing data follow different distributions, which is commonly known as the domain shift problem [1,2]. To tackle this issue, unsupervised domain adaptation (UDA) [3] transfer knowledge from source domain(s) to an unlabeled target domain. However, the target data may not be available in many practical applications, such as deploying a face recognition model in real-world

B. Luo et al. (Eds.): ICONIP 2023, CCIS 1963, pp. 245–257, 2024.
https://doi.org/10.1007/978-981-99-8138-0_20

scenarios. In such cases, traditional UDA methods become impractical and alternative approaches are required. So enhancing generalization without the need for target data becomes crucial for deploying models in new domains. To address this challenge, the emerging field of domain generalization (DG) [4–6] has gained significant attention.

Fig. 1. Example images form the DG benchmark PACS. Each row contains object images of the same class but from different domains.

The previous methods only focus on feature manipulation to learn features that are transferable to target domains. We propose a new perspective to solve the DG problem based on the attention mechanism. As shown in Fig. 1, the appearances of features in different domains are diverse. Therefore, we aim to train an attention module that can adaptively identify the most transferable features within the source domains. This allows the model to dynamically focus on the most discriminative features in the unseen domain. The traditional channel attention module is limited to learning 1D attention weights, which are employed to balance the weights of features in the channel dimension. However, we believe global structure information is vital for enhancing DG performance. Therefore, we design a position-sensitive attention module to effectively capture the spatial information of samples. To identify the most remarkable features in different category appearances in each domain, we set domain-specific attention modules that exclusively pass samples of one domain. Furthermore, to reduce the influence of domain shift, we also present a Test-time Style Align (TSA) method, which aligns the style of the target domain to each source-domain style at test time. Then, the calibrated sample by TSA is passed through the classifier. Finally, we identify the samples by integrating the output from different attention modules.

To reduce computation overhead, we additionally introduce knowledge distillation formulation to train a new lightweight student model that can match the attention capabilities of the original model (teacher). For each sample, we align the attention weights of the student model to the specific one that corresponds to the input domain. The attention module achieves excellent generalization performance by processing samples from multiple domains under the guidance of the teacher model.

Our contributions are summarized as follows: (1) We introduce a new approach to solve the DG problem based on the attention mechanism, which enables the model to focus on the most discriminative features on unseen domains. (2) A new attention module is designed that integrates spatial information compared to the traditional channel attention module. It allows the model to capture the global structure features that are crucial for DG performance. (3) We propose an attention distillation paradigm to ensure the consistency of attention between the teacher and student. It can enhance the generalization performance with diverse guidance from the teacher while reducing the computational overhead. (4) We evaluate our method on three DG benchmarks and demonstrate its superior performance, surpassing the existing state-of-the-art.

2 Related Works

2.1 Attention-Based Network

Attention has remained a prominent area of research in computer vision tasks and has achieved significant advancements. Liu. *et al.* [7] investigate the attention mechanism for fine-grained visual recognition by region localization and feature learning. Deng. *et al.* [8] propose a domain attention consistency network to solve DA problem. They aim to identify transferable features between source and target domains. However, samples on target domains are not available in the DG setting. So, we devote to learning the attention on most transferable and discriminative features among domains, which can benefit the DG performance on the target domain.

2.2 Domain Generalization

DG focuses on the problem of generalizing to out-of-distribution data by leveraging knowledge obtained from multiple source domains. In the early stages of DG research, there is a significant emphasis on the development of methods for learning domain-invariant representations [4,9]. These representations are considered robust as they aim to effectively suppress the inherent domain shift among multiple source domains. Besides, domain augmentation has emerged as another popular approach to tackle the challenges of DG. This approach involves the creation of synthetic samples by applying gradient-based adversarial perturbations [5] or leveraging adversarially trained image generators [10,11]. Meta-learning has emerged as another popular approach for tackling the challenges of DG [12,13]. It involves splitting the source domains into meta-train and meta-test domains, simulating domain shifts during training. Other DG methods utilize self-supervised learning [14], ensemble learning [15], and regularization techniques [16,17] to enhance their performance. In contrast to the methods above, our approach introduces a novel attention-based perspective for addressing the DG problem.

2.3 Knowledge Distillation

Knowledge distillation (KD) is proposed for model compression [18,19] initially. It involves training a smaller target model (referred to as the student) with the guidance of larger pre-trained teacher model. To make the student model perform similarly to the teacher, a common approach is to minimize the consistency loss between their output. In this study, we utilize KD formulation to make the student model match the attention ability of the teacher.

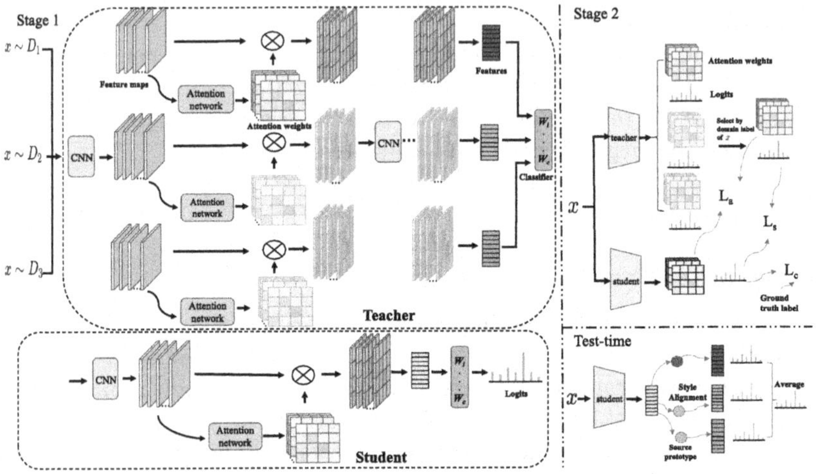

Fig. 2. Framework of our EADG achieved by a two-stage training procedure. In the first stage, we train a teacher network with domain-specific attention modules to learn the most transferable and discriminative features in each source domain. In the second stage, we follow the KD formulation to match the attention ability of the teacher by aligning the attention weights of the student to the specific one of the teacher which corresponds to the input's domain. At test time, we calibrate the style of target-domain samples to each source-domain prototypes and then utilize the average output from multiple calibrated samples as predictions.

3 Method

3.1 Problem Formulation

In multi-source domain generalization, we are provided with a training set composed of multiple labeled source domains $\mathcal{D}_S = \{\mathcal{D}_1, \ldots, \mathcal{D}_K\}$ with N_k training pairs $\{(x_i^k, y_i^k)\}_{i=1}^{N_k}$ in the k-th domain $\mathcal{D}_k$, where x_i^k is an input sample and $y_i^k \in \{1, \ldots, C\}$ is its label. DG aims at learning a domain-agnostic model $f(\cdot; \theta)$ on multiple source domains to perform well on the remaining target domain $\mathcal{D}_T$.

3.2 Overview

To solve the problem of DG, we propose a two-stage framework called *Efficient Attention for Domain Generalization* (EADG). The motivation behind EADG is to focus on the most discriminative features in diverse appearances of categories from different domains. Therefore, we design a position-sensitive attention module to weight the contribution of features in both the spatial and channel dimensions to the discrimination. To reduce the computational-cost, we also introduce attention distillation to train a new lightweight student model. The overview of our framework is shown in Fig. 2.

3.3 Attention Module

Traditional Channel Attention. As shown in Fig. 3, traditional channel attention module consists of a Average-pool, MLP and a Sigmoid function. However, the channel weights is only 1D shape and ignore the spatial attention weights related to global structure information, which is vital for DG performance.

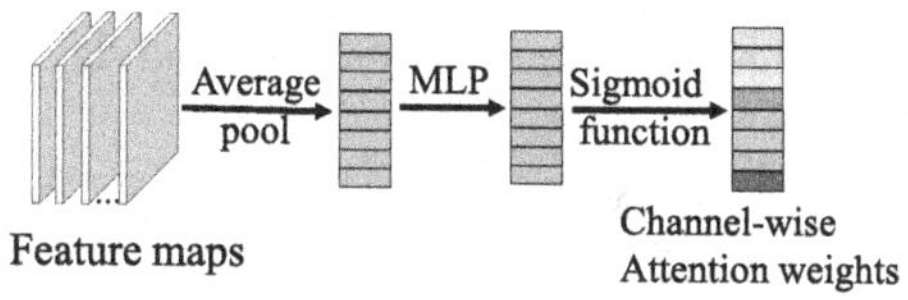

Fig. 3. Traditional channel attention.

Global Structure Information Investigation. Due to the network structure of CNN, it tends to focus on the local features of samples. Follow coordinate attention mechanism [20], we let attention weights be related to the spatial position in feature maps. Besides, except for Average Pool layer, we also use Max Pool layer to calculate attention weights to learn the most discriminative features in each source domain. We add the two attention weights as the final result. The detailed structure is shown in Fig. 4.

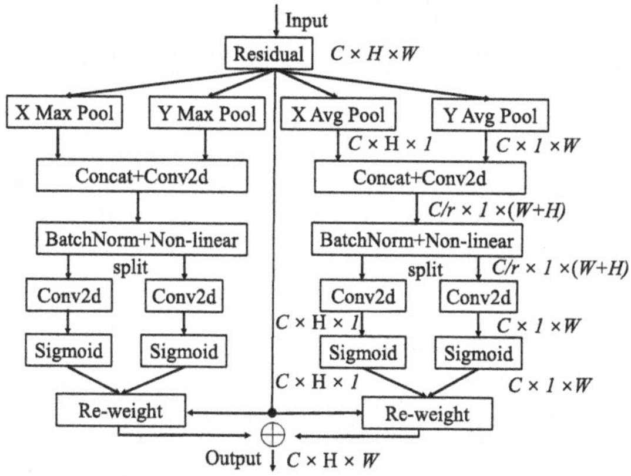

Fig. 4. Framework of our attention module, deigned to integrate global structure information into attention map.

3.4 Efficient Attention for Each Domain

Domain-specific attention learning. In source domains, the appearances of the same category are diverse. Therefore, in the first stage, we deploy domain-specific attention modules to make model focus on the most remarkable features in each domain, as shown in Fig. 2. Cross entropy is introduced as training loss to train the teacher model.

Attention Distillation. To reduce the number of parameters of model, we employ distillation paradigm to train a student model that merges the domain-specific attention knowledge to a single attention module as

$$L_a = \frac{1}{B}\sum_{i=1}^{B}\sum_{k=1}^{K} I(d(x) = k)\left|m_T^k - m_S\right| , \tag{1}$$

where m_t^k denotes the k-th channel weights of teacher for input x. m_s represents the one of student. B is the number of batch size, and $d(x)$ denotes the domain label of input x. $I(\cdot)$ is the indicator function.

For the features that acquire the same attention, the similar prediction is also crucial between the teacher and student. Therefore, we also align the logits the two model as

$$L^s = -\tau^2 \mathbb{E}_{x,y\sim D_k} \sum_{i=1}^{C} p_i^t \log p_i^s , \tag{2}$$

where p_i^t and p_i^s are probabilities of class i predicted by the teacher and student, respectively. Here, softmax function $\sigma(\cdot;\tau)$ is used to transfer the logits outputs to probabilities, where temperature τ is introduced as hyper-parameters to soften the vanilla distribution. τ is set to 5.0 in this paper.

For training the distillation CNN model, we combine the losses aforementioned as

$$L^s = L_c + \lambda_a L_a + \lambda_s L_s , \tag{3}$$

where λ_a and λ_s are balancing weights. They are set to 0.2 and 0.1 in the all experiments, respectively. The final CNN model trained with Eq. 3 is called EADG.

3.5 Test-Time Style Alignment

Although the most discriminative feature has been located in each domain, the domain style divergence still affects the DG performance. To reduce this effect, we propose a module named Test-time Style Alignment (TSA) to match the style of samples on the target domain to each source domain at test time. At first, we need calculate style prototype $\mathbf{Pt_{S_k}} \in \mathbb{R}^{1\times 2}$ of each source domain by exponential moving average in each mini-patch as

$$Pt_{S_k} = \alpha \mathbf{Pt_{S_k}} + (1-\alpha)\frac{1}{B}\sum_{i=1}^{B}(\mu_i, \sigma_i) , \tag{4}$$

where B is the number of samples in mini-batch. μ_i and σ_i is the mean and variety of i-th samples' latent feature $F_i \in \mathbb{R}^{1\times C}$ from the penultimate layer. α is set to 0.99.

Then we leverage the $\mathcal{G}$ function to align the target style $\mathcal{S}(z_T^i)$ to k-th source prototype $\mathbf{Pt_{S_k}}$ when evaluation as

$$\mathcal{G}(\mathbf{Pt_{S_k}}, \mathcal{S}(z_t^i)) = \gamma\mathbf{Pt_{S_k}} + (1-\gamma)\mathcal{S}(z_T^i)\,, \tag{5}$$

where the γ is the hyperparameter to manipulate the degree of alignment and is set to 0.5 as default. When testing, we average the outputs of K style-aligned features as the final prediction.

4 Experiments

4.1 Datasets and Experimental Settings

We examine the performance of our method based on three benchmark datasets including PACS, Office-Home and VLCS. (1)**PACS** is a commonly-used DG dataset, which consists four domains of Photo, Art painting, Carton and Sketch. PACS encompasses 7 distinct classes and 9991 images with significant discrepancy in image styles observed across domains. The split of original train-validation provided by [1] is employed in this paper. (2)**Office-Home** is composed of 65 groups and 15500 graphics in four domains, *i.e.*, Artistic, Clipart, Product, and Real-word. We follow the evaluation setting in [21], where each source domain are divided into 90% for training and 10% for validation purposes. (3)**VLCS** comprises of four domains including VOC2007, LabelMe, Caltech and Sun. We use the original train-validation split provided by [6], where source domain data are randomly split into 80% for training and 20% for validation.

4.2 Implementation Details

The training process of EADG consists of two stages: domain-specific attention training and attention distillation. To ensure a fair comparison with previous methods, we adopt ResNet-18 and ResNet-50 [22] as backbones, following the implementation in [16]. Both the teacher and student models are trained using the SGD optimizer with a total of 60 training epochs. The initial learning rate is set to 0.001, the batch size is set to 30, and a weight decay of 5e-4 is applied. During the training, we adopt a standard augmentation protocol as described in [16], including random resized cropping, horizontal flipping, and color jittering techniques. The learning rate decay is implemented using the cosine annealing rule. In addition, the ResNet model is initialized with parameters pre-trained on the ImageNet dataset. All experiments are conducted following the leave-one-domain-out protocol. Our model is trained using the training set of the source domains, while the hyper-parameters are selected based on the validation set. We evaluate the selected model on the hold-out target domain. The results are reported based on the classification accuracy.

Table 1. Leave-one-domain-out results on PACS using ResNet-18 and ResNet-50 backbones. The best and second-best results are bolded and underlined, respectively.

Method	A	C	P	S	Avg
			ResNet18		
DeepAll	79.35	74.53	95.51	66.61	79.00
CORAL [23]	83.80	76.80	95.50	77.90	83.50
MetaReg [12]	83.70	77.20	95.50	70.30	81.70
JiGen [16]	79.42	75.25	96.03	71.35	80.51
Mixstyle [24]	<u>84.10</u>	78.80	<u>96.10</u>	75.90	<u>83.70</u>
DDAIG [2]	84.20	78.10	95.30	74.70	83.10
MASF [13]	80.29	77.17	94.99	71.69	81.04
DAEL [15]	**84.60**	74.40	95.60	78.90	83.40
DIR-GAN [25]	82.56	76.37	95.65	<u>79.89</u>	83.62
EADG(Ours)	83.00	**79.63**	**96.20**	**81.21**	**85.01**
			ResNet50		
DeepAll	82.80	80.42	95.03	72.33	82.65
MMD-AAE [4]	86.10	79.40	96.60	76.50	84.70
CORAL [23]	85.10	<u>82.30</u>	96.20	76.60	85.00
MetaReg [12]	<u>87.20</u>	79.20	97.60	70.30	83.60
EISNet [26]	86.64	81.53	97.11	78.07	85.84
DMG [27]	82.57	78.11	94.49	78.32	83.37
BatchFomer [28]	**89.00**	80.10	**98.00**	79.80	<u>86.80</u>
MIRO [17]	88.84	78.73	<u>97.61</u>	75.16	85.08
EADG(Ours)	86.89	**83.43**	97.25	**84.09**	**87.92**

4.3 Evaluation on PACS

As Table 1 shows, our EADG achieves the best performance among all the competitors, improving the baseline DeepAll by a large margin. Compared to the methods based on feature manipulation [23, 24, 27], our EADG outperforms them by a large margin. Further, our method also beats competitors based on meta-learning [12, 13], which need a complicated training procedure. Notably, Table 1 shows that our method also has absolute superiority on the *carton* and the *sketch* domain using ResNet-18 and ResNet-50. It is reasonable that global structure information is extremely essential for discrimination, which is properly enhanced by our designed attention mechanism. Our EADG also surpasses the method based on adversarial network including DDAIG [2] and DIR-GAN [25] by a large margin, demonstrating the effectiveness of our method.

4.4 Evaluation on OfficeHome

As shown in Table 2, our EADG performs at a state-of-the-art level overall. OfficeHome is a large-scale dataset with diverse semantic cues, so some

Table 2. Leave-one-domain-out results on Office-Home using ResNet-18 and ResNet-50 backbones. The best and second-best results are bolded and underlined, respectively.

Method	A	C	P	R	Avg
ResNet18					
DeepAll	59.13	48.59	73.60	75.49	64.20
CCSA [9]	59.90	49.90	74.10	<u>75.70</u>	64.90
MMD-AAE [4]	56.50	47.30	72.10	74.80	62.70
Mixstyle [24]	58.70	<u>53.40</u>	<u>74.20</u>	**75.90**	65.50
CrossGrad [5]	58.40	49.40	73.90	75.80	64.40
Jigen [16]	53.04	47.51	71.47	72.79	61.20
DSON [29]	59.37	45.70	71.84	74.68	62.90
LRDG [30]	**61.73**	52.43	72.96	75.89	<u>65.75</u>
DIR-GAN [25]	56.69	50.49	71.32	74.23	63.18
EADG(Ours)	<u>61.22</u>	**56.21**	**74.27**	73.44	**66.28**
ResNet50					
DeepAll	61.28	52.42	75.91	76.11	66.43
MMD-AAE [4]	60.40	53.30	74.30	77.40	66.40
SagNet [31]	63.40	<u>54.80</u>	75.80	78.30	68.10
SelfReg [14]	63.60	53.10	<u>76.90</u>	78.10	67.90
mDSDI [32]	<u>68.10</u>	52.10	76.00	**80.40**	<u>69.20</u>
EADG(Ours)	**68.18**	**56.33**	**78.09**	<u>80.32</u>	**70.73**

methods based on learning domain-invariant features perform unremarkable like DSON [29], CCSA [9], MMD-AAE [4] and the latest method LRDG [30]. Results of DSON and MMD-AAE are even worse than the strong baseline DeepAll. Besides, EADG also surpasses augmentation-based methods [24,25,31] and regularization-based methods [16]. Moreover, our method achieves the best results on the *clipart* and *product* domain and the average performance, further demonstrating the efficacy of our approach.

4.5 Evaluation on VLCS

As shown in Table 3, our EADG performs best compared to other methods. For this dataset, DeepAll is also competitive due to the limited discrepancy among domains and the resemblance to Image-Net. Nevertheless, our EADG achieves overall improvement. We beat the methods based on feature manipulation, such as MMD, RSC by a margin of 1%∼2% with backbone ResNet-50. We also surpass the newest DG competitor BatchFormer by a small margin using ResNet-50. These results again demonstrate the superiority of EADG.

Table 3. Leave-one-domain-out results on VLCS. The best and second-best results are bolded and underlined respectively.

Method	C	L	S	V	Avg
DeepAll	95.27	<u>63.06</u>	69.74	72.51	75.15
BatchFormer [28]	**97.20**	61.30	<u>71.70</u>	**77.40**	<u>76.90</u>
EADG(Ours)	<u>96.80</u>	**63.51**	**72.00**	<u>76.42</u>	**77.18**
ResNet50					
DeepAll	97.22	**64.30**	<u>73.43</u>	74.69	77.41
RSC [33]	<u>97.90</u>	62.50	72.30	<u>75.60</u>	77.10
MMD [4]	97.70	<u>64.00</u>	72.80	75.30	<u>77.50</u>
EADG(Ours)	**98.22**	63.76	**74.92**	**78.16**	**78.76**

4.6 Ablation Study

We conduct an extensive ablation study on PACS to investigate the impact of main components in our EADG as shown in Table 4 and Table 5. All models are trained with the same training parameters. Overall, our final model EADG improves the baseline by a large margin of 6.01%.

Significance of attention mechanism ($\mathcal{A}_t$ and $\mathcal{A}_s$). We apply multiple domain attention modules $\mathcal{A}_t$ in teacher network and one $\mathcal{A}_s$ in the distilled networks. In Table 4, one can see that Model 1 (teacher) and Model 3 (corresponding student) both improve baseline DeepAll by margins of 3%∼4% with additional attention module. Compared to Model 3, Model 1 perform better because the domain-specific attention modules ($\mathcal{A}_t$) can focus on the most discriminate features in each domain.

Importance of L_a. To reduce parameter cost, we introduce KD formulation to learn a new student model possessing the similar attention ability. Therefore, in Model 4, L_a is employed to align its attention weights to the specific one that corresponds to the domain of the input from Model 1. Results show that it achieves a significant improvement to Model 3. Moreover, Model 4 even performs better than its teacher Model 1. This may be because that attention module in student Model 4 has access to more samples, empowering it to adaptively attend to the most discriminative features in each domain with the guidance of teacher.

Table 4. Ablation study on PACS with backbone ResNet-18. $\mathcal{A}_t$: domain-specific attention modules of teacher network. $\mathcal{A}_s$: attention module of the student network. The alignment targets of L_a and L_s are both from Model 1.

#	Method	Avg
DeepAll	L_c	79.00
Model 1	$L_c + \mathcal{A}_t$	83.01
Model 2	$L_c + \mathcal{A}_t + \mathcal{S}$	83.52
	Distillation	**Avg**
DeepAll	L_c	79.00
Model 3	$L_c + \mathcal{A}_s$	82.69
Model 4	$L_c + \mathcal{A}_s + L_a$	84.26
Model 5	$L_c + \mathcal{A}_s + L_a + L_s$	84.77
Model 6	$L_c + \mathcal{A}_s + L_a + L_s + \mathcal{S}$	**85.01**

Effect of L_s. We also hope that the student model has similar prediction output while ensuring the model focuses on the same features. So, L_s is employed in Model 5 to match the soft label outputs of Model 1. It can be observed that it supplies a small improvement by the margin of 0.51% compared to Model 4.

Impact of $\mathcal{S}$. To mitigate the effect of domain shift, we also introduce the TSA module to align the feature style on the target domain to the source domains. Results show that TSA improves the performance of both teacher (Model 1) and student (Model 5), proving this module's effectiveness.

Table 5. Ablation study on where to apply the attention modules on PACS using backbone ResNet-18. The locations of attention modules in teacher and student networks are consistent here.

Method	Apply $\mathcal{A}$ in	Avg
EADG	All four residual blocks	68.80
	Firt three residual blocks	74.33
	Firt two residual blocks	82.92
	Firt residual blocks (Ours)	**85.01**

Where to apply the attention modules? We evaluate four variants of EADG in Table 5, where the attention module is applied in different numbers of residual blocks. The summaries are as follows. 1) Applying the attention module in the first residual blocks perform best. 2) Applying the module in deeper layers worsens the performance. This is as expected: the earlier layer in CNN learn the information of texture and shape, where attention module can help locate the most discriminative features.

5 Conclusion

In conclusion, our approach based on attention mechanism addresses the challenge from domain shift in DNNs. By introducing a domain-specific attention module, we enable the model to learn transferable features from source domains and focus on most discriminative features in unseen domains. Besides, our designed attention module incorporates spatial information into the model, which can capture the global structure of samples and then improve generalization performance. We further introduce a knowledge distillation formulation to train a lightweight student model with similar attention capabilities. We conduct extensive experiments and achieve the state-of-the-art performance on three DG benchmark datasets, demonstrating the superiority of our approach over existing methods.

References

1. Li, D., Yang, Y., Song, Y.-Z., Hospedales, T.M.: Deeper, broader and artier domain generalization. In: Proceedings of the IEEE International Conference on Computer Vision, pp. 5542–5550 (2017)
2. Zhou, K., Yang, Y., Hospedales, T., Xiang, T.: Deep domain-adversarial image generation for domain generalisation. In: Proceedings of the AAAI Conference on Artificial Intelligence. vol. 34, pp. 13025–13032 (2020)
3. Xu, R., Chen, Z., Zuo, W., Yan, J., Lin, L.: Deep cocktail network: multi-source unsupervised domain adaptation with category shift. In: Proceedings of the IEEE Conference on Computer Vision and Pattern Recognition, pp. 3964–3973 (2018)

4. Li, H., Pan, S.J., Wang, S., Kot, A.C.: Domain generalization with adversarial feature learning. In: Proceedings of the IEEE Conference on Computer Vision and Pattern Recognition, pp. 5400–5409 (2018)
5. Shankar, S., Piratla, V., Chakrabarti, S., Chaudhuri, S., Jyothi, P., Sarawagi, S.: Generalizing across domains via cross-gradient training. arXiv preprint arXiv:1804.10745 (2018)
6. Gulrajani, I., Lopez-Paz, D.: In search of lost domain generalization. arXiv preprint arXiv:2007.01434 (2020)
7. Liu, H., Li, J., Li, D., See, J., Lin, W.: Learning scale-consistent attention part network for fine-grained image recognition. IEEE Trans. Multimedia **24**, 2902–2913 (2021)
8. Deng, Z., Zhou, K., Yang, Y., Xiang, T.: Domain attention consistency for multi-source domain adaptation. arXiv preprint arXiv:2111.03911 (2021)
9. Motiian, S., Piccirilli, M., Adjeroh, D.A., Doretto, G.: Unified deep supervised domain adaptation and generalization. In: Proceedings of the IEEE International Conference on Computer Vision, pp. 5715–5725 (2017)
10. Ganin, Y., et al.: Domain-adversarial training of neural networks. J. Mach. Learn. Res. **17**(1), 2096–2030 (2016)
11. Li, Y., Gong, M., Tian, X., Liu, T., Tao, D.: Domain generalization via conditional invariant representations. In: Proceedings of the AAAI Conference on Artificial Intelligence, vol. 32 (2018)
12. Balaji, Y., Sankaranarayanan, S., Chellappa, R., Balaji, Y.: MetaReg: towards domain generalization using meta-regularization. In: Advances in Neural Information Processing Systems, vol. 31 (2018)
13. Dou, Q., de Castro, D.C., Kamnitsas, K., Glocker, B.: Domain generalization via model-agnostic learning of semantic features. In: Advances in Neural Information Processing Systems, vol. 32 (2019)
14. Kim, D., Yoo, Y., Park, S., Kim, J., Lee, J.: SelfReg: self-supervised contrastive regularization for domain generalization. In: Proceedings of the IEEE/CVF International Conference on Computer Vision, pp. 9619–9628 (2021)
15. Zhou, K., Yang, Y., Qiao, Y., Xiang, T.: Domain adaptive ensemble learning. IEEE Trans. Image Process. **30**, 8008–8018 (2021)
16. Carlucci, F.M., D'Innocente, A., Bucci, S., Caputo, B., Tommasi, T.: Domain generalization by solving jigsaw puzzles. In: Proceedings of the IEEE/CVF Conference on Computer Vision and Pattern Recognition, pp. 2229–2238 (2019)
17. Cha, J., Lee, K., Park, S., Chun, S.: Domain Generalization by Mutual-Information Regularization with Pre-trained Models. In: Avidan, S., Brostow, G., Cissé, M., Farinella, G.M., Hassner, T. (eds.) Computer Vision – ECCV 2022. ECCV 2022. LNCS, vol. 13683. Springer, Cham (2022). https://doi.org/10.1007/978-3-031-20050-2_26
18. Luo, P., Zhu, Z., Liu, Z., Wang, X., Tang, X.: Face model compression by distilling knowledge from neurons. In: Thirtieth AAAI Conference on Artificial Intelligence (2016)
19. Polino, A., Pascanu, R., Alistarh, D.: Model compression via distillation and quantization. arXiv preprint arXiv:1802.05668 (2018)
20. Hou, Q., Zhou, D., Feng, J.: Coordinate attention for efficient mobile network design. In: Proceedings of the IEEE/CVF Conference on Computer Vision and Pattern Recognition (CVPR), pp. 13713–13722 (2021)
21. Venkateswara, H., Eusebio, J., Chakraborty, S., Panchanathan, S.: Deep hashing network for unsupervised domain adaptation. In: Proceedings of the IEEE Conference on Computer Vision and Pattern Recognition, pp. 5018–5027 (2017)

22. He, K., Zhang, X., Ren, S., Sun, J.: Deep residual learning for image recognition. In: Proceedings of the IEEE Conference on Computer Vision and Pattern Recognition, pp. 770–778 (2016)
23. Sun, B., Saenko, K.: Deep CORAL: correlation alignment for deep domain adaptation. In: Hua, G., Jégou, H. (eds.) ECCV 2016. LNCS, vol. 9915, pp. 443–450. Springer, Cham (2016). https://doi.org/10.1007/978-3-319-49409-8_35
24. Zhou, K., Yang, Y., Qiao, Y., Xiang, T.: Domain generalization with mixstyle. arXiv preprint arXiv:2104.02008 (2021)
25. Nguyen, A.T., Tran, T., Gal, Y., Baydin, A.G.: Domain invariant representation learning with domain density transformations. Adv. Neural Inf. Process. Syst. **34**, 5264–5275 (2021)
26. Wang, S., Yu, L., Li, C., Fu, C.-W., Heng, P.-A.: Learning from extrinsic and intrinsic supervisions for domain generalization. In: Vedaldi, A., Bischof, H., Brox, T., Frahm, J.-M. (eds.) ECCV 2020. LNCS, vol. 12354, pp. 159–176. Springer, Cham (2020). https://doi.org/10.1007/978-3-030-58545-7_10
27. Chattopadhyay, P., Balaji, Y., Hoffman, J.: Learning to balance specificity and invariance for in and out of domain generalization. In: Vedaldi, A., Bischof, H., Brox, T., Frahm, J.-M. (eds.) ECCV 2020. LNCS, vol. 12354, pp. 301–318. Springer, Cham (2020). https://doi.org/10.1007/978-3-030-58545-7_18
28. Hou, Z., Yu, B., Tao, D., Hou, Z., Yu, B., Tao, D.: BatchFormer: Learning to explore sample relationships for robust representation learning. arXiv preprint arXiv:2203.01522 (2022)
29. Seo, S., Suh, Y., Kim, D., Kim, G., Han, J., Han, B.: Learning to optimize domain specific normalization for domain generalization. In: Vedaldi, A., Bischof, H., Brox, T., Frahm, J.-M. (eds.) ECCV 2020. LNCS, vol. 12367, pp. 68–83. Springer, Cham (2020). https://doi.org/10.1007/978-3-030-58542-6_5
30. Ding, Y., Wang, L., Liang, B., Liang, S., Wang, Y., Chen, F.: Domain generalization by learning and removing domain-specific features. In: Advances in Neural Information Processing Systems (2022)
31. Nam, H., Lee, H., Park, J., Yoon, W., Yoo, D.: Reducing domain gap by reducing style bias. In: Proceedings of the IEEE/CVF Conference on Computer Vision and Pattern Recognition, pp. 8690–8699 (2021)
32. Bui, M.-H., Tran, T., Tran, A., Phung, D.: Exploiting domain-specific features to enhance domain generalization. Adv. Neural. Inf. Process. Syst. **34**, 21189–21201 (2021)
33. Huang, Z., Wang, H., Xing, E.P., Huang, D.: Self-challenging improves cross-domain generalization. In: Vedaldi, A., Bischof, H., Brox, T., Frahm, J.-M. (eds.) ECCV 2020. LNCS, vol. 12347, pp. 124–140. Springer, Cham (2020). https://doi.org/10.1007/978-3-030-58536-5_8

Adaptive Accelerated Gradient Algorithm for Training Fully Complex-Valued Dendritic Neuron Model

Yuelin Wang and He Huang(✉)

School of Electronics and Information Engineering, Soochow University, Suzhou 215006, People's Republic of China
hhuang@suda.edu.cn

Abstract. This paper presents an adaptive complex-valued Nesterov accelerated gradient (ACNAG) algorithm for the training of fully complex-valued dendritic neural model (FCVDNM). Firstly, based on the complex-valued Nesterov accelerated gradient (CNAG) algorithm, an adaptive stepsize update method is introduced by using local curvature information. Secondly, the obtained adaptive stepsize is further constrained by scaling the Malitsky-Mishchenko criterion. Experimental results demonstrate the superior convergence and efficiency of the proposed algorithm compared to CNAG for the training of FCVDNM.

Keywords: Complex-valued accelerated gradient algorithm · Fully complex-valued dendritic neuron model · Adaptive stepsize · Curvature · Malitsky-Mishchenko criterion

1 Introduction

Artificial neural networks (ANNs) are computational models designed to mimic the function of human brains. They emulate the information processing and transmission observed in real neurons. The McCulloch-Pitts neuron model, a fundamental component of ANNs, is established based on the principles of neural message transmission. Through training, the weights and thresholds of these artificial neurons are adjusted, enabling ANNs to tackle complex problems. However, the McCulloch-Pitts neuron model has been criticized for its oversimplified representation of real neurons and its limited ability to handle continuous-valued inputs [6]. As a result, alternative models have been proposed to address these shortcomings and improve the performance of ANNs.

The dendritic neuron model (DNM) [19] has emerged as a promising alternative to address the limitations of the McCulloch-Pitts neuron model. The DNM takes inspiration from the intricate structure and functionality of biological dendrites in real neurons, which involves the synaptic, dendritic, membrane and soma layers. Ongoing research on DNM mainly focuses on understanding its mechanisms, developing efficient learning algorithms and exploring diverse applications. Investigating synaptic plasticity and network architecture is crucial

B. Luo et al. (Eds.): ICONIP 2023, CCIS 1963, pp. 258–269, 2024.
https://doi.org/10.1007/978-981-99-8138-0_21

for unraveling the mechanisms underlying DNM and its computational capabilities. Simultaneously, advancing the development of learning algorithms for DNM enhances our understanding of its characteristics and properties, while also broadening its potential applications. Currently, DNM has demonstrated successful applications in pattern recognition [2], medical diagnosis [5,14] and time series prediction systems [4,13,15,20].

A recent study has expanded the application of DNM from the real domain to the complex domain [3]. The complex-valued gradient descent (CGD) algorithm in this study is used to optimize the parameters of fully complex-valued DNM (FCVDNM). By leveraging CGD, the study enhances FCVDNM's performance in processing complex-valued signals. The optimization of FCVDNM through CGD confirms its effectiveness in tasks such as CXOR, channel equalization and wind prediction.

However, the gradient descent algorithm, which relies solely on the current gradient information for parameter updates, exhibits limitations in terms of its convergence rate and ability to navigate steep or narrow regions of the optimization landscape. This can result in slow convergence and suboptimal solutions [8]. Currently, several complex-valued first-order accelerated algorithms, such as complex-valued Nesterov accelerated gradient (CNAG) [18] and complex-valued Barzilai-Borwein method (CBBM) [17], as well as second-order algorithms including complex-valued GaussNewton (CGN), complex-valued Levenberg-Marquard (CLM) [1] and complex-valued L-BFGS (CL-BFGS) [16], have been proposed to enhance the optimization performance and convergence speed for complex-valued optimization problems.

The NAG algorithm [9], as a refinement of the heavy-ball (HB) method [10], incorporates the concept of momentum from previous iterations to adjust parameter updates. By introducing momentum terms and extrapolation steps, the algorithm effectively improves the performance of optimization and leads to a faster convergence. And in complex domain, in order for CNAG to dynamically respond to local curvature information of the objective function, CNAG needs to be designed with an adaptive stepsize. [17] extends the Barzilai-Borwein method (BBM) to the complex domain, derives several complex-valued adaptive stepsizes that approximating the information in the inverse of complex Hessian matrix for the training of complex-valued neural networks. In [7], two rules were proposed to ensure effective control of the adaptive stepsize in gradient descent. The first rule advises against increasing the stepsize too rapidly and the second rule emphasizes not exceeding one-half times the reciprocal of the local curvature.

Building upon these discussions, this paper introduces an adaptive Nesterov accelerated gradient algorithm for the training of FCVDNM. The algorithm adapts the stepsize based on local curvature information with the inspiration from CBBM, and further constrains the stepsize by incorporating the scaled Malitsky-Mishchenko criterion. Consequently, the proposed algorithm leads to a faster convergence compared to CNAG, significantly improving the training efficiency for FCVDNM. Experimental results confirm the effectiveness of the proposed algorithm for FCVDNM.

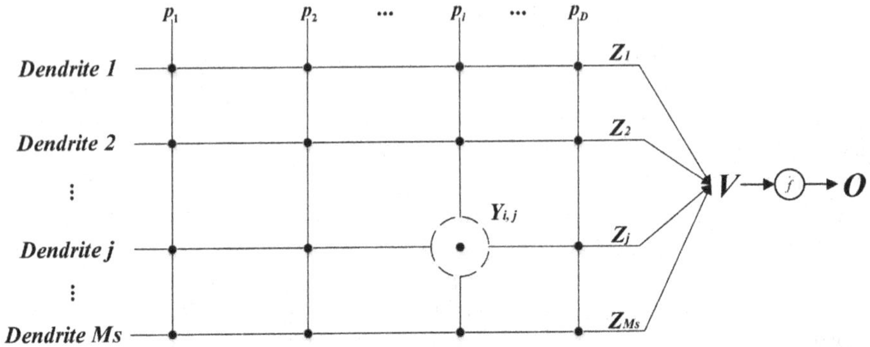

Fig. 1. Structure of FCVDNM

2 Preliminaries

2.1 Fully Complex-Valued Dendritic Neuron Model

Both FCVDNM and DNM exhibit a similar architecture, comprising synaptic, dendritic, membrane and soma layers. The scheme of FCVDNM is provided in Fig. 1.

The first layer of FCVDNM, known as the synaptic layer, connects the individual ith element p_i $(i = 1, 2, \ldots, D)$ of the input sample to the jth dendrite $(j = 1, 2, \ldots, M_s)$ in the dendritic layer using the activation function $\tilde{f}$. Here, D denotes the dimensionality of the input data and M_s represents the total number of dendrites in the dendritic layer. The synaptic layer generates an output matrix Y of size $D \times M_s$. Each element $Y_{i,j}$ of the matrix is computed by

$$Y_{i,j} = \tilde{f}\left(\omega_{i,j} p_i - \theta_{i,j}\right), \tag{1}$$

where $\omega_{i,j}$ and $\theta_{i,j}$ denote the complex-valued weight and bias parameters.

The second layer, known as the dendritic layer, processes the input signals from the previous layer and employs multiplication as its nonlinear activation function. The output Z_j of the jth dendrite is computed by

$$Z_j = \prod_{i=1}^{D} Y_{i,j}. \tag{2}$$

The third layer, known as the membrane layer, computes the summation of the outputs from all dendritic branches. The mathematical formula of this layer can be expressed by

$$V = \sum_{j=1}^{M_s} Z_j. \tag{3}$$

The soma layer, serving as the final layer, determines the output O of FCVDNM. The definition of O is presented as

$$O = \mathring{f}(V), \tag{4}$$

where $\mathring{f}$ is the fully complex-valued activation function used in the soma layer.

The objective function of FCVDNM, encompassing the entire sample set of size N, is defined by

$$L = \frac{1}{N} \sum_{n=1}^{N} L_n, \tag{5}$$

where L_n is the loss corresponding to the nth training sample, which is defined by

$$L_n = \frac{1}{2} e_n \overline{e_n}, \tag{6}$$

$$e_n = O_n - T_n, \tag{7}$$

where e_n is the error between the acutual output O_n and the target T_n for the nth training sample, and $\overline{e_n}$ is the complex conjugate of e_n.

2.2 Nesterov Accelerated Gradient Method

NAG utilizes a fixed step-size and a momentum term that integrates the previous gradient information with the current learning process during solution updates. It achieves a faster convergence rate of $O(1/k^2)$ compared to gradient descent for convex optimization problems, where k represents the number of iterations.

For a unconstrained optimization problem

$$\min_{x \in \mathbb{R}^{\mathbb{D}}} L(x), \tag{8}$$

where $L(\cdot) : \mathbb{R}^{\mathbb{D}} \to \mathbb{R}$ is a real-valued loss function, NAG updates the parameter x using the following pair of equations

$$\begin{cases} y_{k+1} = x_k - \alpha \nabla L(x_k), \\ x_{k+1} = y_{k+1} + \beta(y_{k+1} - y_k), \end{cases} \tag{9}$$

where $x_0 = y_0 \in \mathbb{R}^{\mathbb{D}}$, $\alpha > 0$ is the fixed step-size and $\beta > 0$ is the momentum coefficient. In the absence of the momentum term ($\beta = 0$), NAG simplifies to the traditional gradient descent approach.

Simultaneously, by setting $v_k = y_k - y_{k-1}$ with $v_0 = 0$, (9) can be alternatively expressed by

$$\begin{cases} v_{k+1} = \beta v_k - \alpha \nabla L(y_k + \beta v_k), \\ y_{k+1} = y_k + v_{k+1}. \end{cases} \tag{10}$$

By simplifying (10), the NAG method can be exclusively represented in a univariate form

$$y_{k+1} = y_k + \beta(y_k - y_{k-1}) - \alpha \nabla L(y_k) - \alpha\beta(\nabla L(y_k) - \nabla L(y_{k-1})), \tag{11}$$

The term $\nabla L(y_k) - \nabla L(y_{k-1})$, defined as the gradient correction in [12], plays a crucial role in NAG method. This gradient correction term distinguishes NAG from the heavy-ball method [10] and rectifies the update direction by contrasting the gradients at consecutive iterations. Within the framework of high-resolution ordinary differential equations introduced in [12], the gradient correction term $\nabla L(y_k) - \nabla L(y_{k-1})$ can be approximated as the Hessian-driven gradient correction $\nabla^2 L(y_k)(y_k - y_{k-1})$.

2.3 Complex-Valued Barzilai-Borwein Method

The Barzilai-Borwein method (BBM) introduced in [11] has been extended to the complex domain under the name CBBM [17]. CBBM determines the adaptive stepsize by solving the least squares problem derived from the secant equation. Simultaneously, the adaptive stepsize can be categorized into complex-valued and real-valued variations.

In complex domain, the update under the loss function L is described by

$$x_{k+1} = x_k - \alpha_k \nabla_{x^*} L(x_k), \tag{12}$$

where $x \in \mathbb{C}^{\mathbb{D}}$ represents the parameter to be learned and x^* is the complex conjugate of x. $L(\cdot) : \mathbb{C}^{\mathbb{D}} \to \mathbb{R}$ is the loss function. α_k represents the adaptive stepsize corresponding to the number of iterations k.

By solving the least squares problem of quasi-Newton equation, α_k can be obtained as

$$\alpha_k = \begin{cases} \dfrac{y^{(k)H} s^{(k)}}{y^{(k)H} y^{(k)}}, & \alpha_k \in \mathbb{C} \\ \dfrac{\Re(y^{(k)H} s^{(k)})}{y^{(k)H} y^{(k)}}, & \alpha_k \in \mathbb{R} \end{cases} \tag{13}$$

or

$$\alpha_k = \begin{cases} \dfrac{s^{(k)H} s^{(k)}}{s^{(k)H} y^{(k)}}, & \alpha_k \in \mathbb{C} \\ \dfrac{s^{(k)H} s^{(k)}}{\Re(s^{(k)H} y^{(k)})}, & \alpha_k \in \mathbb{R} \end{cases} \tag{14}$$

where $s^{(k)} = x_k - x_{k-1}$, $y^{(k)} = \nabla_{x^*} L(x_k) - \nabla_{x^*} L(x_{k-1})$ and the notation $\Re(\cdot)$ signifies taking the real part of the complex number.

CBBM provides an effective adaptive complex-valued stepsize selection method for complex-valued networks. By expanding the search space from a half-line to a half-plane, it significantly mitigates the risk of getting stuck in saddle points, thus improving the optimization performance.

3 The Proposed Method

According to (1), the complex-valued weight and bias parameters, $\omega = (\omega_{1,1}, \ldots, \omega_{i,j}, \ldots, \omega_{D,M_s})^T$ and $\theta = (\theta_{1,1}, \ldots, \theta_{i,j}, \ldots, \theta_{D,M_s})^T$ of FCVDNM need to be updated. For convenience, let $x = \left(\omega^T, \theta^T\right)^T$. The complex-valued NAG algorithm is initially used to update the parameter x by

$$\begin{cases} v_{k+1} = \beta v_k - \alpha \nabla_{x^*} L(x_k + \beta v_k), \\ x_{k+1} = x_k + v_{k+1}, \end{cases} \tag{15}$$

In CNAG, the gradient $\nabla_{x^*} L(\cdot)$ signifies the direction of the steepest ascent with respect to x, which is distinct from the gradient $\nabla L(\cdot)$ utilized in NAG, as illustrated in (9). Concurrently, taking inspiration from CBBM [17], there is an increasing urgency to achieve adaptive changes in α during the iterative process to expedite convergence.

Through simplification, (15) can be exclusively expressed in a univariate form

$$x_{k+1} = x_k + \beta v_k - \alpha \nabla_{x^*} L(x_k + \beta v_k). \tag{16}$$

Define $\overset{+}{x}_k = x_k + \beta v_k$

$$x_{k+1} = \overset{+}{x}_k - \alpha \nabla_{x^*} L(\overset{+}{x}_k), \tag{17}$$

which is nearly close to the Newton method $x_{k+1} = \overset{+}{x}_k - \left(H^{(k)}\right)^{-1} \nabla_{x^*} L(\overset{+}{x}_k)$, where $H^{(k)}$ is the complex Hessian matrix of $L(x)$ at the point $x = \overset{+}{x}_k$. Therefore, the representation of α_k (α in the kth iteration) is able to be defined by the inverse of complex Hessian matrix $H^{(k)}$.

Similar to [17], the approximate matrix of $H^{(k)}$ can be obtained by solving the least squares problem associated with the quasi-Newton equation

$$B^{(k)} s^{(k)} = y^{(k)}, \tag{18}$$

where $s^{(k)} = \overset{+}{x}_k - \overset{+}{x}_{k-1}$, $y^{(k)} = \nabla_{x^*} L(\overset{+}{x}_k) - \nabla_{x^*} L(\overset{+}{x}_{k-1})$ and $B^{(k)}$ represents the approximation for the complex Hessian matrix $H^{(k)}$. Therefore, the representation of adaptive stepsize α_k can be obtained as described in (13) and (14), with the choice of $\alpha_k = \frac{{y^{(k)}}^H s^{(k)}}{{y^{(k)}}^H y^{(k)}}$.

Next, the stepsize α_k inspired by CBBM can be further regulated by incorporating the adaptive stepsize strategy [7] proposed by Malitsky and Mishchenko. In [7], an additional sequence γ_k is introduced, where γ_1 is initially set to $+\infty$. The strategy is originally defined by

$$\begin{cases} \alpha_k^2 \leq (1 + \gamma_{k-1}) \alpha_{k-1}^2, \\ \alpha_k \leq \frac{\|s^{(k)}\|}{2\|y^{(k)}\|}, \end{cases} \tag{19}$$

where $\gamma_k = \frac{\alpha_k}{\alpha_{k-1}}$ is the stepsize adjustment factor. When $y^{(k)} = \nabla_{x^*} L(\overset{+}{x}_k) - \nabla_{x^*} L(\overset{+}{x}_{k-1})$ approaches zero, the second formula in (19) can be ignored. These inequalities are derived from the theoretical findings presented in [7].

By comparing the stepsize $\alpha_k = \frac{y^{(k)^H} s^{(k)}}{y^{(k)^H} y^{(k)}}$ inspired by CBBM with the stepsize calculated using formula 2 in (19)

$$\alpha_k \leq \frac{\|y^{(k)^H} s^{(k)}\|}{2\|y^{(k)^H} y^{(k)}\|} \leq \frac{\|y^{(k)}\|\|s^{(k)}\|}{2\|y^{(k)}\|\|y^{(k)}\|} = \frac{\|s^{(k)}\|}{2\|y^{(k)}\|}, \tag{20}$$

According to (20), $\alpha_k \leq \frac{\|s^{(k)}\|}{2\|y^{(k)}\|}$ in (19) is refined into $\alpha_k \leq \frac{\|y^{(k)^H} s^{(k)}\|}{2\|y^{(k)^H} y^{(k)}\|}$ in complex domain. Consequently, the stepsize α_k can be alternatively expressed as

$$\alpha_k = min\{\sqrt{1+\gamma_{k-1}}\alpha_{k-1}, \frac{\|y^{(k)^H} s^{(k)}\|}{2\|y^{(k)^H} y^{(k)}\|}\}. \tag{21}$$

The flowchart of the proposed ACNAG algorithm is available as Algorithm 1.

Algorithm 1. ACNAG

Require:

The initial parameter $x_0 = \left(\omega_0^T, \theta_0^T\right)^T$

The initial stepsize: α_0

The momentum parameter: β

The initial stepsize adjustment factor: γ_0

The iteration counter: k

The total iteration number: K

The parameter $v_0 = 0$

Ensure:

$k \leftarrow 1$

repeat:

$\overset{+}{x}_k = x_k + \beta v_k$

$g_k = \nabla_{x^*} L(\overset{+}{x}_k)$

$s^{(k)} = \overset{+}{x}_k - \overset{+}{x}_{k-1}$, $y^{(k)} = g_k - g_{k-1}$

$\alpha_k = min\{\sqrt{1+\gamma_{k-1}}\alpha_{k-1}, \frac{\|y^{(k)^H} s^{(k)}\|}{2\|y^{(k)^H} y^{(k)}\|}\}$

$v_{k+1} = \beta v_k - \alpha_k g_k$

$x_{k+1} = x_k + v_{k+1}$

$\gamma_k = \frac{\alpha_k}{\alpha_{k-1}}$

$k \leftarrow k+1$

until $k \equiv K$

return $\boldsymbol{x_k}$

---

4 Experiments

The proposed ACNAG algorithm is evaluated through two distinct complex-valued experiments. To demonstrate the superiority of ACNAG, the CGD [3], CNAG [18] and CBBM [17] algorithms are used for comparison.

4.1 Wind Forecast

Wind forecasting plays a vital and multifaceted role in practical applications spanning various domains. It is pivotal for optimizing wind power generation, enhancing power grid operations and resource allocation. Furthermore, wind forecasting contributes significantly to our comprehension of weather patterns and climate dynamics, enabling precise weather forecasting, climate modeling and comprehensive environmental assessments. In transportation and aviation, accurate wind forecasts assume critical importance for ensuring secure and efficient operations, encompassing vital aspects such as route planning and air traffic management.

Wind forecast mainly includes the forecast of wind speed and direction, both of which affect wind energy simultaneously. Wind forecasting in the context of FCVDNM has been extensively discussed in [3,4,13]. This model utilizes a vector-based approach to analyze wind signals by incorporating both wind speed and wind direction.

In this study, the performance of prediction algorithms on FCVDNM are assessed using wind speed and direction data obtained from LaGuardia Airport in NYC. A dataset consisting of 5,000 continuous hours of data from January 1 to June 21, 2022 is collected. Ten hours of consecutive data are used to predict the next hour. The training set consists of the first 4,500 h of data, while the remaining 500 h are designated as the test set for evaluating the effectiveness of the algorithms. For FCVDNM, the experiment iterates 150 times with four different algorithms. In CGD, the learning rate is set to a fixed value of 0.3. Similarly, in CBBM, CNAG and ACNAG, the initial stepsize α_0 is also set to 0.3. Additionally, the momentum parameter in CNAG and ACNAG is fixed at 0.9.

Meanwhile, the coefficient of determination r^2 is employed as the performance indicator, which is defined as

$$r^2 = 1 - \frac{\sum_{n=1}^{N} |T_n - O_n|^2}{\sum_{n=1}^{N} |T_n - T_{\text{mean}}|^2}, \tag{22}$$

Figure 2 illustrates the loss function (5) of the four algorithms (ACNAG, CGD, CBBM and CNAG) over 150 iterations on FCVDNM. The results demonstrate the superiority of the proposed ACNAG algorithm, characterized by faster convergence and a smaller loss value compared to CGD, CBBM and CNAG. Table 1 presents the training loss, testing loss and r^2 coefficient of each algorithm after 150 iterations. Notably, ACNAG demonstrates superior performance compared to the other three algorithms. Additionally, Fig. 3 displays the fitting curve of ACNAG for wind forecasting, illustrating accurate predictions of both wind speed and direction. These results further validate the effectiveness of our proposed algorithm for wind forecasting.

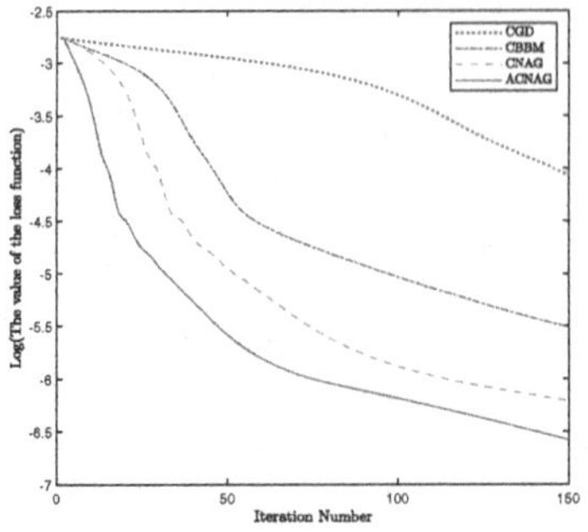

Fig. 2. Loss function of the wind forecasting problem.

Table 1. Performance of different algorithms on the wind forecasting problem

	CGD	CBBM	CNAG	ACNAG
Training loss	0.0171	0.0040	0.0020	**0.0014**
Testing loss	0.0111	0.0040	0.0021	**0.0014**
r^2	0.9520	0.9776	0.9842	**0.9895**

4.2 Stock Price Forecast

Stock price prediction is of paramount importance in financial markets and investment decision-making. Its accuracy empowers investors to anticipate market trends, optimize portfolio allocations and effectively manage risks. Additionally, it supports financial institutions in asset valuation, credit risk assessment and regulatory compliance. Leveraging advanced algorithms and data analytics, stock price prediction enables informed decision-making and enhances comprehension of intricate market dynamics within the multifaceted realm of finance. Recently, FCVDNM has gained significant popularity in the field of stock price forecasting [15,20] due to its ability to effectively extract nonlinear information from historical stock prices.

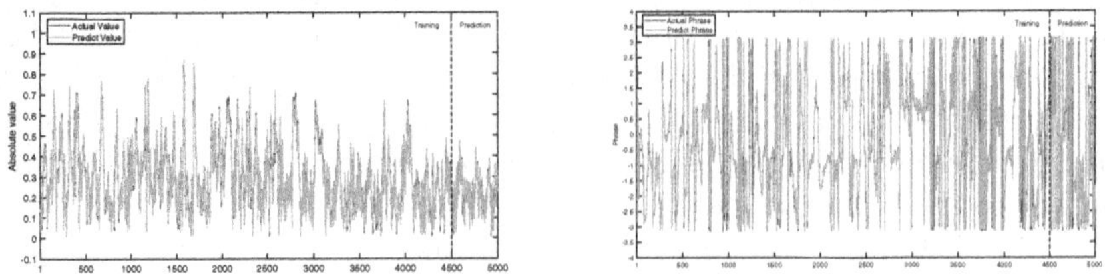

(a) (b)

Fig. 3. Fitting curve of wind forecasting by ACNAG: (a) Normalized absolute wind speed, (b) Wind direction

This section evaluates the performance of different algorithms using stock indexes from four different markets, denoted as Deutscher Aktien Index (DAX), Nasdaq Composite Index (IXIC), Shenzhen Stock Exchange Composite Index (SZSE) and Shanghai Stock Exchange Composite Index (SSE). The prediction model utilizes the closing index data from five consecutive days to forecast the closing index for the next day. Data from different time periods are collected for each index on their respective opening days. The DAX index included closing prices from July 24, 2018, to April 21, 2023. The IXIC index consisted of closing prices from July 10, 2018, to April 21, 2023. The SZSE and SSE indices encompassed closing prices from May 8, 2018, to April 21, 2023. For each of these four datasets, 1200 sets of continuous data are obtained for training and testing respectively.

For FCVDNM, above mentioned four algorithms are evaluated through 200 iterations on four different datasets. In CGD, the learning rate is set to a fixed value of 0.3. Similarly, in CBBM, CNAG and ACNAG, the initial stepsize α_0 is also set to 0.3. Additionally, the momentum parameter in CNAG and ACNAG is fixed at 0.9.

Table 2 presents the training and testing loss obtained by four algorithms on different datasets. The results illustrate the superior performance of the proposed ACNAG algorithm compared to the other three algorithms in terms of stock price forecast, as indicated by lower training and testing loss.

Table 2. Performance of different algorithms on the stock price forecast problem

Dataset	Algorithm	Training loss	Testing loss
DAX	CGD	0.0030	0.0029
	CBBM	0.0015	0.0014
	CNAG	5.2646e-04	4.0877e-04
	ACNAG	**4.5457e-04**	**3.4830e-04**
IXIC	CGD	0.0028	0.0033
	CBBM	8.1764e-04	8.0831e-04
	CNAG	4.1995e-04	3.4393e-04
	ACNAG	**3.4102e-04**	**2.9278e-04**
SZSE	CGD	0.0030	0.0038
	CBBM	0.0011	0.0010
	CNAG	5.4564e-04	2.8447e-04
	ACNAG	**4.4329e-04**	**2.1175e-04**
SSE	CGD	0.0034	0.0026
	CBBM	0.0017	0.0013
	CNAG	8.1588e-04	4.4885e-04
	ACNAG	**6.9484e-04**	**3.7408e-04**

5 Conclusion

This paper has presented the ACNAG algorithm for the efficient training of FCVDNM. The integration of an adaptive stepsize update formula and stepsize limitation rules in CNAG facilitates effective stepsize adaptation. Actually, the adaptive stepsize update formula utilized in the proposed approach incorporates second-order information from approximating the inverse of complex Hessian matrix. This enables accurate adaptation of stepsize using local curvature information, speeding up the convergence process. Moreover, the stepsize is additionally constrained by scaled Malitsky-Mishchenko criterion. Through performance comparison with CGD, CBBM and CNAG algorithms in wind forecast and stock price forecast tasks, it can be concluded that ACNAG has provided a promising way for the training of FCVDNM.

References

1. Amin, M.F., Amin, M.I., Al-Nuaimi, A.Y.H., Murase, K.: Wirtinger calculus based gradient descent and levenberg-marquardt learning algorithms in complex-valued neural networks. In: ICONIP, pp. 550–559 (2011)
2. Gao, S., Zhou, M., Wang, Y., Cheng, J., Yachi, H., Wang, J.: Dendritic neuron model with effective learning algorithms for classification, approximation, and prediction. IEEE Trans. Neural Networks Learn. Syst. **30**(2), 601–614 (2019)
3. Gao, S., et al.: Fully complex-valued dendritic neuron model. IEEE Trans. Neural Networks Learn. Syst. **34**(4), 2105–2118 (2023)
4. Ji, J., Dong, M., Lin, Q., Tan, K.C.: Forecasting wind speed time series via dendritic neural regression. IEEE Comput. Intell. Mag. **16**(3), 50–66 (2021)
5. Ji, J., Dong, M., Lin, Q., Tan, K.C.: Noninvasive cuffless blood pressure estimation with dendritic neural regression. IEEE Trans. Cybern. **53**(7), 4162–4174 (2023)
6. Li, X., et al.: Power-efficient neural network with artificial dendrites. Nat. Nanotechnol. **15**(9), 776–782 (2020)
7. Malitsky, Y., Mishchenko, K.: Adaptive gradient descent without descent. arXiv:1910.09529 (2019)
8. Mustapha, A., Mohamed, L., Ali, K.: An overview of gradient descent algorithm optimization in machine learning: application in the ophthalmology field. In: SADASC, pp. 349–359 (2020)
9. Nesterov, Y.E.: A method of solving a convex programming problem with convergence rate $O(\frac{1}{k^2})$. Dokl. Akad. Nauk. **269**, 543–547 (1983)
10. Polyak, B.T.: Some methods of speeding up the convergence of iteration methods. USSR Comput. Math. Math. Phys. **4**(5), 1–17 (1964)
11. Raydan, M.: The Barzilai and Borwein gradient method for the large scale unconstrained minimization problem. SIAM J. Optim. **7**(1), 26–33 (1997)
12. Shi, B., Du, S.S., Jordan, M.I., Su, W.J.: Understanding the acceleration phenomenon via high-resolution differential equations. Math. Program. **195**, 79–148 (2021)
13. Song, Z., Tang, Y., Ji, J., Todo, Y.: Evaluating a dendritic neuron model for wind speed forecasting. Knowledge-Based Syst. **201**, 106052 (2020)
14. Tang, C., Ji, J., Tang, Y., Gao, S., Tang, Z., Todo, Y.: A novel machine learning technique for computer-aided diagnosis. Eng. Appl. Artif. Intell. **92**, 103627 (2020)

15. Tang, Y., Song, Z., Zhu, Y., Hou, M., Tang, C., Ji, J.: Adopting a dendritic neural model for predicting stock price index movement. Expert Syst. Appl. **205**, 117637 (2022)
16. Wu, R., Huang, H., Qian, X., Huang, T.: A L-BFGS based learning algorithm for complex-valued feedforward neural networks. Neural Process. Lett. **47**, 1271–1284 (2018)
17. Zhang, H., Mandic, D.P.: Is a complex-valued stepsize advantageous in complex-valued gradient learning algorithms? IEEE Trans. Neural Networks Learn. Syst. **27**(12), 2730–2735 (2015)
18. Zhang, S., Xia, Y.: Two fast complex-valued algorithms for solving complex quadratic programming problems. IEEE Trans. Cybern. **46**(12), 2837–2847 (2015)
19. Zheng, T., Tamura, H., Kuratu, M., Ishizuka, O., Tanno, K.: A model of the neuron based on dendrite mechanisms. Electron. Commun. Jpn. **84**, 11–24 (2001)
20. Zhou, T., Gao, S., Wang, J., Chu, C., Todo, Y., Tang, Z.: Financial time series prediction using a dendritic neuron model. Knowledge-Based Syst. **105**, 214–224 (2016)

Interpreting Decision Process in Offline Reinforcement Learning for Interactive Recommendation Systems

Zoya Volovikova[1,2(✉)], Petr Kuderov[1,2], and Aleksandr I. Panov[1,3]

[1] AIRI, Moscow, Russia
zoya.v@ya.ru
[2] Moscow Institute of Physics and Technology, Dolgoprudny, Russia
[3] Federal Research Center "Computer Science and Control" of the Russian Academy of Sciences FRC CSC RAS, Moscow, Russia
https://airi.net/

Abstract. Recommendation systems, which predict relevant and appealing items for users on web platforms, often rely on static user interests, resulting in limited interactivity and adaptability. Reinforcement Learning (RL), while providing a dynamic and adaptive approach, brings its unique challenges in this context. Interpreting the behavior of an RL agent within recommendation systems is complex due to factors such as the vast and continuously evolving state and action spaces, non-stationary user preferences, and implicit, delayed rewards often associated with long-term user satisfaction.

Addressing the inherent complexities of applying RL in recommendation systems, we propose a framework that includes innovative metrics and a synthetic environment. The metrics aim to assess the real-time adaptability of an RL agent to dynamic user preferences. We apply this framework to LastFM datasets to interpret metric outcomes and test hypotheses regarding MDP setups and algorithm choices by adjusting dataset parameters within the synthetic environment. This approach illustrates potential applications of our framework, while highlighting the necessity for further research in this area.

1 Introduction

The task of providing recommendations in machine learning involves predicting the most relevant and interesting items for users on a web resource [1]. Different approaches to prediction exist, each with its own optimization objective in online settings. Traditional methods often build a static vector of user interests based on known interactions and make predictions accordingly [8,15]. However, this approach lacks interactivity and adaptability, as users' interests change over time and some historical data may become irrelevant.

A more effective approach is to view recommendations as a sequence modeling task using the next-item-prediction framework [19,20]. This framework

B. Luo et al. (Eds.): ICONIP 2023, CCIS 1963, pp. 270–286, 2024.
https://doi.org/10.1007/978-981-99-8138-0_22

interactively makes decisions about recommendations, taking into account an encoded sequence of user actions and history. [10] showed that by considering recommendation as a sequence modeling task, it can be observed that the rating given by a website visitor for a particular item is correlated with the items that were recommended prior to it. This suggests the possibility of implementing an algorithm with a policy that optimizes long-term rewards.

By employing Reinforcement Learning (RL) algorithms [9], which are designed to learn from interactions with the environment, it is possible to create more accurate, flexible, and responsive recommendation systems that can keep pace with users' changing interests and needs. The application of RL in recommendation systems presents unique challenges [5,6,10], especially in understanding the behaviour of the RL agent and its potential for performance in real-time situations [3,13]. Existing evaluation methods tend to focus on short-term predictive success, not thoroughly addressing the nuanced implications of RL in real-world scenarios.

To address this, our study provides a controlled, synthetic environment to observe and evaluate RL agents. We perform a series of experiments that examine the behaviour of these agents under different conditions, revealing their adaptability and focus on long-term optimization, and the divergence of their learned strategies from the original dataset. We use our experimental environment to scrutinize the RL agents' behaviour, with particular attention to their adaptability to changing user preferences. Furthermore, we introduce a new set of interactive metrics, giving us deeper insights into the decision-making process of the agent. By applying these tools to the real-world LastFM datasets, we demonstrate how our proposed methods can guide the interpretation of RL agent performance.

2 Related Work

In traditional settings, recommendation systems can be largely reduced to a matrix factorization problem [8]. This approach has been shown to be reliable, and under certain configurations, can compete with more complex methods such as sequence modeling [12]. However, the effectiveness of these systems is often gauged using conventional evaluation metrics such as NDCG, HitRate, etc. The adequacy of these metrics has come under scrutiny, with several studies pointing out their inconsistency and proposing alternative measures [4]. For instance, there are suggestions to compute Negative and Positive Hit Rates separately, showcasing that an algorithm might perform well on conventional metrics but still recommend a significant number of undesired items to the user.

Understanding the online performance of these systems, based on these metrics, is a challenging aspect explored in various research studies [2,3,11]. Some propose the creation of online environments to simulate user behavior and monitor algorithm behavior during online training. For instance, the RecoGym [14] and model is designed to simulate changes in user interests through the parameterization of the hidden space dimensionality. Another study [22] proposes a user simulator based on a Generative Adversarial Network (GAN), enabling the evaluation of the quality of recommendation algorithms. The authors of RecSim [7] propose a configurable platform for authoring simulation environments for recommender systems that supports sequential interaction with users. The goal is to create simulations that mirror specific aspects of user behavior found in real systems to serve as a controlled environment for developing, evaluating, and comparing recommender models and algorithms. Real world based environments provide a more realistic setting for evaluating and training RL algorithms by simulating actual user-item interactions from online platforms.

Several researchers aim to enhance the interpretability of reinforcement learning (RL) agents in recommendation systems. Studies by [3,10], and [6] suggest approaches like counterfactual evaluation and simulation of the online behavior of RL agents. Others, like [18], recommend an initial examination of datasets to assess their suitability for long-term optimization. Simultaneously, [21] and [13] advocate for the SHAP (SHapley Additive exPlanations) approach for interpreting RL agent behavior.

3 Recommendation System as Markov Decision Problem

In order to frame Recommendation System optimization problem as Reinforcement Learning problem (see Sect. C of Appendix for the RL problem definition), we define the following MDP components: actions, states, and rewards, using the notations common in recommender systems literature.

Actions (Items): All items available for selection on a web resource constitute the action space. Formally, the action space A is defined as the set of all possible items a user can choose: $A = \{i_1, i_2, ..., i_n\}$.

States (User Interaction History): The state space comprises the user's last N consecutive interactions, such as browsing history or items added to a shopping cart. Formally, the state space S is defined as the set of all possible user interaction histories: $S = \{h_1, h_2, ..., h_m\}$, where each history h_i contains the last N items interacted with by the user.

Rewards (User Ratings): The reward is the quantitative feedback or evaluation (e.g., rating) assigned by the user to an item. Formally, the reward function R is defined as a mapping from a state-item-state tuple (h, i, h') to a real-valued reward: $R(h, i, h') = r$. Here, h represents the user interaction history, i is the selected item, and h' is the updated interaction history after selecting item i.

The goal of an MDP in the context of recommender systems is to find a policy $\pi : S \rightarrow A$ that maximizes the expected cumulative discounted reward:

$$\pi^* = \underset{\pi}{\operatorname{argmax}} \mathbb{E}\left[\sum_{t=0}^{\infty} \gamma^t R(h_t, i_t, h_{t+1}) | \pi, h_0\right], \tag{1}$$

where h_t is the user interaction history at time t, i_t is the item selected at time t according to policy π, h_{t+1} is the updated user interaction history, and $\gamma \in [0, 1)$ is the discount factor.

4 Reinforcement Learning Interactive Evaluation

Our research introduces two interpretive tools for reinforcement learning (RL) in interactive recommendation systems: a controlled environment simulating user behavior, and two new metrics - Interactive Hit Rate (IHitRate) and Preference Correlation. The environment uses hidden vector embeddings to model users and items, with a dynamic 'satiation' vector reflecting user preference variability. This allows for comprehensive control, insight into internal states, and testing algorithm adaptability, long-term reward focus, and improvement potential.

The IHitRate metric captures real-time user interests, making it suitable for tasks requiring immediate response like music recommendations. The Preference Correlation metric measures the RL-based recommender system's understanding of user preferences by comparing actual user preferences with predictions from the RL agent's Q-function. Both metrics provide more accurate real-time algorithm performance measurement.

4.1 Controllable Environment

The proposed environment for reinforcement learning (RL) captures the dynamics of user behavior in a web-based system, focusing on recommendation algorithms. User preferences are represented by two components: static long-term interests ("tastes") and dynamic short-term interests ("satiation"). By capturing the intricate dynamics of user behavior, this model provides a realistic and challenging environment for training RL agents.

We denote user tastes as a vector $\mathbf{T}$ in the same latent space as items' embeddings. It reflects the user's long-term preferences to the items (defined by an arbitrary selected distance/similarity metric). The tastes are randomly generated with configurable random seed and remain static within a single simulation run, providing a stable user behavior baseline.

In contrast, the satiation is dynamic, evolving over time as the user interacts with the system. We denote satiation as a vector $\mathbf{S}$, where each element represents the current level of interest or satiety for a particular category. As the user interacts with an item, the user's satiation updates, reflecting changes in short-term interests. For instance, if a user frequently interacts with items from a certain category, the satiation for the corresponding category increases, reflecting a decrease in short-term interest due to overexposure. Additionally, the satiation to the similar categories is slightly increased too.

The "relevance" of an item to the user is based on the static user tastes and their volatile satiation as following:

$$\text{relevance} = \text{similarity}(\mathbf{T}, \mathbf{V}_{\text{item}}) \cdot \text{boosting}(\mathbf{A}_{\text{sat}}), \tag{2}$$

where $\mathbf{A}_{\text{sat}}$ represents the "aggregate item satiation". The boosting function modifies the relevance score based on the aggregate satiation for the item's categories, adding a dynamic component to the item's relevance.

To update satiation, we calculate an item's relation to the categories (clusters) with soft classification based on their pairwise relevance first:

$$\mathbf{R}_{\text{item-cluster}} = \text{softmax}\left[\text{similarity}(\mathbf{V}_{\text{item}}, \mathbf{C})\right], \tag{3}$$

where $\mathbf{C}$ represents the clusters' embeddings and $\mathbf{V}_{\text{item}}$ represents the item's embedding vector. The user satiation is then updated based on this item-to-categories relation and the category-specific satiation rate:

$$\mathbf{S} \leftarrow \mathbf{S} \cdot \left[1.0 + \mathbf{R}_{\text{item-cluster}} \cdot \mathbf{V}_{\text{sat-speed}}\right], \tag{4}$$

where $\mathbf{V}_{\text{sat-speed}}$ is a vector representing the rate of change of satiation for each category.

4.2 Metrics

In order to quantify the performance of reinforcement learning (RL) based recommender systems in an interactive setting, we introduce two key metrics: the Interactive Hit Rate (IHitRate) and the Preference Correlation (PC).

The IHitRate is an adaptation of the traditional Hit Rate formula, designed to capture the dynamics of a user's real-time interests. Unlike the traditional Hit Rate, which assesses the overall proportion of relevant recommendations, the IHitRate measures the proportion of relevant recommendations within each episode of a user's real-time interaction. An episode is formed based on the time gaps in the user interaction history. Mathematically, the IHitRate is defined as:

$$\text{IHitRate} = \frac{\text{Number of relevant items in episode}}{\text{Total number of recommended items in episode}}. \tag{5}$$

Moving onto the Preference Correlation metric, its objective is to assess the extent to which the RL agent comprehends a user's relative preferences at any given moment. Given the inherent limitation of not knowing the user's preferences for all items at all times, this metric focuses on the user's relative preferences for a specific group of items.

The essential idea behind Preference Correlation is the comparison of the actual immediate rewards R_t (representing the user's preferences) and the expected future rewards $Q(s_t, a_t)$ predicted by the RL agent's Q-function. If the relative preferences are conserved between the actual rewards and the predicted rewards, we can assert that the RL algorithm understands the dynamics of the user's preferences at that specific time step.

Mathematically, given a set of item evaluations R_t provided by the user (the immediate rewards) and a set of item evaluations $Q(s_t, a_t)$ predicted by the RL agent's Q-function (the expected future rewards), we sort R_t and $Q(s_t, a_t)$ in increasing order of R_t. Let's denote the sorted vectors as R_{ts} and $Q_{ts}(s_t, a_t)$ respectively.

The Preference Correlation metric is then defined as the Pearson correlation coefficient between R_{ts} and $Q_{ts}(s_t, a_t)$:

$$\text{PC} = \frac{\text{cov}(R_{ts}, Q_{ts}(s_t, a_t))}{\sigma_{R_{ts}} \sigma_{Q_{ts}(s_t, a_t)}}, \tag{6}$$

where $\text{cov}(R_{ts}, Q_{ts}(s_t, a_t))$ is the covariance of R_{ts} and $Q_{ts}(s_t, a_t)$, and $\sigma_{R_{ts}}$ and $\sigma_{Q_{ts}(s_t,a_t)}$ are the standard deviations of R_{ts} and $Q_{ts}(s_t, a_t)$ respectively. A high Preference Correlation indicates that the RL agent is capable of accurately modelling the dynamics of the user's preferences, thereby demonstrating its effectiveness.

4.3 Visualizations

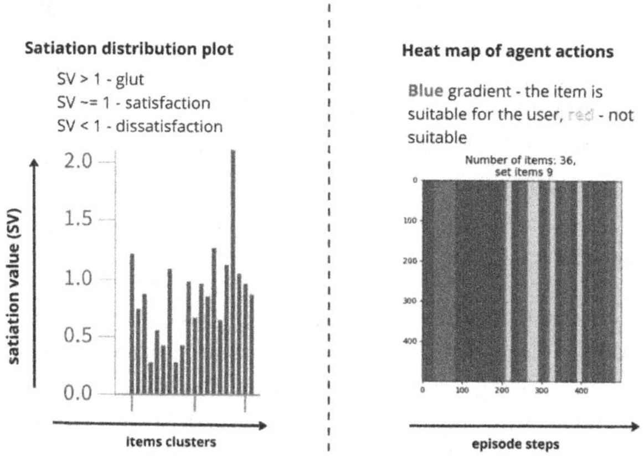

Fig. 1. Left: a user's satiation for each item category. Right: the recommender system's actions relevance heatmap throughout the episode. (Color figure online)

We propose a series of visualizations of the agent's performance throughout a single episode: a user's satiation for each category and the user's action heatmap (see example in Fig. 1). These visualizations provide valuable insights into the agent's behavior and decision making within that specific timeframe, and help assess RL algorithms in our recommender system.

A satiation plot show a user's engagement and interaction. Values > 1 imply oversaturation, < 1 imply dissatisfaction, and near 1 indicate fulfilled interests. A satiation histogram can help detect overfitting (uneven values) or insufficient specificity (uniform distribution) of the agent's actions.

The user's action heatmap indicates relevant (blue) and irrelevant (red) actions. A predominantly blue heatmap suggests an effective adaptation to user interests. Otherwise, a predominantly red heatmap may indicate ineffective response to satiation changes.

4.4 Offline-Datases Action Sampling Strategies

Capitalizing on the state dynamics of our simulator, we've designed an innovative action sampling strategy for offline dataset creation. The array of actions, ranging from highly strategic to suboptimal, are sorted by their theoretical relevance, generating a wide range of possible scenarios. This diversity is critical in the context of offline reinforcement learning (RL) environments for recommendation systems, where real-time adjustments are not feasible.

This approach allows us to comprehensively study RL agents' behavior in offline settings, investigating their adaptability and generalization capabilities. When faced with suboptimal actions, the agents are tasked with extracting beneficial information, showcasing their resilience. Meanwhile, the presence of optimal actions allows us to measure agents' effectiveness in exploiting successful strategies and evaluate potential overfitting risks. Thus, through this varied action sampling, we gain valuable insights into the robustness and adaptability of RL agents in offline recommendation systems.

5 Experiments

In this chapter, we conduct rigorous training of various algorithms on data derived from a controlled environment using the d3rlpy library [16]. Our primary objectives are to analyze how the training datasets influence agents' online behavior and to determine the correlation between offline testing metrics and actual online performance. The study encompasses several algorithms, particularly focusing on their unique learning characteristics such as prediction accuracy and adaptability to user behavior changes.

5.1 Algorithms

Discrete Conservative Q-Learning (CQL) is an offline reinforcement learning algorithm that mitigates overestimation bias in Q-learning. By minimizing an upper bound on the expected value function under the learned policy relative to its value under the behavior policy, it ensures conservative policy improvement.

The principle of entropy maximization drives **Discrete Soft Actor-Critic (SAC)**. This off-policy algorithm optimizes a balance between expected return and entropy, facilitating efficient exploration in discrete action spaces.

BERT4Rec is a sequential recommendation algorithm that extends the transformer model, BERT, to the recommendation domain. By treating the sequence of interactions as a sequence of tokens in a language model, BERT4Rec captures complex sequential patterns in user behavior, thus generating personalized recommendations.

Behavioral Cloning (BC) is a method in imitation learning where the agent aims to mimic the behavior of an expert by direct supervision. The model is trained to replicate the expert's decisions in similar states, thereby creating a policy that is expected to perform comparably in the same environment. BC assumes access to a dataset of high-quality expert demonstrations and is often used as a starting point for more sophisticated imitation learning algorithms.

5.2 Environment Settings

In our study, we explore two settings for our simulator. The first setting assumes static user interests within a single episode, that is optimal actions come from approximately two fixed interest clusters. We define this setting in our environment using the parameter saturation speed [0.0001, 0.0004].

Conversely, the second setting assumes that user interests dynamically change. Thus, within a single episode, the clusters of items that interest the user can vary. We set the saturation speed parameter to [0.1, 0.4] to reflect this dynamic nature.

By employing these different settings, we aim to assess the extent to which agents can adapt to dynamically changing environment parameters or, in the context of recommender systems, adapt to the user's mood.

5.3 Synthetic Datasets Generation

In our study, we generated two distinct dataset variants to explore different scenarios and strategies within recommender systems. The first dataset variant represents a relatively high level of optimality, comprising uniformly distributed actions from the first 10% of optimal actions, the first 25% of optimal actions, and the last 10% of suboptimal actions. The second dataset variant represented a lower level of optimality, featuring a uniform distribution of actions from the first 30% of optimal actions and the last 25% of suboptimal actions.

By employing these datasets, we aim to evaluate the performance of various algorithms, including CQL, SAC, Bert4Rec, and BC (behavioral cloning). Specifically, we seek to determine whether agents can discover strategies that surpass the suboptimal action set and identify which algorithms exhibit superior performance on each dataset variant.

6 Results

How interactive metrics and classic metrics forecast agent behavior online?

We compared the dynamics of metric changes on two datasets of varying quality: one of low quality (Fig. 5) and another of high quality (Fig. 6). This comparison led us to draw certain conclusions about the interpretation of metrics during the training of an agent.

In the case of lower-quality datasets, which contain a high number of negative reviews, the metric that shows the best correlation is Preference Correlation. This is because Preference Correlation is designed to assess the agent's understanding of user needs at a specific step, rather than simply evaluating the relationship between the predicted items and those present in the test set. However, it's important to note that Preference Correlation can exhibit high values if the dataset includes many high ratings and the agent's coverage is low. In such situations, the agent only samples a small number of actions it has learned, and there is a high likelihood that these actions coincide with the positive ratings in the dataset.

Based on our tests, for datasets of higher quality—meaning those with a large number of positive responses—the I-HitRate metric correlates with online evaluations, but is also sensitive to the agent's coverage. When comparing agents with equal accuracy and different coverage, the I-HitRate metric will favor an agent with higher coverage.

On the other hand, the results of the NDCG metric did not seem informative in the context of an interactive setup for recommendations. This is because the dynamic changes of this metric did not correlate with the rewards obtained in the environment. Hence, we deduce that this metric may not be very useful in an interactive recommendation task.

Thus, based on our findings, we propose that in studies involving real-world data, preference should be given to those algorithms that, while demonstrating the highest Preference Correlation (PC), also show a sufficiently high I-HitRate. This combination of metrics is indicative of an algorithm's ability to not only comprehend the user's needs at each interaction but also ensure broad coverage of relevant items. Higher I-HitRate suggests that the algorithm is capable of recommending an assortment of items, thereby accommodating diverse user preferences, while a high PC implies that the algorithm is adept at capturing the user's needs at a specific stage. This combination could yield a more effective recommendation system that satisfies users' immediate and diverse needs, which is crucial in the context of real-world environments with varying data quality.

Can agents learn a better strategy than the data?

We compared the behavior of agents in an environment trained on data with a large number of optimal actions, and on data with a small number of optimal actions. When analyzing the behavior, we need to consider that Reinforcement Learning (RL) agents strive to find an optimal policy—a strategy that defines the best possible action in every state to maximize the cumulative reward.

Figure 2 shows that when the agents are trained on data with a larger number of optimal values, they do not reach the strategy by which this data was generated. This may be due to the inherent complexity of exploring a vast action space and learning to generalize optimal decisions from it. Conversely, when trained on suboptimal data, the agent's behavior surpasses the behavior of the agent that generated the data. This might seem counterintuitive but it's possible, especially if the agent training process manages to find a more efficient or diverse set of actions, compared to the less varied, suboptimal data it was initially trained on.

Moreover, the comparison of the Fig. 2 plots reveals that the Conservative Q-Learning (CQL) algorithm, when trained on suboptimal data, manages to learn a strategy superior to that obtained when trained on optimal data. CQL, a type of Reinforcement Learning algorithm, is specifically designed to handle issues related to overestimation of Q-values and to provide more robust and stable learning from off-policy data. In this case, it seems that it's able to better exploit the suboptimal data by driving exploration in less densely covered action spaces, thereby achieving higher performance.

How do agents adjust to changing user moods?

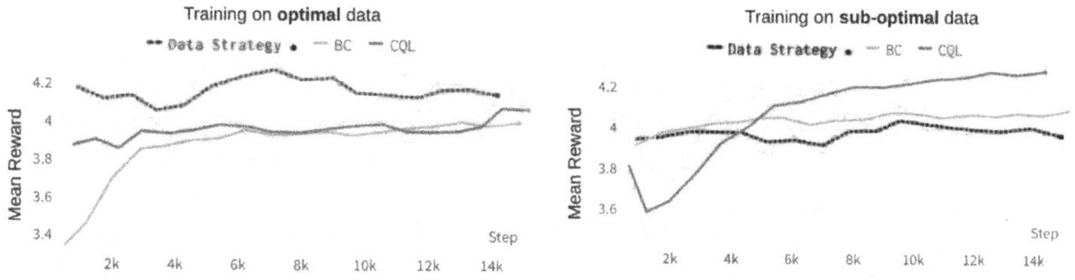

Fig. 2. Learning curves of the CQL and BC algorithms on data obtained by optimal and sub-optimal strategies.

We compared the behavior of agents in an environment with high and low dynamics of user interests (see Table 1). Among the algorithms tested, CQL shows the best results in the high-dynamic environment. However, an analysis of heat map visualizations of agent behavior (see Fig. 3) in episodes reveals that none of the algorithms were able to sustain user interest throughout the entire episode. There is a gradient from blue to red, indicating a decline in interest over time. On the other hand, when comparing visualizations for the low-dynamic environment, it can be observed that both CQL and BC are able to maintain user interest. While SAC and Bert4Rec show decent results in the low-dynamic environment, they perform poorly in the high-dynamic environment. In the case of SAC, this may be attributed to the overestimation of the value function, as SAC lacks any regularizers unlike CQL. As for the Bert4Rec algorithm, the low results in a dynamic environment can be attributed to the fact that the architecture of the algorithm itself does not effectively capture dynamic patterns from the data.

Is it possible to use a framework to improve results on real datasets?

Our research, primarily conducted on synthetic data, aims to elucidate the practicality of behavior analysis within a controlled environment for enhancing reinforcement learning agents' performance on real-world data. In this regard, we simulate key parameters of actual datasets, such as episode lengths, user count, and item count, within our environment.

A subset of the LastFM dataset [17] served as our case study. Predominantly, the dataset comprises items believed to be liked by the user, a conclusion drawn from the observation that tracks were listened to in their entirety. Nonetheless, it's important to note the absence of direct user ratings in the dataset, thus ruling out the possibility of training the agent using rewards mirroring the user's item rating.

To mimic this dataset in our environment, we incorporated available data: the dataset is heavily skewed towards user-relevant actions; it included data from 30 users and 2,000 items, culminating in a total of 10,000 data points. As LastFM is a music service, we inferred user preferences to be highly dynamic, an assumption reflected in our data modeling.

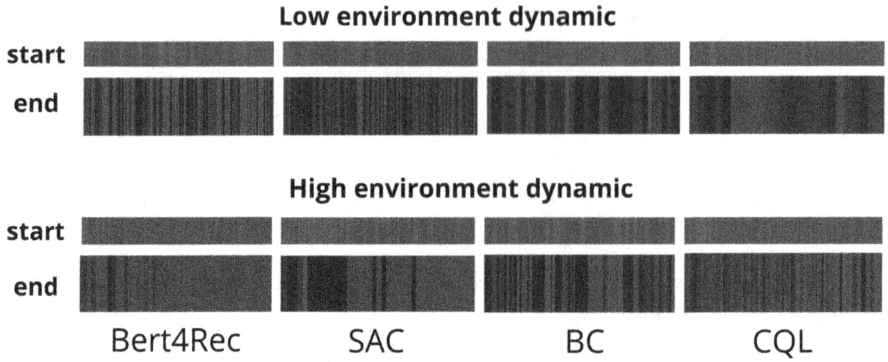

Fig. 3. The pre-training (start) and post-training (end) behavior of BC and CQL agents in environments with varying dynamics. In low dynamic environments, the blue gradient signifies effective adaptation to user preferences. A shift from blue towards red in high dynamic environments indicates difficulty in keeping up with rapidly changing user preferences. (Color figure online)

Our investigation centers on determining whether an agent can formulate a behavior strategy devoid of user rating labels. Specifically, we examine the agent's behavior when given a dense reward, consistently valued at 1. The pivotal question here is - how does this task configuration influence the agent's behavior?

Our analysis of the agent's reward acquisition within the environment revealed that an agent trained with a dense reward always set to 1 learns at a slower pace than when rewards are derived from user ratings (see Fig. 4). However, by the culmination of the training, the agent's strategy converges to a model of rewards that is close to the one based on user ratings. This evidence corroborates our hypothesis that a Markov Decision Process (MDP) framework with a dense reward always equal to 1 can be utilized for training on LastFM data.

Moreover, we discern a correlation between the growth dynamic of the Interactive Hit Rate (I-HitRate) metric within our environment and during training on real-world data. Given our assumption that the data largely encapsulates information about user-relevant actions, we can infer that an increase in I-HitRate should correlate with a rise in reward scores in the actual environment, as supported by our correlation analysis of metrics and rewards (Appendix A.2).

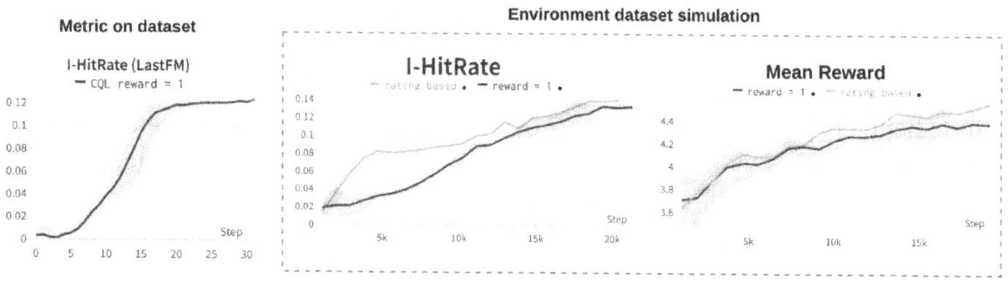

Fig. 4. Left: the agent's training results on real data, measured by the I-HitRate metric. Right: the agent's performance in the environment using simulated LastFM data, where the parameters are modeled based on the known information.

7 Conclusion

Our research presents a new environment and task setup aimed at improving our understanding of reinforcement learning agents' behavior in recommender systems. This setup uses unique metrics such as the interactive hit-rate, which measures an agent's prediction accuracy within simulated user-web interactions, and the newly introduced Preference Correlation metric, assessing the agent's understanding of user's relative preferences at any moment.

Our controlled environment provides a means to effectively analyze the strategies and adaptability of reinforcement learning agents, offering insights into how well they adjust to changes in user moods over time and the reliability of their predictions. Rather than replicating the entire real world within our model, we concentrate on key aspects that affect an agent's performance, including user mood shifts and various dynamic environmental factors. By focusing on these elements, we can better understand why an agent behaves in a certain way and identify possible improvements. In summary, our setup offers a valuable tool for deepening our understanding of agent behavior. It provides a platform for hypothesis testing and analyzing agent behavior, contributing to efforts to refine reinforcement learning agents for real-world applications.

One of the limitations of this work is the assumption that the simulated scenarios reflect the real ones. That is, they set tasks for the recommender system and therefore evaluate exactly those properties that will be reflected in the corresponding real scenarios. The proposed framework is not proposed to be used for direct training of models, but can be used for an approximate assessment of the influence of hyperparameters or for ablative studies of individual parts of the methods.

Acknowledgments. The authors extend their heartfelt gratitude to Evgeny Frolov and Alexey Skrynnyk for their insightful feedback, guidance on key focus areas for this research.

A Classic Recommendation Algorithms

Recommendation systems have been extensively studied, and several classic algorithms have emerged as popular approaches for generating recommendations. In this section, we provide a brief overview of some of these algorithms, such as matrix factorization and other notable examples.

A.1 Algorithms

Matrix factorization is a widely used technique for collaborative filtering in recommendation systems. The basic idea is to decompose the user-item interaction matrix into two lower-dimensional matrices, representing latent factors for users and items. The interaction between users and items can then be approximated by the product of these latent factors. Singular Value Decomposition (SVD) and Alternating Least Squares (ALS) are common methods for performing matrix factorization. The objective function for matrix factorization can be written as:

$$\min_{U,V} \sum_{(i,j)\in\Omega} (R_{ij} - U_i^T V_j)^2 + \lambda(||U_i||^2 + ||V_j||^2), \tag{7}$$

where R_{ij} is the observed interaction between user i and item j, U_i and V_j are the latent factors for user i and item j, respectively, Ω is the set of observed user-item interactions, and λ is a regularization parameter to prevent overfitting.

Besides matrix factorization, other classic recommendation algorithms include:

- **User-based Collaborative Filtering:** This approach finds users who are similar to the target user and recommends items that these similar users have liked or interacted with. The similarity between users can be computed using metrics such as Pearson correlation or cosine similarity.
- **Item-based Collaborative Filtering:** Instead of focusing on user similarity, this method computes the similarity between items and recommends items that are similar to those the target user has liked or interacted with.
- **Content-based Filtering:** This approach utilizes features of items and user profiles to generate recommendations, assuming that users are more likely to be interested in items that are similar to their previous interactions.

A.2 Evaluation Metrics for Recommendation Algorithms

Several evaluation metrics are commonly used to assess the performance of recommendation algorithms. We provide a brief description with equations for the metrics, which we use in our work.

Normalized Discounted Cumulative Gain (NDCG) used for measuring the effectiveness of ranking algorithms, takes into account the position of relevant items in the ranked list. First, DCG is calculated:

$$\text{DCG@}k = \sum_{i=1}^{k} \frac{\text{rel}_i}{\log_2(i+1)}, \tag{8}$$

where k is the number of top recommendations considered, and rel_i is the relevance score of the item at position i in the ranked list. Then DCG value is normalized with the ideal DCG (IDCG), which represents the highest possible DCG value:

$$\text{NDCG@}k = \frac{\text{DCG@}k}{\text{IDCG@}k}. \tag{9}$$

Coverage measures the fraction of recommended items to the total number of items:

$$\text{Coverage} = \frac{|\text{Recommended Items}|}{|\text{Total Items}|}. \tag{10}$$

High coverage indicates that the algorithm can recommend a diverse set of items, while low coverage implies that it is limited to a narrow subset. Hit Rate calculates the proportion of relevant recommendations out of the total recommendations provided:

$$\text{Hit Rate} = \frac{\text{Number of Hits}}{\text{Total Number of Recommendations}}, \tag{11}$$

where a "hit" occurs when a recommended item is considered relevant or of interest to the user. A higher hit rate indicates better performance.

B Experiments in Dynamic Environment

Experiments in a high-dynamic environment for agents trained on optimal and sub-optimal data. Preference Correlation is the best metric for evaluating lower-quality datasets with negative reviews, as it assesses the agent's understanding

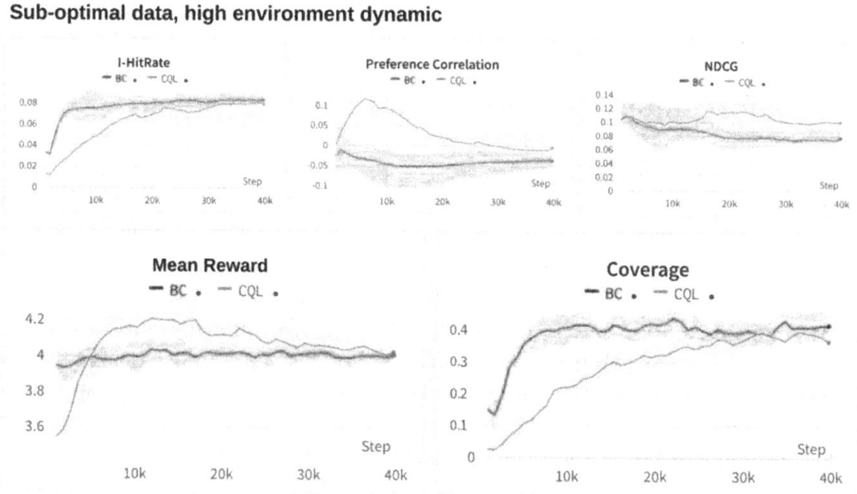

Fig. 5. The results of an experiment in a highly dynamic environment when trained on suboptimal data.

of user needs. However, high values can occur if the dataset has many positive ratings and the agent's coverage is low. For higher-quality datasets with positive responses, the I-HitRate metric correlates with online evaluations but is sensitive to the agent's coverage.

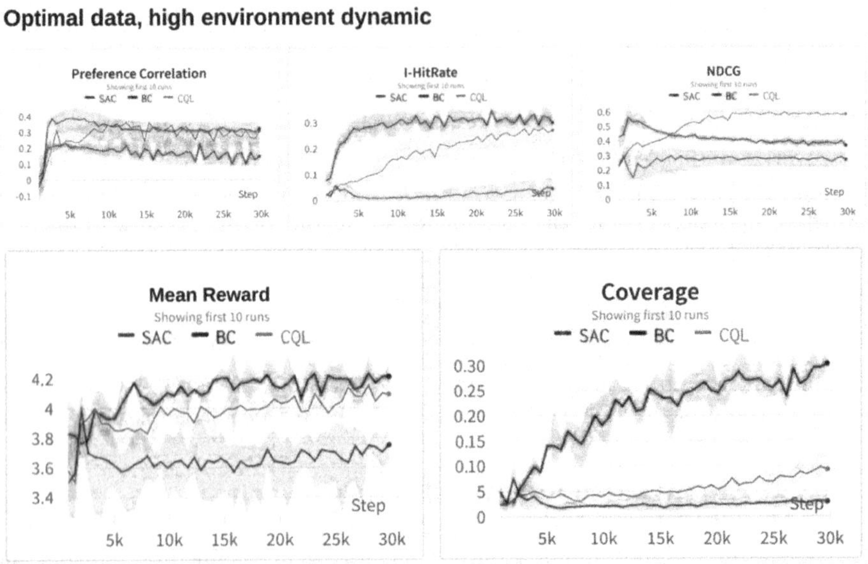

Fig. 6. The results of an experiment in a highly dynamic environment when trained on optimal data.

C Reinforcement Learning as an MDP Problem

Reinforcement learning (RL) addresses the problem of learning optimal behaviors by interacting with an environment. A fundamental concept in RL is the Markov Decision Process (MDP), which models the decision-making problem as a tuple (S, A, P, R, γ). In this framework, S represents the state space, A is the action space, P is the state transition probability function, R denotes the reward function, and γ is the discount factor. By formulating the recommendation problem as an MDP, RL algorithms can learn to make decisions that optimize long-term rewards. The MDP framework provides a solid foundation for designing and evaluating RL agents in recommendation systems, allowing for the development of more adaptive and effective algorithms.

D Comparing Algorithms Behavior on High Dynamic Environment

Table 1. Comparison of CQL, SAC, BC, Dert4Rec algorithms in an environment with high dynamics of user mood changes.

	CQL	SAC	BC	Bert4Rec
I-HitRate	0.1	0.05	**0.11**	0.01
PC	0.1	**0.3**	−0.05	0.01
NDCG	**0.11**	0.1	0.08	0.1
Mean Reward	**4.2**	3.7	4	3.7

References

1. Bobadilla, J., Ortega, F., Hernando, A., Gutiérrez, A.: Recommender systems survey. Knowl.-Based Syst. **46**, 109–132 (2013)
2. Dacrema, M.F., Boglio, S., Cremonesi, P., Jannach, D.: A troubling analysis of reproducibility and progress in recommender systems research. ACM Transactions on Information Systems **39**(2), 1–49 (2021). https://doi.org/10.1145/3434185, arXiv:1911.07698 [cs]
3. Deffayet, R., et al.: Offline evaluation for reinforcement learning-based recommendation: a critical issue and some alternatives. arXiv preprint arXiv:2301.00993 (2023)
4. Frolov, E., Oseledets, I.: Fifty shades of ratings: how to benefit from a negative feedback in top-n recommendations tasks. In: Proceedings of the 10th ACM Conference on Recommender Systems, pp. 91–98 (2016). https://doi.org/10.1145/2959100.2959170, http://arxiv.org/abs/1607.04228, arXiv:1607.04228 [cs, stat]
5. Grishanov, A., Ianina, A., Vorontsov, K.: Multiobjective evaluation of reinforcement learning based recommender systems. In: Proceedings of the 16th ACM Conference on Recommender Systems, pp. 622–627 (2022)
6. Hou, Y., et al.: A deep reinforcement learning real-time recommendation model based on long and short-term preference. Int. J. Comput. Intell. Syst. **16**(1), 4 (2023)
7. Ie, E., et al.: RECSIM: a configurable simulation platform for recommender systems (arXiv:1909.04847) (2019). http://arxiv.org/abs/1909.04847, arXiv:1909.04847 [cs, stat]
8. Koren, Y., Bell, R., Volinsky, C.: Matrix factorization techniques for recommender systems. Computer **42**(8), 30–37 (2009)
9. Levine, S., Kumar, A., Tucker, G., Fu, J.: Offline reinforcement learning: tutorial, review, and perspectives on open problems. arXiv preprint arXiv:2005.01643 (2020)
10. Liu, F., et al.: Deep reinforcement learning based recommendation with explicit user-item interactions modeling. arXiv preprint arXiv:1810.12027 (2018)
11. Lundberg, S.M., Lee, S.I.: A unified approach to interpreting model predictions. In: Advances in Neural Information Processing Systems, vol. 30 (2017)

12. Marin, N., Makhneva, E., Lysyuk, M., Chernyy, V., Oseledets, I., Frolov, E.: Tensor-based collaborative filtering with smooth ratings scale. arXiv preprint arXiv:2205.05070 (2022)
13. Meggetto, F., et al.: Why people skip music? on predicting music skips using deep reinforcement learning. arXiv preprint arXiv:2301.03881 (2023)
14. Rohde, D., Bonner, S., Dunlop, T., Vasile, F., Karatzoglou, A.: RecoGym: a reinforcement learning environment for the problem of product recommendation in online advertising. arXiv preprint arXiv:1808.00720 (2018)
15. Schafer, J.B., Frankowski, D., Herlocker, J., Sen, S.: Collaborative filtering recommender systems. In: Brusilovsky, P., Kobsa, A., Nejdl, W. (eds.) The Adaptive Web. LNCS, vol. 4321, pp. 291–324. Springer, Heidelberg (2007). https://doi.org/10.1007/978-3-540-72079-9_9
16. Seno, T., Imai, M.: d3rlpy: an offline deep reinforcement learning library. J. Mach. Learn. Res. **23**(1), 14205–14224 (2022)
17. Turrin, R., Quadrana, M., Condorelli, A., Pagano, R., Cremonesi, P.: 30music listening and playlists dataset. In: Castells, P. (ed.) Poster Proceedings of the 9th ACM Conference on Recommender Systems, RecSys 2015, Vienna, Austria, September 16, 2015. CEUR Workshop Proceedings, vol. 1441. CEUR-WS.org (2015). https://ceur-ws.org/Vol-1441/recsys2015_poster13.pdf
18. Wang, K., et al.: Rl4rs: A real-world benchmark for reinforcement learning based recommender system. arXiv preprint arXiv:2110.11073 (2021)
19. Wang, S., Cao, L., Wang, Y., Sheng, Q.Z., Orgun, M.A., Lian, D.: A survey on session-based recommender systems. ACM Comput. Surv. (CSUR) **54**(7), 1–38 (2021)
20. Wang, S., Hu, L., Wang, Y., Cao, L., Sheng, Q.Z., Orgun, M.: Sequential recommender systems: challenges, progress and prospects. arXiv preprint arXiv:2001.04830 (2019)
21. Wang, X., et al.: A reinforcement learning framework for explainable recommendation. In: 2018 IEEE International Conference on Data Mining (ICDM), pp. 587–596. IEEE (2018)
22. Zhao, X., Xia, L., Zou, L., Yin, D., Tang, J.: Toward simulating environments in reinforcement learning based recommendations. arXiv preprint arXiv:1906.11462 (2019)

A Novel Framework for Forecasting Mental Stress Levels Based on Physiological Signals

Yifan Li[1], Binghua Li[1], Jinhong Ding[2], Yuan Feng[1], Ming Ma[3], Zerui Han[1], Yehan Xu[1], and Likun Xia[1](✉)

[1] Laboratory of Neural Computing and Intelligent Perception (NCIP), College of Information Engineering, Capital Normal University, Beijing 100048, China
xlk@cnu.edu.cn

[2] School of Psychology, Capital Normal University, Beijing 100048, China

[3] Department of Computer Science, Winona State University, Winona, MN 55987, USA

Abstract. Mental stress may negatively affect individual health, life and work. Early intervention of such disease may help improve the quality of their lives and avoid major accidents caused by stress. Unfortunately, detection of stress levels at the present time can hardly meet the requirements such as stress regulations, which is due to the lack of anticipation of future stress changes, thus it is necessary to forecast the state of mental stress. In this study, we propose a supervised learning framework to forecast mental stress levels. Firstly, we extract a series of features of physiological signals including electroencephalography (EEG) and electrocardiography (ECG); secondly, we apply various autoregressive (AR) models to forecast stress features based on the extracted features; finally, the forecasted features are fed into several conventional machine learning based classification models to achieve forecasting of mental stress levels at subsequent time steps. We compare the effectiveness of the proposed framework on three competitive methods using three different datasets. The experimental results demonstrate that our proposed method outperforms those three methods and achieve a better forecasting accuracy of 89.65%. In addition, we present a positive correlation between mental state changes and forecasting result on theta spectrum at the frontal region.

Keywords: Mental stress · Electroencephalography (EEG) · Electrocardiography (ECG) · Autoregressive · Feature forecasting

1 Introduction

Mental stress in daily life tends to be common while chronic stress can cause some serious consequences to our health, e.g., depression [1], heart attack [2] and stroke

Supported by Beijing Natural Science Foundation (4202011).

Y. Li and B. Li — These authors contributed equally to the work.

B. Luo et al. (Eds.): ICONIP 2023, CCIS 1963, pp. 287–297, 2024.
https://doi.org/10.1007/978-981-99-8138-0_23

[3]. Mental stress affects not only physical health but also work performance [4,5]. According to related researches, excessive work-related stress has been proven to increase the possibility of errors, incidents, and injuries since mental stress at job negatively affects safety behavior in terms of safety compliance [6]. Therefore, early detection of mental stress is an essential step to manage mental stress.

Previous studies highlight that majority of the psychological activities and cognitive behavior can be indicated by psychological signals such as EEG, ECG, electromyography (EMG) and electrodermal activity (EDA) [7]. Among them, EEG and ECG are most investigated signals to detect mental stress, since EEG is closely related to brain activities and ECG is able to reflect emotional state of individual [8–10]. Therefore, discriminative features are essential for mental stress detection. Spectral feature is considered as one of the most effective EEG features which reveals individual changes corresponding to the intensity of the stress response [11–14] such as relative power of beta and alpha [15,16]. Meanwhile, existing researches [17,18] indicate that mental stress can cause changes in Heart Rate (HR) and Heart Rate Variability (HRV) because they are main biomarkers of autonomic nervous system.

Despite a large amount of research to detect mental stress levels at current time, forecasting mental stress based on physiological signals has not yet been well studied. Several studies proposed stress emotion forecasting models based on data from participants' behavior logs and cell phone logs [19–21]. However, behavior observation is less effective than somatic and physiological changes. Moreover, the above forecasting models require a large amount of long-term historical data as support to improve the accuracy. In contrast, by collecting physiological signals, the stress state of subjects can be detected in a relatively shorter term.

Unlike existing mental stress detection methods, which directly identify stress levels from recorded physiological signals, we initially forecast stress features and then identify stress levels based on forecasted features, so as to allow early intervention.

In this study, we develop a framework based on physiological signal for forecasting mental stress. The main contribution is three-fold:

- The proposed framework forecasts mental stress levels using autoregressive (AR) model to estimate stress-related physiological features in advance.
- We find a positive association of mental state changes with forecasting result on theta spectrum at the frontal region.
- The proposed framework achieves desirable performance and is beneficial to the early interventions for mental stress regulation in future research and practice.

2 Methodology

In this section, we detail the proposed mental stress forecasting framework as illustrated in Fig. 1. It mainly includes three phases: (a) data preprocessing, (b) feature extraction and selection, and (c) feature forecasting.

In phase (a), the raw data recorded during the experiment is preprocessed to remove the noise. Phase (b) initially extracts EEG and HRV features from the preprocessed data, and then utilizes principal component analysis (PCA) to realize feature selection and dimension reduction. Afterwards, the features at subsequent time steps are forecasted by AR model followed by the identification of mental stress using several well-known classifiers.

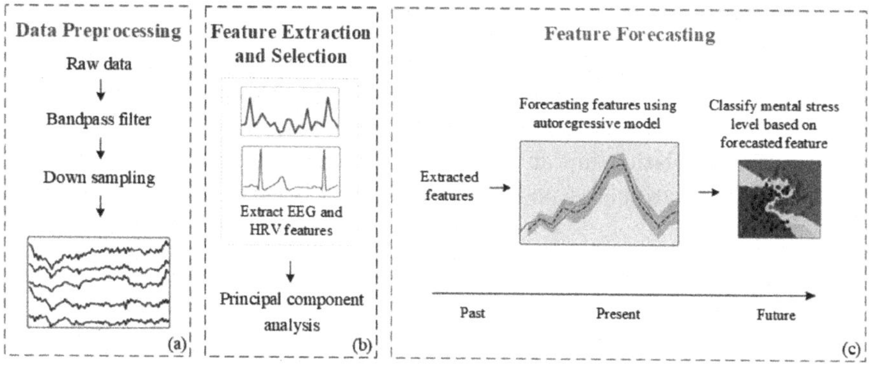

Fig. 1. The proposed framework

2.1 Data Preprocessing

The raw EEG signals are passed through a 0.1 Hz high-pass filter and 50 Hz notch filter to remove the artifacts and down sampled to 128 samples/s. In addition, the R-R intervals of ECG signals are segmented and power spectral analysis is applied to calculate HRV for subsequent feature calculation.

2.2 Feature Extraction and Selection

Feature Extraction. Frequency features of EEG and HRV features of ECG are adopted in this work. The specific details are as follows.

Absolute Power. For all the frequency bands ((delta (0.5–4 Hz), theta (4–8 Hz), alpha (8–13 Hz), beta (13–29 Hz) and gamma (¿29 Hz)), the EEG absolute power is estimated by converting the EEG signal to frequency domain using Fast Fourier Transform (FFT) for each channel.

Relative Power. It reveals the rhythmical changes in the EEG signals. In a specific frequency EEG band B_1, relative power is defined as absolute power of B_1 divided by total power of all bands. Mathematically, it is represented in equation (1):

$$RP_{B1} = \frac{P(B_1)}{P(Total_b)} \times 100\% \tag{1}$$

where $P(\cdot)$ is the absolute power, and $Total_b$ indicates all bands.

Relative Power Ratio. It shows the dominance of one frequency band over another band. Across each electrode, six ratios (delta/theta, delta/alpha, delta/beta, theta/alpha, theta/beta, and beta/alpha) between the powers of four most widely studied frequency bands (delta, theta, alpha and beta) are calculated as shown in Eq. (2), where RP_{B_i} indicates relative power of selected band B_i.

$$RP_{ratio} = \frac{RP_{B_1}}{RP_{B_2}} \tag{2}$$

Coherence. It captures the coupling at specific frequency bands and has been widely used to analyze synchrony-defined cortical neuronal assemblies in the frequency domain. In this study, we adopt frequency-specific coherence which represents linear relationship at a specific frequency between two signals. Considering two EEG signals, a and b, the coherence between them is defined by equation (3):

$$C_{xy} = \frac{|D_{ab}|^2}{D_{aa}D_{bb}} \tag{3}$$

Where D_{ab} is the cross-spectral density between a and b, and D_{aa} and D_{bb} present the power-spectral density of a and b, respectively.

HRV features. A total of seven HRV features are computed in this work, including VLF, LF, HF, LFnu, HFnu, LF/HF and HR. The power spectrum of the HRV is distributed into the very-low-frequency (VLF) (<0.05 Hz), low-frequency (LF) (0.04–0.15 Hz), and high-frequency (HF) (0.15–0.4 Hz) spectrums. The normalized units (nu) represent the relative value of LF and HF components in proportion to the sum of the components $LFnu = LF/(LF+HF)$ and $HFnu = HF/(LF+HF)$. LF/HF is presented as a ratio of the sympathetic to parasympathetic distribution or a reflection of the sympathetic attribution.

Feature Selection. In the application, PCA is applied to compress the extracted features onto a lower-dimensional feature subspace aiming to maintain most of the relevant information. Given a d-dimensional feature set $F = \{x_1, x_2, \cdots, x_d\}$ and dimensionality of the new feature subspace d', firstly, estimate the covariance matrix XX^T of the feature set; secondly, eigenvalue decomposition (EVD) is applied to XX^T to compute eigenvalues and the corresponding eigenvectors; thirdly, construct a projection matrix $W = \{w_1, w_2, \ldots, w_d\}$ from the sorted top d' eigenvectors; finally, transform the d-dimensional feature set F using the projection matrix W to obtain the new d' dimensional feature subspace.

2.3 Feature Forecasting

Autoregressive Model. It is a regression model that utilizes the combination of random variables at a certain time in the previous period to describe the random variables at a certain time in the future. Its parameter identification is relatively simple and has favorable real-time performance. Therefore, we apply AR model to forecast the features of subsequent time steps based on the existing features extracted at several time steps in the previous period. Given a time series variable χ, an autoregressive model of order p can be written as equation (4):

$$\chi_t = \phi_0 + \phi_1\chi_{t-1} + \phi_2\chi_{t-2} + \cdots + \phi_p\chi_{t-p} + \xi_t \tag{4}$$

Where ξ_t is white noise and ϕ is the parameters of the model.

After feature forecasting, the support vector machine (SVM) [9–12] is used to classify the mental stress levels based on the forecasted features, due to its advantages in solving the classification of nonlinear, small samples and high-dimensional objects. Other classifiers have also been applied such as k-nearest neighbors (k-NN), random forest and multilayer perceptron (MLP).

3 Experimental Results

3.1 Dataset and System Environment

The proposed framework was evaluated on three different datasets, namely, a private dataset - MIST [9] and two publicly available datasets - DMAT [22] and STEW [23].

MIST dataset. It was collected from 20 subjects (20 males, mean age of 22.54±1.53 years) during the Montreal Imaging Stress Task (MIST) experiment [24] for four classes. The data was obtained using 23 EEG channels with a 500 Hz sample rate. The recording for each subject included 20 min (4 task levels × 5 min/ level) during tasks and 20 min during tasks with time restriction.

DMAT dataset. It was collected from 36 subjects (27 females and 9 males, mean age of 18.60±0.87 years) during mental arithmetic task performance for two classes. The data was obtained using 23 EEG channels with a 500 Hz sample rate. Each subject's recording contained 3 min at rest and 1 min during mental counting.

STEW dataset. It was collected from 48 subjects (48 males) during a multitasking workload experiment for two classes. The data was obtained using 14 EEG channels with a 128 Hz sample rate. The recording of each subject consisted of 2.5 min at rest state and 2.5 min during task. The ratings of subject's perceived mental workload (on a scale of 1 to 9) are also provided.

System Environment. The proposed framework was implemented on Python environment with AMD Ryzen 9 3950X 16-Core Processor and Quadro RTX 6000 GPU.

3.2 Evaluation of Feature Forecasting Performance

Two evaluation metrics are selected including mean absolute error (MAE) and root mean square error (RMSE), which indicate the actual errors and enlarger the gap between errors, respectively, as shown in Eq. (5) and (6), to measure the difference between actual features and forecasted features.

$$MAE(X,Y) = \frac{1}{N}\sum_{i=1}^{N}|X_i - Y_i| \tag{5}$$

$$MAE(X,Y) = \frac{1}{N}\sum_{i=1}^{N}|X_i - Y_i| RMSE(X,Y) = \sqrt{\frac{1}{N}\sum_{i=1}^{N}|X_i - Y_i|} \quad (6)$$

Where X represents actual features, Y denotes forecasting features and N indicates the number of total samples.

To evaluate the feature forecasting performance of AR, we compare it with Auto Regressive Integrated Moving Average (ARIMA) model and Long Short-Term Memory (LSTM). The comparison of two metrics among different methods is shown in Table 1.

Table 1. Feature forecasting performance using three forecasting models

Dataset	MIST	
	MAE	RMSE
AR	**0.2157 ± 0.2608**	**0.2456 ± 0.3498**
ARIMA	0.2622 ± 0.2892	0.3205 ± 0.3459
LSTM	0.2531±0.2654	0.3099 ± 0.3263
Dataset	DMAT	
	MAE	RMSE
AR	**0.3156 ± 0.3498**	**0.3641 ± 0.4056**
ARIMA	0.3727 ± 0.3166	0.4326 ± 0.3646
LSTM	0.6563 ± 0.8408	0.7379 ± 0.8876
Dataset	STEW	
	MAE	RMSE
AR	**0.1853 ± 0.2169**	**0.2447 ± 0.3344**
ARIMA	0.2529 ± 0.3566	0.3148 ± 0.4457
LSTM	0.2236 ± 0.2930	0.2876 ± 0.3848

It is observed that AR outperforms other models across all evaluation metrics on three datasets, indicating favorable effectiveness for forecasting features. Specifically, on MIST dataset, the MAE of AR is 0.2157, which is about 17% lower than ARIMA and 14% lower than LSTM, while the RMSE values obtained from AR are less than ARIMA by 23% and lower than LSTM by 20%. Meanwhile, in terms of standard deviation of all metrics, AR is lower than other models in most cases indicating better robustness of AR. Taking a close observation on DMAT dataset the MAE and RMSE of LSTM is nearly twice higher than AR, which may be due to insufficient model training caused by small datasets. Moreover, LSTM requires longer training time than other two models. Therefore, in terms of all metrics, among three models, AR has superior performance on feature forecasting.

3.3 Evaluation of Classification Performance

For better observation and objective evaluation of the classification method, we calculated the following metrics: accuracy (Acc) = (TP+TN)/(TP+FP+TN+FN), precision (Pre) = TP/(TP+FP), recall (Rec) = TP/(TP + FN), F1 Score (F1) = 2*TP/(2*TP+FN+FP) and AUC = $\Sigma I(P_{positive}, P_{negative})/(M * N)$, where M denotes the number of positive samples and N indicates the number of negative samples.

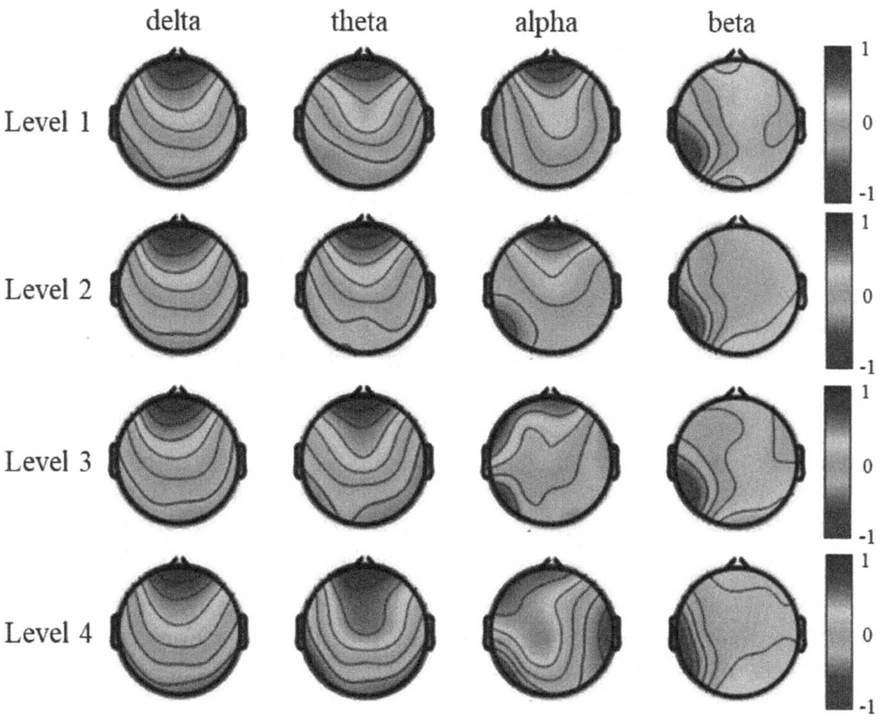

Fig. 2. The topoplots illustrate the averaged EEG spectral power for stress levels (states). Each row presents the averaged spectra powers associated with different spectra; each column highlights the averaged spectra powers related to different stress levels

The results are illustrated in Table 2 with four classifiers including SVM-RBF, k-NN, random forest and MLP for mental stress levels detection at subsequent time steps.

It is observed that across all evaluation metrics SVM-RBF outperforms other classifiers, which indicates better effectiveness for multiclass classification, so as to detect mental stress levels at subsequent time steps based on forecasted features. In details, SVM-RBF provides superior performance as compared to random forest (with 7% higher in Acc, Rec, F1 and AUC, 9% higher in Pre on MIST dataset, about 3% higher on DMAT dataset and 2% higher on STEW

Table 2. Comparison of various classifiers on three datasets

Dataset	MIST				
	Acc(%)	Pre(%)	Rec(%)	F1(%)	AUC(%)
SVM-RBF	**79.79 ± 0.63**	**81.14 ± 8.98**	**79.93 ± 8.68**	**79.69 ± 3.56**	**79.72 ± 3.73**
k-NN	75.83 ± 0.80	77.48 ± 11.59	75.85 ± 7.10	76.06 ± 6.77	76.87 ± 5.18
Random Forest	72.25 ± 1.01	72.29 ± 7.41	72.14 ± 7.65	72.11 ± 7.03	72.53 ± 4.17
MLP	78.90 ± 0.72	80.18 ± 9.29	79.11 ± 6.85	79.48 ± 5.29	79.44 ± 3.70
Dataset	DMAT				
	Acc(%)	Pre(%)	Rec(%)	F1(%)	AUC(%)
SVM-RBF	**79.85 ± 2.66**	**80.32 ± 4.47**	**79.58 ± 8.24**	**79.53 ± 3.92**	**79.58 ± 2.76**
k-NN	77.11 ± 4.02	76.81 ± 7.90	77.02 ± 6.73	76.62 ± 5.44	76.91 ± 3.74
Random Forest	76.38 ± 2.92	76.28 ± 3.87	76.08 ± 4.99	76.11 ± 3.77	76.89 ± 3.07
MLP	78.83 ± 2.18	78.89 ± 3.77	78.69 ± 5.35	78.63 ± 3.13	78.69 ± 2.33
Dataset	STEW				
	Acc(%)	Pre(%)	Rec(%)	F1(%)	AUC(%)
SVM-RBF	**89.65 ± 2.44**	**90.02 ± 4.89**	**89.72 ± 5.71**	**89.61 ± 2.53**	**89.72 ± 2.36**
k-NN	85.35 ± 5.11	86.41 ± 8.38	85.43 ± 10.84	85.09 ± 5.69	85.43 ± 5.11
Random Forest	87.13 ± 2.90	87.10 ± 5.38	87.30 ± 4.74	87.02 ± 3.10	87.30 ± 3.00
MLP	88.88 ± 4.13	88.97 ± 5.01	88.90 ± 5.11	8.83 ±3 4.14	88.90 ± 4.03

dataset) and k-NN (with 4% higher in Acc, Pre and Rec, 3% higher in F1 and AUC on MIST dataset, about 2% higher on DMAT dataset and 4% higher on STEW dataset). In conclusion, SVM-RBF achieves the best classification performance among three classification models based on forecasted features.

3.4 Visualization and Evaluation on Mental States Forecasting Based on Spectral Characteristics

To demonstrate the effectiveness of the proposed forecasting work, we visualize the classification results associated with the stress level changes. This is implemented by comparing the EEG spectrums among all the stress levels (states) from the MIST dataset. Figure 2 illustrates the averaged EEG spectral power for stress levels in different frequency bands. Across subjects and sessions, the state of stress is characterized by an increase in the theta power at the frontal-parietal sites, a decrease in the theta power at the occipital region, and an increase in the alpha power at the left temporal site. Such phenomenon is consistent with the outcome during rest-state [25]. Moreover, a significant suppression of the high-frequency spectra (beta band) can be found over the left temporal region. Taking a close observation on the same spectrum at each stress level, we observe that the changes are insignificant. On the contrary, the spectral power of theta band at stress level 3 and level 4 increases over the parietal lobe. This implies that more difficult task has been performed in the region, which verifies our hypothesis, i.e., stress level changes are associated with task difficulty, thus improving the effectiveness of forecasting.

Since both theta and alpha bands exhibit more significant information than others, we take one channel of theta band at frontal region (F4) to further investigate the accuracy of forecasting. This is implemented by comparing the forecasted data with actual data based on the features of absolute power of theta band with time duration of 30 secs. The results shown in Fig. 3 indicate that both the actual data and the forecasted one are similar in terms of shape and magnitude, which indicates that the forecasting model can predict the mental stress accurately.

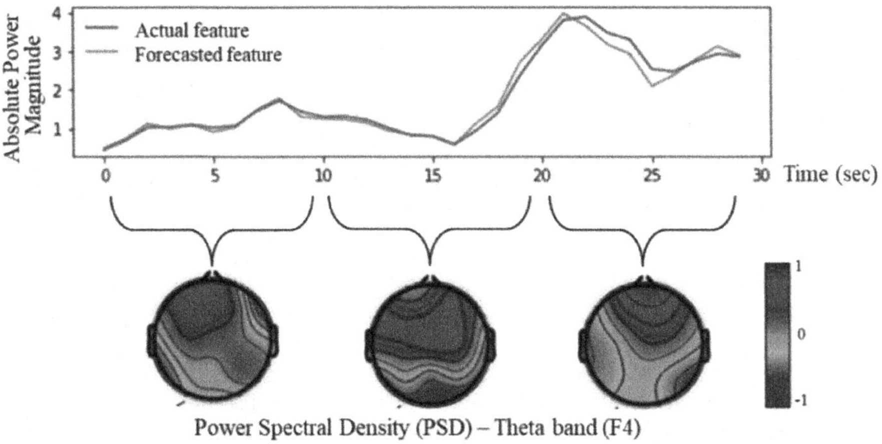

Fig. 3. Evaluation on forecasting capability associated with Power Spectral Density in theta band at frontal lobe (F4). The x axis presents the time in sec, and y axis denotes the magnitude of absolute power

Moreover, we investigate whether the changes of magnitude for stress level changes are associated with the spectral power. The forecasting time is divided into three equivalent parts, each of which reveals the positive correlation between spectral power and the changes of the signal. During 0-10 secs, the magnitude increases gradually, and the corresponding spectral power at F4 increases; while the magnitude decreases during about 11–20 secs as the corresponding spectral power decreases; and the magnitude increases sharply between 21-28 secs, although there is decrease at about 22–25 secs, the corresponding spectral power shows a stronger activity in average than other two parts. This may indicate the higher mental stress level detected via activities in this region. Hence, the proposed work can forecast the mental stress level by taking the absolute power magnitude in F4 region into account.

4 Conclusion

In this study, a framework is presented for forecasting mental stress levels. Initially, we extract a set of effective features from EEG and ECG, and then we apply AR model to forecast mental stress features after comparison of three existing forecasting models including AR, ARIMA and LSTM. Finally, four representative classifiers are employed to detect the mental stress levels at subsequent time steps. Based on three datasets, our results demonstrate that the proposed framework has achieved desirable performance in the forecasting of mental stress levels, and that the mental state correlated to the forecasted outcome of theta spectrum at the frontal region, indicating the presence of the EEG marker that can be applied for evaluating mental stress. However, there is no significant difference of high spectral power in the whole brain region at all stress levels, implying the minimum impact of such power on forecasting of mental states. The proposed framework can be applied to mental stress assessment and regulation.

Acknowledgments. This work is supported by Beijing Natural Science Foundation (4202011). There is no other conflicts of interest.).

References

1. Wang, F., Yang, J., Pan, F., Bourgeois, J.A., Huang, J.H.: Early life stress and depression. Front. Psych. **10**, 964 (2019)
2. Song, H., Fang, F., Arnberg, F.K., et al.: Stress related disorders and risk of cardiovascular disease: population based, sibling controlled cohort study. BMJ. Br. Med. J. **365**, l1255 (2019)
3. Kronenberg, G., Schöner, J., Nolte, C., Heinz, A., Endres, M., Gertz, K.: Charting the perfect storm: emerging biological interfaces between stress and stroke. Eur. Arch. Psychiatry Clin. Neurosci. **267**(6), 487–494 (2017)
4. Day, A.J., Brasher, K., Bridger, R.S.: Accident proneness revisited: the role of psychological stress and cognitive failure. Accid. Anal. Prev. **49**(6), 532–535 (2012)
5. Lu, C.S., Kuo, S.Y.: The effect of job stress on self-reported safety behaviour in container terminal operations: the moderating role of emotional intelligence. Transport. Res. F: Traffic Psychol. Behav. **37**, 10–26 (2016)
6. Leung, M.Y., Liang, Q., Olomolaiye, P.: Impact of job stressors and stress on the safety behavior and accidents of construction workers. J. Manag. Eng. **32**(1), 04015019 (2016)
7. Lehmann, D.: EEG assessment of brain activity: spatial aspects, segmentation and imaging. Int. J. Psychophysiol. **1**(3), 267–276 (1984)
8. Al-Shargie, F.M., Tang, T.B., Kiguchi, M.: Mental stress quantification using EEG signals. In 2015 International Conference for Innovation in Biomedical Engineering and Life Sciences (IFMBE), Singapore: Springer Singapore, 15-19 (2016)
9. Xia, L., Malik, A.S., Subhani, A.R.: A physiological signal-based method for early mental-stress detection. Biomed. Signal Process. Control **46**, 18–32 (2018)
10. Arsalan, A., Majid, M., Butt, A.R., Anwar, S.M.: Classification of perceived mental stress using a commercially available EEG headband. IEEE J. Biomed. Health Inform. **23**(6), 2257–2264 (2019)

11. Jun, G., Smitha K.G.: EEG based stress level identification. IEEE International Conference on Systems, Man, and Cybernetics (SMC), 003270-003274 (2016)
12. Subhani, A.R., Mumtaz, W., Saad, M., Kamel, N., Malik, A.S.: Machine learning framework for the detection of mental stress at multiple levels. IEEE Access **5**, 13545–13556 (2017)
13. Asif, A., Majid, M., Anwar, S.M.: Human stress classification using EEG signals in response to music tracks. Comput. Biol. Med. **107**, 182–196 (2019)
14. Ahirwal, M.K.: Analysis and identification of EEG features for mental stress. In Evolution in Computational Intelligence, Singapore: Springer Singapore, 201–209 (2021)
15. Norhazman, H., Zaini, N., Taib, M.N., Jailani, R., Latip, M.: Alpha and Beta Sub-waves Patterns when Evoked by External Stressor and Entrained by Binaural Beats Tone. 2019 IEEE 7th Conference on Systems, Process and Control (ICSPC), 112-117 (2019)
16. Chang, H.Y., Stevenson, C.E., Jung, T.P., Ko, L.W.: Stress-induced effects in resting EEG spectra predict the performance of SSVEP-based BCI. IEEE Trans. Neural Syst. Rehabil. Eng. **28**(8), 1771–1780 (2020)
17. Taelman, J., Vandeput, S., Spaepen, A., Van Huffel, S.: Influence of mental stress on heart rate and heart rate variability. In 4th European conference of the international federation for medical and biological engineering, 1366-1369 (2009)
18. Pereira, T., Almeida, P.R., Cunha, J.P., Aguiar, A.: Heart rate variability metrics for fine-grained stress level assessment. Comput. Methods Programs Biomed. **148**, 71–80 (2017)
19. Suhara, Y., Xu, Y., Pentland, A.: DeepMood: Forecasting depressed mood based on self-reported histories via recurrent neural networks. the 26th International Conference on International World Wide Web Conferences Steering Committee, 17 (2017)
20. Taylor, S.A., Jaques, N., Nosakhare, E., Sano, A., Picard, R.: Personalized multitask learning for predicting tomorrow's mood, stress, and health. IEEE Trans. Affect. Comput. **11**(2), 200–213 (2020)
21. Umematsu, T., Sano, A., Taylor, S., Picard, R.W.: Improving students' daily life stress forecasting using LSTM neural networks. In 2019 IEEE EMBS International Conference on Biomedical & Health Informatics (BHI), 1-4 (2019)
22. Zyma, I., Tukaev, S., Seleznov, I., Kiyono, K., Popov, A., Chernykh, M., Shpenkov, O.: Electroencephalograms during mental arithmetic task performance. Data **4**(1), 14 (2019)
23. Lim, W.L., Sourina, O., Wang, L.P.: STEW: simultaneous task EEG workload data set. IEEE Trans. Neural Syst. Rehabil. Eng. **26**(11), 2106–2114 (2018)
24. Dedovic, K., Renwick, R., Pruessner, J.C.: The Montreal Imaging Stress Task: using functional imaging to investigate the effects of perceiving and processing psychosocial stress in the human brain. J. Psychiatry Neurosci. **30**(5), 319 (2005)
25. Komarov, O., Ko, L.W., Jung, T.P.: Associations among emotional state, sleep quality, and resting-state EEG spectra: a longitudinal study in graduate students. IEEE Trans. Neural Syst. Rehabil. Eng. **28**(4), 795–804 (2020)

Correlation-Distance Graph Learning for Treatment Response Prediction from rs-fMRI

Francis Xiatian Zhang[3], Sisi Zheng[1,2], Hubert P. H. Shum[3(✉)], Haozheng Zhang[3], Nan Song[1,2], Mingkang Song[1,2], and Hongxiao Jia[1,2(✉)]

[1] Beijing Key Laboratory of Mental Disorders, National Clinical Research Center for Mental Disorders and National Center for Mental Disorders, Beijing Anding Hospital, Capital Medical University, Beijing, China
{zhengsisi,jhxlj}@ccmu.edu.cn, songmed003@mail.ccmu.edu.cn

[2] Advanced Innovation Center for Human Brain Protection, Beijing Anding Hospital, Capital Medical University, Beijing, China

[3] Department of Computer Science, Durham University, Durham, UK
{xiatian.zhang,hubert.shum,haozheng.zhang}@durham.ac.uk

Abstract. Resting-state fMRI (rs-fMRI) functional connectivity (FC) analysis provides valuable insights into the relationships between different brain regions and their potential implications for neurological or psychiatric disorders. However, specific design efforts to predict treatment response from rs-fMRI remain limited due to difficulties in understanding the current brain state and the underlying mechanisms driving the observed patterns, which limited the clinical application of rs-fMRI. To overcome that, we propose a graph learning framework that captures comprehensive features by integrating both correlation and distance-based similarity measures under a contrastive loss. This approach results in a more expressive framework that captures brain dynamic features at different scales and enables more accurate prediction of treatment response. Our experiments on the chronic pain and depersonalization disorder datasets demonstrate that our proposed method outperforms current methods in different scenarios. To the best of our knowledge, we are the first to explore the integration of distance-based and correlation-based neural similarity into graph learning for treatment response prediction.

Keywords: Graph Learning · Functional Connectivity · rs-fMRI

This research is supported in part by the Beijing Hospitals Authority Youth Program (ref: QML20191901), Beijing Hospitals Authority Clinical Medicine Development of Special Funding (ref: ZYLX202129), Beijing Hospitals Authority's Ascent Plan (ref: DFL20191901), Training Plan for High Level Public Health Technical Talents Construction Project (ref: TTL-02-40), Research Cultivation Program of Beijing Municipal Hospital (ref: PZ2023032), EPSRC NortHFutures project (ref: EP/X031012/1).
The original version of this paper has been revised. The name of the first author has been corrected. A correction to this chapter can be found at
https://doi.org/10.1007/978-981-99-8138-0_45

B. Luo et al. (Eds.): ICONIP 2023, CCIS 1963, pp. 298–312, 2024.
https://doi.org/10.1007/978-981-99-8138-0_24

1 Introduction

Resting-state functional magnetic resonance imaging (rs-fMRI) detects spontaneous fluctuations in the blood-oxygen-level dependent (BOLD) signal of the brain [1]. It provides insights into functional connectivity (FC) between different brain regions, aiding the understanding of neurological disorders [2]. However, the dynamic nature of neural patterns may restrict their use in identifying treatment biomarkers and predicting responses [3]. Leveraging deep learning to infer treatment outcomes from rs-fMRI is a potential solution [4], and establishing a framework that considers the current brain states and their underlying mechanisms could increase the clinical relevance of rs-fMRI.

Traditional FC analysis uses Pearson correlation to represent neural similarity, a key component in understanding brain region interactions [1]. However, recent studies suggest that both correlation and distance-based similarity metrics offer distinct advantages in representing neural similarity across different tasks and brain states [5]. In particular, distance-based measures have been shown to enhance machine learning performance in rs-fMRI analysis [6]. That reveals the potential value of integrating distance-based similarity measures for deciphering complex neural mechanisms underlying observed patterns, which may also potentially improve the treatment response prediction.

FC naturally exhibits a graph structure [7], making graph representation and graph neural networks (GNNs) effective for capturing complex relationships in brain networks [8]. Recent studies have introduced graph representation and GNNs into FC analysis, grounded on two perspectives of FC [9]: static FC graph (SG) [10], which assumes constant FC within a scan, and dynamic FC graph (DG) [11–13], which assumes varying FC within a scan. With the non-stationary nature of rs-fMRI FC [14], DG approaches may be potentially more suitable for FC analysis in rs-fMRI. A recent exemplar is the Spatio-Temporal Attention Graph Isomorphism Network (STAGIN) proposed by Kim et al. [13], which applies a Graph Isomorphism Network (GIN) [15] with an attention mechanism to model the spatio-temporal interplay between ROIs and their divided time windows. However, these studies have primarily concentrated on diagnosis rather than predicting treatment responses, and their graph representations exclusively depend on correlation-based similarity measures.

Several studies have utilized machine learning techniques to predict treatment responses from rs-fMRI FC data. Cao et al. [16] employed support vector machines (SVMs) [17] to predict the treatment response of schizophrenia. Similarly, Kong et al. [4] utilized spatial-temporal graph convolutions to predict treatment response for major depressive disorder, outperforming conventional machine learning methods. However, these approaches are still solely based on the correlation-based FC for rs-fMRI and cannot infer the distinctive temporal signal segment associated with the treatment response, thereby providing an incomplete understanding and interpretability of the complex spatiotemporal FC dynamics underlying treatment response in rs-fMRI.

In this work, we aim to handle the challenge of predicting treatment response in rs-fMRI which needs comprehending more intricate underlying mechanisms compared to other status-understanding tasks. We present a framework utilizing both correlation-based and distance-based similarities to capture complex brain dynamics, beyond conventional Pearson correlation-based FC methods. We introduce a dynamic Correlation-Distance Graph Isomorphism Network (CD-GIN) with contrastive loss, enhancing spatio-temporal feature learning across varied time points. Further, we incorporate a Convolutional Block Attention Module (CBAM) [18] to highlight critical graph representations, aiding precise inference. Our method, evaluated on a chronic knee osteoarthritis (OS) clinical trial dataset [19] and a real-world depersonalization disorder (DPD) treatment dataset, demonstrates the robustness and generalizability of our proposed framework. Source code is available on https://github.com/summerwings/CDGIN.

The contributions of this work are summarized as follows:

1. To the best of our knowledge, we are the first to explore the integration of distance-based and correlation-based neural similarity into graph learning for treatment response prediction, finding a complementary relationship between the two streams from our experiments on two datasets.
2. We present the Correlation-Distance Graph Isomorphism Network (CD-GIN) with a contrastive loss function to dynamically learn unique spatio-temporal features from different similarity measures, improving the modeling of complex brain dynamics.
3. We integrate the Convolutional Block Attention Module (CBAM) to identify and interpret key graphs and time windows in rs-fMRI, enhancing predictive accuracy and uncovering significant predictors of treatment response.
4. We provide a new open benchmark for predicting treatment response in chronic OS, using clinical trial data, and construct a real-world dataset for DPD treatment, demonstrating our approach's generalizability.

2 Related Work

2.1 Distance-Based Neural Similarity

Several studies have leveraged distance-based neural similarity in machine learning for rs-fMRI analysis. Xiao et al. introduced distance measures to capture voxel-wise time course spatial relations within ROIs, leading to distance-based covariance descriptors used for correlation computation and age prediction [6]. Ma et al. extended this work using distance-based and Pearson correlations to predict age [20]. Despite these advancements, such studies primarily used correlation analysis to establish FC, without fully utilizing distance measures. This may overlook regional signal dissimilarities and non-linear ROI relationships. They also considered only static FC for single rs-fMRI scans, possibly neglecting FC temporal dynamics. These intricacies, vital for predicting treatment responses [21], inspire our work to directly employ distance measures to dynamically represent FC, aiming to capture complex spatio-temporal ROI relationships.

2.2 Dynamic Graph Learning for fMRI

Several studies have explored dynamic graph representation for FC in machine-learning-based fMRI analysis. For example, Gadgil et al. [12] utilized Spatio-Temporal Graph Convolutional Networks (STGCN) [22] to extract spatial-temporal features, and Kim et al. [13] introduced STAGIN, combining a GIN [15] with an attention mechanism. Although these studies were primarily focused on sex classification, their graph representations align with the dynamic nature of neural activity [14], a property that is also crucial for treatment response prediction. However, these methods solely relied on correlation-based similarity measures and could miss key distance-based and non-linear ROI information. In contrast, our work integrates distance-based FC into dynamic learning, uniquely positioned to capture both linear and non-linear brain dynamics, enhancing its potential for treatment response prediction.

3 Methodology

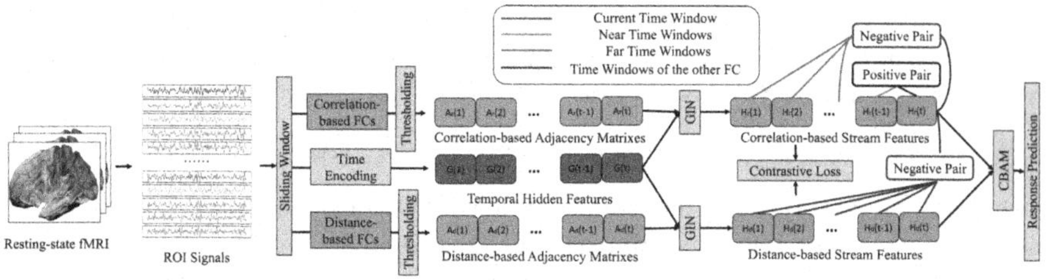

Fig. 1. The Framework of Correlation-Distance Graph Learning: The displayed contrastive pairing pertains to one example time window. However, in practical implementation, the contrastive loss is actually paired and calculated for each time window

In our proposed framework (Fig. 1), signals from predefined ROIs in rs-fMRI are extracted to capture regional activity. The dynamic FC graph is then constructed using both correlation and distance-based similarity measures, representing complex brain dynamics. The CD-GIN, incorporating a contrastive loss, models spatio-temporal features and learns distinctive information between different FC and time windows. Finally, the CBAM is employed to weigh key graphs and time windows to get the probability for prediction, which refines inference and enhances interpretability.

3.1 Dynamic Functional Connectivity Graph Based on Correlation and Distance-Based Similarity

Task Formulation. We employ a pre-defined 3D atlas [23] to partition the brain into distinct ROIs, extracting ROI signal sequences $X \in \mathbb{R}^{T \times M}$ from the input sequence. T is the fMRI data length and M is the atlas-specified ROI number. We formulate treatment response prediction as a classification task, aiming to predict response y, a binary variable determined by the main clinical efficacy. It assesses the reduction rate of relevant symptom rating scores to determine whether the treatment is effective for the corresponding diseases [19]. Our proposed network infers the treatment response $\hat{y}$ from the ROI signal sequences $X(t)$ over different time windows t, where t denotes a sub-sequence extracted from X to capture evolving patterns in rs-fMRI [14].

Dual-Stream Neural Similarity for Dynamic Functional Connectivity. We propose an integration of distance-based and traditional correlation-based measures for fMRI FC analysis, aiming to capture both linear and non-linear interactions among brain regions [24]. The integration of distance-based measures arises from the limitations of correlation-based FC, which often overlooks essential non-linear interactions that influence treatment efficacy. This oversight is partially attributed to a bias induced by stimuli causing large activations [25]. Distance-based similarities, by their nature, assume that entities closer within a metric space share higher similarity, making them suitable for revealing these overlooked non-linear relationships. To align with correlation-based measures, where higher values typically indicate greater similarity, we represent similarity using negative distances. Thus, entities closer in proximity have higher similarity values. Our defined similarity measure between time series $X_i(t)$ and $X_j(t)$, denoted as $s(X_i(t), X_j(t))$, is therefore expressed as the negative distance:

$$s(X_i(t), X_j(t)) = -d(X_i(t), X_j(t)) = -f(X_i(t), X_j(t)), \tag{1}$$

where $f(X_i(t), X_j(t)))$ signifies the distance function between $X_i(t)$ and $X_j(t)$. This transformation into a similarity measure enables intuitive interpretation and comparison among different pairs of time series. Our framework employs a common distance measure, such as Manhattan, Euclidean, or Mahalanobis distance [26], each offering distinct advantages. Manhattan distance, with its simplicity, is beneficial in fMRI analysis where uncorrelated dimensions frequently appear, especially following component decomposition [1]. Euclidean distance calculates the direct path, useful for mapping direct signal interactions in complex brain networks [6]. Mahalanobis distance measures multidimensional distances considering covariance, beneficial for fMRI scans where neural interactions are intrinsically multidimensional [1].

To improve the effectiveness of our representation for predicting treatment responses, we propose integrating distance-based similarity measures with traditional correlation-based similarity measures. This integration provides complementary perspectives on brain dynamics, addressing limitations such as overlooking connectivity and scaling differences in correlation-based measures [6]. Distance-based measures capture these differences, although they may be sensitive to outliers. By combining both types, we aim to achieve a more comprehensive understanding of the data. In our approach, we represent the FC at a given time point t using two matrices: (1) the Pearson correlation coefficient (PCC) matrix $r(t)$ and (2) the negative pairwise distance matrix $d(t)$ proposed in Eq. 1. These matrices are computed as follows:

$$r_{ij}(t) = \frac{Cov(X_i(t), X_j(t))}{\sigma_{X_i(t)}\sigma_{X_j(t)}} \in \mathbb{R}^{N \times N}, \tag{2}$$

$$d_{ij}(t) = -d(X_i(t), X_j(t)), \tag{3}$$

where $Cov(X_i(t), X_j(t))$ denote the covariance between the i-th and j-th ROI at t, while $\sigma_{X_i(t)}$ and $\sigma_{X_j(t)}$ denote the corresponding standard deviations.

Dynamic Correlation-Distance Graph Representation. We introduce a dynamic graph representation design to encapsulate correlation and distance-based similarity that characterize the dynamic FC within a single brain scan. In contrast to prior work that employed static FC representations using distance-based similarity, we utilize a dynamic FC graph representation for rs-fMRI analysis [6,20]. The construction of this dynamic graph enables us to elucidate varying associations and connections among brain regions and subregions, thereby representing the topological relationships of ROIs within the broader brain network [27]. It is particularly effective in decoding brain activities considering the non-stationary dynamic fluctuations inherent to the brain [14,28].

Specifically, we employ a sliding window approach to extract sub-sequences $X(t)$ from the original signal sequence X [29], where the window setting is adjusted to different scenarios and overlapping is allowed. This approximates evolving FC patterns and preserves neural correlates at shorter scales [28]. We assemble a brain FC network graph $G(V, E)$ [7], using ROIs as nodes V and FC between ROIs as edges E. This graph captures the FC temporal evolution at specific time windows t as $G(t)(V(t), E(t))$, where $V(t)$ and $E(t)$ respectively represent the vertices and edges at a time window t. Based on previous experimental study [30], binary adjacency matrices $A_r(t) \in \{0, 1\}$ and $A_d(t) \in \{0, 1\}$ are derived from the correlation and distance-based neural similarity by thresholding the top 30% percentile values of the FC matrices $r(t)$ and $d(t)$, resulting in sparse graphs that provide a clear FC representation for modeling.

To capture the temporal patterns of brain activity within $X(t)$, we derive hidden features from the ROI signals of $X(t)$ for our graph. These features are then concatenated with the identity matrix $e_V \in \{0, 1\}^{M \times M}$, providing a one-hot encoding for each node. This approach introduces a node-aware constraint on feature modeling, thereby enhancing the depiction of temporal patterns in the dynamic FC [13]. The temporal hidden feature is formalized as:

$$G(t) = W_M[e_V || LSTM(X(t))], \tag{4}$$

where $LSTM(X(t)) \in \mathbb{R}^D$ signifies a Long Short-Term Memory (LSTM) [31] unit that processes the encoded ROI signals up to the endpoint of each $X(t)$ as input. Here, $W_M \in \mathbb{R}^{D \times (M+D)}$ denotes a learnable parameter matrix that is utilized for transforming node features.

3.2 Correlation-Distance Graph Isomorphism Network

Graph Isomorphism Network Incorporating Contrastive Loss. We present the CD-GIN with contrastive loss for enhanced treatment response prediction. GINs, efficient in capturing global and local graph features, are utilized for learning correlation and distance-based FC graph features [30]. Our method diversifies GIN expressive power by incorporating contrastive loss, pushing the model to discern unique features across FC graph streams over time windows. Contrasting with prior fMRI contrastive learning designs [32], which primarily focus on learning unique features for each patient, our work emphasizes the differentiation of underlying characteristics within a single patient. It aligns our graph hidden features with the natural structure of these neural and hemodynamic sources, reflecting their typical temporal autocorrelation relationships [33]. This design enhances the interpretability and performance of our contrastive design.

In detail, we assume a perfect graph learning model $f(X(t), A(t))$ to capture characteristic features that can infer treatment response. Here, $A(t)$ denotes the FC between ROIs within $X(t)$, which is either $A_r(t)$ or $A_d(t)$ for correlation-based or distance-based FC streams, respectively. To simulate the autocorrelation pattern in fMRI [33], the feature representations of $P(f(X(t)), A(t))$ and $P(f(X(t \pm \delta)), A(t \pm \delta))$ should be more similar for the same FC $A(t)$ within near time windows t and $t \pm \delta$, where P is a Multi-Layer Perception (MLP) projection with shared weights. Conversely, considering the unique feature representation of temporal patterns, for non-near time windows t and $t + \Delta$, where $\Delta > \delta$, the feature representations of patterns are dissimilar between $P(f(X(t)), A(t))$ and $P(f(X(t \pm \Delta)), A(t \pm \Delta))$. Considering the unique representation between $A(t)$ and different FC $\overline{A(t)}$, the feature representations of patterns should be also dissimilar between $P(f(X(t)), A(t))$ and $P(f(X(t)), \overline{A(t)})$, $P(f(X(t \pm \delta)), \overline{A(t \pm \delta)})$ or $P(f(X(t \pm \Delta)), \overline{A(t \pm \Delta)})$. To achieve this ideal $f(X(t), A(t))$ during training, we treat hidden features from nearby time points within the same graph representation as a positive pair, and those from distant time points or different graph representations as a negative pair. The loss function L_{info} that encourages this assumption can be formulated as follows [34]:

$$L_{\text{info}} = -\frac{1}{N} \sum_{i=1}^{N} \left[\log \frac{\exp(\text{sim}(z_i, z_{i \pm \delta}))}{\sum_j^N \exp(\text{sim}(z_i, \overline{z_j})) + \sum_{j \neq i \pm \delta}^N \exp(\text{sim}(z_i, z_j))} \right] \quad (5)$$

where N is the number of time windows, $sim(\cdot)$ denotes the cosine similarity, $z_i = P(f(X(t_i), A(t_i))$ is the projection of the output of our graph learning model for time point t_i, $z_j = P(f(X(t \pm \Delta), A(t \pm \Delta))$ is the projection of the output of our graph learning model for far time point t_j and $\overline{z_j} = P(f(X(t_j), \overline{A(t_j)})$ is the projection of the output of our model for other FC graph representation.

During the execution of our network, we treat each FC graph stream as $A(t)$, while the alternate stream is considered as $\overline{A(t)}$. Both streams are subject to regulation by our proposed contrastive loss. The hidden features $H_r(t)$ and $H_d(t)$ for the correlation and distance streams of our ideal model are updated according to the GIN process as follows:

$$H_r(t) = R_r(MLP_r((\epsilon_r \cdot I + A_r(t))H_r^{input}(t)W_r), \quad (6)$$

$$H_d(t) = R_d(MLP_d((\epsilon_d \cdot I + A_d(t))H_d^{input}(t)W_d), \quad (7)$$

where R represents an attention-based function for computing the graph's overall representation [13], ϵ is an initially zero parameter that can be learned, I is the identity matrix, and W represents the network weights of the MLP. The subscripts r and d denote the correlation and distance streams, respectively. This structure allows us to capture more non-linear characteristics of functional connectivity for each stream [15], thereby increasing the capacity of our network.

Convolutional Block Attention Module. We integrated the CBAM [18] into our model to enhance its comprehension at each time point by assigning weights to different graph features corresponding to different time windows. Unlike conventional graph integration methods such as later fusion by average pooling [35], the CBAM module enables our model to dynamically adjust its attention focus on critical FC graph representations. Compared to conventional attention, CBAM provides more precise control over the importance assignment to correlation and distance-based FC measures, leading to improved model performance [18]. This adaptive mechanism captures temporal fluctuations in neural activity and selects informative graph representations, improving accuracy in predicting treatment response in rs-fMRI. It also offers insights into the specific time window and FC graph that contribute to the treatment outcome, valuable for further research. The implementation details are described below:

$$Attn_{Stream} = \sigma(W_f(MaxPool(H_f(t))) + W_f(AveragePool(H_f(t)))) \tag{8}$$

$$Attn_{Temporal} = \sigma(Conv1D(MaxPool(H_f(t)))+Conv1D(AveragePool(H_f(t)))), \tag{9}$$

where $H_f(t)$ is $[H_r(t)||H_d(t)]$, σ is an activation function, W_f denotes the network weights of the two-layer MLP for max pooling and average pooling output. The attended hidden features $H_a(t)$ are obtained by:

$$H_a(t) = H_f(t) \times Attn_{Stream} \cdot Attn_{Temporal} \tag{10}$$

The final probability of treatment response prediction is inferred from a two-layer MLP based on the concatenation of k layers attended hidden features. The training loss L is a binary cross-entropy with the contrastive loss:

$$L = -y^T log(\hat{y}) + \alpha L_{info}, \tag{11}$$

where α denotes a hyper-parameter to tune the contrastive loss.

4 Experiment and Results

4.1 Experimental Design

Dataset. Our study used a public clinical trial dataset (https://openfmri.org/s3-browser/?prefix=ds000208, [19]) and a custom dataset, DPD45. The public dataset has pre-treatment fMRI data from 54 OS patients (8 responders and 9 non-responders to duloxetine; 18 responders and 19 non-responders to placebo). The DPD45 dataset, built upon our previous study [36], has pre-treatment fMRI

Table 1. Hyper-parameters for Models in Our Experiment. BS: Batch Size; LR: Learning Rate; WD: Weight Decay Rate; WS: Window Size; SS: Stride Size. To determine the optimal values for these hyperparameters, we used Weight & Bias to select best configurations

	Duloxetine						Placebo						DPD Treatment					
Method	Layer	BS	LR	WD	WS	SS	Layer	BS	LR	WD	WS	SS	Layer	BS	LR	WD	WS	SS
LSTM [31]	2	2	3e-4	1.5e-4	–	–	2	4	3e-4	3e-5	–	–	2	2	4e-4	4e-5	–	–
GCN [41]	2	2	2e-4	1e-5	–	–	2	2	1e-4	1e-5	–	–	2	2	3e-4	1.5e-4	–	–
GIN [15]	4	2	5e-4	2.5e-4	–	–	3	2	5e-4	5e-5	–	–	3	2	4e-4	2e-4	–	–
STGCN [12]	2	2	1e-4	5e-5	50	30	2	2	3e-4	3e-5	30	10	2	2	2e-4	2e-5	40	9
STAGIN [13]	2	2	4e-4	2e-4	25	5	4	2	2e-4	2e-5	50	5	4	2	5e-4	2.5e-4	50	7
Ours	2	4	4e-4	2e-4	35	25	4	2	4e-4	4e-5	50	5	2	2	4e-4	4e-6	35	20

data from 45 DPD patients (19 responders, 26 non-responders, screened by [37]) collected before routine treatment. Data preprocessing was conducted with the DPABI toolbox [38], and the HCP-MMP1 and Harvard-Oxford atlases [39,40] were used for ROI signal extraction. All procedures were approved by our institutional ethics committee, and all participants provided informed consent.

Benchmark Models. Our benchmark models primarily include state-of-the-art dynamic learning methods for fMRI analysis and other conventional methods used in previous machine learning-based fMRI analyses [1]. Models such as LSTM [31], GCN [41], GIN [15], and the latest DG methods including STGCN [12], and STAGIN [13] were used for comparative analysis.

Training Configuration. The experiments used the Weight & Bias tool for hyperparameter optimization and the PyTorch 1.9.0 framework on a GTX 2080 Ti GPU server. We tested our model performance with different distance measures for the negative distance neural similarity: Manhattan distance, Euclidean distance, and Mahalanobis distance. The α and δ parameters were set to 0.1 and 1, respectively, in our framework. Each model was trained using the Adam optimizer, and all other hyperparameters are provided in Table 1. We stratified the dataset into 80% training and 20% testing. Each model was appraised via 4-fold stratified cross-validation, maintaining identical hyperparameters.

Table 2. Performance with OS Dataset. PCC: Pearson Correlation Coefficient stream; MD: Manhattan Distance stream; ED: Euclidean Distance stream; MaD: Mahalanobis Distance stream

		Duloxetine				Placebo			
Method		AUC	ACC	SE	SP	AUC	ACC	SE	SP
LSTM [31]		0.62 ± 0.12	0.56 ± 0.10	0.75 ± 0.25	0.37 ± 0.21	0.67 ± 0.05	0.59 ± 0.10	0.37 ± 0.41	$\mathbf{0.81 \pm 0.32}$
GCN[41]		0.56 ± 0.27	0.56 ± 0.10	0.25 ± 0.43	0.87 ± 0.21	0.60 ± 0.05	0.53 ± 0.05	0.68 ± 0.40	0.37 ± 0.41
GIN[15]		0.68 ± 0.10	0.68 ± 0.10	0.75 ± 0.43	0.62 ± 0.21	0.64 ± 0.09	0.65 ± 0.10	0.50 ± 0.30	$\mathbf{0.81 \pm 0.10}$
STGCN[12]		0.62 ± 0.27	0.68 ± 0.20	0.62 ± 0.41	0.75 ± 0.43	0.62 ± 0.22	0.59 ± 0.13	0.68 ± 0.27	0.43 ± 0.27
STAGIN[13]		0.75 ± 0.17	0.75 ± 0.17	0.62 ± 0.41	$\mathbf{0.87 \pm 0.21}$	0.70 ± 0.17	0.62 ± 0.08	$\mathbf{0.75 \pm 0.30}$	0.50 ± 0.39
Ours	PCC+MD	0.75 ± 0.43	0.62 ± 0.27	0.75 ± 0.25	0.50 ± 0.50	0.65 ± 0.09	0.50 ± 0.08	0.56 ± 0.44	0.43 ± 0.44
	PCC+ED	$\mathbf{0.93 \pm 0.10}$	$\mathbf{0.81 \pm 0.10}$	$\mathbf{0.87 \pm 0.21}$	0.75 ± 0.25	$\mathbf{0.76 \pm 0.17}$	$\mathbf{0.71 \pm 0.13}$	0.68 ± 0.10	0.75 ± 0.17
	PCC+MaD	0.87 ± 0.12	0.68 ± 0.20	0.75 ± 0.43	0.62 ± 0.41	0.71 ± 0.10	0.56 ± 0.13	0.56 ± 0.36	0.56 ± 0.27

Table 3. Performance with DPD Dataset. PCC: Pearson Correlation Coefficient stream; MD: Manhattan Distance stream; ED: Euclidean Distance stream; MaD: Mahalanobis Distance stream

Method		AUC	ACC	SE	SP
LSTM [31]		0.66 ± 0.05	0.58 ± 0.04	0.68 ± 0.20	0.50 ± 0.22
GCN [41]		0.71 ± 0.16	0.61 ± 0.09	0.25 ± 0.43	$\mathbf{0.90 \pm 0.17}$
GIN [15]		0.58 ± 0.09	0.61 ± 0.09	0.37 ± 0.12	0.80 ± 0.14
STGCN [12]		0.70 ± 0.18	0.58 ± 0.09	0.50 ± 0.39	0.65 ± 0.21
STAGIN [13]		0.65 ± 0.09	0.61 ± 0.12	0.56 ± 0.36	0.65 ± 0.25
Ours	PCC+MD	0.58 ± 0.08	0.58 ± 0.09	0.81 ± 0.20	0.40 ± 0.24
	PCC+ED	$\mathbf{0.71 \pm 0.04}$	$\mathbf{0.72 \pm 0.05}$	0.81 ± 0.20	0.65 ± 0.16
	PCC+MaD	0.68 ± 0.07	0.63 ± 0.09	$\mathbf{0.87 \pm 0.12}$	0.45 ± 0.21

Table 4. Ablation Study with OS Dataset. PCC: Pearson Correlation Coefficient stream; ED: Euclidean Distance stream; CL: Contrastive Loss

Method	Duloxetine				Placebo			
	AUC	ACC	SE	SP	AUC	ACC	SE	SP
Full Model	$\mathbf{0.93 \pm 0.10}$	$\mathbf{0.81 \pm 0.10}$	$\mathbf{0.87 \pm 0.21}$	0.75 ± 0.25	$\mathbf{0.76 \pm 0.17}$	$\mathbf{0.71 \pm 0.13}$	0.68 ± 0.10	$\mathbf{0.75 \pm 0.17}$
w/o PCC	0.50 ± 0.35	0.50 ± 0.12	0.37 ± 0.41	0.37 ± 0.41	0.62 ± 0.13	0.59 ± 0.10	0.68 ± 0.40	0.50 ± 0.35
w/o ED	0.75 ± 0.30	0.62 ± 0.12	0.62 ± 0.41	0.62 ± 0.41	0.60 ± 0.20	0.56 ± 0.13	0.62 ± 0.37	0.50 ± 0.35
w/o CL	0.87 ± 0.21	0.68 ± 0.10	0.50 ± 0.35	$\mathbf{0.87 \pm 0.21}$	0.59 ± 0.03	0.53 ± 0.05	$\mathbf{0.81 \pm 0.20}$	0.25 ± 0.25
w/o CBAM	$\mathbf{0.93 \pm 0.10}$	0.75 ± 0.25	0.75 ± 0.43	0.75 ± 0.43	0.68 ± 0.22	0.53 ± 0.10	0.75 ± 0.25	0.31 ± 0.10

Table 5. Ablation Study with DPD Dataset. PCC: Pearson Correlation Coefficient stream; ED: Euclidean Distance stream; CL: Contrastive Loss

Method	AUC	ACC	SE	SP
Full Model	$\mathbf{0.71 \pm 0.04}$	$\mathbf{0.72 \pm 0.05}$	$\mathbf{0.81 \pm 0.20}$	$\mathbf{0.65 \pm 0.16}$
w/o PCC	0.46 ± 0.06	0.50 ± 0.05	0.56 ± 0.27	0.45 ± 0.16
w/o ED	0.55 ± 0.08	0.52 ± 0.04	0.56 ± 0.10	0.50 ± 0.10
w/o CL	0.58 ± 0.04	0.52 ± 0.04	0.62 ± 0.21	0.45 ± 0.16
w/o CBAM	0.55 ± 0.10	0.58 ± 0.04	0.75 ± 0.17	0.45 ± 0.08

Metrics. Performance was evaluated by various metrics, including the area under the ROC curve (AUC), accuracy (ACC), sensitivity (SE), and specificity (SP), according to previous studies for treatment response prediction [42].

4.2 Results and Discussion

Quantitative Analysis. Our model exceeds baseline methods in predicting treatment outcomes as shown in Table 2 and Table 3. Though sensitivity or specificity may appear lower in some scenarios, it is crucial to note the data imbalance and treatment complexities. For instance, placebo administration and

Table 6. Sensitive Analysis of α and δ on Accuracy

Parameter	Duloxetine	Placebo	DPD Routine Treatment
$\alpha = 0.1, \delta = 1$	**0.81 ± 0.10**	**0.71 ± 0.13**	**0.72 ± 0.05**
$\alpha = 0.1, \delta = 2$	0.75 ± 0.17	0.40 ± 0.10	0.58 ± 0.04
$\alpha = 0.1, \delta = 3$	0.65 ± 0.12	0.53 ± 0.13	0.55 ± 0.07
$\alpha = 0.5, \delta = 1$	**0.81 ± 0.10**	0.50 ± 0.08	0.55 ± 0.07
$\alpha = 0.5, \delta = 2$	0.75 ± 0.17	0.59 ± 0.16	0.61 ± 0.09
$\alpha = 0.5, \delta = 3$	**0.81 ± 0.10**	0.62 ± 0.15	0.55 ± 0.05
$\alpha = 1.0, \delta = 1$	0.65 ± 0.27	0.53 ± 0.10	0.58 ± 0.04
$\alpha = 1.0, \delta = 2$	0.68 ± 0.10	0.50 ± 0.00	0.52 ± 0.04
$\alpha = 1.0, \delta = 3$	**0.81 ± 0.20**	0.50 ± 0.00	0.58 ± 0.04

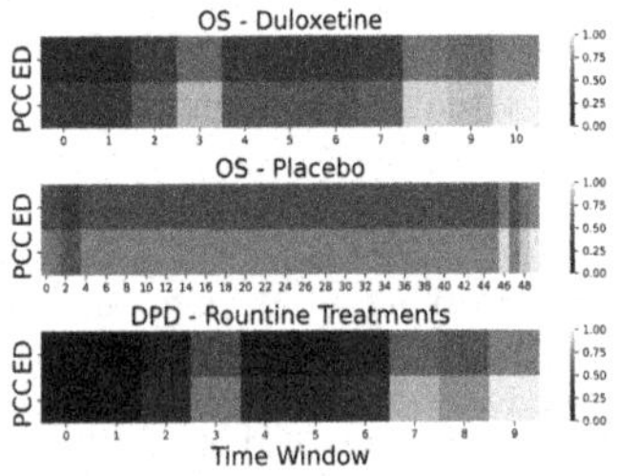

Fig. 2. Attention Map ($Attn_{Stream} \cdot Attn_{Temporal}$) of PCC vs. ED by Time Windows

DPD routine treatments vary in duration and are individually tailored, unlike the consistent duloxetine treatment. Nonetheless, our findings highlight the wide applicability of our model in diverse scenarios, including clinical trials and real-world clinical practice. Throughout our experimentation with different distance measures, we consistently observe that the Euclidean distance-based similarity achieves superior results. This can be attributed to its direct assessment of overall dissimilarity between ROIs, without relying on specific assumptions. This finding suggests that Euclidean distance is the most suitable metric for constructing distance-based neural similarity within our framework.

In our ablation study, we used the Euclidean distance as the basis for distance-based neural similarity due to its superior performance. The results, shown in Tables 4 and 5, confirm that our model, including all components, achieves balanced performance in both OS and DPD datasets. Notably, the AUC and ACC metrics highlight the model effectiveness. That supports our hypothesis regarding the complementary relationship between correlation and distance-based streams in predicting treatment responses. Importantly, excluding either the correlation or distance stream resulted in varying degrees of performance decline, suggesting the influence of dataset complexity. This demonstrates the adaptability of our approach across different data domains and provides evidence that incorporating both correlation and distance-based streams enhances performance.

Our sensitivity analysis (Table 6) investigates the impact of parameters α and δ on model accuracy. The model maintains strong performance across different α and δ values for duloxetine, indicating robustness. However, in placebo and DPD routine treatments, optimal accuracy is achieved at $\alpha = 0.1, \delta = 1$, with a decrease as δ rises, demonstrating the model sensitivity to δ in these cases. This shows the model adaptability to situations with consistent interventions, although in more complex cases, hyperparameter fine-tuning may be required.

Qualitative Analysis. Fig. 2 shows the attention values of graphs derived from PCC and ED streams across various time windows in our test set. The attention values indicate the importance of each stream at different time points. Despite PCC stream having generally higher values, the ED stream also plays a significant role. This validates our ablation study results, suggesting complementary roles of PCC and ED in treatment prediction. Moreover, differences in time windows and attention distribution between duloxetine and placebo show the varied brain dynamics between treatments, highlighting the interpretability and dynamic nature of our method.

5 Conclusion

In this work, we introduce a graph learning framework combining correlation-based and distance-based neural similarity to enhance treatment response prediction understanding. Our validation on two datasets highlights Euclidean distance as particularly effective for distance-based FC, showing the potential of our model for real-world clinical application. Future research will refine our theoretical basis through clinical trials, examining the impact of atlas and FC threshold choices. Additionally, we will also explore advanced methodologies like diffusion models to infer deeper brain mechanisms for treatment response [43], and investigate the applicability of our framework to other clinical contexts requiring complex graph representation, such as surgery [44].

References

1. Khosla, M., Jamison, K., Ngo, G.H., Kuceyeski, A., Sabuncu, M.R.: Machine learning in resting-state fMRI analysis. Magn. Reson. Imaging **64**, 101–121 (2019)
2. Du, Y., Fu, Z., Calhoun, V.D.: Classification and prediction of brain disorders using functional connectivity: promising but challenging. Front. Neurosci. **12**, 525 (2018)
3. Taylor, J.J., Kurt, H.G., Anand, A.: Resting state functional connectivity biomarkers of treatment response in mood disorders: a review. Front. Psych. **12**, 565136 (2021)
4. Kong, Y., et al.: Spatio-temporal graph convolutional network for diagnosis and treatment response prediction of major depressive disorder from functional connectivity. Hum. Brain Map. **42**(12), 3922–3933 (2021)

5. Bobadilla-Suarez, S., Ahlheim, C., Mehrotra, A., Panos, A., Love, B.C.: Measures of neural similarity. Comput. Brain Behav. **3**, 369–383 (2020)
6. Xiao, L., et al.: Distance correlation-based brain functional connectivity estimation and non-convex multi-task learning for developmental fMRI studies. IEEE Trans. Biomed. Eng. **69**(10), 3039–3050 (2022)
7. Wang, J., Zuo, X., He, Y.: Graph-based network analysis of resting-state functional MRI. Front. Syst. Neurosci. **4**, 1419 (2010)
8. Zhou, J., Cui, G., Hu, S., et al.: Graph neural networks: a review of methods and applications. AI Open. **1**, 57–81 (2020)
9. Yu, Q., et al.: Application of graph theory to assess static and dynamic brain connectivity: approaches for building brain graphs. Proc. IEEE **106**(5), 886–906 (2018)
10. Kan, X., Dai, W., Cui, H., Zhang, Z., Guo, Y., Yang, C.: Brain network transformer. In: Advances in Neural Information Processing Systems, vol. 35, pp. 25586–25599 (2022)
11. Dahan, S., Williams, L.Z.J., Rueckert, D., Robinson, E.C.: Improving phenotype prediction using long-range spatio-temporal dynamics of functional connectivity. In: Abdulkadir, A., et al. (eds.) MLCN 2021. LNCS, vol. 13001, pp. 145–154. Springer, Cham (2021). https://doi.org/10.1007/978-3-030-87586-2_15
12. Gadgil, S., Zhao, Q., Pfefferbaum, A., Sullivan, E.V., Adeli, E., Pohl, K.M.: Spatio-temporal graph convolution for resting-state fMRI analysis. In: Martel, A.L., et al. (eds.) MICCAI 2020, Part VII. LNCS, vol. 12267, pp. 528–538. Springer, Cham (2020). https://doi.org/10.1007/978-3-030-59728-3_52
13. Kim, B.H., Ye, J.C., Kim, J.J.: Learning dynamic graph representation of brain connectome with spatio-temporal attention. In: Advances in Neural Information Processing Systems, vol. 34, pp. 4314–4327 (2021)
14. Chang, C., Glover, G.H.: Time-frequency dynamics of resting-state brain connectivity measured with fMRI. Neuroimage **50**(1), 81–98 (2010)
15. Xu, K., Hu, W., Leskovec, J., Jegelka, S. How powerful are graph neural networks?. arXiv preprint arXiv:1810.00826 (2018)
16. Cao, B., et al.: Treatment response prediction and individualized identification of first-episode drug-naive schizophrenia using brain functional connectivity. Mol. Psychiatry **25**(4), 906–913 (2020)
17. Cortes, C., Vapnik, V.: Support-vector networks. Mach. Learn. **20**, 273–297 (1995)
18. Woo, S., Park, J., Lee, J. Y., Kweon, I. S. CBAM: convolutional block attention module. In: Proceedings of the European Conference on Computer Vision, pp. 3–19 (2018)
19. Tétreault, P., Mansour, A., Vachon-Presseau, E., Schnitzer, T.J., Apkarian, A.V., Baliki, M.N.: Brain connectivity predicts placebo response across chronic pain clinical trials. PLoS Biol. **14**(10), e1002570 (2016)
20. Ma, H., Wu, F., Guan, Y., Xu, L., Liu, J., Tian, L.: BrainNet with connectivity attention for individualized predictions based on multi-facet connections extracted from resting-state fMRI data. Cognit. Comput. **15**, 1–15 (2023). https://doi.org/10.1007/s12559-023-10133-8
21. Del Fabro, L., Bondi, E., Serio, F., Maggioni, E., D'Agostino, A., Brambilla, P.: Machine learning methods to predict outcomes of pharmacological treatment in psychosis. Transl. Psychiatry **13**(1), 75 (2023)
22. Yan, S., Xiong, Y., Lin, D.: Spatial temporal graph convolutional networks for action recognition. In: Proceedings of the AAAI Conference on Artificial Intelligence, vol. 32, no. 1 (2018)

23. Faria, A.V., et al.: Atlas-based analysis of resting-state functional connectivity: evaluation for reproducibility and multi-modal anatomy-function correlation studies. Neuroimage **61**(3), 613–621 (2012)
24. Janse, R.J., et al.: Conducting correlation analysis: important limitations and pitfalls. Clin. Kidney J. **14**(11), 2332–2337 (2021)
25. Walther, A., Nili, H., Ejaz, N., Alink, A., Kriegeskorte, N., Diedrichsen, J.: Reliability of dissimilarity measures for multi-voxel pattern analysis. Neuroimage **137**, 188–200 (2016)
26. Perlibakas, V.: Distance measures for PCA-based face recognition. Pattern Recogn. Lett. **25**(6), 711–724 (2004)
27. Smitha, K.A., et al.: Resting state fMRI: a review on methods in resting state connectivity analysis and resting state networks. Neuroradiol. J. **30**(4), 305–317 (2017)
28. Thompson, G.J.: Neural and metabolic basis of dynamic resting state fMRI. Neuroimage **180**, 448–462 (2018)
29. Babcock, B., Datar, M., Motwani, R.: Sampling from a moving window over streaming data. In: Proceedings of the Thirteenth Annual ACM-SIAM Symposium on Discrete Algorithms, pp. 633–634 (2002)
30. Kim, B.H., Ye, J.C.: Understanding graph isomorphism network for RS-fMRI functional connectivity analysis. Front. Neurosci. **14**, 630 (2020)
31. Hochreiter, S., Schmidhuber, J.: Long short-term memory. Neural Comput. **9**(8), 1735–1780 (1997)
32. Wang, X., Yao, L., Rekik, I., Zhang, Y.: Contrastive Functional Connectivity Graph Learning for Population-based fMRI Classification. In: Wang, L., Dou, Q., Fletcher, P.T., Speidel, S., Li, S. (eds.) Medical Image Computing and Computer Assisted Intervention - MICCAI 2022. MICCAI 2022. LNCS, Part I, vol. 13431, pp. 221–230. Springer, Cham (2022). https://doi.org/10.1007/978-3-031-16431-6_21
33. Olszowy, W., Aston, J., Rua, C., Williams, G.B.: Accurate autocorrelation modeling substantially improves fMRI reliability. Nat. Commun. **10**(1), 1220 (2019)
34. Chen, T., Kornblith, S., Norouzi, M., Hinton, G.: A simple framework for contrastive learning of visual representations. In: International Conference on Machine Learning, pp. 1597-1607. PMLR (2020)
35. Dwivedi, C., Nofallah, S., Pouryahya, M., et al.: Multi stain graph fusion for multimodal integration in pathology. In: Proceedings of the IEEE/CVF Conference on Computer Vision and Pattern Recognition, pp. 1835-1845 (2022)
36. Zheng, S., et al.: Potential targets for noninvasive brain stimulation on depersonalization-derealization disorder. Brain Sci. **12**(8), 1112 (2022)
37. Sierra, M., Berrios, G.E.: The Cambridge depersonalisation scale: a new instrument for the measurement of depersonalisation. Psychiatry Res. **93**(2), 153–164 (2000)
38. Yan, C.G., Wang, X.D., Zuo, X.N., Zang, Y.F.: DPABI: data processing & analysis for (resting-state) brain imaging. Neuroinformatics **14**, 339–351 (2016)
39. Glasser, M.F., et al.: A multi-modal parcellation of human cerebral cortex. Nature **536**(7615), 171–178 (2016)
40. Jenkinson, M., Beckmann, C.F., Behrens, T.E., Woolrich, M.W., Smith, S.M.: Fsl. Neuroimage **62**(2), 782–790 (2012)
41. Kipf, T.N., Welling, M.: Semi-supervised classification with graph convolutional networks. arXiv preprint arXiv:1609.02907 (2016)
42. Kesler, S.R., Rao, A., Blayney, D.W., Oakley-Girvan, I.A., Karuturi, M., Palesh, O.: Predicting long-term cognitive outcome following breast cancer with pretreatment resting state fMRI and random forest machine learning. Front. Hum. Neurosci. **11**, 555 (2017)

43. Chang, Z., Koulieris, G.A., Shum, H.P.: On the design fundamentals of diffusion models: a survey. arXiv preprint arXiv:2306.04542 (2023)
44. Zhang, X., Al Moubayed, N., Shum, H.P.: Towards graph representation learning based surgical workflow anticipation. In: 2022 IEEE-EMBS International Conference on Biomedical and Health Informatics, pp. 01–04 (2022)

Measuring Cognitive Load: Leveraging fNIRS and Machine Learning for Classification of Workload Levels

Mehshan Ahmed Khan[1](✉), Houshyar Asadi[1], Thuong Hoang[2], Chee Peng Lim[1], and Saeid Nahavandi[1]

[1] Institute for Intelligent Systems Research and Innovation Geelong, Deakin University, Victoria, Australia
{mehshan.khan,houshyar.asadi,chee.lim}@deakin.edu.au, saeid.nahavandi@ieee.org

[2] Faculty of Sci Eng and Built Env, School of Info Technology, Deakin University, Geelong, VIC, Australia
thuong.hoang@deakin.edu.au

Abstract. Measuring cognitive load, a subjective construct that reflects the mental effort required for a given task, remains a challenging endeavor. While Functional Near-Infrared Spectroscopy (fNIRS) has been utilized in the field of neurology to assess cognitive load, there are limited studies that have specifically focused on high cognitive load scenarios. Previous research in the field of cognitive workload assessment using fNIRS has primarily focused on differentiating between two levels of mental workload. These studies have explored the classification of low and high levels of cognitive load, or easy and difficult tasks, using various Machine Learning (ML) and Deep Learning (DL) models. However, there is a need to further investigate the detection of multiple levels of cognitive load to provide more fine-grained information about the mental state of an individual. This study aims to classify four mental workload levels using classical ML techniques, specifically random forests, with fNIRS data. It assesses the effectiveness of ML algorithms with fNIRS data, provides insights into classification features and patterns, and contributes to understanding neural mechanisms in cognitive processing. ML algorithms used for classification include Naïve Bayes, k-Nearest Neighbors (k-NN), Decision Trees, Random Forests, and Nearest Centroid. Random Forests achieved a promising accuracy and Area Under Curve (AUC) of around 99.99%. The findings of this study highlight the potential of utilizing fNIRS and ML algorithms for accurately classifying cognitive workload levels. The use of multiple features extracted from fNIRS data may contribute to a more robust and reliable classification approach.

Keywords: Functional near-infrared spectroscopy (fNIRS) · machine learning · cognitive load · n-back tasks · classification

B. Luo et al. (Eds.): ICONIP 2023, CCIS 1963, pp. 313–325, 2024.
https://doi.org/10.1007/978-981-99-8138-0_25

1 Introduction

The cognitive workload is a quantitative measure of the mental effort required to perform a particular task. This measure is often used to evaluate cognitive activities' complexity and impact on cognitive resources [1]. Cognitive neuroscience has extensively studied the cognitive workload in various contexts, especially in tasks that require working memory (WM) capabilities. Among these tasks, the n-back task has gained significant attention and has become one of the most widely used paradigms to assess WM capacity [2]. The n-back [3] task is a cognitive test that involves monitoring a sequence of stimuli and detecting when a current stimulus matches one presented n trials earlier. This task requires constant attention, active monitoring, and updating of information in WM, making it an excellent tool for measuring cognitive workload. Researchers have used this task in various settings, such as assessing cognitive load during complex decision-making [4], cognitive control [5], and multitasking [6].

Measuring mental workload (MWL) is an important aspect of studying human cognition, as it helps researchers to understand the cognitive demands of various tasks. Traditionally, Electroencephalography (EEG) has been the primary method used to record brain activity to measure MWL [7]. However, in recent years, functional Near Infrared Spectroscopy (fNIRS) has emerged as a promising alternative measure of the electrical activity of the brain using electrodes placed on the scalp [8]. While EEG has been widely used in the past, it has some limitations, such as low spatial resolution and susceptibility to artefacts caused by eye movement, head movements or muscle activity [9]. In contrast, fNIRS uses light to measure changes in oxygen levels in the brain. This non-invasive technique provides a higher spatial resolution compared to EEG, making it easier to localize brain activity. Additionally, fNIRS is more robust for certain artefacts, such as eye movement or typing [10]. Furthermore, fNIRS optodes do not require the use of gel, which means that an fNIRS headset can be mounted more quickly compared to EEG electrodes. This feature makes fNIRS a more convenient and practical method for measuring MWL in various contexts, such as in real-world environments or during prolonged task performance.

Over the past few years, there has been a noticeable trend towards the use of Machine learning (ML) and Deep learning (DL) techniques in a wide range of fields, including image processing [11], signal processing [12], and remote sensing [13]. Among the various ML algorithms, Support Vector Machines (SVM) [14], Linear Discriminant Analysis (LDA) [15], Random Forests [16], and Naive Bayes [17] are particularly popular for analyzing fNIRS signals and extracting information about cognitive processes and workload. The ability of these algorithms to identify complex patterns in large datasets sets them apart from traditional statistical methods such as Generalized Linear Model (GLM), wavelet analysis, ANOVA, regression analysis and t-test which are often limited in their ability to handle such data. With the help of ML techniques, researchers can now extract meaningful insights from massive amounts of data that would have been nearly impossible to process using traditional statistical approaches [18]. In particular, fNIRS signals, which measure changes in blood oxygenation levels in the brain, have become an increasingly popular area of research for ML applications [19]. By analyzing these signals with ML algorithms, researchers have been able to extract valuable information about cognitive processes and workload, including mental states such as attention, concentration, and fatigue. For instance, researchers have used ML techniques to analyze fNIRS signals during tasks such as driving [20], hazard perception [21], and cognitive tests [22], with promising results. Le et al. [23] employed the auditory n-back task to classify fNIRS signals related to various levels of mental workload while driving a car at around 40km/h speed and found that Random Forests outperformed other methods in terms of accuracy when data from all channels were used for classification. Additionally, Le et al. [24] investigated the mental state of senior drivers and discovered significant changes in relaxed driving, trail driving, and parking tasks, using Random Forests again proving more effective compared to other methods. Cakir et al. [25] evaluated the mental workload of eight pilots at three levels and observed that a model trained on the data of one pilot could be generalized to evaluate the mental workload of other pilots. However, the accuracy of the model in predicting high levels of workload was lower due to frequent head movements. Zhou et al. [26] analyzed hazard perception tasks in a lab setting and found that Linear Discriminant Analysis of fNIRS data achieved 70% accuracy when trained on features obtained from the left prefrontal cortex. Fisher criteria [27] were used to select the top 5 optimal features, indicating that the left PFC is more involved in hazard perception tasks than other brain regions. These studies demonstrate the successful application of ML techniques in extracting meaningful information about cognitive processes and workload from fNIRS signals, outperforming traditional statistical methods in identifying complex patterns in large datasets across different scenarios such as n-back tasks, driving, hazard perception, and simulated flight environments.

In recent years, there has been also a growing interest in utilizing deep learning models, such as Convolutional Neural Networks (CNN) [36], Deep Belief Networks (DBN) [28], and Long Short-Term Memory (LSTM) [29] models, for workload classification. Previous studies have mainly focused on using these models to classify only two levels of workload and have reported high accuracy. However, it should be noted that CNN models can be computationally expensive and challenging to fine-tune, which may limit their practicality in certain applications. In contrast, the current study employs a classical machine learning technique, namely random forests, to classify all four levels of workload using a similar dataset. The main reason for selecting classical machine learning algorithms is their ease of handling and effectiveness in real-time functional Near-Infrared Spectroscopy (fNIRS) signal classification. Random forests are known for their ability to handle large datasets, handle non-linear relationships, and provide robust performance with minimal hyperparameter tuning. Moreover, classical ML algorithms can provide insights into the features and patterns that are driving the classification, which can help in better understanding the underlying physiological and cognitive processes associated with the workload. Furthermore, classical ML algorithms are effective in handling fNIRS signals, which are often characterized by high-dimensional and noisy data. These algorithms can effectively extract relevant features from fNIRS signals, such as changes in oxygenated and deoxygenated hemoglobin concentrations, and use them to accurately classify different levels of workload.

The primary objective of this study is to classify four distinct levels of mental workload by analyzing the publicly available Tufts fNIRS dataset [30]. Previous research studies on this dataset have focused solely on differentiating between two levels of mental workload. Therefore, this study aims to expand on previous findings by utilizing classical ML techniques to identify and classify a wider range of mental workload levels. To accomplish this objective, we will conduct a comprehensive analysis of the TUFTS fNIRS dataset, which contains a vast amount of neuroimaging data collected from participants performing cognitive tasks with various levels of mental workloads. The dataset includes measurements of the hemodynamic responses of the brain, which are indicative of neural activity in response to cognitive tasks. The remainder of the paper is as follows.

Section 2 provides the proposed methodology for fNIRS measurements, followed by Sect. 3 which outlines the results, and validation, and the subsequent discussion is presented in the same section. Finally, Sect. 4 concludes this study.

2 Methodology

The proposed method aims to classify different levels of mental workload through the n-back task. Figure 1 depicts the block diagram of the method, comprising three stages: pre-processing, ML/DL classification, and model evaluation for accurate classification. By combining these three stages, the proposed method provides a comprehensive framework for accurately classifying different levels of mental workloads using the n-back task. In the following sub-section, we will provide a detailed explanation of each stage, including the techniques and methods used, and discuss the potential implications.

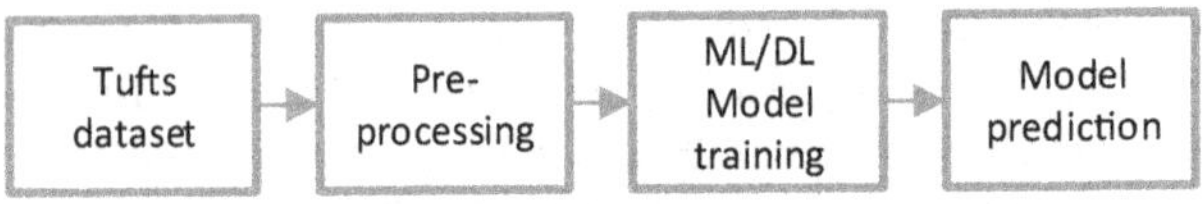

Fig. 1. The complete framework for ML/DL-based cognitive load classification.

2.1 Dataset

To evaluate the proposed models for classifying mental workload, the Tufts fNIRS open-access dataset [30] was used. This dataset contains fNIRS data collected from 68 participants who performed a series of controlled n-back tasks, including 0-back, 1-back, 2-back, and 3-back. The raw fNIRS measurements consist of temporal traces of alternating current intensity and changes in phase at two different wavelengths (690 and 830 nm), using a 110 MHz modulation frequency. From these intensity and phase measurements, time traces of the dynamic changes in oxyhemoglobin (HbO) and deoxyhemoglobin (HbR) were obtained by calculating the spatial dependence of the optical measurements. These dynamic changes in HbO and HbR are known to reflect changes in neural activity associated with cognitive processes. To ensure the removal of respiration, heartbeat, and drift artefacts, each univariate time series was subjected to bandpass filtering using a 3rd-order zero-phase Butterworth filter, retaining only the frequencies between 0.001 and 0.2 Hz. After eliminating noise, the classification process focused on eight specific features, namely: "A-I-O," "AB-PHI-O," "AB-I-DO," "AB-PHI-DO," "CD-I-O," "CD-PHI-O," "CD-I-DO," and "CD-PHI-DO". The measurement of hemoglobin concentration in the blood flowing through the brain is recorded in each column. Moreover, overlapping windows of size 30 s, with a stride of 0.6 s, were used to segment the fNIRS data. The choice of window size and the stride was based on the findings of the dataset authors, who reported that 30-s windows yield the highest accuracy.

2.2 Pre-Processing

Data pre-processing is an essential step in preparing data for ML applications. One common technique used in data pre-processing is the standard scalar function. This function scales and normalizes the data to unit variance, making it easier for ML algorithms to process. The standard scalar function computes the standard score of a sample x using Eq. 1.

$$k = \frac{x - \mu}{s} \tag{1}$$

where k is the scaled value, μ is the mean of the training samples, and s is the standard deviation of the training samples. This equation ensures that the feature values are in the range of -1 to 1, making it easier for algorithms to handle and reducing the potential impact of outliers in the data.

2.3 Classification

In our study, we utilized the Random Forest (RF) [31] algorithm as a classifier model to effectively differentiate between the four distinct states of cognitive fatigue such as 0-back, 1-back, 2-back and 3-back. RF is an advanced supervised learning algorithm that harnesses the power of decision trees to generate more accurate and reliable results. Gini impurity is a criterion used in decision tree algorithms, including Random Forest, to measure the impurity of a node or a split. Gini impurity is a measure of how "impure" the node or split is, in terms of the class labels of the data points it contains. A node or split with low Gini impurity means that the majority of the data points belong to a single class, while a node or split with high Gini impurity means that the data points are evenly distributed across multiple classes. The Gini (G) impurity is calculated using the following formula:

$$G = 1 - \sum\nolimits_{i=0}^{n} p(i)^2 \tag{2}$$

In Eq. 2 $p(i)$ represents the probability of a randomly selected data point in the node or split belonging to class i. . The Gini impurity ranges from 0 (pure node or split) to 1 (impure node or split). The Gini impurity, ranging from 0 (pure node) to 1 (impure node), is minimized by selecting the best split at each node of individual decision trees. This is achieved by randomly selecting a subset of features for each split, introducing randomness. The Random Forest algorithm improves accuracy and robustness by averaging predictions of multiple decision trees trained on random subsets of data points and features. This combination of randomness and averaging reduces overfitting, enhances overall performance, and increases generalization capabilities. The final decision of the RF classifier is determined by aggregating the outputs of each decision tree using a majority-voting system. Table 1 summarizes the default hyperparameters for Random Forests in Scikit-learn:

Table 1. Hyperparameters of Random Forests.

Hyperparameters	Values
No of estimators	100
Criterion	"Gini"
Minimum impurity decrease	0
Max Depth	None
Minimum samples split	2
Min samples leaf	1
Maximum features	"Auto"
Bootstrap	TRUE

The default values are carefully chosen based on considerations such as commonly used values, generalization and performance, and efficiency and scalability. For example, the default value for no of estimators is set to 100, which is a commonly used value for the number of decision trees in a Random Forest. The default value for Maximum depth is set to None, which allows the decision trees to grow deep and split even with a small number of samples, potentially leading to overfitting. Additionally, the default value for maximum features is set to "auto", which means that the square root of the total number of features is used to determine the number of features to consider for the best split, balancing the need for accuracy with computational efficiency. This allows the RF algorithm to make more precise and reliable predictions compared to a single decision tree. In addition to Random Forest, we also evaluated other classification algorithms such as Decision Trees, k-Nearest Centroid, Naive Bayes, and Nearest Centroid to compare the performance of each method. By employing these various algorithms, we were able to thoroughly assess the efficiency and accuracy of the RF algorithm in differentiating between the four states of cognitive fatigue.

3 Results and Discussion

In this section, we provide a comprehensive overview of our implementation of the ML algorithms, including a detailed description and analysis of their classification results. Our algorithms are all coded in Python, leveraging the powerful Scikit-learn [32] module for their implementations. We carefully examine the performance of each algorithm, considering various metrics and techniques to assess their effectiveness and reliability. In this section, we will delve into the specifics of our implementation, highlighting the key steps and methodologies employed in each algorithm.

In our experimental setup, we employed an 80–20 training-testing data split, where 80% of the data was used for training the proposed models and the remaining 20% was reserved for validation during testing. This approach ensured that the models were trained on a substantial portion of the data while also providing a separate set of data for unbiased evaluation. The performance of the proposed random forest classifier has been thoroughly evaluated and compared with other classifiers, as illustrated in Fig. 2. From the results shown in Fig. 2, it is evident that the nearest centroid and naïve Bayes classifiers exhibit the poorest performance, with an accuracy of less than 35%. On the other hand, the k-NN, decision trees, and random forests classifiers demonstrate comparable performance, with slight differences in accuracy. As shown in Fig. 2, the random forests classifier outperforms the other classifiers, achieving an accuracy of 99.99%. This highlights the superior performance of the random forests algorithm in accurately classifying workload levels, which is critical for decision-making and resource allocation in various applications, based on the evaluated dataset. The high accuracy achieved by the random forests classifier suggests that it is a promising choice for the given task and may be well-suited for real-world applications.

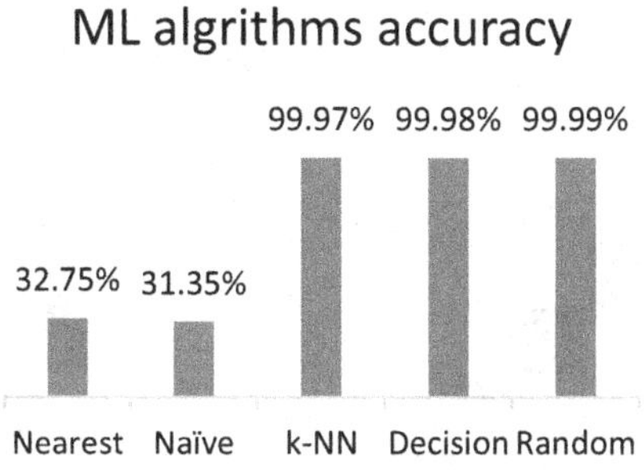

Fig. 2. The bar plot for the accuracy performance of various ML algorithms.

Figure 3 presents the confusion matrix obtained from the classification results of the test data using the random forest classifier. The confusion matrix provides a comprehensive overview of the performance of the classifier in predicting the target labels for each class. From the confusion matrix, it is evident that the majority of the classes have been predicted correctly by the model, with only a few exceptions. The diagonal elements of the matrix represent the correctly predicted instances for each class, while the off-diagonal elements represent the misclassified instances. The higher the values on the diagonal, the better the model's performance in accurately predicting the respective class labels. The small number of misclassified instances in the confusion matrix indicates the overall high accuracy of the random forest classifier. The few misclassifications could be due to various factors, such as overlapping features or similarities between certain classes, imbalanced data distribution, or noise in the dataset. These misclassifications can be further analyzed and addressed to potentially improve the performance of the classifier.

In comparison, Fig. 4 showcases the confusion matrices of other algorithms, which are being compared with the Random Forest algorithm. Upon observing the confusion

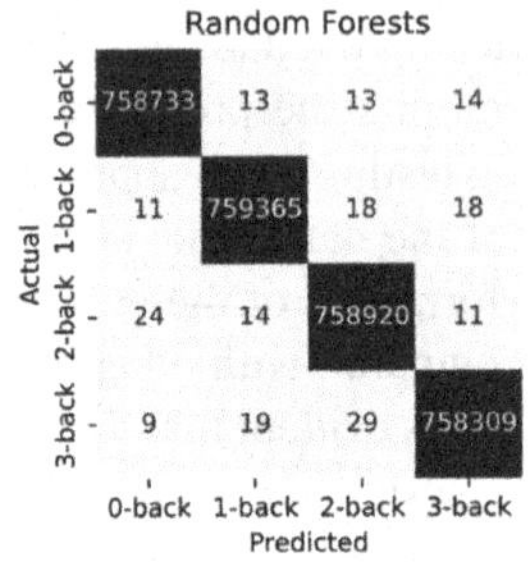

Fig. 3. Confusion matrix of Random Forest algorithm depicting the classification performance of the model on test data.

matrices, it is apparent that some of the other algorithms, such as Naïve Bayes and Nearest Centroid, exhibit lower performance in terms of accuracy. These algorithms display higher numbers of misclassifications, as evidenced by the larger number of entries in the off-diagonal cells of their respective confusion matrices. This shows that they may not be as effective as the Random Forest algorithm in accurately predicting the target classes. On the other hand, algorithms such as k-NN and Decision Trees showcase relatively better performance compared to Naïve Bayes and Nearest Centroid, with fewer misclassifications.

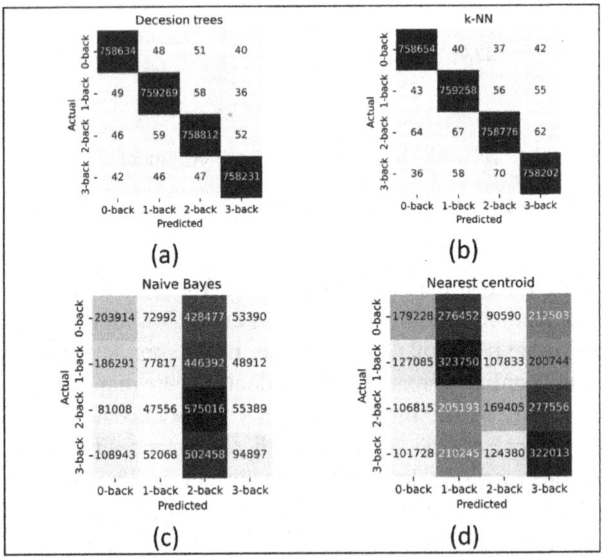

Fig. 4. Comparison of confusion matrices for various classification algorithms: (a) Decision Trees, (b) k-Nearest Neighbor (k-NN), (c) Naive Bayes, and (d) Nearest Centroid on test data.

The results of the quantitative evaluation for different classifiers, namely Naïve Bayes, Nearest Centroid, k-NN, Decision Trees, and Random Forests, are displayed in Table 2. The evaluation metrics used include Precision, Recall, F1-score, and Area

Under Curve (AUC). These metrics provide insights into the performance of each classifier in terms of their ability to accurately predict the target labels. The results in Table 2 highlight that k-NN, Decision Trees, and Random Forests classifiers achieved remarkably high values for F1-score, Precision, Recall and AUC, all close to 1, indicating high performance. This suggests that these classifiers are highly effective in accurately predicting the target labels, making them promising choices for classifying cognitive load levels at multiple levels. In contrast, Naïve Bayes and Nearest Centroid classifiers exhibit lower values for these metrics, indicating comparatively poorer performance in terms of accuracy. These classifiers may not be as suitable for the specific task at hand. The superior performance of k-NN, Decision Trees, and Random Forests classifiers, as demonstrated by the high values of the evaluation metrics, suggests their effectiveness in accurately predicting the target labels for the given dataset.

Table 2. Performance evaluation metrics include precision, recall, F1-score, and AUC of various classification algorithms.

Classifiers	Evaluation metrics			
	F1-score	Precision	Recall	AUC
Naïve Bayes	0.2676	0.3332	0.3135	0.5586
Nearest centroid	0.3201	0.3322	0.3275	0.5727
k-NN	0.9997	0.9997	0.9997	0.9998
Decision trees	0.9998	0.9998	0.9998	0.9998
Random forests	0.9999	0.9999	0.9999	0.9999

The training times (in seconds) for each classifier are listed in Table 3. It can be observed that Nearest Centroid and Naïve Bayes classifiers have relatively shorter training times. Among the ML classifiers, Random Forests outperforms the other ML algorithms with an accuracy of 99.99%. However, it takes almost 25 min to train, which is a significant amount of time compared to the other classifiers. On the other hand, k-NN and Decision Tree classifiers show similar performance with shorter training times. The choice of classifier would depend on the specific requirements of the application, where accuracy, computational complexity, and other factors need to be considered.

The evaluation of ML algorithms in this study has provided valuable insights into the performance of different classifiers in accurately predicting workload levels. The random forest classifier demonstrated high accuracy in predicting classes, with the majority of classes predicted correctly and only a few misclassifications. In comparison, Naïve Bayes and Nearest Centroid algorithms exhibited lower performance with higher misclassification rates. This suggests that the random forests algorithm is a promising choice for real-world applications that require accurate classification.

Table 3. Training time for different classification algorithms.

ML techniques	Training time
Nearest centroid	1.3 s
Naïve Bayes	2.05 s
k-NN	17.2 s
Decision trees	63 s
Random forests	1522 s

4 Conclusion and Future Work

This research work delves into the analysis of functional near-infrared spectroscopy (fNIRS) data to detect different levels of cognitive load. The dataset used in this study is based on four different cognitive activities, namely 0-back, 1-back, 2-back, and 3-back tasks. This study builds upon previous research that has primarily focused on deep learning models for classifying only two levels of workload using fNIRS data. By leveraging classical ML classifiers such as Random forests, k-NN, Decision trees, Nearest centroid, and Naïve Bayes the study aims to evaluate their effectiveness in handling high-dimensional and noisy fNIRS data, provide insights into the underlying features and patterns driving the classification, and compare their performance with other algorithms, such as Naïve Bayes and Nearest Centroid, which may have limitations in handling complex datasets. Various performance metrics, including accuracy, precision, recall, F1-score, and Area Under Curve (AUC), were calculated to evaluate the classification of different cognitive load groups. Notably, the results obtained from the proposed Random Forests classifier demonstrated a remarkable accuracy of 99.99%, outperforming the other ML classifiers. This study presents a promising approach for cognitive load detection using fNIRS signals and ML classifiers. The findings highlight the potential of the proposed method for accurately identifying cognitive load levels and pave the way for future research to explore additional factors that may impact cognitive load in diverse populations.

The current study has a limitation in that it does not consider cohort effects, which could be an important area for investigation in future research. Cohort effects, which refer to the influence of demographic factors such as age, gender and skin colour on fNIRS signals, are crucial to understanding to ensure the validity and generalizability of the findings [30]. Different cohorts may exhibit different patterns of brain activity, and accounting for these effects can enhance the robustness and reliability of the research findings. In addition to addressing cohort effects, further analysis and interpretation of the results are necessary to make informed decisions about the most appropriate classifier for the specific task. Factors such as computational complexity, interpretability, and domain-specific requirements should also be taken into consideration when selecting the optimal classifier for the task at hand. Choosing the right classifier is critical as it can greatly impact the accuracy and reliability of the cognitive load assessment using fNIRS signals. Furthermore, future research plans to integrate n-back tasks with driving [33] and flying [34] simulators to gain further insights into cognitive load in real-world settings. By incorporating more extensive experimental data and accounting for variables such as task engagement motivation, and individual cognitive abilities, a more comprehensive understanding of the complex interplay between cognitive load and individual differences can be achieved [34, 35]. This approach has the potential to provide valuable insights into the nuances of cognitive load assessment and its applications in real-world scenarios [36], where cognitive load plays a crucial role in performance outcomes. To address the limitations mentioned above, future research should aim to conduct a comprehensive evaluation of different classifiers for fNIRS signals analysis. This may involve comparing the performance of various classifiers using appropriate evaluation metrics and statistical analyses and considering the trade-offs between factors such as computational complexity, interpretability, and applicability to the specific requirements of the task. Such a thorough evaluation can aid in identifying the most suitable classifier for the specific application of fNIRS signals analysis and contribute to the advancement of cognitive load assessment research.

References

1. Howie, E.E., et al.: Cognitive load management: an invaluable tool for safe and effective surgical training. Journal of Surgical Education (2023)
2. Ren, M., et al.: Neural signatures for the n-back task with different loads: an event-related potential study. Biological Psychology, 108485 (2023)
3. Kirchner, W.K.: Age differences in short-term retention of rapidly changing information. J. Exp. Psychol. **55**(4), 352 (1958)
4. Pedersen, M.L., et al.: Computational modeling of the N-Back task in the ABCD study: associations of drift diffusion model parameters to polygenic scores of mental disorders and cardiometabolic diseases. Biological Psychiatry: Cognitive Neuroscience and Neuroimaging **8**(3), 290–299 (2023)
5. Kerver, G.A., Engel, S.G., Gunstad, J., Crosby, R.D., Steffen, K.J.: Deficits in cognitive control during alcohol consumption after bariatric surgery. Surgery for Obesity and Related Diseases **19**(4), 344–349 (2023)
6. Kaul, R., Jipp, M.: Influence of cognitive processes on driver decision-making in dilemma zone. Transportation Research Interdisciplinary Perspectives **19**, 100805 (2023)
7. Gountas, J., Gountas, S., Ciorciari, J., Sharma, P.: Looking beyond traditional measures of advertising impact: using neuroscientific methods to evaluate social marketing messages. J. Bus. Res. **105**, 121–135 (2019)
8. Qu, Y., et al.: Methodological issues of the central mechanism of two classic acupuncture manipulations based on fNIRS: suggestions for a pilot study. Frontiers in Human Neuroscience (2023)
9. Wu, J., Srinivasan, R., Burke Quinlan, E., Solodkin, A., Small, S.L., Cramer, S.C.: Utility of EEG measures of brain function in patients with acute stroke. Journal of Neurophysiology **115**(5), 2399–2405 (2016)
10. Meidenbauer, K.L., Choe, K.W., Cardenas-Iniguez, C., Huppert, T.J., Berman, M.G.: Load-dependent relationships between frontal fNIRS activity and performance: a data-driven PLS approach. Neuroimage **230**, 117795 (2021)
11. Khan, M.A., et al.: Gastrointestinal diseases segmentation and classification based on duo-deep architectures. Pattern Recogn. Lett. **131**, 193–204 (2020)
12. Ghandorh, H., et al.: An ICU admission predictive model for COVID-19 patients in Saudi Arabia. International Journal of Advanced Computer Science and Applications **12**(7) (2021)

13. Boulila, W., Ghandorh, H., Khan, M.A., Ahmed, F., Ahmad, J.: A novel CNN-LSTM-based approach to predict urban expansion. Eco. Inform. **64**, 101325 (2021)
14. Chen, L., et al.: Classification of schizophrenia using general linear model and support vector machine via fNIRS. Physical and Engineering Sciences in Medicine **43**, 1151–1160 (2020)
15. Gemignani, J.: Classification of fNIRS data with LDA and SVM: a proof-of-concept for application in infant studies. In: 2021 43rd Annual International Conference of the IEEE Engineering in Medicine & Biology Society (EMBC), IEEE, pp. 824–827 (2021)
16. Oku, A.Y.A., Sato, J.R.: Predicting student performance using machine learning in fNIRS data. Front. Hum. Neurosci. **15**, 622224 (2021)
17. Hasan, M.Z., Islam, S.M.R.: Suitibility Investigation of the different classifiers in fNIRS signal classification. In: 2020 IEEE Region 10 Symposium (TENSYMP), IEEE, pp. 1656–1659 (2020)
18. Eastmond, C., Subedi, A., De, S., Intes, X.: Deep learning in fNIRS: a review. Neurophotonics **9**(4), 041411 (2022)
19. Guevara, E., et al.: Prediction of epileptic seizures using fNIRS and machine learning. J. Intelligent & Fuzzy Systems **38**(2), 2055–2068 (2020)
20. Izzetoglu, M., Jiao, X., Park, S.: Understanding driving behavior using fNIRS and machine learning. International Conference on Transportation and Development **2021**, 367–377 (2021)
21. Hu, M., Shealy, T., Hallowell, M., Hardison, D.: Advancing construction hazard recognition through neuroscience: measuring cognitive response to hazards using functional near infrared spectroscopy. Construction Research Congress **2018**, 134–143 (2018)
22. Perpetuini, D., et al.: Working memory decline in Alzheimer's disease is detected by complexity analysis of multimodal EEG-fNIRS. Entropy **22**(12), 1380 (2020)
23. Le, A.S., Aoki, H., Murase, F., Ishida, K.: A novel method for classifying driver mental workload under naturalistic conditions with information from near-infrared spectroscopy. Front. Hum. Neurosci. **12**, 431 (2018)
24. Le, A.S., Xuan, N.H., Aoki, H.: Assessment of senior drivers' internal state in the event of simulated unexpected vehicle motion based on near-infrared spectroscopy. Traffic Injury Prevention, pp. 1–5 (2022)
25. Çakır, M.P., Vural, M., Koç, S.Ö., Toktaş, A.: Real-time monitoring of cognitive workload of airline pilots in a flight simulator with fNIR optical brain imaging technology. In: International Conference on Augmented Cognition, pp. 147–158. Springer (2016)
26. Zhou, X., Hu, Y., Liao, P.-C., Zhang, D.: Hazard differentiation embedded in the brain: a near-infrared spectroscopy-based study. Autom. Constr. **122**, 103473 (2021)
27. Loog, M., Duin, R.P.W., Haeb-Umbach, R.: Multiclass linear dimension reduction by weighted pairwise Fisher criteria. IEEE Trans. Pattern Anal. Mach. Intell. **23**(7), 762–766 (2001)
28. Ho, T.K.K., Gwak, J., Park, C.M., Song, J.-I.: Discrimination of mental workload levels from multi-channel fNIRS using deep leaning-based approaches. Ieee Access **7**, 24392–24403 (2019)
29. Kang, M.-K., Hong, K.-S.: Application of deep learning techniques to diagnose mild cognitive impairment: functional near-infrared spectroscopy study. In: 2021 21st International Conference on Control, Automation and Systems (ICCAS), IEEE, pp. 2036–2042 (2021)
30. Huang, Z., et al.: The Tufts fNIRS Mental Workload Dataset & Benchmark for Brain-Computer Interfaces that Generalize
31. Breiman, L.: Random forests. Mach. Learn. **45**, 5–32 (2001)
32. Pedregosa, F., et al.: Scikit-learn: machine learning in Python. J. Machine Learning Research **12**, 2825–2830 (2011)
33. Asadi, H., Mohamed, S., Nelson, K., Nahavandi, S., Oladazimi, M.: An optimal washout filter based on genetic algorithm compensators for improving simulator driver perception. In: DSC

2015: Proceedings of the Driving Simulation Conference & Exhibition, 2015: Max Planck Institute for the Advancement of Science, pp. 1–10
34. Asadi, H., Mohammadi, A., Mohamed, S., Nahavandi, S.: Adaptive translational cueing motion algorithm using fuzzy based tilt coordination. In: Neural Information Processing: 21st International Conference, ICONIP 2014, Kuching, Malaysia, November 3–6, 2014. Proceedings, Part III 21, Springer, pp. 474–482 (2014)
35. Asadi, H., Mohammadi, A., Mohamed, S., Rahim Zadeh, D., Nahavandi, S.: Adaptive washout algorithm based fuzzy tuning for improving human perception. In: Neural Information Processing: 21st International Conference, ICONIP 2014, Kuching, Malaysia, November 3–6, 2014. Proceedings, Part III 21, Springer, pp. 483–492 (2014)
36. Asadi, H., Bellmann, T., Qazani, M.C., Mohamed, S., Lim, C.P., Nahavandi, S.: A Novel Decoupled Model Predictive Control-based Motion Cueing Algorithm for Driving Simulators (2023)

Enhanced Motor Imagery Based Brain-Computer Interface via Vibration Stimulation and Robotic Glove for Post-Stroke Rehabilitation

Jianqiang Su[1,2], Jiaxing Wang[2(✉)], Weiqun Wang[2], Yihan Wang[1,2], and Zeng-Guang Hou[1,2(✉)]

[1] School of Artificial Intelligence, University of Chinese Academy of Sciences, Beijing 100049, China
{sujianqiang2021,wangyihan2022,zengguang.hou}@ia.ac.cn
[2] State Key Laboratory of Multimodal Artificial Intelligence Systems, Institute of Automation, Chinese Academy of Science, Beijing 100190, China
{jiaxing.wang,weiqun.wang}@ia.ac.cn

Abstract. Motor imagery based brain-computer interface (MI-BCI) has been extensively researched as a potential intervention to enhance motor function for post-stroke patients. However, the difficulties in performing imagery tasks and the constrained spatial resolution of electroencephalography complicate the decoding of fine motor imagery (MI). To overcome the limitation, an enhanced MI-BCI rehabilitation system based on vibration stimulation and robotic glove is proposed in this paper. First, a virtual scene involving object-oriented palmar grasping and pinching actions, is designed to enhance subjects' engagement in performing MI tasks by providing straightforward and specific goals. Then, vibration stimulation, which can offer proprioceptive feedback, is introduced to help subjects better switch their attention to the corresponding MI limbs. Finally, the self-designed pneumatic manipulator control module is developed for motion execution based on the MI classification results. Seven healthy individuals were recruited to validate the feasibility of the system in improving subjects MI abilities. The results show that the classification accuracy of three-class fine MI can be improved to 65.67%, which is significantly higher than the state-of-the art studies. This demonstrates the great potential of the proposed system in the application of post-stroke rehabilitation training.

Keywords: Brain computer interface · Enhanced motor imagery · Pneumatic exoskeleton · Hand rehabilitation

1 Introduction

Stroke is considered the primary cause of disability among survivors, which can result in functional impairments such as motor, cognitive, language, and speech-related deficits [11]. Typically, most patients cannot fully recover from the use of

B. Luo et al. (Eds.): ICONIP 2023, CCIS 1963, pp. 326–337, 2024.
https://doi.org/10.1007/978-981-99-8138-0_26

their upper limbs and hands after stroke, which leads to a significant impact on their ability to live independently in daily activities [4]. Recently, motor imagery based brain-computer interface (MI-BCI) has presented an alternative approach to rehabilitation. Brain-computer interface (BCI) establishes a direct communication pathway between the brain and the external environment by decoding intentions from their brain activation patterns to control devices [20]. Motor imagery (MI) involves the cognitive simulation of motor actions without overt movement. It has been suggested that MI elicits similar patterns of neural activity in the motor cortex as actual physical movement [13]. Multiple studies have shown that MI-BCI obtains positive results in clinical stroke rehabilitation studies [2,5].

Hand motion, which accounts for approximately 90% of upper limb function, activates a broad region of the posterior precentral gyrus [3,17]. Therefore, rehabilitation training for the hands is crucial for stroke patients to improve neuroplasticity and regain their ability to perform daily activities. Nevertheless, the limited decoding accuracy of imagery hand gestures significantly hinders the clinical application of BCI. For example, Ofner *et al.* [14] adopted time domain information of low-frequency electroencephalography (EEG) signals to classify six kinds of upper limb movements (elbow flexion/extension, forearm supination/pronation, hand close/open), with 55% average accuracy for motor execution (ME) and 27% average accuracy for MI. Yong *et al.* [28] proposed a three-class (hand grasping, elbow movements, and rest) BCI system. The average classification accuracy for two-class MI tasks classification was 66.9% and the three-class average classification accuracy was 60.7%. Jochumsen *et al.* [7] used a single channel of EEG signals to classify motor execution/imagery of four types of palmar grasp for 15 healthy subjects and ultimately achieved 45% accuracy.

There are three crucial reasons for the low accuracy of decoding fine motor intentions. On the one hand, there are a few abnormal brain activation patterns for post-stroke patients, such as reduced cortical activity, slowing of EEG rhythms, and reduced sensorimotor rhythm power, which make EEG signals weaker [11]. On the other hand, the combination of low spatial resolution and signal-to-noise ratio (SNR), as well as poor spatial separability of the activated cerebral cortex in the same hand during MI, limit the decoding performance in imagery hand gestures [21]. Moreover, imagining the tasks associated with different hand gestures accurately is an inherent problem and there are even some individuals who are considered "BCI blind". Hence, how to improve the quality of EEG signals generated by users is a key factor affecting the accuracy of recognition performance.

Many works have been devoted to stimulus paradigm design and decoding model optimization to address these challenges [8,16,18]. In the stimulus paradigm design, virtual reality (VR) technology, which can provide more intuitive visual stimuli compared to the arrows or words, has been applied in BCI [1,24]. Liang *et al.* [9] indicated that the utilization of object-directed movement as visual guidance could enhance the discriminability of imagined patterns. The average accuracy based on object-oriented scene was improved by 7%. Functional electrical stimulation (FES) can activate the sensorimotor areas of cortex and provide humans with proprioceptive feedback. The work conducted by Ren *et al.* [16] demonstrated that the FES can enhance the precision of MI-BCI classification as well as the activation of the motor cortex of participants. In terms of decoding algorithm, multiple state-of-the-art feature extraction and classification methods have been used for EEG decoding, such as common spatial pattern (CSP), filter-bank CSP (FBCSP) [26], support vector machine (SVM) and linear discriminant analysis (LDA) [12]. Additionally, deep learning methods have been increasingly used in the field of processing EEG [10,21].

Inspired by the previous researches, in this paper, we proposed an enhanced BCI-based stroke rehabilitation system that integrated fine MI with a pneumatic exoskeleton. In the design of the MI paradigm, the adoption of object-oriented gestures and proprioceptive feedback allowed patients to perform simulating mental gestures better. In consideration of security, vibration was employed in lieu of FES to provide proprioceptive feedback [29]. Visual and mechanical feedback from the VR scenarios and the pneumatic exoskeleton, respectively, could potentially amplify their sense of engagement and elicit advantageous neural plasticity effects. An analysis of EEG data obtained from seven healthy individuals showed that MI of distinct gestures in the same limb could be distinguished.

2 System Composition

A BCI-based stroke rehabilitation manipulator system is designed in this study. The system is composed of three main components, which are the EEG acquisition equipment, the pneumatic manipulator with vibrating motors, and the VR platform, as is shown in Fig. 1.

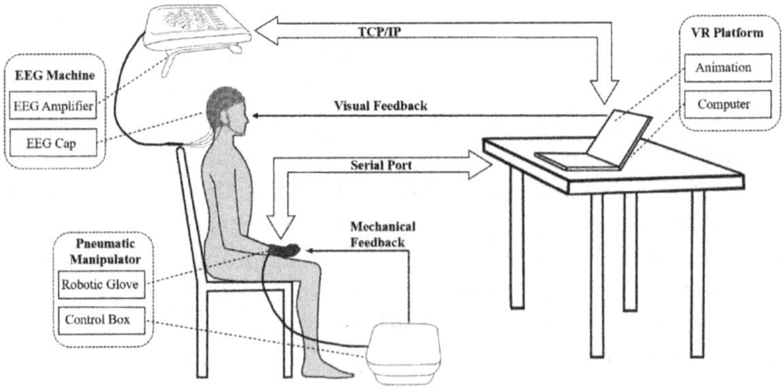

Fig. 1. The system diagram of experimental apparatus. The rehabilitation system is comprised of three main components: EEG acquisition system, pneumatic manipulator, and VR platform

Specially, EEG acquisition equipment collects the EEG data during MI tasks; the pneumatic manipulator is applied to provide assistance for the hand movement and five vibrating motors attached to the glove are designed to stimulate each finger; the VR platform offers the object-oriented cue of MI tasks by playing correlative animations. In a BCI trial, a prompt indicating their intended motor action through both vibratory stimulus and object-oriented visual guidance is given first. Subjects are supposed to do corresponding MI tasks after receiving the cue. Following this, the robotic glove drives the subject to perform the hand gesture required. Meanwhile, animations of successfully executed tasks are displayed on the VR platform.

The pneumatic manipulator part mainly consists of two parts, one contains the control unit and the drive unit, the other is a robotic glove that can be worn on the hand. The back of each finger sleeve of the exoskeleton glove is equipped with the pneumatic telescopic regulating mechanism, which is composed of an air chamber and the corresponding gas-guide tube. The outer end of the gas-guide tube is connected to the solenoid valve, which joins the air pump port. Vibration motors are placed on the inside of each finger sleeve of the glove to generate stimulation. The core of the control unit is an STM32F103 micro-controller (MCU) communicating with the computer through USB to serial port chip CH340G. The drive unit includes a current amplifying circuit, two air pumps, and six solenoid valves. According to the instructions issued by the computer, the MCU regulates the air pumps responsible for inflating and deflating as well as controlling the on/off state of the solenoid valves, which results in the flexion and extension of each finger sleeve of the robotic glove. Ultimately, the pneumatic manipulator drives the patients' hand to grasp, stretch, pinch, and other different rehabilitation training gestures.

3 Method

3.1 Data Acquisition

Seven subjects (right-handed, mean age 24), all of whom were in good physical and mental condition, participated in this experiment. The subjects sat about 50 cm away from the computer screen, kept relaxed. The EEG data were acquired using the Grael 4K EEG amplifier (NeuroScan, Australia), along with a 32-channel (Fig. 2a) cap that was equipped with Ag/AgCI active scalp electrodes. The electrodes were positioned in accordance with the 10–20 standard system. The ground electrode was located in Afz and the reference electrode was placed between Cz and CPz. During the experiment, the impedance of the electrode was kept below 15KΩ. The EEG was acquired at a sampling rate of 256 Hz and pre-filtered within the frequency range of 1 to 60 Hz while the power line interference was eliminated by using a notch filter at 50 Hz.

3.2 Experiment Protocol

Despite the significant number of degrees of freedom present in the human hand, many daily life tasks can be accomplished by using several primary grasping patterns [19]. In this study, we selected two grasping motions from the right hand,

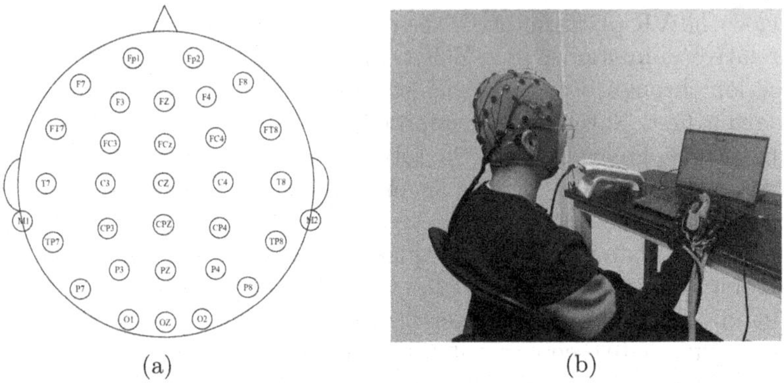

Fig. 2. (a) The channel distribution of the Quick-cap. (b) A subject was performing one MI task

which comprised five-finger palmar grasping and two-finger pinching. Preceding the experiment, all the subjects were required to wear the exoskeleton glove on their right hand and touch the vibration motor. During the experiment, participants were instructed to perform MI tasks by mentally simulating palmar grasping, pinching, or maintaining a state of rest for a specific duration without performing any corresponding physical movements. Each participant kept their hands relaxed without actively exerting any effort as is shown in Fig. 2b.

The experimental paradigm for a single trial is shown in Fig. 3. At second 0, the subjects were presented with a red cross located in the center of the screen. After 2 s, the fingers were randomly subjected to three distinct modes of stimulation that persisted for a duration of 2 s. These modes include five-finger vibration, two-finger vibration, and no vibration, corresponding to three distinct MI states: palmar grasping, pinching, and rest. The subsequent 1 s interval was given to eliminate the effect of the steady-state somatosensory evoked potential brought by the vibration stimulus. After that, a 6 s long animation associated with the vibration stimulus was then looped in the red box at the upper middle part of the screen.

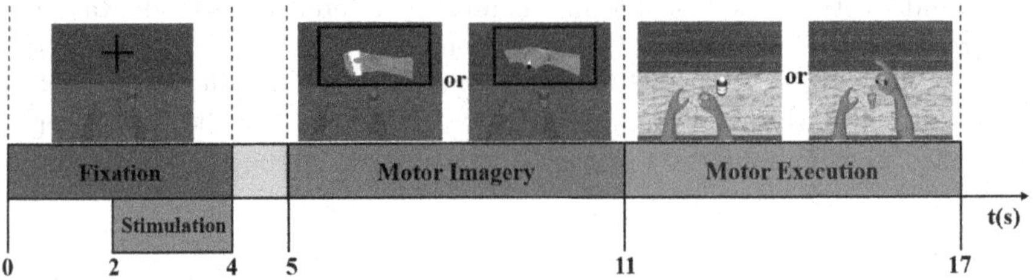

Fig. 3. The experiment paradigms of one trial

In this period, the volunteers executed sustained MI tasks with their right hand, which involved imagining holding a water cup and pinching a capsule. Subsequently, the VR platform presented visual feedback to the subject and showed an animation of performing the task. The robotic glove drove the subject to perform the hand posture required at the same time. Each session consists of 15 trials (5 for each of the three classes). Each subject did 1 session for training and 6 sessions for the experiment. To avoid mental fatigue, there is 3 min of rest between sessions. For each participant, a total of 90 EEG trials were recorded, corresponding to 30 trials with 6 s per MI corresponding

3.3 EEG Preprocessing

Generally, the collected EEG signals contain various noises and artifacts such as eye blinks and muscle activity [23]. To improve the SNR of the EEG signals, the common average reference (CAR) method, which can reduce the influence of common noise sources that are present in all electrodes, such as physiological artifacts, was used in the preprocessing. During the MI period, the effective signals were primarily found within the alpha (8-13 Hz) and beta (14-30 Hz) rhythms [15]. Therefore, a fifth-order Butterworth filter with a cutoff frequency of 8 Hz and 30 Hz was employed to emphasize frequency components of interest while simultaneously reducing artifacts. In order to improve the temporal resolution of EEG trials and augment the dataset size, the EEG data were segmented using a window length of 2 s and step size of 1 s. To ensure result consistency, each classification process was repeated ten times. For each repetition, two-thirds of randomly selected trial samples were allocated as the training set, while the remaining one-third of the data was used as the test set. To avoid data leakage, the dividing of the data set was performed before data cropping.

3.4 Feature Extraction and Classification

CSP is an extension of principal component analysis (PCA) for feature extraction, which has achieved a wide range of applications in EEG signals decoding [23], especially in extracting sensorimotor rhythms evoked by MI [16,24]. The main idea of the CSP algorithm is to maximize the difference between different categories by projecting the multi-channel EEG data into a low-dimensional spatial subspace [25]. Mathematically, the spatial filters w are the stationary points of the following optimization problem [26]:

$$\max_{\mathbf{w}} J(\mathbf{w}) = \frac{\mathbf{w}^T C_1 \mathbf{w}}{\mathbf{w}^T C_2 \mathbf{w}} \qquad s.t. \quad ||\mathbf{w}||_2 = 1 \tag{1}$$

where $\mathbf{w}$ denotes a spatial filter, and C_k denotes the estimated spatial covariance matrix for condition k. Since $J(\mathbf{w})$ is a Rayleigh quotient, the stationary points can be obtained from the equation: $C_1 w = \lambda C_2 w$, where λ denotes the eigenvalue associated with $\mathbf{w}$.

The current study involved three distinct classes. In order to implement the CSP algorithm for this task, the one-vs-all strategy was employed in this paper.

To be more specific, three CSP filters were trained, each of which corresponded to one category. For each filter, samples of the corresponding category were taken as positive examples and samples of the other categories were taken as negative examples. Subsequently, the filters that had undergone training were concatenated to yield a comprehensive filter. After projecting the CSP matrix using the filter, the logarithm of the variance of the EEG signals in the selected band was extracted as the features [25]. Then SVM algorithm was used to classify the obtained features. During the training process, the Gaussian kernel function was selected as the kernel function type. To avoid over-fitting, the grid search algorithm was not employed to optimize the parameters due to the small size of the dataset and the default values were used for parameters.

4 Result

4.1 Classification

This experiment investigated the performance of the model across varying numbers of filter pairs because the number of filter pairs can be adjusted to capture a varying range of spatial information and enhance feature discriminability [26].

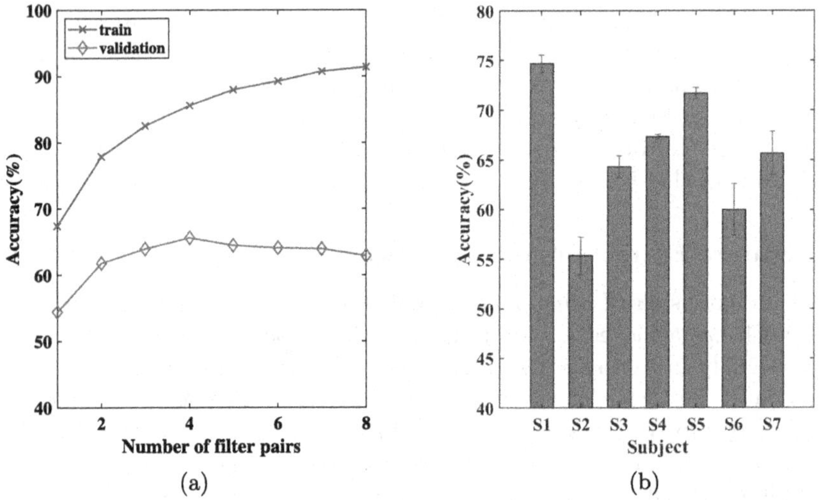

Fig. 4. (a) The mean classification accuracy for various numbers of filter pairs across all subjects during the training and validation phases. (b) The average classification accuracy of three MI tasks for 7 subjects

As depicted in the Fig. 4a, initially, both the training accuracy and the validation accuracy improved with the increase of the number of filter pairs, which indicated an improving fitting capability of the model. As the number continued to increase, the training accuracy continued to rise while the validation accuracy started to decline, suggesting the state of overfitting. The validation accuracy was highest with four filter pairs, therefore, four filter pairs were employed in the final classification model.

Figure 4b shows the average decoding accuracy of three MI tasks for each one. The average accuracy of the seven subjects was 65.67%. Subject S3 obtained the highest classification accuracy among all subjects, reaching a maximum of 74.67%. Notably, two subjects (S1 and S5) exhibited classification accuracy exceeding 70%. Conversely, one subject (S2) achieved accuracy below 60%, possibly due to the phenomenon known as "BCI Illiteracy" as described in [16].

In order to evaluate the prediction accuracy of the classifier on different categories, the confusion matrix for MI tasks was depicted in Fig. 5, which demonstrated that tasks involving MI and rest state were better distinguishable than tasks between two gesture [19]. This observation can be attributed to the fact that the MI tasks involving movements of the same hand activate neighboring areas within the sensorimotor cortex, resulting in overlapping neural patterns and reducing the discriminability of the corresponding EEG signals [17].

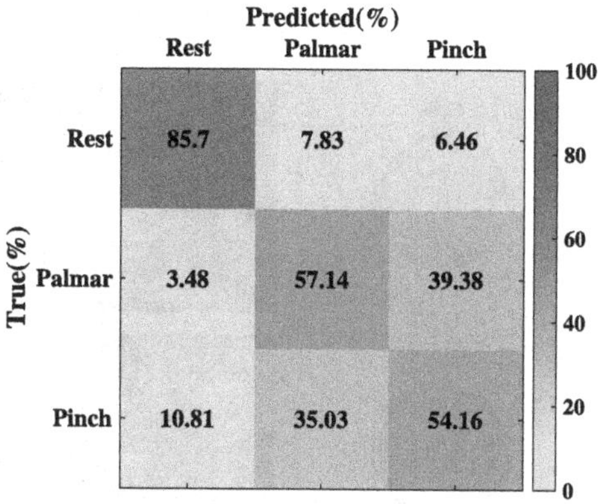

Fig. 5. The confusion matrix for three-class MI tasks with average correct classification and misclassification across all participants

Table 1 presents a overview of studies focusing on the classification of MI within the same limb, employing traditional machine learning methods. It can be seen that the proposed BCI system in this paper can achieve superior classification result, which demonstrates the feasibility of the proposed MI-BCI system in EEG-based classification accuracy improvement.

4.2 ERD/ERS Analysis

To examine the topographical distribution of EEG signals on the scalp during different paradigms, the power values for alpha and beta frequency bands were computed, and the average power topographic maps were generated, as

Table 1. BCI studies that classify different same limb movements

Bibliography	BCI Classes	Accuracy
Costa *et al.* [6]	3-class: Palmar grasp, Pinch, Elbow flexion	47.4%
Ofner *et al.* [14]	6-class: Elbow flexion/extension, Forearm supination/pronation, Hand close/open	27.0%
Tidare *et al.* [22]	2-class: Hand close, Hand open	60.4%
Yang *et al.* [27]	2-class: Arm-reaching, Wrist-twisting	59.4%
Yong *et al.* [28]	3-class: Hand grasp, Elbow flexion, Rest	60.7%
Ours	3-class: Palmar grasp, Pinch, Rest	65.67%

depicted in Fig. 6a. The figure effectively displayed the discernible fluctuations in the intensity of motor cortex activation during different MI states. In the rest state, the power of the alpha band exhibited higher intensity in the occipital lobes compared to other cerebral regions. Furthermore, there were no significant differences in the distribution of power between the two hemispheres observed in either alpha or beta frequency bands. After baseline correction, no ERD/ERS phenomenon was observed in the event-related spectral perturbation (ERSP) time-frequency maps of the C3 electrode location, as the Fig. 6b shown.

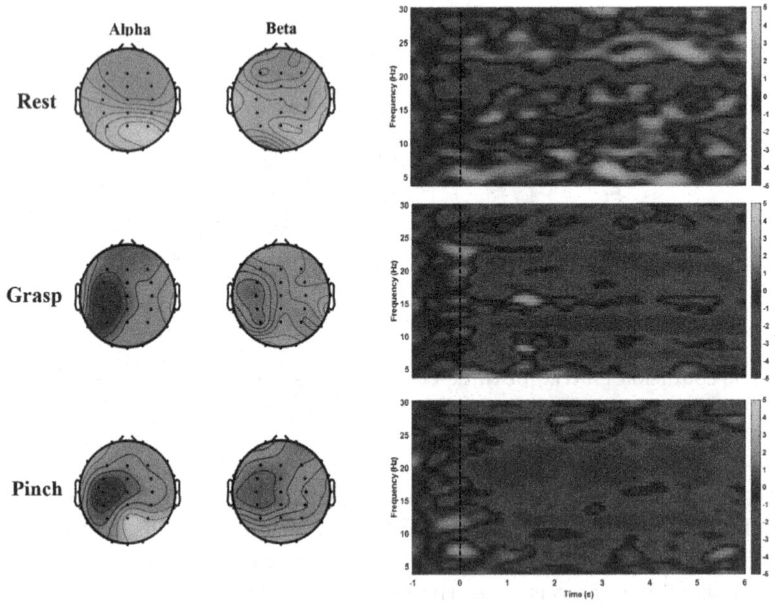

(a) (b)

Fig. 6. (a) Averaged topographical distribution of power in the alpha band (8-13 Hz) and beta band (14-30 Hz) for three MI tasks. (b) The time-frequency maps at the C3 electrode location during three MI tasks

When it comes to imagining grasping movements, a notable ERD phenomenon was identified in the parietal lobe. Specifically, an asymmetrical spatial distribution of EEG power was observed within the motion-related frequency, with a pronounced decline in EEG power localized to the left hemisphere. The alpha band exhibited a pronounced decrease in power compared to the beta band. Additionally, pinching elicited a more prominent ERD phenomenon in the beta band compared to palmar grasping, as evidenced by the time-frequency maps. During MI, the EEG energy associated with palmar grasping predominantly concentrated within the alpha band, whereas the EEG energy reduction induced by pinching exhibited a more widespread distribution across the entire motion-related frequency, as the Fig. 6b shown.

Figure 6 provides physiological evidence supporting the differentiation of distinct gestures, which have significant implications for facilitating the intuitive control of neuroprosthetic devices. Moreover, more efficient decoding algorithms need to be developed to improve the accuracy of classification in the future.

5 Conclusions

In this study, an enhanced MI-BCI system based on vibration stimulation and robotic glove was developed to improve classification accuracy of fine MI. Compared with previous works, our rehabilitation system had the advantages of being lightweight, compact size, and easy to wear. In addition, vibration stimulation instead of FES ensured the security of the system. The feasibility of distinguishing imagined grasping actions commonly used in daily life was validated through the experiment with seven subjects. This system holds promising potential for assisting stroke patients in enhancing hand function during rehabilitation.

Acknowledgments. This work is supported in part by the National Natural Science Foundation of China (Grants 1720106012, U1913601, and 62203440), Beijing Sci&Tech Program (Grant Z211100007921021), Beijing Natural Science Foundation (Grant 4202074), and ANSO Collaborative Research Project (Grant ANSO-CR-PP-2020-03).

References

1. Achanccaray, D., Izumi, S.I., Hayashibe, M.: Visual-electrotactile stimulation feedback to improve immersive brain-computer interface based on hand motor imagery. Comput. Intell. Neurosci. **2021**, e8832686 (2021). https://doi.org/10.1155/2021/8832686
2. Biasiucci, A., et al.: Brain-actuated functional electrical stimulation elicits lasting arm motor recovery after stroke. Nat. Commun. **9**(1), 2421 (2018). https://doi.org/10.1038/s41467-018-04673-z
3. Chen, W., et al.: Soft exoskeleton with fully actuated thumb movements for grasping assistance. IEEE Trans. Rob. **38**(4), 2194–2207 (2022). https://doi.org/10.1109/TRO.2022.3148909

4. Cheng, N., et al.: Brain-computer interface-based soft robotic glove rehabilitation for stroke. IEEE Trans. Biomed. Eng. **67**(12), 3339–3351 (2020)
5. Cheng, N., et al.: Brain-computer interface-based soft robotic glove rehabilitation for stroke. IEEE Trans. Biomed. Eng. **67**(12), 3339–3351 (2020). https://doi.org/10.1109/TBME.2020.2984003
6. Costa, A.P., Møller, J.S., Iversen, H.K., Puthusserypady, S.: An adaptive CSP filter to investigate user independence in a 3-class MI-BCI paradigm. Comput. Biol. Med. **103**, 24–33 (2018)
7. Jochumsen, M., Niazi, I.K., Taylor, D., Farina, D., Dremstrup, K.: Detecting and classifying movement-related cortical potentials associated with hand movements in healthy subjects and stroke patients from single-electrode, single-trial EEG. J. Neural Eng. **12**(5), 056013 (2015)
8. Lawhern, V.J., Solon, A.J., Waytowich, N.R., Gordon, S.M., Hung, C.P., Lance, B.J.: EEGNet: a compact convolutional neural network for EEG-based brain–computer interfaces. J. Neural Eng. **15**(5), 056013 (2018). https://doi.org/10.1088/1741-2552/aace8c
9. Liang, S., Choi, K.S., Qin, J., Pang, W.M., Wang, Q., Heng, P.A.: Improving the discrimination of hand motor imagery via virtual reality based visual guidance. Comput. Methods Programs Biomed. **132**, 63–74 (2016). https://doi.org/10.1016/j.cmpb.2016.04.023
10. Lu, B., Ge, S., Wang, H.: EEG-based classification of lower limb motor imagery with STFT and CNN. In: Mantoro, T., Lee, M., Ayu, M.A., Wong, K.W., Hidayanto, A.N. (eds.) ICONIP 2021. CCIS, vol. 1517, pp. 397–404. Springer, Cham (2021). https://doi.org/10.1007/978-3-030-92310-5_46
11. Mane, R., Wu, Z., Wang, D.: Poststroke motor, cognitive and speech rehabilitation with brain-computer interface: A perspective review. Stroke Vasc. Neurol. **7**(6), 541–549 (2022). https://doi.org/10.1136/svn-2022-001506
12. Miao, M., Zeng, H., Wang, A.: Composite and multiple kernel learning for brain computer interface. In: Liu, D., Xie, S., Li, Y., Zhao, D., El-Alfy, ES. (eds.) Neural Information Processing. ICONIP 2017. Lecture Notes in Computer Science, vol. 10635, pp. 803–810. Springer, Cham (2017). https://doi.org/10.1007/978-3-319-70096-0_82
13. Nojima, I., Sugata, H., Takeuchi, H., Mima, T.: Brain–computer interface training based on brain activity can induce motor recovery in patients with stroke: A meta-analysis. Neurorehabil. Neural Repair **36**(2), 83–96 (2022). https://doi.org/10.1177/15459683211062895
14. Ofner, P., Schwarz, A., Pereira, J., Müller-Putz, G.R.: Upper limb movements can be decoded from the time-domain of low-frequency EEG. PLoS ONE **12**(8), e0182578 (2017). https://doi.org/10.1371/journal.pone.0182578
15. Pfurtscheller, G., Brunner, C., Schlögl, A., Da Silva, F.L.: Mu rhythm (de) synchronization and EEG single-trial classification of different motor imagery tasks. Neuroimage **31**(1), 153–159 (2006)
16. Ren, S., Wang, W., Hou, Z.G., Liang, X., Wang, J., Shi, W.: Enhanced motor imagery based brain-computer interface via FES and VR for lower limbs. IEEE Trans. Neural Syst. Rehabil. Eng. **28**(8), 1846–1855 (2020). https://doi.org/10.1109/TNSRE.2020.3001990
17. Sanes, J.N., Donoghue, J.P., Thangaraj, V., Edelman, R.R., Warach, S.: Shared neural substrates controlling hand movements in human motor cortex. Science **268**(5218), 1775–1777 (1995). https://doi.org/10.1126/science.7792606

18. Schwarz, A., Höller, M.K., Pereira, J., Ofner, P., Müller-Putz, G.R.: Decoding hand movements from human EEG to control a robotic arm in a simulation environment. J. Neural Eng. **17**(3), 036010 (2020). https://doi.org/10.1088/1741-2552/ab882e
19. Schwarz, A., Ofner, P., Pereira, J., Sburlea, A.I., Müller-Putz, G.R.: Decoding natural reach-and-grasp actions from human EEG. J. Neural Eng. **15**(1), 016005 (2018). https://doi.org/10.1088/1741-2552/aa8911
20. Shih, J.J., Krusienski, D.J., Wolpaw, J.R.: Brain-computer interfaces in medicine. Mayo Clin. Proc. **87**(3), 268–279 (2012)
21. Tao, Y., et al.: Decoding multi-class EEG signals of hand movement using multivariate empirical mode decomposition and convolutional neural network. IEEE Trans. Neural Syst. Rehabil. Eng. **30**, 2754–2763 (2022). https://doi.org/10.1109/TNSRE.2022.3208710
22. Tidare, J., Leon, M., Xiong, N., Astrand, E.: Discriminating EEG spectral power related to mental imagery of closing and opening of hand. In: 2019 9th International IEEE/EMBS Conference on Neural Engineering (NER), pp. 307–310. IEEE (2019)
23. Urigüen, J.A., Garcia-Zapirain, B.: EEG artifact removal-state-of-the-art and guidelines. J. Neural Eng. **12**(3), 031001 (2015)
24. Vourvopoulos, A., Bermúdez i Badia, S.: Motor priming in virtual reality can augment motor-imagery training efficacy in restorative brain-computer interaction: a within-subject analysis. J. Neuroeng. Rehabil. **13**(1), 1–14 (2016)
25. Wang, Y., Gao, S., Gao, X.: Common spatial pattern method for channel selection in motor imagery based brain-computer interface. In: 2005 IEEE Engineering in Medicine and Biology 27th Annual Conference, pp. 5392–5395 (2005). https://doi.org/10.1109/IEMBS.2005.1615701
26. Wu, W., Chen, Z., Gao, X., Li, Y., Brown, E.N., Gao, S.: Probabilistic common spatial patterns for multichannel EEG analysis. IEEE Trans. Pattern Anal. Mach. Intell. **37**(3), 639–653 (2015). https://doi.org/10.1109/TPAMI.2014.2330598
27. Yang, B., Ma, J., Qiu, W., Zhu, Y., Meng, X.: A new 2-class unilateral upper limb motor imagery tasks for stroke rehabilitation training. Med. Novel Technol. Devices **13**, 100100 (2022)
28. Yong, X., Menon, C.: EEG classification of different imaginary movements within the same limb. PLoS ONE **10**(4), e0121896 (2015). https://doi.org/10.1371/journal.pone.0121896
29. Zhang, W., Song, A., Lai, J.: Motor imagery BCI-based online control soft glove rehabilitation system with vibrotactile stimulation. In: Tanveer, M., Agarwal, S., Ozawa, S., Ekbal, A., Jatowt, A. (eds.) Neural Information Processing. ICONIP 2022. Communications in Computer and Information Science, vol. 1792, pp. 456–466. Springer, Singapore (2023). https://doi.org/10.1007/978-981-99-1642-9_39

MTSAN-MI: Multiscale Temporal-Spatial Convolutional Self-attention Network for Motor Imagery Classification

Junkongshuai Wang[1], Yangjie Luo[1], Lu Wang[1], Lihua Zhang[1,2], and Xiaoyang Kang[1,2,3,4](✉)

[1] Laboratory for Neural Interface and Brain Computer Interface, Engineering Research Center of AI & Robotics, Ministry of Education, Shanghai Engineering Research Center of AI & Robotics, MOE Frontiers Center for Brain Science, State Key Laboratory of Medical Neurobiology, Institute of AI & Robotics, Institute of Meta-Medical, Academy for Engineering & Technology, Fudan University, Shanghai, China
{jkswang21,yjluo23,lwang22}@m.fudan.edu.cn, {lihuazhang, xiaoyang_kang}@fudan.edu.cn

[2] Ji Hua Laboratory, Foshan 528200, Guangdong Province, China

[3] Yiwu Research Institute of Fudan University, Yiwu City 322000, Zhejiang, China

[4] Research Center for Intelligent Sensing, Zhejiang Lab, Hangzhou 311121, China

Abstract. EEG signals are widely utilized in brain-computer interfaces, where motor imagery (MI) data plays a crucial role. The effective alignment of MI-based EEG signals for feature extraction, decoding, and classification has always been a significant challenge. Decoding methods based on convolution neural networks often encounter the issue of selecting the optimal receptive field, while convolution in the spatial domain cannot fully utilize the rich spatial topological information contained within EEG signals. In this paper, we propose a multiscale temporal-spatial convolutional self-attention network for motor imagery classification (MTSAN-MI). The proposed model starts with a multiscale temporal-spatial convolution module, in which temporal convolutional layers of varying scales across three different branches can extract corresponding features based on their receptive fields respectively, and graph convolution networks are better equipped to leverage the intrinsic relationships between channels. The multi-head self-attention module is directly connected to capture global dependencies within the temporal-spatial features. Evaluation experiments are conducted on two MI-based EEG datasets, which show that the state-of-the-art is achieved on one dataset, and the result is comparable to the best method on the other dataset. The ablation study also proves the importance of each component of the framework.

Keyword: Motor imagery (MI) · Multiscale temporal-spatial convolution · Graph convolution network (GCN) · Self-attention

B. Luo et al. (Eds.): ICONIP 2023, CCIS 1963, pp. 338–349, 2024.
https://doi.org/10.1007/978-981-99-8138-0_27

1 Introduction

Electroencephalography (EEG) plays an important role in the field of brain-computer interfaces (BCI) and is highly desirable for widespread public adoption due to its non-invasive, low-cost, and user-friendly nature [1]. Motor imagery (MI) is one of the basic BCI systems, which refers to the subjects' cognitive process of motor activity without physical execution. The low signal-to-noise ratio, instability, and inefficient properties of MI-based EEG make how to extract effective features and improve the decoding performance become a major challenge [2].

In the current MI-based EEG system, electrodes are arranged on the cerebral cortex according to different distributions, and collect potential changes simultaneously, making EEG data have two dimensions of channel and time [3]. This 2-D representation is well suited for using a CNN-based method for feature extraction or prediction [4]. Lawhern et al. proposed EEGNet [5], a CNN with depthwise and separable convolutions used to create a lightweight EEG-specific model. Mouad et al. proposed MI-EEGNET [6], based on Inception and Xception architectures that extract temporal and spatial features through convolutional layers. However, these methods have also confronted the challenge of determining the optimal receptive field.

Meanwhile, the EEG storage structure typically adopts a multidimensional time series, with electrode channels arranged in a specific order that may not reflect their actual spatial relationship [7]. As a result, the actual topology between electrodes can be easily overlooked. Some past studies using CNNs to extract spatial information from data suffered from this problem. In [5, 6, 8], the spatial features are extracted by using 1-D CNNs along the electrode dimension. Due to the structure of the convolutional layer [4], only the k closest neighboring channels are convolved, however, channels with real topological relationships cannot be computed with each other, resulting in ineffective learning of spatial information or physiological correlations.

Since the electrodes are distributed in various regions of the non-Euclidean cerebral cortex, the graph structure will be the most appropriate data representation for revealing interconnections among different signal channels [7]. In such a data structure, the nodes represent the EEG channel while the edges represent the spatial relations [9] or correlations [10] among channels. The graph neural network (GNN) learns spatial patterns in EEG by aggregating the features of different nodes. EEG decoding applications based on GNN have several sceneries, such as sleep stage classification [11], emotion recognition [3, 12], and motor imagery [9, 13]. Wei et al. proposed a multi-level spatial-temporal adaptation network [9], that captures more discriminative and domain-invariant spatial-temporal features to improve performance in cross-domain MI classification. Biao et al. proposed an end-to-end EEG channel active inference neural network [13], which consists of CNN-based temporal feature extraction module and GCN-based channel active reasoning module. In [11], they proposed an adaptive spatial-temporal graph convolutional networks for sleep stage classification, named GraphSleepNet, which contains an adaptive sleep graph learning mechanism. The adaptive graph learning mechanism achieved better performance than predefined graph structure.

Although the temporal-spatial convolution can extract local features and inter-channel connections from the studies mentioned above, it is still a key problem to further extract global dependencies within the higher dimensional features. The self-attention mechanism has achieved great performance in image processing [14] and EEG decoding [8]. Methods based on self-attention mechanism can perceive dependencies between features on different dimensions. In [8], they proposed EEG Conformer which combines CNN and Transformer straightforwardly for end-to-end EEG classification.

To solve the issues mentioned above, also inspired by the above studies, we propose a multiscale temporal-spatial convolutional self-attention network for motor imagery EEG signals decoding (MTSAN-MI), which mainly consists of a multiscale temporal-spatial convolution module, a multi-head self-attention mechanism module, and an MLP-based motor imagery tasks classifier. The multiscale temporal-spatial convolution module includes different scales of temporal convolution and each convolution layer will be immediately followed by a graph convolution layer. This enables the extraction of the temporal information at different scales and the topological relationships between node pairs. These temporal-spatial features are concatenated and fed into the multi-head self-attention module which consists of a self-attention layer and a feed-forward layer. The final classifier is utilized for the prediction of the motor imagery category. This method achieves remarkable results on two public datasets, while the ablation study further confirms the effectiveness of each component.

2 Methods

2.1 Problem Definition

Before giving a detailed description of the methodological part, we first briefly describe the motor imagery experiment process and the pipeline of motor imagery classification based on graph representation that defines EEG signals. The paradigm of a typical motor imagery experiment is shown in Fig. 1. Each trial starts with a beep and visual cue indicating one of four MI tasks, followed by subjects performing the corresponding motor imagery. The duration of MI is of research interest.

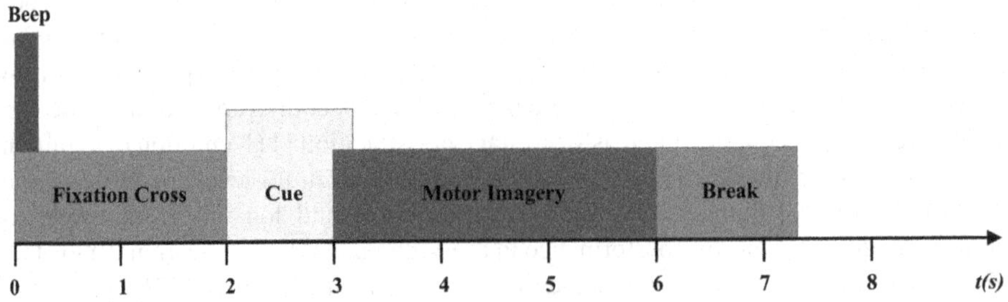

Fig. 1. An example of the paradigm of BCI Competition IV 2a dataset.

Figure 2 illustrates how graph representation is used to decode MI-based EEG signals. By transforming the original 2D EEG signal into a graph construction that reflects the inter-channel spatial topology or physiological correlation, we can obtain corresponding features and classify signals using graph theory operations such as edge computation and node feature aggregation.

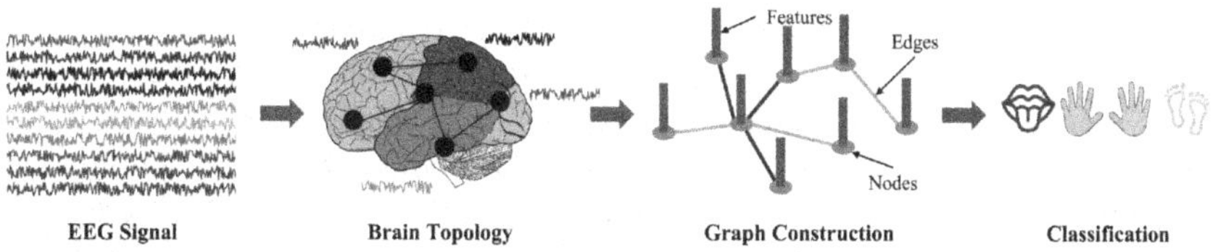

Fig. 2. The flowchart of MI-based EEG signals classification based on graph representation.

Given an MI-based EEG trial representation $x_i \in \mathbb{R}^{C \times T}$, $i \in [1, N]$, in which C represents EEG channels and T represents input sample points, the MI-based EEG dataset contains N trials with corresponding MI task labels. The MI-based EEG decoding problem can be framed as a supervised classification task, in which a classification model is trained on the labeled training dataset to predict the intention of the test dataset.

2.2 Network Overview

The overall framework shown in Fig. 3 consists of three components. The first module contains three branches. Each branch incorporates a temporal convolution network and a spatial convolution network. The multi-head self-attention module comprises three linear transformation layers, a self-attention layer, and a feed-forward layer. The last module is a classifier based on a multilayer perceptron. The numbers before and after "@" are the size of the convolution channel and the convolution kernel. " +" represents the element-wise addition operation, and "C" represents the concatenate operation.

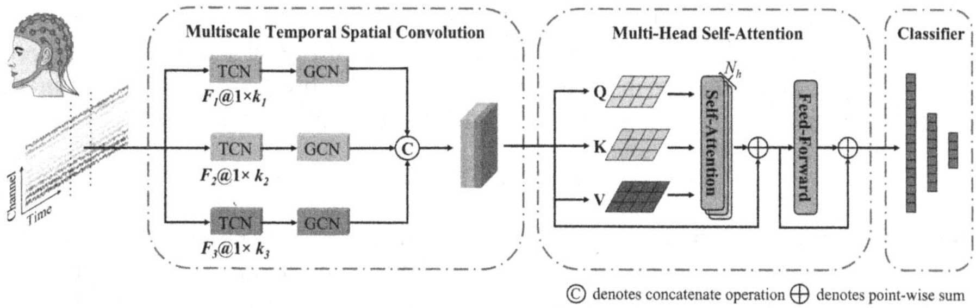

Fig. 3. The network structure of MTSAN-MI. It includes a multiscale temporal-spatial convolution module, a multi-head self-attention module, and a classifier module.

2.3 Multiscale Temporal Spatial Module

The multiscale temporal-spatial module can extract features in both time and space dimensions. As shown in the left part of Fig. 3, each parallel branch is composed of a temporal convolution layer and a spatial convolution layer based on GCN. The temporal layer can dynamically acquire the EEG representation in the time domain, and the spatial layer can learn complex relationships between different channels dynamically and update them continuously through the trainable adjacency matrix.

Temporal Convolution Layer. [15] denotes that using temporal convolution can extract rich motor imagery features. However, selecting an appropriate size for the convolution kernel is a crucial issue as it directly determines its receptive field and consequently affects the features extracted from EEG signals. The method of multiscale time domain convolution can be employed [7], where multiple convolution kernels of different scales are arranged in parallel to extract features from input EEG signals. This enables the acquisition of effective information at various scales. The scale, i.e., the length of the temporal convolution kernel, is set in different proportions of the sampling rate *fs*. Specifically, each temporal convolution layer has a 1-D kernel of its size. The ratio factor can be expressed as $\alpha^{l-1} \bullet r$, where r is the initial ratio value, α is the coefficient for different layers, and l is the level of the layer. The values of l range from 1 to 3, while α is set to 0.5 and r is set to 0.1. Therefore, the convolution kernel size of each temporal convolution layer can be expressed as

$$k^l = (1, \alpha^{l-1} \cdot r \cdot fs), l \in [1, 2, 3] \tag{1}$$

For an input signal sample $x \in \mathbb{R}^{C \times T}$, it is first replicated into three copies and then fed into the multiscale temporal convolution to generate a corresponding feature map $S_l \in \mathbb{R}^{F_l \times C \times P_l}$ for each scale model. The dimension of the channel is preserved, and the original temporal feature space $\mathbb{R}^T$ is mapped into $\mathbb{R}^{P_l}$. Each temporal convolution layer possesses its feature channel, denoted as F_l. To facilitate subsequent computation and concatenation, all three F_l are set to the same. The average pooling operation is also set to all layers to preserve the salient features of the input feature map.

Each temporal convolution layer is followed by a Batch normalization layer and a Leaky rectified linear unit (ReLU) activation layer which can enhance the learning process and optimize performance. Then feature map $S_l \in \mathbb{R}^{F_l \times C \times P_l}$ is permuted and reshaped into $\overline{S}_l \in \mathbb{R}^{C \times F_l * P_l}$, where the features from different feature channels are mapped into the node attribute in graph representation for subsequent graph convolution. The entire temporal convolution process is shown in the following formula

$$\overline{S}_l = \mathcal{F}_{Reshape}(\Phi_{ReLU}(\mathcal{F}_{BN}(\mathcal{F}_{AP}(\mathcal{F}_{Tconv}(x, k^l))))) \tag{2}$$

Spatial Convolution Layer. The spatial convolution layers based on GCN [10] aim to learn spatial features from temporal features. The brief pipeline of MI-based EEG classification based on graph representation is introduced in Sect. 2.1. Each node v_i in graph G represents the EEG channel and is with a feature vector $f_i \in \mathbb{R}^{F_l * P_l}$ learned after the l-level temporal convolution layer.

Based on the graph convolution method, node features are updated by aggregating information from neighboring nodes, thereby emphasizing inter-node relationships within the features. To explore more intricate connectivity patterns, we employed the trainable adjacency matrix [13]. The trainable adjacency matrix can dynamically reflect the interconnections among different nodes. The adjacency matrix is initialized as $A \in \mathbb{R}^{C*C}$ based on the spatial relationship. The degree of each node is obtained by summing the adjacency matrix in the row direction, and these degrees can form a degree matrix, i.e., $D = \sum_j A_{ij}$. The normalized adjacency matrix $\tilde{A}$ can be calculated by

$$\tilde{A} = D^{-\frac{1}{2}} A D^{-\frac{1}{2}} \tag{3}$$

so that the characteristics of the node cannot be affected by the degree of the subsequent aggregation operation. ReLU operation is used to ensure that the elements in matrix A are non-negative. In the backpropagation at the end of each training epoch, the loss will be utilized to perform the gradient descent calculation and update each element of the adjacency matrix. The process of updating elements in the adjacency matrix is shown in the following two equations

$$\frac{\partial \text{Loss}}{\partial A} = \begin{pmatrix} \frac{\partial Loss}{\partial A_{11}} & \cdots & \frac{\partial Loss}{\partial A_{1C}} \\ \vdots & \vdots & \vdots \\ \frac{\partial Loss}{\partial A_{C1}} & \cdots & \frac{\partial Loss}{\partial A_{CC}}, \end{pmatrix} \tag{4}$$

$$A = (1 - \rho)A + \rho \frac{\partial Loss}{\partial A} \tag{5}$$

where $Loss$ is the whole training loss which will be introduced in Sect. 2.5 and ρ is the learning rate of the model.

To enable each spatial convolution layer in different levels to output features of the same size, a feature map is used to transform the input temporal features into the embedding feature space. $W \in \mathbb{R}^{F_l * P_l \times GE}$ is a level-specific trainable weight matrix and $b \in \mathbb{R}^{1 \times GE}$ is the trainable bias vector, where GE is the length of the graph hidden embedding. The output of the spatial convolution of one parallel branch, i.e., $H_l \in \mathbb{R}^{C \times GE}$, can be expressed as

$$H_l = \Phi_{ReLU}(\tilde{A}(\overline{S}_l \cdot W + b)) \tag{6}$$

Concatenation and Projection. The outputs of all parallel branches are concatenated along the feature dimension, followed by a projection layer that maps the output features to a common higher-dimensional space. The output of the whole multiscale temporal-spatial convolution module can be set as $Z \in \mathbb{R}^{EM \times 3 * GE}$, where EM is the number of feature channels in high dimensional space. The specific expression is as follows:

$$Z = \mathcal{F}_{Proj}([H_1 \| H_2 \| H_3]) \tag{7}$$

2.4 Multi-Head Self-Attention Mechanism

After previous feature extractions, there still exist valid connections among the high-dimensional feature channels. Therefore, we employ an attention-based mechanism to weigh the different feature channels. The multi-head self-attention (MHSA) layer is inspired by the Transformer model [14], which consists of a multi-head self-attention network and a feed-forward network. The self-attention mechanism contains three elements called query (Q), key (K), and value (V), which are calculated from input Z by linear transformations. We compute the weighted attention features as:

$$\text{Attention}(Q, K, V) = Softmax(\frac{QK^T}{\sqrt{d_k}})V \tag{8}$$

where the dot products of Q and K^T that reflect the correlation between different input channels are divided by a scaled factor $\sqrt{d_k}$. d_k is the dimension of the input matrix K. The result is then normalized by a *Softmax* function and multiplied by V so that the input and output sizes remain the same.

For the multi-head mechanism, the input is mapped into h different subspaces, each of which has an independent set of attention weights to compute the output. The specific equations are given:

$$head_i = \text{Attention}(Q_i, K_i, V_i) \tag{9}$$

$$\text{MSHA}(Q, K, V) = [head_1; \cdots ; head_h] \tag{10}$$

The second part is a feed-forward layer which consists of two linear transformation layers and a Gaussian Error Linear Unit (GeLU) activation layer, which can enhance the fitting ability. The overall process of multi-head self-attention computation is repeated once in total, as shown in Fig. 3.

2.5 Classifier

The features are then flattened and passed to three fully-connected layers for prediction. The output is an M-dimensional vector after *Softmax* calculation. The whole training process used the cross-entropy loss with a regular penalty. The formula is as follows:

$$Loss = -\frac{1}{N_{bs}}\sum_{i=1}^{N_{bs}}\sum_{c=1}^{M} y \log y_p + \beta\|\Theta\| \tag{11}$$

where y and y_p represent the true label of the data and the probability of model prediction, M means the number of categories, and N_{bs} is the batch size. Θ contains the parameters of the whole model, and β is the regular penalty.

3 Experiments and Results

3.1 Dataset and Preprocessing

We investigate the proposed approach on two MI datasets, including the dataset 2a in BCI Competition IV (BCIC-IV-2a) [16] and the High Gamma Dataset (HGD) [17].

BCIC-IV-2a. BCIC-IV-2a is a cue-based motor imagery paradigm with four-class tasks including left hand, right hand, feet, and tongue recorded in 22 Ag/AgCl EEG electrodes and recorded with a sampling rate of 250 Hz with nine subjects (called A01-A09). The subjects completed six runs of 12 trials for each of the four classes, either in a training or evaluation session, resulting in a total of 576 trials. The raw signals are filtered by third-order Butterworth lowpass filter into 0-38Hz [18].

HGD. The HGD was gathered from 14 participants (S01-S14). Each participant completed three distinct motor imagery tasks (left hand, right hand, foot), along with a stationary task. Data collection was conducted across two sessions, yielding approximately 880 trials and 160 trials for the training set and test set. Scalp EEG recordings were obtained using 128 electrodes and a sampling rate of 500 Hz. To reduce computational complexity, we selectively utilized 44 channels out of the initial 128 that are more relevant to motor imagery tasks. Then the signals are resampled into 250 Hz and filtered into 0125 Hz [17].

Preprocessing. In two datasets, signals for each trial were obtained using data between [−0.5 s, 4 s] relative to the onset of the cue. The training dataset is used to calculate the mean and standard deviation. Based on these, the training and test dataset are then normalized.

Experiment Settings. We adopt the subject-dependent training setting which refers to the method that trains a separate model for each subject and apply 10-fold cross-validation to better verify the generalization ability of the model. All the models are implemented by PyTorch deep learning framework and trained on Tesla V100 GPU in a fully-supervised manner. The number of the heads in multi-head self-attention module is set to 4 and the number of MHSA block is set to 2. Adam with a learning rate of 0.001 is used as the optimizer. For both datasets, the mini-batch size is set to 64.

3.2 Metrics

In this paper, accuracy and kappa score are used as the performance metrics, which can be denoted:

$$ACC = \frac{\sum_{i=1}^{n} TP_i}{n} \tag{12}$$

$$kappa = \frac{1}{n}\sum_{a=1}^{n} \frac{A_a - A_e}{1 - A_e} \tag{13}$$

where n is the number of categories of motor imagery tasks, TP_i is the number of true positive samples for the i-th category, A_a is the classification accuracy, and A_e is the random classification accuracy.

3.3 Results and Discussion

In this section, the performances of the proposed method are compared with some state-of-the-art methods on two public datasets mentioned above. The method based on hand-crafted features includes FBCSP [19]. The CNN-based methods include EEGNet [5], TS-SEFFNet [18], MI-EEGNET [6], Shallow ConvNet [18], and DeepConvNet [18]. The attention-based method includes Conformer [8]. The GNN-based method includes MSTAN [9]. The best results of each model on the two datasets from the reported literatures are summarized in Table. 1.

Table 1. Subject-dependent classification accuracy and kappa score on BCIC-IV-2a and HGD.

Methods	BCI-IV-2a		HGD	
	Accuracy (%)	Kappa	Accuracy (%)	Kappa
FBCSP [19]	67.75	0.5700	90.09	0.8790
DeepConvNet [18]	71.99	0.6270	89.51	0.8600
Shallow ConvNet [18]	72.92	0.6390	88.90	0.8520
EEGNet [5]	74.50	0.6600	86.36	-
TS-SEFFNet [18]	74.71	0.6630	**93.25**	**0.9100**
MI-EEGNET [6]	77.49	0.6620	92.49	-
Conformer [8]	78.66	0.7155	87.71	0.8320
MSTAN [9]	79.20	0.7200	-	-
Ours	**82.33**	**0.7542**	90.57	0.8586
Ours w/o Attention	80.89	0.7378	88.14	0.8464
Ours w/o GCN	78.33	0.7089	87.57	0.8186

Comparison results show that our proposed method achieves the best performance, reaching 82.33% and 0.7542 in accuracy and kappa score on the BCIC-IV-2a dataset. The proposed method generally outperformed CNN-based methods, indicating that some valuable information may be lost when only considering the nearest channels or some adjacent features on this dataset. Compared with the attention-based method Conformer, the proposed method achieves a 3.67% improvement in accuracy. The use of the self-attention mechanism is effective in extracting features' importance, but incorporating multiscale temporal-spatial features can better capture the potential information within the MI-EEG signals and enhance the performance of the attention mechanism. The GNN-based method MSTAN [9] achieves the second highest accuracy, indicating that the method based on the graph neural network can effectively extract inter-channel relationships while increasing the task-specific differences for improved decoding and classification performance, which also verifies the superiority of the proposed model.

Although on the HGD dataset our method is not the best, the results it has achieved can also have some validity and persuasiveness, which reaches 90.57% in accuracy and 0.8586 in kappa score. The results of TS-SEFFNet and MI-EEGNET are better than our method to some extent. Their methods mainly consist of CNNs with different depths. The effectiveness of this approach on the HGD has been well demonstrated [17]. Our method employs a one-layer CNN structure only in the temporal convolution, and other parts are composed of the spatial convolution and the self-attention mechanism. By incorporating more intricate and interpretable modules, we significantly enhanced the accuracy of the BCIC-IV-2a dataset while upholding the accuracy of the HGD dataset. The overall impact of our method is undeniably superior to that of other methods.

3.4 Ablation Study

To further verify the role of GCN-based spatial convolution and the self-attention mechanism in performance improvement, ablation studies are carried out on our proposed model by examining the contributions of these two parts to the specific training process. The results of the metrics on the two datasets are respectively shown at the bottom of Table. 1, where "w/o" means without, both indicating that the overall architecture of our proposed model achieves the best performance on the two datasets in the ablation study.

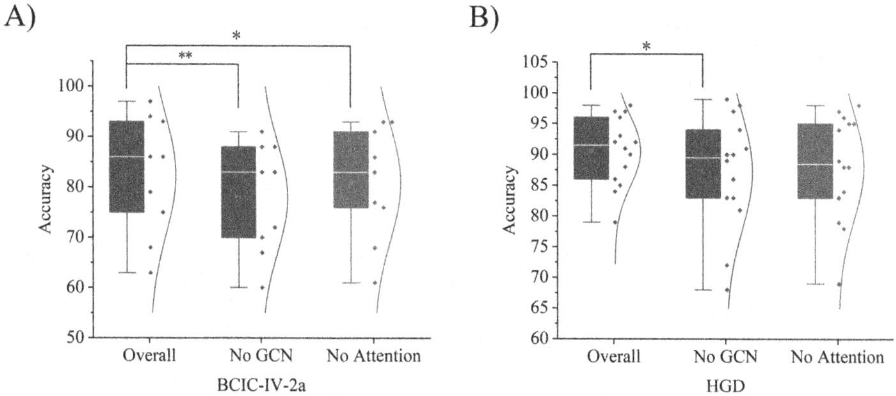

Fig. 4. The classification accuracy of ablation study on the spatial convolution layer and self-attention module for both datasets.

According to whether the two groups of data have normal distribution at the same time, the paired sample t-test or Wilcoxon signed rank test is used to statistically analyze the statistical significance. The results of both datasets are shown in Fig. 4 and the samples for each case are derived from the 10-fold cross-validation results. *No GCN* means the model without the GCN-based spatial convolution layer and *No Attention* means the model without the self-attention module. The accuracy of the overall architecture is significantly better than *No GCN* ($p < 0.01$) and *No Attention* ($p < 0.05$) on the BCIC-IV-2a dataset, and is significantly better than *No GCN* ($p < 0.05$) on the HGD dataset. This result also indicates that the GCN-based spatial convolution layer has a greater impact on MTSAN-MI than the self-attention module.

4 Conclusion

In this paper, we proposed a multiscale temporal-spatial convolutional self-attention network for MI-based EEG decoding called MTSAN-MI. The multiscale temporal-spatial module can reveal salient features in both the time and space domains, in which multiscale temporal convolutions are utilized to capture the complex EEG patterns, and the trainable adjacency matrix can dynamically reflect inter-channel relationships. It can solve the adverse effects caused by the limited receptive field and can extract the neglected interconnections among different channels. The experiment results show that our model exhibits outstanding performance on two datasets. Additionally, the ablation study demonstrates that the overall architecture plays a crucial role in achieving optimal performance and validates the contribution of each module.

Acknowledgments. Research supported by the National Key R&D Program of China, grant no. 2021YFC0122700; National Natural Science Foundation of China, grant no. 61904038 and no. U1913216; Shanghai Sailing Program, grant no. 19YF1403600; Shanghai Municipal Science and Technology Commission, grant no. 19441907600; Opening Project of Zhejiang Lab, grant no. 2021MC0AB01; Fudan University-CIOMP Joint Fund, grant no.FC2019–002; Opening Project of Shanghai Robot R&D and Transformation Functional Platform, grant no. KEH2310024; Ji Hua Laboratory, grant no. X190021TB190 and no.X190021TB193; Shanghai Municipal Science and Technology Major Project, grant no. 2021SHZDZX0103 and no. 2018SHZDZX01; ZJ Lab, and Shanghai Center for Brain Science and Brain-Inspired Technology.

References

1. He, B., Yuan, H., Meng, J., Gao, S.: Brain–computer interfaces. In: He, B. (ed.) Neural Engineering, pp. 131–183. Springer International Publishing, Cham (2020). https://doi.org/10.1007/978-3-030-43395-6_4
2. Pfurtscheller, G., Neuper, C.: Motor imagery and direct brain-computer communication. Proc. IEEE **89**, 1123–1134 (2001). https://doi.org/10.1109/5.939829
3. Ding, Y., Robinson, N., Tong, C., Zeng, Q., Guan, C.: LGGNet: learning from local-global-graph representations for brain–computer interface. IEEE Transactions on Neural Networks and Learning Systems, pp. 1–14 (2023). https://doi.org/10.1109/TNNLS.2023.3236635
4. O'Shea, K., Nash, R.: An Introduction to Convolutional Neural Networks. http://arxiv.org/abs/1511.08458. (2015)
5. Lawhern, V.J., Solon, A.J., Waytowich, N.R., Gordon, S.M., Hung, C.P., Lance, B.J.: EEGNet: a compact convolutional neural network for EEG-based brain–computer interfaces. J. Neural Eng. **15**, 056013 (2018). https://doi.org/10.1088/1741-2552/aace8c
6. Riyad, M., Khalil, M., Adib, A.: MI-EEGNET: a novel convolutional neural network for motor imagery classification. J. Neurosci. Methods **353**, 109037 (2021). https://doi.org/10.1016/j.jneumeth.2020.109037
7. Wang, H., Xu, L., Bezerianos, A., Chen, C., Zhang, Z.: Linking attention-based multiscale CNN With dynamical GCN for driving fatigue detection. IEEE Trans. Instrum. Meas. **70**, 1–11 (2021). https://doi.org/10.1109/TIM.2020.3047502
8. Song, Y., Zheng, Q., Liu, B., Gao, X.: EEG Conformer: convolutional transformer for EEG decoding and visualization. IEEE Trans. Neural Syst. Rehabil. Eng. **31**, 710–719 (2023). https://doi.org/10.1109/TNSRE.2022.3230250
9. Xu, W., Wang, J., Jia, Z., Hong, Z., Li, Y., Lin, Y.: Multi-Level spatial-temporal adaptation network for motor imagery classification. In: ICASSP 2022 - 2022 IEEE International Conference on Acoustics, Speech and Signal Processing (ICASSP), pp. 1251–1255 (2022). https://doi.org/10.1109/ICASSP43922.2022.9746123
10. Hou, Y., et al.: GCNs-Net: a graph convolutional neural network approach for decoding time-resolved EEG motor imagery signals. IEEE Transactions on Neural Networks and Learning Systems, pp. 1–12 (2022). https://doi.org/10.1109/TNNLS.2022.3202569
11. Jia, Z., et al.: GraphSleepNet: Adaptive Spatial-Temporal Graph Convolutional Networks for Sleep Stage Classification (2020). https://doi.org/10.24963/ijcai.2020/184

12. Zhong, P., Wang, D., Miao, C.: EEG-based emotion recognition using regularized graph neural networks. IEEE Trans. Affect. Comput. **13**, 1290–1301 (2022). https://doi.org/10.1109/TAFFC.2020.2994159
13. Sun, B., Liu, Z., Wu, Z., Mu, C., Li, T.: Graph Convolution Neural Network based End-to-end Channel Selection and Classification for Motor Imagery Brain-computer Interfaces. IEEE Transactions on Industrial Informatics, pp. 1–10 (2022). https://doi.org/10.1109/TII.2022.3227736
14. Vaswani, A., et al.: Attention is All you Need. In: Advances in Neural Information Processing Systems. Curran Associates, Inc. (2017)
15. Musallam, Y.K., et al.: Electroencephalography-based motor imagery classification using temporal convolutional network fusion. Biomed. Signal Process. Control **69**, 102826 (2021). https://doi.org/10.1016/j.bspc.2021.102826
16. Tangermann, M., et al.: Review of the BCI Competition IV. Front. Neurosci. **6** (2012). https://doi.org/10.3389/fnins.2012.00055
17. Schirrmeister, R.T., et al.: Deep learning with convolutional neural networks for EEG decoding and visualization. Hum. Brain Mapp. **38**, 5391–5420 (2017). https://doi.org/10.1002/hbm.23730
18. Li, Y., Guo, L., Liu, Y., Liu, J., Meng, F.: A temporal-spectral-based squeeze-and- excitation feature fusion network for motor imagery EEG decoding. IEEE Trans. Neural Syst. Rehabil. Eng. **29**, 1534–1545 (2021). https://doi.org/10.1109/TNSRE.2021.3099908
19. Ang, K.K., Chin, Z.Y., Wang, C., Guan, C., Zhang, H.: Filter bank common spatial pattern algorithm on BCI competition IV datasets 2a and 2b. Front. Neurosci. **6** (2012). https://doi.org/10.3389/fnins.2012.00039

How Do Native and Non-native Listeners Differ? Investigation with Dominant Frequency Bands in Auditory Evoked Potential

Yifan Zhou[1], Md Rakibul Hasan[2], Md Mahbub Hasan[3], Ali Zia[1], and Md Zakir Hossain[1,2](✉)

[1] Australian National University, Canberra, ACT 2600, Australia
{Ali.Zia,Zakir.Hossain}@anu.edu.au
[2] Curtin University, Perth, WA 6102, Australia
{Rakibul.Hasan,Zakir.Hossain1}@curtin.edu.au
[3] Khulna University of Engineering & Technology, Khulna 9203, Bangladesh
mahbub01@eee.kuet.ac.bd

Abstract. EEG signal provides valuable insights into cortical responses to specific exogenous stimuli, including auditory and visual stimuli. This study investigates the evoked potential in EEG signals and dominant frequency bands for native and non-native subjects. Songs in different languages are played to subjects using conventional in-ear phones or bone-conducting devices. Time-frequency analysis is performed to characterise induced and evoked responses in the EEG signal, focusing on the phase synchronisation level of the evoked potential as a significant feature. Phase locking value (PLV) and weighted phase lag index (WPLI) are used to assess the phase synchrony between the EEG signal and sound signal, while the frequency-dependent effective gain is analysed to understand its impact. The results demonstrate that native subjects experience higher levels of evoked potential, indicating more complex cognitive neural processes compared to non-native subjects. Dominant frequency windows associated with higher levels of evoked potential are identified using a peak-picking algorithm. Interestingly, the choice of playing device has minimal influence on the evoked potential, suggesting similar outcomes with both in-ear phones and bone-conducting devices. This study provides valuable insights into the neural processing differences between native and non-native subjects and highlights the potential impact of playing devices on the evoked potential.

Keywords: EEG · phase locking value · weighted phase lag index · dominant frequency

Y. Zhou and M. R. Hasan—These authors contributed equally to this work.

B. Luo et al. (Eds.): ICONIP 2023, CCIS 1963, pp. 350–361, 2024.
https://doi.org/10.1007/978-981-99-8138-0_28

1 Introduction

The human brain possesses a remarkable capacity to perceive external stimuli, such as auditory and visual cues. This perceptual process manifests itself through discernible patterns of neural activity, which can be effectively characterised through electroencephalogram (EEG) signals [5]. EEG recordings capture cortical oscillatory activities, which can be classified into evoked, induced and ongoing responses [6,16]. Among these, auditory evoked potentials have emerged as a valuable tool for scrutinising the intricacies of the human auditory system, particularly within the hearing-impaired population [9]. These potentials provide insights into the auditory process by analysing EEG signals in response to auditory stimuli. Auditory evoked potentials can be categorised as early, middle and late latency responses, which are determined by the temporal relationship between the cessation of auditory stimuli and the onset of corresponding brain waves. It has been observed that early and middle latency responses are devoid of cognitive involvement, whereas late responses exhibit engagement in cognitive processes such as discriminating pure tones from complex acoustic signals like speech [1]. In addition to evoked potentials, induced potentials that capture more intricate cognitive processes within the brain can also be discerned through EEG recordings. The utilisation of mathematical methodologies to analyse EEG signals plays a crucial role in exploring neural activities within the human brain in response to external stimuli [17].

The characterisation of evoked and induced responses within EEG signals involves comparing the amplitude at a specific time relative to the baseline amplitude. An evoked response is identified when the amplitude surpasses the baseline level subsequent to the stimulus presentation and consistently exhibits a fixed latency delay relative to the stimulus onset. In contrast, induced responses are distinguished by segments of time following the stimulus that display higher amplitudes compared to the baseline without adhering to a fixed temporal latency. It is crucial to account for different frequency bands as evoked and induced responses exhibit more pronounced manifestations within specific frequency ranges, contingent upon the nature of the stimulus. Thus, time-frequency analysis [14] assumes a critical role as an initial step in identifying and examining evoked and induced responses. This paper leverages time-frequency analysis for subsequent signal processing and to determine the frequency bands associated with evoked responses.

Investigating the differences in neural activity patterns between native and non-native listeners has garnered significant attention in neuroscience and has implications for developing machine-brain interfaces [11,19]. This study analyse EEG signals recorded from subjects while playing various sound signals, along with the corresponding sound signal itself. We employ mathematical analysis methods, including calculating frequency based on phase locking value (PLV) [3], weighted phase lag index (WPLI) [10], and effective gain, to identify the dominant frequency band for evoked potentials in native listeners listening to either native or non-native songs. The PLV quantifies the phase difference variability between two signals at a specific time and frequency, enabling the measurement of synchronicity. It is computed by evaluating the phase angle difference derived from the Short-Time Fourier Transform (STFT) outcomes of the two signals. WPLI assess the asymmetry of the phase difference distribution between the signals. The frequency resolution of the STFT is contingent upon the length of the sampling window, necessitating appropriate adjustments to attain the desired resolution. Consequently, a *downsampling factor* is computed based on the desired and modified frequency resolutions, ensuring that the modified window length for both EEG and sound data is an integer value. An algorithm is proposed to determine the optimal *downsampling factor.*

2 Materials and Methods

2.1 Sound and EEG Data

We utilised a publicly available dataset [2], sourced from PhysioNet [7]. The dataset consists of more than 240 two-minute EEG recordings obtained from 20 subjects. The recordings were conducted using an OpenBCI Ganglion Board, where four channels (T7, F8, Cz and P4) were sampled at a rate of 200 Hz.

The EEG signals were recorded in two different conditions: resting state and auditory stimuli. During the resting state, the EEG signals were captured with both eyes open and eyes closed. For auditory stimuli, the study consisted of two types of auditory devices (in-ear phones and bone-conducting devices) and three categories of songs (native, non-native and neutral). Accordingly, EEG signals were recorded with six experimental conditions involving the combinations of these devices and songs. For the non-native song specification, Italian subjects listened to Arabic songs, while non-Italian subjects listened to Italian songs.

2.2 Data Processing and Analysis

The data processing for this study involves several steps conducted using Python. In Step 1, STFT is applied to both the EEG signal and the corresponding sound signal for all experimental conditions and trials, resulting in time-frequency distribution maps. In Step 2, the phase is extracted from these maps at each time-frequency point for each condition and trial. Step 3 involves calculating the phase difference at each time-frequency point. Step 4 computes the normalised effective gain, PLV and WPLI using the phase difference and values from the time-frequency distribution. Moving to Step 5, the metrics obtained in Step 4 are averaged over all time points to obtain the frequency-dependent effective gain, PLV and WPLI for each experimental condition. In Step 6, these frequency-dependent metrics are averaged over all trials for each condition. Finally, Step 7 identifies the dominant frequency bands with higher effective gain and extracts the peak values of PLV and WPLI at these bands for each experimental condition. Through these steps, the study enables the analysis and comparison of frequency-dependent metrics and the identification of dominant frequency bands in the EEG and sound signals across different experimental conditions.

Algorithm 1. The algorithm for choosing the *downsampling factor*

Input: desired frequency resolution, sampling frequency for EEG signal, sampling frequency for sound signal
Output: downsampling factor
downsampling factor = 1
while True **do**
  sound time window length = $\frac{\text{sound sampling frequency}}{\text{modified frequency resolution}}$
  EEG time window length = $\frac{\text{EEG sampling frequency}}{\text{modified frequency resolution}}$
  if sound and EEG time window length are integers **then**
    break
  else
    downsampling factor = downsampling factor + 1
  end if
end while

Downsampling Factor. The frequency resolution of STFT is determined by the frequency resolution of the Fast Fourier Transform corresponding to each window size. To achieve a desired frequency resolution lower than the acquired resolution, the sample window length is increased to maintain an integer ratio between the desired and modified resolutions. This ratio, referred to as the *downsampling factor*, remains the same for both sound and EEG data. Consequently, the modified frequency resolution and sampling window length should also be integers. The *downsampling factor* is determined through Algorithm 1.

Effective Gain, PLV and WPLI. STFT generates a 2D time-frequency map where each time-frequency point corresponds to a complex value, representing the phase angle [4]. The STFT of a speech signal $x(t)$ and an EEG signal $y(t)$ can be expressed as:

$$X(\tau,\omega) = \sum_{t=0}^{L} x(t)W(t-\tau)e^{-i\frac{\omega t}{L}} \qquad Y(\tau,\omega) = \sum_{t=0}^{L} y(t)W(t-\tau)e^{-i\frac{\omega t}{L}} \quad (1)$$

where τ is the centre of the window, W is the window function, L is the length of the window and ω is the frequency.

The wrapped phase of the sound and EEG can be expressed as:

$$\Delta\phi_x = \tan^{-1}\left[\frac{\text{imag}(X(\tau,\omega))}{\text{real}(X(\tau,\omega))}\right] \qquad \Delta\phi_y = \tan^{-1}\left[\frac{\text{imag}(Y(\tau,\omega))}{\text{real}(Y(\tau,\omega))}\right] \quad (2)$$

Both of these phases are unwrapped to get the continuous phases:

$$\phi_x(\tau) = \sum_{t=0}^{\tau} \Delta\phi_x(t) \qquad \phi_y(\tau) = \sum_{t=0}^{\tau} \Delta\phi_y(t) \quad (3)$$

Now, the effective gain at time τ can be expressed as:

$$|H_{eff}(\tau,\omega)| = \frac{|Y(\tau,\omega)|}{|X(\tau,\omega)|} \quad (4)$$

A higher effective gain value is indicative of a stronger perception of the EEG signal in response to the sound stimulus. The time-average effective gain:

$$|H_{eff}(\omega)| = \frac{1}{N}\sum_{\tau=0}^{N-1}|H_{eff}(\tau,\omega)| \tag{5}$$

where N is the total number of samples.

PLV calculation requires the phase angle of the same frequency component from two different signals at different times:

$$\text{PLV}_{xy}(\omega) = \frac{1}{N}\sum_{\tau=0}^{N-1}\exp(i\theta(\tau,\omega)) \tag{6}$$

where $\theta(\tau,\omega) = \phi_x(\tau,\omega) - \phi_y(\tau,\omega)$

PLV serves as a valuable metric for assessing the inter-trial variability of the phase difference between two signals at a specific time and frequency [4,12]. By calculating the average over multiple trials, the PLV provides insights into the degree of phase synchronisation between the signals. The PLV ranges between 0 and 1. A PLV of 1 signifies a constant phase difference consistently present at $\frac{\pi}{2}$ or $\frac{3\pi}{2}$ radians. Conversely, a PLV of 0 indicates that the phase difference is uniformly distributed across the range of $[0,\ 2\pi]$ radians, suggesting a lack of synchronisation between the signals.

WPLI shares a similar purpose to the PLV, as it evaluates the distribution patterns of phase differences between $x(t)$ and $y(t)$ at a specific time and frequency. Its primary focus lies in quantifying the asymmetry within the distribution of phase differences, providing valuable insights into the synchronisation characteristics of the signals. Assuming $S(\omega) = X(\tau,\omega)Y^*(\tau,\omega)$, where $*$ represent complex-conjugate, WPLI can be expressed as:

$$\text{WPLI}(\omega) = \frac{|\mathbb{E}\{|S(\omega)|\,\text{sgn}(S(\omega))\}|}{\mathbb{E}\{|S(\omega)|\}} \tag{7}$$

where $\mathbb{E}$ is the expectation operator and sgn is the sign operator. When the phase differences of all trials share the same sign, the WPLI value becomes 1. As a result, a WPLI value closer to 0 indicates a more symmetric distribution of phase differences.

The evoked potential of EEG signals is characterised by both time-locked and phase-locked responses to external stimuli. In general, higher values of WPLI and PLV indicate a more concentrated distribution of phase differences, reflecting increased phase synchronisation at specific time-frequency points. However, solely relying on elevated PLV and WPLI values at a particular time-frequency point is insufficient for identifying the presence of evoked potentials. It is also crucial to consider the corresponding effective gain value. Generally, evoked potentials are more likely to be located at time-frequency points with relatively higher values of all three metrics: effective gain, PLV and WPLI [12,18].

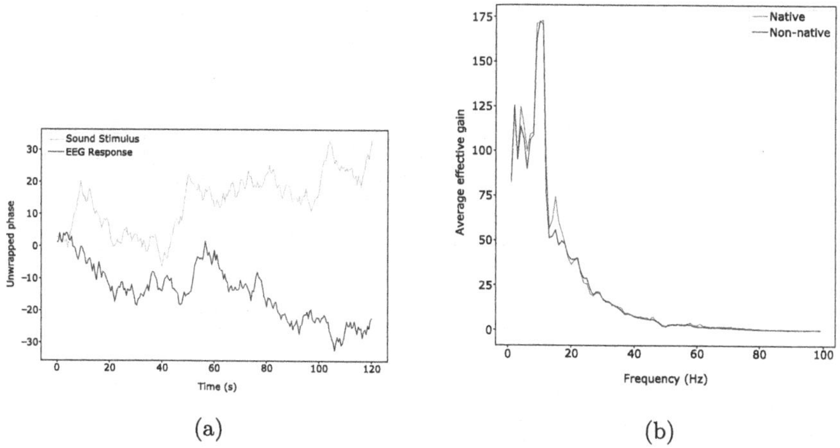

Fig. 1. (a) Phase waveform of a typical sound and an EEG signal at 20 Hz and (b) Normalised effective gain for native and non-native cases using conventional in-ear phones (bone-conducting devices also result in a very similar trend).

3 Results and Discussion

3.1 Unwrapped Phase Waveform

The phase of both EEG and sound signals is obtained by extracting the phase values at each time-frequency point from the STFT results of respective signals. Figure 1a illustrates the phase values at a frequency of 20 Hz for both the EEG and sound signals over a time span of 0 to 120 s. The EEG signal consistently exhibits lower phase values compared to the sound signal. Notably, the phase difference between the two signals demonstrates an increasing trend, starting from approximately 0 at 0 s and reaching approximately 60 at 120 s.

3.2 Frequency-Dependent Normalised Effective Gain

The time-frequency distribution involves applying STFT in each time window to obtain the time-specified spectrum. The sampling rate of the EEG signal is 200 Hz. So, according to the Nyquist-Shannon sampling theorem, a maximum frequency of 100 Hz is acquired in the time-frequency distribution. In the case of the sound signal, a higher sampling frequency (>10,000 Hz) results from a wider frequency range in the extracted spectrum. However, to align with the frequency range of interest in the corresponding EEG signal, only components within 0 to 100 Hz are considered from the sound signal's time-frequency distribution.

The frequency-dependent normalised effective gain for native and non-native cases using conventional in-ear phones is depicted in Fig. 1b. The trends are similar for both native and non-native cases. It is evident that the frequency range from 0 Hz to 40 Hz has relatively higher values compared to other frequencies.

Table 1. Dominant frequency bands for auditory evoked potential characterised by phase locking value (PLV). Here, $H(j\omega)$ is the normalised effective gain.

Device	Native			Non-native		
	Freq. (Hz)	$H(j\omega)$	Peak PLV	Freq. (Hz)	$H(j\omega)$	Peak PLV
In-ear	8–10	**171.608**	0.066	6–8	109.982	0.071
	17–19	44.540	0.063	8–10	**171.608**	0.068
	21–23	36.004	**0.070**	10–12	132.346	0.064
	28–30	18.892	0.070	15–17	57.704	**0.080**
	30–32	15.941	0.067	17–19	44.540	0.068
				19–21	37.749	0.070
				30–32	15.941	0.067
				33–35	13.225	0.072
				38–40	8.281	0.071
Bone-conducting	1–3	109.952	0.068	2–4	110.126	0.078
	7–9	140.798	0.076	6–8	109.982	0.066
	9–11	**172.493**	**0.076**	9–11	**172.493**	0.061
	11–13	74.092	0.066	12–14	59.001	0.072
	15–17	57.704	0.068	16–18	51.726	0.072
	25–27	19.576	0.063	19–21	37.749	**0.080**
	27–29	20.429	0.063	22–24	29.038	0.066
	32–34	13.632	0.064	24–26	23.050	0.068
				33–35	13.225	0.060
				36–38	9.413	0.064

As our focus is on the dominant frequency bands that normally exhibit higher levels of evoked potential, the frequency range of 0 Hz to 40 Hz is selected for subsequent analysis.

3.3 Dominant Frequency Characterised by Phase Locking Value

Table 1 summarises dominant frequency bands and corresponding normalised effective gain and phase locking value (PLV) peaks for both native and non-native experiments.

Native Listener. There are five PLV peaks observed in the in-ear experiment, while the bone-conducting experiment exhibits eight peaks in the frequency range of 0 Hz to 40 Hz (Table 1). The maximum value of average normalised effective gain (171.608) for the in-ear experiment is obtained in the frequency band of 8 Hz to 10 Hz, and the corresponding peak PLV is 0.066. The highest peak PLV value (0.070) is observed in the frequency band of 21 Hz to 23 Hz, with a corresponding average normalised effective gain of 36.004. For the bone-conducting experiment, the maximum value of average normalised effective gain (172.493) is observed in the frequency band of 9 Hz to 11 Hz. The corresponding peak PLV, which is also the highest value, is 0.076.

A higher value of PLV implies a lower level of phase difference between the EEG and sound signals. Additionally, a higher value of effective gain represents a higher ratio of EEG amplitude to sound amplitude, indicating a stronger

EEG response at frequencies where evoked potentials occur. Compared to those who used conventional in-ear phones, the bone-conducting group shows a higher number of peaking PLV values, larger peak PLV values and larger effective gain corresponding to the highest peak PLV. This suggests that native subjects using bone-conducting devices experience a relatively higher level of evoked potential compared to those using conventional in-ear phones.

Non-native Listener. In the case of non-native experiments with conventional in-ear phones, the average normalised effective gain is highest (171.608) in the frequency band of 8 Hz to 10 Hz, with a corresponding peak PLV of 0.068 (Table 1). The highest peak PLV value (0.080) occurs at 15 Hz to 17 Hz, with an average normalised effective gain of 57.704.

In the case of bone-conducting devices, the maximum average normalised effective gain is obtained in the frequency band of 9 Hz to 11 Hz, with a corresponding peak PLV of 0.061, which is slightly lower than that observed in conventional in-ear phones. The highest peak PLV value (0.080) is achieved at 19 Hz to 21 Hz, with an average normalised effective gain of 37.749. Comparing the data with in-ear phones, we observe an additional peaking PLV. However, the largest peaking PLV is very similar to that observed in conventional in-ear phones.

Comparison Between Native and Non-native Listeners. Compared to native subjects using conventional in-ear phones, the non-native subjects exhibit higher peaking PLV, higher maximum peaking PLV and corresponding effective gain values. Additionally, the peaking value corresponding to the highest normalised effective gain for non-native subjects using conventional in-ear phones is higher than that observed in native subjects.

The non-native subjects using either conventional in-ear phones or bone-conducting devices have higher peaking PLV values across most sub-frequency bands compared to native subjects, especially where the effective gain reaches its maximum. This suggests that non-native subjects experience relatively higher evoked potentials (characterised by PLV) within the dominant frequency range. Since the bone-conducting devices exhibit a very similar level of evoked potential to conventional devices for both native and non-native experiments, it can be said that the choice of different playing devices has minimal impact on the evoked potential. With the higher peaking PLV value, it can be inferred that subjects tend to focus on a broader range of frequencies when listening to non-native songs compared to native subjects, regardless of the playing device used.

3.4 Dominant Frequency Characterised by Weighted Phase Lag Index

Table 2 summarises the dominant frequency bands characterised by weighted phase lag index (WPLI), with corresponding normalised effective gain and WPLI peaks for both native and non-native experiments.

Table 2. Dominant frequency bands for auditory evoked potential characterised by weighted phase lag index (WPLI). Here, $H(j\omega)$ is the normalised effective gain.

Device	Native			Non-native		
	Freq. (Hz)	$H(j\omega)$	Peak WPLI	Freq. (Hz)	$H(j\omega)$	Peak WPLI
In-ear	2–4	110.126	0.121	2–4	110.126	0.106
	4–6	107.132	0.113	5–7	104.708	0.107
	7–9	**140.798**	0.093	7–9	140.798	0.109
	10–12	132.346	0.124	9–11	**172.493**	0.109
	14–16	67.360	0.116	12–14	59.001	0.096
	16–18	51.726	0.103	15–17	57.704	0.151
	19–21	37.749	0.095	19–21	37.749	0.112
	22–24	29.038	0.125	21–23	36.004	0.109
	24–26	23.050	0.135	23–25	25.956	0.121
	28–30	18.892	0.135	25–27	19.576	0.111
	30–32	15.941	0.097	31–33	14.715	0.130
	33–35	13.225	0.133	33–35	13.225	**0.158**
	36–38	9.413	**0.140**	37–39	9.055	0.134
Bone-conducting	1–3	109.952	0.130	5–7	104.708	0.092
	7–9	**140.798**	**0.159**	7–9	140.798	0.105
	12–14	59.001	0.088	9–11	**172.493**	**0.143**
	15–17	57.704	0.102	13–15	67.903	0.106
	18–20	38.607	0.102	15–17	57.704	0.086
	20–22	39.490	0.100	19–21	37.749	0.098
	22–24	29.038	0.132	21–23	36.004	0.101
	25–27	19.576	0.104	27–29	20.429	0.114
	28–30	18.892	0.094	31–33	14.715	**0.143**
	33–35	13.225	0.135	34–36	12.558	0.101
	36–38	9.413	0.124	37–39	9.055	0.119

Native Listener. In the conventional in-ear experiment, a total of 13 WPLI peaks are observed, and the largest peaking value (0.140) occurs in the sub-frequency band of 36 Hz to 38 Hz, with a corresponding average normalised effective gain of 9.413 (Table 2). Conversely, the lowest peaking value of the WPLI (0.093) is observed in the sub-frequency band of 7 Hz to 9 Hz, with a corresponding average normalised effective gain of 140.798 (the highest value observed). In the case of bone-conducting devices, a total of 11 WPLI peaks are observed, and the largest peaking value (0.159) occurs in the sub-frequency band of 7 Hz to 9 Hz, with a corresponding average normalised effective gain of 140.798 (the highest value observed).

Compared with conventional in-ear phones, the bone-conducting devices exhibit a higher value for the largest peaking WPLI. The WPLI at the frequency bands where the largest normalised effective gain is obtained is also higher than the value observed in conventional in-ear phones. Additionally, there are two fewer peaking WPLIs observed in the dominant frequency range. The higher value of the peaking WPLI and the corresponding average normalised effective gain indicate that the evoked potential (characterised by WPLI) is at a higher level for bone-conducting devices compared to conventional in-ear phones.

Non-native Listener. In conventional in-ear phones, a total of 13 peaking values in the WPLI are observed (Table 2). The highest peaking WPLI (0.158) is obtained in the sub-frequency band of 33 Hz to 35 Hz, and the second-highest peaking WPLI (0.151) is observed in the sub-frequency band of 15 Hz to 17 Hz, with a corresponding average normalised effective gain of 57.704. In the case of bone-conducting devices, a total of 11 peaking WPLI values are observed. The maximum peaking WPLI (0.143) is obtained in the sub-frequency band of 7 Hz to 9 Hz, with a corresponding average normalised effective gain of 172.493 (also the largest value of average normalised effective gain).

Comparison Between Native and Non-native Listeners. For subjects using conventional in-ear phones, both the top two peaking values of the WPLI for native cases are higher than non-native cases. However, at other sub-frequency bands within the dominant frequency range, higher values of peaking WPLI are observed for native cases. These findings indicate that the number of sub-frequency bands where non-native subjects experience a higher level of evoked potential, characterised by the WPLI, compared to native subjects is reduced compared to the case of evoked potential characterised by the PLV. This discrepancy arises due to the different mathematical definitions between them.

According to the mathematical definition, a higher PLV implies a lower phase difference between two signals averaged over the trials. However, such a lower phase difference does not necessarily result in a higher value of the WPLI. The WPLI incorporates the absolute value and sign of the imaginary part of the cross-spectrum between two signals. Therefore, a larger WPLI indicates that the phase difference tends to be densely distributed either in the range from 0 to π or from π to 2π, or the phase difference across all trials tends to have the same sign. Consequently, the phase difference for non-native subjects tends to be distributed around 0 for all trials, whereas the phase difference for native subjects tends to have the same sign but not as close to 0.

These findings indicate that native subjects using in-ear phones experience a lower level of evoked potential with a higher latency value, while non-native subjects using in-ear phones experience a higher level of evoked potential with a lower latency value. A larger latency value of the evoked potential often reflects a more complex cognitive brain process following external stimuli [13]. Hence, the acquired data is consistent with the hypothesis that native subjects experience a more complex cognitive brain process than non-native subjects.

The maximum peak WPLI for non-native subjects using bone-conducting devices is lower than that of native subjects (Table 2). Only at the sub-frequency bands from approximately 8 to 15 Hz and from 28 to 33 Hz, the peaking WPLI is higher for non-native cases. This comparison is similar to the case of conventional in-ear phones. Thus, the choice of the playing device has minimal impact on the evoked potential level experienced by native and non-native subjects.

The native subjects experience a higher evoked potential characterised by the WPLI at more frequencies from 0 Hz to 40 Hz. However, as discussed earlier, non-native subjects experience a higher evoked potential characterised by the PLV than native subjects. This supports the hypothesis that native subjects undergo more complex cognitive brain processes, as reflected by the evoked potential with a larger latency. In contrast, non-native subjects experience more intense evoked stimuli, indicated by the presence of the evoked potential with a lower latency.

4 Conclusion

This study analyses EEG and sound signals from participants using in-ear phones or bone-conducting devices while listening to both native and non-native songs. The data was subjected to time-frequency analysis to extract phase information and calculate normalised effective gain, phase locking value (PLV) and weighted phase lag index (WPLI). The results indicate that the frequency range of 0 Hz to 40 Hz exhibits a dominant band for higher normalised effective gain. A comparison of PLV and WPLI values reveals that non-native subjects generally exhibit higher PLV but lower WPLI across most sub-frequencies. This implies that non-native subjects experience a higher level of evoked potential with a lower phase difference, whereas native subjects display a lower evoked potential level with a higher phase difference. The outcome is useful for detecting music genres [15] or speech recognition [8] considering native and non-native listeners. Our findings also support to the hypothesis that native subjects engage in more complex cognitive processes. Furthermore, the choice of playing device was found to have a limited impact on the evoked potential level. Future research endeavours could expand the sample size by including a larger number of participants and investigating induced potentials to identify the dominant frequency bands associated with high induced potential levels across different experimental conditions.

References

1. Alain, C., Roye, A., Arnott, S.R.: Chapter 9 - middle- and long-latency auditory evoked potentials: what are they telling us on central auditory disorders? In: Celesia, G.G. (ed.) Disorders of Peripheral and Central Auditory Processing, Handbook of Clinical Neurophysiology, vol. 10, pp. 177–199. Elsevier (2013). https://doi.org/10.1016/B978-0-7020-5310-8.00009-0
2. Alzahab, N.A., et al.: Auditory evoked potential EEG-biometric dataset (2021). https://doi.org/10.13026/ps31-fc50. version 1.0.0
3. Aydore, S., Pantazis, D., Leahy, R.M.: A note on the phase locking value and its properties. Neuroimage **74**, 231–244 (2013). https://doi.org/10.1016/j.neuroimage.2013.02.008
4. Cohen, M.X.: Analyzing Neural Time Series Data: Theory and Practice. MIT Press, Cambridge (2014)
5. Dehaene-Lambertz, G.: Electrophysiological correlates of categorical phoneme perception in adults. NeuroReport **8**(4), 919–924 (1997). https://doi.org/10.1097/00001756-199703030-00021

6. Galambos, R.: A comparison of certain gamma band (40-hz) brain rhythms in cat and man. In: Başar, E., Bullock, T.H. (eds.) Induced Rhythms in the Brain, pp. 201–216. Birkhäuser Boston, Boston (1992). https://doi.org/10.1007/978-1-4757-1281-0_11
7. Goldberger, A.L., et al.: PhysioBank, PhysioToolkit, and PhysioNet: components of a new research resource for complex physiologic signals. Circulation **101**(23), e215–e220 (2000). https://doi.org/10.1161/01.CIR.101.23.e215
8. Hasan, M.R., Hasan, M.M., Hossain, M.Z.: Effect of vocal tract dynamics on neural network-based speech recognition: a Bengali language-based study. Expert. Syst. **39**(9), e13045 (2022). https://doi.org/10.1111/exsy.13045
9. Ibrahim, I.A., Ting, H.N., Moghavvemi, M.: The effects of audio stimuli on auditory-evoked potential in normal hearing Malay adults. Int. J. Health Sci. **12**(5), 25 (2018)
10. Imperatori, L.S., et al.: EEG functional connectivity metrics wPLI and wSMI account for distinct types of brain functional interactions. Sci. Rep. **9**(1), 8894 (2019). https://doi.org/10.1038/s41598-019-45289-7
11. Jagiello, R., Pomper, U., Yoneya, M., Zhao, S., Chait, M.: Rapid brain responses to familiar vs. unfamiliar music-an EEG and pupillometry study. Sci. Rep. 9(1), 15570 (2019). https://doi.org/10.1038/s41598-019-51759-9
12. Lachaux, J., Rodriguez, E., Martinerie, J., Varela, F.J.: Measuring phase synchrony in brain signals. Hum. Brain Mapp. **8**(4), 194–208 (1999). https://doi.org/10.1002/(SICI)1097-0193(1999)8:4$⟨$194::AID-HBM4$⟩$3.0.CO;2-C
13. Michalopoulos, K., Iordanidou, V., Giannakakis, G.A., Nikita, K.S., Zervakis, M.: Characterization of evoked and induced activity in EEG and assessment of intertrial variability. In: 2011 10th International Workshop on Biomedical Engineering, pp. 1–4. IEEE (2011). https://doi.org/10.1109/IWBE.2011.6079037
14. Morales, S., Bowers, M.E.: Time-frequency analysis methods and their application in developmental EEG data. Dev. Cogn. Neurosci. **54**, 101067 (2022). https://doi.org/10.1016/j.dcn.2022.101067
15. Rahman, J.S., Gedeon, T., Caldwell, S., Jones, R., Hossain, M.Z., Zhu, X.: Melodious micro-frissons: detecting music genres from skin response. In: 2019 International Joint Conference on Neural Networks (IJCNN), pp. 1–8. IEEE (2019). https://doi.org/10.1109/IJCNN.2019.8852318
16. Tallon-Baudry, C., Bertrand, O.: Oscillatory gamma activity in humans and its role in object representation. Trends Cogn. Sci. **3**(4), 151–162 (1999). https://doi.org/10.1016/S1364-6613(99)01299-1
17. Vialatte, F.B., Dauwels, J., Musha, T., Cichocki, A.: Audio representations of multi-channel EEG: a new tool for diagnosis of brain disorders. Am. J. Neurodegener. Dis. **1**(3), 292–304 (2012)
18. Vinck, M., Oostenveld, R., Van Wingerden, M., Battaglia, F., Pennartz, C.M.: An improved index of phase-synchronization for electrophysiological data in the presence of volume-conduction, noise and sample-size bias. Neuroimage **55**(4), 1548–1565 (2011). https://doi.org/10.1016/j.neuroimage.2011.01.055
19. Wagner, M., Ortiz-Mantilla, S., Rusiniak, M., Benasich, A.A., Shafer, V.L., Steinschneider, M.: Acoustic-level and language-specific processing of native and non-native phonological sequence onsets in the low gamma and theta-frequency bands. Sci. Rep. **12**(1), 314 (2022). https://doi.org/10.1038/s41598-021-03611-2

A Stealth Security Hardening Method Based on SSD Firmware Function Extension

Xiao Yu[1], Zhao Li[1], Xu Qiao[2], Yuan Tan[2], Yuanzhang Li[2], and Li Zhang[3](✉)

[1] School of Computer Science and Technology, Shandong University of Technology, Zibo, Shandong, China

[2] School of Computer Science and Technology, Beijing Institute of Technology, Beijing, China

[3] Department of Media Engineering, Communication University of Zhejiang, Hangzhou, China

nythhsg@163.com

Abstract. In recent years, issues related to information security have received increasing attention. Expanding the security-related functionality of SSD firmware can provide an additional method for implementing security features in the host system while taking advantage of the excellent performance of the SSD controller. This paper proposes a stealth security hardening method based on SSD Firmware Function Extension. By reverse engineering the firmware program and inserting jump instructions at specific locations, the firmware program can jump to and execute the extension program inserted into the original unused space of the firmware. This can be done without affecting the normal use of the SSD, realizing the functional expansion of the firmware, which mainly includes executing remote code sent by the host, invoking timers, direct read and write flash memory, and self-destruction under specific circumstances. The availability of extended functions and the change in read and write performance after the expansion were experimentally tested.

Keywords: Functional extension · SSD firmware · Security hardening

1 Introduction

At present, flash-based Solid State Drives (SSDs) have become the mainstream choice in the personal consumer market, and they are increasingly favored by enterprise users. Cloud service providers such as Tencent, Alibaba, and Amazon have also begun to choose SSDs for building storage servers in order to address the performance challenges in current network applications.

A typical SSD usually consists of three main components: a host interface, a controller chip, and flash memory units. The controller chip generally contains a general-purpose processor for running firmware programs, SRAM for storing firmware programs and related data, and controllers for managing different interfaces mapped to memory addresses. Some high-performance SSDs are also equipped with DRAM chips for buffering read and write data to improve the overall read and write performance of the hard drive. Since the controller chip contains a complete computer system, it is possible to

B. Luo et al. (Eds.): ICONIP 2023, CCIS 1963, pp. 362–375, 2024.
https://doi.org/10.1007/978-981-99-8138-0_29

modify the firmware so that the controller chip executes a program pre-written into the firmware when detecting specific commands, thereby implementing more complex functions or assisting the host in computation.

This paper mainly focuses on the functional extension of SSD firmware. Through reverse engineering analysis of SSD firmware, we modified the Flash Translation Layer (FTL) of the SSD firmware so that when the SSD receives read and write commands for specific logical addresses, it executes the pre-set program in the firmware, thereby achieving functional expansion of the SSD firmware. The specific research content and contributions of this paper can be summarized as follows:

(1) Reverse engineered the firmware of a specific SSD model, analyzed its primary operating processes, and identified the program used for processing read and write commands, laying the foundation for further modifications in subsequent research;
(2) Expanded the functionality of the firmware for the specified SSD model, allowing it to execute pre-set or remotely sent code from the host (referred to as remote code hereafter) when receiving read and write commands for specific Logical Block Addresses (LBAs), thus completing specific tasks and assisting the host in computation;
(3) Implemented an SSD self-destruction feature, where if an attacker attempts to read the specific logical address mentioned earlier using software like WinHex, the SSD firmware will permanently disable the extended functionality, preventing further debugging by the attacker and protecting the pre-set program within the firmware.

2 Related Work

The Buffer Management Layer (BML) is one of the essential components in SSD firmware. Its primary responsibility is to store frequently accessed data in the buffer, thus reducing the direct read and write times of flash memory and improving the read-write performance and service life of SSD. Yang et al. proposed a BML algorithm called Clean-First and Dirty-Redundant-Write (CFDRW), which aims to improve the performance of SSD-based database applications [7]. Wang et al. presented a method for evaluating hot data using particle swarm optimization, and the data replacement strategy developed through this approach significantly outperforms the traditional LRU method in terms of performance [8]. Bao proposed a hot data identification algorithm based on Bloom Filters and temporal locality, which utilizes Bloom Filters for preliminary identification of hot data and utilizes LRU algorithm for data replacement. This results in a noticeable improvement in performance when compared to traditional BML algorithms [9].

The Flash Translation Layer (FTL) is another critical component of SSD firmware. Due to the characteristic of SSDs that they cannot be updated in place, traditional file systems designed for mechanical hard drives cannot be directly applied to SSDs. To maintain compatibility with existing file systems, the FTL in SSD firmware hides the differences between SSDs and mechanical hard drives at a higher level, while establishing a mapping relationship between logical addresses and physical addresses at a lower level, converting the read and write methods initially designed for mechanical hard drives into those suitable for SSDs. Additionally, considering the lifespan issue of flash memory,

FTL also needs to be responsible for wear leveling and garbage collection of flash-memory. Xie et al. proposed an Adaptive Separation-Aware FTL (ASA-FTL) algorithm, which employs a data clustering approach to distinguish between cold and hot data. The FTL efficiency can be improved [10]. Zhou et al. proposed the Correlation-Aware Page-level FTL (CPFTL) method, which optimizes the FTL algorithm by leveraging the semantic correlation in I/O operations, significantly reducing SSD response time [11]. Tang proposed a Superblock-based FTL that performs block-level mapping for ordinary blocks and page-level mapping within superblocks, balancing the time and space overhead of mapping [12].

In addition to research on SSD firmware performance, the current mainstream research on embedded device firmware mainly focuses on detecting vulnerabilities in firmware and securely updating embedded device firmware. In terms of firmware vulnerability detection, FIRM-AFL proposed by Zheng et al. is a fuzzing method for IoT device firmware, which solves compatibility issues and performance bottlenecks of general emulation methods [14]. Fuzzware, proposed by Scharnowski et al., is also a fuzzing method that generates models to improve the effectiveness of the method by determining how values generated by hardware are used [15]. The FIT method, proposed by Liang et al., uses neural network techniques to analyze firmware semantics to find hidden security vulnerabilities [16].

Regarding firmware secure updates, Falas et al. proposed a firmware secure update method based on hardware primitives and encryption modules [17]. This method can be flexibly adjusted according to the hardware resources of embedded devices. The RapidPatch method proposed by He et al. is a firmware hot update method that can update firmware while the device is performing other tasks, thereby improving the update efficiency [18].

3 SSD Firmware Function Expansion

3.1 Overall Design of Function Expansion

The SSD used in this paper is the Vertex Plus 128 GB version produced by OCZ. This SSD adopts SATA2 interface, uses Barefoot chip from Indilinx as the main control chip, is equipped with a 64 MB DRAM chip for buffering read and write data, and uses 16 MLC NAND flash chips with a single capacity of 64 GB produced by Micron as storage units.

There are two parts of space available in the firmware. The first part starts from 0x48BC to 0x54AF, with a total space of 3060 bytes, which is used to store the firmware patch program jump table and the main function extension programs. The second part starts from 0x16800 to 0x16FFF, with a total space of 2048 bytes, which is used to store remote code and larger data needed during runtime.

The method of firmware function expansion and the overall structure after expansion are shown in Fig. 1. Firstly, the compiled firmware function expansion patch is placed in the corresponding free position in the firmware. Then, at a specific position in the original firmware code, the original instruction is changed to an instruction to jump to a certain entry in the program jump table-using the program jump table is for convenient management of the addresses of various patch programs without recalculating the address of each program after modifying the patch program. The jump table entry is also a jump instruction, which allows the firmware to jump to the specified patch program for execution. It should be noted that in addition to using the stack to protect the firmware context in the patch program, it is also necessary to add the original instruction covered by the jump instruction to prevent errors caused by missing logic in the original firmware functions.

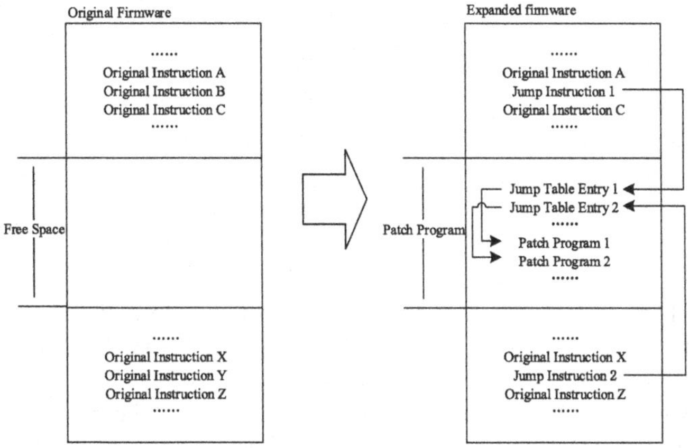

Fig. 1. The method of firmware function expansion

All modifications to the firmware are shown in Table 1. The write command processing flow has been modified so that the firmware can save the written data as remote code to SRAM when detecting a write command to a specific LBA. The read command processing flow has also been modified, including cases where the buffer does not contain the data of the read LBA and cases where it does, allowing the firmware to execute the remote code when detecting a read command to a specific LBA and place the result in the corresponding buffer to return to the host. The SSD initialization process has also been modified to hide some available sectors from the host and initialize timer 3. An interrupt handler for timer 3 is set up in the IRQ interrupt handler program.

3.2 Function Expansion of Write Command Processing Flow

As shown in Table 1, the function expansion for the write command processing flow includes two parts, modifying the addresses at 0x21410 and 0x244A0.

The address 0x244A0 is located in the main function. After the main function retrieves a write command from the read-write command queue, it will execute up to this point. Subsequently, the firmware will put parameters such as the starting LBA and the number of sectors to be written into memory, and in the following main function loop, call the write processing program to write the data in the buffer to the flash memory according to this information. An instruction at 0x244A0 is inserted that jumps to the patch program. The LBA and the number of sectors to be written, used as patch program parameters, have already been saved at the top of the stack by the original firmware.

Table 1. Firmware modification content

Address	Original Instruction	Extended Function
0x55AC	LDR R0, [R0]	Set the interrupt handling program for timer 3
0x7F38	LDR R0, [R6]	Initialize Timer 3 and set the number of readable sectors After power on
0x20C40	ADD R0, R6, #0x43000000	Modify the read command processing program and return the host sending program result
0x21410	MOV R5, #0x00270000	Modify the write command processing program and get the program from host
0x244A0	LDMFD SP, {R0,R1}	Set flag bit when writing specific LBA
0x244AC	CMP R0, #0	update the contents of the buffer when reading data existing in the buffer

When using them, they can be loaded into registers directly from the stack. In the patch program, first, judge the state of the SSD. If the SSD has already been destroyed or the extended function hasn't been enabled, exit the patch program; otherwise, determine whether the written LBA is the same as the preset value. If it is the same, it is considered that the transferred data is the remote code sent by the host, and set a flag to true in memory to inform subsequent patch programs that the transferred data should be transferred to SRAM.

Algorithm 1. Write command processing patch part 2

```
store context
if self-erased or not enabled
   recover context
   return
if writing writeLBA flag set
   copy writeBuffer[bufferID] to &remoteCode
   sectorOffset++
else if writeBuffer[bufferID] == selfEraseMassege
   call selfErase
recover context
return
```

The address 0x21410 is located at the beginning of the write command processing program. We insert an instruction here to jump to the patch program as shown in Algorithm 1. The buffer number, sector offset in the logical page, and other variables used as patch program parameters have already been placed in registers by the original firmware and can be used as needed. In the patch program, similarly, first judge the state of the SSD. Under normal conditions, the patch program will check the flag set earlier. If the flag is found to be set to true, the memcpy function in the firmware is used to copy the remote code in the buffer to SRAM for later execution, and the flag is reset to false. Otherwise, the write data will be checked, and if it contains specific content, the self-destruction program is triggered. After copying the remote code to the specified location, the patch program also modifies the sector offset and other parameters so that the remote code will not be written to the flash memory in the end. This prevents attackers from obtaining the remote code by reading the flash memory content and analyzing the extended firmware.

3.3 Functional Expansion of Read Command Processing Procedure

As shown in Table 1, the functional expansion of the read command processing procedure also includes two parts. We have made modifications at addresses 0x20C40 and 0x244AC, targeting situations where the data of the read LBA does not exist in the buffer and where the LBA data exists in the buffer, respectively.

Algorithm 2. Read command processing patch part 1

```
store context
if self-erased
   recover context
   return
if LBA == enableLBA
   set enable flag
else LBA == executeLBA
   // Save BufferID for executing LBA
   exeLBABufferID = bufferID
   if enabled
      call remoteCode
      copy result to readBuffer[bufferID]
recover context
return
```

The address 0x20C40 is located within the read command processing program. At this point, the firmware has already broken down the read command into a process of reading multiple logical pages. Subsequently, it will calculate the physical address from the logical address, fetch the required data from flash memory, and place it into the buffer. Here, we insert an instruction to jump to the patch program as shown in Algorithm 2. The buffer address and the number of sectors to be read, which serve as patch program parameters, have been placed into registers. Although the LBA has not been directly placed into a register, it can be calculated from the logical page number and sector offset that have been placed into registers. The patch program first checks whether the SSD has executed the self-destruction process; if not, it checks whether the read LBA is equal to the pre-set LBA used to enable other extended functions. If equal, a flag is set to indicate that the current SSD extended function has been enabled, and the program jumps back to the normal read process within the original firmware. If the LBA is not equal, the patch continues to check if the read LBA is equal to another pre-set LBA used for executing remote code and returning execution results. If unequal, the patch program returns to the original firmware execution; if equal, the patch program first records the buffer number of this read command in SRAM, then checks if the extended function has been enabled. The reason for prioritizing the saving of the buffer number will be explained later. When the extended function is confirmed to be enabled, the patch program executes the remote code and copies the contents of the pre-agreed memory space for storing the execution results to the buffer corresponding to this read operation.

Algorithm 3. Read command processing patch part 2

```
store context
if self-erased or not enabled
    recover context
    return
if LBA == executeLBA and cmd == read
    call remoteCode
    //Update the execution results to the corresponding buffer
    bufferID = exeLBABufferID
    copy result to readBuffer[bufferID]
recover context
return
```

Due to the fact that when the data of the read LBA exists in the read buffer, the firmware will directly return the existing data in the buffer to reduce the read latency caused by flash operations. Therefore, only modifying the read command processing program mentioned earlier may result in the same value being always returned even when the result should change due to some random factors during continuous remote code execution. To solve this problem, we have also modified the program at address 0x244AC. This address is located within the main function and runs when the firmware detects that the content of the read LBA exists in the buffer. Here, we insert an instruction to jump to the patch program as shown in Algorithm 3. The parameters of the patch program, such as the read/write command type, starting LBA, and the number of sectors to be read, are saved in the stack by the original firmware and can be accessed and used through memory access instructions. The patch program also first checks the SSD's status, including whether it has self-destructed or whether the extended functions have been enabled. In situations where the SSD has not self-destructed and the extended functions have been enabled, the patch program checks if the command type, LBA, and the number of sectors to be read are equal to the pre-set values for triggering remote code execution.

3.4 Other Function Extensions

(1) Self-destruction Function

There are currently two self-destruction functions in the extended firmware. The first one is triggered when the host writes a specific string to a specific LBA, and the second one is triggered when the host reads from a specific LBA. When self-destructing, the patch program first sets the flag indicating the self-destruct state to true, then transfers this flag to the address 0xED24 in memory. This location is an idle byte in the area used to store various SSD settings. After the SSD is powered on, the firmware will read these settings from the flash memory and save them in the internal memory. Subsequently, the patch program calls the function within the firmware that saves the settings in memory to flash memory, saving the self-destruct flag to flash memory as well. In addition, to copy the self-destruct flag saved in memory after power-on to the specified location, we have also added a patch program at the address 0x7F38 in the firmware. Address 0x7F38 is located in the function used by the firmware to process ATA identify commands, so it will inevitably be executed after the SSD is powered on.

(2) Timer settings

In the patch program mentioned in the part of self-destruction function, we have added settings for timer 3. The value of this timer starts at 0xFFFFFFFF and decreases by 1 in each clock cycle. The timer will generate an interrupt after overflow and will automatically reload the initial timer value and run again after the interrupt is processed. At the same time, we have also set a flag in this patch program to inform subsequent patch programs that the current timer setting is made by the extended patch, so the modified interrupt handling program should be executed when an interrupt is generated. We have added a patch program at address 0x55AC in the firmware to modify the interrupt handling for timer 3. This patch program first increments an integer value used to record the number of timer 3 interrupts by 1, and then checks the flag mentioned earlier - if the current timer is configured by the patch program, it directly returns to the position where the original firmware's interrupt handling program ends. Otherwise, it returns to the original address and continues to execute the original interrupt handling program in the firmware.

3.5 Selection of Specified LBAs for Function Extension

In this study, we hope to not affect the original data storage function of SSDs while extending their functionality. The extensions to read and write functions mentioned earlier, as well as the self-destruction function, require firmware to check read and write commands for specified LBAs to trigger extended features. However, taking the LBA used to obtain remote code execution results in the read command function extension as an example, if this LBA is set in a position that might be used for normal data storage in the SSD, it may cause the operating system to obtain the content that is actually the result of remote code execution when reading hard disk data, resulting in problems with the normal use of SSDs. In addition, in the Windows operating system, if an application directly writes to a specific LBA on a formatted and in-use hard drive, the operating system will block the operation to maintain file system integrity. This also implies that if the LBA used for writing remote code execution results is placed arbitrarily in some position in the SSD, it may lead to the inability to use this function properly.

Therefore, we placed the specified LBAs used in function extensions in the unallocated space of the SSD after formatting. Taking the current SSD with firmware extensions as an example, after initializing the hard drive using the Master Boot Record (MBR) partition structure in the Windows operating system, there are 114175 MB of available space, but only 114173 MB can be formatted and used normally, leaving the remaining

2 MB reserved and unable to be formatted into any partition. The corresponding LBAs of these spaces are from 0x0DEFF000 to 0x0DEFFFFF, of which the last 33 sectors are used by MBR-related content and are also not available as specified LBAs for function extension. Finally, we selected some LBAs between 0x0DEFF600 and 0x0DEFF900 as the specified LBAs for function extension. This approach isolates the space used for normal data storage in the SSD from the space used to activate function extensions, allowing the SSD to be used normally after function extension.

4 Experiments

4.1 Experimental Setup

Since the experiment is conducted in a Windows operating system, the host-side application utilizes some Windows APIs related to device input and output.

The host-side application uses the CreateFile function to open the SSD and obtain its handle for subsequent read and write operations. In this case, "\\.\PhysicalDrive" is used as the value for the lpFileName parameter, which represents the file or physical input/output device name, indicating that the opened file is a physical drive. It should be noted that the physical drive number needs to be added at the end of this file name, which can be viewed in Disk Management within the Windows system. The dwFlagsAndAttributes parameter in the function is also set to FILE_FLAG_NO_BUFFERING to prevent the operating system from buffering some of the written data, resulting in an inability to correctly obtain the results of remote code execution in certain cases. Additionally, after the program has finished running, the CloseHandle function must be used to close the SSD's handle, preventing unnecessary occupation of system resources.

Subsequently, the host-side application uses the DeviceIoControl function to perform read and write operations on the SSD. To accurately control the LBA and the number of sectors for read and write operations, thereby precisely triggering the extended functionality, the control code of this function employs the IOCTL_SCSI_PASS_THROUGH_DIRECT macro definition. With this approach, the function directly sends the SCSI command, which is provided as a parameter, to the SSD for processing. Constructing the SCSI command requires the use of the SCSI_PASS_THROUGH_DIRECT structure, which includes members such as data transfer direction, transfer data length, response timeout duration, and the direct SCSI command CDB. By utilizing this function, this study constructs functions in the host-side application that activate the SSD's extended functionality, transmit remote code, and execute remote code while obtaining runtime results.

It is important to note that copying the remote code to SRAM takes a certain amount of time. If an SCSI command to execute the remote code and return the result is sent immediately after using DeviceIoControl to transfer the remote code to the SSD, it may result in receiving data that is not the actual execution result of the remote code but the genuine data stored in the flash memory instead. Therefore, this study adds a 1-ms delay at the end of the function for transmitting remote code, providing sufficient time for the SSD to copy the remote code to the correct location.

```
D:\src2023-g>host
18 56 3F F3 00 00 00 00  00 00 00 00 00 00 00 00
0E A8 72 C9 00 00 00 00  00 00 00 00 00 00 00 00
E3 27 AC 9F 00 00 00 00  00 00 00 00 00 00 00 00
DC D0 E5 75 00 00 00 00  00 00 00 00 00 00 00 00
55 59 11 4C 00 00 00 00  00 00 00 00 00 00 00 00
89 49 4E 22 00 00 00 00  00 00 00 00 00 00 00 00
BA 8E 86 F8 01 00 00 00  00 00 00 00 00 00 00 00
46 17 B3 CE 01 00 00 00  00 00 00 00 00 00 00 00
```

Fig. 2. The execution results of the timer remote code

4.2 Experimental Results and Analysis

(1) The Execution Results of the Timer Remote Code

We made a simple modification to the host-side application so that after sending the remote code, it sends an execute remote command to the SSD every 8 s, for a total of 8 times, and prints out the first 16 bytes of each execution result. The obtained results are shown in Fig. 2. Each row represents one execution result. The first four bytes of each row represent the little-endian stored timer value, while the following four bytes represent the little-endian stored timer interrupt count. It can be seen that the timer value gradually decreases as expected and triggers an interrupt after overflow.

```
D:\src2023-g>host
Original Data in LBA 0x0DF00040:
        AF 00 00 EB FF 5F BD E8  18 E0 9F E5 1E FF 2F E1
New Data written to LBA 0x0DF00040:
        16 00 00 EA 48 00 00 EA  5C 00 00 EA 84 00 00 EA
Data in LBA 0x0DF00040 now:
        16 00 00 EA 48 00 00 EA  5C 00 00 EA 84 00 00 EA
```

Fig. 3. The execution results of read and write remote code

(2) The Execution Results of Read and Write Remote Code

We also made modifications to the host-side application, enabling it to send the read remote code and execute first, obtaining the original data in the specified LBA; then send the write remote code and execute, writing a portion of the firmware code as designated data into the specified LBA, and retrieving this data; finally, resend the read remote code and execute again, obtaining the current data in the specified LBA. By comparing the three sets of data, we can determine whether the direct read and write programs are usable. The host-side application execution results are shown in Fig. 3. It can be seen that the current data in the specified LBA is different from the original data but identical to the written data, indicating that the extended direct read and write programs are usable.

(3) The Test of Self-destruction Feature

We opened the hard drive using WinHex and read the sector used to trigger firmware self-destruction. Then, we used the host-side program and remote code designed for testing the timer functionality to test the SSD after self-destruction. The results are

```
D:\src2023-g>host
FF FF FF FF FF FF FF FF  FF FF FF FF FF FF FF FF
FF FF FF FF FF FF FF FF  FF FF FF FF FF FF FF FF
FF FF FF FF FF FF FF FF  FF FF FF FF FF FF FF FF
FF FF FF FF FF FF FF FF  FF FF FF FF FF FF FF FF
FF FF FF FF FF FF FF FF  FF FF FF FF FF FF FF FF
FF FF FF FF FF FF FF FF  FF FF FF FF FF FF FF FF
FF FF FF FF FF FF FF FF  FF FF FF FF FF FF FF FF
FF FF FF FF FF FF FF FF  FF FF FF FF FF FF FF FF
```

Fig. 4. Timer remote code execution results after self-destruction

shown in Fig. 4. It can be seen that the returned content is all the original data in the sectors rather than the remote code execution results. After multiple power cycles, the same results were obtained, indicating that the self-destruct feature has persistence. We also triggered another self-destruct condition of the firmware by sending specific content to a specific sector and tested the self-destruct effect in the same manner as described above. The results were consistent with Fig. 4. This demonstrates that both self-destruct methods in the SSD functionality extension are usable.

Table 2. Crystal Disk Mark test results

	SEQ1M Q8T1	SEQ1M Q1T1	RND4K Q32T16	RND4K Q1T1
Read 16MB before expansion	250.83 MB/s	225.42 MB/s	46.80 MB/s	16.89 MB/s
Read 16MB after expansion	251.63 MB/s	213.03 MB/s	46.88 MB/s	16.83 MB/s
Performance comparison	+0.32%	**-5.05%**	+0.17%	-0.35%
Write 16MB before expansion	264.56 MB/s	241.04 MB/s	18.22 MB/s	18.25 MB/s
Write 16MB after expansion	262.38 MB/s	228.43 MB/s	18.31 MB/s	18.33 MB/s
Performance comparison	-0.82%	**-5.23%**	+0.38%	+0.44%
Read 8GB before expansion	253.38 MB/s	230.07 MB/s	51.67 MB/s	16.56 MB/s
Read 8GB after expansion	254.33 MB/s	228.23 MB/s	51.75 MB/s	16.51 MB/s
Performance comparison	+0.37%	-0.80%	+0.15%	-0.30%
Write 8GB before expansion	165.73 MB/s	166.97 MB/s	21.03 MB/s	21.86 MB/s
Write 8GB after expansion	170.51 MB/s	170.26 MB/s	21.51 MB/s	22.33 MB/s
Performance comparison	+2.88%	+1.97%	+2.28%	+2.15%

(4) The Test of SSD Read and Write Performance

Table 2 shows the Crystal Disk Mark test results. The SEQ1M Q8T1 item represents sequential read and write test results for an SSD with a depth of 1024K, 1 thread, and a queue depth of 8 per thread. The SEQ1M Q1T1 item represents sequential read and write test results for an SSD with a depth of 1024K, 1 thread, and a queue depth of 1 per thread. Both items reflect the SSD's sequential read and write performance for large files. The RND4K Q32T16 item represents random read and write test results for an SSD with a depth of 10244K, 16 threads, and a queue depth of 32 per thread. The RND4K Q1T1 item represents random read and write test results for an SSD with a depth of 10244K, 1 thread, and a queue depth of 1 per thread. The number of threads refers to the number of threads simultaneously accessing the hard drive for reading and writing, while the queue depth represents the capacity of a single thread's read and write queue. These two items reflect the SSD's random read and write performance for a large number of small files. From the information in the table, it can be seen that except for the 1-thread 1-queue depth sequential read and write test for 16 MB files, where the extended firmware SSD showed a performance decrease of 5.05% and 5.23%, the performance changes in other projects can be considered within the margin of error.

Table 3. AS SSD Benckmark test results

	Seq	4K	4K-64Thrd	AccTime
Read 1GB before expansion	235.08 MB/s	13.30 MB/s	46.97 MB/s	0.160 ms
Read 1GB after expansion	235.72 MB/s	13.24 MB/s	46.95 MB/s	0.183 ms
Performance comparison	+0.27%	-0.45%	-0.04%	**+14.38%**
Write 1GB before expansion	162.04 MB/s	15.75 MB/s	16.44 MB/s	0.239 ms
Write 1GB after expansion	162.14 MB/s	16.14 MB/s	16.65 MB/s	0.234 ms
Performance comparison	+0.06%	+2.48%	+1.28%	-2.09%
Read 10GB before expansion	234.68 MB/s	13.46 MB/s	47.10 MB/s	0.187 ms
Read 10GB after expansion	234.94 MB/s	13.28 MB/s	47.14 MB/s	0.200 ms
Performance comparison	+0.11%	-1.34%	+0.08%	**+6.95%**
Write 10GB before expansion	155.37 MB/s	16.68 MB/s	16.74 MB/s	2.377 ms
Write 10GB after expansion	158.09 MB/s	17.00 MB/s	17.01 MB/s	2.342 ms
Performance comparison	+1.75%	+1.92%	+1.62%	-1.47%

Table 3 shows the AS SSD Benchmark test results. The Seq item represents the sequential read and write test results of the SSD; the 4K item and the 4K-64Thrd represent the 4K random read and write test results of the SSD under different thread counts; the AccTime item represents the read and write response times of the SSD, where the read response time is tested by randomly reading contents within the full disk address range in 4K bytes, and the write response time is tested by randomly writing content into a specified address range in 512-byte units. From the information in the table, it can be seen that except for the read response times in both experiments having relatively noticeable increases of 14.38% and 6.95%, other performance changes can basically be considered within the margin of error. Moreover, since the read response time itself is already a very small value, such an increase does not have a perceptible impact on the normal use of the SSD.

The slight performance degradation exhibited by the extended SSD may be related to the additional jump instructions and memory access instructions in the firmware. However, compared to the original number of instructions in the firmware, the increase in instruction count is almost negligible when the extension functionality is not triggered.

At the same time, triggering the extension functionality requires reading and writing specific addresses, which is almost impossible during normal use. Therefore, in most cases, there will be no significant performance degradation, which is consistent with our requirement that the functionality extension should not affect the normal use of the SSD.

5 Conclusion

This paper proposes a method for extending the functionality of existing SSD firmware binary files. The implemented SSD firmware extension functions mainly include executing remote code from the host, configuring and using timers, and self-destruct functionality under specific circumstances. By simulating host-side applications, we call the extended functionality of the SSD and verify its usability. The results show that the remote code execution function and timer-related functions can all be used normally. Both types of self-destruct functions under specific conditions can also be triggered normally, causing the extension functions in the firmware to fail and not recover even after restarting the SSD. In addition, commonly used SSD testing software such as Crystal Disk Mark was used to compare and analyze the read and write performance of the SSD before and after the firmware functionality extension. The results show that all of the SSD performance changes are within the permissible margin of error.

Acknowledgments. This study was supported by the National Key Research and Development Program of China (2020YFB1005704).

References

1. IBMArchives, 20th century disk storage chronology. https://www.ibm.com/ibm/history/exhibits/storage/storage_chrono20.html. Accessed 10 Jun 2023
2. Schultz: Research on Key Technologies of Solid State Disk Storage System. National University of Defense Technology, Hunan (2019)
3. Zhou, J.: Research on Firmware Optimization Algorithm in Solid State Disk. Hangzhou University of Electronic Science and Technology, Zhejiang (2019)
4. Luo, Y., Ghose, S., Cai, Y., et al.: HeatWatch: improving 3D NAND flash memory device reliability by exploiting self-recovery and temperature awareness. In: 2018 IEEE International Symposium on High Performance Computer Architecture (HPCA) (2018)
5. Mingbo, Y.: Research on Firmware Optimization of Solid State Disk. Hangzhou University of Electronic Science and Technology, Zhejiang (2020)
6. Vladimirov, S., Pirmagomedov, R., Kirichek, R., et al.: Unique degradation of flash memory as an identifier of ICT device. IEEE Access **7**, 107626–107634 (2019)
7. Yang, C., Jin, P., Yue, L., et al.: Efficient buffer management for tree indexes on solid state drives. Int. J. Parallel Prog. **44**, 5–25 (2016)
8. Wang, Y.L., Kim, K.T., Lee, B., et al.: A novel buffer management scheme based on particle swarm optimization for SSD. J. Supercomput. **74**, 141–159 (2018)
9. Jiecheng, B.: Optimization Design of Solid State Disk Firmware Based on Thermal Data Identification. Hangzhou University of Electronic Science and Technology, Zhejiang (2022)
10. Xie, W., Chen, Y., Roth, P.C.: ASA-FTL: an adaptive separation aware flash translation layer for solid state drives. Parallel Comput. **61**, 3–17 (2017)

11. Zhou, J., Han, D., Wang, J., et al.: A correlation-aware page-level FTL to exploit semantic links in workloads. IEEE Trans. Parallel Distrib. Syst. **30**(4), 723–737 (2018)
12. Tang, H.: Research and Design of Management Algorithm for Super Block Flash Conversion Layer. Zhejiang University, Zhejiang (2020)
13. Zhang, J., Kwon, M., Swift, M., et al.: Scalable parallel flash firmware for many-core architectures. In: Proceedings of the 18th USENIX Conference on File and Storage Technologies, pp. 121–136 (2020)
14. Zheng, Y., Davanian, A., Yin, H., et al.: FIRM-AFL: high-throughput greybox fuzzing of IoT firmware via augmented process emulation. In: 28th USENIX Security Symposium, pp. 1099–1114 (2019)
15. Scharnowski, T., Bars, N., Schloegel, M., et al.: Fuzzware: using precise {MMIO} modeling for effective firmware fuzzing. In: 31st USENIX Security Symposium, pp. 1239–1256 (2022)
16. Liang, H., Xie, Z., Chen, Y., et al.: FIT: inspect vulnerabilities in cross-architecture firmware by deep learning and bipartite matching. Comput. Secur. **99**, 102032 (2020)
17. Falas, S., Konstantinou, C., Michael, M.K.: A modular end-to-end framework for secure firmware updates on embedded systems. ACM J. Emerg. Technol. Comput. Syst. (JETC) **18**(1), 1–19 (2021)
18. He Y., Zou Z., Sun K. et al.: {RapidPatch}: firmware hotpatching for {real-time} embedded devices. In: 31st USENIX Security Symposium, pp. 2225–2242 (2022)

Attention-Based Deep Convolutional Network for Speech Recognition Under Multi-scene Noise Environment

Chuanwu Yang[1], Shuo Ye[1], Zhishu Lin[2], Qinmu Peng[1,4(✉)], Jiamiao Xu[3], Peipei Yuan[1], Yuetian Wang[1], and Xinge You[1,4]

[1] School of Electronic Information and Communications, Huazhong University of Science and Technology, Wuhan 430074, China
[2] Communication and Information systems, Wuhan Research Institute of Posts and Telecommunications, Wuhan 430074, China
[3] Department of Deep Learning, Deeproute Co. Ltd, Shenzhen 518000, China

[4] Shenzhen Research Institute, Huazhong University of Science and Technology, Shenzhen 518000, China
pengqinmu@hust.edu.cn

Abstract. One goal of Automatic Speech Recognition (ASR) is to convert the command of human speech into computer-readable input, but noise interference is an important yet challenging problem. Capturing speech context, deep neural networks have demonstrated to be superior in learning networks for identifying specific command words. Existing deep neural networks generally rely on the two-layer structure, different layer is used to identify the noisy environment and speech respectively, which makes the model large and complex. In addition, their performance generally drops dramatically in unknown noisy environments, which restricts the generalization of the method. In this paper, we propose a novel deep framework, named Adaptive-Attention and Joint Supervision (AJS) to circumvent the above challenge. Specifically, we use the spectrogram as the input. Adaptive attention is employed to refine the features from the noise environment and get rid of the limitation of the noisy scene. Furthermore, a combination of coarse-to-fine losses are adopted to process difficult words step by step. Extensive experiments on four public datasets demonstrate the robustness of our method to various noise environments and its superiority for ASR in terms of accuracy. Codes are available at: https://github.com/zhishulin/bajs.

Keywords: Speech recognition · Multi-scene noise environment · Attention model · Joint supervision · Acoustic model

1 Introduction

Automatic speech recognition (ASR) plays a key role in the area of human-computer interaction. Although some progress has been made, low accuracy

B. Luo et al. (Eds.): ICONIP 2023, CCIS 1963, pp. 376–388, 2024.
https://doi.org/10.1007/978-981-99-8138-0_30

of ASR is still a challenge in the low signal-noise ratio (SNR) environment. Various methods have been used to reduce noise interference [2,6,12]. Feature-based methods that aim to reduce the interference of noise by purification of the characteristics, such as subspace filtering [4,13,19], wavelet transform [9], dictionary learning algorithms [1]. Recently, GAN is used to compensate the multi-path effect on both target speech and interference [15,35]. However, the effect of this approach is obvious only under the condition of long reverberation time. Besides, finding robust features is also a feasible solution, such as GFCC feature [10,26,36]. Compared to the previous MFCC feature [17,31], it can better fit the frequency-selective characteristics of the human ear [18], unfortunately, it only achieves good performance with moderate noise. Although the above method can resist the effects of the noise to a certain extent, it also has an impact on useful signals. Note that, this impact becomes more and more serious with the reduction of SNR, which brings challenges to ASR in a low SNR environment.

To circumvent the difficulty of identification caused by low SNR, model-based methods are proposed to fit the distribution of features. Hidden Markov Model (HMM) is one of the most classic methods [11,22,24]. However, acoustic characteristics have a strong continuity correlation that does not obey the assumptions of HMM. To cope with this, artificial neural networks based models are introduced, such as Tandem Architectures (TA), Phone-based Background Model (PBM), and Universal Attribute Model (UAM) [25,27]. To enhance the generalization ability of the model-based method, multiple models, such as Two-layer model [3,41], were trained to adapt to different noise scenarios. To take full advantage of long-term contextual information, a dual-stream architecture based on GMM and Long Short-Term Memory (LSTM) is proposed to enable the model to learn a long-term time background [5,30], for enhancing the robustness to noise and reverberation. Benefiting from the joint learning which can obtain the characteristics of objects from different aspects, joint training framework for speech separation and recognition [37] concatenates a DNN based speech separation frontend and a DNN based acoustic model to build a larger neural network, combination models [29] are proposed to incorporate SRE into the framework of ASR. Unfortunately, it also brings about the problem of over parameter tuning. As the complexity of ASR technology increases, this process becomes increasingly difficult and time-consuming [20]. Although a mutual-learning sequence-level knowledge distillation with distinct student structures approach is proposed to reduce the model size and computation cost [16], it also decreases the recognition accuracy. Overall speaking, much of the research has examined how to deal with the recognition under a multi-scene noise environment. However, their coverage of noise types is restricted by the training data, and it is difficult to apply to unknown scenarios [1].

This paper aims at solving the problem of insufficient generalization ability of speech recognition model in low SNR environment. To consider the discriminant information of spectrogram, different from previous work, this paper proposes a novel AJS algorithm. Specifically, attention is incorporated into the fundamental framework of ASR. It can be self-adapting to different types of noise, thus enhancing the expansibility and robustness of the model. Besides, AJS design a loss function, which effectively reduces the training complexity. To summarize, our main contributions are as follows:

1. The attention model is used to focus on the difference among spectrograms and reduce the interference of noise. Moreover, compared with the previous noise reduce counterparts, our method can be applied to more scenes.
2. A combination of coarse-to-fine losses are proposed, which simultaneously calculates the cross-entropy loss on category level and center loss on sample level during optimization procedure.
3. Experiments on four benchmark datasets demonstrate that our method can effectively improve the recognition performance in multi-scene noise environment.

The rest of this paper is organized as follows. In Sect. 2, we briefly introduce the attention-based framework for speech recognition and the details of joint supervision. Experimental results and some limitations of this method also be described in Sect. 3. At last, we conclude the paper and mentions future work in Sect. 4.

2 The Proposed Method

In this section, we first detail the general idea of our AJS algorithm and then present relevant discussion and optimization.

2.1 Spectrogram Extraction and Representation

The presence of noise in speech data poses a challenge in maintaining the clarity and features of the data. This requires the simultaneous handling of both noise discrepancy and discriminability. While deep learning has demonstrated its ability in feature extraction, its direct application to speech signal processing has not produced the desired results [33]. Hence, manual feature MFCC [8] is still commonly used in the field of automatic speech recognition.

Suppose the speech signal $x(t)$ is given, the extraction of the spectrogram can be roughly divided into four steps: sampling, framing, windowing, and Fourier transform (FT). The purpose of sampling is to discretize continuous speech, and the sampling rate is set as 16kHz in this letter. Since the speech signal is non-stationary, it is not applicable to FT. To solve this problem, a common method is framing to achieve a short-time signal stabilization. To reduce the intensity of the side lobe after FT and obtain a higher quality spectrum [44], each frame of the signal will multiply a smooth window function to ensure that the end of the frame could smoothly attenuate to zero. The formula is as follows:

$$X(n,k) = \sum_{m=0}^{N-1} x_n(m) \exp(j\frac{2\pi km}{N}) \qquad k \in [0, N-1]. \tag{1}$$

Due to the human perception of sound is logarithmic, we need to send spectrogram to the Mel-filter bank to compress the signal range. After the Discrete Cosine Transform (DCT), one frame of MFCC can be obtained. We spliced each frame to obtain the MFCC Feature spectrogram. The spectrogram describes the spectral information of the speech. Specifically, abscissa and ordinate are speech length and frequency, and the corresponding values are the energy of the speech.

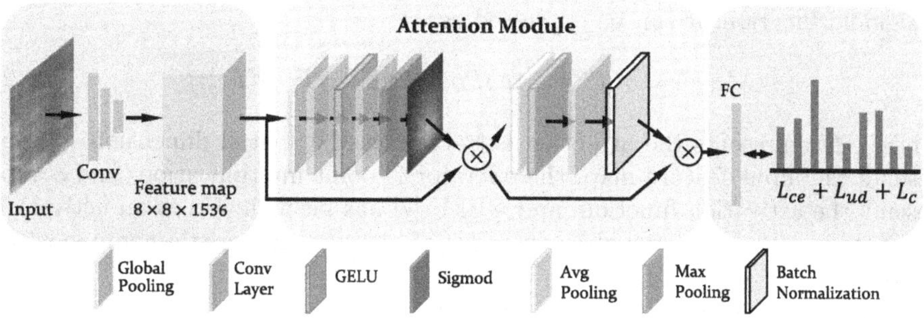

Fig. 1. Framework of our attention modules. It uses inception_v4 to extract spectrogram features. The upper branch is the attention module, which uses global pooling first to make global information available to the lower layers, after that we use average pooling and maximum pooling, and a layer of standardized operation to avoid the gradient disappearing.

2.2 Attention for Spectrogram

Deep learning-based methods are typically designed under the assumption that noise effects are not severe or that the environment scene is relatively fixed [20,29,42]. However, strong noise interference is common in real-world scenarios, often accompanied by variations in the scene, resulting in a texture fuzzy problem in the spectrogram.

To overcome the limitations of CNN, where each feature map cannot use the context information of other feature maps, we incorporate a global average pooling method [21,43] into our framework instead of pooling on local receptive fields. The input feature map size is $8 \times 8 \times 1536$, and after global pooling, a sequence of $1 \times 1 \times 1536$ is obtained. This approach has two benefits: the global average pooling layer does not require parameters, avoiding overfitting, and it enables the lower layers of the network to utilize global information, which is more robust to the spatial variation of the input by summing the spatial information. The global average pooling formula is applied to multiple feature maps and can be interpreted as a collection of local descriptors, whose statistical information is expressive for the whole image.

To make full use of the information between different feature layers, a convolutional layer is used to transform the feature sequence into $1 \times 1 \times 96$. After activation and up-sampling, we add this information to the original feature map to improve discriminability. To further enhance the feature learning and address the impact of complicated noise, AJS incorporates attention mechanisms [40] into the model. AJS includes two pooling layers: maximum pooling highlights the texture information of the spectrogram, while average pooling minimizes the role of features corrupted by noise.

After getting the attention weights, we conduct a multiplication between the attention weight matrix and feature map, the final weights are generated through a sigmoid function. It can be expressed as:

$$M_s(F) = \sigma(f^{8\times8}([AvgPool(F), MaxPool(F)])), \tag{2}$$

where M_s represents the attention extraction in the spatial dimension, F represents the input feature map, the operator is point multiplication, and σ represents the activation function, and GELU [7] has been chosen as an activation function in our model to take advantage of every characteristic component of speech.

Figure 2 visualizes the speech signals in this paper. It contains four sets of images, including the display for spectrogram, feature map, attention heat map, and noise. (a) contains two pictures, the top is a spectrogram of a pure speech signal in dataset, the bottom is the feature map obtained after passing the MFCC filter; (b) is the same speech mixed with white noise, signal-noise ratio is 5. As can be seen, noise obscures part of the texture feature; (c) is the heat map of attention, where the brightness represents the weight; (d) visualizes the noise. We can see that the spectrogram of different noises has obvious differences.

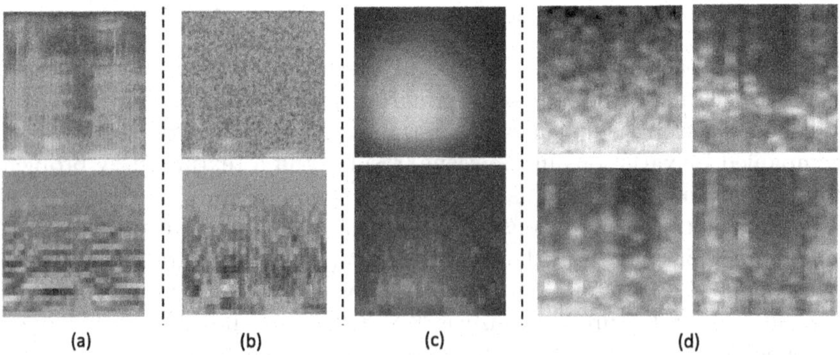

Fig. 2. Visualization of the signals, including (a) spectrogram, (b) feature map, (c) attention heat map, and (d) noise.

2.3 Loss Function for Joint Supervision

In this section, we discuss the loss function in details. In order to identify speech categories, we utilize the following correlation entropy loss:

$$L_{ce} = \sum_{i=1}^{n} y_i \cdot log(p_i) \tag{3}$$

where n is the total number of samples, y_i is true label of sample x_i, and p_i is prediction label. The aim of (3) is to fit the original samples rather than the noise samples, as complex noise is often applied to the original samples and noise models may fail to work even if the noise obeys predefined distributions. This is due to the fact that the distribution of different noises can vary greatly. To address this issue, we introduce an extra pair of antagonistic losses, L_{ud} and L_c to improve the model's robustness [39].

$$\begin{aligned} L_{ud} &= \sum_{C=1}^{M} \frac{1}{M} \cdot log(p_i), \\ L_c &= \sum_{i=1}^{n} |x_i - C_{yi}|_2^2, \end{aligned} \tag{4}$$

where M is the number of categories, C_{y_i} is center of the samples in category y_i. L_{ud} is used to learn the uniform distribution of data, which can decrease the confidence of the sample x identified to one class, thus reducing the influence of the model on the classification errors of difficult samples. L_c constrains the boundaries of each category of data. This can gradually shrink the classification boundary and reduce the classification difficulty of difficult samples. The total loss function is:

$$Loss = \alpha L_{ce} + \beta L_{ud} + \gamma L_c. \tag{5}$$

where α, β, γ are the weights.

3 Experiments

In this section, four benchmark databases for speech data are used to evaluate the performance of our method including THCHS-30 [34], NoiseX-92 [32], Speech Commands dataset [38] and HI-MIA [23], where speech signals from these datasets are corrupted by common noises.

3.1 Datasets and Experimental Setting

Databases for speech data are used to evaluate the performance of our method are introduced as follow. **Speech Commands dataset** [38] contains 65,000 pieces of data in 30 categories, each lasting one second. **HI-MIA** [23] is a newer voice dataset that can be used for speaker verification and keyword awakening research. It contains 86 speakers, two sampling rates, and provides data on multiple speech rates and far-field speech. The noise signals we adopt are Cafe Noise and White Noise in the THCHS-30 dataset [34], and that with Speech Babble Noise, Factory Floor Noise, Machine Gun Noise and Volvo Noise in the NoiseX-92 [32] are selected in our experiment. Since the data samples are balanced, all experiments in this letter use the Top-1 precision to measure the recognition performance of each model. The accuracy rate calculation formula is as follows:

$$acc(f; D) = \frac{1}{M} \sum_{i}^{M} F(f(x_i) = y_i), \tag{6}$$

where acc represents the accuracy rate, D is the sample set, M is the number of samples in the current sample set, and F is an illustrative function. To make readers easier to understand our experimental results, the definition of Signal-Noise Ratio (SNR) is provided here. SNR is an important parameter to measure signal quality. It is expressed as the logarithm of the ratio of useful signal power to noise power. The measurement unit of the ratio is dB. The larger the value is, the lower the noise will be.

Our network is optimized with Stochastic Gradient Descent (SGD), and its relevant setting are as follows: learning_rate = 0.001, epochs = 50, batch_size = 16. The computer used in the experiment of this paper is configured as: Core i7-9700K processor, 32G memory, Linux operating system, Nvidia 1080Ti and the framework is TensorFlow.

3.2 Illustration of Baseline

In our experiments, the noise data used is obtained by randomly cutting out the noise segment from the pure noise environment and mixing it with pure speech data by -5 dB, 0 dB, 5 dB, 10 dB, 20 dB. Four methods have been chosen as the baseline for comparison, including the traditional MFCC feature, GFCC feature, LSTM [5], two-layer model S-DH [3], and typical deep convolutional neural network named inception-v4 [28]. Among them, MFCC and GFCC based models are established by HMM [14]. For a two-layer model, we train a Support Vector Machine (SVM) at first to distinguish noise types, and then, train a DNN-HMM in the current noise environment. We extract spectrograms for inception-v4 and our method AJS. In all convolution operations, the convolution kernel has a move step size of 1. For all datasets, the AJS takes input images with a size of 299 × 299 pixels. To make fair comparisons with other methods, all the layers in our model are randomly initialized. Our pre-processing follows the commonly used configurations. Specifically, horizontal flipping is adopted for data augmentation in the training phase. To prevent over-fitting, a 50% dropout is set in the network. For a rigorous comparison, this data set is further divided into training set, validation set, and test set at a ratio of 3:1:1.

Table 1. Comparison results (%) on Speech Commands Dataset

Noise Type	white					cafe				
	–5 dB	0 dB	5 dB	10 dB	20 dB	–5 dB	0 dB	5 dB	10 dB	20 dB
MFCC	0.2318	0.3723	0.5243	0.6538	0.7974	0.5414	0.6872	0.7382	0.8002	0.8908
GFCC	0.2616	0.3014	0.4685	0.6667	0.7936	0.5730	0.7424	0.7936	0.8224	0.8944
LSTM	0.2470	0.2676	0.3573	0.5516	0.7875	0.5436	0.7428	0.7901	0.7922	0.8609
S-DH	**0.4874**	**0.5374**	**0.7188**	0.8203	0.8606	0.5624	0.7499	0.8657	0.8635	0.9011
Inc-v4	0.2644	0.3215	0.4164	0.5776	0.7903	0.5828	0.7351	0.8208	0.8635	0.9004
OURS	0.4148	0.4815	0.6185	**0.8230**	**0.8824**	**0.6068**	**0.7627**	**0.8741**	**0.8984**	**0.9335**
	babble					factory1				
	–5 dB	0 dB	5 dB	10 dB	20 dB	–5 dB	0 dB	5 dB	10 dB	20 dB
MFCC	0.3012	0.4140	0.5367	0.6673	0.7944	0.3820	0.5858	0.6540	0.8034	0.8843
GFCC	0.3601	0.4434	0.5229	0.7082	0.8080	0.3875	0.6245	0.7050	0.8248	0.8650
LSTM	0.3172	0.4488	0.6471	0.7304	0.8186	0.3107	0.4267	0.6177	0.8130	0.9151
S-DH	**0.4643**	**0.6228**	0.7542	0.7604	0.8816	**0.5219**	0.6217	0.8034	0.8440	0.9423
Inc-v4	0.2798	0.3903	0.6052	0.7409	0.8654	0.4110	0.6279	0.7721	0.8412	0.9002
OURS	0.3211	0.4549	**0.7738**	**0.8031**	**0.9079**	0.4431	**0.6303**	**0.8135**	**0.8627**	**0.9440**

3.3 Comparison Results

Speech Commands Dataset. The experiment to evaluate the performance of AJS on Speech Commands Dataset and its identification results are shown in Table 1. As can be seen, although GFCC is a more robust feature, it achieves similar performance to MFCC. Moreover, MFCC, GFCC and LSTM methods all perform badly when the SNR drops sharply. One possible reason is that the noise seriously damages the context characteristics of the signal, whereas the above methods cannot effectively decrease it. By contrast, S-DH achieves a promising performance among the competitors, since it acquires context characteristics of the signal under noise during the training. As expected, our method achieves a remarkable improvement in various noise environments, which can be attributed to the effective elimination of noise interference by the attention model. However, since the spectrogram information is seriously damaged, this method fails to provide a satisfactory result when the SNR is extremely low. This indicates that two-layer model presents advantages when dealing with a known noise type.

Further experiments are conducted to verify the generalization ability of the algorithm proposed in this letter. Two types of noise that not in the template have been added: Machine Gun and Volvo. The identification results are shown in Table 2. As can be seen, our model is trained using pure speech, S-DH model still uses previous noisy template. Experimental results on the unknown noise environment still demonstrate the superiority of our method according to the overall performance.

HI-MIA. To further evaluate the performance of our method, experiments on HI-MIA data set were carried out. Compared to the previous data set, HI-MIA data set provides a wealth of choices in speech rates, which enables us to study whether the speech speed will affect the recognition of our model. Experimental results are presented in Table 3. As can be seen, our method is much more insensitive to these noises than their corresponding baselines.

Table 2. Comparison results (%) of three methods in unknown noise environment

Noise Type	Machine Gun					Volvo				
	−5 dB	0 dB	5 dB	10 dB	20 dB	−5 dB	0 dB	5 dB	10 dB	20 dB
LSTM	0.6110	0.7005	0.7726	0.7919	0.8373	0.7301	0.7927	0.8331	0.8546	0.8755
S-DH	0.7650	0.8034	0.8336	0.8547	0.8959	0.8461	0.8312	0.8479	0.8841	0.9008
Inc-v4	0.7342	0.7818	0.8199	0.8499	0.8847	0.8761	0.8871	0.8979	0.9085	0.9103
OURS	**0.7828**	**0.8332**	**0.8728**	**0.8953**	**0.9309**	**0.8987**	**0.9124**	**0.9250**	**0.9343**	**0.9399**

Table 3. Comparison results (%) of three methods on HI-MIA

Noise Type	white					cafe				
	−5 dB	0 dB	5 dB	10 dB	20 dB	−5 dB	0 dB	5 dB	10 dB	20 dB
S-DH	**0.7852**	**0.8312**	0.8657	0.8544	0.8824	**0.8257**	0.8630	0.8977	0.8843	0.9093
Inc-v4	0.5137	0.5558	0.8526	0.9410	0.9686	0.7666	0.8826	0.9491	0.9817	0.9948
OURS	0.5642	0.7298	**0.9529**	**0.9750**	**0.9892**	0.8052	**0.9142**	**0.9683**	**0.9913**	**0.9980**
	babble					factory1				
	−5 dB	0 dB	5 dB	10 dB	20 dB	−5 dB	0 dB	5 dB	10 dB	20 dB
S-DH	**0.7411**	**0.8101**	0.8425	0.8988	0.9421	**0.8073**	0.8280	0.8958	0.9193	0.9205
Inc-v4	0.5648	0.6980	0.8456	0.9352	0.9939	0.6218	0.7340	0.8709	0.9500	0.9834
OURS	0.5622	0.6875	**0.8500**	**0.9541**	**0.9974**	0.7311	**0.8288**	**0.9430**	**0.9770**	**0.9942**

3.4 Ablation Study

Table 4 shows the quantitative comparison for loss function in terms of classification accuracy. Note that we obtain four noise types and five SNR for the HI-MIA, respectively. It can be observed that 1) learning the uniform distribution (L_{ud}) plays an important role in improving the recognition accuracy for AJS. It reduces the influence of the model on the classification errors of difficult samples. And 2) using center loss (L_c) single will fail to boost accuracy. When L_{ud} and L_c are used in combination, the boundary can be effectively shrinked and the identification accuracy can be further improved. Note that, our method can effectively adapt to different noise environments and SNR, further improve the accuracy of model recognition. The weight values used in this paper are α =0.9, β= 0.1 and γ =0.0005 respectively. In different tasks, fine-tuning the weight of loss can get the effect improvement.

Table 4. Ablation study OF loss function

Noise Type	white					cafe				
	-5 dB	0 dB	5 dB	10 dB	20 dB	-5 dB	0 dB	5 dB	10 dB	20 dB
L_{ce}	**0.5738**	0.7241	0.8235	0.8730	0.9433	0.7677	0.8672	0.9241	0.9602	0.9878
$L_{ce} + L_{ud}$	0.5493	0.6806	0.9317	0.9675	0.9828	**0.8324**	**0.9321**	**0.9739**	0.9907	0.9968
$L_{ce} + L_c$	0.5599	0.6994	0.8881	0.9154	0.9500	0.7666	0.8552	0.9067	0.9483	0.9898
Total	0.5642	**0.7298**	**0.9529**	**0.9750**	**0.9892**	0.8052	0.9142	0.9683	**0.9913**	**0.9980**
	babble					factory1				
	-5 dB	0 dB	5 dB	10 dB	20 dB	-5 dB	0 dB	5 dB	10 dB	20 dB
L_{ce}	**0.6250**	0.7573	0.8462	0.9125	0.9791	0.6797	0.7709	0.8703	0.9119	0.9608
$L_{ce} + L_{ud}$	0.6229	**0.7741**	**0.8974**	**0.9578**	0.9915	0.6339	0.7733	0.9251	0.9758	0.9911
$L_{ce} + L_c$	0.6119	0.7465	0.8442	0.9218	0.9799	0.6390	0.7532	0.8680	0.9116	0.9657
Total	0.5622	0.6875	0.8500	0.9541	**0.9974**	**0.7311**	**0.8288**	**0.9430**	**0.9770**	**0.9942**

3.5 Discussion

As can be seen, the performance S-DH does not suffer from the change of known noise environment compared with MFCC and GFCC based methods, whereas inception-v4 and our AJS achieve satisfactory accuracy in unknown noise environment. One should note that two-layer methods have good performance in the case there the noise type is known, but we also need to see the limitations. Experiment demonstrates the impact of different templates on recognition. It can be seen that the effect is not ideal when a template with high SNR speech training is used to identify low SNR speech. By contrast to S-DH, our AJS is insensitive to the change of noise environment. This nice property can benefit real applications. Despite the satisfactory results achieved by our AJS, more effects need to be taken to reduce the impact of human speech noise. According to current results, the recognition rate in the Speech Babble environment is lower than others. Command speech and interference speech have roughly the same distribution of features, which makes ASR difficult.

4 Conclusion

In this paper, inspired by attention and joint supervision, we present a novel attention-based method for automatic speech recognition in various noise environments. Extensive experiments are conducted on four commonly used datasets. Compared with the other approaches, our method has a wider range of application scenarios and can effectively reduce noise impact, thus significantly improving the classification accuracy. Moreover, ablation study indicates the positive efficacy of our loss function on handling complicated noise.

Despite the satisfactory results achieved by our AJS, more effects need to be taken to reduce interference of human voice. According to current results, our method relaxes the processing of noise to enhance the generalization ability and thus leads to the difficulty to separate human speech. One feasible method to handle this problem is source separation.

Acknowledgements. This work was supported in part by the National Key R&D Program (2022YFF0712300 and 2022YFC3301004), in part by the Fundamental Research Funds for the Central Universities, HUST: 2023JYCXJJ031.

References

1. Baby, D., Virtanen, T., Gemmeke, J.F., et al.: Coupled dictionaries for exemplar-based speech enhancement and automatic speech recognition. IEEE-ACM Trans. Audio Speech **23**(11), 1788–1799 (2015)
2. Boll, S.: Suppression of acoustic noise in speech using spectral subtraction. IEEE-ACM Trans. Audio Speech **27**(2), 113–120 (1979)
3. Cao, J., Xu, J., Shao, S.: Research on multi-noise-robust auto speech recognition. Comput. Appl. 1790–1794 (2018)
4. Ephraim, Y., Van Trees, H.L.: A signal subspace approach for speech enhancement. IEEE Trans. Speech Audio Process. **3**(4), 251–266 (1995)
5. Geiger, J.T., Weninger, F., Gemmeke, J.F., Wöllmer, M., Schuller, B., Rigoll, G.: Memory-enhanced neural networks and NMF for robust ASR. IEEE-ACM T Audio Speech **22**(6), 1037–1046 (2014)
6. Grancharov, V., Samuelsson, J., Kleijn, B.: On causal algorithms for speech enhancement. IEEE-ACM Trans. Audio Speech **14**(3), 764–773 (2006)
7. Hendrycks, D., Gimpel, K.: Gaussian error linear units (GELUS). arXiv preprint arXiv:1606.08415 (2016)
8. Hu, J., Shen, L., Sun, G.: Squeeze-and-excitation networks, pp. 7132–7141 (2018)
9. Hu, Y., Loizou, P.C.: Speech enhancement based on wavelet thresholding the multitaper spectrum. IEEE Trans. Speech Audio Process. **12**(1), 59–67 (2004)
10. Islam, M.: GFCC-based robust gender detection. In: ICISET, pp. 1–4. IEEE (2016)
11. Juang, B.H., Hou, W., Lee, C.H.: Minimum classification error rate methods for speech recognition. IEEE Trans. Speech Audio Process. **5**(3), 257–265 (1997)
12. Kamath, S., Loizou, P., et al.: A multi-band spectral subtraction method for enhancing speech corrupted by colored noise. In: ICASSP, vol. 4, pp. 44164–44164. Citeseer (2002)
13. Lev-Ari, H., Ephraim, Y.: Extension of the signal subspace speech enhancement approach to colored noise. IEEE Signal Process. Lett. **10**(4), 104–106 (2003)
14. Li, X., Wang, Z.: A hmm-based mandarin Chinese singing voice synthesis system. JAS **3**(2), 192–202 (2016)
15. Li, Y., Zhang, W.T., Lou, S.T.: Generative adversarial networks for single channel separation of convolutive mixed speech signals. Neurocomputing **438**, 63–71 (2021)
16. Li, Z., Ming, Y., Yang, L., Xue, J.: Mutual-learning sequence-level knowledge distillation for automatic speech recognition. Neurocomputing **428**, 259–267 (2021)
17. Liu, L., Li, W., Wu, X., Zhou, B.X.: Infant cry language analysis and recognition: an experimental approach. JAS **6**(3), 778–788 (2019)
18. Meriem, F., Farid, H., Messaoud, B., Abderrahmene, A.: Robust speaker verification using a new front end based on multitaper and gammatone filters. In: SITIS, pp. 99–103. IEEE (2014)
19. Mittal, U., Phamdo, N.: Signal/noise KLT based approach for enhancing speech degraded by colored noise. IEEE Trans. Speech Audio Process. **8**(2), 159–167 (2000)
20. Moriya, T., Tanaka, T., Shinozaki, T., Watanabe, S., Duh, K.: Evolution-strategy-based automation of system development for high-performance speech recognition. IEEE-ACM Trans. Audio Speech **27**(1), 77–88 (2018)

21. Nilufar, S., Ray, N., Molla, M.I., Hirose, K.: Spectrogram based features selection using multiple kernel learning for speech/music discrimination, pp. 501–504 (2012)
22. Povey, D., Woodland, P.C.: Minimum phone error and i-smoothing for improved discriminative training. In: ICASSP, vol. 1, pp. I-105. IEEE (2002)
23. Qin, X., Bu, H., Li, M.: Hi-MIA: a far-field text-dependent speaker verification database and the baselines, pp. 7609–7613 (2020)
24. Rabiner, L.R.: A tutorial on hidden Markov models and selected applications in speech recognition. Proc. IEEE **77**(2), 257–286 (1989)
25. Schwarz, P., Matejka, P., Cernocky, J.: Hierarchical structures of neural networks for phoneme recognition. In: ICASSP, vol. 1, pp. I-I. IEEE (2006)
26. Shi, X., Yang, H., Zhou, P.: Robust speaker recognition based on improved GFCC. In: IEEE INFOCOM, pp. 1927–1931. IEEE (2016)
27. Siniscalchi, S.M., Reed, J., Svendsen, T., Lee, C.H.: Universal attribute characterization of spoken languages for automatic spoken language recognition. Comput. Speech Lang. **27**(1), 209–227 (2013)
28. Szegedy, C., Ioffe, S., Vanhoucke, V., Alemi, A.: Inception-v4, inception-resnet and the impact of residual connections on learning. arXiv preprint arXiv:1602.07261 (2016)
29. Tang, Z., Li, L., Wang, D., Vipperla, R.: Collaborative joint training with multitask recurrent model for speech and speaker recognition. IEEE-ACM Trans Audio Speech **25**(3), 493–504 (2016)
30. Tu, Y.H., Du, J., Lee, C.H.: Speech enhancement based on teacher-student deep learning using improved speech presence probability for noise-robust speech recognition. IEEE-ACM Trans Audio Speech **27**(12), 2080–2091 (2019)
31. Umesh, S., Sinha, R.: A study of filter bank smoothing in MFCC features for recognition of children's speech. IEEE-ACM Trans Audio Speech **15**(8), 2418–2430 (2007)
32. Varga, A., Steeneken, H.J.: Assessment for automatic speech recognition: Ii. noisex-92: a database and an experiment to study the effect of additive noise on speech recognition systems. Speech Commun. **12**(3), 247–251 (1993)
33. Variani, E., Sainath, T.N., Shafran, I., Bacchiani, M.: Complex linear projection (CLP): a discriminative approach to joint feature extraction and acoustic modeling (2016)
34. Wang, D., Zhang, X.: THCHS-30: a free Chinese speech corpus. arXiv preprint arXiv:1512.01882 (2015)
35. Wang, K., Gou, C., Duan, Y., Lin, Y., Zheng, X., Wang, F.Y.: Generative adversarial networks: introduction and outlook. JAS **4**(4), 588–598 (2017)
36. Wang, Q., Du, J., Dai, L.R., Lee, C.H.: A multiobjective learning and ensembling approach to high-performance speech enhancement with compact neural network architectures. IEEE-ACM Trans. Audio Speech **26**(7), 1185–1197 (2018)
37. Wang, Z.Q., Wang, D.: A joint training framework for robust automatic speech recognition. IEEE-ACM Trans. Audio Speech **24**(4), 796–806 (2016)
38. Warden, P.: Speech commands: a dataset for limited-vocabulary speech recognition. arXiv preprint arXiv:1804.03209 (2018)
39. Wen, Y., Zhang, K., Li, Z., Qiao, Y.: A discriminative feature learning approach for deep face recognition, pp. 499–515 (2016)
40. Woo, S., Park, J., Lee, J.Y., So Kweon, I.: CBAM: convolutional block attention module, pp. 3–19 (2018)
41. Xiang, B., Jing, X., Yang, H.: Vehicular speech recognition based on noise classification and compensation. Comput. Eng. (3), 37 (2017)

42. Ye, S., et al.: Discriminative suprasphere embedding for fine-grained visual categorization. IEEE Trans. Neural Netw. Learn. Syst. (2022)
43. Ye, S., Wang, Y., Peng, Q., You, X., Chen, C.P.: The image data and backbone in weakly supervised fine-grained visual categorization: A revisit and further thinking. IEEE Trans. Circ. Syst. Video Technol. (2023)
44. Yu, G., Slotine, J.J.: Audio classification from time-frequency texture, pp. 1677–1680 (2009)

Discourse-Aware Causal Emotion Entailment

Dexin Kong[1], Nan Yu[1], Yun Yuan[1], Xin Shi[1], Chen Gong[1,2], and Guohong Fu[1,2](✉)

[1] School of Computer Science and Technology, Soochow University, Suzhou, China
{nyu,yyuanwind,shixin}@stu.suda.edu.cn, {gongchen18,ghfu}@suda.edu.cn
[2] Institute of Artificial Intelligence, Soochow University, Suzhou, China

Abstract. Causal Emotion Entailment (CEE) aims to identify the corresponding causal utterances for a target emotional utterance in conversations. Most previous research has focused on the use of sequential encoding to model conversational contexts, without fully considering the interaction effects between different utterances. In this paper, we explore the significance of discourse parsing in addressing these interactions, and propose a new model called discourse-aware model (DAM) to tackle the CEE task. Concretely, we use a multi-task learning framework to jointly model CEE and discourse parsing to fuse rich discourse information. In addition, we use a graph neural network to further enhance our CEE model by explicitly encoding discourse and other discourse-related structure features. The results on the benchmark corpus show that the DAM outperforms the state-of-the-art systems in the literature. This suggests that the discourse structure may contain a potential link between emotional utterances and their corresponding cause expressions. We will release the codes of this paper to facilitate future research (https://github.com/Sakurakdx/DAM).

Keywords: Causal emotion entailment · Discourse · Utterance interaction

1 Introduction

Causal Emotion Entailment (CEE) is an important task in the field of conversation sentiment analysis. It has received increasing attention [21] with the open conversational data deluge on social media platforms. Similar to text-level emotion cause extraction (ECE) [14], this task aims to extract which utterances are responsible for the non-neutral emotion in a conversational utterance. The Fig. 1 shows an example of the CEE. The emotion of the target utterance "*That sounds wonderful. Will there be anyone there that I know?*" is "happiness". The cause utterance spans are "*it would be fun*", and "*It will give my wife a chance to dress up*", CEE is designed to extract the utterances where these spans are located. This suggests that the explicit emotion causes aforementioned could be helpful for many downstream tasks, such as opinion mining [6].

B. Luo et al. (Eds.): ICONIP 2023, CCIS 1963, pp. 389–401, 2024.
https://doi.org/10.1007/978-981-99-8138-0_31

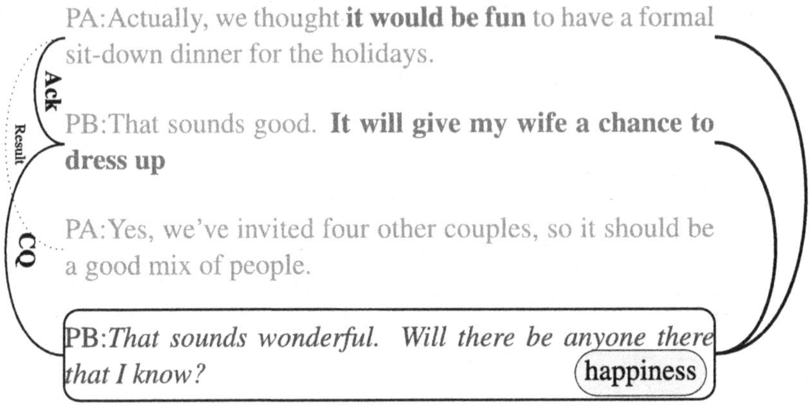

Fig. 1. An example of CEE. Left arcs are predicted discourse structures. Right arcs are emotion-cause pair annotations.

ECE has been investigated intensively since early research [4,14]. It can be treated as a classification task that requires the contexts of a document as inputs and determines which clause is a cause. Poria et al. [21] extend the ECE task to conversation named CEE, and apply neural approaches of text-level emotion cause extraction [8,23]. CEE is considered more challenging than ECE due to the complexity of conversations and the fact that the cause utterances can be far away from the target utterance.

Poria et al. [21] consider CEE as emotion-cause utterances pair classification problems with historical conversation information. Bhat et al. [1] and Zhang et al. [25] think that the correlations between utterances and emotional information are important for CEE. They respectively use a multi-task framework and a dual-stream attention mechanism to enhance the performance of the model by encoding this information. These approaches ignore the discourse structure interaction information between utterances, and other discourse-related structural features. It is a serious problem when performing the CEE task specifically in long conversations.

The discourse structures in a conversation represent the interactive relationships between utterances. Intuitively, these structures contain the potential links between emotional utterances and their corresponding cause expressions. As shown by the solid arc on the left in Fig. 1, most of the emotion-cause pairs are linked with the target utterance by discourse relations. In addition, several findings in previous studies on CEE support our point of view. Chakrabarty et al. [2] and Poria et al. [21] believe that discourse structures play a very important role in the reasoning steps pertaining to the extracted causes.

In this paper, we propose a discourse-aware model (DAM) for CEE. Concretely, we jointly model CEE and discourse parsing in conversations. It uses a shared pre-trained language model (PLM) to represent conversational contexts. The discourse parsing task can implicitly integrate rich utterance interaction information into the shared PLM. Besides, we use a graph neural network (GNN) [18] to explicitly encode discourse structures that generated by the discourse parser. In addition, we follow Wang et al. [22], exploiting a GNN to further integrate discourse-related structural features such as the relative utterance distance and speakers into our model. Both of them can help the model integrate rich utterance interaction information and mitigate the long-distance dependency problem.

We conduct the experiments on the standard benchmark dataset of the CEE task to verify our DAM model. Experiments show that the utterance interaction features are effective for this task.

In summary, we make three main contributions as follows:

- We propose a discourse-aware model (DAM) that integrates rich discourse interaction features for CEE through both implicit and explicit ways.
- We further exploit a GNN to capture discourse-related structural features such as the relative utterance distance and speakers for CEE.
- We advance the performance of the SOTA models for CEE.

2 Related Work

Emotion Cause Extraction (ECE). The ECE task is proposed for the first time by Lee et al. [14]. Early research adopts linguistic rules [4,14] and traditional machine learning [10,11] to identify the clause in the text that reflects the real reason for the emotion. In recent years, several neural network models are introduced to the ECE task with different granularities, such as clause-level [7,13] and span-level [16,17]. Compared with ECE, the CEE task is more challenging due to the informality of conversations and the fact that the cause utterances can be far away from the target utterance.

Causal Emotion Entailment (CEE). Poria et al. [21] extend ECE to the conversation domain and introduce the CEE task. They consider CEE as an emotion-cause utterances pair classification problem. Bhat and Modi et al. [1] and Zhang et al. [25] think that the correlations between utterances and emotional information are important for CEE. Bhat and Modi et al. [1] introduce a multi-task learning framework to use sentiment classification as an auxiliary task to fuse sentiment information. Zhang et al. [25] uses a dual-stream attention mechanism to separately encode the speaker's behavior and emotional state to enhance model performance. However, most of the existing CEE works do not consider the influence of the interaction and discourse relation between utterances. Intuitively, utterance structure interaction information such as discourse structures are promising for the CEE task. Chakrabarty et al. [2] and Poria et al. [21] believe that discourse structures play a very important role in the cause extraction. In this paper, we employ discourse parsing as an auxiliary task by multi-task learning framework [9,12] to implicitly capture utterance interaction information. To better model the discourse structure, we refer to Wang et al.; Li et al. [18,22], and use GNN to encode discourse-related structural information such as relative distance and speakers. Both of them can help the model integrate rich utterance interaction information and mitigate the long-distance dependency problem.

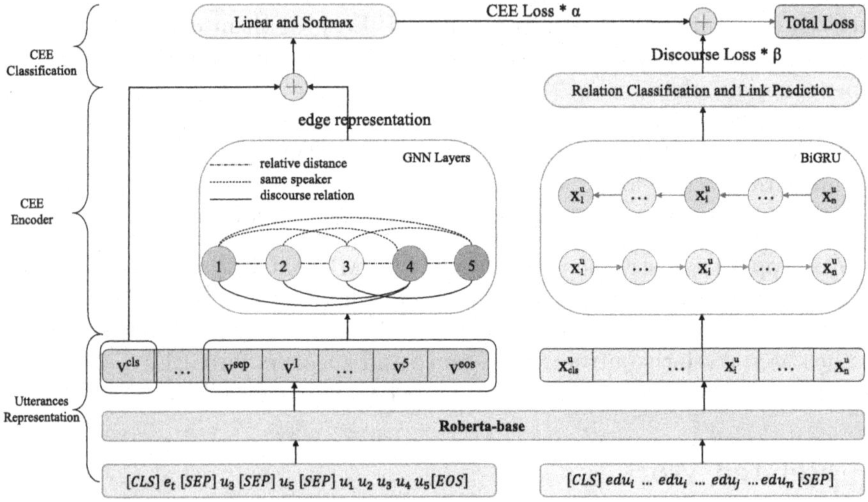

Fig. 2. Framework of our model. In this example, the model judges whether u_3 is the cause utterance of emotion e_t in u_5.

3 Our Proposed Model

In this section, we present the proposed discourse-aware model (DAM) for CEE. As shown in Fig. 2, we use a multi-task learning (MTL) framework to model the CEE task and discourse parsing jointly. It can implicitly integrate rich utterance interaction information into our conversational context representations. Furthermore, we use a graph neural network (GNN) to encode discourse-related structural features, such as relative distance and speakers between utterances, to help the model capture the discourse structure information.

3.1 Basic CEE Model

Following Poria et al. [21], we consider CEE as the emotion-cause utterances pair classification problem with historical conversation information. Formally, given a conversation with n utterances $\{u_1, u_2, ..., u_n\}$, for each target utterance u_t with emotion e_t and its historical utterances set $H(u_t)$, the CEE task aims to extract the cause utterances set $C(u_t)$ of the emotion expressed by u_t. Therefore, this task can be regarded as a triplet classification. For each utterance u_i in $H(u_t)$, the tuple $(u_t, u_i, H(u_t))$ is taken as a positive case if $u_i \in C(u_t)$. Otherwise, it is classified as a negative case.

Utterance Representations. As shown in Fig. 2, the input of RoBERTa is formatted as "[CLS]e_t[SEP]u_t[SEP]u_i[SEP]$H(u_t)$[EOS]", where e_t represents the emotion of the target utterance u_t. We extract the hidden states $\mathbf{v}_{cls}$ of the [CLS] token as well as the hidden states $\mathbf{v}^i$ of all utterances belonging to $H(u_t)$ for later use.

CEE Encoder. In the CEE Encoder, we build a fully connected graph according to the discourse relations, speaker, and relative utterance distance. Then, we use the GNN layer to encode inter-utterance representations and obtain hidden states $\mathbf{h}_i$. See more details in Sect. 3.3.

CEE Classification. After obtaining the output hidden states $\mathbf{h}_i$ of the encoder, we use softmax function to get the probability distribution $\mathbf{p}_i$ through a linear layer and generate the prediction $\widehat{y}_i$ of inputs as follows:

$$\mathbf{p}_i = \mathrm{SoftMax}(\mathbf{W}_s\mathbf{h}_i + \mathbf{b}_s) \tag{1}$$

$$\widehat{y}_i = \arg\max(\mathbf{p}_i) \tag{2}$$

where $\mathbf{W}_s$ and $\mathbf{b}_s$ are the weight matrix and bias of the linear before softmax, respectively.

Loss Function. To train the model, we calculate the negative log-likelihood of train data as loss of the main task L_{cee} as follows:

$$L_{cee} = -\sum_{v=1}^{V}\sum_{z=1}^{Z} \mathbf{Y}_{vz} \log \mathbf{P}_{vz} \tag{3}$$

where $\mathbf{Y}$ is the label indication matrix, and $\mathbf{P}$ is the corresponding probability matrix.

3.2 Multi-task Learning Framework

We employ discourse parsing as an auxiliary task by the MTL framework to capture discourse structure interaction information. Specifically, we use hard parameter sharing to combine these two tasks, and use RoBERTa [19] to obtain the shared conversational context representations.

Discourse Parser. Given a conversation with n elementary discourse units (EDU) $\{edu_1, ..., edu_n\}$, discourse parsing aims to predict dependency links and the corresponding relation types $\{(edu_j, edu_i, r_{ji}) | j \neq i\}$ between the EDUs, where (edu_j, edu_i, r_{ji}) stands for a link of relation type r_{ji} from edu_j to edu_i. We employ the model proposed by Yu et al. [24] as the discourse parser module in our framework, as shown by the right part of Fig. 2.

Discourse Representations. For the ith EDU edu_i, the [CLS] token embedding is taken as the representation of edu_i, denoted as $\mathbf{x}_i^u$. We take $\{\mathbf{x}_1^u, \mathbf{x}_2^u, ..., \mathbf{x}_n^u\}$ as the input of Bi-directional GRU (BiGRU) [5] to get hidden states $\{\mathbf{h}_1^u, \mathbf{h}_2^u, ..., \mathbf{h}_n^u\}$ as EDUs representations.

Link Prediction and Relation Classification. The link predictor predicts the parent of edu_i. The relation classifier is responsible for predicting the relationship type r_{ji} between edu_i and edu_j. The link predictor and relation classifier have similar structures. They first convert the input vector $\mathbf{h}_{i,j}(j < i)$ into a hidden representation through a linear layer:

$$\mathbf{L}_{i,j}^{\mathcal{D}} = \tanh(\mathbf{W}_{\mathcal{D}}\mathbf{h}_{i,j} + \mathbf{b}_{\mathcal{D}}) \tag{4}$$

where $\mathbf{W}$ and $\mathbf{b}$ are learnable weights, and $\mathcal{D} \in \{link, rel\}$ represent the weights of the link predictor and relation classifier respectively.

Loss Function. We add the negative log-likelihood loss of link prediction and relation classification together as the final loss.

$$L_{dp} = \text{CE}(label_{\mathcal{D}}, \text{SoftMax}(\mathbf{W'}_{\mathcal{D}}\mathbf{L}_{i,j}^{\mathcal{D}} + \mathbf{b'}_{\mathcal{D}})) \tag{5}$$

where CE is cross-entropy function and $label_{\mathcal{D}}$ is the golden label of link and relation.

Training. During training, two models are learned simultaneously for the two sub-tasks (CEE and discourse parsing). The bottom RoBERTa$_{base}$ encoding layer is shared by the above two tasks. The total training objective is defined as:

$$L = \alpha L_{cee} + \beta L_{dp} \tag{6}$$

where α and β are hyperparameters.

3.3 Graph Neural Network Module

Inspired by Wang et al. [22] and Li et al. [18], we explicitly encode conversational discourse structures and discourse-related structural features by GNN. This can not only enhance the discourse awareness of the model, but also enrich other structural interaction information.

Graph Construction. We construct a fully connected graph as input to the GNN, where each utterance in historical utterances set $H(u_t)$ is regarded as a node of the graph. The construction rules are as follows:

Discourse Relation Graph. In order to explicitly encode the interaction between discourse structures, we constructed a discourse relation graph based on the discourse relations between utterances. Specifically, we first trained a discourse parser to obtain the discourse relations between utterances:

$$\{(i, j, r_{ij}), ...\} = \texttt{Parser}(u_1, ..., u_n) \tag{7}$$

where (i, j, r_{ij}) denotes a link of relation type r_{ij} from u_i to u_j. We connect an edge if there is a discourse relation between any two nodes, and set different weights to different discourse relation types.

Speaker Graph. Speaker interactions are important cues of dialogue discourse. Based on speaker interaction information, we constructed a speaker graph that connects utterances with edges when they have the same speaker. This can further assist the model in integrating the interaction information between utterances.

Relative Distance Graph. For the CEE task, it is very important to establish the long-distance dependency between utterances. In order to better model this dependency, we add relative utterance distance features. There is an edge between any two nodes and a self-loop edge on each node. It represents the relative utterance distance between its linked nodes.

GNN. Each node and edge in the graph represents a learnable vector for the GNN. We directly use the utterance level representation $\mathbf{v}^i$ to initialize the vector representation of the corresponding node. For the edge between any two nodes, we initialize it by a learnable vector $\mathbf{e}_{ij}$ as follows.

$$\mathbf{e}_{ij} = [\mathbf{s}_{ij}, \mathbf{d}_{ij}, \mathbf{r}_{ij}] \tag{8}$$

where $\mathbf{s}_{ij}$, $\mathbf{d}_{ij}$ and $\mathbf{r}_{ij}$ respectively denote the speaker, relative utterance distance, and discourse relation type vectors between utterance u_i and utterance u_j.

After that, we conduct structure-aware scaled dot-product attention to update the hidden state of nodes. Due to space limitations, more details can be seen in Wang et al. [22]. We iterate T times to update the above hidden states. Eventually, we concatenate the hidden state vectors of two directions $\mathbf{e}_{ij}^T and \mathbf{e}_{ji}^T$ in the top layer:

$$\widehat{\mathbf{E}}_{i,j} = [\mathbf{e}_{ij}^T; \mathbf{e}_{ji}^T] \tag{9}$$

where the subscripts i and j represent the ith and j^{th} utterance, respectively. The GNN encoding representation $\widehat{\mathbf{E}}_{i,j}$ and the hidden representation $\mathbf{v}_{cls}$ of [CLS] in the previous stage are concatenated together as the final representation $\mathbf{h}_i$:

$$\mathbf{h}_i = [\mathbf{v}_{cls}; \widehat{\mathbf{E}}_{i,j}] \tag{10}$$

The final representation $\mathbf{h}_i$ will be used for CEE classification.

4 Experiment Settings

4.1 Datasets

To verify the effectiveness of utterance interaction information, we conduct experiments on the RECCON [21] dataset. The data in this dataset are divided into three sub-datasets, named Fold1, Fold2, and Fold3, which are constructed by using different strategies to generate negative examples. We follow Zhang et al. [25], using Fold1-DD to validate our proposed model. Training, validation, and testing data are 27,915, 1,185, and 7,224 samples respectively.

Moreover, we use the Molweni [15] dataset to help complete the training of auxiliary task discourse parsing. There are 10,000 conversations with 88,303 EDUs and 78,245 discourse relations in Molweni.

Table 1. The results of our model in comparison to other models. Bold font denotes the best performance. w/o CC means ignoring the conversation history.

Model	w/o CC			w/ CC		
	Pos.F1	Neg.F1	Macro F1	Pos.F1	Neg.F1	Macro F1
SIMP_{base}	56.64	85.13	70.88	64.28	88.74	76.51
MuTEC	59.18	84.20	**71.69**	69.20	85.90	77.55
JOINT	–	–	–	66.61	89.11	77.86
TSAM	–	–	–	68.59	89.75	79.17
DAM	–	–	–	69.32	89.35	**79.34**

4.2 Evaluation

For fair comparison, we use Pos.F1, Neg.F1 and macro-averaged F1 to evaluate our proposed model, the same as Poria et al. [21]. Pos.F1 and Neg.F1 represent the F1-score on the positive and negative examples, respectively. The overall macro-averaged F1 is based on the Pos.F1 and Neg.F1.

4.3 Implementation Details

We use RoBERTa_{base} to encode the conversational contexts, and adopt an AdamW [20] algorithm to optimize the parameters of our model. We adopt the grid search strategy to determine our hyperparameters. The initial learning rate is set to 1e-5. The batch size of the CEE and discourse parsing are set to 16 and 2 respectively. We use the warm-up strategy, and the radio is set to 0.06. In the global loss function, α and β are set to 1 and 0.5, respectively. The dimensions for the edge and the node states in GNN are both 768. The dimensions for $\mathbf{s}_{ij}, \mathbf{d}_{ij}$ and $\mathbf{r}_{ij}$ are set to 192, 384 and 192, respectively. We train 10 epochs on the training set and save the best model according to the macro-averaged F1 on the validation set. Based on the best model, we test the performance on the test set.

4.4 Compared Methods

To validate the effectiveness of our model, we compare it with several baseline models. These methods use RoBERTa-base as an encoder to facilitate a fair comparison.

1. SIMP_{base} [21]: It tackles CEE in emotion-cause utterances pair classification framework with historical conversation information. We use it as our baseline model.
2. MuTEC [1]: It is an end-to-end MTL framework, where emotions are predicted as auxiliary task and CEE is the main task;
3. JOINT [25]: It tackles CEE in a joint classification framework. It aims to capture the correlations between contextual utterances in a global view;

4. TSAM [25]: Based on JOINT, it proposes a two-stream attention module to effectively model the speaker's emotional influences in the conversational history.

5 Results and Analysis

5.1 Main Results

The results are shown in Table 1, since the conversational history is naturally occurring, we followed the experimental settings of TSAM [25] and did not report results under the w/o CC setting. We can see that our method has achieved a significant improvement compared to the baseline model, from 76.51 to 79.34. This shows that discourse structure information and discourse-related structural features are beneficial to the CEE task.

By further comparison, we can find that encoding conversation history information can bring a huge performance boost to the task (comparing two different settings of w/o CC and w/ CC in Table 1). This shows the importance of interaction information in conversation history for this task. The DAM model significantly outperforms the JOINT model by above 1 points of macro-F1 score (comparing DAM with JOINT in Table 1). There may be two main factors: 1). More interaction information is beneficial for the CEE task; 2). The potential links between emotional utterances and their corresponding cause expressions are contained in the discourse structures.

5.2 Ablation Study

Previous main results show that the effectiveness of our model. As shown in Table 2, the performance of removing auxiliary task decreases by 1.71% over the DAM. It indicates that multi-task framework can help our model capture discourse structures information. This information improves our model's ability to establish relationships between utterances. The second row shows the model without GNN decreases by 1.27% over the DAM, showing the help of encoding conversation context information for understanding conversation. We also verify the influence of speakers and relative utterance distance features respectively, the performance of them all decrease. It demonstrates that these structure information containing conversational features have a significant impact on the model.

Table 2. Ablation results. "w/o" represents "without".

Model	Pos.F1	Neg.F1	MacroF1
DAM	**69.32**	**89.35**	**79.34**
w/o multi-task	66.76	88.51	77.63
w/o gated-gnn	67.76	88.37	78.07
w/o speaker	68.13	89.38	78.75
w/o distance	67.44	88.31	77.87
w/o discourse	67.77	89.47	78.62

Table 3. Result of different fusion methods of discourse relations.

Model	Pos.F1	Neg.F1	MacroF1
SIMP_{base}	64.28	88.74	76.51
DAM-mtl	67.76	88.37	78.07
DAM-cat	68.92	88.47	78.64
DAM-gnn	65.74	88.90	77.32
DAM-mtl-gnn	66.75	88.04	77.40
DAM-arc	68.50	89.27	78.89
DAM	**69.32**	**89.35**	**79.34**

5.3 Effect of Different Fusion Methods

In this section, we explore the impact of different ways of integrating discourse relations on the model. Chen et al. [3] and Ding et al. [8] use graph convolutional network or joint learning to model the dependency relations between clauses. In this paper, we design and experiment with six alternative methods for incorporating discourse relations. Table 3 shows the performance of these methods, and we can draw the following conclusions. First, we can see that discourse structure information can be strengthened by three strategies (DAM-mtl/cat/gnn), where "mtl/gnn" means using only MTL/GNN strategy, and "cat" means concatenating the last hidden parameters of the discourse parser with our corresponding model parameters. Therefore, bringing better performance than the SIMP_{base} that initialized by RoBERTa_{base}. However, the improvement of DAM-cat is slight. It may due to that it suffers from the serious error propagation directly using not exactly correct hidden states. When we combine the DAM-mtl and DAM-gnn (DAM-mtl-gnn), the performance degrades. It may be due to the duplicated discourse information or error propagation. We add the gate module (DAM) on the basis of DAM-mtl-gnn that effectively alleviates the above problems and performs best. So we choose this fusion method as final method. Moreover, we also try to not use the specific discourse relation types and instead focus on whether they have discourse relations (DAM-arc) or not, but it still does not comparable with DAM.

5.4 Long-Distance Dependency Analysis

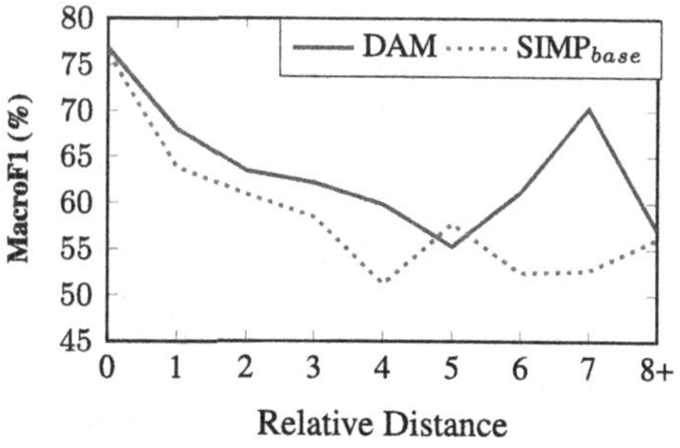

Fig. 3. Performance of different relative utterance distance between u_t and u_i.

To verify whether our model can alleviate the long-distance dependency problem, we present a set of results of different relative utterance distances between u_t and u_i in Fig. 3. We can find that as the relative distance increases, the performance of the two models tends to decrease. This shows that both models suffer from long-distance dependency problem. Moreover, we can find DAM performs better than previous SIMP$_{base}$ model in most cases. That means our model is more robust in terms of distance by combining the utterance interaction features. We find that our model performs abnormally at the distance of 7. This could be due to the shock caused by the small number of samples in this case. In the case of longer distances, our model's performance is similar to the SIMP$_{base}$ model. It suggests that although our approach can help to ameliorate the problem, deep reasoning needs to be further explored in the case of overextended turns or distance.

6 Conclusion

In this paper, we introduce a discourse-aware model for causal emotion entailment task. Our approach incorporates the discourse parsing task using a multi-task framework. By leveraging shared pre-trained language models, this framework enables implicit learning of discourse structure representation in conversations. Furthermore, we employ a graph neural network to explicitly encode discourse and discourse-related structural features. Experimental results on a benchmark dataset demonstrate the effectiveness of both approaches in integrating rich information from utterance interactions, thereby benefiting CEE task.

Acknowledgments. We thank the anonymous reviewers for their valuable comments. This work was supported by the National Natural Science Foundation of China (No.62076173, No. 62176174), the High-level Entrepreneurship and Innovation Plan of Jiangsu Province (No. JSSCRC2021524), and the Project Funded by the Priority Academic Program Development of Jiangsu Higher Education Institutions.

References

1. Bhat, A., Modi, A.: Multi-task learning framework for extracting emotion cause span and entailment in conversations. In: Transfer Learning for Natural Language Processing Workshop, pp. 33–51. PMLR (2023)

2. Chakrabarty, T., Hidey, C., Muresan, S., McKeown, K., Hwang, A.: Ampersand: argument mining for persuasive online discussions. arXiv preprint arXiv:2004.14677 (2020)
3. Chen, Y., Hou, W., Li, S., Wu, C., Zhang, X.: End-to-end emotion-cause pair extraction with graph convolutional network. In: Proceedings of the 28th International Conference on Computational Linguistics, pp. 198–207 (2020)
4. Chen, Y., Lee, S.Y.M., Li, S., Huang, C.R.: Emotion cause detection with linguistic constructions. In: Proceedings of the 23rd International Conference on Computational Linguistics (Coling 2010), pp. 179–187 (2010)
5. Cho, K., et al.: Learning phrase representations using RNN encoder-decoder for statistical machine translation. arXiv preprint arXiv:1406.1078 (2014)
6. Das, D., Bandyopadhyay, S.: Finding emotion holder from bengali blog texts—an unsupervised syntactic approach. In: Proceedings of the 24th Pacific Asia Conference on Language, Information and Computation. pp. 621–628 (2010)
7. Ding, J., Kejriwal, M.: An experimental study of the effects of position bias on emotion cause extraction. arXiv preprint arXiv:2007.15066 (2020)
8. Ding, Z., Xia, R., Yu, J.: Ecpe-2d: Emotion-cause pair extraction based on joint two-dimensional representation, interaction and prediction. In: Proceedings of the 58th Annual Meeting of the Association for Computational Linguistics, pp. 3161–3170 (2020)
9. Fan, C., Yuan, C., Gui, L., Zhang, Y., Xu, R.: Multi-task sequence tagging for emotion-cause pair extraction via tag distribution refinement. IEEE/ACM Trans. Audio Speech Lang. Process. **29**, 2339–2350 (2021)
10. Gui, L., Xu, R., Lu, Q., Wu, D., Zhou, Y.: Emotion cause extraction, a challenging task with corpus construction. In: Li, Y., Xiang, G., Lin, H., Wang, M. (eds.) SMP 2016. CCIS, vol. 669, pp. 98–109. Springer, Singapore (2016). https://doi.org/10.1007/978-981-10-2993-6_8
11. Gui, L., Yuan, L., Xu, R., Liu, B., Lu, Q., Zhou, Y.: Emotion cause detection with linguistic construction in Chinese Weibo text. In: Zong, C., Nie, J.-Y., Zhao, D., Feng, Y. (eds.) NLPCC 2014. CCIS, vol. 496, pp. 457–464. Springer, Heidelberg (2014). https://doi.org/10.1007/978-3-662-45924-9_42
12. He, Y., Zhang, Z., Zhao, H.: Multi-tasking dialogue comprehension with discourse parsing. arXiv preprint arXiv:2110.03269 (2021)
13. Hu, G., Lu, G., Zhao, Y.: Bidirectional hierarchical attention networks based on document-level context for emotion cause extraction. In: Findings of the Association for Computational Linguistics (EMNLP 2021), pp. 558–568 (2021)
14. Lee, S.Y.M., Chen, Y., Huang, C.R.: A text-driven rule-based system for emotion cause detection. In: Proceedings of the NAACL HLT 2010 Workshop on Computational Approaches to Analysis and Generation of Emotion in Text, pp. 45–53 (2010)
15. Li, J., et al.: Molweni: a challenge multiparty dialogues-based machine reading comprehension dataset with discourse structure. arXiv preprint arXiv:2004.05080 (2020)
16. Li, X., Gao, W., Feng, S., Wang, D., Joty, S.: Span-level emotion cause analysis with neural sequence tagging. In: Proceedings of the 30th ACM International Conference on Information & Knowledge Management, pp. 3227–3231 (2021)
17. Li, X., Gao, W., Feng, S., Zhang, Y., Wang, D.: Boundary detection with bert for span-level emotion cause analysis. In: Findings of the Association for Computational Linguistics (ACL-IJCNLP 2021), pp. 676–682 (2021)
18. Li, Y., Tarlow, D., Brockschmidt, M., Zemel, R.: Gated graph sequence neural networks. arXiv preprint arXiv:1511.05493 (2015)

19. Liu, Y., et al.: Roberta: a robustly optimized bert pretraining approach. arXiv preprint arXiv:1907.11692 (2019)
20. Loshchilov, I., Hutter, F.: Fixing weight decay regularization in adam. arXiv preprint arXiv:1711.05101 (2017)
21. Poria, S., et al.: Recognizing emotion cause in conversations. Cogn. Comput. **13**, 1317–1332 (2021)
22. Wang, A., et al.: A structure self-aware model for discourse parsing on multi-party dialogues. In: Zhou, Z. (ed.) Proceedings of the Thirtieth International Joint Conference on Artificial Intelligence (IJCAI 2021), Virtual Event, Montreal, 19–27 August 2021, pp. 3943–3949. ijcai.org (2021).https://doi.org/10.24963/ijcai.2021/543
23. Wei, P., Zhao, J., Mao, W.: Effective inter-clause modeling for end-to-end emotion-cause pair extraction. In: Proceedings of the 58th Annual Meeting of the Association for Computational Linguistics, pp. 3171–3181 (2020)
24. Yu, N., Fu, G., Zhang, M.: Speaker-aware discourse parsing on multi-party dialogues. In: Proceedings of the 29th International Conference on Computational Linguistics, pp. 5372–5382. International Committee on Computational Linguistics, Gyeongju (2022). https://aclanthology.org/2022.coling-1.477
25. Zhang, D., Yang, Z., Meng, F., Chen, X., Zhou, J.: Tsam: a two-stream attention model for causal emotion entailment. arXiv preprint arXiv:2203.00819 (2022)

DAformer: Transformer with Domain Adversarial Adaptation for EEG-Based Emotion Recognition with Live-Oil Paintings

Zhong-Wei Jin[1], Jia-Wen Liu[1], Wei-Long Zheng[1,2], and Bao-Liang Lu[1,2,3,4](✉)

[1] Department of Computer Science and Engineering, Shanghai Jiao Tong University, Shanghai 200240, China
{izzie_kim,ljw_venn,weilong,bllu}@sjtu.edu.cn

[2] Key Laboratory of Shanghai Commission for Intelligent Interaction and Cognitive Engineering, Shanghai Jiao Tong University, Shanghai 200240, China

[3] Center for Brain-Machine Interface and Neuromodulation, RuiJin Hospital, Shanghai Jiao Tong University School of Medicine, Shanghai 200020, China

[4] RuiJin-Mihoyo Laboratory, RuiJin Hospital, Shanghai Jiao Tong University School of Medicine, Shanghai 200020, China

Abstract. The emergence of domain adaptation has brought remarkable advancement to EEG-based emotion recognition by reducing subject variability thus increasing the accuracy of cross-subject tasks. A wide variety of materials have been employed to elicit emotions in experiments, however, artistic works that aim to evoke emotional resonance of observers are relatively less frequently utilized. Previous research has shown promising results in electroencephalogram(EEG)-based emotion recognition on static oil paintings. As video clips are widely recognized as the most commonly used and effective stimuli, we adopted animated live oil paintings, a novel set of emotional stimuli in the live form which are essentially a type of video clip while possessing fewer potential influencing factors for EEG signals compared to traditional video clips, such as abrupt switches on background sound, contrast, and color tones. Moreover, previous studies on static oil paintings focused primarily on the subject-dependent task, and further research involving cross-subject analysis remains to be investigated. In this paper, we proposed a novel DAformer model which combines the advantages of Transformer and adversarial learning. In order to enhance the evocative performance of oil paintings, we introduced a type of innovative emotional stimuli by transforming static oil paintings into animated live forms. We developed a new emotion dataset SEED-LOP (SJTU EEG Emotion Dataset-Live Oil Painting) and constructed DAformer to verify the effectiveness of SEED-LOP. The results demonstrated higher accuracies in three-class emotion recognition tasks when watching live oil paintings, with a subject-dependent accuracy achieving 61.73% and a cross-subject accuracy reaching 54.12%.

B. Luo et al. (Eds.): ICONIP 2023, CCIS 1963, pp. 402–414, 2024.
https://doi.org/10.1007/978-981-99-8138-0_32

Keywords: Emotion recognition · Transfer learning · Electroencephalogram · Oil paintings

1 Introduction

Affective computing aims at identifying, analyzing, and interpreting human emotional states by employing various techniques such as natural language processing, data mining, and machine learning to recognize and understand emotional information effectively, among which electroencephalogram(EEG)-based emotion recognition draws the most attention for its neural pattern stability in emotion [1,2]. In recent years, a diverse range of emotion elicitation materials has been employed in studies focusing on emotion recognition based on EEG signals: images [3], video clips [4], musical segments [5], etc. However, artistic works which aim to evoke the emotional resonance of observers are relatively less frequently utilized. Luo *et al.* innovatively utilized oil paintings as emotional stimuli and demonstrated their effectiveness in a subject-dependent EEG-based emotion recognition task [6]. Subsequently, Lan *et al.* employed Transformer to conduct subject-dependent emotion recognition on the oil painting dataset and achieved promising performance [7].

In recent years, cross-subject emotion recognition has emerged as a research focus due to its compatibility with the practical requirements of real-world applications. With the development of transfer learning, remarkable advancement has been brought to cross-subject emotion recognition tasks by reducing subject variability. Domain adaptation (DA) is a highly important branch of transfer learning [8] whose primary objective is to effectively map features from various source domains into a unified feature space, with the goal of eliminating the domain discrepancy between the source and target domains, thus increasing the accuracy of the target domain. Methods based on domain adversarial neural networks (DANNs) were proposed to identify common representations of EEG signals among all subjects, thereby improving the cross-subject performance [9]. Apart from adversarial methods, subdomain adaption (DSAN) has also yielded promising results. DSAN partitions similar samples within a domain into subdomains by certain criteria and aligns these subdomains rather than the global domain alignment [10].

In addition to advanced classification methods, more neural network architectures for EEG signals to extract efficient representations in various domains have also garnered significant attention in the field of affective computing. Wang *et al.* employed an attention mechanism to fuse EEG and eye signals [11]. Spectral-spatial-temporal adaptive graph convolutional neural network (SST-AGCN) was designed to extract EEG features from spectral, temporal, and spatial domains based on a graph convolutional neural network and achieved outstanding performance in a subject-dependent confidence estimation task [12].

As mentioned before, previous studies of oil painting datasets focused mainly on the subject-dependent tasks [6,7]. However, the effectiveness of this paradigm has not yet been validated across subjects which lays the foundation for practical cross-subject applications. This paper aims to address this issue by introducing a novel DAformer model. By combining the advantage of the attention mechanism in feature extraction of EEG signals with the effectiveness of domain adversarial methods for the cross-subject task, we employed DAformer to recognize three different classes of emotions (negative, neutral, positive) from EEG signals. Since the emotional film clip is one of the most popular stimuli which has been proven to be effective [13], we fabricated a novel set of emotional stimuli of animated live oil paintings in order to enhance the emotion-inducing effects. We conducted experiments to collect EEG signals and eye movements of participants under the new stimuli and developed a new dataset: SEED-LOP (SJTU Emotion EEG Dataset-Live Oil Painting)[1]. Finally, we demonstrated the superior performance of the DAformer model as well as the feasibility of the SEED-LOP dataset. The main contribution of this new dataset lies in its revolutionary implementation of 2D live art pieces as stimuli, which has established a more streamlined and lightweight experimental paradigm compared to traditional emotional video clips and provided innovative methods in the field of affective computing.

2 Methods

We constructed the DAformer model to verify the effectiveness of SEED-LOP (Fig. 1). DAformer is composed of a feature extractor G_f, an emotion predictor G_y, and a domain discriminator G_d. We employed a multi-layer encoder based on the multi-head self-attention mechanism as the feature extractor, a fully connected feed-forward network for emotion prediction, together with a domain adversarial module to eliminate the domain distribution discrepancy across different subjects. The gradient reversal layer propagates the gradients from the domain discriminator reversely to the feature extractor during backward propagation, thereby enabling the feature extractor to extract invariant EEG features in different domains. The training dataset included EEG signals from both labeled data in the source domain and unlabeled data in the target domain. All data are labeled by their corresponding domains to train the domain discriminator, while only source data and its emotion labels are utilized to train the emotion predictor to predict three types of emotions (0: negative, 1: neutral, 2: positive). The test dataset consists of target data and their emotion labels to evaluate the performance of the feature extractor and emotion predictor.

2.1 Data Preprocessing

The raw EEG signals were first downsampled to 200 Hz and processed with a 1–70 Hz bandpass filter and a 50 Hz notch filter. We extracted the differential entropy (DE) features of EEG signals as it is proven to have excellent performance in previous studies [11,12,14]. The preprocessed EEG data underwent the short-time Fourier transform (STFT) using a one-second Hanning window. This

[1] https://bcmi.sjtu.edu.cn/home/seed/.

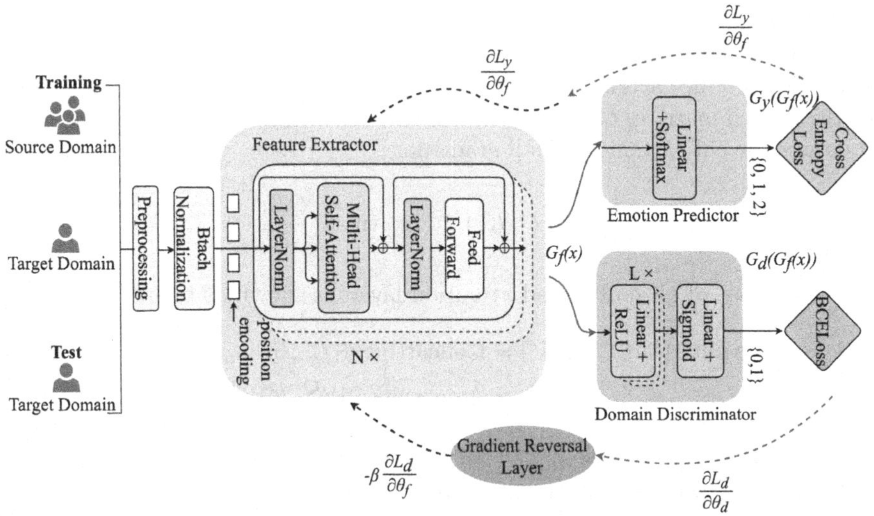

Fig. 1. The overall architecture of the DAformer. Solid arrows represent the forward propagation and dashed arrows represent the backward propagation. The gradient reversal layer propagates the gradient reversely to the feature extractor during backward propagation, thereby enabling the feature extractor to learn invariant representations in the source and target domains.

transformation was performed on each epoch to extract the differential entropy (DE) features from five distinct frequency bands: delta (1–3 Hz), theta (4–7 Hz), alpha (8–13 Hz), beta (14–30 Hz), and gamma (31–50 Hz). The extracted DE features of EEG signals are defined as $X \in \mathbb{R}^{N \times F \times C}$ where N stands for the total number of samples, F represents five frequency bands and C denotes the EEG channels.

2.2 DAformer

Feature Extractor. As EEG signals are temporal signals, we employed an encoder based on the multi-head self-attention mechanism as the feature extractor. A batch normalization layer is applied at the beginning to avoid over-fitting and accelerate the training process. After batch normalization and position encoding of a sequence length L_{seq}, the representations of EEG signals are fed into the encoder which consists of N identical layers, each of the layers contains two sub-layers: a multi-head attention layer and a fully connected feed-forward network. Both sub-layer is preceded by a layer normalization and applied with a residual connection.

A multi-head attention layer consists of a parallel concatenation of N scaled dot product attention layers. First of all, the original input $X \in \mathbb{R}^{B \times T \times D}$ where B denotes the batch size, T the overlapping window, and D the feature dimension of EEG signals after preprocessing. After a linear transformation, a query of dimension d_k, a key of dimension d_k, and a value of dimension d_v are generated to compose the input of the scaled dot-product attention. The output of a single scaled dot-product attention is calculated as below, where Q, K, and V stand for the packed matrices of queries, keys, values, and d_k denotes the scaling factor in order to avoid extremely small gradients:

$$\text{Attention}(Q, K, V) = \text{softmax}(\frac{QK^T}{\sqrt{d_k}})V. \tag{1}$$

The output of all multi-head attention layers is shown as below:

$$\begin{aligned} \text{MultiHead}(Q, K, V) &= \text{Concat}(\text{head}_1, ..., \text{head}_\text{h})W^O, \\ \text{head}_i &= \text{Attention}(QW_i^Q, KW_i^K, VW_i^V), \end{aligned} \tag{2}$$

where the weight matrices $W_i^Q \in \mathbb{R}^{d_{en} \times d_k}$, $W_i^K \in \mathbb{R}^{d_{en} \times d_k}$, $W_i^V \in \mathbb{R}^{d_{en} \times d_v}$, $W_i^O \in \mathbb{R}^{hd_v \times d_{en}}$ and d_{en} denotes the output dimension of the encoder.

The output is then fed into a residual connection before the fully connected feed-forward network which comprises two linear transformations connected by a ReLU activation function:

$$\begin{aligned} x &= \text{LayerNorm}(x + \text{Sublayer}(x)), \\ \text{FFN}(x) &= \max(0, xW_1 + b_1)W_2 + b_2. \end{aligned} \tag{3}$$

Emotion Predictor. We utilized a single layer fully connected feed-forward network as the emotion predictor.

Domain Discriminator. The domain discriminator is composed of L layers where each is a single linear transformation with a ReLU activation function. The output of the last linear transformation is activated by a Sigomoid activation function. Between feature extractor G_f and domain discriminator G_d, a gradient reversal layer is employed to propagate the gradient reversely, thereby fostering an adversarial training process between G_f and G_d. As G_y and G_d gain increasing precision in classifying emotion types and domain types, the reversed gradient is intended to guide G_f in extracting features that are indistinguishable for G_d.

We denoted $x_{d,i}$ the ith input of the model, $d \in \{s, t\}$ which represents the domain of $x_{d,i}$ (s for source domain and t for target domain), the feature extractor produces its output $G_f(x_{d,i})$. The extracted features of source domain data are then fed into the emotion predictor and domain discriminator while those of the target domain are only applied to the domain discriminator. The emotion predictor predicts the emotion label $G_y(G_f(x_{s,i}))$ only for source data, while the domain discriminator makes domain prediction $G_d(G_f(x_{d,i}))$ for both source data and target data.

We utilized the cross entropy loss for the emotion predictor. The loss function can be presented as:

$$L_y = -\frac{1}{|s|}\sum_{i=1}^{|s|} y_{s,i} \log \hat{y}_{s,i}, \tag{4}$$

where $\hat{y}_{s,i}$ denotes the output of the emotion predictor: $\hat{y}_{s,i} = G_y(G_f(x_{s,i}))$ and $y_{s,i}$ denotes the true emotion label of $x_{s,i}$.

We applied the binary cross entropy loss for the domain discriminator. The loss functions for the source domain and target domain are expressed as:

$$L_s = -\frac{1}{|s|}\sum_{i=1}^{|s|}(d_{s,i}\log \hat{d}_{s,i} + (1-d_{s,i}) * \log \hat{d}_{s,i}), \tag{5}$$

$$L_t = -\frac{1}{|t|}\sum_{i=1}^{|t|}(d_{t,i}\log \hat{d}_{t,i} + (1-d_{t,i}) * \log \hat{d}_{t,i}), \tag{6}$$

where $\hat{d}_{s,i}$ and $\hat{d}_{t,i}$ denote the predictions of the domain discriminator: $\hat{d}_{s,i} = G_d(G_f(x_{s,i}))$, $\hat{d}_{t,i} = G_d(G_f(x_{t,i}))$, respectively, and $d_{s,i}$, $d_{t,i}$ denote the true domain label of $x_{s,i}$, $x_{t,i}$, respectively.

The total loss of DAformer is composed as below:

$$L = L_y + \beta(L_s + L_t), \tag{7}$$

where β is a hyperparameter that balances the weight between the emotion class loss and the domain loss.

We optimized the parameters of the feature extractor and the emotion predictor by minimizing the total loss function, and we updated the parameters of the domain discriminator by maximizing the total loss function:

$$\begin{aligned} \hat{\theta}_f, \hat{\theta}_y &= argmin_{\theta_f, \theta_y}\mathcal{L}(\theta_f, \theta_y, \hat{\theta}_d), \\ \hat{\theta}_d &= argmax_{\theta_d}\mathcal{L}(\hat{\theta}_f, \hat{\theta}_y, \theta_d). \end{aligned} \tag{8}$$

3 Experiments

We designed a novel emotional experiment paradigm using animated live oil paintings as stimuli. The value of this innovative dataset lies in its groundbreaking use of 2D live art pieces as stimuli, which has achieved a more lightweight experimental paradigm compared to video clips and yielded effective classification results. A total of 40 healthy, right-handed participants (20 males, 20 females) aged between 17 and 29 years (mean age: 21.6 years, standard deviation: 2.95) were recruited from Shanghai Jiao Tong University for the experiment. Each participant viewed 60 oil paintings (20 positive, 20 neutral, 20 negative) during the experiments, and both their EEG signals and eye movements were recorded during the observation of paintings. As previous studies proved that the power of the theta band and the alpha band power of EEG signals differ between artists and non-artists in response to abstract and representational paintings [15], our participants were selected intentionally as half artistically experienced and half artistically naive under the results of the art experience questionnaire [16]. We successfully collected EEG signals and eye movements of 40 participants with a statistic distribution of 28% negative emotion samples, 34% neutral emotion samples, and 38% positive emotion samples.

3.1 Dataset

Stimuli. Luo *et al.* employed a total of 114 world-renowned paintings spanning from the mid-16th century to the 19th century [6]. The paintings encompassed a diverse range of genres, including portraiture, animal depiction, still life, landscape, and cityscape, which were representative of most major art styles. Questionnaires were distributed among students of Shanghai Jiao Tong University and China Academy of Art on the perception of emotion types (negative, neutral, positive) of the 60 paintings as well as their level of intensity. We selected 60 paintings (20 of each type of emotion) that have the highest intensity ratings and put them into our stimuli set. Before the animation of the oil paintings, we conducted an experiment to collect the eye movement heatmaps during the observations of these paintings in order to locate the parts that draw the most attention. 40 healthy participants were recruited (9 males, 31 females, age: 25.35 ± 3.71) to observe the 60 oil paintings selected. Each painting was presented for 20 s with a one-second interval of focus time between each display. The eye movements of the participants during the observation were recorded by a Tobii Pro X3-120 screen-based eye tracker at a sampling rate of 120 Hz. We produced an attention heatmap for each painting by calculating the average gaze point of all participants (Fig. 2). Based on the attention heatmaps, we successfully transformed 60 static oil paintings into animated live GIFs using Photoshop software and Cartoon Animator 4 software.

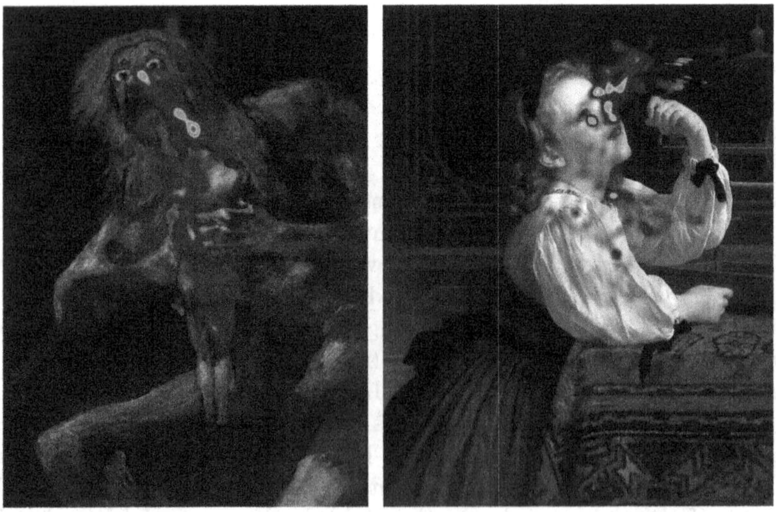

Fig. 2. Examples of attention heatmaps of the average eye movements of 40 participants during observations of the oil paintings from our stimuli set. The transition from static oil paintings to animation form was based on the attention heatmaps.

Procedure. The experimental protocol is presented in Fig. 3 to illustrate the details of our main experiment. In the experiment for each participant, 60 animated oil paintings were organized into 5 groups. Each group comprised 3 randomly decided different emotion type batches, with each batch consisting of 4 paintings randomly selected with the same emotion class in order to minimize the occurrence of frequent emotional transformation within a short timeframe. Before the experiment, the participants were briefed on the experimental protocol and shown a tutorial of three example paintings representing negative, neutral, and positive emotions as a reference level for rating. Each painting was preceded by a one-second focus time of a black screen with a white + symbol in the middle. The duration of display of each painting was fixed at 20 s, during which the participants were asked to intently observe and perceive the painting. After a one-second focus session that followed, participants were instructed to report both their emotional perception of the emotion type (negative, neutral, and positive) and the intensity (1–9) at the rating session with no time limit.

During the experiments, the EEG signals of subjects were collected by an ESI NeuroScan System with a 62-channel module arranged according to the international 10–20 system at a sampling rate of 1000 Hz. Eye movements were recorded by a Tobii Pro X3-120 screen-based eye tracker at a sampling rate of 120 Hz.

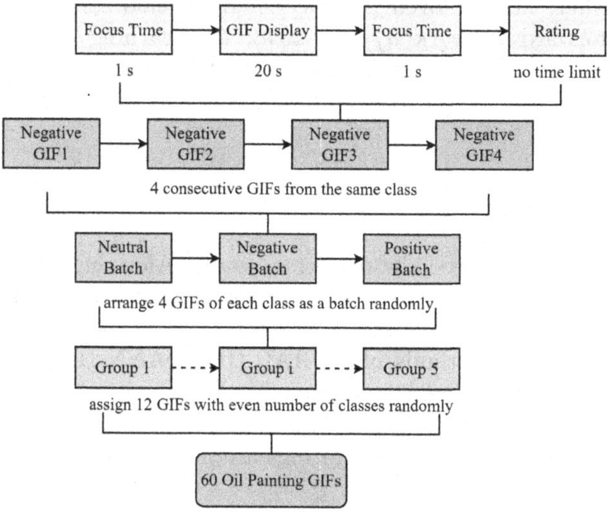

Fig. 3. Illustration of the experimental protocol.

3.2 Implementation Details

During the experiment, the reported emotion categories of the participants (0: negative emotion, 1: negative emotion, 2: positive emotion) were used as classification labels to investigate the effectiveness of this paradigm in emotion classification tasks based on EEG signals. We utilized EEG features in a total of five frequency bands. In the subject-dependent experiment, all models were trained on each participant using three-fold cross-validation, while the data under the observation of the same oil painting did not appear in both the training sets and the testing sets simultaneously. For the cross-subject experiment, we employed Leave-One-Out-Cross-Validation. To evaluate the performance of DAformer, we compared our model with four other classifiers, including support vector machine (SVM), multilayer perceptron (MLP), Transformer [17], and spectral-spatial-temporal adaptive graph convolutional neural network (SST-AGCN) [12], both in the subject-dependent task and the cross-subject task, as well as the comparison of four different domain adaptation methods: correlation alignment (CORAL) [18], domain-adversarial neural networks (DANN) [19], dynamic adversarial adaptation network (DAAN) [20] and deep subdomain adaptation network (DSAN) [10] additionally for the cross-subject task.

In our experiments, we set batch size $B = 128$, window size $= 4$ s thus window number $T = 5$. The number of channels of EEG signals was 62 for our experiment, and together with the 5 frequency bands, the feature dimension D of EEG signals added to 310. For both the subject-dependent task and the cross-subject task, the SVM classifiers were applied with the RBF kernel with the range of parameter C is $2^{[-8:8]}$.

For DAformer, we employed the encoder dimension $d_{en} = 512$, the dimension of feed-forward network $d_{ff} = 2048$, the learning rate in a range of $[1e-3, 5e-4, 1e-4]$, the domain loss weight $\beta = 10$, the layer number of domain discriminator $L = 1$ with 256 dimensions and the sequence length L_{seq} within a range of $[1, 2, 4]$.

3.3 Results Analysis

In this section, we compared the performance of DAformer with the other four classifiers, including SVM, MLP, Transformer [17], and SST-AGCN [12], both in the subject-dependent task and the cross-subject task, with the comparison of 4 different DA methods of CORAL [18], DANN [19], DAAN [20] and DSAN [10] in order to recognize emotions under stimuli of active oil paintings for the cross-subject task. To further analyze the performance of SEED-LOP, the neural patterns under live oil paintings are also investigated.

Table 1. Results of the subject-dependent task on SEED-LOP

Model	SVM	MLP	SST-AGCN	Transformer
mean	53.67	56.47	56.58	**61.73**
std	7.85	8.86	4.46	7.65

Emotion Recognition Performance of DAformer. The mean accuracies and standard deviation of SVM, MLP, SST-AGCN, and Transformer under the subject-dependent task are listed in Table 1 while Table 2 presents the results of SVM, MLP, SST-AGCN, Transformer, DANN, CORAL, DAAN, DSAN and DAformer. The experimental results indicate that for the subject-dependent task, Transformer achieved the best performance with an accuracy of 61.73%/7.65%, while among the nine models for the cross-subject task, DAformer exhibited superior performance with the highest classification accuracy of 54.12%/6.89%.

Table 2. Results of the cross-subject task on SEED-LOP

Model	SVM	MLP	SST-AGCN	Transformer
mean	39.07	47.54	50.59	50.09
std.	8.47	9.86	6.88	6.92

Model	DANN	CORAL	DAAN	DSAN	DAformer
mean	49.23	44.29	46.08	45.54	**54.12**
std.	6.85	8.31	9.31	8.61	6.89

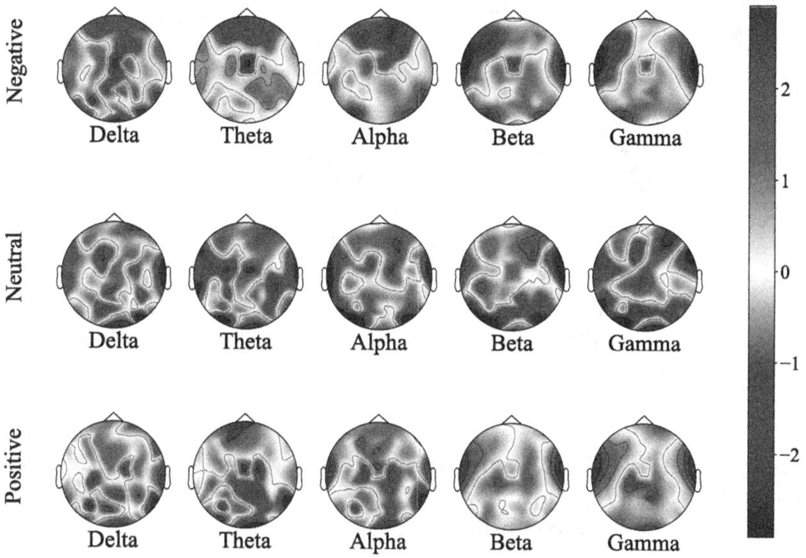

Fig. 4. The average topographic maps of the 40 subjects for three emotion classes with the five frequency bands. The row denotes the different emotion classes and the column denotes the different frequency bands.

Visualization of the Brain Topographic Maps. The neural patterns of the five frequency bands are depicted in Fig. 4. These patterns were derived by averaging the DE features across all 40 participants for each EEG channel. It demonstrates that the lateral temporal areas are more active in the beta and gamma band and the prefrontal sites have higher theta and alpha responses for positive emotions than negative emotions. While the neural patterns of neutral emotions show a greater activation at prefrontal sites for all bands and a stronger gamma response at the occipital sites. We also observed that the negative emotion patterns have significantly higher activation in all bands at the occipital sites and parietal sites. These observations highlight the potential existence of neural patterns associated with the stimuli of animated oil paintings.

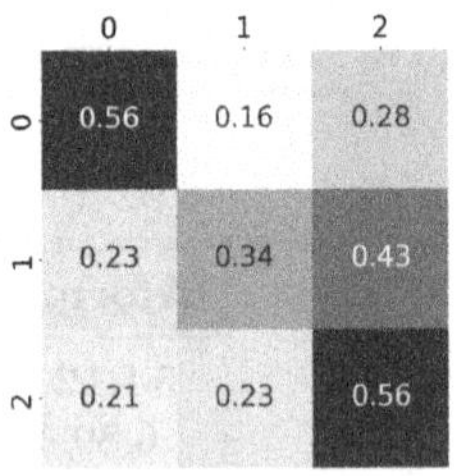

(a) SVM

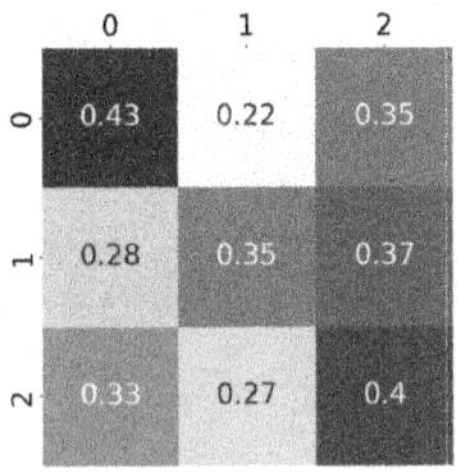

(b) SST-AGCN

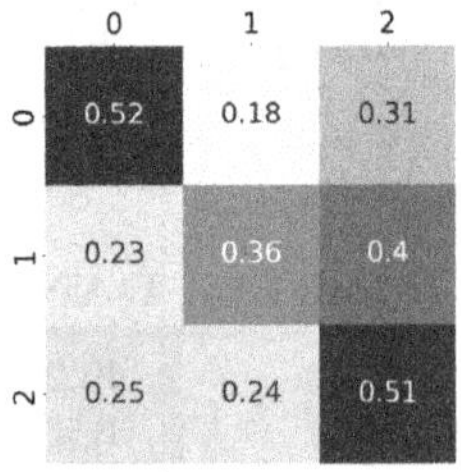

(c) Transformer

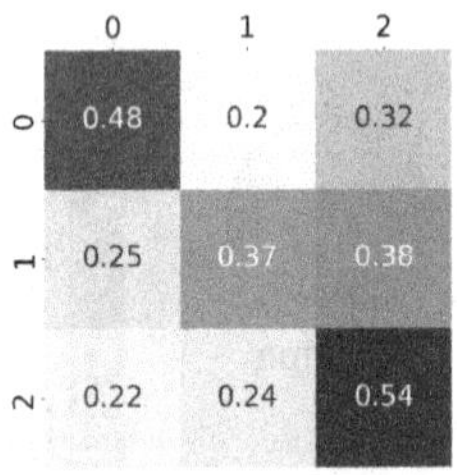

(d) MLP

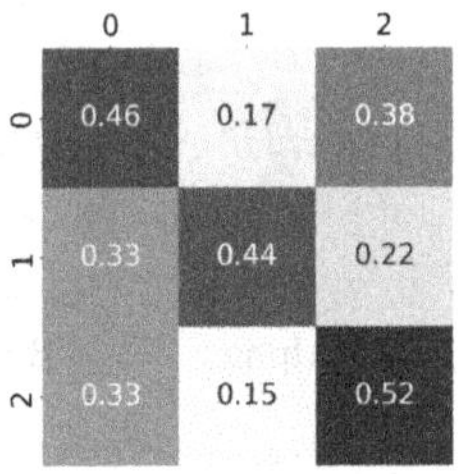

(e) SST-AGCN

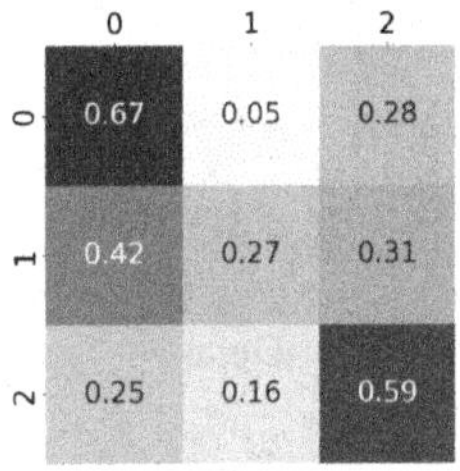

(f) DAformer

Fig. 5. Confusion matrices of experiments. 0: negative, 1: neutral, 2: positive; (a), (b), (c): SVM, SST-AGCN, Transformer in the subject-dependent task, (d), (e), (f): MLP, SST-AGCN, DAformer in the cross-subject task.

Confusion Matrix of the Experiments. To further analyze the experimental outcomes of emotion recognition on SEED-LOP, we present the confusion matrices in Fig. 5. From these confusion matrices, we observe that negative emotion and positive emotion are more readily distinguishable by the model with higher accuracies while there exists a tendency for neutral emotions to be more easily predicted as positive emotion, which aligns with the neural pattern observed in the average topographic map. The topographic map indicates a greater similarity in the neural pattern between neutral and positive emotions as they both show higher activation in prefrontal sites at the theta and alpha bands. This phenomenon also corresponds to the reports of some participants that they believed to have observed a smaller number of neutral emotion paintings in comparison to paintings of the other two emotions. However, we have utilized an equal number of neutral emotion paintings compared to the other two categories in the experiments. This finding suggests that the criteria for rating neutral emotions may be ambiguous for different participants, leading to confusion in the classification task of the models.

4 Conclusion

In this paper, we proposed a novel EEG-based emotion recognition experimental paradigm by utilizing animated oil paintings as stimuli and created the dataset SEED-LOP. The significance of this new dataset lies in its revolutionary utilization of 2D live art pieces as stimuli, which has established a more streamlined and lightweight experimental paradigm in comparison to traditional emotional video clips. We employed Transformer for subject-dependent tasks and introduced DAformer by combining the effectiveness of Transformer with domain adversarial methods for cross-subject tasks to test our dataset SEED-LOP. The experimental results demonstrate that Transformer performs an outstanding accuracy of 61.73% on the subject-dependent task and an improved accuracy of 54.12% of DAformer on the cross-subject task which both serves as an evidence for the effectiveness of the SEED-LOP.

Acknowledgements. This work was supported in part by grants from National Natural Science Foundation of China (Grant No. 61976135), STI 2030-Major Projects+2022ZD0208500, Shanghai Municipal Science and Technology Major Project (Grant No. 2021SHZD ZX), Shanghai Pujiang Program (Grant No. 22PJ1408600), Medical-Engineering Interdisciplinary Research Foundation of Shanghai Jiao Tong University "Jiao Tong Star" Program (YG2023ZD25), and GuangCi Professorship Program of RuiJin Hospital Shanghai Jiao Tong University School of Medicine.

References

1. Alarcao, S.M., Fonseca, M.J.: Emotions recognition using EEG signals: a survey. IEEE Trans. Affect. Comput. **10**(3), 374–393 (2017)
2. Zheng, W.-L., Zhu, J.-Y., Lu, B.-L.: Identifying stable patterns over time for emotion recognition from EEG. IEEE Trans. Affect. Comput. **10**(3), 417–429 (2019)
3. Schaaff, K., Schultz, T.: Towards emotion recognition from electroencephalographic signals. In: 3rd International Conference on Affective Computing and Intelligent Interaction and Workshops, pp. 1-6. IEEE (2009)
4. Koelstra, S., Muhl, C., Soleymani, M.: Deap: a database for emotion analysis; using physiological signals. IEEE Trans. Affect. Comput. **3**(1), 18–31 (2011)
5. Lin, Y.P., et al.: EEG-based emotion recognition in music listening. IEEE Trans. Biomed. Eng. **57**(7), 1798–1806 (2010)

6. Luo, S., Lan, Y.T., Peng, D., Li, Z., Zheng, W.L., Lu, B.L.: Multimodal emotion recognition in response to oil paintings. In: 44th Annual International Conference of the IEEE Engineering in Medicine & Biology Society (EMBC), pp. 4167-4170. IEEE (2022)
7. Lan, Y.T., Li, Z.C., Peng, D., Zheng, W.L., Lu, B.L.: Identifying artistic expertise difference in emotion recognition in response to oil paintings. In: 11th International IEEE/EMBS Conference on Neural Engineering (NER), pp. 1–4. IEEE (2023)
8. Ben-David, S., Blitzer, J., Crammer, K., Kulesza, A., Pereira, F., Vaughan, J.W.: A theory of learning from different domains. Mach. Learn. **79**, 151–175 (2010)
9. Li, H., Jin, Y.-M., Zheng, W.-L., Lu, B.-L.: Cross-subject emotion recognition using deep adaptation networks. In: Cheng, L., Leung, A.C.S., Ozawa, S. (eds.) ICONIP 2018. LNCS, vol. 11305, pp. 403–413. Springer, Cham (2018). https://doi.org/10.1007/978-3-030-04221-9_36
10. Zhu, Y., et al.: Deep subdomain adaptation network for image classification. IEEE Trans. Neural Networks Learn. Syst. **32**(4), 1713–1722 (2021)
11. Wang, Y., Jiang, W. B., Li, R., Lu, B.L.: Emotion transformer fusion: complementary representation properties of EEG and eye movements on recognizing anger and surprise. In: 2021 IEEE International Conference on Bioinformatics and Biomedicine (BIBM), pp. 1575–1578. IEEE (2021)
12. Li, R., Wang, Y., Lu, B.L.: Measuring decision confidence levels from EEG using a spectral-spatial-temporal adaptive graph convolutional neural network. In: Tanveer, M., Agarwal, S., Ozawa, S., Ekbal, A., Jatowt, A. (eds.) ICONIP 2022. LNCS, pp. 395–406. Springer, Cham (2022). https://doi.org/10.1007/978-981-99-1642-9_34
13. Schaefer, A., et al.: Assessing the effectiveness of a large database of emotion-eliciting films: a new tool for emotion researchers. Cogn. Emot. **24**(7), 1153–1172 (2010)
14. Zheng, W.L., Lu, B.L.: Personalizing EEG-based affective models with transfer learning. In: International Joint Conference on Artificial Intelligence, pp. 2732–2738. AAAI Press, New York (2016)
15. Batt, R., Palmiero, M., Nakatani, C., van Leeuwen, C.: Style and spectral power: processing of abstract and representational art in artists and non-artists. Perception **39**(12), 1659–1671 (2010)
16. Chatterjee, A., Widick, P., Sternschein, R., Smith, W.B., Bromberger, B.: The assessment of art attributes. Empir. Stud. Arts **28**(2), 207–222 (2010)
17. Vaswani, A., et al.: Attention is all you need. In: Advances in Neural Information Processing Systems, vol. 30 (2017)
18. Sun, B., Saenko, K.: Deep CORAL: correlation alignment for deep domain adaptation. In: Hua, G., Jégou, H. (eds.) ECCV 2016. LNCS, vol. 9915, pp. 443–450. Springer, Cham (2016). https://doi.org/10.1007/978-3-319-49409-8_35
19. Ganin, Y., et al.: Domain-adversarial training of neural networks. J. Mach. Learn. Res. **17**(1), 2030–2096 (2016)
20. Wang, J., Feng, W., Chen, Y., Yu, H., Huang, M., Yu, P.S.: Visual domain adaptation with manifold embedded distribution alignment. In: Proceedings of the 26th ACM International Conference on Multimedia, pp. 402-410 (2018)

Time-Frequency Transformer: A Novel Time Frequency Joint Learning Method for Speech Emotion Recognition

Yong Wang[1,3], Cheng Lu[1,2,4(✉)], Yuan Zong[1,2,4(✉)], Hailun Lian[1,3], Yan Zhao[1,3], and Sunan Li[1,3]

[1] Key Laboratory of Child Development and Learning Science of Ministry of Education, Nanjing 210096, China
[2] School of Biological Science and Medical Engineering, Southeast University, Nanjing 210096, China
{cheng.lu,xhzongyuan}@seu.edu.cn
[3] School of Information Science and Engineering, Southeast University, Nanjing 210096, China
[4] Pazhou Lab, Guangzhou 510320, China

Abstract. In this paper, we propose a novel time-frequency joint learning method for speech emotion recognition, called Time-Frequency Transformer. Its advantage is that the Time-Frequency Transformer can excavate global emotion patterns in the time-frequency domain of speech signal while modeling the local emotional correlations in the time domain and frequency domain respectively. For the purpose, we first design a Time Transformer and Frequency Transformer to capture the local emotion patterns between frames and inside frequency bands respectively, so as to ensure the integrity of the emotion information modeling in both time and frequency domains. Then, a Time-Frequency Transformer is proposed to mine the time-frequency emotional correlations through the local time-domain and frequency-domain emotion features for learning more discriminative global speech emotion representation. The whole process is a time-frequency joint learning process implemented by a series of Transformer models. Experiments on IEMOCAP and CASIA databases indicate that our proposed method outdoes the state-of-the-art methods.

Keywords: Speech emotion recognition · Time-frequency domain · Transformer

1 Introduction

Speech Emotion Recognition (SER) aims to automatically recognize the emotional states in human speech [1], which has become a hotspot in many fields,

Y. Wang and C. Lu—Equal Contributions.

B. Luo et al. (Eds.): ICONIP 2023, CCIS 1963, pp. 415–427, 2024.
https://doi.org/10.1007/978-981-99-8138-0_33

e.g., affective computing and Human-Computer Interaction (HCI) [2]. For SER task, how to obtain a discriminative emotion features is a key to realize the superior performance [3,4].

In order to recognition speech emotions well, early SER works mainly combined some low-level descriptor (LLD) features or their combinations [5], e.g., Mel-Frequency Cepstral Coefficients (MFCC), Zero-Crossing Rate, and Pitch, with classifiers [6], e.g., K-Nearest Neighbor (KNN) and Support Vector Machine (SVM), for emotion prediction. Furthermore, with the rapid development of deep learning, high-dimensional speech emotion features generated by deep neural networks (DNN), e.g., Convolutional Neural Network (CNN) [7] and Recurrent Neural Network (RNN) [8], have emerged on SER and achieved superior performance. Currently, the input features of DNNs are mainly based on spectrogram, e.g., magnitude spectrogram [9] and Mel-spectrogram [10], which are the time-frequency representations of speech signals.

The time and frequency domains of the spectrogram contain rich emotional information. To excavate them, a practical approach is joint time-frequency modeling strategy with the input features of spectrograms [3,10]. Among these methods, combining CNN and RNN structure, i.e., CNN+Long Short-Term Memory (LSTM), is a classic method, which utilizes CNN and RNN to encode the time-frequency information. For instance, Satt et al. [11] combined CNN with a special RNN (i.e., LSTM) to model the time-frequency domain of emotional speech. Wu et al. [12] proposed a recurrent capsules network to extract time-frequency emotional information from the spectrogram.

Although current time-frequency joint learning methods have achieved certain success on SER, they still suffer from two issues. The first one is that they usually shared the modeling both in time and frequency domains, ignoring the specificality of the respective domains. For instance, the time-frequency domain shares a uniform size convolution kernel in CNN [9] and a uniform-scale feature map is performed on the time-frequency domain in RNN [11]. Therefore, separate modeling of time-domain and frequency-domain information should be considered to ensure the specificity and integrity of the encoding of time-frequency domain information. The other issue is that only some low-level feature fusion operations (e.g., splicing and weighting) are adopted in time frequency joint learning process, leading to poor discriminativeness of fusion features [10]. This indicates that the effective fusion of emotional information in the time-frequency domain is also the key for time-frequency joint learning.

To cope with the above issues, we propose a novel time frequency domain joint learning method for SER, called Time-Frequency Transformer, which consisting of three modules, i.e., the Transformer modules of Time, Frequency, and Time-Frequency, as shown in Fig. 1. Firstly, the Time and Frequency Transformer modules are designed to model the local emotion correlations between frames and inside frequency bands respectively, ensuring the integrity of the emotion information in the time-frequency domain. Then, we also propose a Time-Frequency Transformer module to excavate the time-frequency emotional correlations through the local time- frequency emotion features for learning more discriminative global speech emotion representation.

Overall, our contributions in this paper can be summarized as the following three points: *(1) We propose a novel time frequency joint learning method based on Transformers (i.e., Time-Frequency Transformer), which can effectively excavate the local emotion information both in time frames and frequency bands to aggregate global speech emotion representations. (2) We propose a Time Transformer and Frequency Transformer to ensure the integrity of modeling time-frequency local emotion representations. (3) Our proposed Time-Frequency Transformer outperform on the state-of-the art methods on IEMOCAP database and CASIA database.*

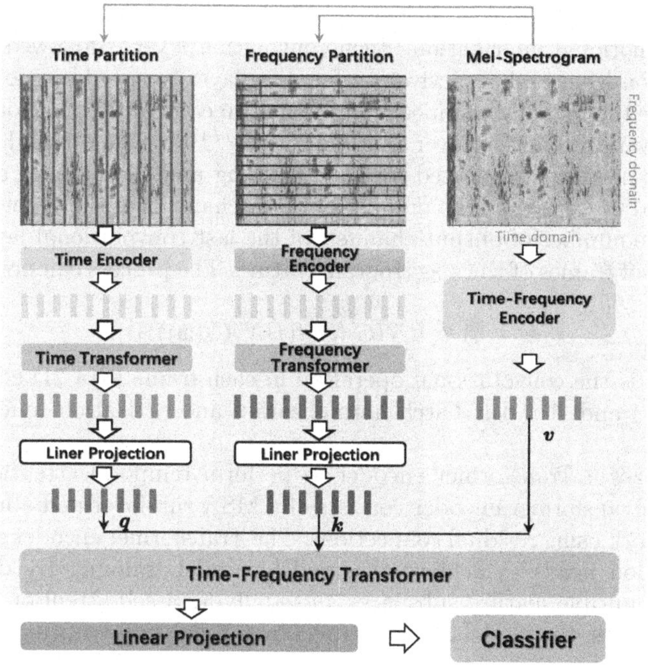

Fig. 1. Overview of Time-Frequency Transformer for speech emotion recognition. It mainly consists of three modules, i.e., Time Transformer module (light orange background), Frequency Transformer module (light blue background), and Time-Frequency Transformer module (light green background). (Color figure online)

2 Proposed Method

In this section, we will introduce our proposed Time-Frequency Transformer shown in Fig. 1, including three modules, i.e., Time Transformer module, Frequency Transformer module and Time-Frequency Transformer module.

2.1 Time Transformer Module

The role of Time Transformer is to capture the local emotion correlations across time frames for time domain encoding of emotional speech. This module utilizes the time encoder to reduce the dimensionality of the input feature $\boldsymbol{x}$ of the model and extract part of the emotional information in time-domain of $\boldsymbol{x}$. Then, Multi-head Self-Attention (MSA) mechanism of the Transformer encoder is used to calculate the emotional correlation between frames, making the model focus on emotion related frames and reducing the impact of emotion unrelated frames on speech emotion recognition.

Basically, the time encoder consists of a 2D convolutional layer, a 2D batch normalization layer, an activation function layer, a 2D convolutional layer, a 2D batch normalization layer and an activation function layer in sequence. We convolute the input spectrogram feature $\boldsymbol{x} \in \mathbb{R}^{b \times c \times f \times d}$ frame by frame and extract temporal emotional information. Each convolution layer is followed by a batch normalization layer and an activation function layer to speed up model training and improve the model's nonlinear representation capabilities. The output of the second activation function layer is $\hat{\boldsymbol{x}}' \in \mathbb{R}^{b \times c1 \times (f/4) \times d}$, where b and f represent the number of samples selected for each training and the number of Mel-filter banks, c and *c1* represent the number of input channels of the spectrogram feature and the number of output channels of the last convolutional layer and d is the number of frames of the spectrogram feature. The process can be represented as

$$\hat{\boldsymbol{x}}' = Act(BN(C_t(Act(BN(C_t(\boldsymbol{x})))))), \tag{1}$$

where $C_t(\cdot)$ is the convolutional operation in each frame by a 2D convolutional layer, $BN(\cdot)$ and $Act(\cdot)$ is batch normalization and activation function operations.

We utilizes a Transformer encoder to perform temporal attention focusing on $\hat{\boldsymbol{x}}'$. The Transformer encoder consists of a MSA sub-layer and a feed-forward neural network using residual connections. The Transformer encoder applies multiple attention heads to achieve the model parallel training. By dividing the input into multiple feature subspaces and applying a self-attention mechanism in the subspaces, the model can be trained in parallel while capturing emotional information.

We take the mean value of the convolution output channel dimension of $\hat{\boldsymbol{x}}'$ to get $\boldsymbol{s}' \in \mathbb{R}^{b \times (f/4) \times d}$ and then transpose $\boldsymbol{s}'$ to get $\hat{\boldsymbol{s}}' \in \mathbb{R}^{b \times d \times (f/4)}$. The feature $\hat{\boldsymbol{s}}' \in \mathbb{R}^{b \times d \times (f/4)}$ is used as the input of the time-domain Transformer encoder. We can represent the process as

$$\boldsymbol{m} = LN(MSA(\hat{\boldsymbol{s}}') + \hat{\boldsymbol{s}}'), \tag{2}$$

$$\hat{\boldsymbol{q}} = LN(MLP(\boldsymbol{m}) + \boldsymbol{m}), \tag{3}$$

where $MSA(\cdot)$, $LN(\cdot)$ and $MLP(\cdot)$ is MSA, layer normalization and feed-forward neural network respectively in a Transformer encoder. Besides, $\boldsymbol{m}$ is the output of $\hat{\boldsymbol{s}}'$ after the MSA and layer normalization of a Transformer encoder and $\hat{\boldsymbol{q}} \in \mathbb{R}^{b \times d \times (f/4)}$ is the output of a Transformer encoder. The output feature

$\hat{\boldsymbol{q}}$ will be used as one of the inputs of time-frequency Transformer module after linear mapping.

2.2 Frequency Transformer Module

This module has a similar structure to the time Transformer module, both containing a Transformer encoder. The difference between the two modules is that this module performs 2D convolutional operation on the input features $\boldsymbol{x}$ frequency band by frequency band, which is used for reducing dimension of feature and extracting frequency-domain emotional information. We use 2D convolutional operation $C_f(\cdot)$ on each frequency band of feature $\boldsymbol{x}$. Then, we normalize the feature and activate the feature using the activation function. Finally get the output $\hat{\boldsymbol{x}}'' \in \mathbb{R}^{b \times c1 \times f \times (d/4)}$ of the last activation function layer. The operations can be represented as

$$\hat{\boldsymbol{x}}'' = Act(BN(C_f(Act(BN(C_f(\boldsymbol{x})))))). \tag{4}$$

We take the mean value of the convolution output channel dimension of $\hat{\boldsymbol{x}}''$ to get $\hat{\boldsymbol{s}}'' \in \mathbb{R}^{b \times f \times (d/4)}$. Then, we use a Transformer encoder to calculate the emotional correlation between frequency bands for $\hat{\boldsymbol{s}}''$, so that the model focuses on the emotional related parts in frequency domain, reducing the impact of emotional unrelated frequency bands on speech emotion recognition. The operations can be represented as

$$\boldsymbol{n} = LN(MSA(\hat{\boldsymbol{s}}'') + \hat{\boldsymbol{s}}''), \tag{5}$$

$$\hat{\boldsymbol{k}} = LN(MLP(\boldsymbol{n}) + \boldsymbol{n}), \tag{6}$$

where $\boldsymbol{n}$ is the output of $\hat{\boldsymbol{s}}''$ after the MSA and layer normalization of the Transformer encoder and $\hat{\boldsymbol{k}} \in \mathbb{R}^{b \times f \times (d/4)}$ is the output of the Transformer encoder. After linear mapping, $\hat{\boldsymbol{k}}$ will be used as one of the inputs to the time-frequency Transformer module.

2.3 Time-Frequency Transformer Module

Time-Frequency Transformer Module aims to aggregate the local emotion encoding in time and frequency domains generated by the Time and Frequency Transformers into the global emotion representation in time-frequency domain. Therefore, the main function of this module is to use the time-frequency features to further weight so that the model pays more attention to the emotion-related segments in time-frequency domain, thereby improving accuracy of speech emotion recognition.

First, We utilize the 2D convolution operation $C(\cdot)$ to encode $\boldsymbol{x}$ in the time-frequency domain. After that, we normalize and activate the feature to obtain $\hat{\boldsymbol{x}} \in \mathbb{R}^{b \times c1 \times (f/4) \times (d/4)}$. The process can be represented as

$$\hat{\boldsymbol{x}} = Act(BN(C(Act(BN(C(\boldsymbol{x})))))). \tag{7}$$

Then, we take the mean value of $\hat{\boldsymbol{x}}$ in the convolution output channel dimension to get $\boldsymbol{v} \in \mathbb{R}^{b\times(f/4)\times(d/4)}$. We linearly map $\hat{\boldsymbol{q}}$ and $\hat{\boldsymbol{k}}$ to get $\boldsymbol{q} \in \mathbb{R}^{b\times(f/4)\times(d/4)}$ and $\boldsymbol{k} \in \mathbb{R}^{b\times(f/4)\times(d/4)}$ respectively. The $\boldsymbol{q}$, $\boldsymbol{k}$ and $\boldsymbol{v}$ are used as the input of the Time-Frequency Transformer encoder. Compared with an original Transformer encoder, we replace the MSA sub-layer in the encoder with a MA sub-layer to obtain a Time-Frequency Transformer encoder. A Time-Frequency Transformer encoder is composed of a MA sub-layer and a feed-forward neural network using residual connection. The main difference between MSA and MA is that when doing attention calculations, the inputs $\boldsymbol{Q}$ (Query Vector), $\boldsymbol{K}$ (Keyword Vector), and $\boldsymbol{V}$ (Value Vector) of MSA are the same, but the inputs $\boldsymbol{Q}$, $\boldsymbol{K}$, and $\boldsymbol{V}$ of MA are different. The process can be denoted as

$$\boldsymbol{p} = LN(MA(\boldsymbol{q}, \boldsymbol{k}, \boldsymbol{v}) + \boldsymbol{v}), \tag{8}$$

$$\boldsymbol{y} = LN(MLP(\boldsymbol{p}) + \boldsymbol{p}), \tag{9}$$

where $MA(\cdot)$ is MA and $\boldsymbol{p}$ is the output of the input features after the MA and layer normalization of the Time-Frequency Transformer encoder. Besides, $\boldsymbol{y} \in \mathbb{R}^{b\times(f/4)\times(d/4)}$ is the output of the Time-Frequency Transformer encoder. The obtained $\boldsymbol{y}$ is the input of the classifier and finally obtain the predicted emotional category.

The classifier consists of a pooling layer and a fully connected layer. The main function of the pooling layer is to reduce the feature dimension. The pooling layer takes the mean and standard deviation in the frequency-domain dimension of $\boldsymbol{y}$ and concatenates them to get $\hat{\boldsymbol{y}} \in \mathbb{R}^{b\times(d/2)}$, which can be represented as

$$\hat{\boldsymbol{y}} = Pool(\boldsymbol{y}). \tag{10}$$

We calculate the prediction probability of all emotions through a fully connected layer and ultimately obtain $\hat{\boldsymbol{y}}' \in \mathbb{R}^{b\times c}$, where c is the number of emotion categories of the corpus. We take the emotion with the largest prediction probability as the predicted emotion of the model and optimize the model by reducing the cross-entropy loss $Loss$ between the predicted emotion label $\hat{\boldsymbol{y}}'' \in \mathbb{R}^{b\times c}$ and the true emotion label $\boldsymbol{z} \in \mathbb{R}^{b\times c}$. The operations can be represented as

$$\hat{\boldsymbol{y}}' = FC(\hat{\boldsymbol{y}}), \tag{11}$$

$$\hat{\boldsymbol{y}}'' = Softmax(\hat{\boldsymbol{y}}'), \tag{12}$$

$$Loss = CrossEntropyLoss(\hat{\boldsymbol{y}}'', \boldsymbol{z})), \tag{13}$$

where $FC(\cdot)$, $Softmax(\cdot)$ and $CrossEntropyLoss(\cdot)$ is the fully connected layer operation, Softmax function [13] and Cross-Entropy loss function [14], respectively.

3 Experiments

3.1 Experimental Databases

Extensive experiments are conducted on two well-known speech emotion databases, *i.e.*, **IEMOCAP** [15] and **CASIA** [16]. **IEMOCAP** is an audio-visual database, consisting of five dyadic sessions, and each session is performed by a male actor and a female actor in improvised and scripted scenarios to obtain various emotions (angry, happy, sad, neutral, frustrated, excited, fearful, surprised, disgusted, and others). We select the audio samples in improvised scenario, including 2280 sentences belonging to four emotions (angry, happy, sad, neutral) for experiments. **CASIA** is a mandarin database collected by the Institute of Automation of the Chinese Academy of Sciences. Four speakers are required to perform six different emotions, *e.g.*, happiness, anger, fear, sadness, neutrality and surprise. A total of 1200 sentences are utilized in this experiment. The sampling rate of both databases is 16 kHz.

3.2 Experimental Protocol

In the experiment, we follow the same protocol of the previous research [10] and adopt the Leave-One-Speaker-Out (LOSO) cross-validation for evaluation. Specifically, for CASIA and IEMOCAP, when one speaker's samples are served as the testing data, the remaining three speakers' samples are used for training. Moreover, since IEMOCAP contains 5 sessions, the LOSO protocol is also a common way for evaluation [17]. Therefore, we also choose some methods using this protocol for comprehensive comparison. Also, we choose the weighted average recall (WAR) [6] and the unweighted average recall (UAR) [5], which are widely-used SER evaluation indicators, to effectively measure the performance of the proposed method.

Table 1. Experimental Results on IEMOCAP and CASIA

Database	Protocol	Comparison Method	Accuracy(%)	
			WAR	UAR
IEMOCAP	LOSO (5 Sessions or 10 Speakers)	MSA-AL [17]	72.34	58.31
		CNN-LSTM [11]	68.80	59.40
		STC-Attention [21]	61.32	60.43
		DNN-SALi [22]	62.28	58.02
		GRU-CNN-SeqCap [12]	72.73	59.71
		DNN-BN [23]	59.70	61.40
		Ours	**74.43**	**62.90**
CASIA	LOSO (4 Speakers)	GA-BEL [24]	39.50	39.50
		ELM-DNN [25]	41.17	41.17
		LoHu [26]	43.50	43.50
		DCNN-DTPM [27]	45.42	45.42
		ATFNN [10]	48.75	48.75
		Ours	**53.17**	**53.17**

3.3 Experimental Setting

Before the feature extraction, we preprocess the audio samples by dividing them into small segments of 80 frames (20 ms per frame). With this operation, samples are not only augmented, but also maintain the integrity of speech emotions. After that, we pre-emphasize the speech segments and the pre-emphasis coefficient is 0.97. Then, we use a 20 ms Hamming window with a frame shift of 10 ms to extract log-Mel-spectrogram, where the number of points of Fast Fourier Transform (FFT) and bands of Mel-filter are 512 and 80 respectively. Finally, the model input features of $b = 64$, $c = 1$, $f = 80$, $d = 80$ are obtained.

Besides, the parameters of $C_t(\cdot)$, $C_f(\cdot)$, $C(\cdot)$ are shown in Table 1. The $BN(\cdot)$ and $Act(\cdot)$ denote the BatchNorm function [18] and the ReLU function [13], respectively. The parameters of the Transformer encoder used in our model are shown in Table 1. The proposed method is implemented by Pytorch [19] with NVIDIA A10 Tensor Core GPUs, which is trained from scratch with 1000 epochs and optimized by Adam optimizer [20] with the initialized learning rate of 0.001.

3.4 Experimental Results and Analysis

Results on IEMOCAP. We selected some state-of-the-art methods for performance comparison with the proposed method, *i.e.*, a model based on MSA that fuses acoustic and linguistic features (MSA-AL) [17], a model that combines CNN with Long Short Term Memory (LSTM) and uses spectrogram as the input features (CNN-LSTM) [11], spectro-temporal and CNN with attention model(STC-Attention) [21], a Deep Neural Network (DNN) method combining Gaussian Mixture Models (GMM) and Hidden Markov Models (HMM) using subspace alignment strategy (DNN-SAli) [22], a method of using Gated Recurrent Unit (GRU) in CNN layers and combining with sequential Capsules (GRU-CNN-SeqCap) [12], and a DNN method with Bottleneck features (DNN-BN) [23].

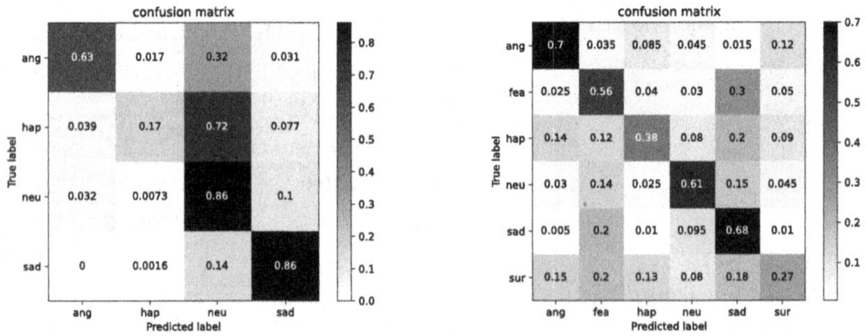

(a) Confusion matrix on IEMOCAP (b) Confusion matrix on CASIA

Fig. 2. Confusion matrices on IEMOCAP and CASIA.

The experimental results are shown in Table 1. From the results, we can find something interesting. Firstly, our proposed Time-Frequency Transformer achieves the best performance on both WAR (74.43%) and UAR (62.90%) compared to other mentioned methods. Moreover, compared to all methods using

Leave-One-Speaker-Out protocol, our proposed method achieves a promising increase over 1.7% in term of WAR and 1.5% in UAR.

The confusion matrix of IEMOCAP is shown in Fig. 2(a). What we can observe first is that the proposed method exhibits a excellent performance in classifying specific emotions, *e.g.*, *angry*, *neutral* and *sad*. However, it is difficult for the proposed model to correctly recognize the emotion *happy*. As shown in the figure, 72% *happy* samples are misclassified as *neutral* and only 17% are correctly classified. Obviously, it cannot be caused by the reason that *happy* is more close to *neural* than other emotions (negative emotions: *anger* and *sad*) since the possibility of *neutral* samples being misclassified as *happy* is only 0.73%. This situation lead us to consider the other reason which is the unbalanced sample size. Since the number of *happy* samples in IEMOCAP is only 284 which is the smallest among all emotions, the model cannot learn the unique emotional characteristics of *happy* well. It may lead to this situation that the *happy* samples are more likely to be mistaken for *neutral*.

Results on CASIA. Some state-of-the-art methods that also use the LOSO protocol are used for comparison with the proposed method, including Genetic Algorithm (GA) combined with Brain Emotional Learning (BEL) model (GA-BEL) [24], Extreme Learning Machine (ELM) combined with DNN (ELM-DNN) [25], weighted spectral features based on Local Hu moments (LoHu) [26], Deep CNN (DCNN) combined with a Discriminant Temporal Pyramid Matching (DTPM) strategy (DCNN-DTPM) [27] and an Attentive Time-Frequency Neural Network (ATFNN) [10].

The results on the CASIA database is shown in Table 1. It is obvious that our method achieves state-of-the-art performance among all algorithms. Specifically, our method obtains the best result on WAR (53.17%) and UAR (53.17%) than all comparison methods. Since the sample numbers of the 6 emotions of CASIA used in the experiment are balanced, WAR and UAR are equal. Besides, our results are not only the best, but also far superior to other methods. Even compared to ATFNN which is the second best method, the proposed method still obtain an over 4% performance increase.

Table 2. Ablation experiments of different architectures for our model, where '✓' or '✗' represents the network with or without the module. 'T-Trans', 'F-Trans', and 'TF-Trans' are the modules of Time, Frequency, and Time-frequency Transformers, respectively.

Architecture	Ablation Experiments			IEMOCAP(%)		CASIA(%)	
	T-Trans	F-Trans	TF-Trans	WAR	UAR	WAR	UAR
T+F	✓	✓	✗	70.12	58.77	40.32	40.32
T+TF	✓	✗	✓	70.96	60.34	48.91	48.91
F+TF	✗	✓	✓	71.47	60.59	49.26	49.26
T+F+TF (ours)	✓	✓	✓	**74.43**	**62.90**	**53.17**	**53.17**

From the confusion matrix of CASIA in Fig. 2(b), it is obvious that the proposed method has a high recognition rate in four types of emotions (*angry*, *fear*, *neutral*, *sad*), but the recognition effect on *happy* and *surprise* is poor. Since *happy* is easily misclassified as *sad*, it may be caused by the pendulum effect [28] in psychology. Human emotions are characterized by multiplicity and bipolarity under the influence of external stimuli. Beside that, *surprise* is always confused with *fear*. Due to the similar arousal [29] of the two emotions, it may lead to them inducing each other.

Ablation Experiments. We verified the effectiveness of our method by removing some modules of the proposed method. The experimental results are shown in Table 2, where 'T-Trans', 'F-Trans', and 'TF-Trans' are the modules of Time Transformer, Frequency Transformer, and Time-frequency Transformer, respectively. According to the results of the ablation experiments, the TF-Trans can effectively make the model focus on the emotion-related segments in the time-frequency domain and improve the emotional discrimination of the model. In addition, the effect of removing T-Trans is better than removing F-Trans, indicating that the frequency domain information of speech is of great significance for emotion recognition. Moreover, it is easy to observe that T+F model achieves the worst result, particular on CASIA. Compared to T+TF model and F+TF model, T+F model has a significant performance degradation over 8% of both WAR and UAR on CASIA. This phenomenon indicates the effectiveness of the Time-Frequency Transformer module. If we remove this part, the local time-domain and frequency-domain emotion features are not fully utilized to mine the time-frequency emotional correlations. Thus, the model cannot learn more discriminative global acoustic emotional feature representations.

Visualization of Attention. In order to further investigate whether the proposed method focuses on frequency bands with specific energy activations and emotional key speech frames in the speech signal, we visualize the attention of Time Transformer, Frequency Transformer and Time-Frequency Transformer to the log-Mel-spectrogram, as shown in Fig. 3. Form the visualization of Time Transformer in Fig. 3(b), we can observe that there are strong activation values in $15^{th}-20^{th}$ frames and $50^{th}-60^{th}$ frames, which indicates that the emotion correlations between these frames is important to represent speech emotions. And these frames correspond to the positions with richer semantic information in Fig. 3(a). The visualization of Frequency Transformer in Fig. 3(c) shows that the activation of the middle and low frequency bands is more obvious, demonstrating that the middle and low frequency bands are key for the *sad* emotion representation. From the results of time-frequency attention in Fig. 3(d), we can see that the larger activation value (i.e., the salient patches) corresponds to the regions where the semantic information is more concentrated in Fig. 3(a). Therefore, our proposed Time-Frequency Transformer can fully capture the time-frequency regions highly correlated with emotions while ensuring the complete modeling

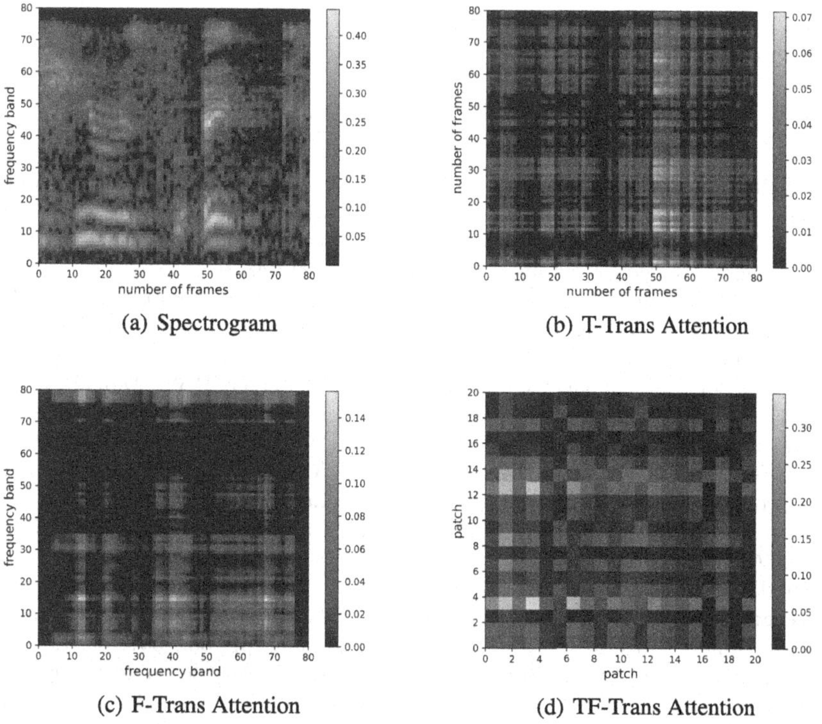

Fig. 3. Visualization of log-Mel-spectrogram, T-Trans. Attention, F-Trans. Attention, and TF-Trans. Attention (taking *sad* of IEMOCAP as an example).

of local emotion information in the time and frequency domains to obtain discriminative speech emotion features.

4 Conclusion

In this paper, we propose a novel Transformer-based time frequency domain joint learning method for SER, i.e., Time-Frequency Transformer. It can effectively model local emotion correlations between frames and frequency bands through Time Frequency and Frequency Transformer. Then these local emotion features are aggregated into more discriminative global emotion representations by a Time-Frequency Transformer. However, the MSA operation in Transformer is aiming at model global long-range discrepancy, which is easily disturbed by noisy frames or frequency bands in speech. Therefore, our Future research will focus on sparse MSA for speech emotion representations.

Acknowledgements. This work was supported in part by the National Key R & D Project under the Grant 2022YFC2405600, in part by the National Natural Science Foundation of China under the Grant U2003207, in part by the Jiangsu Frontier Technology Basic Research Project under the Grant BK20192004, in part by the Zhishan

Young Scholarship of Southeast University, and in part by Jiangsu Province Excellent Postdoctoral Program.

References

1. Schuller, B., Batliner, A.: Computational Paralinguistics: Emotion, Affect and Personality in Speech and Language Processing. John Wiley & Sons (2013)
2. Schuller, B.W.: Speech emotion recognition: two decades in a nutshell, benchmarks, and ongoing trends. Commun. ACM **61**(5), 90–99 (2018)
3. Akçay, M.B., Oğuz, K.: Speech emotion recognition: emotional models, databases, features, preprocessing methods, supporting modalities, and classifiers. Speech Commun. **116**, 56–76 (2020)
4. Lu, C., Zong, Y., Zheng, W., Li, Y., Tang, C., Schuller, B.W.: Domain invariant feature learning for speaker-independent speech emotion recognition. IEEE/ACM Trans. Audio Speech Lang. Process. **30**, 2217–2230 (2022)
5. Stuhlsatz, A., Meyer, C., Eyben, F., Zielke, T., Meier, G., Schuller, B.: Deep neural networks for acoustic emotion recognition: raising the benchmarks. In: 2011 IEEE International Conference on Acoustics, Speech and Signal Processing (ICASSP), pp. 5688–5691. IEEE (2011)
6. Schuller, B., Vlasenko, B., Eyben, F., Rigoll, G., Wendemuth, A.: Acoustic emotion recognition: a benchmark comparison of performances. In: 2009 IEEE Workshop on Automatic Speech Recognition & Understanding, pp. 552–557. IEEE (2009)
7. Abbaschian, B.J., Sierra-Sosa, D., Elmaghraby, A.: Deep learning techniques for speech emotion recognition, from databases to models. Sensors **21**(4), 1249 (2021)
8. Wang, J., Xue, M., Culhane, R., Diao, E., Ding, J., Tarokh, V.: Speech emotion recognition with dual-sequence LSTM architecture. In: 2020 IEEE International Conference on Acoustics, Speech and Signal Processing (ICASSP 2020), pp. 6474–6478. IEEE (2020)
9. Mao, Q., Dong, M., Huang, Z., Zhan, Y.: Learning salient features for speech emotion recognition using convolutional neural networks. IEEE Trans. Multimedia **16**(8), 2203–2213 (2014)
10. Lu, C., et al.: Speech emotion recognition via an attentive time–frequency neural network. IEEE Trans. Computat. Soc. Syst. (2022)
11. Satt, A., et al.: Efficient emotion recognition from speech using deep learning on spectrograms. In: Interspeech, pp. 1089–1093 (2017)
12. Wu, X., et al.: Speech emotion recognition using capsule networks. In: ICASSP 2019-2019 IEEE International Conference on Acoustics, Speech and Signal Processing (ICASSP), pp. 6695–6699. IEEE (2019)
13. Dubey, S.R., Singh, S.K., Chaudhuri, B.B.: Activation functions in deep learning: a comprehensive survey and benchmark. Neurocomputing (2022)
14. Zhang, Z., Sabuncu, M.: Generalized cross entropy loss for training deep neural networks with noisy labels. Adv. Neural Inf. Process. Syst. **31** (2018)
15. Busso, C., et al.: Iemocap: interactive emotional dyadic motion capture database. Lang. Resour. Eval. **42**, 335–359 (2008)
16. Zhang, J., Jia, H.: Design of speech corpus for mandarin text to speech. In: The Blizzard Challenge 2008 Workshop (2008)
17. Bhosale, S., Chakraborty, R., Kopparapu, S.K.: Deep encoded linguistic and acoustic cues for attention based end to end speech emotion recognition. In: 2020 IEEE International Conference on Acoustics, Speech and Signal Processing (ICASSP 2020), pp. 7189–7193. IEEE (2020)

18. Ioffe, S., Szegedy, C.: Batch normalization: accelerating deep network training by reducing internal covariate shift. In: International Conference on Machine Learning, pp. 448–456. PMLR (2015)
19. Paszke, A., et al.: Pytorch: an imperative style, high-performance deep learning library. Adv. Neural Inf. Process. Syst. **32** (2019)
20. Adam, K.D.B.J., et al.: A method for stochastic optimization. arXiv preprint arXiv:1412.6980 (2014)
21. Guo, L., Wang, L., Xu, C., Dang, J., Chang, E.S., Li, H.: Representation learning with spectro-temporal-channel attention for speech emotion recognition. In: 2021 IEEE International Conference on Acoustics, Speech and Signal Processing (ICASSP 2021), pp. 6304–6308. IEEE (2021)
22. Mao, S., Tao, D., Zhang, G., Ching, P., Lee, T.: Revisiting hidden Markov models for speech emotion recognition. In: 2019 IEEE International Conference on Acoustics, Speech and Signal Processing (ICASSP 2019), pp. 6715–6719. IEEE (2019)
23. Kim, E., Shin, J.W.: DNN-based emotion recognition based on bottleneck acoustic features and lexical features. In: 2019 IEEE International Conference on Acoustics, Speech and Signal Processing (ICASSP 2019), pp. 6720–6724. IEEE (2019)
24. Liu, Z.T., Xie, Q., Wu, M., Cao, W.H., Mei, Y., Mao, J.W.: Speech emotion recognition based on an improved brain emotion learning model. Neurocomputing **309**, 145–156 (2018)
25. Han, K., Yu, D., Tashev, I.: Speech emotion recognition using deep neural network and extreme learning machine. In: Interspeech 2014 (2014)
26. Sun, Y., Wen, G., Wang, J.: Weighted spectral features based on local hu moments for speech emotion recognition. Biomed. Signal Process. Control **18**, 80–90 (2015)
27. Zhang, S., Zhang, S., Huang, T., Gao, W.: Speech emotion recognition using deep convolutional neural network and discriminant temporal pyramid matching. IEEE Trans. Multimedia **20**(6), 1576–1590 (2017)
28. Wegner, D.M., Ansfield, M., Pilloff, D.: The putt and the pendulum: ironic effects of the mental control of action. Psychol. Sci. **9**(3), 196–199 (1998)
29. Hanjalic, A., Xu, L.Q.: Affective video content representation and modeling. IEEE Trans. Multimedia **7**(1), 143–154 (2005)

Asymptotic Spatiotemporal Averaging of the Power of EEG Signals for Schizophrenia Diagnostics

Włodzisław Duch[1(✉)], Krzysztof Tołpa[1], Ewa Ratajczak[2], Marcin Hajnowski[2], Łukasz Furman[1], and Luís A. Alexandre[3]

[1] Department of Informatics, Institute of Engineering and Technology, Faculty of Physics, Astronomy and Informatics, Nicolaus Copernicus University, Toruń, Poland
wduch@umk.pl

[2] Institute of Psychology, Faculty of Philosophy and Social Sciences, Nicolaus Copernicus University, Toruń, Poland

[3] Universidade da Beira Interior, and NOVA LINCS, Covilhã, Portugal

Abstract. Although many sophisticated EEG analysis methods have been developed, they are rarely used in clinical practice. Individual differences in brain bioelectrical activity are quite substantial, therefore simple methods that can provide stable results reflecting the basic characteristics of individual neurodynamics are very important. Here, we explore the potential for brain disorder classification based on patterns extracted from the asymptotic spatial power distributions, and compare it with 4–20 microstates, providing information about the dynamics of clustered global power patterns. Applied to the 16-channel EEG data such methods gave discrimination between adolescent schizophrenia patients and a healthy control group at the level of 86–100%.

Keywords: EEG · power spectra · STFT · microstates · neurodynamics · schizophrenia diagnostics · machine learning

1 Introduction

Bioelectrical brain activity is frequently measured using electroencephalography (EEG) or magnetoencephalography (MEG). Such signals may have a high sampling rate, with temporal resolution below a millisecond. On the other hand, functional magnetic resonance (fMRI) measuring blood-oxygen-level dependent (BOLD) hemodynamic signals provides information about metabolic demands with a much lower temporal resolution of the order of one second. Both methods are used to diagnose mental disorders, neurofeedback, and many other applications.

Although many sophisticated approaches to EEG analysis have been developed, they are rarely used in clinical practice. Bioelectrical brain activity is non-stationary, even during short periods. Individual differences are quite large. Neurodynamics, especially in the resting state, strongly depends on hundreds of confounds [1]. Methods tested on a

B. Luo et al. (Eds.): ICONIP 2023, CCIS 1963, pp. 428–439, 2024.
https://doi.org/10.1007/978-981-99-8138-0_34

small number of samples give good results for favorably tuned parameters but in real life do not generalize well. Good biomarkers for objective diagnosis of brain disorders are still unknown [2]. Neural and genetic fingerprints of brain disorders may belong to large clusters, reflecting the fundamental character of individual genetic or neurodynamic processes. Still, rare cases may be impossible to diagnose without similar patterns in the dataset used for training. Methods that reach 100% accuracy on small datasets have little chance of being useful.

EEG recordings involve a spatial distribution of power and temporal dynamics. The human brain contains about 2–4 million cortical columns, each containing tens of thousands of neurons, generating oscillations with frequencies reaching hundreds of hertz. EEG measurements are characterized by a spatial resolution of one centimeter and are usually analyzed in the 0–50 Hz range (intracranial iEEG may include 500 Hz ripples). The activity of large-scale neural networks can be used for biometric identification [3]. Specific patterns (fingerprints) of brain activity have diagnostic value. EEG recordings may detect the activity of more than ten large-scale networks linked to specific information processing (visual, auditory, sensorimotor, salience, dorsal and ventral attention, or default mode). This requires co-registration with the fMRI signal, which is technically very difficult [4].

In this paper, we have used a data-driven phenomenology that reflects the actual physiological processes. First, the limits of both spatial averaging of power in the narrow frequency bands and temporal characterization of brain processes using a large number of microstates, are explored. Tests were made on a typical, small EEG dataset of 45 schizophrenic adolescents [5]. Many papers were tested on even smaller EEG datasets, of 14 schizophrenia cases. Second, we investigate why several papers (see [6] and the review in [15]) can get 100% accuracy on such datasets. Classification may be successful if cases of similar structure are in the training set. Perfect classification accuracy is reached when features are selected on the whole data, before training a classifier (which is a common practice). Feature selection on the cross-validation training partition may lead to errors, identifying rare cases that should be inspected separately.

The methods applied in our study are described in the next section, followed by the description of the real data used for testing in the third section, and the results obtained in section four; the paper ends with the final discussion.

2 Methods

2.1 Microstates

EEG microstate analysis is a very popular method that may be used to generate features useful for classification. Global field potential (GFP) is calculated as the variance of potential $V_i(t)$ on the scalp:

$$GFP(t) = \sqrt{\frac{1}{N}\sum_{i=1}^{N}\left(V_i(t) - \overline{V}\right)^2}$$

Local maxima of GFP represent quasi-stable attractor states, transient activity patterns across all electrodes lasting from milliseconds to seconds. Such metastable field potential patterns are clustered for groups of subjects to identify classes of microstate topologies. Two clusterization methods are most frequently used: the k-means approach and the "topographic atomize and agglomerate hierarchical clustering" (TAAHC, [7]). The whole time series is divided into windows assigned to a given microstate class based on the spatial similarity metric between each consecutive EEG sample and each microstate class. Initially, only four stable microstates were distinguished [8], and rarely more than ten classes were used for analysis.

Despite its great popularity, the microstate approach is burdened by some serious methodological limitations [9], drastically oversimplifying complex EEG signals. Ascribing various brain states to only a few clusters makes using statistical methods and symbolic dynamics techniques feasible, but a lot of information is lost. EEG oscillations show complex dynamics in different frequency bands between GFP peaks. Spatial power distribution patterns of microstates are too simple to accurately reflect the activity of large-scale brain networks.

We have performed MS analysis using the 'global maps strategy' [8–10] where one common set of k global maps is identified for all recordings, producing a set of common prototypes for both study groups. The analysis was carried out using freely available toolboxes for the MATLAB environment - the Microstate Toolbox EEGLAB plug-in [7] and the +microstate stand-alone package [11]. The initial steps included re-referencing to average reference and aggregating all EEG data fragments without normalization by average channel standard deviation. For each participant 1000 randomly selected EEG maps of the highest GFP at a minimum map distance of 10 ms were selected, discarding maps with GFP values exceeding one standard deviation of all maps' GFPs. Following normalization across the whole dataset, the selected maps were subjected to clustering using modified k-means (50 repetitions, 1000 maximum iterations, 10^{-6} threshold, cross-validation criterion measure of fitness). Maps were clustered selecting 4 to 20 microstate classes, and sorted by the global explained variance (GEV). Even number of microstate prototypes was used for microstate segmentation and backfitting to the data; sample maps were labeled based on the maximum similarity to the prototypes assessed with the global map dissimilarity (GMD) measure. The resulting microstate label syntax was subjected to temporal smoothing by rejecting small fragments below 30 ms. The following microstate statistics were calculated for each participant and used as features for classification: occurrence, duration, coverage, GEV, global field potential, mean spatial correlation, and the transition probabilities between microstate classes.

2.2 Spatial Distribution of Power

Our ToFFi toolbox for frequency-based fingerprinting of brain signals allows for the identification of specific frequencies ("fingerprints") arising in local brain regions, depending on the subnetwork that engages it in its activity [12]. Asymptotic average power distribution maps are more complex than MS maps and are good candidates for prototype states characterizing brain neurodynamics. To create such maps, we calculated short-time Fourier spectra (STFT) in 1-s sliding time windows. The STFT spectra were generated in windows starting in consecutive EEG samples, providing cumulative and average power estimations at discrete frequencies f for each electrode on the scalp. Given a raw

EEG data matrix $\mathbf{U}_k = (u_{ik}) = (u_k(t_i))$, where the index $i = 1\ldots N$ enumerates time-series samples and index $k = 1\ldots N_c$ refers to electrode number (N_c data streams, input channels), the algorithm is summarized as follows:

- For each subject, given EEG data matrix $\mathbf{U}_k = (u_{ik})$
- Segment the data into time windows with τ samples, $\boldsymbol{w}_k(t_i) = [u_{ik}, u_{ik+\tau}]$.
- For each time window $i = 1.. N-\tau$ calculate STFT power spectra $\boldsymbol{S}_k(t_i, f)$.
- Sum all $\boldsymbol{S}_k(t_i, f)$ over time windows to get local cumulative power $\boldsymbol{S}_k(f)$ in channel k at frequency f.
- Calculate the average power $\boldsymbol{R}_k(f) = \boldsymbol{S}_k(f)/(N-\tau)$ in each channel.
- Sum $\boldsymbol{R}_k(f)$ over selected frequency ranges to estimate power in each band.

If the sampling frequency is high, windows may be shifted by several samples to speed up the calculations. This procedure creates for each subject a vector with the number of N_c components estimating average power in different frequency ranges. We have found that these averages stabilize after about 60 s. Average power maps are relatively stable for each individual but do not contain any information about the dynamics or frequency. This kind of information may be added by dividing the whole frequency range of the STFT spectra into typical frequency bands (δ, θ, α, β, γ) or by focusing on several narrow few-Hz bands to capture power peaks that arise at the same time in several channels, reflecting synchronized processes.

The avPP analysis can be extended in many ways. The global EEG signal at a given time moment $\mathbf{P}_\mathrm{M}(t)$ may be decomposed using the basis calculated by asymptotic spatial averaging. The cumulative average power values and variances were calculated separately for each electrode for the broadband spectrum (0.5–60 Hz). After calculation of the STFT spectra, it is trivial to calculate the average power for classical EEG bands: delta (0.5–4 Hz), theta (4–8 Hz), alpha (8–12 Hz), beta (12–30 Hz), gamma (30–60 Hz). This will create vectors with $5 \times N_c$ components. We have also calculated power in a narrow 1 Hz band to see which frequencies lead to the most characteristic patterns, creating vectors with $60 \times N_c$ components. Attractor states may be defined by monitoring the time windows in which the power decreases below a certain threshold. We can combine that with our approach to recurrence quantification analysis [13], estimating distances to our asymptotic distributions instead of self-similarity. To capture the flexibility of brain subnetworks [14] we can look at various functional correlation measures and average them in longer recordings.

3 Dataset and Related Papers

Diagnosis of schizophrenia based on EEG has been a popular subject, with over 40 methods mentioned in the summary by Khare, Bajaj, and Acharya [15]. They have tested their SchizoNET approach on EEG recordings of 45 adolescent boys (10–14 years old) with schizophrenia (schizotypical and schizoaffective disorders), described in Borisov et al. paper [5]. The control group of similar age consisted of 39 healthy schoolboys. EEG recordings were made in a wakeful, relaxed state with the eyes closed, using 16 electrodes placed in the standard 10–20 system at O1, O2, P3, P4, Pz, T5, T6, C3, C4, Cz, T3, T4, F3, F4, F7, and F8. The sampling rate was 128 Hz, and only artifact-free EEG segments of the recordings were used for analysis.

Analysis performed in the original paper [5] was based on the index of structural synchrony (ISS), the synchronization frequency between 120 pairs of electrodes during quasi-stationary segments of the EEG signals, free from random coincidences. Resulting graphs showing connections between pairs of electrodes with high ISS have a little chance of being stable. We have made a similar investigation and in each cross-validation fold such graphs differ significantly. The overall conclusion from this paper is that schizophrenics have less synchronicity between electrodes that are far apart, but more for electrodes that are adjacent.

This data has been analyzed in many other papers, therefore we have a comparison with different methods of analysis. Several convolutional neural networks were applied to this data. SchizoNET [15] combined the Margenau–Hill time-frequency distribution (MH-TFD) and convolutional neural network (CNN) with only five layers. The time-frequency amplitude is converted to two-dimensional plots and fed as an image to the CNN model. Using 5-fold cross-validation (5 CV) this model achieved a very high accuracy of 97.4%. Phang et al. [16] developed a deep convolutional neural network (CNN) framework for the classification of electroencephalogram (EEG)-derived brain connectome in schizophrenia. They have used a combination of 3 methods: connectivity features based on a vector autoregressive model, partial directed coherence, and complex network measures of network topology. They have employed different fusion strategies a parallel ensemble of 1D and 2D CNNs to integrate the features from various domains, and analysis of dynamic brain connectivity using the recurrent neural networks. In the 5 $\times$ CV tests, their fusion-based models CNNs outperform the SVM classifier (90.4% accuracy), achieving the highest accuracy of 91.7 $\pm$ 4.6%. These models use full connectivity matrices in 5 bands, requiring vectors with 1280 components, and the full fusion models use vectors with 2730 dimensions. Aslan and Akin [17] converted EEG signals to 2D images using continuous wavelet transform and trained VGG16 deep learning network architecture on such images. They created vectors from 5-s sequences, dimension 10240, and transformed them into 224 $\times$ 224 scalograms. They claim an accuracy of 98% but have used a single partition of 20–80%, so it cannot be compared to cross-validation. Shen and colleagues [7] used dynamic functional connectivity analysis and 3D deep convolutional neural networks. A time-frequency domain functional connectivity analysis was used to extract the features in the alpha band only with a cross-mutual information algorithm. This gave 97.7 $\pm$ 1.2% accuracy and some differences between the connectivity of temporal lobe areas in both the right and left side of the brain, between the schizophrenia and control subjects.

All these results are almost perfect, but none of these methods have the chance to be useful in a clinical setting. First, the "black box" approach provides no insight into the brain processes that may characterize schizophrenia. Second, feature selection on the whole dataset leads to high accuracy that significantly drops when selection is done only on the training partition. This is illustrated below.

4 Results

4.1 CNN Calculations

We have tested a typical convolutional neural network, considering the data from each subject as an image. The images were created by building a $N_c \times N_c$ distance matrix from the $N_c = 16$ channel data, where the (i, j) pixel represents distance $d(\mathrm{X}_i, \mathrm{X}_j)$. Here X_i is a N_c-dimensional vector with the value of each EEG channel at time instant $i = 1,.., N$, the number of EEG samples $N = 7680$ (128 Hz $\times$ 60 s), and the distance function was Euclidean. This matrix was subsampled to a final size of 240 $\times$ 240 pixels. An example of such a matrix can be seen in Fig. 1. Our CNN had 3 convolutional layers, with 16 3 $\times$ 3 filters in the first layer, 32 3 $\times$ 3 filters in the second and 64 3 $\times$ 3 filters in the third convolutional layer, a ReLU layer after all convolutional layers, and two fully connected maxpooling layers after the ReLU layers of the convolutional blocks. Batch normalization layers are after the second and third convolutional layers, the output layer is followed by softmax nodes, with a total of 945170 parameters. This network was trained with a dropout of 0.25 before the first fully connected layer, using the Adadelta optimizer, for 200 epochs, and the best cross-entropy validation loss weights were used for testing. No information about the test partition has been used at any stage of calculations.

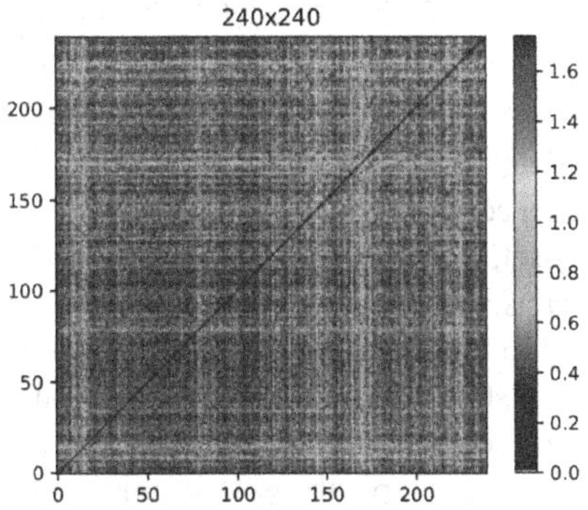

Fig. 1. An example of a subsampled distance matrix used in the experiments with CNNs.

5-fold cross-validation gave 79.8 $\pm$ 5.9% average accuracy. This is more realistic than accuracies that are close to 100%. Such impressive results may come from a common error, performing feature selection on the whole dataset instead of only on the training partition within cross-validation folds. Sophisticated neural models can achieve high classification accuracy, but do not help to understand EEG data. The use of fMRI data for schizophrenia diagnosis seems to be about as accurate as EEG, for example, a model for automated schizophrenia diagnosis with fMRI features had accuracy below 80% in all cases [18]. Unfortunately, we do not have EEG and fMRI data for the same group of patients, therefore direct comparison of the two approaches is impossible.

Classification results presented below were made using the linear SVM method implemented in a Scikit-learn Python library, with the stratified 5-fold cross-validation.

One advantage of such an approach is that the model is simple, and we can identify combinations of electrodes and frequency bands providing specific spatial information about localized oscillatory brain activity.

4.2 Microstate-Based Calculations

We have performed calculations with up to 20 microstates, generating features for the LSVM classifier. Detecting a large number of microstates should create smaller and more specific clusters. However, microstate algorithms cluster many states around the peak global power, so all maps resemble variations on similar patterns (Fig. 2). Additional microstates do not show complex patterns that could represent the activation of large-scale brain networks. The results of the classification are in Table 1 and Fig. 3.

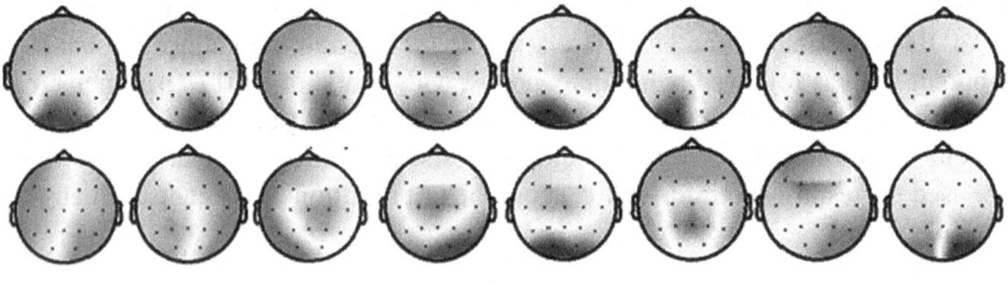

Fig. 2. Maps of 16 microstate.

Results of classification based on 14 to 20 microstate features derived from the whole dataset were surprisingly high, reaching 100% in the 5-fold cross-validation. For these calculations, we used a common set of features selected before training the LSVM classifier on the whole dataset, using recursive feature elimination with cross-validation (RFECV). With a small dataset and a large number of parameters (the number of transition probabilities is equal to the square of the number of microstates), we can always find some parameters that distinguish a single case from all others. Feature selection within cross-validation done separately on the training partition will not discover such prominent features, and therefore atypical cases are misclassified.

Using a common set of features 100% accuracy is reached for 14 or more microstates, with the variance of results quickly decreasing to zero. Most useful features are based on transition probabilities (Fig. 4), showing the importance of dynamics. Performing feature selection within the training partition gives much worse results. Accuracy reaches 83.5 ± 13,7% for 16 microstates, using 185 features. Almost as high accuracy 82.1 ± 15.2% is reached for 12 microstates with only six features. Among the additional 179 features, some are useful only for a single, very specific case (a single test error contributes about 6% to accuracy). Cross-validation may put such cases in the test partition, leading to a large variation in the number of errors.

To find unusual cases in the schizophrenia dataset we have performed the leave-one-out tests, checking which cases/features are responsible for errors. For 16 microstates accuracy grows to 86.8%, with about 2–3 cases misclassified. Reliable classifiers should estimate the confidence of their predictions distinguishing a group of typical cases for a given class and designating all others for more detailed evaluation.

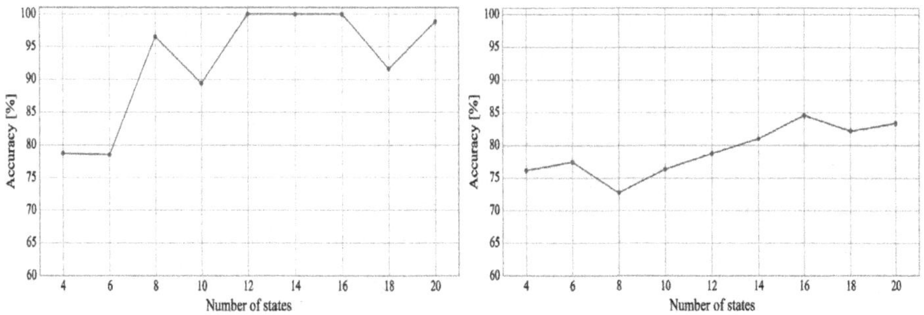

Fig. 3. 5 × CV accuracy dependence on the number of microstates; left - recursive feature selection on all data; right - feature selection performed separately within each training fold.

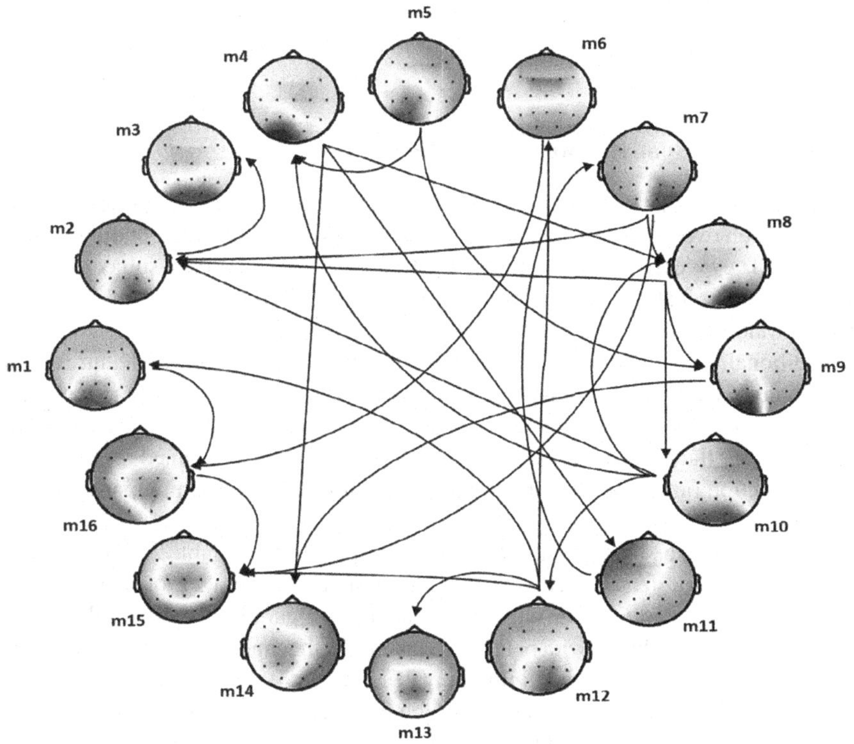

Fig. 4. Most important transition probabilities between 16 microstates used as features for the LSVM classifier that works with 100% accuracy.

Table 1. Classification results using parameters derived from microstates.

Microstates		Selection on all data		Selection on training only		
N states	Type	N dim	Acc ± Var %	Type	Ndim	Acc ± Var %
4	TAAHC	4	78.7 ± 17.2	TAAHC	4	76.1 ± 18.6
6	TAAHC	52	78.5 ± 17.7	TAAHC	52	77.4 ± 17.7
8	TAAHC	17	96.5 ± 3.4	TAAHC	17	72.7 ± 20.2
10	TAAHC	93	89.4 ± 9.4	TAAHC	93	76.3 ± 18.5
12	K-means	55	**100**	K-means	55	78.7 ± 17.5
14	K-means	90	**100**	K-means	90	81.0 ± 15.9
16	TAAHC	42	**100**	K-means	17	**84.6** ± 13.0
18	TAAHC	281	91.6 ± 7.8	TAAHC	281	82.2 ± 14.0
20	TAAHC	221	98.8 ± 1.2	TAAHC	221	83.4 ± 14.3

4.3 Asymptotic Spatial Power Distribution

For the purpose of the classification, the last cumulative average power values (following the summation over the whole signal; in our case, 60 s) and the corresponding variances were used. They were normalized within the frequency range for each subject separately and used for the feature matrix. We have tested several combinations of feature sets, selecting values separately for different bands (see Table 2).

Additionally, to exclude irrelevant features, we have used a recursive feature elimination technique with 5-fold cross-validation taken from the Scikit-learn function. During cross-validation, the algorithm performs step-wise removal of data components based on the value of the SVM coefficients. We also present results obtained using recursive feature selection on the whole dataset and selection within the training partition in 5 $\times$ CV.

A summary of the best classification results is presented in Table 2. The best results are obtained for the feature set consisting of cumulative average power values using the delta band in combination with theta, alpha, and gamma. Good results are also achieved for theta combined with beta bands taken together and reduced to 42 dimensions following the recursive feature elimination (RFECV).

Power plots in Fig. 5 are averaged over the five frequency bands. Power distribution patterns differ for each band and have a more complex structure than the microstates. Compared to the control group, schizophrenia patients show larger regions of high activity, as well as lower activity in the left and higher in the right temporal lobe. This confirms the observations reported by Shen et al. [7]. The number of features was optimized and fixed for all folds, using the RFECV scikit function.

Table 2. Summary of classification results for features obtained from asymptotic spatial averaging in selected bands. The window size was 256 samples.

EEG bands	Selection on all data		Selection on training	
	N dim	Acc ± Var %	N dim	Acc ± Var %
broadband	10	72.5 ± 20.6	10	68.0 ± 22.8
β+γ	3	73.8 ± 19.6	3	72.7 ± 20.5
θ+β	23	74.9 ± 18.4	23	65.4 ± 23.0
δ+θ	19	68.9 ± 21.0	19	65.6 ± 23.6
δ+θ+α	19	90.5 ± 8.6	19	76.2 ± 19.2
δ+θ+α+β	21	**95.2 ± 4.6**	21	78.5 ± 17.8
δ+θ+α+β+γ	71	79.5 ± 15.5	71	**79.8 ± 17.1**

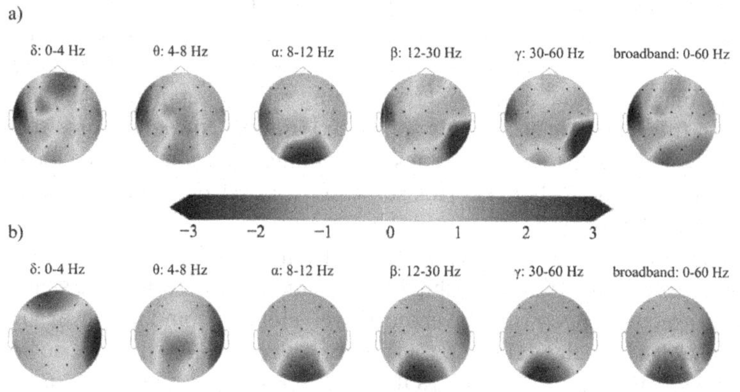

Fig. 5. Example of asymptotic power distributions. a) S27w1 schizophrenia, b) 719w1 healthy.

The electrodes and frequency bands that contribute the most to the theta-beta SVM classification based on the maximum absolute values of the weights are shown below. Starting with the most important combinations, the top 10 include:

T4 α	F8 α	P3 β	C3 β	O2 α	Pz α	F8 θ	P4 θ	T4 θ	T3 θ

5 Discussion

The dynamics of the brain states is characterized by a sequence of metastable states. The key problem is to find the optimal complexity of representation, capturing sufficient information about spatial patterns and their dynamics. We have introduced here frequency-dependent asymptotic spatial averaging, creating reference patterns more complex than those provided by microstates. In our tests, these spatial states show high discriminatory power. Splitting and optimizing frequency bands, selecting channels/sources and specific bands, and adding statistical information about the dynamics of these states, in the same way as done in the microstate analysis, should improve the results further. The microstate approach maximizes global explained variance, creating a few maps (usually 4 to 10). Michel and Koenig [9] have summarized microstate parameter changes (Dur, Occ, Cov, GEV) calculated for subjects suffering from neuropsychiatric diseases. This is very crude and not sufficient for precise diagnosis. Here, we have performed microstate analysis with up to 20 states, focusing on generating features useful for classification. Transition probabilities between microstates (Fig. 4) provide especially useful features. Selecting a subset of these features on the whole dataset gives a set of features that give 100% correct classification using linear SVM in 5-fold cross-validation. Although no information about the test partition is used in the classifier training, such selection performed strictly on the training data strongly influences the results even in the leave-one-out procedure.

Small EEG datasets can always contain a few unique cases. All papers should clearly state whether feature selection has been performed separately in each cross-validation partition, or on the whole dataset. Rahman and colleagues [19] proposed an approach for the analysis of a small complex fMRI dataset called MILC (Mutual Information Local to the whole Context). A self-supervised pre-training scheme captured potentially relevant information from large datasets. This approach is similar to foundational models, pre-trained on large datasets to enable context embedding and trained on the local data. We see a similar phenomenon here but lack sufficiently large EEG data collections to create such foundational models.

Recurrence analysis is very well suited to the analysis of time series. Further work is needed to understand the relationship between microstates, recurrence states and their transitions, spectral fingerprints, average power plots, decomposition of signals into template models, and flexibility of transitions between different large-scale network states, graphs of transitions between ROIs, motifs derived from hidden Markov models, and subnetworks. All these issues require deeper investigation.

Acknowledgments. This work was supported by the Polish National Science Center grant UMO-2016/20/W/NZ4/00354, and NOVA LINCS (UIDB/04516/2020) with the financial support of FCT.IP, Portugal.

References

1. Van De Ville, D., Farouj, Y., Preti, M.G., Liégeois, R., Amico, E.: When makes you unique: temporality of the human brain fingerprint. Sci. Adv. **7**, eabj0751 (2021)
2. Abi-Dargham, A., Moeller, S.J., Ali, F., et al.: Candidate biomarkers in psychiatric disorders: state of the field. World Psychiatry **22**, 236–262 (2023)
3. Finn, E.S., et al.: Functional connectome fingerprinting: identifying individuals using patterns of brain connectivity. Nat. Neurosci. **18**, 1664–1671 (2015)
4. Abreu, R., Leal, A., Figueiredo, P.: EEG-Informed fMRI: a review of data analysis methods. Front. Hum. Neurosci. **12**, 29 (2018)

5. Borisov, S.V., Kaplan, A.Y., Gorbachevskaya, N.L., Kozlova, I.A.: Analysis of EEG structural synchrony in adolescents with schizophrenic disorders. Hum. Physiol. **31**(3), 255–261 (2005). https://doi.org/10.1007/s10747-005-0042-z
6. Shen, M., Wen, P., Song, B., Li, Y.: Automatic identification of schizophrenia based on EEG signals using dynamic functional connectivity analysis and 3D convolutional neural network. Comput. Biol. Med. **160**, 107022 (2023)
7. Poulsen, A.T., Pedroni, A., Langer, N., Hansen, L.K.: Microstate EEGlab toolbox: an introductory guide. bioRxiv, 289850 (2018)
8. Michel, C.M., Koenig, T.: EEG microstates as a tool for studying the temporal dynamics of whole-brain neuronal networks: a review. Neuroimage **180**, 577–593 (2018)
9. Shaw, S.B., Dhindsa, K., Reilly, J.P., Becker, S.: Capturing the forest but missing the trees: microstates inadequate for characterizing shorter-scale EEG dynamics. Neural Comput.put. **31**, 2177–2211 (2019)
10. Khanna, A., Pascual-Leone, A., Farzan, F.: Reliability of resting-state microstate features in electroencephalography. PLoS ONE **9**, e114163 (2014)
11. Tait, L., Zhang, J.: +microstate: a MATLAB toolbox for brain microstate analysis in sensor and cortical EEG/MEG. Neuroimage **258**, 119346 (2022)
12. Komorowski, M.K., et al.: ToFFi – toolbox for frequency-based fingerprinting of brain signals. Neurocomputing **544**, 126236 (2023)
13. Furman, Ł, Duch, W., Minati, L., Tołpa, K.: Short-time Fourier transform and embedding method for recurrence quantification analysis of EEG time series. Eur. Phys. J. Spec. Top. **232**, 135–149 (2023). https://doi.org/10.1140/epjs/s11734-022-00683-7
14. Chinichian, N., et al.: A fast and intuitive method for calculating dynamic network reconfiguration and node flexibility. Front. Neurosci. **17**(2023), 1025428 (2023)
15. Khare, S.K., Bajaj, V., Acharya, U.R.: SchizoNET: a robust and accurate Margenau-Hill time-frequency distribution based deep neural network model for schizophrenia detection using EEG signals. Physiol. Meas. **44**, 035005 (2023)
16. Phang, C.-R., Noman, F., Hussain, H., Ting, C.-M., Ombao, H.: A multi-domain connectome convolutional neural network for identifying schizophrenia from EEG connectivity patterns. IEEE J. Biomed. Health Inform. **24**, 1333–1343 (2020)
17. Aslan, Z., Akin, M.: A deep learning approach in automated detection of schizophrenia using scalogram images of EEG signals. Phys Eng Sci Med. **45**, 83–96 (2022)
18. Ellis, C.A., Miller, R.L., Calhoun, V.D.: Towards greater neuroimaging classification transparency via the integration of explainability methods and confidence estimation approaches. Inform. Med. Unlocked **37**, 101176 (2023)
19. Rahman, M.M., et al.: Interpreting models interpreting brain dynamics. Sci. Rep. **12**, 12023 (2022)

Human Centred Computing

Non-contact Respiratory Flow Extraction from Infrared Images Using Balanced Data Classification

Ali Roozbehi[1(✉)], Mahsa Mohaghegh[2], and Vahid Reza Nafisi[3]

[1] Amirkabir University of Technology, Tehran, Iran
Roozbehi_ali@yahoo.com

[2] Auckland University of Technology, 55 Wellesley Street East, Auckland City, New Zealand
mahsa.mohaghegh@aut.ac.nz

[3] Biomedical Engineering Group, Iranian Research Organization for Science and Technology, Tehran, Iran
vrnafisi@yahoo.com

Abstract. The COVID-19 pandemic has emphasized the need for non-contact ways of measuring vital signs. However, collecting respiratory signals can be challenging due to the transmission risk and physical discomfort of spirometry devices. This is problematic in places like schools and workplaces where monitoring health is crucial. Infrared fever meters are not accurate enough since fever is not the only symptom of these diseases. The objective of our study was to develop a non-contact method for obtaining Respiratory Flow (RF) from infrared images. We recorded infrared images of three subjects at a distance of 1 m while they breathed through a spirometry device. We proposed a method called Balanced Data Classification to distribute frames equally into several classes and then used the DenseNet-121 Convolutional Neural Network Model to predict RF signals from the infrared images. Our results showed a high correlation of 97% and a RMSE of 5%, which are significant compared to other studies. Our method is fully non-contact and involves standing at a distance of 1 m from the subjects. In conclusion, our study demonstrates the feasibility of using infrared images to extract RF.

Keywords: Biomedical infrared imaging · Spirometry · Biomedical measurement · CNN

1 Introduction

Respiratory Flow (RF), Respiratory Volume (RV), and Respiratory Rate (RR) are vital physiological indicators. RV is derived from the integration of the RF signal [1], while RR is based on the frequency domain of the RF signal [2]. Traditional methods of measuring these signals involve physical connections to patients, leading to inconvenience and discomfort [3–5]. The COVID-19 pandemic underscored the significance of non-contact methods for measuring vital signs, given the transmission risks. This study aims to develop a non-contact method to extract RF from infrared images, providing particular value in pandemic situations.

B. Luo et al. (Eds.): ICONIP 2023, CCIS 1963, pp. 443–454, 2024.
https://doi.org/10.1007/978-981-99-8138-0_35

Abbas et al. [6] utilized thermal imaging to extract RR using the Continuous Wavelet technique. Lewis et al. [7] employed video tracking algorithms to obtain breath-by-breath measures of RR and relative tidal volume. Pereira et al. [8] devised an algorithm to extract a more precise region of interest (ROI) by detecting nostril holes, subsequently using signal processing methods to extract RR. Elphick et al. [9] formulated algorithms to capture images, identify the face's location within each image, and obtain RR in real-time using signal processing techniques. Jagadev et al. [10] proposed a non-contact respiration monitoring method that functions without the need for visible facial landmarks and suits uncontrolled sleeping postures in the dark. Lyra et al. [11] harnessed object detection and deep learning to derive RR from thermal images. Mozafari et al. [12] implemented tensor decomposition on the entire facial image to extract RR. Transue et al. [13] introduced a non-contact respiratory evaluation technique utilizing Thermal-CO2 imaging for exhale flow and volume estimation, aligning closely with our work. This method investigates natural breathing patterns without necessitating participant effort.

In spite of comprehensive research, a discernible gap persists in extracting the RF signal, with the majority of studies centering on RR. Capturing RF facilitates the determination of supplementary respiratory parameters, such as RV and RR. This research endeavors to capture the RF signal by directly processing infrared images of the oral regions through neural networks. To this end, we introduced an innovative data classification approach to tackle the challenges posed by imbalanced datasets. Furthermore, we integrated various CNN models and pinpointed the most fitting model for the precise prediction of RF values from infrared images. A holistic overview of the project's workflow and its prospective applications is illustrated in Fig. 1.

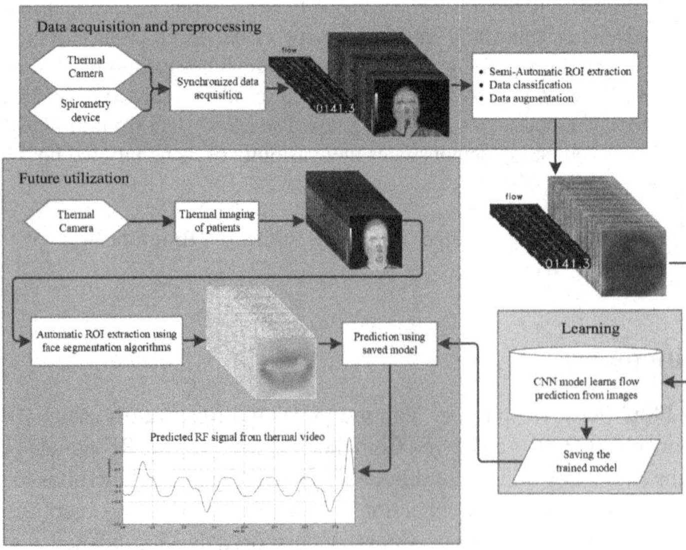

Fig. 1. Illustration of the project workflow and future applications

2 Experimental Setup

In our study, we utilized a Lenovo Legion Y540 laptop, a Satir Hotfind-S infrared camera, and an NTF-9501 model ventilator analyzer spirometer from Novin Teb Fanavaran Company. The infrared camera was set to capture temperatures ranging from 20 to 36 °C, with a resolution of 240 × 320 pixels and a sampling frequency of 20 fps. Although the spirometer was capable of measuring various respiratory parameters, only the RF was recorded for the purposes of this research. Thermal camera videos and spirometer readings were simultaneously recorded for the participants. The camera was strategically positioned 1 m from the subjects, ensuring that the respiratory flow was captured in real-time directly from their mouths (Fig. 2).

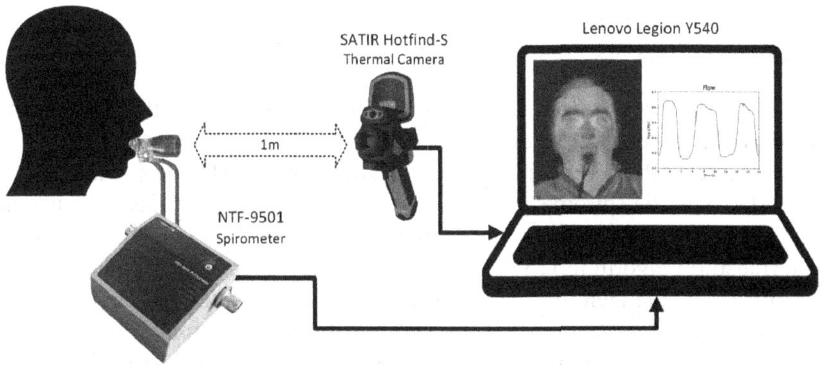

Fig. 2. Experimental setup schematic

During recording, subjects performed six breathing maneuvers (shallow-slow, shallow-normal, shallow-rapid, deep-slow, deep-normal, and deep-rapid) for 20 s each, totaling 2 min per subject. We used hand gestures in front of the camera to separate maneuvers. The study included two males and one female, averaging 22.3 years, 178 cm, and 74.6 kg in age, height, and weight, respectively.

3 Method

During data recording, hand gestures were employed to mark the beginning of each breathing maneuver in the videos. We edited the thermal videos to commence from the instant a hand gesture was detected, signaling a new breathing maneuver, and let them run for an additional 20 s (as illustrated in Fig. 3 for one of the participants). This approach produced 18 videos, each 20 s in duration, for all three participants.

Convolutional Neural Networks (CNNs) are adept at automatically extracting features [14]. However, when dealing with intricate features—like the Region of Interest (ROI) in this case, which represents the mouthpiece—large datasets are often required. To mitigate the complexities of learning, we designed a semi-automatic algorithm to identify the ROI in each frame. It's crucial to note that we trialed both techniques and

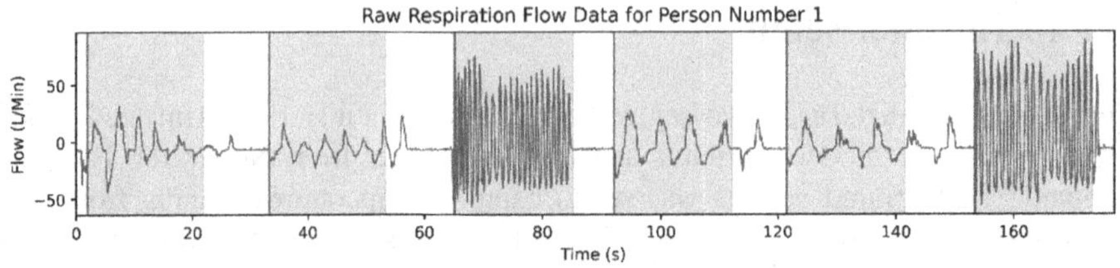

Fig. 3. Visualization of subject's breathing maneuvers

discovered that isolating the ROI using our method and solely feeding the ROI to the CNN resulted in better validation accuracy, even when data augmentation was applied. Nonetheless, this observation is based on a limited sample of only three participants.

3.1 ROI Extraction

Our study aims to detect respiratory changes based on temperature fluctuations around the mouth during the respiratory cycle. As per [6], the mouth area serves as an appropriate region of interest (ROI) for this objective. Extracting the ROI was challenging due to its overlap with the spirometry mouthpiece, rendering prevalent Face Segmentation Algorithms impractical. Instead, we proposed a semi-automatic technique employing the Hough Transform algorithm [15] to detect circles resembling the Mouthpiece Region (MR). This technique identifies circles near the manually marked circle from the first frame, assuming a consistent distance between subjects and minimal movement. The identified circles had an approximate radius of 8 pixels, which we rounded up to 10, cropping them into 20 × 20-pixel squares as the ROI (as shown in Fig. 4).

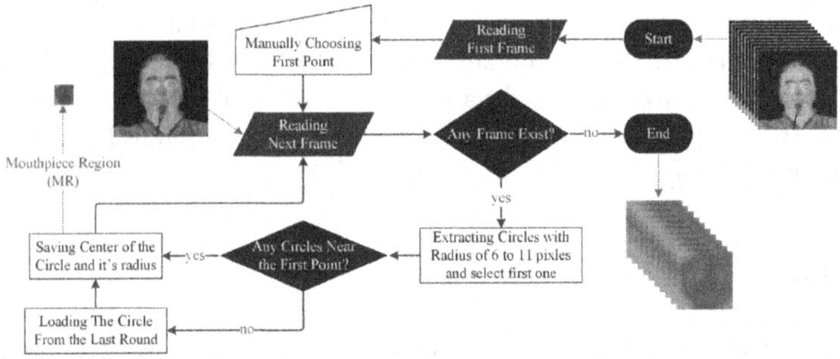

Fig. 4. Schematic of MR area extraction algorithm

The algorithm presents two primary limitations: the requirement for manual pinpointing of the MR center in the initial video frame, and the assumption that subjects remain largely static. In real-world scenarios, subjects may move during thermal imaging, and the mouthpiece might not always be present. To overcome these challenges, one can employ face segmentation algorithms, such as the MediaPipe Face Mesh algorithm [16], to extract the mouth area of patients.

3.2 Preprocessing

After collecting RF signals and infrared videos, followed by trimming and extracting ROIs, several preprocessing steps were undertaken to prepare the data for the neural network. These include data normalization, synchronizing these signals with the corresponding videos, filtering out noisy data, selecting test data, classifying frames, and data augmentation.

Data Normalization. Normalizing inputs and outputs is advised for convolutional neural networks [17]. In this study, the input consists of images with 8-bit pixels, having values that range from 0 to 255. These were normalized to fall between zero and one. The output, conversely, is the RF signal. Considering the global maximum and minimum values of RF signals were 193.55 and −135.82 respectively, they were normalized to range between 1 and 0 across all 18 signals.

Noisy Data Removal. Post the normalization of RF signals, we evaluated their synchrony with the videos. Four out of the 18 breathing maneuvers presented significant distortions (as depicted in Fig. 5) and were consequently discarded. This left us with 14 data sets, each containing a 20-s video, accompanied by its respective RF signal and mouthpiece coordinates.

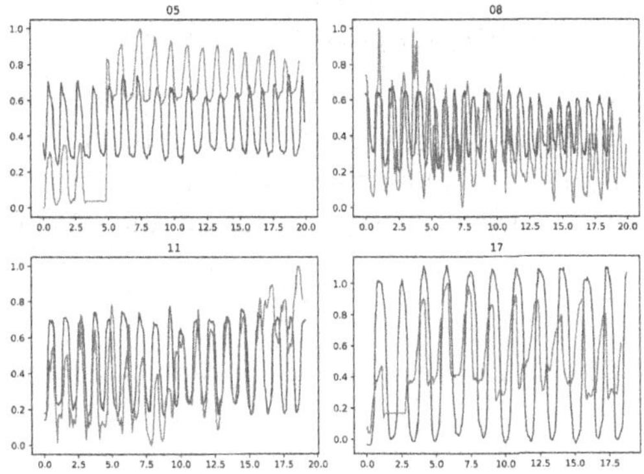

Fig. 5. Excluded noisy data – averaged ROI region (Orange) and respiration flow (Blue) signals for removed subjects (Color figure online)

Videos and RF Data Synchronization. Synchronization was executed both manually and through automated means. The primary phase necessitated the alignment of the initial discernible local extremum in both the RF signal and the video, illustrated in (Fig. 6 – b). The wireless protocol adopted by the spirometry device occasionally led to a form of 'elastic' desynchronization. As a remedy, we identified peaks within both signals, deduced the temporal disparities between these peaks, and subsequently synchronized the distances between the peaks of the RF signal and the video ROI signal via either up-sampling or down-sampling methods (See Fig. 6 - c).

Test Videos Selection. Out of the 14 available videos, three (videos 1, 7, and 13) were earmarked for testing, with the remaining 11 set aside for training purposes. The 20-s

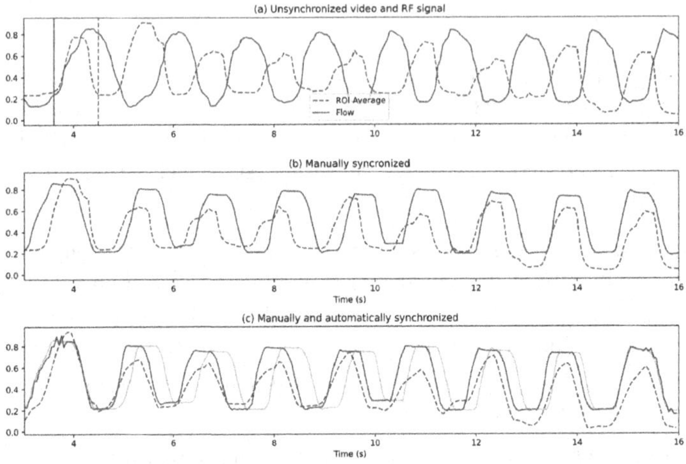

Fig. 6. Manual and automatic synchronization of RF signals and corresponding video

breathing maneuvers, captured at a frequency of 20 Hz, got segmented into 20×20 pixel frames. Once the aforementioned pre-processing protocols were instituted, we were left with video clips containing anywhere between 326 to 400 frames. This culminated in 4075 images for training and 1078 images designated for testing.

Data Classification. While predicting continuous RF signals from thermal images can be viewed as a regression task, we approached it as a classification challenge by dividing normalized RF signals into 10 distinct flow boundaries, as outlined in Table 1. The marked variance in the data distribution rendered it unsuitable for neural network applications.

Table 1. Data distribution across classes after classification using the normal method

Class num	1	2	3	4	5	6	7	8	9	10	CV
Test	6	86	159	159	320	90	126	45	79	2	82%
Train	36	165	460	836	1399	443	347	221	119	34	99%

The distribution of data in Table 1 deviates significantly from uniformity, exhibiting a pronounced coefficient of variation (CV) [18]. This non-uniformity, when processed using a CNN, culminated in an accuracy of less than 10%. To rectify this, we conceived a novel classification technique dubbed the Balanced Data Classification (BDC).

Balanced Data Classification (BDC). BDC is a strategy that refines classification boundaries to foster a semi-uniform distribution across classes, as illustrated in Fig. 7.

As depicted in Fig. 8, the BDC method contracts the boundaries for classes with higher populations, resulting in a more equitable data distribution, detailed in Table 2.

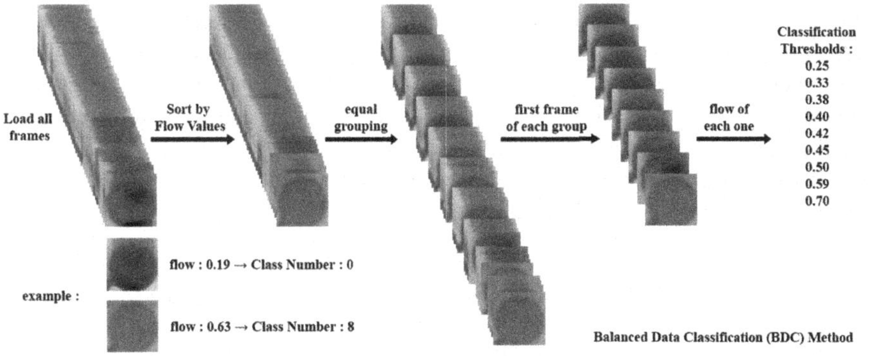

Fig. 7. Schematic of the Balanced Data Classification (BDC) method

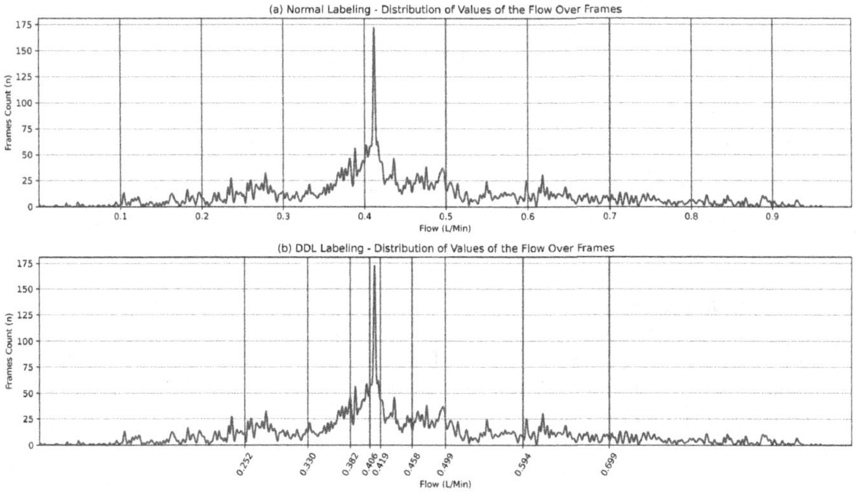

Fig. 8. Histogram and thresholds for normal method (a) and BDC method (b) of RF data

Table 2. Data distribution across classes after classification using the BDC method

Class num	1	2	3	4	5	6	7	8	9	10	CV
Test	132	147	56	108	116	141	28	81	143	126	35%
Train	383	369	460	408	395	373	487	431	372	397	9%

By leveraging BDC, the data becomes apt for neural network training, with each training class containing approximately 400 data points. This ensures the network comprehends all classes uniformly. Efficaciously, the BDC technique addressed the dataset's imbalances, paving the way for more accurate predictions.

Data Augmentation. Our research entailed the application of several data augmentation techniques to enhance both the size and diversity of the dataset. The adopted augmentation strategies included Gaussian distortion, random distortion, random erasing, random rotation, and random flipping. We established the optimal augmentation rate as five-fold the size of the most populated class, leading to a final count of 2435 images per class in the final dataset. A comprehensive breakdown of the image count for each class post-augmentation is detailed in Table 3.

Table 3. Data distribution across classes after processing steps

Class num	1	2	3	4	5	6	7	8	9	10
Test	128	147	56	110	105	156	33	72	144	127
Train	2435	2435	2435	2435	2435	2435	2435	2435	2435	2435

3.3 Model Input

As alluded to earlier, the chosen MRs were depicted using 20x20-pixel squares, a representation evident in Fig. 4.

3.4 Evaluation Metrics

Utilizing 5-fold stratified cross-validation [19], we assessed the model's efficacy based on three metrics: the coefficient of determination (R-squared) [20], the root mean squared error (RMSE) [21], and the correlation coefficient [22].

4 Results

In our quest to pinpoint the most effective neural network architecture for image processing, we experimented with multiple structures. The performance of these models, gauged by the average f1-score [23] during a 5-fold cross-validation [19], is encapsulated in Table 4. With an exemplary average f1-score of 92%, DenseNet-121 emerged as the leader. To curtail the potential for overfitting, we incorporated dropout [24] at a rate of 25% and applied Batch Normalization [25] within the fully-connected layers. It's pertinent to note that specific networks demand a minimum input size; hence, all our Mouthpiece Region (MR) images underwent reshaping to dimensions of $(3 \times 75 \times 75)$.

Table 4. Highest training accuracy attained by selected prominent CNN models

Model	DenseNet121 [26]	InceptionV3 [27]	EfficientNetB0 [28]
Avg f1-score (%)	92.9	68.3	57.2

5. Matthews, G., Sudduth, B., Burrow, M.: A non-contact vital signs monitor. In: Critical Reviews in Biomedical Engineering, pp. 173–178. Begell House Inc. (2000). https://doi.org/10.1615/CritRevBiomedEng.v28.i12.290
6. Abbas, A.K., Heimann, K., Jergus, K., Orlikowsky, T., Leonhardt, S.: Neonatal non-contact respiratory monitoring based on real-time infrared thermography. Biomed. Eng. Online **10**, 93 (2011). https://doi.org/10.1186/1475-925X-10-93
7. Lewis, G.F., Gatto, R.G., Porges, S.W.: A novel method for extracting respiration rate and relative tidal volume from infrared thermography. Psychophysiology **48**, 877–887 (2011). https://doi.org/10.1111/j.1469-8986.2010.01167.x
8. Pereira, C.B., Yu, X., Czaplik, M., Rossaint, R., Blazek, V., Leonhardt, S.: Remote monitoring of breathing dynamics using infrared thermography. Biomed. Opt. Express **6**, 4378 (2015). https://doi.org/10.1364/boe.6.004378
9. Elphick, H., Alkali, A., Kingshott, R., Burke, D., Saatchi, R.: Thermal imaging method for measurement of respiratory rate. Eur. Respir. J., PA1260 (2015). https://doi.org/10.1183/13993003.congress-2015.pa1260
10. Jagadev, P., Giri, L.I.: Non-contact monitoring of human respiration using infrared thermography and machine learning. Infrared Phys. Technol. **104**, 103117 (2020). https://doi.org/10.1016/j.infrared.2019.103117
11. Lyra, S., et al.: A deep learning-based camera approach for vital sign monitoring using thermography images for ICU patients. Sensors **21**, 1–18 (2021). https://doi.org/10.3390/s21041495
12. Mozafari, M., Law, A.J., Green, J.R., Goubran, R.A.: Respiration rate estimation from thermal video of masked and unmasked individuals using tensor decomposition. In: Conference Record - IEEE Instrumentation and Measurement Technology Conference. Institute of Electrical and Electronics Engineers Inc. (2022). https://doi.org/10.1109/I2MTC48687.2022.9806557
13. Transue, S., Min, S.D., Choi, M.-H.: Expiratory flow and volume estimation through thermal-CO_{2} imaging. IEEE Trans. Biomed. Eng., 1–10 (2023). https://doi.org/10.1109/TBME.2023.3236597
14. Deep Learning and the Future of Machine Learning | AltexSoft. https://www.altexsoft.com/blog/deep-learning/. Accessed 22 Jan 2023
15. Duda, R.O., Hart, P.E.: Use of the Hough transformation to detect lines and curves in pictures. Commun. ACM **15**, 11–15 (1972). https://doi.org/10.1145/361237.361242
16. Face Mesh | mediapipe. https://google.github.io/mediapipe/solutions/face_mesh.html. Accessed 22 Jan 2023
17. Lecun, Y., Bengio, Y., Hinton, G.: Deep learning. Nature **521**(7553), 436–444 (2015). https://doi.org/10.1038/nature14539
18. Co-efficient of Variation Meaning and How to Use It. https://www.investopedia.com/terms/c/coefficientofvariation.asp. Accessed 18 Aug 2023
19. A Gentle Introduction to k-fold Cross-Validation - MachineLearningMastery.com. https://machinelearningmastery.com/k-fold-cross-validation/. Accessed 18 Aug 2023
20. Coefficient of Determination (R^2) | Calculation & Interpretation. https://www.scribbr.com/statistics/coefficient-of-determination/. Accessed 18 Aug 2023
21. Root Mean Square Error (RMSE) - Statistics By Jim. https://statisticsbyjim.com/regression/root-mean-square-error-rmse/. Accessed 18 Aug 2023
22. The Correlation Coefficient: What It Is, What It Tells Investors. https://www.investopedia.com/terms/c/correlationcoefficient.asp. Accessed 18 Aug 2023
23. F-score – Wikipedia. https://en.wikipedia.org/wiki/F-score. Accessed 18 Aug 2023
24. A Gentle Introduction to Dropout for Regularizing Deep Neural Networks. https://machinelearningmastery.com/dropout-for-regularizing-deep-neural-networks/. Accessed 14 Sept 2022

25. Ioffe, S., Szegedy, C.: Batch normalization: accelerating deep network training by reducing internal covariate shift (2015). https://doi.org/10.48550/arxiv.1502.03167
26. Huang, G., Liu, Z., van der Maaten, L., Weinberger, K.Q.: Densely connected convolutional networks (2016). https://doi.org/10.48550/arxiv.1608.06993
27. Szegedy, C., Vanhoucke, V., Ioffe, S., Shlens, J., Wojna, Z.: Rethinking the inception architecture for computer vision (2015). https://doi.org/10.48550/arxiv.1512.00567
28. Tan, M., Le, Q.V.: EfficientNet: rethinking model scaling for convolutional neural networks (2019)

The Construction of DNA Coding Sets by an Intelligent Optimization Algorithm: TMOL-TSO

Yongxu Yan, Wentao Wang, Zhihui Fu, and Jun Tian(✉)

College of Software, NanKai University, Tianjin 300350, China
{wangwt,fuzhihui}@mail.nankai.edu.cn, jtian@nankai.edu.cn

Abstract. DNA computing has a natural advantage in solving NP-complete problems due to its high concurrency and low energy consumption. With the development of DNA computing, the preliminary formation of DNA logic circuit architectures further illustrates the potential of this field. Additionally, DNA holds great potential for storage due to its high density, large capacity, and long-term stability. It is suitable for database construction and data storage. However, non-specific hybridization of DNA molecules may cause unexpected outcomes during computation or storage processes. Therefore, it is crucial to apply pressure constraints to DNA coding to ensure stability. Designing a DNA coding set that satisfies constraints is a primary challenge. In this paper, we propose a Tuna Swarm Optimization (TSO) algorithm that employs random opposition-based learning strategy and Two-Swarm Merge strategy. This algorithm has stronger global exploration capabilities. Experimental results demonstrate that this algorithm can find a better coding set in some cases.

Keywords: DNA Design · TSO Algorithm · Two-Swarm Merge

1 Introduction

The relationship between DNA and computers has long been a subject of interest in the fields of computer science and bioinformatics. Over the years, research in DNA information science has given rise to several branches, including DNA computing, DNA storage, DNA nanotechnology, DNA quantum computing, and more.

DNA computing is a novel molecular biological computing method that uses DNA and related enzymes as basic materials and utilizes specific biochemical reactions to perform calculations. The method encodes information using DNA's double helix structure and base pairing rules, maps computational data onto DNA chains, and generates data pools using biological enzymes. Then, according to specific rules, we map the highly parallel computational process of the original problem's data onto the controllable biochemical process of DNA molecular

B. Luo et al. (Eds.): ICONIP 2023, CCIS 1963, pp. 455–469, 2024.
https://doi.org/10.1007/978-981-99-8138-0_36

chains. Finally, the required computational results are detected using molecular biology techniques [1]. DNA computing has many advantages [1,6], such as 1) high parallelism; 2) high storage density and large capacity; 3) low energy consumption. Under the same computational conditions, a molecular computer requires only one billionth of the energy an electronic computer needs.

Adleman [1] first proposed using DNA molecules for computing, using the clipping sequences of DNA molecules as path encoding, and using standard protocols and enzymes to execute "operations" during the computing process to solve the Hamiltonian path problem. The high parallelism of DNA computing demonstrated its enormous potential. However, the inflexibility of its operations has become a bottleneck in DNA computing. Lipton et al. [16] extended Adleman's biological computing method for solving the Hamiltonian path problem, and proposed a framework for biological computing to solve NP-complete problems for the first time. However, the operations in biological information computing methods still need improvement, and the critical problem is how to solve errors. Liu et al. [17] introduced a scalable and automated method, DNA computing on surfaces, with potential for automation. Later, Wu [30] proposed a cost-effective, shorter operation time, reusable surface, and simpler method for DNA computing, where he proposed the steps of DNA computing, converting the SAT problem into a problem solved through optical computing steps. In subsequent research, Benenson [3] described a programmable finite automaton that autonomously solves computational problems using DNA and DNA operating enzymes. Sakakibara [23] proposed a novel DNA chip with executable logical operations and developed methods to represent and evaluate Boolean functions on DNA chains, which solved the "AND' and "OR" logic operations of gene expression. By combining DNA coding methods, they created a DNA chip that detects gene expression and finds logical formulas for gene expression. The development of DNA computing has moved closer to DNA computers and is no longer limited to solving NP-complete problems. Qian [19,20] proposed a simple DNA scaffold architecture for large-scale circuits that can synthesize thousands of gates and demonstrated several digital logic circuits based on a simple DNA reaction mechanism of reversible strand displacement, forming a four-bit square root circuit containing 130 DNA strands. In addition, this design incorporates other important components for large-scale circuits, including a debugging tool that can be used universally, support for parallel circuits, and an abstraction hierarchy that is facilitated by an automatic circuit compiler. Building on previous research, Qian [21] designed a DNA molecular-level system based on the self-assembly and self-catalysis properties of DNA molecules, demonstrating that DNA strand displacement cascades can be used to enable autonomous systems to recognize patterns of molecular events, make decisions and respond to their environment, thereby achieving neural network computation. In recent years, Song [25] proposed a DNA logic circuit architecture based on single-strand logic gates using chain replacement DNA polymerases, improving computing speed and reducing architecture complexity. The research and development of DNA computing have become more three-dimensional and mature under the previous research.

In addition to research on DNA computing, DNA storage has attracted considerable research attention due to its advantages such as high capacity, high density, and long-term storage [29]. DNA is formed by connecting four types of deoxyribonucleotides, and the pairing of A-T and C-G forms the stable double-stranded structure. The DNA base sequence can be mapped to binary data for storage, whether paired double-stranded or unpaired single-stranded DNA.

The methods for storing information in DNA include DNA encoding technology [18] and DNA decoding technology [24]. The DNA data storage architecture was first proposed by Baum [2], who used containers containing DNA as memory and wrote appropriate DNA chains into the container as memory writes. This model can be used to construct DNA memory with a large capacity. The length of DNA sequences is arbitrary but limited, so when writing to memory, they are typically divided into small blocks and then reassembled into the original data. During the reassembly process, an index is added to each block [7,10]. According to a specific algorithm, the DNA coding sequence is processed into information encoding, which is then written into DNA molecules [18]. For reading DNA data, DNA sequencing methods such as PCR are used to obtain the base sequence, which is then converted into binary data to obtain the original data [24]. After completing the basic research on DNA storage, research on error correction and database construction for DNA storage received attention. Yamamoto et al. [12,13,33] constructed a database with a storage capacity of 16.8MB using a nested PCR-based storage method. Stewart et al. [26] proposed an image DNA storage database based on associative search. Heckel et al. [11] provided a quantitative and qualitative understanding of DNA data storage channels. Their research is helpful for designing a DNA data storage system.

Both DNA computing and DNA storage may encounter errors due to non-specific hybridization of DNA during the "operation" process of DNA-specific hybridization or the "search" process of sequencing [1,4,27]. Therefore, designing a set of robust DNA coding sequences is an important issue. The way to improve the robustness of the DNA coding set is to add related constraints to optimize the coding sequence. Construction methods, intelligent algorithms, neighborhood search algorithms, and random search algorithms are currently used to design an excellent coding set. In recent years, Yin et al. [34] proposed an NOL-HHO algorithm improved by a nonlinear control strategy. Experimental results showed that a better lower bound for DNA storage was obtained. Cao et al. [5] proposed a KMVO algorithm superior to the MOV algorithm in satisfying Hamming distance constraint, GC content constraint, and No-runlength constraint. Li et al. [14] proposed an improved DTW distance constraint and used the ROEAO algorithm to construct a coding set with higher lower bounds and better coding quality under traditional constraints and enhanced constraints. Rasool [22] proposed an MFOS evolutionary algorithm based on Levy flight cooperative moth-flame optimizer and opponent learning to obtain better coding sets than before. S.T. [8] proposed a new reversible code construction method based on group ring-derived composite matrices to construct DNA coding sets that satisfy Hamming constraint, reverse constraint, reverse complement constraint, and GC constraint. Xie [32] proposed an improved arithmetic optimization of the billiards algorithm, introduced an excellent initial point set, expanded the global search range with the billiard hitting strategy, and added a random lens opponent learning mechanism. Experimental results showed that the DNA sequences designed by this method have higher quality.

Through the efforts of previous researchers, evolutionary algorithms have gradually evolved to improve the lower bound of DNA coding sets by combining different evolutionary algorithms with optimization strategies. However, many of the optimization evolutionary algorithms proposed by previous researchers are single-objective strategies with a narrow search space, and their algorithmic performance could be better. This article proposes a strategy that combines multi-objective optimization and merging algorithms. The strategy is based on the Tuna Swarm algorithm and involves expanding the search space using a target dispersion and merging strategy. Optimization algorithms with random opponent learning after merging are used to improve the DNA coding set under combination optimization. We name it the Two-Merge and random Opposition-base Learning TSO algorithm(TMOL-TSO).

The structure of the paper is as follows: Sect. 2 discusses the constraints on DNA coding. Section 3 introduces the Tuna Swarm Optimization (TSO) algorithm. Section 4 presents the algorithmic process of the improved Tuna Swarm Optimization algorithm and the application of TMOL-TSO in DNA coding optimization. Section 5 shows the experimental results and compares them with other optimization algorithms. Finally, Sect. 6 summarizes this article.

2 Constructions on DNA Code

The coding problem is the foundation of DNA computing and DNA Storage. The purpose of the coding problem is to design a coding set that meets the requirements of the problem and can be effectively used in solving the problem. A high-quality coding design can effectively reduce the space complexity of DNA computing, improve the feasibility of biochemical operations in DNA computing, and store data more reliably in DNA. Therefore, different constraints must be applied to the DNA coding set to enhance the robustness of the DNA sequence.

2.1 Hamming Distance Constraint

The Hamming distance between two DNA sequences, denoted as H(x, y), is defined as the number of positions where the corresponding bases differ between the two sequences [9]. Satisfying the Hamming distance implies that H(x, y)≥ d. Let X and Y be two DNA sequences of equal length, respectively: $X = 5' - x_1x_2...x_n - 3'$ and $Y = 5' - y_1y_2...y_n - 3'$. The Hamming distance is calculated using the following Eq. (1):

$$H(X,Y) = \sum_{i=1}^{n} h(x_i, y_i), h(x_i, y_i) = \begin{cases} 0, x_i = y_i \\ 1, x_i \neq y_i \end{cases} \tag{1}$$

Hamming distance is a measure of the dissimilarity between two DNA sequences. It indicates the number of positions where the bases of two sequences differ. The larger the Hamming distance between two sequences X and Y, the fewer bases are identical between them. This means that if the reverse complement sequences of X and Y have fewer complementary base pairs, they are less likely to hybridize, resulting in greater stability between the two sequences.

2.2 No-Runlength Constraint

In DNA storage, a No-runlength (NL) constraint can be added to the DNA sequence to avoid an increased error rate during decoding. NL means that when constructing the DNA base sequence, it is required that no two adjacent bases can be the same. Otherwise, errors are likely to occur during the sequencing and synthesis processes. For a DNA sequence such as $X = 5' - x_1 x_2 ... x_n - 3'$, this constraint is defined by Eq. (2) as follows:

$$x_i \neq x_{i-1}, i \in [2, n] \tag{2}$$

2.3 GC-Content Constraint

The GC content constraint in DNA represents the total content of C and G bases in the DNA sequence, expressed as a percentage of the full length of the DNA sequence. The GC content constraint is commonly employed to regulate the melting temperature and stability of DNA molecules, as the hydrogen bonds between G and C bases in a DNA molecule are more stable than those between A and T bases. Thus, higher GC content results in higher melting temperature and greater stability of DNA molecules. For a DNA sequence of length n, GC content is calculated using the following Eq. (3):

$$GC(n) = \frac{|G| + |C|}{|n|} \tag{3}$$

2.4 Reverse Complement Hamming Distance Constraint

The complementary bases for A, G, C, and T are T, C, G, and A, respectively. For a DNA sequence $X = 5' - x_1 x_2 x_3 .. x_n - 3'$, its reverse complement DNA sequence $X^{RC} = 5' - \hat{x}_n \hat{x}_{n-1} \hat{x}_{n-2} ... \hat{x}_1 - 3'$ The inverse Hamming distance constraint describes the similarity between X and Y^{RC}.

The purpose of the reverse complement Hamming distance constraint is to prevent the formation of secondary structures between the reverse complement sequences of a DNA strand, which can affect the stability and reliability of DNA molecules. The larger the reverse complement Hamming distance, the better, as it indicates the degree of dissimilarity between the two sequences. A larger reverse complement Hamming distance implies that the similarity between the two sequences is smaller, which reduces the likelihood of forming secondary structures and increases the stability and reliability of DNA molecules. The calculation formula of Reverse complement Hamming distance H(X,Y^{RC}) is Eq. (4):

$$H(X,Y^{RC}) = \sum_{i=1}^{n} h(x_i, \hat{y}_{n-i+1}), h(x_i, \hat{y}_{n-i+1}) = \begin{cases} 0, x_i = \hat{y}_{n-i+1} \\ 1, x_i \neq \hat{y}_{n-i+1} \end{cases} \quad i \in [1, n] \tag{4}$$

3 TSO Algorithm

The Tuna Swarm Optimization (TSO) algorithm is a metaheuristic algorithm based on swarm intelligence, inspired by the cooperative foraging behavior of tuna schools. Although tuna swim very fast, they are less agile than smaller fish regarding reaction speed, so they have developed various effective predation strategies. One such strategy is spiral foraging, where tuna form a spiral queue to drive prey towards shallow waters, where they are more vulnerable to attack. Another strategy is parabolic foraging, where tuna swim behind each other in a parabolic shape to encircle prey. The TSO algorithm models these two foraging behaviors of tuna schools to create an efficient metaheuristic algorithm.

3.1 Spiral Foraging Strategy

When a small school of fish encounters a predator, they gather into a queue and swim in a constantly changing direction to avoid the predator. In this situation, a tuna school needs to form a tight spiral shape to chase after prey. In the beginning, most of the school members do not have a sense of direction, but when a small number of tuna determine a direction, nearby members follow them, forming a large group with a clear goal. Information exchange between adjacent tuna also enables the school to follow the target. Based on the above information, the model is as follows:

$$X_i^{t+1} = \begin{cases} \alpha_1 \cdot (X_{best}^t + \beta \cdot |X_{best}^t - X_i^t|) + \alpha_2 \cdot X_i^t, & i = 1 \\ \alpha_1 \cdot (X_{best}^t + \beta \cdot |X_{best}^t - X_i^t|) + \alpha_2 \cdot X_{i-1}^t, & i = 2, 3...N \end{cases} \tag{5}$$

Tuna exhibit good exploration ability when foraging in a spiral formation around food. However, blindly following the optimal individual is not beneficial for the group's foraging when the optimal individual fails to locate food. Therefore, it is necessary to allow some individuals to break away from following the optimal individual at the appropriate time and search a wider area. This gives the TSO algorithm global search capability. The mathematical model for TSO is as follows:

$$X_i^{t+1} = \begin{cases} \alpha_1 \cdot (X_{rand}^t + \beta \cdot |X_{rand}^t - X_i^t|) + \alpha_2 \cdot X_i^t, & i = 1 \\ \alpha_1 \cdot (X_{rand}^t + \beta \cdot |X_{rand}^t - X_i^t|) + \alpha_2 \cdot X_{i-1}^t, & i = 2, 3...N \end{cases} \tag{6}$$

3.2 Parabolic Foraging Strategy

In addition to spiral foraging, tuna schools use parabolic cooperative foraging around a reference point, forming a parabolic structure. Additionally, tuna search for food by moving randomly in the search space. The selection probability of these two foraging strategies is 50%. The mathematical model for this is described as follows:

$$X_i^{t+1} = \begin{cases} X_{best}^t + rand \cdot (X_{best}^t - X_i^t) + TF \cdot p^2 \cdot (X_{best}^t - X_i^t), \\ \qquad \text{if } rand < 0.5 \\ TF \cdot p^2 \cdot X_i^t, \\ \qquad \text{if } rand \geq 0.5 \end{cases} \tag{7}$$

In the model, X_{best}^t, X_i^t, and X_{rand}^t represent the current best position obtained by the tuna school, the position of the i-th individual, and a random position. α_1 and α_2 are weight parameters used to control the shape and density of the spiral. TF is a random number that can be either 1 or −1. Other parameters are detailed in the original paper.

4 The Improved Algorithm

The Tuna Swarm Optimization (TSO) algorithm switches between two foraging modes: spiral and parabolic. The two modes are switched with equal probability. The spiral foraging mode searches around the current best solution in a spiral pattern, expanding the search range and possessing global exploration capability. The parabolic foraging mode is a local search strategy that utilizes information exchange and cooperation among tuna schools. When a group of tuna discovers the optimal foraging point, other individuals quickly gather around the point. However, both foraging modes are prone to falling into local optima.

To improve the performance of the TSO algorithm and enhance its ability to escape from local optima and perform global search, we considered two optimization strategies: changing search direction and reducing search constraints to expand the search target.

4.1 Random Opposition-Based Learning

Swarm Intelligence algorithms inevitably tend to converge in a certain direction. When the solution set of the problem to be solved is too large, or when the parameter settings of the Swarm Intelligence algorithms are unreasonable, the algorithm is likely to be trapped in oscillations around a local solution. Therefore, it is necessary to introduce some optimization strategies to improve these situations. Reverse Opposite Learning [28] generates an opposite solution set for the current solution set. For x, in the interval [a, b], the offset of x relative to a equals the offset of $\hat{x}$ relative to b.

Using the opposite strategy during iteration can increase the probability of individuals searching for the global optimal solution. However, the initial optimal solution during iteration may be far from solving the global optimal solution, so there is a possibility that both the current individual and the generated opposite individual are far from the individual representing the global optimal solution. Therefore, to improve the effectiveness of reverse learning, we use a random reverse opposite learning strategy.

$$\hat{X} = a + b - rand \cdot X, X \in [a, b] \tag{8}$$

4.2 Two-Swarm Merge

Merge sort is a highly efficient and stable sorting algorithm, and its key algorithmic idea is the binary merge. Merge sort divides the sequence to be sorted into two parts, recursively processes the left and right parts, and then merges the ordered sequences of the left and right parts into a single ordered sequence. The split and merge operations of merge sort provide us with good inspiration for optimizing swarm intelligence algorithms. For swarm intelligence algorithms, we can adopt a binary strategy for the initial population, dividing it into two parts and implementing different search strategies or evaluations for each part. After each iteration, we can merge the populations according to certain rules or constraints. This expands the search space of swarm intelligence algorithms and speeds up the search process.

4.3 TMOL-TSO Algorithm's Description

The process of using TMOL-TSO to solve the DNA coding problem is as follows. First, we initialize a constrained DNA set, DNASet, which contains a DNA chain that satisfies the constraints we set. Next, we initialize the parameters of the TSO algorithm and two populations, named LeftTunas and RightTunas. Taking the combination constraints of DNA set satisfying HL constraint, GC(x) = $\lfloor n/2 \rfloor$, and HD(X,Y) $\leq$ d as an example, we set LeftTunas to satisfy the HL constraint and GC constraint, and RightTunas to satisfy the HD constraint and GC constraint. By setting different constraints for the two populations, we expand the search space and speed up the search process with the same population size.

We need a fitness value to evaluate the fitness of individuals in the population. For the four constraints, we set their corresponding constraint fitness values as follows:

$$GCfitness(x) = abs(|C| + |G| - \lfloor n/2 \rfloor) \tag{9}$$

$$HDfitness(x, y) = max(0, d - H(x, y)) \tag{10}$$

$$HLfitness(x) = HL(x) \tag{11}$$

$$RCfitness(x, y) = max(0, H(x, y^{RC}) \tag{12}$$

An individual's fitness value is related to each DNA sequence in DNASet. Therefore, when an individual has HD or RC constraints, its fitness must be calculated for each DNA sequence in a loop.

$$Fitness(LeftTunas_i) = \sum_{i=1}^{n} HDfitness(LeftTunas_i, seq_i) + GCfitness(LeftTunas_i), seq_i \in DNASet \quad (13)$$

$$Fitness(RightTunas_i) = GCfitness(RightTunas_i) + HLfitness(RightTunas_i) \quad (14)$$

It is evident that when the fitness value of an individual is 0, it is a DNA sequence that satisfies the constraints.

In each iteration, we first decide whether to randomly initialize an individual by comparing a random number and parameter z. If the random number is less than z, we initialize the individual randomly according to Eq. (15). Otherwise, we compare the random number with 0.5. If it is less than 0.5, we update the individual using the spiral foraging Strategy, selecting Eq. (5) or Eq. (6) based on the number of iterations. Otherwise, we update the population using the parabolic foraging strategy Eq. (7).

$$X_i^{int} = rand \cdot (ub - lb) + lb, i = 1, 2...N \quad (15)$$

After each iteration, we perform random opposite learning on LeftTunas and RightTunas to generate two opposite populations. Then, we evaluate the four populations to check whether the individuals in the population satisfy the constraints. If an individual in the population satisfies the constraints with DNASet, we add it to DNASet. For individual A, assuming that all individuals in the DNASet collection are compatible except for individual B, we remove the incompatible DNA sequence B from the DNASet and add individual A to the DNASet.

After these operations, we need to update the populations by selecting the top N best individuals from the opposite and original populations for the next iteration.

The description and flow chart of the TMOL-TSO algorithm for solving the DNA coding set are shown below. Algorithm 1 shows the pseudo-code of the TMOL-TSO algorithm. The flow chart is shown in Fig. 1.

5 Experimental Results and Analysis

In DNA coding sets, NL, GC, and HD constraints are typically coding constraints for DNA storage, while GC, HD, and RC constraints are typically coding constraints for DNA computation. The quality of a DNA coding set can be evaluated from two perspectives: one is to find a higher-quality DNA coding set that meets the constraints, and the other is to maximize the number of DNA sequences in the coding set while maintaining a certain level of quality. The algorithm proposed in this paper focuses on the latter case. We define $A^{GC,HL}(n, d, w)$ as the number of DNA coding sets that satisfy the Hamming distance $H(x_i, x_j) \geq d$, the GC content constraint with a length of w, and the HL constraint. $A^{GC,RC}(n, d, w)$ is defined as the number of DNA coding sets

Algorithm 1. TMOL-TSO algorithm

```
Initialize a series of parameters.
Initialize the random population LeftTunas_i(i = 1, 2...N).
Initialize the random population RightTunas_i(i = 1, 2...N).
Initalize the result DNASet.
while t ≤ T do
  Calculate separately the fitness values of LeftTunas,RightTunas.
  Update LeftBestT,RightBestT.
  for i from 1 to N do
    update α1α2, p
    if rand < z then
      Update the position LeftTunas_i^{t+1}, RightTunas_i^{t+1} using equation (15).
    else if rand ≥ z then
      if rand < 0.5 then
        if t/t_max < rand then
          Update the position LeftTunas_i^{t+1}, RightTunas_i^{t+1}using equation (6)
        else if t/t_max ≥ rand then
          Update the position LeftTunas_i^{t+1}, RightTunas_i^{t+1} using equation (5)
        end if
      else if rand ≥ 0.5 then
        Update the position LeftTunas_i^{t+1}, RightTunas_i^{t+1} using Equation (7)
      end if
    end if
  end for
  Execute Random Opposition-Based learning based on Equation(x)
  Get oppsition population OpLeftTunas,OpRightTunas.
  for i from 1 to N do
    for s in OpLeftTunas_i, OpRightTunas_i, LeftTunas_i, Right_tunas_i do
      if s incompatible with set DNASet then
        if just one DNA Sequence in DNASet incompatible with s then
          Remove the Sequence which incompatible with s from DNASet.
          Add the s to DNASet.
        end if
      else
        Add the s to DNASet.
      end if
    end for
  end for
  Calculate the fitness values LeftFitval of LeftTunas,OpLeftTunas.
  NewLeftTunas equal the top N tunas in LeftTunas,OpLeftTunas with the maxi-
  mum fitness values.
  Do the same for NewRightTunas.
end while
```

that satisfy the Hamming distance $H(x_i, x_j) \geq d$, the reverse complement Hamming distance $H(x_i, \hat{x_j}) \geq d$, and the GC content constraint with a length of w. In both experiments, w is set to $\lfloor n/2 \rfloor$, which means the proportion of GC content is 0.5.

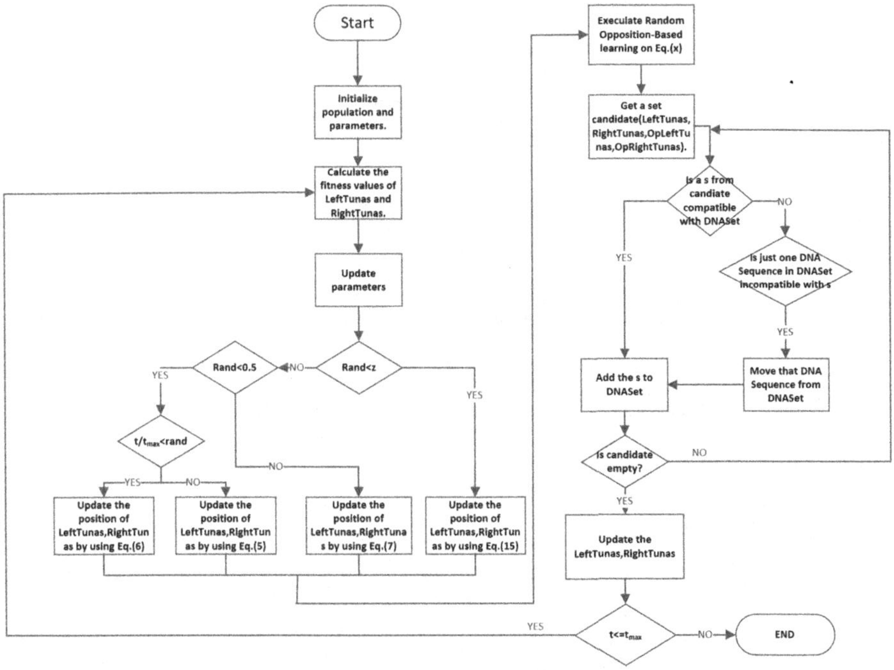

Fig. 1. TMOL-TSO algorithm flow chart

Table 2 shows the meaning of the data indices in Table 1 and Table 3. Table 1 shows the experimental results of $A^{GC,RC}(n,d,w)$, and Table 3 shows the experimental results of $A^{GC,HL}(n,d,w)$.

Table 1. The maximum number of sequences obtained by $A^{GC,RC}(n,d,w)$ set.

n\d	3	4	5	6	7	8	9
4	$\mathbf{6}^{C}$ $\mathbf{6}^{T}$						
5	$\mathbf{15}^{C}$ $\mathbf{15}^{T}$	$\mathbf{3}^{C}$ $\mathbf{3}^{T}$					
6	$\mathbf{42}^{C}$ $\mathbf{42}^{T}$	13^{C} $\mathbf{14}^{T}$	$\mathbf{4}^{A}$ $\mathbf{4}^{T}$				
7	$\mathbf{129}^{C}$ 128^{T}	34^{C} $\mathbf{35}^{T}$	$\mathbf{11}^{C}$ 10^{T}	$\mathbf{2}^{C}$ $\mathbf{2}^{T}$			
8	$\mathbf{434}^{C}$ 339^{T}	$\mathbf{101}^{C}$ $\mathbf{101}^{T}$	23^{C} $\mathbf{26}^{T}$	8^{C} $\mathbf{11}^{T}$	$\mathbf{2}^{C}$ $\mathbf{2}^{T}$		
9	1298^{C} $\mathbf{1319}^{T}$	$\mathbf{280}^{C}$ 271^{T}	$\mathbf{65}^{C}$ 62^{T}	$\mathbf{18}^{C}$ $\mathbf{18}^{T}$	$\mathbf{7}^{C}$ $\mathbf{7}^{T}$	$\mathbf{2}^{C}$ $\mathbf{2}^{T}$	
10	$\mathbf{4552}^{C}$ 4395^{T}	$\mathbf{865}^{C}$ 832^{T}	167^{A} $\mathbf{174}^{T}$	$\mathbf{55}^{A}$ 46^{T}	14^{A} $\mathbf{15}^{T}$	$\mathbf{6}^{A}$ $\mathbf{6}^{T}$	$\mathbf{2}^{C}$ $\mathbf{2}^{T}$

Table 2. The meaning of Superscript.

Superscript	Meaning
A	Altruistic algorithm [15]
N	NOL-HHO [34]
E	EORS [31]
R	ROEAO [14]
C	ACO [35]
T	TMOL-TSO

Table 3. The maximum number of sequences obtained by $A^{GC,HL}(n,d,w)$ set.

n\d	3	4	5	6	7	8	9
4	11^{A}						
	12^{N}						
	12^{E}						
	12^{R}						
	10^{T}						
5	17^{A}	7^{A}					
	20^{N}	**8^{N}**					
	20^{E}	**8^{E}**					
	20^{R}	**8^{R}**					
	20^{T}	**8^{T}**					
6	44^{A}	16^{A}	6^{A}				
	55^{N}	23^{N}	**8^{N}**				
	55^{E}	21^{E}	**8^{E}**				
	60^{R}	27^{R}	**8^{R}**				
	58^{T}	**28^{T}**	**8^{T}**				
7	110^{A}	36^{A}	11^{A}	4^{A}			
	121^{N}	42^{N}	14^{A}	**7^{N}**			
	125^{E}	46^{E}	16^{A}	6^{E}			
	127^{R}	**47^{R}**	**17^{R}**	**7^{R}**			
	132^{T}	43^{T}	**17^{T}**	**7^{T}**			
8	289^{A}	86^{A}	29^{A}	9^{A}	4^{A}		
	339^{N}	108^{N}	35^{A}	13^{N}	**5^{N}**		
	364^{E}	110^{E}	38^{A}	15^{E}	**5^{E}**		
	327^{R}	110^{R}	36^{R}	14^{R}	**5^{R}**		
	346^{T}	**114^{T}**	**39^{T}**	**16^{T}**	**5^{T}**		
9	662^{A}	199^{A}	59^{A}	15^{A}	8^{A}	4^{A}	
	705^{N}	216^{N}	69^{A}	22^{N}	**11^{N}**	4^{N}	
	737^{E}	226^{E}	**71^{A}**	26^{E}	**11^{E}**	**5^{E}**	
	786^{R}	228^{R}	**71^{R}**	**27^{R}**	**11^{R}**	**5^{R}**	
	752^{T}	**229^{T}**	69^{T}	**27^{T}**	**11^{T}**	**5^{T}**	
10	1810^{A}	525^{A}	141^{A}	43^{A}	7^{A}	5^{A}	4^{A}
	1796^{N}	546^{N}	148^{A}	51^{N}	20^{N}	9^{N}	4^{N}
	1856^{E}	546^{E}	153^{A}	53^{E}	22^{E}	9^{E}	**5^{E}**
	1964^{R}	**581^{R}**	**157^{R}**	**57^{R}**	21^{R}	10^{R}	**5^{R}**
	1884^{T}	550^{T}	156^{T}	56^{T}	**24^{T}**	**11^{T}**	**5^{T}**

From Table 1, it can be seen that in the experimental results on $A^{GC,RC}(n, d, w)$, such as $A^{GC,RC}(10, 3, 5)$ and $A^{GC,RC}(10, 4, 5)$, there are some distances between the results and the available experimental results. Because the result is already optimal, some results are unchanged, such as $A^{GC,RC}(9, 6, 4) = 18$ and $A^{GC,RC}(6, 3, 3) = 42$. In addition, our algorithm also obtains some more excellent results, such as $A^{GC,RC}(10, 5, 5) = 174$ and $A^{GC,RC}(8, 5, 4) = 26$.

The results shown in Table 3 are similar to those in Table 1, and the analysis method is the same as that used for Table 2. As can be seen from the data in the table, compared to the combined constraints of GC, HD and RC, the algorithm obtains more optimal solutions under the combined constraints of GC, HD and HL.

Experimental data shows that in certain cases, particularly when n is fixed and d is relatively large, the algorithm is capable of discovering a greater number of DNA codings that satisfy the given constraints. This applies to combined constraints involving GC, HD, RC, as well as combined constraints involving GC, HD, HL. These results demonstrate the efficacy of the algorithm in the construction of the DNA coding sets, exploring a larger search space and confirming the feasibility of the algorithm. Since the global optimum has already been achieved for some identical results, further optimization is not possible. For the remaining results, the proposed algorithm in this paper and the algorithm developed by the predecessor complement each other. They can provide a complete set of DNA codes for DNA computing and storage.

6 Conclusions

In this paper, we constructed corresponding coding sets for DNA computing and storage using an improved TSO algorithm. We combined the original TSO algorithm with a learning strategy based on random opposition and a Two-Swarm merge strategy and named it TMOL-TSO. DNA coding is a crucial element in both DNA computing and DNA storage. To limit the non-specific hybridization of DNA during DNA computing or when encrypting/decrypting data for storage, we proposed an algorithm presented in this paper to address this issue.

Our experiments have shown that the algorithm proposed in this paper can identify a DNA coding set with better size when n is fixed and d is relatively large. This holds true for both the combination constraints of DNA computing and the combination constraints of DNA storage, as compared to existing results. This demonstrates the algorithm's advantage and supports future research on DNA coding.

References

1. Adleman, L.M.: Molecular computation of solutions to combinatorial problems. Science **266**(5187), 1021–1024 (1994)
2. Baum, E.B.: Building an associative memory vastly larger than the brain. Science **268**(5210), 583–585 (1995)

3. Benenson, Y., Paz-Elizur, T., Adar, R., Keinan, E., Livneh, Z., Shapiro, E.: Programmable and autonomous computing machine made of biomolecules. Nature **414**(6862), 430–434 (2001)
4. Bornholt, J., Lopez, R., Carmean, D.M., Ceze, L., Seelig, G., Strauss, K.: A DNA-based archival storage system. In: Proceedings of the Twenty-First International Conference on Architectural Support for Programming Languages and Operating Systems, pp. 637–649 (2016)
5. Cao, B., Zhao, S., Li, X., Wang, B.: K-means multi-verse optimizer (KMVO) algorithm to construct DNA storage codes. Ieee Access **8**, 29547–29556 (2020)
6. Chang, W.L., Guo, M., Ho, M.H.: Fast parallel molecular algorithms for DNA-based computation: factoring integers. IEEE Trans. Nanobiosci. **4**(2), 149–163 (2005)
7. Church, G.M., Gao, Y., Kosuri, S.: Next-generation digital information storage in DNA. Science **337**(6102), 1628–1628 (2012)
8. Dougherty, S.T., Korban, A., Şahinkaya, S., Ustun, D.: Construction of DNA Codes from composite matrices and a bio-inspired optimization algorithm. IEEE Trans. Inf. Theory **69**(3), 1588–1603 (2022)
9. Frutos, A.G., et al.: Demonstration of a word design strategy for DNA computing on surfaces. Nucleic Acids Res. **25**(23), 4748–4757 (1997)
10. Grass, R.N., Heckel, R., Puddu, M., Paunescu, D., Stark, W.J.: Robust chemical preservation of digital information on DNA in silica with error-correcting codes. Angew. Chem. Int. Ed. **54**(8), 2552–2555 (2015)
11. Heckel, R., Mikutis, G., Grass, R.N.: A characterization of the DNA data storage channel. Sci. Rep. **9**(1), 1–12 (2019)
12. Kashiwamura, S., Yamamoto, M., Kameda, A., Ohuchi, A.: Experimental challenge of scaled-up hierarchical DNA memory expressing a 10, 000-address space. In: Preliminary Proceeding of 11th International Meeting on DNA based Computers, London, UK (2005)
13. Kashiwamura, S., Yamamoto, M., Kameda, A., Shiba, T., Ohuchi, A.: Potential for enlarging DNA memory: the validity of experimental operations of scaled-up nested primer molecular memory. Biosystems **80**(1), 99–112 (2005)
14. Li, X., Zhou, S., Zou, L.: Design of DNA storage coding with enhanced constraints. Entropy **24**(8), 1151 (2022)
15. Limbachiya, D., Gupta, M.K., Aggarwal, V.: Family of constrained codes for archival DNA data storage. IEEE Commun. Lett. **22**(10), 1972–1975 (2018)
16. Lipton, R.J.: DNA solution of hard computational problems. Science **268**(5210), 542–545 (1995)
17. Liu, Q., Wang, L., Frutos, A.G., Condon, A.E., Corn, R.M., Smith, L.M.: DNA computing on surfaces. Nature **403**(6766), 175–179 (2000)
18. Palluk, S., et al.: De novo DNA synthesis using polymerase-nucleotide conjugates. Nat. Biotechnol. **36**(7), 645–650 (2018)
19. Qian, L., Winfree, E.: A simple DNA gate motif for synthesizing large-scale circuits. J. R. Soc. Interface **8**(62), 1281–1297 (2011)
20. Qian, L., Winfree, E.: Scaling up digital circuit computation with DNA strand displacement cascades. Science **332**(6034), 1196–1201 (2011)
21. Qian, L., Winfree, E., Bruck, J.: Neural network computation with DNA strand displacement cascades. Nature **475**(7356), 368–372 (2011)
22. Rasool, A., Jiang, Q., Wang, Y., Huang, X., Qu, Q., Dai, J.: Evolutionary approach to construct robust codes for DNA-based data storage. Front. Genet. **14**, 415 (2023)
23. Sakakibara, Y., Suyama, A.: Intelligent DNA chips logical operation of gene expression profiles on DNA computers. Genome Inform. **11**, 33–42 (2000)

24. Shendure, J., et al.: DNA sequencing at 40: past, present and future. Nature **550**(7676), 345–353 (2017)
25. Song, T., et al.: Fast and compact DNA logic circuits based on single-stranded gates using strand-displacing polymerase. Nat. Nanotechnol. **14**(11), 1075–1081 (2019)
26. Stewart, K., et al.: A content-addressable DNA database with learned sequence encodings. In: Doty, D., Dietz, H. (eds.) DNA 2018. LNCS, vol. 11145, pp. 55–70. Springer, Cham (2018). https://doi.org/10.1007/978-3-030-00030-1_4
27. Tabatabaei Yazdi, S.H., Yuan, Y., Ma, J., Zhao, H., Milenkovic, O.: A rewritable, random-access DNA-based storage system. Sci. Rep. **5**(1), 14138 (2015)
28. Tizhoosh, H.R.: Opposition-based learning: a new scheme for machine intelligence. In: International Conference on Computational Intelligence for Modelling, Control and Automation and International Conference on Intelligent Agents, Web Technologies and Internet Commerce (CIMCA-IAWTIC 2006), vol. 1, pp. 695–701. IEEE (2005)
29. Wang, Y., Noor-A-Rahim, M., Gunawan, E., Guan, Y.L., Poh, C.L.: Construction of bio-constrained code for DNA data storage. IEEE Commun. Lett. **23**(6), 963–966 (2019)
30. Wu, H.: An improved surface-based method for DNA computation. Biosystems **59**(1), 1–5 (2001)
31. Xiaoru, L., Ling, G.: Combinatorial constraint coding based on the EORS algorithm in DNA storage. PLoS ONE **16**(7), e0255376 (2021)
32. Xie, L., Wang, S., Zhu, D., Hu, G., Zhou, C.: DNA sequence optimization design of arithmetic optimization algorithm based on billiard hitting strategy. In: Interdisciplinary Sciences: Computational Life Science, pp. 1–18 (2023)
33. Yamamoto, M., Kashiwamura, S., Ohuchi, A., Furukawa, M.: Large-scale DNA memory based on the nested PCR. Nat. Comput. **7**, 335–346 (2008)
34. Yin, Q., Cao, B., Li, X., Wang, B., Zhang, Q., Wei, X.: An intelligent optimization algorithm for constructing a DNA storage code: NOL-HHO. Int. J. Mol. Sci. **21**(6), 2191 (2020)
35. Zhou, Q., Wang, X., Zhou, C.: DNA design based on improved ant colony optimization algorithm with bloch sphere. IEEE Access **9**, 104513–104521 (2021)

Heterogeneous Graph Fusion with Adversarial Learning for Recommendation Service

Jiaxi Wang, Tong Mo, and Weiping Li(✉)

School of Software and Microelectronics, Peking University, Beijing, China
wangjxi@pku.edu.cn, {motong,wpli}@ss.pku.edu.cn

Abstract. Social recommendation, in particular, relies on modeling social information to provide high-order information beyond user-item interaction in recommendation service. However, current approaches rely on graph neural network-based embedding, which can result in an over-smoothing problem. Additionally, graph diffusion, which encodes high-order features, can add noise to the model. Previous research has not adequately addressed the latent influence of social relations. In this work, we introduce a new recommendation framework named Heterogeneous Information Network Fusion with Adversarial Learning (HIN-FusionGAN), which inherently fuses adversarial learning-enhanced social networks with the fused graph between the user-item interaction graph and user-user social graph. We propose a heterogeneous information network that fuses social and interaction graphs into a unified heterogeneous graph, explicitly encoding high-order collaborative signals. We employ user embeddings using both interaction information and adversarial learning-enhanced social networks, which are efficiently fused by the feature fusion model. To address the issue of over-smoothing and uncover latent feature representation, we use the structure of an adversarial network in social relation graph. Comprehensive experiments on three real-world datasets demonstrate the superiority of our proposed model.

Keywords: Social Recommendation · Heterogeneous Information Network · Adversarial Learning

1 Introduction

The development and evolution of recommender systems have effectively addressed the pervasive issue of information overload. However, traditional recommender systems encounter a significant obstacle in that a vast majority of users only engage with a limited selection of items, resulting in inferior recommendation accuracy due to the data sparsity problem.

In recent years, Graph Neural Networks (GNNs) have gained significant popularity as advanced techniques for graph representation learning, as highlighted in their comprehensive study [12]. Recommender systems often leverage graphs

B. Luo et al. (Eds.): ICONIP 2023, CCIS 1963, pp. 470–482, 2024.
https://doi.org/10.1007/978-981-99-8138-0_37

to represent user-item interactions, with nodes representing users and items, and edges representing the interactions between them. GNN-based recommender systems utilize graph convolution operations to learn user and item embeddings by propagating messages across graph nodes in a layered fashion. Nevertheless, due to data sparsity, recommender systems relying solely on user-item interaction history suffer from limited performance, particularly for users with scarce behavior data. Fortunately, social networks have become increasingly prevalent, and research has shown that modeling social connections as crucial side information can enhance user and item embeddings. Social recommendation, supported by the theory of social influence [1], has emerged as a promising research direction.

The HIN based model enables the expression of complex relationships that exist between nodes in social networks through a variety of link types. Past studies [6,17] have analyzed the relationships between users and items using HIN and have yielded promising results, demonstrating the efficacy of HIN in the realm of recommendation. Nonetheless, these approaches rely heavily on the availability of explicit data and lack the capacity for in-depth mining and analysis of social relations between users and items via the HIN network embedding method.

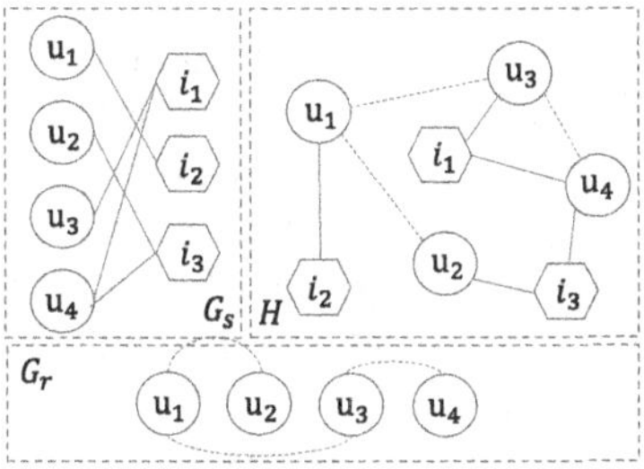

Fig. 1. The Heterogeneous Information Network (HIN) is constructed using both the user-item interaction graph and the social network.

This research paper focuses on social relations and proposes a model, named HIN-FusionGAN, that explicitly models high-order collaborative signals by integrating social relations into the recommendation system using Heterogeneous Information Network (HIN). Specifically, the user-user social graph and the user-item interaction graph are merged into a social information enhanced HIN, which addresses the incomplete exploration of indirect and explicit information. This model represents the first attempt to adopt HIN embedding with different types of social relations for social recommendation.

The primary contributions of this paper are as follows:

- We propose a novel social recommendation model, called Heterogeneous Information Network Fusion with Adversarial Learning (HIN-FusionGAN), To the best of our knowledge, we are the first to inherently fuses adversarial learning-enhanced social networks with the fused graph between the user-item interaction graph and user-user social graph.

- We propose an adversarial learning enhanced social network to improve the social recommendation predictive performance and address the issue of over-smoothing and uncover latent feature representation.
- We perform extensive experiments on three real-world datasets to demonstrate the superiority of our proposed model.

2 Related Work

In this section, we mainly review social recommendation, HIN-based social recommendation and GNN-based recommendation and adversarial learning in GNN-based recommender system.

Non-deep learning social recommendation methods frequently employ co-factorization techniques, with SoRec [9] using a probability matrix to perform co-factorization on both user-item rating matrix and user-user social relation matrix. GraphRec [2] is a GNN-based model for rating prediction that aggregates representations for items and users from their linked neighbors. MHCN [15] utilizes hierarchical mutual information maximization to retrieve connection information by incorporating self-supervised learning into the training of the hypergraph convolutional network.

Several HIN-based models have been proposed to address recommendation tasks from the perspective of heterogeneous graphs. For instance, HteRec [16] focuses on implicit feedback and uses different user preference matrices based on various meta-paths, followed by matrix factorization to perform recommendations. The IF-BPR [13] model uses a meta-path approach to learn user embedding representations and identifies potential friends through an adaptive approach.

Adversarial learning has emerged as a highly successful approach in various domains, including recommendation systems. DASO [3] employs two adversarial learning components to model interactions between two domains. RSGAN [14] presents an end-to-end social recommendation framework based on Generative Adversarial Nets (GANs). APR [5] is another adversarial learning-based model that improves the pairwise ranking technique Bayesian Personalized Ranking (BPR).

3 Methodology

In this section, we propose the HIN-FusionGAN method for social recommendation. Its framework mainly consists of three components, including graph convergence by HIN, social relation modeling by adversarial learning and fusion approach with optimized loss function. First of all, the user-user social graph and the user-item interaction graph will be fused by the structure of HIN and we design multiple meta-path to obtain the high-order relation in the merged network. Then we utilize generative adversarial network rather than relying on influence-based graph neural network (GNN) architecture for more accurate graph node embedding. A social relation adversarial network can be employed to create more precise embeddings that avoid the common issue of over-smoothing in node representations. Last but not least, to make predictions, we utilize a multi-layer perceptron (MLP) in combination with an optimized loss function that identifies important features from the two aforementioned modules (Fig. 2).

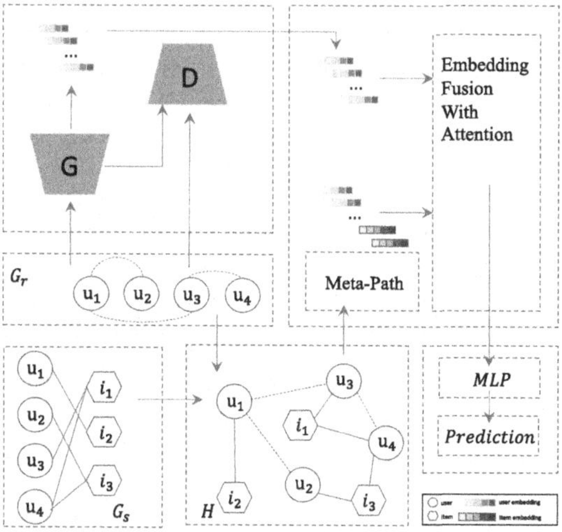

Fig. 2. The Overview of HIN-FusionGAN model, including three modules, including graph convergence by HIN, social relation modeling by adversarial learning and fusion approach with optimized loss function.

3.1 Graph Convergence by HIN

In the social recommendation task, we denote the user set as U and the item set as I. The user-item bipartite network and the user-user social network are represented by Gr and Gs respectively. The network Gr $= (V_r, E_r)$ contains two types of nodes: user and item, where $(u, i) \in E_r$ indicates that the user u purchased/rated the item i. The network Gs $= (V_s, E_s)$ contains only one type of node: user, where $(u_1, u_2) \in E_s$ indicates that the user u_1 trusts the user u_2. The relationships between users are asymmetric, and the edge (u_1, u_2) is different from the edge (u_2, u_1).

We consider concatenating Gr and Gs as a whole to form a heterogeneous information network H as Fig. 1, which enables us to capture rich information shared across Gr and Gs. By doing so, we can effectively quantify user similarity for social recommendation purposes.

When dealing with large-scale real-world recommender systems modeled by heterogeneous information networks (HINs), computational efficiency becomes a crucial challenge in identifying implicit friends while preserving the original network's embedded information. To address this challenge, we take inspiration from successful network embedding models and propose using meta-paths to generate biased meta-path-based random walks to solve the problem.

A meta-path schema is a composite relation between the start type and the end type of a path, characterized by a series of relation types. In this work, we define six carefully designed meta-paths to capture user similarities across the user-item bipartite network and the user-user social network. These meta-paths help identify similar entities that may be distant from each other on the networks.

We exploit meta-paths to guide the random walks, resulting in a social corpus of node sequences. However, social relations in HINs can be noisy, and thus it is essential to identify reliable sequences. We use UUU, UUI, UIU, UUUI for HIN meta-path investigation (Table 1).

Table 1. Meta-paths Description

Path	Schema	Detailed Information
#P1	UUU	Friend's friend, two-hop connection
#P2	UUI	short-term similar preference relationship
#P3	UIU	Users who have similar preference
#P4	UUUI	long-term similar preference relationship

3.2 Social Relation Modeling by Adversarial Learning

To improve the performance of recommendations by incorporating social relation features, it is essential to fully explore user embeddings containing social relationship information. Graph neural network (GNN) and graph convolutional network (GCN) methods are suitable for graph embedding tasks. However, they are not capable of addressing the data sparsity and noise issues in social networks. To mitigate these problems, we propose a social relation GAN model to discover latent representations of each user. The proposed model consists of a generator, which generates latent friends of a given user by fitting the connectivity pattern distribution in the social relation network, and a discriminator, which engages in a minimax game during training to progressively enhance its capability. Drawing inspiration from [18], we implement the discriminator using a sigmoid function of the inner product, as follows:

$$D\left(u;\boldsymbol{\theta}_D\right)=\frac{1}{1+\exp\left(-\phi_D\left(\hat{u}_k;\boldsymbol{\theta}_D\right)^{\top}\phi_D\left(u_i;\boldsymbol{\theta}_D\right)\right)} \tag{1}$$

where $\phi_D\left(\cdot;\boldsymbol{\theta}_D\right)$ denotes the embedding function in discriminator, we utilize graph softmax to implement it as follows:

$$G\left(u \mid u_i;\theta_G\right)=\left(\prod_{k=1}^{L_p} p_r\left(u_k^{\rho} \mid u_{k-1}^{\rho}\right)\right)\times\left(p_r\left(u_{L_p-1}^{\rho} \mid u_{L_p}^{\rho}\right)\right) \tag{2}$$

where L_p is the path length from u_i to $\hat{u}_k$, u^ρ is the user in the path. $p_r(u_j \mid u_i)$ is the conditional probability.

According to the mechanism of adversarial learning, the generator and discriminator play the following two-player minimax game. We minimize the generator (see Eq. 2) and maximize the discriminator (see Eq. 1):

$$\arg\min_{\theta_G}\max_{\theta_D}\mathcal{L}(\boldsymbol{\theta}_G,\boldsymbol{\theta}_D)=\sum_{i=1}^{|U|}\left(\mathbb{E}_{u\sim p_{true}(\cdot|u_i)}\left[\log D(u,u_i;\boldsymbol{\theta}_D)\right]\right.$$
$$\left.+\mathbb{E}_{u\sim p_G(\cdot|u_i)}\left[\log\left(1-D(u,u_i;\boldsymbol{\theta}_D)\right)\right]\right), \tag{3}$$

Where the $p_{true}(\cdot \mid u_i)$ is the ground truth value of users. In the iteration of adversarial learning, discriminator is feed with true value from $p_{true}(\cdot \mid u_i)$ and the fake sample from generator. The generator is updated with the regulation from discriminator. The minimax game will continue until the sample from generator is indistinguishable from the ground truth value.

3.3 Fusion Approach and Optimized Loss Function

The node features in two graphs are combined by an attention mechanism-based multi-graph information fusion model to improve the learning of the vertex representation after being learned from the user-item bipartite graph Gs and the social relation network Gr. An user-item attention network is used to learn a set of weights for all the items associated to each user vector in the user-item graph. A user-user attention network is used to learn a set of weights for each user's neighbors in the social relation network. These two categories of features are combined using a concatenation operation after the weighted features have been created as follow:

$$\mathbf{e}_{ij}=\sigma\left(\boldsymbol{W}_{UI}\times\left(\left(\boldsymbol{W}_R\times\left(\sum\nolimits_{i,j=1}^{|N(u_i,v_j)|}\omega_{ij}\times\boldsymbol{e}(u_i,v_j)\right)+\boldsymbol{\beta}_R\right)+\boldsymbol{\beta}_{UI}\right) \tag{4}$$

$\boldsymbol{W}_{UI}$ and $\boldsymbol{\beta}_{UI}$ represent the parameters of user-item aggregation neural network. $\boldsymbol{W}_R$ and $\boldsymbol{\beta}_R$ represent the parameters of item-rating representation neural network. $\boldsymbol{e}(u_i,v_j))$ denotes the rating representation between u_i and v_j from heterogeneous graph learning. After the training of neural network with attention mechanism we can obtain the representation of user-item pair(u_i,v_j).

Similarly to the user-item attention network, a user-user graph is to learn a set of weights for all the neighbors for each user in social relation network Gr . This network can be formulated as:

$$\mathbf{e}_r=\sigma\left(\boldsymbol{W}_{UU}\times\left(\sum_{k=1}^{|N(u_i)|}\omega_{ik}\times\boldsymbol{e}(u_i,u_k)\right)+\boldsymbol{\beta}_{UU}\right) \tag{5}$$

To improve the representation learning, an information fusion model is used to fuse the representation e_{ij} from user-item bipartite graph Gs and the representation e_r from social relation network Gr together. This model is implemented by an l-layers deep neural network model $z_{ij} = \text{Dnn}\left(e_{ij}, e_r; \theta_{Dnn}\right)$. Specifically, the first layer is a concatenation operator that concatenates e_{ij} and e_r, and the remaining several layers construct a multiple-layer Perceptron. The structure of this DNN can be formulated as

$$\begin{aligned}
\boldsymbol{a}_{ij}^{(1)} &= \mathbf{e}_{ij} \oplus \mathbf{e}_r, \\
a_{ij}^{(2)} &= \sigma\left(\boldsymbol{W}_2 \times \boldsymbol{a}_{ij}^{(1)} + \boldsymbol{\beta}_2\right), \\
&\cdots \\
\boldsymbol{z}_{ij} &= \sigma\left(W_l \times \boldsymbol{a}_{ij}^{(l)} + \boldsymbol{\beta}_l\right),
\end{aligned} \tag{6}$$

where $a^{(j)}$ is the output of j th layer, $j \in [1, \ldots, l]$. The parameter vector is $\theta_{Dnn} = \left(\{\boldsymbol{W}_j\}|_{j=2}^{l}, \{\boldsymbol{\beta}_j\}|_{j=2}^{l}\right)$. Then the fused representations are fed into a linear prediction layer to generate the predicted recommendation as follows:

$$r'_{ij} = \text{Predict}\left(z_{ij}; \theta_P\right) = W \times z_{ij}. \tag{7}$$

Then the loss function of the model can be formulated as:

$$\mathcal{L}_{\text{p}} = \frac{1}{2\Omega|} \sum_{i,j \in O} \left(r'_{ij} - r_{ij}\right)^2 \tag{8}$$

where $|O|$ is the number of user-item pair and r_{ij} is the ground truth value of rating. So the overall loss function can be formulated as:

$$\mathcal{L} = \mathcal{L}_{\text{p}} + \mathcal{L}(\boldsymbol{\theta}_G, \boldsymbol{\theta}_D) \tag{9}$$

4 Experiment

The performance of our proposed algorithm has been rigorously evaluated using three real-world datasets. The overarching objective of our research is to provide insights into the following research questions: **RQ1:** How does the performance of our model compare to other state-of-the-art recommendation algorithms? **RQ2:** What is the influence of the key components and hyperparameter optimization, such as varying the embedding dimension, on the performance of our model? **RQ3:** What is the effectiveness of the fused HIN graph that we have proposed?

4.1 Experimental Setting

Dataset. In order to validate the performance of the HIN-FusionGAN, we conduct experiments on three public datasets, Epinions, Yelp and Ciao. The three datasets are described in detail as follows (Table 2).

Table 2. The statistics of the three datasets

Dataset	Epinions	Yelp	Ciao
#User	22164	17235	7375
#Item	296277	37378	105114
#Ratings	911442	207945	169730
#Relations	574202	169150	57544
#Rating density	0.014%	0.032%	0.035%
#Relation density	0.116%	0.087%	0.313%

- **Epinions:** The dataset comes from a general consumer review site Epinions.com which contains trust relationships interactions and review ratings information.
- **Yelp:** This dataset is a location-based review dataset which also contains social information and ratings.
- **Ciao:** This dataset contains rating information of users given to items, and also contain item category information.

Evaluation Metrics. We make inferences about the item that a user is likely to interact with, and each candidate method generates a ranked list of items for recommendation. To evaluate the performance of the methods, we employ two widely accepted ranking-based metrics, namely, Hit Ratio at rank k (Hit Ratio@k) and Normalized Discounted Cumulative Gain at rank k (NDCG@k). We report the top K (K = 5 and K = 10) items in the ranking list as the recommended set.

Baselines. We compare the proposed model with the following baseline models:

- **BPR** [10]**:** BPR is a state-of-the-art model for traditional item recommendation that only uses user-item interaction information. It optimizes the matrix factorization model using a pairwise ranking loss.
- **SocialMF** [7]**:** Matrix factorization techniques are used in this approach, which incorporates trust propagation mechanism into the model. This reduces cold start problem and leads to more accurate prediction.
- **GraphRec** [2]**:** This approach proposes a novel GNN-based model that integrates user-user social graph with user-item interaction graph. Additionally, an attention network is used to filter out high-influence users and items.
- **Diffnet** [11]**:** The approach uses a layer-wise GNN diffusion architecture to model users' preferences.
- **LightGCN** [4]**:** User and item embeddings are obtained through LightGCN by linearly propagating them on the user-item interaction graph. The final embedding is the weighted sum of embeddings learned at all layers. This model is significantly simpler to implement and train.

- **MHCN** [15]: Hierarchical mutual information maximization is utilized in this approach to retrieve connection information by incorporating self-supervised learning into training of the hypergraph convolutional network.

Parameter Settings. We implement our proposed model by Pytorch. We utilize Adam [8] as the optimize for the gradient descent processes. For the best result of all the baseline model, we search the learning rate in the range of [0.001, 0.005, 0.1, 0.2, 0.5]. In order to avoid over-fitting, dropout has been adopted with the rate of 0.5. The embedding size is [32, 64, 128, 256]. For the diffusion model, we set the layer number with 2 and 3 respectively. In our evaluations, we adopt early stopping for training termination when the performance degrades for 5 continuous epochs on the validation dataset.

4.2 Performance Comparision (RQ1)

To provide a comprehensive evaluation of the performance of the proposed model and compare it with other recommendation methods, we report the results in terms of HR@[5, 10] and NDCG@[5, 10] in Table 3.

This is a table showing the overall comparison of different recommendation models on three datasets (Epinions, Yelp, and Ciao) using different evaluation metrics (H@5, H@10, N@5, N@10). The table includes six baseline models: BPR, SocialMF, GraphRec, Diffnet, LightGCN, and MHCN, as well as the proposed model (Ours). The results indicate that the proposed model performs better than all baseline models on all three datasets and across all evaluation metrics. In particular, the proposed model achieves the highest H@5, H@10, N@5, and N@10 scores on all three datasets, indicating its superior performance in top-K recommendation. The table also shows that LightGCN and MHCN are the two best-performing baseline models, and that the proposed model outperforms both of them by a significant margin.

Table 3. Overall Comparison with Baseline

Dataset	Epinions				Yelp				Ciao			
	H@5	H@10	N@5	N@10	H@5	H@10	N@5	N@10	H@5	H@10	N@5	N@10
BPR	0.417	0.535	0.298	0.379	0.560	0.791	0.391	0.373	0.203	0.367	0.131	0.163
SocialMF	0.411	0.534	0.359	0.385	0.600	0.701	0.420	0.464	0.241	0.336	0.154	0.180
GraphRec	0.430	0.541	0.341	0.388	0.591	0.813	0.450	0.535	0.260	0.384	0.170	0.211
Diffnet	0.460	0.576	0.350	0.391	0.600	0.749	0.441	0.490	0.241	0.365	0.163	0.196
LightGCN	0.485	0.609	0.360	0.391	0.644	0.790	0.474	0.526	0.253	0.390	0.172	0.227
MHCN	0.491	0.611	0.373	0.420	0.667	0.809	0.511	0.557	0.261	0.398	0.170	0.231
Ours	0.501	0.631	0.390	0.445	0.690	0.831	0.551	0.570	0.286	0.410	0.186	0.258

Table 4. Overall Comparison with Different Version Model

Dataset	Epinions				Yelp				Ciao			
	H@5	H@10	N@5	N@10	H@5	H@10	N@5	N@10	H@5	H@10	N@5	N@10
Our-Normal	0.414	0.520	0.310	0.368	0.570	0.784	0.435	0.501	0.213	0.391	0.160	0.216
Our-nHIN	0.460	0.571	0.358	0.403	0.643	0.793	0.503	0.545	0.259	0.391	0.166	0.207
Our-nGAN	0.483	0.621	0.381	0.436	0.684	0.827	0.539	0.551	0.273	0.409	0.177	0.239
MHCN	0.491	0.611	0.373	0.420	0.667	0.809	0.511	0.557	0.261	0.398	0.170	0.231
Ours	0.501	0.631	0.390	0.445	0.690	0.831	0.551	0.570	0.286	0.410	0.186	0.258

4.3 Experiment with Effectiveness of HIN (RQ2)

To assess the contribution of each of the main components of our model, we conducted ablation experiments by simplifying the model into three versions that represent different perspectives.

- **Ours-nHIN:** This version only omits the HIN part of our model instead of the traditional diffusion model.
- **Ours-nGAN:** This version only omits the GAN part of our model and using GAT as the core model to model the social network.
- **Ours-Normal:** This version omits the HIN part and GAN part of our model as the basic baseline in comparsion.

The performance of different version of our model is shown in Table 4. The results in the table show that the proposed algorithm (Ours) outperforms the other algorithms on all three datasets and for all evaluation metrics. In particular, it achieves the highest Hit Ratio (H@5 and H@10) and Normalized Discounted Cumulative Gain (N@5 and N@10) scores, indicating that it is able to recommend relevant items to users. The results also show that the performance of the proposed algorithm improves as more complex models (nHIN and nGAN) are used, indicating the effectiveness of the proposed neural network architectures.

4.4 Hyper-parameter Tuning Analysis (RQ3)

This section explores the impact of embedding dimension and validates the effectiveness of HIN-FusionGAN model. The study comprises three sets of experiments. Firstly, we conduct a study on embedding dimension, where we vary the embedding dimension in the range of [64, 128, 256, 512] to determine the optimal value as shown in Fig. 3. The result shows that the optimal dimension is around 256.

The GAN framework faces challenges with slow model convergence and lengthy training times, particularly when applied to models involving discrete data sampling. Our study presents the learning curves of NDCG@10 and HR@10 for various GAN-based models, including our proposed model, on the Ciao dataset as shown in Fig. 4. The results demonstrate that our model achieves superior performance and converges more quickly compared to other GAN-based models. These findings are consistent across other datasets with larger data amounts, indicating that our model offers better recommendation performance while requiring less time for training.

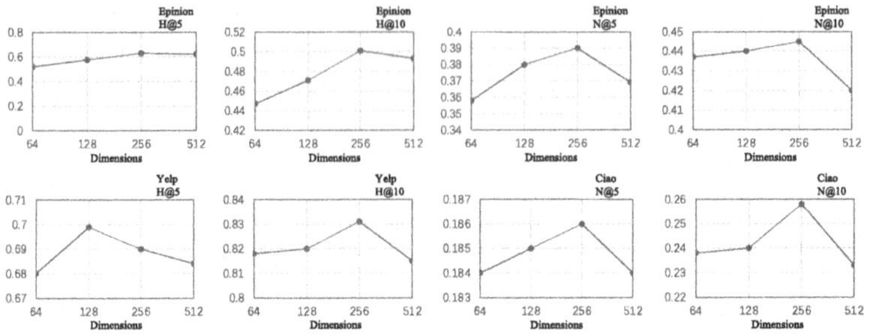

Fig. 3. The Embedding Dimension Analysis of HIN-FusionGAN model

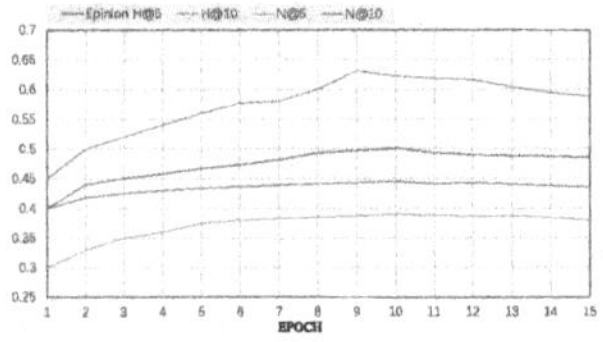

Fig. 4. The Convergence Analysis of HIN-FusionGAN model

5 Conclusions

Our work introduces a novel recommendation framework, named Heterogeneous Information Network Fusion with Adversarial Learning (HIN-FusionGAN), that explicitly models high-order collaborative signals in social recommendation service. We have devised a set of meaningful meta-paths, which are used to create random walks and obtain more accurate embedding feature. Our experimental analysis on three publicly available datasets demonstrates the efficacy of our proposed approach in handling simple heterogeneous information networks. Moving forward, we aim to enhance our model to accommodate complex network structures with additional attribute information. In the future, we consider improving our method to handle more complex graph such as hypergraph.

Acknowledgements. This work was supported by the National Key R&D Program of China [2022YFF0902703].

References

1. Cialdini, R.B., Goldstein, N.J.: Social influence: compliance and conformity. Annu. Rev. Psychol. **55**(1), 591–621 (2004)
2. Fan, W., et al.: Graph neural networks for social recommendation. In: The World Wide Web Conference, pp. 417–426 (2019)
3. Fan, W., Ma, Y., Yin, D., Wang, J., Tang, J., Li, Q.: Deep social collaborative filtering. In: Proceedings of the 13th ACM Conference on Recommender Systems, pp. 305–313 (2019)
4. He, X., Deng, K., Wang, X., Li, Y., Zhang, Y., Wang, M.: LightGCN: simplifying and powering graph convolution network for recommendation. In: Proceedings of the 43rd International ACM SIGIR conference on research and development in Information Retrieval, pp. 639–648 (2020)
5. He, X., He, Z., Du, X., Chua, T.S.: Adversarial personalized ranking for recommendation. In: The 41st International ACM SIGIR Conference on Research & Development in Information Retrieval, pp. 355–364 (2018)
6. Hu, B., Shi, C., Zhao, W.X., Yu, P.S.: Leveraging meta-path based context for top-n recommendation with a neural co-attention model. In: Proceedings of the 24th ACM SIGKDD International Conference on Knowledge Discovery & Data Mining, pp. 1531–1540 (2018)
7. Jamali, M., Ester, M.: A matrix factorization technique with trust propagation for recommendation in social networks. In: Proceedings of the Fourth ACM Conference on Recommender Systems, pp. 135–142 (2010)
8. Kingma, D.P., Ba, J.: Adam: a method for stochastic optimization. arXiv preprint arXiv:1412.6980 (2014)
9. Ma, H., Yang, H., Lyu, M.R., King, I.: SoRec: social recommendation using probabilistic matrix factorization. In: Proceedings of the 17th ACM Conference on Information and Knowledge Management, pp. 931–940 (2008)
10. Rendle, S., Freudenthaler, C., Gantner, Z., Schmidt-Thieme, L.: BPR: bayesian personalized ranking from implicit feedback. arXiv preprint arXiv:1205.2618 (2012)
11. Wu, L., Sun, P., Fu, Y., Hong, R., Wang, X., Wang, M.: A neural influence diffusion model for social recommendation. In: Proceedings of the 42nd international ACM SIGIR Conference on Research and Development in Information Retrieval, pp. 235–244 (2019)
12. Wu, Z., Pan, S., Chen, F., Long, G., Zhang, C., Philip, S.Y.: A comprehensive survey on graph neural networks. IEEE Trans. Neural Netw. Learn. Syst. **32**(1), 4–24 (2020)
13. Yu, J., Gao, M., Li, J., Yin, H., Liu, H.: Adaptive implicit friends identification over heterogeneous network for social recommendation. In: Proceedings of the 27th ACM International Conference on Information and Knowledge Management, pp. 357–366 (2018)
14. Yu, J., Gao, M., Yin, H., Li, J., Gao, C., Wang, Q.: Generating reliable friends via adversarial training to improve social recommendation. In: 2019 IEEE International Conference on Data Mining (ICDM), pp. 768–777. IEEE (2019)
15. Yu, J., Yin, H., Li, J., Wang, Q., Hung, N.Q.V., Zhang, X.: Self-supervised multi-channel hypergraph convolutional network for social recommendation. In: Proceedings of the Web Conference 2021, pp. 413–424 (2021)
16. Yu, X., et al.: Personalized entity recommendation: a heterogeneous information network approach. In: Proceedings of the 7th ACM International Conference on Web Search and Data Mining, pp. 283–292 (2014)

17. Yu, X., et al.: Recommendation in heterogeneous information networks with implicit user feedback. In: Proceedings of the 7th ACM Conference on Recommender Systems, pp. 347–350 (2013)
18. Zhang, C., Wang, Y., Zhu, L., Song, J., Yin, H.: Multi-graph heterogeneous interaction fusion for social recommendation. ACM Trans. Inf. Syst. (TOIS) **40**(2), 1–26 (2021)

SSVEP Data Augmentation Based on Filter Band Masking and Random Phase Erasing

Yudong Pan[1], Ning Li[1], Lianjin Xiong[1], Yiqian Luo[1], and Yangsong Zhang[1,2(✉)]

[1] Southwest University of Science and Technology, Mianyang 621010, China
zhangysacademy@gmail.com
[2] Key Laboratory of Testing Technology for Manufacturing Process, Ministry of Education, Southwest University of Science and Technology, Mianyang 621010, China

Abstract. Steady-state visual evoked potential (SSVEP)-based brain-computer interfaces (BCIs) have shown promising results in various applications, such as controlling prosthetic devices and augmented reality systems. However, current data-driven frequency recognition methods used to build high-performance SSVEP-BCIs often encounter overfitting and poor generalization when training data is limited. To address this issue, in this paper, we propose two potential SSVEP data augmentation methods, namely filter band masking (FBM) and random phase erasing (RPE), based on the inherent features of SSVEPs. These methods can produce high-quality supplementary training data to improve the performance of SSVEP-BCIs without parameter learning, making them easy to implement. To evaluate the proposed methods, two large-scale publicly available datasets (Benchmark and BETA) were used, and the experimental results showed that the proposed methods significantly could enhance the classification performance of baseline classifiers with a limited amount of calibration data. Specifically, evaluated on two methods, FBM increased the average accuracies by 7.40%, 8.55%, and RPE increased the average accuracies by 5.85%, 6.10%, respectively, with as few as two 1-scecond calibration trials on the Benchmark dataset. These findings demonstrate the potential of these data augmentation methods in enhancing the practicality of SSVEP-BCI for real-life scenarios.

Keywords: brain-computer interface (BCI) · steady-state visual evoked potential (SSVEP) · filter band masking (FBM) · random phase erasing (RPE) · data augmentation

1 Introduction

Steady-state visual evoked potential (SSVEP) is a kind of physiological signal mainly generated in occipital-frontal region of the brain which elicited by a fixed frequency of visual stimuli [1]. SSVEP response is frequency-dependent, which

B. Luo et al. (Eds.): ICONIP 2023, CCIS 1963, pp. 483–493, 2024.
https://doi.org/10.1007/978-981-99-8138-0_38

comprised of the fundamental frequency components occur at the frequencies as same as the flicking stimulus and its harmonics. In view of this characteristic, SSVEP has been widely adopted to build brain-computer interface (BCI) system for manipulating the external device consciously via decoding user's neural activities into specific control demands. SSVEP-based BCI systems can achieve a higher information transmission rate (ITR) and require less user training [2,3]. Since entering the millennium, numerous SSVEP-BCIs based applications have been developed, such as mental spellers, prosthetic devices, and augmented reality systems [4–6].

To build high-performance SSVEP-based BCI systems, the most crucial aspect is to design a frequency recognition method which could ensure robust and accurate detection results with the relatively short visual stimulation time [7]. There are mainly two types of frequency recognition methods, namely training-free and data-driven methods. Training-free methods such as canonical correlation analysis (CCA) and multivariate synchronization index (MSI) were once considered the best candidate algorithms for SSVEP-BCI since they did not require long-tedious calibration processes [8,9]. However, recent studies have shown that they are highly susceptible to spontaneous EEG interference, and their performance in identifying short signals and datasets with a large number of stimuli would severely decline [10]. By contrast, data-driven methods manifest superiority in achieving higher and more stable classification performance, particularly under challenging conditions. Among them, spatial filtering-based and deep learning-based methods are two representative research branches. Spatial filtering-based methods utilize the calibration data to estimate the spatial filters for constructing the linear combinations of multi-channel EEG data, then calculate the correlation coefficients for all stimulus frequencies to identify the user's gazing target, such as task-related component analysis (TRCA) [11] and task-discriminant component analysis (TDCA) [12], etc. Deep learning-based methods leverage the existing databases and advanced technologies like attention mechanism or label smooth techniques to provide a powerful end-to-end framework for automatically executing feature extraction and classification, such as SSVEPformer [13] and SSVEPNet [14], etc.

These data-driven methods could achieve satisfactory results even when the data length shorter than 1-second and under the large-scale datasets with up to 40 targets. However, these methods may suffer from overfitting and poor generalization when training data is scarce, which limit their practicality in building a high-performance SSVEP-BCI systems with much comfort for users. To address this issue, recent studies have demonstrated that the data augmentation strategies could be an effective solution [15–17]. In our previous work, we verified the feasibility of using a generative adversarial network (GAN)-based framework, termed as TEGAN, to extend the data length and enhance the classification performance [18]. However, we found that the improvement effect of TEGAN weakened when handling a large amount of targets. Additionally, the TEGAN is also a data-driven model, so its effectiveness is impacted by the amount of training data available.

In this paper, we propose two potential SSVEP data augmentation methods, i.e., filter band masking (FBM) and random phase erasing (RPE), based on the inherent features of SSVEPs. These methods could produce high-quality supplementary training data to improve the performance of SSVEP-BCIs, which require no parameter learning and are easy to implement. To evaluate the proposed methods, we used two large-scale publicly available datasets, i.e., Benchmark dataset and BETA dataset. The results showed that the proposed methods significantly enhanced the classification performance of two baseline methods, especially with a limited amount of calibration data. Specifically, evaluated on two methods, FBM increased the average accuracies by 7.40%, 8.55%, and RPE increased the average accuracies by 5.58%, 6.10%, respectively, with as few as two 1-scecond calibration trials on the Benchmark dataset. These findings demonstrate the potential of these data augmentation methods in enhancing the practicality of SSVEP-BCI for real-life scenarios.

2 Materials and Methods

2.1 Datasets

To evaluate the proposed augmentation methods, two large-scale publicly SSVEP datasets with 40 stimulus targets, namely Benchmark dataset [19] collected in the laboratory setting and BETA dataset [20] recorded outside the laboratory were used in this study, respectively. The 40 targets were encoded utilizing the joint frequency and phase modulation (JFPM) technique [4]. The target stimulation frequencies spanned from 8 Hz to 15.8 Hz, with an increment of 0.2 Hz, while the phases began at 0 and incremented by 0.5 π. Thirty-five and seventy subjects participated in the 6-block and 4-block experiments for Benchmark and BETA dataset, respectively. Each block contained 40 trials corresponding to all 40 stimuli generated in a random order. Each trial lasted for 6 s, which comprised of 0.5 s cuing period, 5 s targeted stimulus, and 0.5 s for stimulus offset before the next trial began, respectively. The more detailed information refers to [19] and [20].

For both datasets, EEG signals from 64 channels were acquired using a Synamps2 EEG system at a sampling rate of 1000 Hz, and then down-sampled to 250 Hz. EEG signals from the nine electrodes covering the occipital-parietal area (Pz, PO5, PO3, POz, PO4, PO6, O1, Oz, and O2) were used in this study as previous studies. All data were band-pass filtered between 7 and 90 Hz via fourth-order forward-backward Butterworth bandpass filter. Visual latency of 0.14 s and 0.13 s after the stimulus onset was considered for Benchmark and BETA dataset to extract data, respectively.

2.2 Baseline Classifiers

In this subsection, we briefly introduce a classical spatial filtering-based method (FBTRCA) and a representative deep learning-based method (SSVEPNet). Both methods were adopted as baseline classifiers to verify the effectiveness of our augmentation strategies.

TRCA is a signal processing method which could extract the task-related components via maximizing the reproducibility during task periods from neuroimaging data [21]. In 2018, Nakanishi et al. used TRCA algorithm as a spatial filtering-based method for SSVEP frequency detection [11]. To further improve classification performance, the filter bank extended TRCA variant, termed as FBTRCA, was proposed by combining the original TRCA and the filter bank technology proposed by Chen et al. [22]. The TRCA algorithm has been validated on multiple SSVEP datasets, and it can still achieve satisfactory results in short time windows and with a large number of stimulus targets. Due to its effectiveness, it has been used as a benchmark method in extensive research in recent years [23,24]. In current study, we implemented the FBTRCA algorithm according to the reference [11], and four filter banks ranged from $[8 \times m, 90]$ Hz were leveraged in the filter bank technology, where $m \in \{1, 2, 3, 4\}$ represents m-th sub-band.

SSVEPNet is an efficient CNN-LSTM network with spectral normalization and label smooth technologies, which aims to implement a high-performance SSVEP-BCI system with limited training data [14]. Among the network components, a convolutional neural network (CNN) was used to extract spatial-temporal features of SSVEP data, while a bidirectional long short-term memory network (BiLSTM) was employed to learn the dependency between them. Finally, the encoded fine-grained features were input into a fully connection network with three dense layers. As for regularization techniques, spectral normalization restricts the updateable weight matrix in a neural network to the Lipschitz condition that makes the training process stable. The label smoothing technique mitigates the adverse effects of interference from non-target flicking stimuli by extending the original single label to multiple labels. SSVEPNet has been evaluated on a 4-class and a 12-class SSVEP dataset, and it can yield about classification accuracy of 88.0% even with a few calibration trials available in the user-dependent (UD) training scenarios. Until now, we find that SSVEPNet has not been validated on large-scale SSVEP datasets with up to 40 stimulus targets, such as Benchmark and BETA. For implementing the SSVEPNet in this study, we simplified the stimulus interface as a 5×8 matrix in both datasets to calculate the attention score for label smooth technology.

2.3 Filter Band Masking

The EEG signals collected from the scalp can be regarded as linear mixed signals of multiple electrodes reflecting brain electrophysiological activities at the cortical level [9]. Although SSVEP response for the same stimulus target varies from trial to trial, we assume that the recorded multi-channel SSVEPs could be derived from the identical sources. According to the classical SSVEP theoretical model [15]:

$$X = S + N \tag{1}$$

where $X \in \mathbb{R}^{N_c \times N_t}$ represents the multi-channel EEG data, $X \in \mathbb{R}^{N_c \times N_t}$ denotes the source component, and $N \in \mathbb{R}^{N_c \times N_t}$ indicates the noise component. N_c and N_t represent number of channels and number of time points, respectively.

Suppose that the source component is invariant, which equivalents to the linear combination of multi-frequency sine-cosine signals. Therefore, the variant that causes the discrepancies in SSVEPs is the background noise which comprised of spontaneous EEG and artificial interference. Once the source component is determinant, the observed SSVEP signal could be simulated by adjusting the noise component. Since the response curve of the SSVEP amplitude with frequency can be roughly divided into three regions: low-frequency (5–15 Hz), medium frequency (15–30 Hz), and high-frequency (30–45 Hz) bands [25]. We consider that the SSVEP signal components in these three frequency bands are source components, while those beyond this frequency band are noise components. Thus, by retaining the information of the source component frequency band and disturbing the information of the noise frequency band component, an artificially synthetic SSVEP data used for enlarge the training dataset could be created.

More specifically, assuming the frequency range of the source component is $[l, h]$, where l is the lower limit frequency and h is the upper limit frequency. First, the discrete time domain data of each channel $x[n]$ is converted into discrete frequency domain data $X[k]$ through fast Fourier transform (FFT):

$$\begin{aligned} X[k] &= \sum_{n=0}^{N-1} x[n] e^{\frac{-2\pi jkn}{N}} \\ &= A[k] + iB[k], k = 0, 1, \ldots, N-1 \end{aligned} \tag{2}$$

Where N is number of sample points used for FFT, Fs is sampling rate, and the discrete frequency is $f = k * Fs/N$. $A[k]$ and $B[k]$ represent the real part and imaginary part, respectively. Then the perturbation that masking noises $\epsilon \sim \mathcal{N}(0, \sigma^2)$ are added on the real part and imaginary part as follows:

$$A[k], B[k] = \epsilon, k = \{0, \ldots, l\} \cup \{h, \ldots, (N+1)/2\} \tag{3}$$

It is noted that in order to obtain more accurate frequency resolution RES, we added $(Fs/RES - N)$ zeros point when calculating with FFT. Hence, the l and h in Eq. (3) would subsequently change to the original $1/RES$. In addition, due to the symmetry of $(N+1)/2$ as the node center in the frequency domain data, the data of the last $N - (N+1)/2$ points in the real and imaginary part data are also equivalent to the first $(N+1)/2$ data after noise masking.

Finally, the frequency domain data masked by noise is restored to discrete time domain data $x'[n]$ through inverse fast Fourier transform (IFFT). And the first N_t points of $x'[n]$ were extracted as the supplementary training data to enlarge the size of training dataset.

$$x'[n] = \frac{1}{N} \sum_{n=0}^{N-1} X[k] e^{\frac{2\pi jkn}{N}}, k = 0, 1, \ldots, N-1 \tag{4}$$

2.4 Random Phase Erasing

Phase is crucial for JFPM encoded stimulus recognition, as it can enhance the distinguishability between SSVEPs with quite narrow frequency ranges [4]. Furthermore, recent studies found that the different segments from the same trial could also be classified into different clusters [26]. This once again emphasizes the impact of phase on SSVEP frequency recognition. Therefore, perturbing the phase to increase the overlap between the training set and the testing set may be a potential SSVEP data augmentation method.

Concretely, we implement the phase perturbation through random phase erasing technology. Similar with filter band masking technology, the raw SSVEP data was converted into frequency domain by FFT as Eq. (2) firstly. And the magnitude and phase were calculated through the following formula:

$$M(k) = \sqrt{A^2(k) + B^2(k)}, k = 0, 1, \ldots, N - 1 \tag{5}$$

$$P(k) = arctan\frac{B(k)}{A(k)}, k = 0, 1, \ldots, N - 1 \tag{6}$$

where $M(k)$ and $P(k)$ represents the magnitude and phase information, respectively. Secondly, the phase of some random frequency points in proportion is randomly reset to the random values of $\eta \sim U(-\pi, \pi)$, i.e., erasing operation is performed.

$$P(r) = \eta, r \subseteq k \tag{7}$$

Finally, the IFFT operation $\mathcal{F}^{-1}()$ utilizing the magnitude and phase information was conducted to restore the frequency domain data to time domain data $x'[n]$. And the first N_t points of $x'[n]$ were extracted as the supplementary training data to enlarge the size of training dataset.

$$x'[n] = \mathcal{F}^{-1}\{A(k)e^{jP(k)}\}, k = 0, 1, \ldots, N - 1 \tag{8}$$

2.5 Performance Evaluation

To verify the effectiveness of FBM and RPE, we evaluate them on Benchmark and BETA dataset in the intra-subject experiments. The source code of SSVEPNet can be accessed at https://github.com/YuDongPan/SSVEPNet. For both datasets, the hyperparameters are set to as follows: mini-batch = 40, max-epochs = 500, learning rate = 0.01, dropout probability = 0.5. The optimizer we choose is Adam(beta1 = 0.9, beta2 = 0.999) combined with cosine annealing to stable training. The remaining hyperparameters of SSVEPNet is identical with the original implementation code. For FBM, the hyperparameters of l, h, and σ are set to 8, 50, and 0.6, respectively. As for RPE, the proportion of phase-erased frequency points is set to 0.1. Given the frequency distribution of stimuli in the Benchmark and BETA datasets, RES is fixed at 0.2 for both augmentation methods. The trained and tested models were running on a server with an Intel

Core I7-10700K CPU and an NVIDIA GeForce GTX 3090 (24 GB Memory) GPU.

In this study, classification accuracy was selected as the evaluation metric. The average classification results across all subjects were presented in the form of mean ± standard deviation. When the training blocks in the training set can be evenly divided, K-fold cross validation would be leveraged to eliminate potential randomness in the experiment. For example, when the training block used is 2, 3-fold cross validation would be performed. The 6 training blocks in the dataset would be evenly divided into three parts, with anyone of the three parts used for training and the remaining two parts used for testing. The average classification results of the final three experiments would be recorded. Five time-window lengths, i.e., 0.2 s to 1 s with interval of 0.2 s were analyzed in the experiments. Paired t-tests were implemented to investigate whether there were significant differences in the classification accuracy between all pairs of methods at each condition.

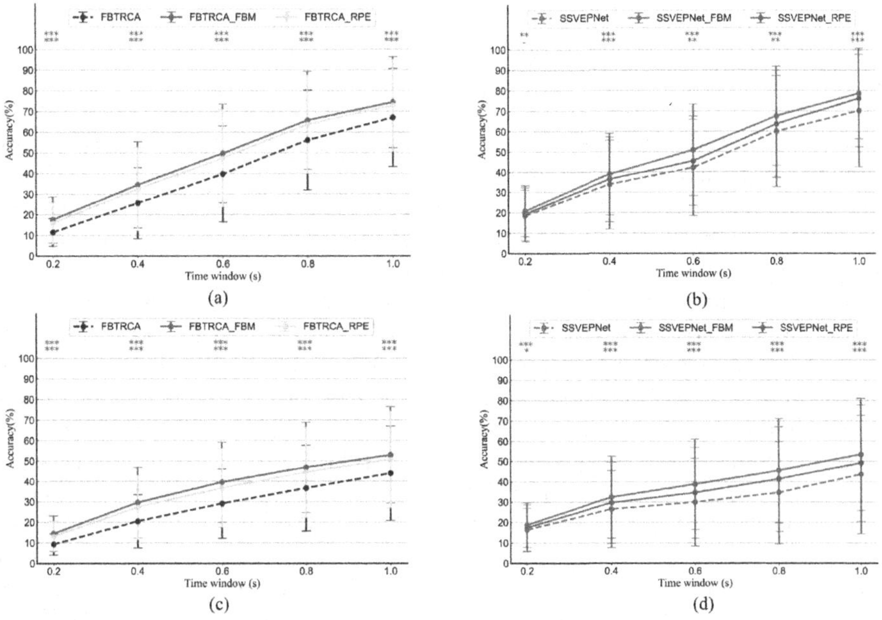

Fig. 1. The classification results of across all subjects obtained by baseline classification methods and augmented methods at various time window lengths. The number of training blocks is set to $N_b = 2$. (a) and (b) are on the Benchmark dataset, (c) and (d) are on the BETA dataset. The red asterisk in two rows (first row: FBM, second row: RPE) indicates significant difference between each pair of methods evaluated by paired t-tests (*$p < 0.05$, **$p < 0.01$, ***$p < 0.001$). (Color figure online)

Table 1. Classification results of using the proposed augmentation methods with different training blocks. The highest accuracy (in bold) indicates the optimal scheme for this setting. The asterisk indicates significant difference between baseline methods and with augmentation methods by paired t-tests (* $p < 0.05$, ** $p < 0.01$, *** $p < 0.001$).

	Method	Number of training blocks			
		Nb = 2	Nb = 3	Nb = 4	Nb = 5
Benchmark	FBTRCA	66.99 ± 23.72	83.54 ± 19.12	88.46 ± 16.82	91.21 ± 15.29
	FBTRCA w/ FBM	**74.39 ± 21.99*****	**85.13 ± 17.96*****	**89.21 ± 16.11***	**91.79 ± 14.10**
	FBTRCA w/ RPE	72.84 ± 22.28***	84.39 ± 18.82***	89.00 ± 15.98	91.57 ± 14.79
	SSVEPNet	70.05 ± 27.69	85.75 ± 21.12	91.07 ± 14.14	92.86 ± 12.61
	SSVEPNet w/ FBM	**78.60 ± 22.22*****	**87.57 ± 16.79***	**91.82 ± 12.78**	93.07 ± 12.00
	SSVEPNet w/ RPE	76.15 ± 23.73***	86.42 ± 20.46	90.75 ± 15.32	**93.21 ± 11.42**
BETA	FBTRCA	43.85 ± 23.04	60.67 ± 24.28	/	/
	FBTRCA w/ FBM	**52.86 ± 23.52*****	**63.76 ± 23.66*****	/	/
	FBTRCA w/ RPE	50.58 ± 23.66***	62.36 ± 24.14***	/	/
	SSVEPNet	43.61 ± 29.39	63.48 ± 27.37	/	/
	SSVEPNet w/ FBM	**53.41 ± 27.63*****	**67.64 ± 25.76*****	/	/
	SSVEPNet w/ RPE	49.13 ± 28.64***	66.73 ± 26.03***	/	/

3 Results

Firstly, we investigated the results how the classification performance of two baseline methods and their variant augmentation methods varied with different time-window lengths on two SSVEP datasets. The number of training blocks are set to 2 for both datasets. Figure 1 illustrates averaged classification results across subjects of all methods at different time-window lengths. Figure 1 (a) and Fig. 1 (b) present the results on the Benchmark dataset, and the results of BETA dataset are depicted in Fig. 1 (c) and Fig. 1 (d). On the Benchmark dataset, we can observe that both FBTRCA and SSVEPNet could achieve better classification results after using the augmentation methods across all time-window lengths. Paired t-tests manifest that there were significant differences between FBTRCA and its corresponding variants (FBTRCA vs. FBTRCA_FBM: all $p < 0.001$, FBTRCA vs. FBTRCA_RPE: all $p < 0.001$), SSVEPNet and its corresponding variants (SSVEPNet vs. SSVEPNet_FBM: all $p < 0.01$, SSVEPNet vs. SSVEPNet_RPE: 0.4 s to 1.0 s $p < 0.001$). On the BETA dataset, the similar tendency was obtained. Paired t-tests show that there were significant differences between FBTRCA, SSVEPNet and their corresponding variants (FBTRCA vs. FBTRCA_FBM, FBTRCA_RPE: all $p < 0.001$, SSVEPNet vs. SSVEPNet_FBM, SSVEPNet_RPE: all $p < 0.05$).

Secondly, the classification performance of using the proposed augmentation methods with different training blocks were investigated. The time-window length for all methods was fixed at 1.0 s. Table 1 listed averaged classification accuracies across all subjects of all methods on two datasets. From the results of Table 1, we could find that improvement effect of the proposed augmentation method was gradually weakened with continuous expansion of training blocks. Using two training blocks on the Benchmark dataset, FBM and RPE could improve the accuracies of FBTRCA about 7.40% and 5.85%, improve the accuracies of SSVEPNet about 8.55% and 6.10%, respectively. While using three training blocks, the improved accuracy of FBM and RPE decreased to 1.59%, 0.85% for FBTRCA, 1.82% and 0.67%, respectively. Moreover, the specific statistical results are summarized in Table 1.

4 Discussion

To promote the deployment of SSVEP-BCI systems in real-life scenarios, reducing user calibration data to improve user experience has become a crucial research topic in recent years [7]. To address this issue, developing the frequency recognition methods that require less calibration data and designing the augmentation methods are both effective strategies. In our previous work, we proposed an efficient CNN-LSTM network termed SSVEPNet aiming to build a high-performance SSVEP-BCI system with even a few calibration trials available. We have evaluated SSVEPNet on a 4-class and a 12-class dataset, achieving significantly better classification performance than representative spatial-filtering method, i.e. FBTRCA. In this study, we subsequently verified SSVEPNet on two publicly datasets with forty categories, which manifest a similar tendency. These results demonstrate the potential of SSVEPNet in enhancing the classification performance with limited training data. Moreover, we also found that the performance of SSVEPNet and FBTRCA could be further improved employing the proposed augmentation methods, i.e. FBM and RPE. Based on the intrinsic characteristics of SSVEPs, FBM and RPE, respectively, disturb the noise components in SSVEPs and reset the phase of some random frequency points to obtain supplementary data to enlarge the size of training dataset. Different from data augmentation methods based on deep learning methods, such as GANs, these two augmentation methods require no parameter learning, making them easy to implement. Besides, due to the fact that the optimization of GANs is essentially a training problem for two neural networks, and the high demand for data in neural networks, FBM and RPE are expected to improve the performance of GAN-based models such as TEGAN on small sample datasets.

5 Conclusion

In this work, two novel data augmentation methods named FBM and RPE are proposed to enhance the performance of SSVEP-BCIs to obtain more accurate classification results with limited calibration data. The intrinsic properties of SSVEPs are taken into account in the design of the proposed augmentation methods. They also require no parameter learning, making them easy to implement. Results on the Benchmark and BETA datasets with forty categories demonstrate the proposed augmentation methods could significantly improve the performance of two high-performance frequency recognition methods, i.e., FBTRCA and SSVEPNet, in target identification tasks. These findings suggest the potential of these data augmentation methods in enhancing the utility of SSVEP-BCI for real-life scenarios.

Acknowledgements. This work was supported in part by the National Natural Science Foundation of China under Grant No.62076209.

References

1. Yuan, X., Sun, Q., Zhang, L., Wang, H.: Enhancing detection of SSVEP-based BCIs via a novel CCA-based method. Biomed. Signal Process. Control **74**, 103482 (2022)
2. Zhang, Y., Xu, P., Liu, T., Hu, J., Zhang, R., Yao, D.: Multiple frequencies sequential coding for SSVEP-based brain-computer interface. PLoS ONE **7**(3), e29519 (2012)
3. Chen, X., Chen, Z., Gao, S., Gao, X.: A high-ITR SSVEP-based BCI speller. Brain-Computer Interfaces **1**(3–4), 181–191 (2014)
4. Chen, X., Wang, Y., Nakanishi, M., Gao, X., Jung, T.P., Gao, S.: High-speed spelling with a noninvasive brain-computer interface. Proc. Natl. Acad. Sci. **112**(44), E6058–E6067 (2015)
5. Kwak, N.S., Müller, K.R., Lee, S.W.: A convolutional neural network for steady state visual evoked potential classification under ambulatory environment. PLoS ONE **12**(2), e0172578 (2017)
6. Zhang, R., et al.: The effect of stimulus number on the recognition accuracy and information transfer rate of SSVEP-BCI in augmented reality. J. Neural Eng. **19**(3), 036010 (2022)
7. Pan, Y., Chen, J., Zhang, Y.: A Survey of deep learning-based classification methods for steady-state visual evoked potentials. Brain-Apparat. Commun. J. Bacomics **2**(1), 2181102 (2023). https://doi.org/10.1080/27706710.2023.2181102
8. Lin, Z., Zhang, C., Wu, W., Gao, X.: Frequency recognition based on canonical correlation analysis for SSVEP-based BCIs. IEEE Trans. Biomed. Eng. **53**(12), 2610–2614 (2006)
9. Zhang, Y., Xu, P., Cheng, K., Yao, D.: Multivariate synchronization index for frequency recognition of SSVEP-based brain-computer interface. J. Neurosci. Methods **221**, 32–40 (2014)
10. Chen, Y., Yang, C., Ye, X., Chen, X., Wang, Y., Gao, X.: Implementing a calibration-free SSVEP-based BCI system with 160 targets. J. Neural Eng. **18**(4), 046094 (2021)
11. Nakanishi, M., Wang, Y., Chen, X., Wang, Y.T., Gao, X., Jung, T.P.: Enhancing detection of SSVEPs for a high-speed brain speller using task-related component analysis. IEEE Trans. Biomed. Eng. **65**(1), 104–112 (2017)
12. Liu, B., Chen, X., Shi, N., Wang, Y., Gao, S., Gao, X.: Improving the performance of individually calibrated SSVEP-BCI by task-discriminant component analysis. IEEE Trans. Neural Syst. Rehabil. Eng. **29**, 1998–2007 (2021)
13. Chen, J., Zhang, Y., Pan, Y., Xu, P., Guan, C.: A Transformer-based deep neural network model for SSVEP classification. arXiv preprint arXiv:2210.04172 (2022)
14. Pan, Y., Chen, J., Zhang, Y., Zhang, Y.: An efficient CNN-LSTM network with spectral normalization and label smoothing technologies for SSVEP frequency recognition. J. Neural Eng. **19**(5), 056014 (2022)
15. Luo, R., Xu, M., Zhou, X., Xiao, X., Jung, T.P., Ming, D.: Data augmentation of SSVEPs using source aliasing matrix estimation for brain-computer interfaces. IEEE Trans. Biomed. Eng. **70**, 1775–1785 (2022)

16. Li, R., Wang, L., Suganthan, P., Sourina, O.: Sample-based data augmentation based on electroencephalogram intrinsic characteristics. IEEE J. Biomed. Health Inform. **26**(10), 4996–5003 (2022)
17. Aznan, N.K.N., Atapour-Abarghouei, A., Bonner, S., Connolly, J.D., Al Moubayed, N., Breckon, T.P.: Simulating brain signals: creating synthetic EEG data via neural-based generative models for improved ssvep classification. In: 2019 International Joint Conference on Neural Networks (IJCNN), pp. 1–8. IEEE, Budapest, Hungary (2019)
18. Pan, Y., Li, N., Zhang, Y.: Short-time SSVEP data extension by a novel generative adversarial networks based framework. arXiv preprint arXiv:2301.05599 (2023)
19. Wang, Y., Chen, X., Gao, X., Gao, S.: A benchmark dataset for SSVEP-based brain-computer interfaces. IEEE Trans. Neural Syst. Rehabil. Eng. **25**(10), 1746–1752 (2016)
20. Liu, B., Huang, X., Wang, Y., Chen, X., Gao, X.: Beta: a large benchmark database toward SSVEP-BCI application. Front. Neurosci. **14**, 627 (2020)
21. Tanaka, H., Katura, T., Sato, H.: Task-related component analysis for functional neuroimaging and application to near-infrared spectroscopy data. Neuroimage **64**, 308–327 (2013)
22. Chen, X., Wang, Y., Gao, S., Jung, T.P., Gao, X.: Filter bank canonical correlation analysis for implementing a high-speed SSVEP-based brain-computer interface. J. Neural Eng. **12**(4), 046008 (2015)
23. Wong, C.M., et al.: Learning across multi-stimulus enhances target recognition methods in SSVEP-based BCIs. J. Neural Eng. **17**(1), 016026 (2020)
24. Sun, Q., Chen, M., Zhang, L., Li, C., Kang, W.: Similarity-constrained task-related component analysis for enhancing SSVEP detection. J. Neural Eng. **18**(4), 046080 (2021)
25. Wang, Y., Wang, R., Gao, X., Hong, B., Gao, S.: A practical VEP-based brain-computer interface. IEEE Trans. Neural Syst. Rehabil. Eng. **14**(2), 234–240 (2006)
26. Waytowich, N., et al.: Compact convolutional neural networks for classification of asynchronous steady-state visual evoked potentials. J. Neural Eng. **15**(6), 066031 (2018)

ONEI: Unveiling Route and Phase of Breathing from Snoring Sounds

Xinhong Li[1], Baoai Han[3], Li Xiao[1], Xiuping Yang[3], Weiping Tu[1,2(✉)], Xiong Chen[3(✉)], Weiyan Yi[1], Jie Lin[1], Yuhong Yang[1], and Yanzhen Ren[4]

[1] National Engineering Research Center for Multimedia Software, School of Computer Science, Wuhan University, Wuhan, China

[2] Hubei Key Laboratory of Multimedia and Network Communication Engineering, Wuhan University, Wuhan, China

tuweiping@whu.edu.cn

[3] Sleep Medicine Centre, Zhongnan Hospital of Wuhan University, Wuhan, China

zn_chenxiong@whu.edu.cn

[4] School of Cyber Science and Engineering, Wuhan University, Wuhan, China

Abstract. Obstructive Sleep Apnea-Hypopnea Syndrome (OSAHS) is a chronic respiratory disorder caused by the obstruction of the upper airway. The treatment approach for OSAHS varies based on the individual patient's breathing route and phase during snoring. Extensive research has been conducted to identify various snoring patterns, including the breathing route and the breathing phase during snoring. However, the identification of breathing routes and phases in snoring sounds is still in the early stages due to the limited availability of comprehensive datasets with scientifically annotated nocturnal snoring sounds. To address this challenge, this study presents ONEI, an innovative dataset designed for recognizing and analyzing snoring patterns. ONEI encompasses 5171 snoring recordings and is annotated with four distinct labels, namely nasal-dominant inspiratory snoring, nasal-dominant expiratory snoring, oral inspiratory snoring, and oral expiratory snoring. Experimental evaluations reveal discernible acoustic features in snoring sounds, which can be effectively utilized for accurately identifying various snoring types in real-world scenarios. The dataset will be made publicly available for access at https://github.com/emleeee/ONEI.

Keywords: OSAHS · Breathing route · Snoring sound

1 Introduction

Obstructive sleep apnea-hypopnea syndrome (OSAHS) is a prevalent medical condition characterized by recurrent upper airway obstruction during sleep, commonly accompanied by heavy snoring. The management of OSAHS encompasses various treatment approaches, such as continuous positive airway pressure (CPAP), oral appliances, surgical interventions, and lifestyle modifications.

B. Han—Equal contribution.

B. Luo et al. (Eds.): ICONIP 2023, CCIS 1963, pp. 494–505, 2024.
https://doi.org/10.1007/978-981-99-8138-0_39

Among these, CPAP stands as the primary treatment modality for patients diagnosed with moderate to severe OSAHS. Identifying the breathing route is important in determining the most suitable CPAP treatment interface prescription [5]. In clinical practice, the snoring route is often determined by the patient's self-description. However, such self-reported information fails to provide an objectively measured breathing route [19], thus highlighting the need for more reliable and accurate methods of determining the actual breathing route.

Snoring is a prevalent clinical characteristic of OSAHS and is associated with upper airway structures. Previous studies, such as Mikami *et al.* [16–18] employed machine learning methods to distinguish breathing routes during snoring. However, these studies relied on simulated snoring sounds collected from fifteen participants, who were instructed to produce snoring sounds using different breathing methods (oral, nasal, and oronasal). The classification strategies in these works included a three-category scheme (oral, nasal, and oronasal snoring) and a two-category scheme (oral and nasal snoring), achieving overall accuracies of 81.7% and 85.91%, respectively. Similarly, Qian *et al.* [23] focused on detecting inspiration-related snoring signals (IRSS) throughout the night and achieved an accuracy of 85.91%. Recent research aimed at detecting mouth breathing utilized commercially available earbuds to capture audio from 30 participants, resulting in an accuracy of 78% in detecting mouth breathing [2].

Snoring route can be classified as either oral or nasal, and snoring cycle can be further categorized into inspiratory and expiratory phases [20]. However, the detection of these snoring breathing routes and phases is currently in its early stages. The existing studies in this field have several notable limitations, which can be summarized as follows:

1) **Reliance on Simulated Snoring Sounds.** In the field of oral/nasal snore detection, previous studies have predominantly utilized simulated snoring sounds rather than actual recordings. These simulations involved participants occluding their nostrils with fingers or closing their mouths to mimic nocturnal rhythmic snoring [18]. However, significant discrepancies exist in the acoustic characteristics between simulated snoring and actual snoring during sleep [12]. Solely relying on simulated snoring sound datasets may hinder the machine learning models' ability to fully capture the characteristics of real snoring sounds, leading to potential inaccuracies and misclassification of relevant snoring events. Such limitations in accurate snoring recognition could have consequential implications in clinical diagnosis, potentially resulting in incorrect assessments of patients' conditions.
2) **The Neglected Importance of Expiratory Snoring.** Current research in the field of detecting breathing phases during snoring primarily focuses on inspiratory snoring, neglecting the significance of expiratory snoring. However, the detection of expiratory snoring holds crucial clinical implications. Woodson *et al.* [28] highlight that the treatment directed at preventing expiratory obstruction is likely critical to successfully managing OSAHS. Additionally, studies suggest that patients displaying expiratory snoring should undergo evaluation for potential pulmonary conditions such as asthma and

chronic obstructive pulmonary disease (COPD) [3]. To the best of our knowledge, there is currently an absence of publicly available datasets containing annotations for snoring phases, leading to a lack of research on expiratory snoring detection methods.

In this paper, a novel snoring sound dataset with breathing route and phase information has been proposed. In contrast to the conventional subjective judgment based approach, the annotation method employed in this study is grounded on Polysomnography (PSG) signals which are considered the gold standard for diagnosing OSAHS in sleep laboratories [21]. Annotating snoring based on PSG signal waveforms provides a more objective and accurate approach to analysis. The establishing of this comprehensive database is significant for future research on the detection of breathing route and phase during snoring. It will provide ENT (ear, nose, and throat) experts with more precise information regarding OSAHS patients' snoring condition, rather than relying solely on self-reported data. As a result, it will facilitate more effective therapeutic interventions for OSAHS patients. Additionally, the research on detecting expiratory snoring will raise awareness among patients about the potential occurrence of pulmonary diseases when they experience frequent expiratory snoring. Moreover, leveraging recent advancements in audio classification, this study employs PSLA, a CNN model as the baseline method for snoring sounds classification. Experimental results demonstrate the significant potential of our work in identifying distinct snoring patterns, which can aid physicians in prescribing appropriate treatments for OSAHS and help individuals avoid the risk of developing COPD and other pulmonary diseases.

The main contributions of this paper are summarized as follows:

1) Introducing ONEI, a comprehensive and natural snoring sound dataset, we have successfully integrated essential information about breathing routes and phases during snoring. This novel dataset provides a solid foundation for the detection and analysis of breathing routes and phases, both of which are crucial in devising effective treatment strategies for OSAHS and self-health monitoring.
2) The annotations in our snoring sound dataset have been derived through a scientific and objective approach, utilizing signals obtained from PSG systems, thereby surpassing the reliance on subjective judgments. This methodology enhances the credibility and robustness of the dataset, laying the groundwork for future research.

The rest of this paper is organized as follows. We first describe our dataset and annotation rules in Sect. 2, then discuss our baseline experimental results in Sect. 3 and finally provide conclusions in Sect. 4.

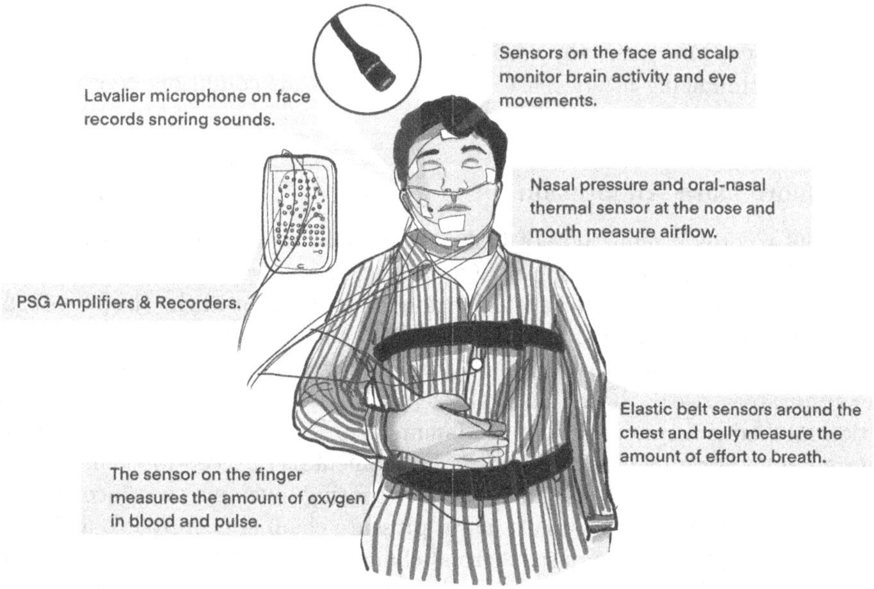

Fig. 1. The detail of microphone and PSG set up. The lavalier microphone placement is on the patient's face between nose and mouth, and the sensors of PSG system are attached to the patient's head and body.

2 Dataset

2.1 Snore Data Acquisition

This study involved 16 adult participants (14 males, 2 females) with an average age of 45.1 ($\pm$22.9) years, a mean body mass index (BMI) of 27.2 ($\pm$4.5), and an average apnea-hypopnea index (AHI) of 61.5 ($\pm$36.6). Participants were recruited from a Sleep Medicine Research Center at a local hospital. The study obtained ethical approval from the local medical ethics committee, and written consent was obtained from each participant, including permission for audio and video recordings during sleep. Personal information was collected and stored anonymously to ensure privacy protection.

During the sleep period, the snoring incidents were recorded using a lavalier microphone, which was synchronized with the PSG signals. The sleep monitoring was performed using a full night PSG system to obtain the AHI. A high-quality cardioid lavalier microphone (Shure MX150/C, USA) connected to an audio interface (Antelope Zen Go Synergy Core, Bulgaria) was securely attached to the subject's face using medical tape, positioned between the nose and mouth as depicted in Fig. 1. The microphone captured sleep-related sounds at a sampling rate of 32 kHz and 16-bit resolution, from a close distance to effectively capture subtle abnormal breathing sounds that might be masked by background noise in typical long-distance recording setups. Acquiring a comprehensive record of all snoring sounds is crucial for the analysis and diagnosis of health conditions such as sleep apnea. Moreover, the cardioid microphone design contributes to noise reduction, minimizing interference from factors like air conditioners or external sources.

2.2 Snore Data Annotation

The PSG records a multitude of biological signals, including electroencephalogram (EEG), electrooculography (EOG), electromyography (EMG), nasal/oral airflow, and blood oxygen saturation throughout the night in the sleep medicine research center [21]. The EEG, EOG, and EMG channels are employed to ensure the patient's sleep stage while the nasal/oral airflow channels are utilized to monitor respiratory status during natural sleep. Oral/nasal-dominant snoring annotation method is inspired by AASM Manual V2.6 [7] and inspiratory/expiratory snoring annotation method is based on Fundamentals of Sleep Medicines [6].

When annotating the breathing route of snoring, our approach incorporates waveform data from two channels: nasal pressure channel (referred to as PFlow channel in Fig. 2) and oral-nasal thermal channel (referred to as TFlow channel in Fig. 2). The PFlow channel specifically records airflow only from the nose and the TFlow channel records airflow from both nose and mouth. When the PFlow signal indicates an absence of airflow and the TFlow signal shows continued airflow, this airflow pattern corresponds to oral breathing [6]. However, defining the absence of airflow in real-world clinical settings can be inherently ambiguous. To address this challenge, we employ the apnea definition rule based on PSG signals to accurately define the absence of airflow within our study. Following the established guidelines outlined in the AASM Manual V2.6, we consider a drop in peak thermal sensor excursion by 90% of the baseline as a reliable indicator of absence [7]. This criterion is subsequently applied to the nasal pressure signal, wherein if the peak of the airflow curve, as monitored by the nasal pressure sensor, decreases by more than 90% from its baseline, we classify the signal as an absence, and the associated snoring during this period is classified as oral snoring, as illustrated in box A of Fig. 2. On the other hand, due to limitations in PSG sensors, we are unable to differentiate between nasal snoring and oronasal snoring when the PFlow and TFlow signals indicate a consistent airflow. Therefore, we categorize this type of snoring as nasal-dominant snoring (box B of Fig. 2). Our approach adheres to rigorous scientific principles and holds practical applicability in real-world scenarios.

The annotation of inspiratory and expiratory snoring primarily relies on waveform data obtained from the oral-nasal thermal channel, with supplementary reference to the nasal pressure channel. This choice is driven by the jagged and spiky nature of pressure waveforms, which can introduce complexities in making accurate judgments. The oral-nasal thermal sensors consist of three devices commonly used in clinical sleep studies to monitor airflow. These sensors detect temperature changes resulting from the flow of air, where cooler air corresponds to inhalation and warmer air corresponds to exhalation [6,24]. Consequently, a decrease in airflow is indicative of inspiratory snoring, while an

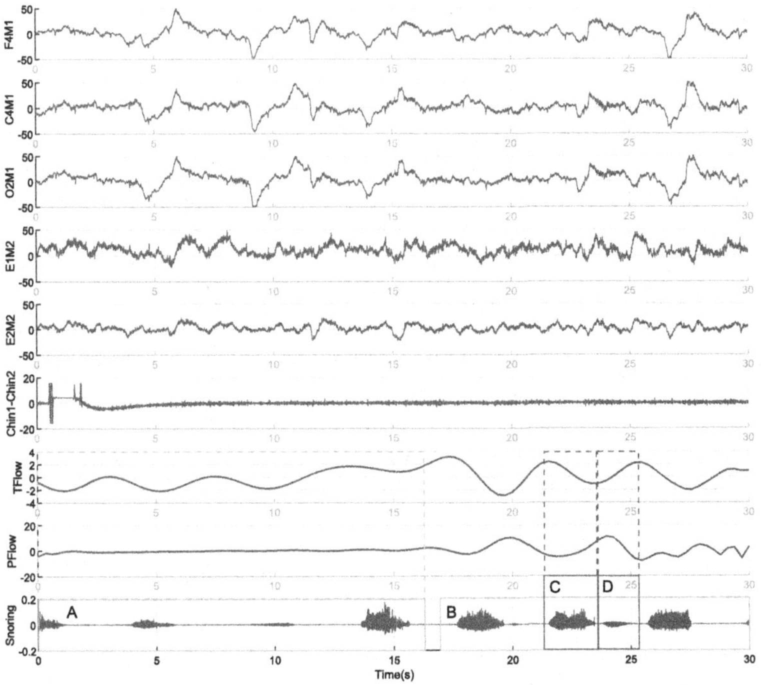

Fig. 2. This figure displays a 30-s segment of PSG data from a full night's sleep. Signals recorded include EEG (F4M1, C4M1, O2M1), EOG (E1M2, E2M2), EMG (Chin1-Chin2), airflow (TFlow, PFlow). Box A shows oral snoring sounds with corresponding airflow waveforms indicating absent PFlow and fluctuating TFlow. Box B indicates nasal-dominant snoring sounds with continuous airflow observed in both PFlow and TFlow. Snoring in box C represents inspiratory snoring with decreasing TFlow and increasing PFlow. Snoring in box D represents expiratory snoring with increasing TFlow and decreasing PFlow.

increase in airflow signifies expiratory snoring. On the other hand, nasal pressure (NP) is measured using a nasal cannula connected to a precise pressure transducer. The measured pressure reflects the pressure drop across the resistance of the nasal inlet associated with nasal airflow [6]. In a complete respiratory cycle, the waveform tends to rise to its peak during inhalation and decrease to valleys during exhalation. Snoring that occurs when the TFlow signal decreases and the PFlow signal increases is annotated as inspiratory snoring (box C in Fig. 2). Conversely, snoring is annotated as expiratory snoring when the TFlow signal increases and the PFlow signal decreases (box D in Fig. 2).

2.3 Snore Data Distribution

Previous research consistently indicates a higher prevalence of expiratory snoring among chronic heavy snorers [10]. Additionally, it has been observed that women generally exhibit less severe OSAHS compared to males, with lower AHI scores and shorter apneas and hypopneas [27]. In order to ensure a well-balanced representation of both inspiratory and expiratory snoring in our study sample, we intentionally included more severe OSAHS patients in the ONEI dataset. As a result, the dataset exhibits an imbalanced gender distribution, with only 2 female and 14 male patients included. Within the ONEI corpus, the distribution of snoring sound samples is relatively balanced among three categories (nasal-dominant inspiratory snoring, nasal-dominant expiratory snoring, oral inspiratory snoring), except for oral expiratory snoring, which is limited by the scarcity of its parent classes, namely, oral snoring and expiratory snoring. The ONEI dataset comprises a total of 5171 snoring samples. Further information on the data split can be found in Table 1.

Table 1. The ONEI dataset comprises labeled snoring clips classified into different snoring patterns: nasal-dominant inspiratory snoring (*na-in*), nasal-dominant expiratory snoring (*na-ex*), oral inspiratory snoring (*or-in*), and oral expiratory snoring (*or-ex*). It should be noted that the number of samples in the oral expiratory snoring class is relatively lower compared to the other classes.

Labels	Patient-based			Snore-based			$\sum$
	Train	Validation	Test	Train	Validation	Test	
na-in	954	250	101	779	270	256	1305
na-ex	1077	88	195	797	275	288	1360
or-in	1054	264	90	864	282	262	1408
or-ex	956	90	52	663	207	228	1098
$\sum$	4041	692	438	3103	1034	1034	5171

3 Baseline Experiments

3.1 Baseline Method

Convolutional neural networks (CNNs) are a type of artificial neural network known for their exceptional performance in processing structured data, such as images, speech, and audio [1,11,14]. By utilizing convolutional operations, CNNs extract high-level features from input data, eliminating the need for manually designed features. This characteristic makes CNNs highly efficient and robust in learning patterns from diverse datasets. In the field of snoring sound analysis, CNNs have shown their effectiveness in tasks such as upper airway obstruction location classification [4] and OSAHS snoring detection [8]. However, similar to other AI applications in medicine, the availability of annotated snore data is limited, and the inherent imbalance in snore characteristics cannot be overlooked [22]. Therefore, addressing the issue of imbalance in snoring labeling is of significant importance.

To address these issues, we employed a PSLA audio classifier as our baseline experiment, which incorporates ImageNet-pretraining, balanced sampling, data augmentation, label enhancement and model aggregation [9]. These techniques have been proposed individually in previous studies, but PSLA is the first CNN-based neural model that combines them simultaneously to improve the performance for audio tagging. While we made efforts to achieve better balance in the ONEI dataset, the incidence of oral expiratory snoring remains infrequent in the corpus, which is consistent with findings from clinical studies. To mitigate the impact of data imbalance, as in PSLA, we employed a balanced sampling algorithm and applied time/frequency masking data augmentation techniques. These approaches were designed to address the imbalance problem inherent in snoring sound data. The processed snoring sound data was then input into a pretrained EfficientNet model [25] and an attention pooling layer for prediction.

3.2 Experiments Settings

Model Inputs. Speech is produced by the vibratory excitation of a hollow anatomic system called the vocal tract. Similarly, snoring is caused by the vibration of anatomical structures in the pharyngeal airway. Previous studies have shown that snoring sounds contain valuable information about the upper airway state, as different breathing routes and respiratory phases generate distinct patterns [15,26]. Given these physiological similarities, we adopt a speech analysis approach to study snoring. Following the methodology of PSLA [9], we transform the snoring sound waveform into a sequence of 128-dimensional log Mel filterbank (fbank) features. These features are computed with a 25 ms Hanning window every 10 ms, and zero padding is applied to ensure that all audio clips have a consistent length of 1056 frames. This process results in a 1056×128 feature vector, which serves as input to our baseline model.

Baseline Experiments. In this study, all labeled snoring sounds in ONEI were utilized to conduct two types of experiments as follows.

1) We employed a snore-based partitioning method to randomly divide the snoring sounds of all patients in ONEI into training, validation, and test sets with a ratio of 6:2:2. The objective of this experiment is to evaluate the potential of utilizing snoring sounds from the ONEI dataset for accurate classification of various snoring patterns.
2) To investigate the individual variability present in the ONEI dataset and enhance alignment with the clinical diagnosis reality, we conducted a supplementary patient-based partitioning experiment. This experiment involved

splitting the snoring sounds according to individual patients. The ONEI dataset comprises 16 patients, with 12 patients included in the training set, and 2 patients each allocated to the validation and test sets.

Training Details. The model has been trained on two NVIDIA GeForce RTX 4090 GPUs with batch size 120 for 50 epochs. The experimental setup mostly follows the PSLA. The model utilizes the Adam optimizer [13], and use binary cross-entropy (BCE) loss. We use a fixed initial learning rate of 1e−3 and cut it in half every 5 epochs after the 35th epoch. We use a linear learning rate warm-up strategy for the first 1,000 iterations. The aim of the ONEI dataset is to identify the different snoring pattern of patients.

Evaluation Metrics. The objective of the ONEI corpus is to identify snoring patterns during sleep, thus *Accuracy* serves as the primary evaluation metric for performance assessment. Additionally, in line with medical practices, *Sensitivity* and *Specificity* are incorporated into baseline experiments.

3.3 Results

As indicated in Table 2, the results achieved an overall accuracy of 54.6% and 88.4% in the test set when we partitioned based on patients and snores, respectively. Moreover, when partitioning data with snoring, there is excellent performance with average sensitivity and specificity values of 92.5% and 94.4%. When partitioning data with patients, the average sensitivity and specificity values are 85.4% and 64.8% respectively, which means the experiment is capable of detecting oral and expiratory snoring, but it may also identify snores that do not belong to these categories. The obtained results indicate that the proposed ONEI dataset can be use to effectively differentiate between different snoring patterns and accurately detect both oral snoring sounds and expiratory snoring sounds. These findings hold significant implications for guiding the treatment of OSAHS and facilitating the inspection of pulmonary diseases. However, it is worth mentioning that the patient-based partitioning method exhibits a significant decrease in all evaluation metrics compared to the snore-based partitioning method. This phenomenon can be attributed to the distinct snoring sound characteristics exhibited among individual patients. This observation aligns with our data annotation experience and leads to a notable decline in classification performance when using patient-based partitioning. Therefore, to improve the performance of snoring classification, it is recommended to expand the dataset and enhance the generalization performance of the algorithms.

Moreover, to investigate the specific category responsible for the observed decline in classification performance, we have presented the corresponding confusion matrices in Fig. 3. These matrices reveals significant differences between the *or-ex* class and the other classes. Specifically, the confusion matrix shows that the recognition accuracy of the *or-ex* class is the worst in our experiment. In both Fig. 3a and Fig. 3b, the number of correctly recognized samples for *or-ex* is the lowest among the four classes. Furthermore, in Fig. 3a, the number of

Table 2. The classification results of patient-based and snore-based partitioning method, including overall accuracy, sensitivity and specificity of oral snoring, expiratory snoring and their average value.

Accuracy	Patient-based			Snore-based		
	54.6%			88.4%		
	Oral	Expiratory	Average	Oral	Expiratory	Average
Sensitivity	87.3%	83.4%	85.4%	90.2%	94.8%	92.5%
Specificity	67.2%	62.3%	64.8%	94.5%	94.3%	94.4%

misjudged samples as *or-in* is nearly 1.5 times higher than the number of correctly classified ones. This discrepancy can be attributed to a limited number of patients and potential limitations in the chosen model architecture or input. To address this, it is recommended to augment the ONEI dataset by including additional snoring sounds from diverse patients and to further explore the incorporation of snoring acoustics to capture more effective snoring features, specifically focusing on the features of *or-ex* snoring. These efforts are expected to enhance the classification performance in future research.

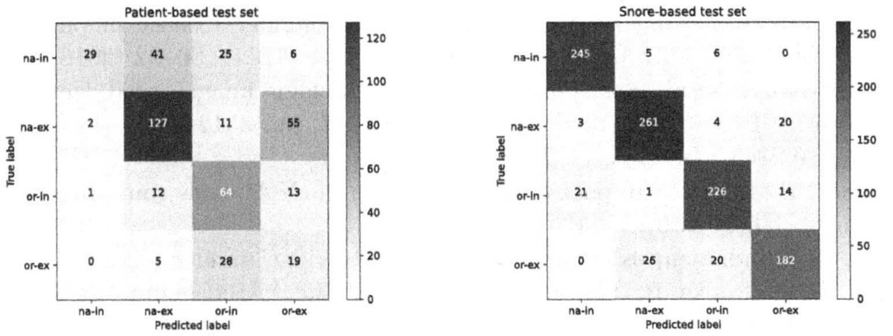

(a) Patient-based (b) Snore-based

Fig. 3. Confusion matrices of two data splitting method.

4 Conclusion

The characteristics of snoring sounds are influenced by the breathing routes and phases of the respiratory cycle. However, there has been insufficient research on methods for detecting oral and expiratory snoring, which are important for determining CPAP treatment interface prescriptions for OASHS patients and preventing pulmonary diseases. To address this gap, this paper aims to construct a snoring sound dataset called ONEI, which includes four snoring classes: nasal-dominant inspiratory snoring, nasal-dominant expiratory snoring, oral inspiratory snoring, and oral expiratory snoring. The ONEI dataset is expected to provide valuable insights for snoring research. Additionally, baseline experiments have shown that automatic classification models based on acoustic properties can successfully differentiate the four mentioned snoring types. However, the accuracy decreases significantly when patients in the test set are not present in the training and validation sets. To improve the accuracy of classifying various snoring sounds, future research can focus on expanding the number of participants in the ONEI database and exploring innovative approaches for snoring sound classification.

Acknowledgements. This work was supported by the Hubei Province Technological Innovation Major Project (No. 2022BCA041).

References

1. Abdel-Hamid, O., Mohamed, A.R., Jiang, H., Deng, L., Penn, G., Yu, D.: Convolutional neural networks for speech recognition. IEEE/ACM Trans. Audio Speech Lang. Process. **22**(10), 1533–1545 (2014)
2. Ahmed, T., Rahman, M.M., Nemati, E., Kuang, J., Gao, A.: Mouth breathing detection using audio captured through earbuds. In: 2023 IEEE International Conference on Acoustics, Speech and Signal Processing (ICASSP), ICASSP 2023, pp. 1–5. IEEE (2023)
3. Alchakaki, A., Riehani, A., Shikh-Hamdon, M., Mina, N., Badr, M.S., Sankari, A.: Expiratory snoring predicts obstructive pulmonary disease in patients with sleep-disordered breathing. Ann. Am. Thorac. Soc. **13**(1), 86–92 (2016)
4. Amiriparian, S., et al.: Snore sound classification using image-based deep spectrum features. In: Proceedings of the Interspeech 2017, pp. 3512–3516 (2017). https://doi.org/10.21437/Interspeech.2017-434
5. Andrade, R.G., et al.: Impact of acute changes in CPAP flow route in sleep apnea treatment. Chest **150**(6), 1194–1201 (2016)
6. Berry, R.: Fundamentals of Sleep Medicine. Elsevier/Saunders (2012)
7. Berry, R.B., Brooks, R., Gamaldo, C.E., et al.: The AASM manual for the scoring of sleep and associated events: rules, terminology and technical specifications, Version 2.6.0. American Academy of Sleep Medicine, Darien, Illinois (AASM) (2020). https://aasm.org/
8. Ding, L., Peng, J., Song, L., Zhang, X.: Automatically detecting apnea-hypopnea snoring signal based on VGG19 + LSTM. Biomed. Sig. Process. Control **80**, 104351 (2023)
9. Gong, Y., Chung, Y.A., Glass, J.: PSLA: improving audio tagging with pretraining, sampling, labeling, and aggregation. IEEE/ACM Trans. Audio Speech Lang. Process. **29**, 3292–3306 (2021)
10. Guilleminault, C., Stoohs, R., Duncan, S.: Snoring (i): daytime sleepiness in regular heavy snorers. Chest **99**(1), 40–48 (1991)
11. He, K., Zhang, X., Ren, S., Sun, J.: Deep residual learning for image recognition. In: Proceedings of the IEEE Conference on Computer Vision and Pattern Recognition, pp. 770–778 (2016)
12. Herzog, M., et al.: Frequency analysis of snoring sounds during simulated and nocturnal snoring. Eur. Arch. Otorhinolaryngol. **265**, 1553–1562 (2008)
13. Kingma, D.P., Ba, J.: Adam: a method for stochastic optimization. arXiv preprint arXiv:1412.6980 (2014)

14. Kong, Q., Cao, Y., Iqbal, T., Wang, Y., Wang, W., Plumbley, M.D.: PANNs: large-scale pretrained audio neural networks for audio pattern recognition. IEEE/ACM Trans. Audio Speech Lang. Process. **28**, 2880–2894 (2020)
15. Liistro, G., Stanescu, D., Veriter, C.: Pattern of simulated snoring is different through mouth and nose. J. Appl. Physiol. **70**(6), 2736–2741 (1991)
16. Mikami, T., Kojima, Y., Yamamoto, M., Furukawa, M.: Recognition of breathing route during snoring for simple monitoring of sleep apnea. In: Proceedings of SICE Annual Conference 2010, pp. 3433–3434. IEEE (2010)
17. Mikami, T., Kojima, Y., Yamamoto, M., Furukawa, M.: Automatic classification of oral/nasal snoring sounds based on the acoustic properties. In: 2012 IEEE International Conference on Acoustics, Speech and Signal Processing (ICASSP), pp. 609–612. IEEE (2012)
18. Mikami, T., Kojima, Y., Yonezawa, K., Yamamoto, M., Furukawa, M.: Spectral classification of oral and nasal snoring sounds using a support vector machine. J. Adv. Comput. Intell. Intell. Inf. **17**(4), 611–621 (2013)
19. Nascimento, J.A., et al.: Predictors of oronasal breathing among obstructive sleep apnea patients and controls. J. Appl. Physiol. **127**(6), 1579–1585 (2019)
20. Pevernagie, D., Aarts, R.M., De Meyer, M.: The acoustics of snoring. Sleep Med. Rev. **14**(2), 131–144 (2010)
21. Punjabi, N.M.: The epidemiology of adult obstructive sleep apnea. Proc. Am. Thorac. Soc. **5**(2), 136–143 (2008)
22. Qian, K., et al.: Can machine learning assist locating the excitation of snore sound? A review. IEEE J. Biomed. Health Inform. **25**(4), 1233–1246 (2020)
23. Qian, K., Xu, Z., Xu, H., Ng, B.P.: Automatic detection of inspiration related snoring signals from original audio recording. In: 2014 IEEE China Summit & International Conference on Signal and Information Processing (ChinaSIP), pp. 95–99. IEEE (2014)
24. Sleep-related breathing disorders in adults: recommendations for syndrome definition and measurement techniques in clinical research. The Report of an American Academy of Sleep Medicine Task Force. Sleep **22**(5), 667–689 (1999)
25. Tan, M., Le, Q.: EfficientNet: rethinking model scaling for convolutional neural networks. In: International Conference on Machine Learning, pp. 6105–6114. PMLR (2019)
26. Whitelaw, W.: Characteristics of the snoring noise in patients with and without occlusive sleep apnea. Am. Rev. Respir. Dis. **147**, 635–644 (1993)
27. Wimms, A., Woehrle, H., Ketheeswaran, S., Ramanan, D., Armitstead, J.: Obstructive sleep apnea in women: specific issues and interventions. Biomed. Res. Int. **2016**, 1764837 (2016)
28. Woodson, B.T.: Expiratory pharyngeal airway obstruction during sleep: a multiple element model. Laryngoscope **113**(9), 1450–1459 (2003)

MVCAL: Multi View Clustering for Active Learning

Yi Fan, Biao Jiang, Di Chen, and Yu-Bin Yang(✉)

State Key Laboratory for Novel Software Technology, Nanjing University, Nanjing 210023, China
fanyiplus@smail.nju.edu.cn, yangyubin@nju.edu.cn

Abstract. Various active learning methods with ingenious sampling strategies have been proposed to solve the lack of labeled samples in supervised learning, but most are designed for specific tasks. In this paper, we propose a simple but task-agnostic active sampling method. We introduce 'multi-view clustering module' to extract multiple feature maps at different levels for unsupervised clustering. According to clustering distribution, we calculate consistency, representativeness and stability to guide sampling and training. Among them, consistency measures the similarity between clustering results of two views, representativeness reflects the distance between a sample and the corresponding cluster center, and stability reflects the model's feature representation and recognition ability for the same sample. Our method does not depend on the specific network, and can be constructed as a two-stage sampling module to supplement the existing sampling algorithm. Experiments results on image classification and object detection tasks show that our method can further enhance the effect of active learning on the basis of baseline methods.

Keywords: Computer Vision · Active Learning · Image Classification · Object Detection · Clustering

1 Introduction

In computer vision, complex networks and tasks require high-quality labeled data. The increasing labeling overhead has hampered access to this data to some extent. Active learning (AL) aims to use as few labeled samples, which may contain more information, as possible to obtain the same effect as fully supervised training. In a classic pool-based active learning scenario with a limited training set, numerous unlabeled samples form a candidate sample pool (called *unlabelpool*). The model continuously selects critical samples from the unlabelpool through a sampling strategy for annotation to expand the training set, so as to optimize the current model iteratively. Existing AL basically follows the above

This work is funded by the Natural Science Foundation of China under Grant No. 62176119.

B. Luo et al. (Eds.): ICONIP 2023, CCIS 1963, pp. 506–517, 2024.
https://doi.org/10.1007/978-981-99-8138-0_40

framework by designing different active sampling strategies. For example, the classical Least Confidence (LC), Margin, and Entropy algorithms in classification tasks [9,14,16] measure the prediction uncertainty of the current model to guide sampling. In object detection tasks, there are both the sampling method for classification branch only [24], which is transferred from the classification task, and the method using the stability predicted by regression box [11] as the sampling index from the perspective of regression branch.

However, the AL sampling strategies in the above methods depend on specific tasks. Although they can be adapted to other tasks after appropriate modifications, they often do not work as well on the new task. In recent years, researchers began to explore and design a task-agnostic AL method, hoping to provide a general sampling strategy. For example, [28] proposes a task-agnostic loss prediction module to predict sample loss directly to guide sampling. [22] proposes a method of active sampling by measuring data distribution called *Coreset*. Unfortunately, the sampling standard of the above methods still has some one-sidedness. [28] only considers the feedback of the model and ignores the characteristics of the data, while [22] only considers the feature distribution of the data on the macro level. As deep learning methods, they do not make full use of powerful feature representation ability of neural networks.

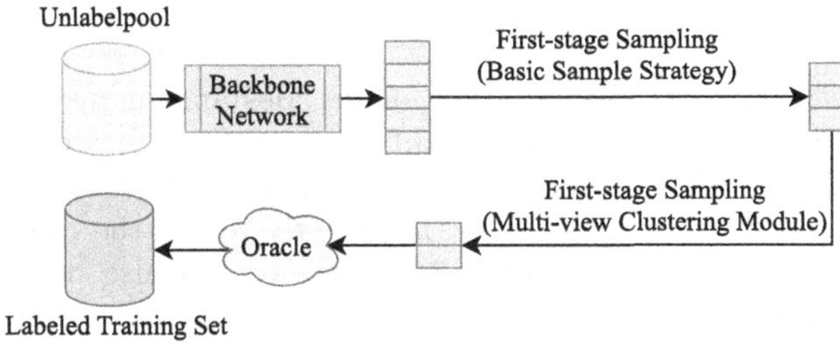

Fig. 1. A basic AL architecture with MVC module. For unlabeled samples, the backbone network is used to obtain features. Then the Basic Sample Strategy and Multi-view Clustering Module are used for First-stage Sampling and Second-stage Sampling respectively. After the two stages of sampling, the obtained samples are labeled as Labeled Training Set.

For most computer vision tasks, backbone networks are used to extract the features of input images. The process of analyzing these features can be independent of specific tasks. Inspired by this, we propose a plug-and-play Multi-view Clustering (MVC) module that can be conveniently embedded in the backbone network to improve the performance of deep learning models. A basic AL architecture with our MVC module is shown in Fig. 1. We believe it can be applied to any task that uses a deep network. In the first-stage sampling, the candidate sampling set is obtained through the existing basic sampling strategy. Then, the MVC module is adopted for the second-stage sampling to further screen out the key samples. Meanwhile, in the training process, we also conduct the MVC of batch data and then calculate the overall consistency of the batch data distribution as consistency loss to optimize the feature extraction capability of the model.

Since the MVC module does not depend on specific task types and can supplement the effect of existing active sampling methods, we believe it is a convenient and effective task-agnostic method. Besides, our method has strong scalability and can be combined with several existing AL methods. Experiments show that in image classification and object detection tasks, taking three existing active sampling methods as the baseline, the model effect of adding the MVC module exceeds the baseline methods.

Our contributions can be summarized as follows:

- We propose a novel active sampling method with the MVC module, which is task-agnostic and can be directly embedded in any tasks with deep networks.
- We use MVC as a supplement to the existing active sampling method. The existing sampling method is used for first-stage sampling, followed by second-stage MVC sampling, which can further improve the effect of active learning.
- We evaluate the proposed method with two learning tasks including image classification and object detection. Experimental results show that the proposed method significantly outperforms baseline methods.

2 Related Work

Active learning (AL) has been studied for decades and many excellent methods have emerged [1,4]. According to different application scenarios, AL can be divided into pool-based, stream-based and Query Synthesis active learning [7]. However, this division method can not clearly reflect the characteristics of different active sampling strategies. Therefore, we can also divide them into uncertainty-based, distribution-based, expected model change and metaheuristic active learning based on different sampling strategies.

2.1 Uncertainty-Based Methods

Uncertainty sampling is one of the most classical sampling strategies in AL. In multi-classification tasks, the uncertainty can be calculated by Least Confidence [14], Margin [9], and Entropy [16] algorithms. In addition, SVMs [26] can also define the distance from the decision boundary as uncertainty. Recently, uncertainty-based methods have been applied to many tasks such as video moment retrieval [8] and image segmentation [10].

2.2 Distribution-Based Methods

Uncertainty-based methods measure the information of samples from the perspective of models, while distribution-based methods mine representative samples in the overall data distribution of unlabeled sample pool. The typical approach is to conduct unsupervised clustering of data [17], and then calculate representative and diversity scores according to the distance between samples, so as to sample diversity but overall representative samples. [6] calculate the distance between the sample and its nearest neighbors, and then samples that can better represent the data features of the current fixed region will be sampled. Coreset [5] further defines active learning as the generation process of candidate sample set, that is, by sampling key samples to form a subset that can represent the characteristics of the whole unlabeled pool, training on this subset can obtain the same task effect.

2.3 Expect Model Change Methods

Deep learning models are usually optimized by gradient descent and minimize prediction losses during training, which inspires researchers to design sampling algorithms from the perspective of model adjustment. [20] predict the gradient descent degree in the training process and samples with larger values will be sampled. LLAL [28] predicts the task loss in the training process, which will be used to guide active sampling. Since expect model change methods can effectively save the consumption of model training, it has been applied in diverse fields such as [13,23,25].

2.4 Metaheuristic Methods

Metaheuristic methods have gained significant attention in training various neural networks due to their ability to optimize complex problems by exploring the problem space efficiently. [19] proposes the distributed wound treatment optimization method for training CNN models. [12] proposes the neuroevolutionary approach to control complex multicoordinate interrelated plants. [29] introduces the concept of simulated annulment in convolutional neural networks, and uses metaheuristics to remove unnecessary connections in the network, simplifying the model and improving its efficiency. [3] proposes a novel convolutional neural network model based on the beetle antennae search optimization algorithm for computerized tomography diagnosis. These studies offer promising solutions for enhancing the performance and efficiency of neural network models in a variety of domains.

The above methods define the key samples from different perspectives, and then derive a variety of sampling strategies. However, single sampling strategy can not avoid the problem of one-sided sampling. In addition, while task-agnostic methods already exist, they all fail to effectively utilize the powerful feature representation ability of neural networks, and room for improvement still exists.

3 Method

In neural networks, the feature map (FM) is a universal feature representation layer available by the combination of convolutions. Different combinations generate various FMs reflecting the characteristics of samples at different views. Based on this, we propose multi-view clustering active learning (MVCAL). The overall framework is shown in Fig. 2. The core of MVCAL, the MVC module, performs clustering by extracting multiple FMs corresponding to multiple views. Based on the clustering results, the representativeness and stability of samples will be calculated as the sampling strategy. Meanwhile, consistency will be calculated as part of the loss function to improve the feature extraction ability of the model in the training process.

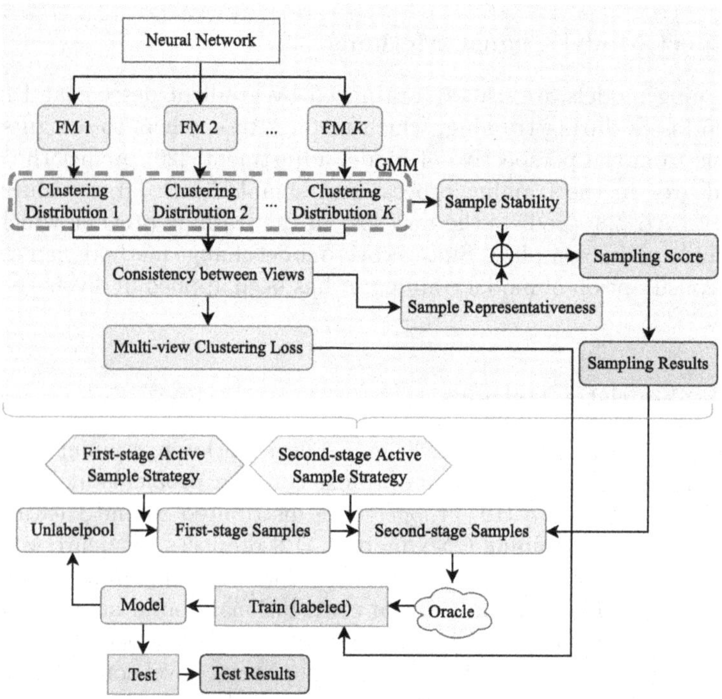

Fig. 2. Our proposed MVCAL. On the bottom is the two-stage sampling process. On the top is the concrete structure of the MVC module: Firstly, FM of different levels (corresponding to multiple views) are extracted from the backbone network as the input of unsupervised clustering. Then, sample stability and sample representativeness are calculated respectively according to the clustering results of multiple views, which are integrated as sampling scores.

The MVC module is task-agnostic and can be easily embedded into existing networks. Therefore, we expand the traditional AL into a two-stage AL based on MVC. By combining MVC with other different sampling strategies, the evaluation index of AL can be more comprehensive, and critical samples can be sampled simultaneously from the perspective of model expectation and data distribution.

3.1 FM Clustering in Different Views

In the MVC module, we first extract multiple FMs for each sample. Then we use one of the most classic clustering models, Gaussian mixture model (GMM) [27], to estimate the distribution and conduct clustering in each view. In GMM, the distribution of samples is expressed by

$$p\left(\boldsymbol{x}|\boldsymbol{\theta}\right)=\sum_{k=1}^{K}\alpha_k\phi\left(\boldsymbol{x}|\boldsymbol{\mu}_k,\boldsymbol{\sigma}_k\right), \tag{1}$$

where $\boldsymbol{x}$ denotes the input sample, K denotes the number of Gaussian model, $\phi\left(\boldsymbol{x}|\boldsymbol{\mu}_k,\boldsymbol{\sigma}_k\right)$ denotes the k-th Gaussian distribution with mean $\boldsymbol{\mu}_k$ and variance $\boldsymbol{\sigma}_k$, α_k denotes the weight coefficient, i.e., the probability that the observation sample belongs to the k-th Gaussian model. So for a total of U views, we obtain U clustering distributions, $p_1\left(\boldsymbol{x}|\boldsymbol{\theta}_1\right)$, $p_2\left(\boldsymbol{x}|\boldsymbol{\theta}_2\right)$, ..., $p_U\left(\boldsymbol{x}|\boldsymbol{\theta}_U\right)$. According to these clustering distributions, we can divide each sample into different classes in each view.

3.2 Consistency Between Views

A well-trained model should be able to extract common features at different levels, so as to make the clustering distribution of each view as consistent as possible. Therefore, the consistency between two views is used to measure the similarity between their clustering results. In this paper, we choose a simple but effective algorithm, Rand statistic [21], to calculate the consistency.

Denote the clustering label of a sample $\boldsymbol{x}_i$ in view V_m as $l_{V_m}\left(\boldsymbol{x}_i\right)$. Then for all the sample pairs in two different views, V_m and V_n, we get $s\left(s-1\right)/2$ sample pairs (s denotes sample size), $\left(\boldsymbol{x}_i,\boldsymbol{x}_j\right)$ $(i\neq j)$. In these sample pairs, we use s_p to denote those who satisfying both $l_{V_m}\left(\boldsymbol{x}_i\right)=l_{V_m}\left(\boldsymbol{x}_j\right)$ and $l_{V_n}\left(\boldsymbol{x}_i\right)=l_{V_n}\left(\boldsymbol{x}_j\right)$, or satisfying both $l_{V_m}\left(\boldsymbol{x}_i\right)\neq l_{V_m}\left(\boldsymbol{x}_j\right)$ and $l_{V_n}\left(\boldsymbol{x}_i\right)\neq l_{V_n}\left(\boldsymbol{x}_j\right)$, and s_n to denote other sample pairs. Then the consistency between V_m and V_n can be calculated as

$$R\left(V_m,V_n\right)=\left\|s_p\right\|/\left(s_p+s_n\right), \tag{2}$$

where $\|\cdot\|$ denotes the number of element in a set.

3.3 Training Strategy

In the training process, the parameters of networks are optimized to perform the specific task better and extract FM better simultaneously. So the loss function is mainly composed of two parts, task loss (TL), $\mathcal{L}_{\text{task}}$, and multi-view clustering loss (MVCL), $\mathcal{L}_{\text{MVC}}$. TL is the loss of a specific task, such as the cross entropy of classification [18]. MVCL is related to consistency, and can be calculated as

$$\mathcal{L}_{\text{MVC}}=\sum_{m=1}^{U}\sum_{n=1}^{U}(1-R(V_m,V_n)). \tag{3}$$

Finally, the total loss is calculated as

$$\mathcal{L} = \mathcal{L}_{\text{task}} + \lambda \cdot \mathcal{L}_{\text{MVC}}, \tag{4}$$

where λ denotes the weight between two items.

3.4 Sampling Strategy

In the sampling process, we use representativeness and stability as indicators to measure the quality of a sample.

Representativeness is a commonly used sampling strategy in the field of AL. [17] proposed clustering as the data preprocessing process and active sampling through representativeness. After that, AGPR [27] also proposes a method of sampling by pixel comparison of the whole image. Different from the existing methods, our method selects the distribution with the highest consistency among all the views to calculate the representativeness, where the consistency of view V_m can be calculated as

$$\text{Cons}\left(V_m\right) = \sum_{n=1, n \neq m}^{U} R\left(V_m, V_n\right). \tag{5}$$

Then the representativeness of sample $\boldsymbol{x}_i$ is just the probability density of the selected distribution, expressed as

$$\begin{aligned} &\text{Rep}\left(\boldsymbol{x}_i\right) = p_o\left(\boldsymbol{x}_i \middle| \boldsymbol{\theta}_o\right), \\ &\text{where } o = \arg\max_m p_m\left(\boldsymbol{x} \middle| \boldsymbol{\theta}_m\right). \end{aligned} \tag{6}$$

By this design, the representativeness can reflect the distance between the sample and the center of the fixed cluster. The larger its value is, the closer it is to the cluster center, i.e., it has better representativeness.

Unlike sample representativeness represents a class of samples with key common features, sample stability measures the stability of distribution in various views, which reflects the model's feature representation and recognition ability for the same sample. Assume the set of samples owning the same cluster label as $\boldsymbol{x}_i$ in V_m is $S_{V_m}\left(\boldsymbol{x}_i\right)$, then the stability of $\boldsymbol{x}_i$ in V_m and V_n is

$$\text{Stab}_{mn}\left(\boldsymbol{x}_i\right) = \frac{S_{V_m}\left(\boldsymbol{x}_i\right) \cap S_{V_n}\left(\boldsymbol{x}_i\right)}{S_{V_m}\left(\boldsymbol{x}_i\right) \cup S_{V_n}\left(\boldsymbol{x}_i\right)}. \tag{7}$$

Finally, the stability of sample $\boldsymbol{x}_i$ can be defined as

$$\text{Stab}\left(\boldsymbol{x}_i\right) = \sum_{m=1}^{U} \sum_{n=m+1}^{U} \left(\text{Stab}_{mn}\left(\boldsymbol{x}_i\right)\right). \tag{8}$$

Now we can calculate the score of sample $\boldsymbol{x}_i$ as

$$S\left(\boldsymbol{x}_i\right) = \text{Rep}\left(\boldsymbol{x}_i\right) + \text{Stab}\left(\boldsymbol{x}_i\right) \tag{9}$$

to decide which samples should be sampled.

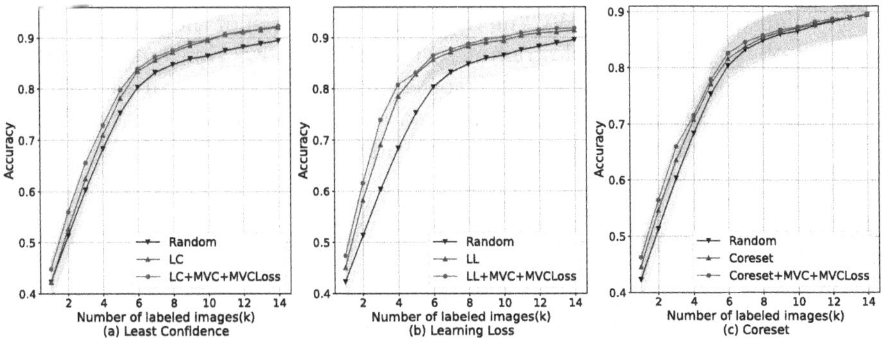

Fig. 3. Results for image classification on CIFAR-10. LC, LL and Coreset mean the one-stage AL using the least confidence, learning loss and coreset strategy, respectively. +MVC means adding the MVC as the second-stage sampling. +MVCLoss means adding $\mathcal{L}_{\mathrm{MVCL}}$ in the training process. We took the average of the three experiments as the final result.

4 Experiments

4.1 Image Classification

For image classification, we use Resnet-18 as the backbone network and CIFAR-10 as dataset. Due to the numerous samples in the unlabelpool at the beginning, it is expensive to use all unlabeled samples for prediction. Therefore, we follow the practice in [2]. In each round of AL, we first select 10 000 images as candidate sets in a random way. Then in the first stage of MVCAL, 2000 images are sampled out of 10 000, and in the second stage, 1000 images are further sampled from 2000.

Experimental Setup. The number of clustering centers is specified as 10. In Resnet-18, FM of the last four convolution layers are taken as views, and their sizes are $64 \times 32 \times 32$, $128 \times 16 \times 16$, $256 \times 8 \times 8$, $512 \times 4 \times 4$ respectively. The learning rate is set to 10^{-3}, and we train 200 epochs each iteration. We use Adam optimizer with $\alpha_1 = 0.9$ and $\alpha_2 = 0.99$.

We use LC [14], learning loss (LL) [11] and Coreset [22] as baseline methods respectively. The results are shown in Fig. 3. Results show that all the methods have better results than the random baseline. After adding MVC and MVCLoss, further improvements are achieved. For the LL-based method, the improvement of MVC and MVCLoss is the most obvious. This is in line with expectations because LL does not evaluate the distribution of the data, and the sampling index is single. For LC-based, the results are similar. For Coreset-based, the improvement of MVC and MVCLoss is not apparent. This may be because the coreset method itself is distribution-based, and the MVC model also measures the distribution characteristics. Nevertheless, our method can still bring improvement, which shows that our method is better than coreset in mining sample distribution information.

The improvement of '+MVC' methods in the first half of the training cycle is the most obvious in the whole training cycle, which shows that our method can effectively accelerate the convergence speed of the model, and can also obtain a weak final effect improvement on LL-based. These demonstrate after adopting our method, we can obtain better classification results.

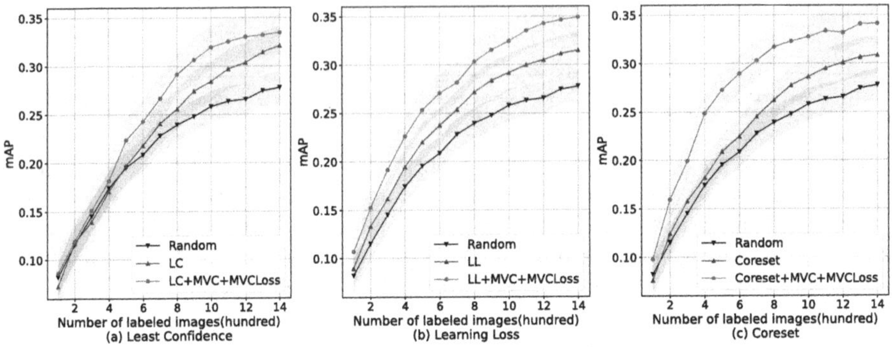

Fig. 4. Results for object detection on Pascal VOC07. Similar to Fig. 3.

4.2 Object Detection

We conduct experiments on object detection to verify the excellent task-agnostic of our method. We use SSD [15] as the backbone network and Pascal VOC2007 as dataset. Since Pascal VOC2007 does not contain too many samples, we no longer build candidate sets for sampling, but actively sample 200 images from the unlabelpool each round.

Experiment Setup. The number of clustering centers is set to 6 and 20 respectively. In SSD, FM for MVC is extracted from layer 4_3, 7, 8_2, 9_2, 10_2, 11_2 [15], same with [28]. We use Adam optimizer with $\alpha_1 = 0.9$ and $\alpha_2 = 0.99$. Each round of AL trains 6 epochs, of which the learning rate of the first 4 epochs is set to 10^{-3} and that of the last 2 epochs is set to 5^{-4}.

We also use LC [14], LL [11] and Coreset [22] as baseline methods respectively. The results are shown in Fig. 4. It can be seen that compared with baseline methods(one-stage), our method shows a significant performance improvement. This indicates that for complex visual tasks such as object detection, the existing one-stage sampling method is ineffective in assessing images' sampling information with multiple candidate instances. Our method measures the stability of clustering results under multiple views by adding the MVC module. The more instances there are, the more significant the impact on the consistency of clustering, so it can achieve a significant improvement.

In addition, our method shows almost the highest performance for all AL cycles. In the last cycle, our method achieves mAPs of 0.3346 (LC-based), 0.3489

(LL-based) and 0.3431 (Coreset-based). The results are 3.34%, 10.8% and 11.7% higher than the LC-based, LL-based and Coreset-based methods respectively.

4.3 Ablation Study

We conduct ablation experiments on CIFAR-10 by removing parts of our method. Results are shown in Fig. 5. We can see that the effect of single-view clustering is even worse than that of random strategy, which indicates that in simple tasks, single-view clustering pays too much attention to representative samples and ignores the impact of others. In contrast, inherent random search characteristics can better avoid overfitting in a random strategy. The MVC sampling, which measures sample stability and cluster consistency simultaneously, can improve the results to a certain extent compared with random. After adding $\mathcal{L}_{\mathrm{MVCL}}$, the results can be further improved, and even better than the existing typical task-agnostic method LL. This fully proves the effectiveness of our proposed MVC module.

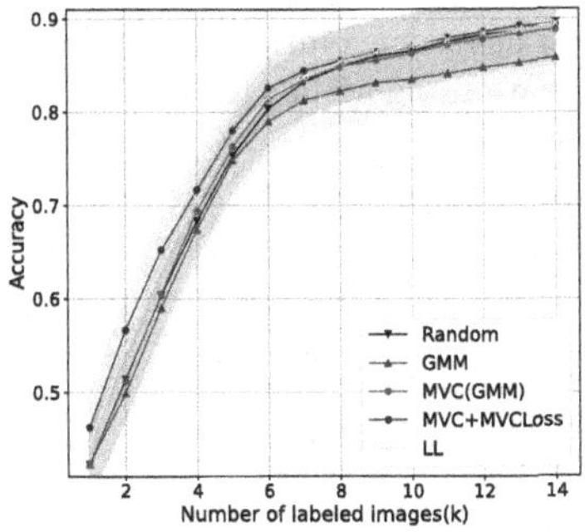

Fig. 5. Ablation study. 'Random' means random sampling. 'GMM' means sampling only according to representativeness. 'MVC' means sampling according to representativeness and stability. Both GMM and MVC do not contain $\mathcal{L}_{\mathrm{MVCL}}$.

5 Conclusion and Further Work

In this paper, we propose a task-agnostic active sampling module, MVC, and further embed it into the existing AL methods to construct a two-stage AL framework. The MVC module plays a critical role in both training and sampling. In the training process, it is used to calculate the overall clustering consistency of batch data and optimize the parameters of networks. In the sampling process, it calculates the stability and representativeness of samples to make up for the deficiency in the one-stage sampling. Extensive experiments on image classification and object detection tasks show that our method outperforms three traditional

ALs. This proves that our method is suitable for different tasks and different baseline AL methods.

In the future, we will verify our method in more tasks such as natural language processing and speech recognition. Moreover, we acknowledge the high computational cost associated with the clustering methods used in our method. To address this limitation, we will explore and develop more efficient clustering methods that can maintain or improve the performance while reducing computational overhead. By optimizing the clustering process, we aim to enhance the scalability and practicality of our method, making it more accessible and feasible for real-world AL scenarios.

References

1. Bachman, P., Sordoni, A., Trischler, A.: Learning algorithms for active learning. In: International Conference on Machine Learning, pp. 301–310. PMLR (2017)
2. Beluch, W.H., Genewein, T., Nürnberger, A., Köhler, J.M.: The power of ensembles for active learning in image classification. In: Proceedings of the IEEE Conference on Computer Vision and Pattern Recognition, pp. 9368–9377 (2018)
3. Chen, D., Li, X., Li, S.: A novel convolutional neural network model based on beetle antennae search optimization algorithm for computerized tomography diagnosis. IEEE Trans. Neural Netw. Learn. Syst. **34**, 1418–1429 (2021)
4. Gal, Y., Islam, R., Ghahramani, Z.: Deep Bayesian active learning with image data. In: International Conference on Machine Learning, pp. 1183–1192. PMLR (2017)
5. Guo, C., Zhao, B., Bai, Y.: DeepCore: a comprehensive library for coreset selection in deep learning. In: Strauss, C., Cuzzocrea, A., Kotsis, G., Tjoa, A.M., Khalil, I. (eds.) Database and Expert Systems Applications, DEXA 2022, Part I. Lecture Notes in Computer Science, vol. 13426, pp. 181–195. Springer, Cham (2022). https://doi.org/10.1007/978-3-031-12423-5_14
6. Hasan, M., Roy-Chowdhury, A.K.: Context aware active learning of activity recognition models. In: Proceedings of the IEEE International Conference on Computer Vision, pp. 4543–4551 (2015)
7. He, T., Jin, X., Ding, G., Yi, L., Yan, C.: Towards better uncertainty sampling: active learning with multiple views for deep convolutional neural network. In: 2019 IEEE International Conference on Multimedia and Expo (ICME), pp. 1360–1365. IEEE (2019)
8. Ji, W., et al.: Are binary annotations sufficient? Video moment retrieval via hierarchical uncertainty-based active learning. In: Proceedings of the IEEE/CVF Conference on Computer Vision and Pattern Recognition, pp. 23013–23022 (2023)
9. Joshi, A.J., Porikli, F., Papanikolopoulos, N.: Multi-class active learning for image classification. In: 2009 IEEE Conference on Computer Vision and Pattern Recognition, pp. 2372–2379. IEEE (2009)
10. Kang, C.J., Peter, W.C.H., Siang, T.P., Jian, T.T., Zhaofeng, L., Yu-Hsing, W.: An active learning framework featured Monte Carlo dropout strategy for deep learning-based semantic segmentation of concrete cracks from images. Struct. Health Monit. **22**, 3320–3337 (2023). https://doi.org/10.1177/14759217221150376
11. Kao, C.-C., Lee, T.-Y., Sen, P., Liu, M.-Y.: Localization-aware active learning for object detection. In: Jawahar, C.V., Li, H., Mori, G., Schindler, K. (eds.) ACCV

2018. LNCS, vol. 11366, pp. 506–522. Springer, Cham (2019). https://doi.org/10.1007/978-3-030-20876-9_32
12. Kondratenko, Y., Kozlov, O., Gerasin, O.: Neuroevolutionary approach to control of complex multicoordinate interrelated plants. Int. J. Comput. **18**(4), 502–514 (2019)
13. Kosugi, S., Yamasaki, T.: Crowd-powered photo enhancement featuring an active learning based local filter. IEEE Trans. Circ. Syst. Video Technol. **33**, 3145–3158 (2023)
14. Lewis, D.D.: A sequential algorithm for training text classifiers: corrigendum and additional data. ACM SIGIR Forum **29**, 13–19 (1995)
15. Liu, W., et al.: SSD: single shot multibox detector. In: Leibe, B., Matas, J., Sebe, N., Welling, M. (eds.) ECCV 2016. LNCS, vol. 9905, pp. 21–37. Springer, Cham (2016). https://doi.org/10.1007/978-3-319-46448-0_2
16. Luo, W., Schwing, A., Urtasun, R.: Latent structured active learning. In: Advances in Neural Information Processing Systems, vol. 26 (2013)
17. Nguyen, H.T., Smeulders, A.: Active learning using pre-clustering. In: Proceedings of the Twenty-First International Conference on Machine Learning, p. 79 (2004)
18. Nielsen, M.A.: Neural Networks and Deep Learning, vol. 25. Determination Press, San Francisco (2015)
19. Ponce, H., Moya-Albor, E., Brieva, J.: Towards the distributed wound treatment optimization method for training CNN models: analysis on the MNIST dataset. In: 2023 IEEE 15th International Symposium on Autonomous Decentralized System (ISADS), pp. 1–6. IEEE (2023)
20. Roy, N., McCallum, A.: Toward optimal active learning through Monte Carlo estimation of error reduction. In: ICML, Williamstown, vol. 2, pp. 441–448 (2001)
21. Schütze, H., Manning, C.D., Raghavan, P.: Introduction to Information Retrieval, vol. 39. Cambridge University Press, Cambridge (2008)
22. Sener, O., Savarese, S.: Active learning for convolutional neural networks: a core-set approach. In: International Conference on Learning Representations (2018)
23. Sivaraman, G., Jackson, N.E.: Coarse-grained density functional theory predictions via deep kernel learning. J. Chem. Theor. Comput. **18**(2), 1129–1141 (2022)
24. Sivaraman, S., Trivedi, M.M.: Active learning for on-road vehicle detection: a comparative study. Mach. Vis. Appl. **25**, 599–611 (2014)
25. Ureel, Y., et al.: Active learning-based exploration of the catalytic pyrolysis of plastic waste. Fuel **328**, 125340 (2022)
26. Vijayanarasimhan, S., Grauman, K.: Large-scale live active learning: training object detectors with crawled data and crowds. Int. J. Comput. Vis. **108**, 97–114 (2014)
27. Wang, H., Gao, X., Zhang, K., Li, J.: Single-image super-resolution using active-sampling gaussian process regression. IEEE Trans. Image Process. **25**(2), 935–948 (2015)
28. Yoo, D., Kweon, I.S.: Learning loss for active learning. In: Proceedings of the IEEE/CVF Conference on Computer Vision and Pattern Recognition, pp. 93–102 (2019)
29. Zhou, C.: Simulated annulling in convolutional neural network. In: 2022 2nd International Symposium on Artificial Intelligence and its Application on Media (ISAIAM), pp. 38–42. IEEE (2022)

Extraction of One Time Point Dynamic Group Features via Tucker Decomposition of Multi-subject FMRI Data: Application to Schizophrenia

Yue Han[1], Qiu-Hua Lin[1(✉)], Li-Dan Kuang[2], Ying-Guang Hao[1], Wei-Xing Li[1], Xiao-Feng Gong[1], and Vince D. Calhoun[3]

[1] School of Information and Communication Engineering, Dalian University of Technology, Dalian 116024, China
qhlin@dlut.edu.cn

[2] School of Computer and Communication Engineering, Changsha University of Science and Technology, Changsha 410114, China

[3] Tri-Institutional Center for Translational Research in Neuroimaging and Data Science (TReNDS), Georgia State University, Georgia Institute of Technology, Emory University, Atlanta, GA, USA

Abstract. Group temporal and spatial features of multi-subject fMRI data are essential for studying mental disorders, especially those exhibiting dynamic properties of brain function. Taking advantages of a low-rank Tucker model in effectively extracting temporally and spatially shared features of multi-subject fMRI data, we propose to extract dynamic group features via Tucker decomposition for identifying patients with schizophrenia (SZs) from healthy controls (HCs). We segment multi-subject fMRI data using sliding-window technique with different lengths and step size of one time point, and analyze amplitude of low frequency fluctuations and voxel features for shared time courses and shared spatial maps obtained by Tucker decomposition of segmented data. Results of two-sample *t*-tests show that HCs have higher amplitudes of low frequency fluctuations within 0.01–0.08 Hz than SZs within window length of 40 s–160 s, and significant HC-SZ activation differences exist in such as the inferior parietal lobule and left part of auditory within 40 s window, providing new evidence for analyzing schizophrenia.

Keywords: Multi-subject fMRI data · One time point dynamic estimates · Group features · Low-rank Tucker decomposition · Schizophrenia

1 Introduction

Temporal and spatial group feature extraction is important in analyses of multi-subject functional magnetic resonance imaging (fMRI) data, which can provide evidence for studying mental disorders such as schizophrenia [1–6]. Blind source separation such as independent component analysis (ICA) has been widely used to extract group features

B. Luo et al. (Eds.): ICONIP 2023, CCIS 1963, pp. 518–527, 2024.
https://doi.org/10.1007/978-981-99-8138-0_41

including time courses (TCs) and spatial maps (SMs) from multi-subject fMRI data. TCs of interest can be used to form functional network connectivity (FNC) based on correlations [1, 2] or to calculate amplitude of low frequency fluctuations [3] for finding group differences. SMs of interest, e.g., the default mode network (DMN), can be directly used to identify spatial variations of SZs using statistical methods (F-test [4] or *t*-test [5]).

Extracting dynamic group features from multi-subject fMRI data can be promising, since human brain is a highly dynamic system [6]. Kiviniemi et al. [7] conducted ICA of fMRI data using sliding window of 60 time points and sliding step of one time point, and calculated correlations between group averaged DMN and its template for each window to find dynamic spatial variances of healthy participants. Dynamic group features can also be used to identify brain abnormalities. Lu et al. [2] obtained individual TCs by group ICA (GICA), used a sliding window with 15 time points and step size of one time point to have segmented TCs. Each window of segmented TCs from nine resting-state networks (e.g., DMN and visual network) were used to construct dynamic FNC to explore the aberrant of neurocognitive performance in patients with acute mild traumatic brain injury. Ma et al. [8] performed independent vector analysis (IVA) on segmented multi-subject fMRI data with 50 time points and 50% overlap and built dynamic FNC of SMs based on mutual information to find differences between health individuals and schizophrenia patients.

Group components with or without individual variability were extracted and utilized via the above-mentioned methods. In practice, group shared temporal and spatial information, simultaneously extracted with individual temporal and spatial information, is essential to fully analyze brain function of multi-subject fMRI data. Among blind source separation methods, Tucker decomposition (TKD), as one of the tensor decomposition algorithms, is characterized by fully utilization of tensor structure information of multi-subject fMRI data and the non-diagonal core tensor that carry out subject-specific information [9–11], so we hypothesize that the difference of static group features extracted by TKD can be largely improved by dynamic analyses. We proposed a sparse and low-rank Tucker model (slcTKD) for decomposing three-way (voxel $\times$ time $\times$ subject) multi-subject fMRI data [11]. The slcTKD model makes full use of tensor structure of multi-subject fMRI data to obtain group shared TCs and SMs, and individual information in the core tensor without compressing the original fMRI data.

Considering that extracting dynamic group shared features via slcTKD could provide new evidence for studying mental disease such as schizophrenia, we propose to extract one time point dynamic group features via slcTKD for identifying differences between patients with schizophrenia (SZs) and healthy controls (HCs). We select DMN and auditory cortex (AUD) as two components of interest since they have been verified to be biomarkers of schizophrenia [4, 5, 8]. First, we segment multi-subject fMRI data of HCs and SZs using sliding window with different lengths (20–140 time points) in steps of one time point. Then, we obtain shared TCs and shared SMs via slcTKD from segmented fMRI data. Next, we extract frequency features within multiple bands from shared TCs for all time windows based on amplitude of low-frequency fluctuations (i.e., averaged square root of the power spectrum within typical band: 0.01–0.08 Hz, slow-5 band: 0.01–0.027 Hz, and slow-4 band: 0.027–0.073 Hz for each shared TCs), and calculate dynamic voxel features by performing dynamic voxel-level analysis on shared SMs. Finally, two-sample t-tests are conducted on the frequency features and voxel features between HCs and SZs to find significant differences with different window lengths. Since dynamic analysis on (voxel × time × subject) fMRI data via TKD has not been explored to our best knowledge, our findings within different window lengths can be promising in finding new evidences for identifying SZs.

To summarize, contributions of this study are three-fold:

(1) We propose to extract one time point dynamic group shared features via slcTKD for three-way (voxel × time × subject) fMRI data, and extract frequency and voxel features for finding significant temporal and spatial differences between HCs and SZs.
(2) We examine HC-SZ differences within a wide range of window lengths (20–140 time points) by using the strategy of one time point dynamic estimates.
(3) New temporal and spatial evidence for identifying SZs from HCs are found, e.g., HCs have higher amplitudes of low frequency fluctuations within 0.01–0.08 Hz for window lengths of 40 s–160 s; exhibit higher activations in left part of auditory (AL) but lower activations in right inferior parietal lobule (RIPL) within a window length of 40 s, compared with SZs.

2 Methods

2.1 The slcTKD Model

For a multi-subject fMRI data $\underline{\mathbf{X}} \in \mathbb{R}^{V \times T \times K}$, where V, T, K denote the number of in-brain voxels, time points, and subjects, the Tucker-2 model is built as follows:

$$\underline{\mathbf{X}} = \underline{\mathbf{G}} \times_1 \mathbf{S} \times_2 \mathbf{B} + \underline{\mathbf{E}} \tag{1}$$

where $\times_n$ denotes the mode-n product; $\mathbf{S} = [\mathbf{s}_1, \mathbf{s}_2, \ldots \mathbf{s}_N] \in \mathbb{R}^{V \times N}$ and $\mathbf{B} = [\mathbf{b}_1, \mathbf{b}_2, \ldots \mathbf{b}_N] \in \mathbb{R}^{T \times N}$ represent shared SM matrix and shared TC matrix; N is the model order; $\underline{\mathbf{G}} \in \mathbb{R}^{N \times N \times K}$ and $\underline{\mathbf{E}} \in \mathbb{R}^{V \times T \times K}$ are the core tensor and residual tensor. In slcTKD algorithm [11], the tensors and matrices are updated based on the following:

$$\min_{\mathbf{S},\mathbf{B},\underline{\mathbf{G}},\underline{\mathbf{E}}} \left\|\underline{\mathbf{X}} - \underline{\mathbf{G}} \times_1 \mathbf{S} \times_2 \mathbf{B} - \underline{\mathbf{E}}\right\|_{\mathbf{F}}^2 + \|\mathbf{S}\|_{\mathbf{F}}^2 + \|\mathbf{B}\|_{\mathbf{F}}^2 + \delta\|\mathbf{S}\|_p + \lambda\left\|\underline{\mathbf{G}}\right\|_1 + \gamma\left\|\underline{\mathbf{E}}\right\|_1 \tag{2}$$

where $\|\cdot\|_F$ denotes low-rank ℓ_F constraint, $\|\cdot\|_1$ and $\|\cdot\|_p$ are ℓ_1 and ℓ_p sparsity constraints ($0 < p \leq 1$), positive parameters δ, λ, and γ control the sparsity constraints effects. The model is solved by alternating direction method of multipliers and half quadratic splitting. More details of the updating of $\mathbf{B}$ and $\mathbf{S}$ can be referred to [11].

2.2 Extraction of Dynamic Group Shared Temporal and Spatial Components via Sliding Window and slcTKD

We denote a component of interest by a subscript of "*" (e.g., $\mathbf{b}_*$), and a component from the m th window by a superscript "m" (e.g., $\mathbf{b}^m$), where $m = 1, 2, ..., M$, and M is the total number of windows.

Given multi-subject fMRI data of HCs and SZs $\underline{\mathbf{X}}$, we divide T time points into M time windows $\left[\underline{\mathbf{X}}^1, \underline{\mathbf{X}}^2, \ldots, \underline{\mathbf{X}}^M\right]$, where $M = T - L + 1$, L is the window length, and $\underline{\mathbf{X}}^m \in \mathbb{R}^{V \times L \times K}$ is defined as $\underline{\mathbf{X}}^m = \underline{\mathbf{X}}(:, m : m + L - 1, :)$, $m = 1, 2, ..., M$. Then, for each $\underline{\mathbf{X}}^m$, we obtain $\mathbf{B}^m \in \mathbb{R}^{L \times N}$ and $\mathbf{S}^m \in \mathbb{R}^{V \times N}$ within each window using the slcTKD algorithm [11]. Next, we extract shared TCs and shared SMs of interest from $\mathbf{B}^m \in \mathbb{R}^{L \times N}$ and $\mathbf{S}^m \in \mathbb{R}^{V \times N}$ for analysis.

2.3 The Proposed Dynamic Group Features Extraction Method

Figure 1 shows extraction of dynamic temporal and spatial group features with DMN as an example. Group shared TCs of M windows are obtained by $\overline{\mathbf{B}} = \left[\mathbf{b}_*^1, \mathbf{b}_*^2, \ldots \mathbf{b}_*^M\right] \in \mathbb{R}^{L \times M}$, where $\mathbf{b}_*^m \in \mathbb{R}^L$ is the DMN component extracted from $\mathbf{B}^m$. Similarly, group shared SMs for DMN of M windows are defined by $\overline{\mathbf{S}} = \left[\mathbf{s}_*^1, \mathbf{s}_*^2, \ldots \mathbf{s}_*^M\right] \in \mathbb{R}^{V \times M}$, where $\mathbf{s}_*^m \in \mathbb{R}^V$ is the DMN component extracted from $\mathbf{S}^m$.

Dynamic Frequency Feature Extraction by Using Amplitudes of Low Frequency Fluctuations. For each $\mathbf{b}_*^m$, power spectrum is obtained by fast Fourier transform as follows:

$$\mathbf{p}^m(f) = \sum_l \mathbf{b}_*^m(l) e^{\frac{-j2\pi lf}{L}} \tag{3}$$

where $\mathbf{p}^m \in \mathbb{C}^L$ is the power spectrum of $\mathbf{b}_*^m$, $l = 1, 2, ..., L$, $j = \sqrt{1}$, $f = 1, 2, ..., L$ denotes the frequency time point. Then the amplitude of low-frequency fluctuation of $\mathbf{b}_*^m$ within a specific frequency band is obtained as follows:

$$\tilde{\mathbf{b}}_*^m = \sum_{\tau(f) \in \Omega} \sqrt{\frac{|\mathbf{p}^m(f)|^2}{L}} \tag{4}$$

where $\tau(f)$ is the frequency in point f, $\tilde{\mathbf{b}}_*^m \in \mathbb{R}^1$ denotes the amplitude of low-frequency fluctuation, $\Omega = 0.01$–0.08 Hz (typical band) [12], 0.01–0.027 Hz (slow-5 band), or 0.027–0.073 Hz (slow-4 band) [13]. The dynamic temporal group features for all the M windows are collected by $\tilde{\mathbf{b}} = \left[\tilde{\mathbf{b}}_*^1, \tilde{\mathbf{b}}_*^2, \ldots, \tilde{\mathbf{b}}_*^M\right] \in \mathbb{R}^M$ for HCs and SZs, respectively, defined as $\tilde{\mathbf{b}}^{HC} \in \mathbb{R}^M$ and $\tilde{\mathbf{b}}^{SZ} \in \mathbb{R}^M$. Then two-sample t-tests are conducted to determine significant HC-SZ differences as follows:

$$\boldsymbol{t_b} = ttest\left(\tilde{\mathbf{b}}^{HC}, \tilde{\mathbf{b}}^{SZ}\right) \tag{5}$$

where $\boldsymbol{t_b} \in \mathbb{R}^1$ is the t-value, $ttest(\cdot)$ denotes two-sample t-test operation between HCs and SZs.

Dynamic Voxel Feature Extraction. The dynamic spatial group features are obtained by the vectorizing the same voxel values for M time windows, defined as $\overline{\mathbf{s}}_{(v)} = [\mathbf{s}_*^1(v), \mathbf{s}_*^2(v), \ldots, \mathbf{s}_*^M(v)] \in \mathbb{R}^M$, where $v = 1, 2, ..., V$. The voxel features extracted from HCs and SZs are defined as $\overline{\mathbf{s}}_{(v)}^{HC} \in \mathbb{R}^M$ and $\overline{\mathbf{s}}_{(v)}^{SZ} \in \mathbb{R}^M$, respectively, then two-sample t-tests are conducted between each voxel vectors to generate the HC-SZ difference t-map $\boldsymbol{t_s} \in \mathbb{R}^V$:

$$\boldsymbol{t_{s(v)}} = \Phi(v) \cdot ttest\left(\overline{\mathbf{s}}_{(v)}^{HC}, \overline{\mathbf{s}}_{(v)}^{SZ}\right) \tag{6}$$

where $\mathbf{\Phi} \in \mathbb{R}^V$ is a binary mask determined by the significant level, defined as:

$$\Phi(v) = \begin{cases} 1, & if \left| ttest\left(\bar{\mathbf{s}}_{(v)}^{HC}, \bar{\mathbf{s}}_{(v)}^{SZ}\right)\right| > t_{th} \\ 0, & otherwise \end{cases} \tag{7}$$

where t_{th} is the threshold of t-value for two-sample t-test.

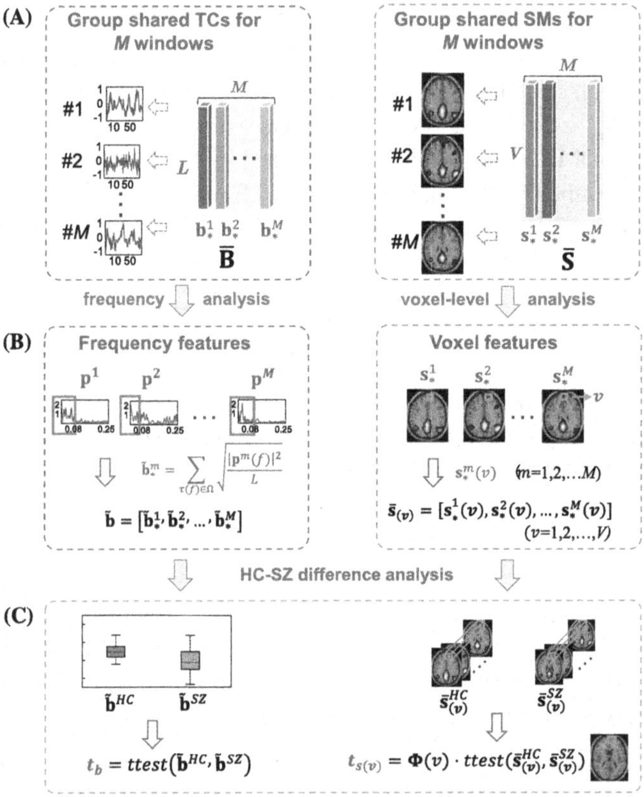

Fig. 1. Dynamic temporal and spatial group feature extraction of the proposed method. (A) Group shared TCs and SMs of M windows. (B) Dynamic frequency and voxel feature extraction. (C) Two-sample t-tests on dynamic features between HCs and SZs.

3 Experimental Methods

3.1 Resting-State fMRI Data

The resting-state fMRI data were collected from 10 HCs and 10 SZs with written subject consent overseen by the University of New Mexico Institutional Review Board. During the scan, all participants were instructed to rest quietly in the scanner, keeping their eyes open without sleeping and not to think of anything in particular. FMRI scans were acquired using a 3.0 T Siemens Allegra scanner, equipped with 40 mT/m gradients and a standard quadrature head coil. The functional scan was acquired using gradient-echo echo-planar imaging with the following parameters: TR = 2 s, TE = 29 ms, field of view = 24 cm, acquisition matrix = 64 × 64, flip angle = 75°, slice thickness = 3.5 mm, slice gap = 1 mm. Data preprocessing was performed using the SPM software package (http://www.fil.ion.ucl.ac.uk/spm). Five scans were excluded due to steady state magnetization effects and 146 resting state scans were used for analysis. After motion correction, the functional images were normalized into Montreal Neurological Institute standard space. Following spatial normalization, the data were slightly sub-sampled to $3 \times 3 \times 3$ mm^3, resulting in $53 \times 63 \times 46$ voxels. Data were then spatially smoothed with an $8 \times 8 \times 8$ mm^3 full-width half-maximum (FWHM) Gaussian kernel. After removing the voxels out of the brain and flattening the volume image data of all time points for each subject, we construct the three-way fMRI data of size $62336 \times 146 \times 10$ for HCs and $62336 \times 146 \times 10$ for SZs, respectively.

3.2 Experimental Methods

In order to explore HC-SZ differences with a wide range of window lengths, we change L from 20 to 140 at an interval of 10. The model order is selected by $N = min(L, 40)$, since the components of interest cannot be well extracted with smaller model orders. Other parameters of slcTKD given in Eq. (2) are selected to be the same with those used in [11]. The references for DMN and AUD provided by Smith et al. [14] are used to select components of interest. Before extracting frequency features, each group shared TCs components are detrended and normalized to the range of 0 to 1. The t-values obtained from two-sample t-tests ($df = 2M - 2, p < 0.05$, corrected by false discovery rate) are utilized to evaluate the performance, and thresholds are changed for different L, ranging from $|t| = 1.969$ ($L = 20, M = 127, df = 252$) to $|t| = 2.179$ ($L = 140, M = 7, df = 12$).

4 Results

4.1 Dynamic Frequency Features of Group Shared TCs

Figure 2A shows multiple bands of results of t-values with different window lengths. We see that positive t-values are consistently obtained for the typical and slow-4 bands, indicating that HCs have higher amplitude of low-frequency fluctuation than SZs, agreeing with previous static analyses on raw fMRI data [15–17]. With the increase of L, the t-values show decreasing trend within the typical and slow-5 bands, while the changes are not obvious in slow-4 band. Significant differences ($p < 0.05$) are obtained within typical band when $L = 20$–80 (40 s–160 s) for both components, which extend the range of previous utilized window lengths in dynamic analyses (commonly 30 s–60 s) [6]. Figure 2B shows boxplots for amplitude of low-frequency fluctuations within the typical band 0.01–0.08 Hz for DMN and AUD when $L = 20$, 50, and 80. We see that the amplitude of low-frequency fluctuations within the typical band show less obvious HC-SZ differences with the increase of L.

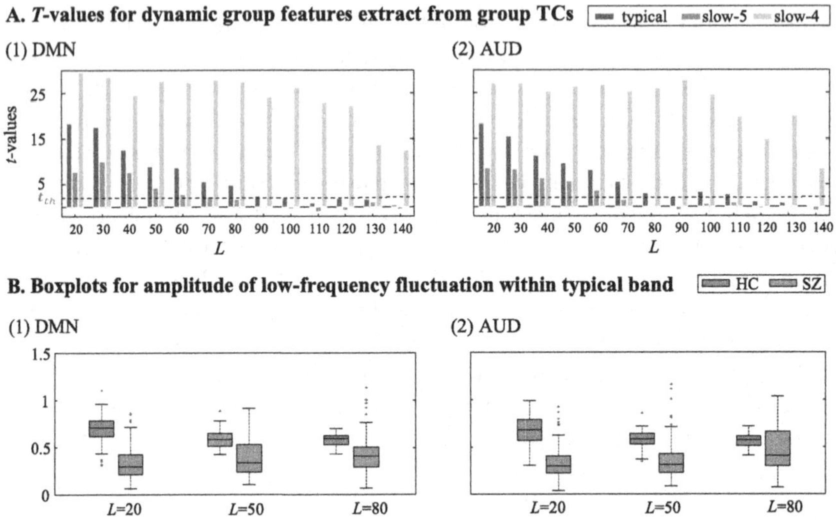

Fig. 2. Results of dynamic frequency features of group shared TCs. (A) Multiple bands of t-values of two sample t-test on dynamic temporal features between HCs and SZs. (B) Boxplots for amplitude of low-frequency fluctuation within the typical band. (1) DMN. (2) AUD. Thresholds of t-values t_{th} ($p < 0.05$) for each window length are marked by dotted lines (typical band: 0.01–0.08 Hz, slow-5 band: 0.01–0.027 Hz, slow-4 band: 0.027–0.073 Hz).

4.2 Dynamic Voxel Features of Group Shared SMs

Figure 3 shows voxel numbers inside the DMN and AUD reference masks with all window lengths and comparison of HC-SZ difference t-maps ($p < 0.05$) when $L = 20$, 50, and 80. Figure 3A1 also shows the voxel numbers inside anterior cingulate cortex (ACC) since it has been verified as a biomarker of SZs [4, 5]. HC-SZ differences exist in DMN when $L = 20$–130 (40 s–260 s), and in ACC when $L = 30$–100 (60 s–200 s). In addition, we also notice that SZs exhibit significantly larger activations in RIPL (see Fig. 3B), especially when $L = 20$ (40 s). Changes of IPL in SZs are also important [18], but previous findings are not exactly consistent, e.g., [19] found increased functional connectivity between RIPL with the right lingual gyrus and inferior occipital gyrus in SZs, while [20] showed decreased regional homogeneity in RIPL for SZs. This study provides new evidence for studying RIPL in SZs for dynamic analyses.

Significant HC-SZ differences are captured in AUD within $L = 20$–120 (40 s–240 s), especially when $L = 20$ in AL, as shown in Fig. 3C, indicating the abnormalities in AL of SZs [4] and verifying that AL is specialized for rapid temporal processing [21]. In addition, we newly find that significant differences exist in right part of auditory (AR) within larger window length (see Fig. 3C3–C4), which may be caused by the increasing activity of right temporal lobe to environmental sounds [22] and auditory hallucinations of SZs. These results indicate that different subcomponents have different sensitivities for different window lengths, i.e., different dynamic characteristics, suggesting the potential of one time point dynamic estimates for detailed investigation of HCs and SZs.

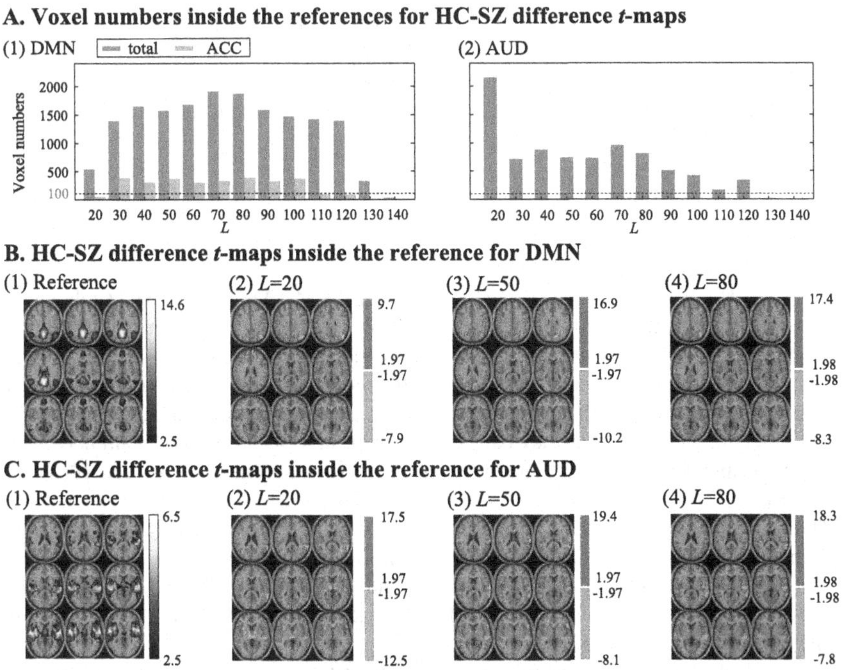

Fig. 3. Results of dynamic frequency features of group shared SMs. (A) Voxel numbers inside the references for HC-SZ difference t-maps. (1) DMN. (2) AUD. (B) HC-SZ difference t-maps inside the reference for DMN. (C) HC-SZ difference t-maps inside the reference for AUD. (1) Reference. (2) $L = 20$. (3) $L = 50$. (4) $L = 80$. Threshold of 100 voxels is marked by dotted lines.

5 Conclusion

In this study, one time point dynamic group shared frequency and voxel features are extracted via slcTKD for finding significant temporal and spatial differences between HCs and SZs, within a wide range of sliding window lengths (20–140 time points). Results show that HCs have significant higher amplitudes of low frequency fluctuations within 0.01–0.08 Hz than SZs for window lengths of 40 s–160 s, and significant HC-SZ activation differences are captured in subcomponents of DMN and AUD within specific lengths of window (e.g., ACC: 60 s–200 s, RIPL: 40 s, AL: 40 s). Compared to the existing group analysis algorithms such as GICA in [6] and IVA in [8], the proposed algorithm has two advantages: (1) TKD utilizes high dimensional structure of multi-subject fMRI data while GICA and IVA do not; (2) TKD can simultaneously extracts group shared spatial and temporal features while the existing algorithm cannot, thus, the group features extracted by TKD are more distinctive than those extracted by the existing algorithms. As a result, the dynamic frequency and voxel-level analyses proposed in this paper are better and more complete. Since slcTKD model can simultaneously provide subject-specific temporal and spatial features [11], we will also further extract dynamic subject-specific features for studying SZs together with shared features. Current results appear to show strong HC-SZ differences, we will evaluate on more subjects later.

Acknowledgement. This work was supported in part by the National Natural Science Foundation of China under Grants 61871067 and 62071082, the NSF under Grant 2112455, the NIH Grant R01MH123610, the Fundamental Research Funds for the Central Universities, China, under Grants DUT20ZD220 and DUT20LAB120, and the Supercomputing Center of Dalian University of Technology.

References

1. Sakoglu, U., Pearlson, G.D., Kiehl, K.A., Wang, Y.M., Michael, A.M., Calhoun, V.D.: A method for evaluating dynamic functional network connectivity and task-modulation: application to schizophrenia. Magn. Reson. Mater. Phys. Biol. Med. **23**(5–6), 351–366 (2010)
2. Lu, L., et al.: Aberrant static and dynamic functional network connectivity in acute mild traumatic brain injury with cognitive impairment. Clin. Neuroradiol. **32**(1), 205–214 (2022)
3. Qi, S., et al.: Multiple frequency bands analysis of large scale intrinsic brain networks and its application in schizotypal personality disorder. Front. Comput. Neurosci. **12**(64), 1–16 (2018)
4. Qiu, Y.: Spatial source phase: a new feature for identifying spatial differences based on complex-valued resting-state fMRI data. Human Brain Mapp. **40**(9), 2662–2676 (2019)
5. Kuang, L.D., Lin, Q.H., Gong, X.F., Cong, F., Sui, J., Calhoun, V.D.: Model order effects on ICA of resting-state complex-valued fMRI data: application to schizophrenia. J. Neurosci. Methods **304**, 24–38 (2018)
6. Fu, Z., et al.: Characterizing dynamic amplitude of low-frequency fluctuation and its relationship with dynamic functional connectivity: an application to schizophrenia. Neuroimage **180**, 619–631 (2018)
7. Kiviniemi, V., et al.: A sliding time-window ICA reveals spatial variability of the default mode network in time. Brain Connectivity **1**(4), 339–347 (2011)
8. Ma, S., Calhoun, V.D., Phlypo, R., Adalı, T.: Dynamic changes of spatial functional network connectivity in healthy individuals and schizophrenia patients using independent vector analysis. Neuroimage **90**, 196–206 (2014)
9. Kolda, T.G., Bader, B.W.: Tensor decompositions and applications. SIAM Rev. **51**(3), 455–500 (2009)
10. Han, Y., Lin, Q.H., Kuang, L.D., Gong, X.F., Cong, F., Calhoun, V.D.: Tucker decomposition for extracting shared and individual spatial maps from multi-subject resting-state fMRI data. In: IEEE International Conference on Acoustics, Speech and Signal Processing (ICASSP), pp. 1110–1114, June 2021
11. Han, Y., et al.: Low-rank Tucker-2 model for multi-subject fMRI data decomposition with spatial sparsity constraint. IEEE Trans. Med. Imaging **41**(3), 667–679 (2022)
12. Zang, Y.F., et al.: Altered baseline brain activity in children with ADHD revealed by resting-state functional MRI. Brain Develop. **29**(2), 83–91 (2007)
13. Zuo, X.N., et al.: The oscillating brain: complex and reliable. Neuroimage **49**(2), 1432–1445 (2010)
14. Smith, S.M., et al.: Correspondence of the brain's functional architecture during activation and rest. Nat. Acad. Sci. United States Am. **106**(31), 13040–13045 (2009)
15. Fryer, S.L., Roach, B.J., Wiley, K., Loewy, R.L., Ford, J.M., Mathalon, D.H.: Reduced amplitude of low-frequency brain oscillations in the psychosis risk syndrome and early illness schizophrenia. Neuropsychopharmacology **41**(9), 2388–2398 (2016)
16. Wang, X., et al.: Frequency-specific alteration of functional connectivity density in antipsychotic-naive adolescents with early-onset schizophrenia. J. Psychiatr. Res. **95**, 68–75 (2017)

17. Chang, M., et al.: Spontaneous low-frequency fluctuations in the neural system for emotional perception in major psychiatric disorders: amplitude similarities and differences across frequency bands. J. Psychiatry Neurosci. **44**(2), 132–141 (2019)
18. Torrey, E.F.: Schizophrenia and the inferior parietal lobule. Schizophr. Res. **97**(1–3), 215–225 (2007)
19. Liu, X., et al.: Selective functional connectivity abnormality of the transition zone of the inferior parietal lobule in schizophrenia, NeuroImage Clin. **11**, 789–795 (2016)
20. Wang, S., et al.: Abnormal regional homogeneity as a potential imaging biomarker for adolescent-onset schizophrenia: a resting-state fMRI study and support vector machine analysis. Schizophr. Res. **192**, 179–184 (2018)
21. Zatorre, R.J., Belin, P.: Spectral and temporal processing in human auditory cortex. Cereb. Cortex **11**(10), 946–953 (2001)
22. Hugdahl, K., Bronnick, K., Kyllingsbaek, S., Law, I., Gade, A., Paulson, O.B.: Brain activation during dichotic presentations of consonant-vowel and musical instrument stimuli: a 15O-PET study. Neuropsychologia **37**(4), 431–440 (1999)

Modeling Both Collaborative and Temporal Information for Sequential Recommendation

Jinyue Dai[1,2], Jie Shao[1,2], Zhiyi Deng[2], Hongcai He[2], and Feiyu Chen[1,2,3](✉)

[1] Sichuan Artificial Intelligence Research Institute, Yibin 644000, China
[2] University of Electronic Science and Technology of China, Chengdu 611731, China
{jydai,zhiyideng,hehongcai}@std.uestc.edu.cn,
{shaojie,chenfeiyu}@uestc.edu.cn
[3] Intelligent Terminal Key Laboratory of Sichuan Province, Yibin 644000, China

Abstract. Sequential recommendation has drawn a lot of attention due to its good performance in recent years. The temporal order of user interactions cannot be ignored in sequential recommendation, and user preferences are constantly changing over time. The application of deep neural network in sequential recommendation has achieved many remarkable results, especially self-attention based methods. However, previous works mainly focused on item-item temporal information of the sequence while ignoring the latent collaborative relations in user-item interactions. Therefore, we propose a new method named Collaborative-Temporal modeling for Sequential Recommendation (CTSR) to learn both collaborative relations and temporal information. The proposed CTSR method introduces a graph convolutional network to learn the user-item collaborative relations while using self-attention to capture item-item temporal information. We apply focal loss to reduce the loss contribution of the easy samples and increase the contribution of the hard samples, to train the model more effectively. We extract a portion of item-item pairs that are most valuable, and then feed these pairs as augmented information into adjacency matrix of the graph neural network. More importantly, it is the first work to encode the relative positions of items into their embeddings in sequential recommendation. The experimental results show that CTSR surpasses previous works on three real-world datasets.

Keywords: Sequential recommendation · Graph convolutional networks · Collaborative-temporal modeling

1 Introduction

To reduce information overload on the web, recommendation systems are widely deployed to perform personalized information filtering. Sequential Recommendation (SR) has attracted a lot of attention from both the academic community

B. Luo et al. (Eds.): ICONIP 2023, CCIS 1963, pp. 528–539, 2024.
https://doi.org/10.1007/978-981-99-8138-0_42

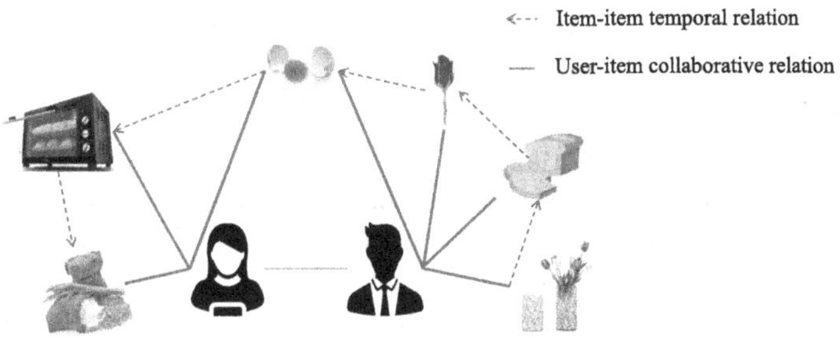

Fig. 1. Collaborative and temporal relations. The two users have the same preference item, i.e., eggs. The female user also interacts with flour and oven, and the male user may also be interested in flour and oven. Therefore, in the latent space, there is a collaborative relation between them.

and industry due to its great success and scalability in recent years [1,23,26]. The user-item interactions are sequential and constantly changing. Sequential recommendation considers these interactions as temporal-order sequences, which can effectively capture users' recent preferences, thus improving the recommendation performance [24,25]. For example, a user who purchased sports items a year ago may recently browse painting tools. Under the circumstances, sequential recommendation tends to capture his/her recent interactions and then recommend painting tools instead of sports items.

Various methods have been used in sequential recommendation, including Markov chains [4] and deep neural network based methods [6,11,21]. Among them, self-attention based models have been widely applied due to its good capability of capturing interrelations [8,11,12,18]. Yet, prior works [8,12] capture only the item-item temporal information, but ignore the user-item collaborative relations (also called collaborative signals) as shown in Fig. 1. Users are similar to each other through items, hence there are collaborative relations between users and items [22]. To address this issue, we propose a novel sequential recommendation model called Collaborative-Temporal modeling for Sequential Recommendation (CTSR), which consists of two important components, a Temporal Information Extraction (TIE) module and a Collaborative Relation Modeling (CRM) module. The first one, TIE, is applied to extract the item-item temporal information using a self-attention based structure. The vanilla self-attention lacks the position of items in the sequence, and existing methods based on absolute position encoding cannot model relative position relations between items, such as predecessor and adjacency relations. Therefore, we use a novel and effective method in order to capture the relative position relations. The second one, CRM, is employed to model the user-item collaborative relations in a non-Euclidean space using a graph neural network. Apart from that, we extract a portion of item-item pairs that are most valuable from noisy interaction sequences and use these pairs as augmented information for adjacency matrix of the graph convolutional network. Finally, to reduce the loss contribution of simple samples and increase the contribution of difficult samples, we explore the application of focal loss [13] in sequential recommendation.

To summarize, this work has the following three contributions:

- We present a novel sequential recommendation model CTSR, using a graph neural network to learn the user-item collaborative relations and self-attention with relative position encoding to learn the item-item temporal information.
- We extract a portion of most valuable item-item pairs from noisy interaction sequences and use these pairs as augmented information for adjacency matrix of the graph convolutional network.
- To reduce the loss contribution of the easy samples and increase the contribution of the hard samples, we explore the application of focal loss in sequential recommendation.

2 Related Work

Sequential recommendation is a task applying temporal information to make recommendation more personalized. Early works use Markov chains [4] to calculate the transfer probabilities in behavior sequences. However, Markov chains assume that the current item depends on the most recent item or items, so it only captures short-term dependency but ignores the long-term. The classic representative is Factorizing Personalized Markov Chains (FPMC) [16].

More recently, due to the development of deep learning techniques, researchers gradually turn to models based on deep neural network. The Convolutional Neural Network (CNN) based models first construct the embedding vector of historical behaviors into a matrix, then use this matrix as an 'image' in time and latent space, finally obtain a short-term representation of the user by convolution operations. The most classic one is the convolutional sequence embedding recommendation model Caser [19]. Recurrent Neural Network (RNN) based sequential recommendations [6] predict the next interaction by modeling the sequential dependencies of user-item historical interactions.

Unlike CNN-based or RNN-based sequential models, Transfomer [20] is purely based on an attention mechanism known as self-attention. It is highly effective in uncovering syntactic and semantic patterns between words in a sentence. Inspired by Transformer, Self-Attentive Sequential Recommendation (SASRec) [8] uses a similar architecture to the encoder part of it. SASRec achieves state-of-the-art performance in sequential recommendation.

However, previous works have ignored the latent collaborative relations in user-item interactions. Unlike them, the proposed CTSR method captures both user-item collaborative relations and item-item temporal information.

3 Methodology

3.1 Problem Definition

Based on the basic setting of sequential recommendation, we are given n users and each user engages with a subset of m items in temporal order. The goal of

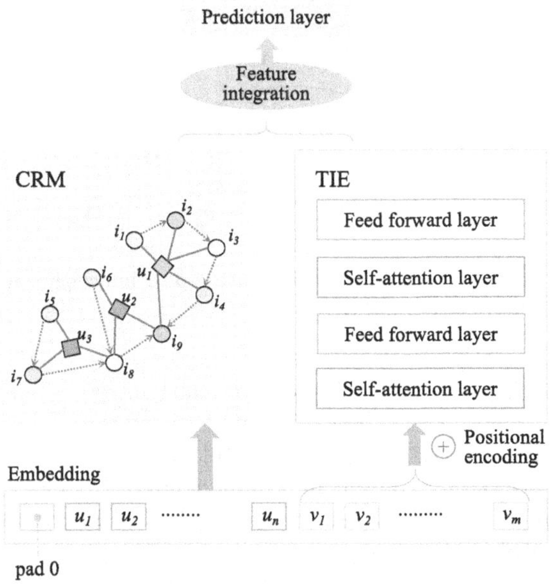

Fig. 2. An illustration of our proposed method.

sequential recommendation is to predict the next item which will be most likely to interact with. We assume that the item interaction sequence of n users is:

$$S_u = (v_1, v_2, ..., v_T), 1 \leq u \leq n, \tag{1}$$

where v is the interacted item. Sequence S_u of length T contains the last T items that user u has interacted with in the temporal order (from old to new). For different users, the sequence lengths can vary.

3.2 CTSR

Previous works mainly focused on item-item temporal information of the sequence while ignoring the latent collaborative relations in user-item interactions. Therefore, it leads to a deficiency of suboptimal embeddings when capturing valuable relations. In order to solve this issue, we obtain embeddings from both collaborative relations and temporal information. This leads to more effective embeddings for recommendation, since the embedding refinement step explicitly injects both collaborative relation and temporal information into embeddings.

To capture item-item temporal information, we use a self-attention based structure [8] to learn the context of user behavior. The vanilla self-attention lacks the position of items in the sequence, and existing methods based on absolute position encoding cannot model relative position relations between items, such as predecessor and adjacency relations. Therefore, we use a novel and effective method in order to capture the relative position relations [17]. At the same

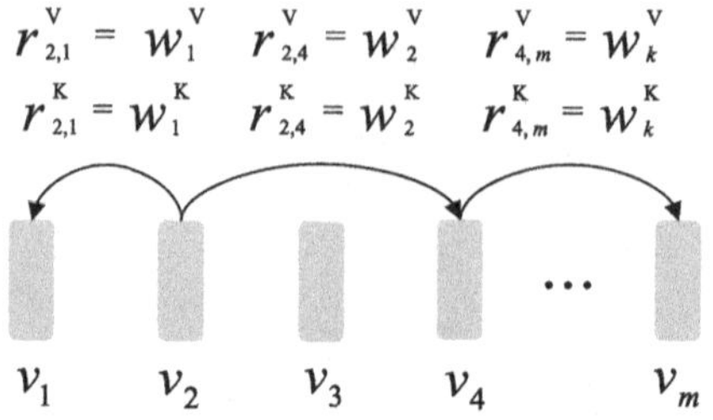

Fig. 3. We learn representations for each relative position within a clipping distance k. The figure assumes $2 \leqslant k \leqslant n - 4$.

time, we capture the user-item collaborative relations by a graph neural network [5]. The overall framework of the CTSR model is shown in Fig. 2. The Temporal Information Extraction (TIE) module and Collaborative Relation Modeling (CRM) module are the two important components of the model. TIE is used to extract the item-item temporal information, and CRM is employed to model the user-item collaborative relations.

Embedding Layer. We consider encoding item-item temporal information and collaborative relations simultaneously in the same feature space, hence we use a shared embedding $\mathbf{E} \in \mathbb{R}^{(n+m+1)\times d}$ where d is the latent dimensionality for both user and item, and 1 is added because the sequence needs to be padded.

$$\mathbf{E} = [E_0, \underbrace{E_{u_1}, ..., E_{u_n}}_{\text{users embeddings}}, \underbrace{E_{v_1}, ..., E_{v_m}}_{\text{items embeddings}}], \tag{2}$$

where E_0 is the padding item 0 (not representing any actual items).

For each user, the longest sequence length that the model can handle is t. If the user's interaction sequence length is longer than t, we include the most recent t actions, i.e., $s_u = (v_{T-t+1}, v_{T-t+2}, ..., v_T), T > t$. If the sequence length is less than t, we pad a constant zero vector 0 (not representing any actual items) to the left until the length is t, i.e., $s_j = (0, 0, ..., v_1, v_2, ..., v_X), X < t$.

Temporal Information Extraction (TIE). After the embedding layer, the item-item temporal information is obtained through this module. In this module, we stack B self-attention blocks (i.e., a self-attention layer $\mathbf{C}$ and a feed-forward network $\mathbf{F}$) to learn more complex feature transformations, where each layer extracts features for each time step based on the previous layer's outputs. Meanwhile, we use vectors $r^{\mathbf{V}}_{ij}, r^{\mathbf{K}}_{ij}$ to represent the information of relative positions between v_i and v_j, as shown in Fig. 3. v_i, v_j are the i-th, j-th item embeddings of the sequence. The self-attention operation takes the embeddings v_i and v_j as input, converts them to three matrices through linear projections, and feeds them into an attention layer:

$$\mathbf{C} = \text{softmax}\left(\frac{(v_i\mathbf{W}^{\mathbf{Q}})(v_j\mathbf{W}^{\mathbf{K}} + r^{\mathbf{K}}_{ij})^T}{\sqrt{d}}\right)(v_j\mathbf{W}^{\mathbf{V}} + r^{\mathbf{V}}_{ij}), \tag{3}$$

where the projection matrices $\mathbf{W^Q}, \mathbf{W^K}, \mathbf{W^V} \in \mathbb{R}^{d\times d}$. The maximum relative position distance is set to $k = 16$, and the relative position beyond k is clipped:

$$\begin{aligned} r_{ij}^{\mathbf{K}} &= w_{\text{clip}(j-i,k)}^{\mathbf{K}}, \\ r_{ij}^{\mathbf{V}} &= w_{\text{clip}(j-i,k)}^{\mathbf{V}}, \\ \text{clip}(x,k) &= \max(-k, \min(k, x)), \end{aligned} \tag{4}$$

where $r_{ij}^{\mathbf{K}}$ and $r_{ij}^{\mathbf{V}}$ are obtained by learning, and the relative position representations $w^{\mathbf{K}} = (w_{-k}^{\mathbf{K}}, \ldots, w_k^{\mathbf{K}})$ and $w^{\mathbf{V}} = (w_{-k}^{\mathbf{V}}, \ldots, w_k^{\mathbf{V}})$, where $w_i^{\mathbf{K}}, w_i^{\mathbf{V}} \in \mathbb{R}^d$.

We apply a point-wise feed-forward network to $\mathbf{C}$ identically to endow the model with nonlinearity:

$$\mathbf{F} = \text{ReLU}\left(\mathbf{CW}^{(1)} + \mathbf{b}^{(1)}\right)\mathbf{W}^{(2)} + \mathbf{b}^{(2)}, \tag{5}$$

where $\mathbf{W}^{(1)}, \mathbf{W}^{(2)} \in \mathbb{R}^{d\times d}$ and $\mathbf{b}^{(1)}, \mathbf{b}^{(2)}$ are d-dimensional vectors. After B self-attention blocks, the user embedding U_{TIE} represented by the sequence of historical interactions is obtained.

Collaborative Relation Modeling (CRM). In addition to extracting item-item temporal information, we also model the collaborative relations between user and item. To model the collaborative relations, general methods used to has a single input and cannot use user-item features as input. Besides, general methods are weak for sparse interaction matrices. Graph convolutional network can solve both of these problems. First, it uses node features as the initial embedding, where user and item features can be added. Second, it can introduce high-order connectivity information. Thus, we apply a graph convolutional network to model the collaborative relations.

Graph Construction. We regard users and items as nodes of the graph, and construct two different types of edges. The first type indicates an user-item interaction. When there is an interaction between user u and item v, the entry in the corresponding position of the adjacency matrix A is set to 1; otherwise 0. The second type of edge represents the contextual relation between item and item. To capture this relation without noise interference, we pick out high-frequency item-item pairs to add to the adjacency matrix. The items in a window size range are formed into pairs, and all item-item pairs are sorted according to the frequency. Through the experiment, it is observed that the experimental results of using different frequency of edges on different datasets largely vary.

Simplified Graph Convolution (SGC). We adopt the simple weighted sum aggregator and abandon the use of feature transformation and nonlinear activation. Propagation rule is defined as:

$$\begin{aligned} e_u^{(k+1)} &= \sum_{v\in\mathcal{N}_u} \frac{1}{\sqrt{|\mathcal{N}_u|}\sqrt{|\mathcal{N}_v|}} e_v^{(k)}, \\ e_v^{(k+1)} &= \sum_{u\in\mathcal{N}_v} \frac{1}{\sqrt{|\mathcal{N}_v|}\sqrt{|\mathcal{N}_u|}} e_u^{(k)}, \end{aligned} \tag{6}$$

where $e_u^{(k)}$ and $e_v^{(k)}$ respectively denote the refined embedding of user u and item v after k layers propagation, $\mathcal{N}_u$ and $\mathcal{N}_v$ represent the first-hop neighbors of user u and item v, and $\frac{1}{\sqrt{|\mathcal{N}_u|}\sqrt{|\mathcal{N}_v|}}$ is the graph Laplacian norm which reflects how much the historical item contributes to the user preference. Matrix form can facilitate implementation and discussion. The matrix equivalent form of SGC is:

$$\mathbf{E}^{(k+1)} = (D^{-\frac{1}{2}}AD^{-\frac{1}{2}})\mathbf{E}^{(k)}, \tag{7}$$

where D is an $(n+m+1)\times(n+m+1)$ diagonal matrix, in which each entry D_{uu} denotes the number of nonzero entries in the u-th row vector of the adjacency matrix A (D is also named as degree matrix). Hence, the final embedding matrix is:

$$\begin{aligned}\mathbf{E} &= \mathbf{E}^{(0)} + \mathbf{E}^{(1)} + \mathbf{E}^{(2)} + ... + \mathbf{E}^{(K)} \\ &= \mathbf{E}^{(0)} + \tilde{A}\mathbf{E}^{(0)} + \tilde{A}^2\mathbf{E}^{(0)} + ... + \tilde{A}^K\mathbf{E}^{(0)},\end{aligned} \tag{8}$$

where $\tilde{A} = D^{-\frac{1}{2}}AD^{-\frac{1}{2}}$ is the symmetrically normalized matrix, and $\mathbf{E}^{(0)} \in \mathbb{R}^{(n+m+1)\times d}$ denotes the 0-th layer embedding matrix. In our experiments, we observed that setting K as 2 leads to good performance. Finally, the user embedding U_{SGC} can be obtained out of $\mathbf{E}$ after SGC:

$$U_{SGC} = \mathbf{E}(u). \tag{9}$$

Feature Integration. The user embeddings respectively obtained by SGC and TIE are merged to obtain the final user embedding F_u for prediction. We use softmax to learn a weight for each dimension of each user embedding. The following U^l is the l-th dimension of user u's embedding:

$$\begin{aligned}U^l &= \frac{e^{\alpha_l}}{e^{\alpha_l} + e^{\beta_l}}U^l_{SGC} + \frac{e^{\beta_l}}{e^{\alpha_l} + e^{\beta_l}}U^l_{TIE}, \\ F_u &= [U^1, U^2, ..., U^d],\end{aligned} \tag{10}$$

where U^l_{TIE} is the l-th dimension of the embedding obtained by TIE, U^l_{SGC} is the l-th dimension of the embedding obtained by SGC, and e^{α_l} and e^{β_l} denote the learnable weights of the l-th dimension.

Prediction Layer. A Matrix Factorization (MF) [10] layer is used to predict the score of item v:

$$r_{v,t} = F_u \cdot E_v^T, \tag{11}$$

where F_u is the integration of user embedding, E_v represents embedding of item v, and $r_{v,t}$ denotes the score of item v as the next item given the first t items.

Focal Loss. We adopt the focal loss [13] as the base loss for measuring the ranking prediction error. It is transformed from cross entropy loss. Lin et al. [13] consider that easily classified negatives comprise the majority of the loss and dominate the gradient. This may prevent the loss function from being optimized to the best. Inspired by this, they proposed the focal loss. Focal loss is a dynamically scaled cross entropy loss, where the scaling factor γ can automatically down-weight the contribution of easy examples during training and rapidly guide the model to focus on hard examples. Focal loss is in the form of:

$$FL(y) = -\alpha(1 - y)^{\gamma} log(y), \tag{12}$$

where the weight factor $\alpha \in [0, 1]$ is used to balance the importance of positive/negative examples, $y \in [0, 1]$ is the model's estimated probability for the positive class, and $(1 - y)^{\gamma}$ indicates the modulating factor. Inspired by this loss function, the objective function L of our model can be formalized as:

$$L = -\alpha(1 - \sigma(r))^{\gamma} log(\sigma(r)), \tag{13}$$

where r is the score obtained by Eq. (11), σ represents the sigmoid function, and $\sigma(r)$ aims to map the score r to $[0, 1]$.

4 Experiments

In this section, we compare our proposed model CTSR with other state-of-the-art models on three real-world datasets. We implement our code in PyTorch and conduct all our experiments on a server with 48-core Intel Xeon Silver 4214R CPU@2.40 GHz, 256G RAM and Nvidia GeForce RTX 3090 GPUs.

4.1 Experimental Setup

Datasets. Since the datasets are sorted by time, it cannot be randomly divided into train, validation and test sets. In chronological order, it must be ensured that the training sets precede the validation sets and the validation sets precede the test sets. We use the last item in each user's interaction sequence as the test set, the penultimate one as the validation set, and the rest as the training set. The three datasets are:

- MovieLens-1M (*ML-1M*): This dataset has been widely used in the previous studies, which consists of 1 million ratings for 3416 items by 6040 users [2].
- *Beauty*: This is a series of product review datasets crawled from Amazon.com by McAuley et al. [14]. They split the data into separate datasets according to the top-level product categories on Amazon. In this work, we adopt the 'Beauty' category.
- *Games*: This dataset is notable for its high sparsity and variability from Amazon [14].

The specific statistics of these three datasets are shown in Table 1.

Table 1. Dataset statistics (after preprocessing).

Dataset	#Users	#Items	Avg. length	#Interactions
ML-1M	6,040	3,416	163.5	1.0M
Beauty	52,024	57,289	7.6	0.4M
Games	31,013	23,715	9.3	0.3M

Baselines. We compare the proposed CTSR with the following baselines to verify the effectiveness of our model. These baselines can be divided into three groups. The first group consists of static recommendation methods that ignore the sequential order of interactions, including BPR [15] and PopRec (basically ranking items according to their popularity). The second group comprises some sequential recommendation methods based on Markov chains: Factorized Markov Chains (FMC), FPMC [16] and TransRec [3]. The third group includes some deep learning methods for sequential recommendations: GRU4Rec [7], GRU4Rec$^+$ [6], Caser [19] and SASRec [8].

Parameter Settings. We use the same datasets as in [8] and apply the same data split method. For the structure of the model, we consider using two TIE blocks ($B = 2$), and summing the two layers of SGC ($K = 2$) at the collaborative relation modeling. Embedding table E in the embedding layer and prediction layer is shared. The parameter γ in the loss function is 2, and α is set to 0.2 on *ML-1M*, 0.3 on *Games* and 0.4 on *Beauty*. We choose the *Adam* optimizer [9] with a learning rate of 0.001, momentum exponential decay rates $\beta_1 = 0.9$ and $\beta_2 = 0.98$. Batch size is set to 2048 on all datasets except *ML-1M* which is 1024. The maximum sequence length t is 200 on *ML-1M* and 50 on the other datasets. The dropout rate is set to 0.2 on *ML-1M* and 0.5 on others. For the construction of high-frequency edges, it is observed that adding the top 60% of item-item pairs leads to the best results on the *Games* dataset, while on the other datasets, the top 80% are required.

Evaluation Metrics. For each user, we rank the prediction scores calculated by Eq. (11), resulting in a list of the top-N recommendations. We use the typical top-N ranking evaluation metrics, Recall@K and Normalized Discounted Cumulative Gain (NDCG@K) [3]. By default, we set $K = 10$. Our evaluation setting follows [8], which is to predict the ratings at the next moment $t + 1$ given the previous t ratings. To speed up the verification, we randomly sample 100 negative items, while always keeping the positive item that would be engaged next. By sorting these 101 items, Recall@10 and NDCG@10 can be evaluated.

4.2 Performance Comparison

We show the performance of all models in Table 2, where CTSR (AP) indicates the replacement of relative position encoding with absolute position encoding

Table 2. Overall performance comparison table. In each column, the best scores are in bold.

Methods	Beauty		Games		ML-1M	
	Recall	NDCG	Recall	NDCG	Recall	NDCG
PopRec	0.4003	0.2277	0.4724	0.2779	0.4239	0.2377
BPR	0.3775	0.2185	0.4853	0.2875	0.5781	0.3287
FMC	0.3771	0.2477	0.6358	0.4456	0.6986	0.4676
FPMC	0.4310	0.2891	0.6802	0.4680	0.7599	0.5176
TransRec	0.4607	0.3020	0.6838	0.4557	0.6413	0.3969
GRU4Rec	0.2125	0.1203	0.2938	0.1837	0.5581	0.3381
GRU4Rec$^+$	0.3949	0.2256	0.6599	0.4759	0.7501	0.5513
Caser	0.4264	0.2547	0.5282	0.3214	0.7886	0.5538
SASRec	0.4680	0.3108	0.7298	0.5270	0.8220	0.5880
CTSR (AP)	0.4889	0.3307	0.7390	0.5410	0.8270	0.5960
CTSR	**0.4957**	**0.3389**	**0.7425**	**0.5451**	**0.8284**	**0.5983**

in our model. After comparison we can conclude that our model outperforms all baselines on all three datasets. Compared with the state-of-the-art method SASRec, our CTSR model outperforms it on all three datasets. The reason for the enhancement is mainly due to the fact that we capture both user-item collaborative relations and item-item temporal information.

4.3 Ablation Analysis

In this section, we conduct ablation experiments on three real-world datasets to explore the effectiveness of TIE and CRM, which are the two important components of CTSR. Apart from that, the effectiveness of using focal loss and extracting certain proportion of edges will be verified. The results are shown in

Table 3. Ablation analysis on three datasets.

Methods		Beauty		Games		ML-1M	
		Recall	NDCG	Recall	NDCG	Recall	NDCG
CTSR		**0.4957**	**0.3389**	**0.7425**	**0.5451**	**0.8284**	**0.5983**
CTSR (AP)		0.4889	0.3307	0.7390	0.5410	0.8270	0.5960
R-CRM		0.4612	0.3083	0.7268	0.5101	0.8177	0.5786
R-TIE		0.4638	0.3012	0.6876	0.4637	0.7621	0.4856
Proportion of high-frequency edges	0%	0.4675	0.3142	0.7302	0.5331	0.8180	0.5859
	30%	0.4722	0.3178	0.7290	0.5297	0.8228	0.5912
	60%	0.4794	0.3257	0.7390	0.5410	0.8183	0.5896
BCE		0.4763	0.3166	0.7371	0.5313	0.8224	0.5852

Table 3. **CTSR (AP)** indicates the replacement of relative position encoding with absolute position encoding in our model. **R-TIE** removes the TIE module described in Sect. 3.2. **R-CRM** removes the CRM module. **Proportion of high-frequency edges** extracts different proportion of edges. **BCE** replaces the focal loss in CTSR with the Binary Cross Entropy (BCE) loss. Except for the changes mentioned above, the other parts of the models and experimental settings remain identical to ensure the fairness of comparison.

We can conclude the following points from Table 3. (1) The removal of either TIE and CRM would reduce the effect. (2) The proper proportion of high-frequency edges leads to significant experimental performance. This indicates that high-frequency edges can improve the collaborative relation modeling performance of graph neural networks. (3) Focal loss function is more helpful for training on two datasets. (4) Relative position encoding is more effective in improving prediction accuracy.

5 Conclusion

In this paper, we propose a model named CTSR which captures both item-item temporal information and user-item collaborative relations, thus enabling the recommendation effect to achieve the state-of-the-art performance. Focal loss is used to make the training of the model more effective by focusing on hard samples. The results demonstrate that our model outperforms the most classical self-attention based sequential recommendation model.

Acknowledgements. This work is supported by China Postdoctoral Science Foundation (No. 2023M730503) and Open Fund of Intelligent Terminal Key Laboratory of Sichuan Province (No. SCITLAB-20008).

References

1. Chang, J., et al.: Sequential recommendation with graph neural networks. In: SIGIR, pp. 378–387 (2021)
2. Harper, F.M., Konstan, J.A.: The MovieLens datasets: history and context. ACM Trans. Interact. Intell. Syst. **5**(4), 19:1–19:19 (2016)
3. He, R., Kang, W., McAuley, J.J.: Translation-based recommendation. In: RecSys, pp. 161–169 (2017)
4. He, R., McAuley, J.J.: Fusing similarity models with Markov chains for sparse sequential recommendation. In: ICDM, pp. 191–200 (2016)
5. He, X., Deng, K., Wang, X., Li, Y., Zhang, Y., Wang, M.: LightGCN: simplifying and powering graph convolution network for recommendation. In: SIGIR, pp. 639–648 (2020)
6. Hidasi, B., Karatzoglou, A.: Recurrent neural networks with top-k gains for session-based recommendations. In: CIKM, pp. 843–852 (2018)
7. Hidasi, B., Karatzoglou, A., Baltrunas, L., Tikk, D.: Session-based recommendations with recurrent neural networks. In: ICLR (2016)
8. Kang, W., McAuley, J.J.: Self-attentive sequential recommendation. In: ICDM, pp. 197–206 (2018)

9. Kingma, D.P., Ba, J.: Adam: a method for stochastic optimization. In: ICLR (2015)
10. Koren, Y., Bell, R.: Advances in collaborative filtering. In: Ricci, F., Rokach, L., Shapira, B. (eds.) Recommender Systems Handbook, pp. 77–118. Springer, Boston, MA (2015). https://doi.org/10.1007/978-1-4899-7637-6_3
11. Li, Y., Chen, T., Zhang, P., Yin, H.: Lightweight self-attentive sequential recommendation. In: CIKM, pp. 967–977 (2021)
12. Lin, J., Pan, W., Ming, Z.: FISSA: fusing item similarity models with self-attention networks for sequential recommendation. In: RecSys, pp. 130–139 (2020)
13. Lin, T., Goyal, P., Girshick, R.B., He, K., Dollár, P.: Focal loss for dense object detection. IEEE Trans. Pattern Anal. Mach. Intell. **42**(2), 318–327 (2020)
14. McAuley, J.J., Targett, C., Shi, Q., van den Hengel, A.: Image-based recommendations on styles and substitutes. In: SIGIR, pp. 43–52 (2015)
15. Rendle, S., Freudenthaler, C., Gantner, Z., Schmidt-Thieme, L.: BPR: Bayesian personalized ranking from implicit feedback. In: UAI, pp. 452–461 (2009)
16. Rendle, S., Freudenthaler, C., Schmidt-Thieme, L.: Factorizing personalized Markov chains for next-basket recommendation. In: WWW, pp. 811–820 (2010)
17. Shaw, P., Uszkoreit, J., Vaswani, A.: Self-attention with relative position representations. In: NAACL-HLT, vol. 2, pp. 464–468 (2018)
18. Tan, Q., et al.: Dynamic memory based attention network for sequential recommendation. In: AAAI, pp. 4384–4392 (2021)
19. Tang, J., Wang, K.: Personalized top-N sequential recommendation via convolutional sequence embedding. In: WSDM, pp. 565–573 (2018)
20. Vaswani, A., et al.: Attention is all you need. In: NIPS, pp. 5998–6008 (2017)
21. Wang, H., Ma, Y., Ding, H., Wang, Y.: Context uncertainty in contextual bandits with applications to recommender systems. In: AAAI, pp. 8539–8547 (2022)
22. Wang, X., He, X., Wang, M., Feng, F., Chua, T.: Neural graph collaborative filtering. In: SIGIR, pp. 165–174 (2019)
23. Wang, Y., Zhang, H., Liu, Z., Yang, L., Yu, P.S.: ContrastVAE: contrastive variational autoencoder for sequential recommendation. In: CIKM, pp. 2056–2066 (2022)
24. Wang, Z., et al.: Counterfactual data-augmented sequential recommendation. In: SIGIR, pp. 347–356 (2021)
25. Zhou, K., et al.: S3-Rec: self-supervised learning for sequential recommendation with mutual information maximization. In: CIKM, pp. 1893–1902 (2020)
26. Zhou, K., Yu, H., Zhao, W.X., Wen, J.: Filter-enhanced MLP is all you need for sequential recommendation. In: WWW, pp. 2388–2399 (2022)

Multi-level Attention Network with Weather Suppression for All-Weather Action Detection in UAV Rescue Scenarios

Yao Liu[1(✉)], Binghao Li[2], Claude Sammut[1], and Lina Yao[1,3]

[1] School of Computer Science and Engineering, University of New South Wales Sydney, NSW, Australia
{yao.liu3,c.sammut,lina.yao}@unsw.edu.au
[2] School of Minerals and Energy Resources Engineering, University of New South Wales, Sydney, NSW, Australia
binghao.li@unsw.edu.au
[3] Data 61, CSIRO, Sydney, NSW, Australia

Abstract. Unmanned Aerial Vehicles (UAVs) possess significant advantages in terms of mobility and range compared to traditional surveillance cameras. Human action detection from UAV images has the potential to assist in various fields, including search and rescue operations. However, UAV images present challenges such as varying heights, angles, and the presence of small objects. Additionally, they can be affected by adverse illumination and weather conditions. In this paper, we propose a Multi-level Attention network with Weather Suppression for all-weather action detection in UAV rescue scenarios. The Weather Suppression module effectively mitigates the impact of illumination and weather, while the Multi-level Attention module enhances the model's performance in detecting small objects. We conducted detection experiments under both normal and synthetic harsh conditions, and the results demonstrate that our model achieves state-of-the-art performance. Furthermore, a comparison of relevant metrics reveals that our model strikes a balance between size and complexity, making it suitable for deployment on UAV platforms. The conducted ablation experiments also highlight the significant contribution of our proposed modules.

Keywords: Unmanned Aerial Vehicles · Human action detection · Weather suppression · Multi-level attention

1 Introduction

With their advantages of high mobility, flexible deployment, and a large surveillance range, Unmanned Aerial Vehicles (UAVs) have progressively demonstrated their utility in surveillance, target tracking, aerial photography, and rescue operations in recent years [2,3,8,39,51]. Particularly during natural disasters, UAVs

B. Luo et al. (Eds.): ICONIP 2023, CCIS 1963, pp. 540–557, 2024.
https://doi.org/10.1007/978-981-99-8138-0_43

can be rapidly deployed to remote areas for swift and extensive scanning, making them highly valuable for search and rescue operations. However, the current application of UAVs in search and rescue often involves capturing aerial photographs, which are subsequently manually analyzed to identify individuals requiring rescue. This heavy reliance on human intervention makes search and rescue operations involving UAVs labor-intensive. Although object detection has been extensively studied for many years, there are still certain limitations in UAV image detection. UAV images exhibit variations in terms of heights, angles, object scales, and backgrounds, compared to conventional camera images [46]. In the field of search and rescue, UAV images are frequently impacted by inclement weather and challenging illumination conditions. Moreover, UAV rescue operations specifically focus on identifying individuals in need of assistance, requiring the capability to differentiate between various human actions to determine the urgency of rescue. The differentiation of human actions helps exclude unrelated individuals and expedite the search and rescue process. However, the subtle variations between human actions pose a greater challenge for distinction compared to the distinctions between categories in general object detection.

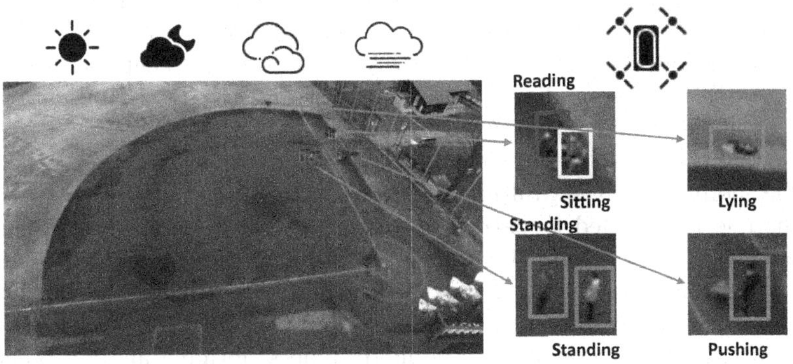

Fig. 1. Human action detection on UAV images. In different illumination and weather conditions, we perform action recognition and localization of humans in UAV-captured images. Strong lighting and nighttime conditions can cause objects to lose textural detail and blend in with the background. Adverse weather conditions can obscure parts of the object and reduce overall visibility.

Apart from manual recognition of UAV images, traditional machine learning approaches often employ a sliding window paradigm and rely on hand-crafted features [28,45]. However, these methods are time-consuming, and their feature robustness is insufficient. In recent years, the advancements in deep learning have led to the emergence of Convolutional Neural Networks (CNNs) [34] and Generative Adversarial Networks (GANs) [14]. These techniques have significantly impacted the field of object detection and have naturally extended their influence to UAV image detection as well. UAV image detection encompasses various tasks, such as object detection [52], dense detection [30], and object counting [5]. However, it currently encounters challenges related to small object detection [20] and handling long-tailed distribution of objects [50]. Hence, UAV image detection cannot be directly adopted from conventional object detection methods, and further improvements are necessary to address its unique challenges and requirements. Indeed, UAV images present unique tasks, especially in human detection, which encompass gesture detection [23] and action detection [1]. In the context of search and rescue, UAVs face the challenge of operating under complex conditions and conducting extensive scanning operations. Due to their inherent limitations, UAVs often cannot maintain steady recording of small areas, resulting in a shorter duration for capturing human activity compared to regular circumstances. Detection methods that rely on video analysis typically require multiple consecutive frames, preferably with a stable background [26,31]. Consequently, detecting humans and recognizing their actions in search and rescue scenarios using a single image poses a significant and practical problem [27].

Our research focuses on human action detection in UAV images under various illumination and weather conditions, as illustrated in Fig. 1. In our study, we utilize UAV data that encompasses diverse angles, altitudes, and backgrounds. Additionally, we incorporate a wide range of challenging conditions, including strong lighting, nighttime scenarios, cloudy weather, and foggy environments. Our objective is to create a human action detection model that can effectively operate in various scenarios using just a single image. This capability is particularly valuable in demanding environments like areas affected by natural disasters or underground mines, where video-based methods may not be feasible or practical. Our model consists of two primary modules: the Weather Suppression module and the Multi-level Attention module. The Weather Suppression module is designed to extract the noise map and illumination map from the backbone network, enabling the acquisition of True Color map features. This separation process helps suppress weather-related noise and improve the quality of the image features. Furthermore, the Weather Suppression module can be seamlessly integrated into the model's neck, forming an end-to-end network. This integration is distinct from the approach of training separate image enhancement networks independently. To achieve more effective cross-level feature aggregation, we employ the Multi-level Attention module. This module intelligently aggregates multi-level features by assigning weights based on the attention mechanism. This enables a rational and adaptive fusion of features from different levels, resulting in improved performance and better representation of the target objects. Through our experiments, we have demonstrated that our model achieves state-of-the-art performance in this task.

Our main contributions are as follows:

- We introduce a novel Multi-level Attention network with Weather Suppression specifically designed for all-weather action detection in UAV rescue scenarios. Additionally, we synthesize a multi-weather human action detection dataset by augmenting an existing dataset, facilitating effective training and accurate detection under various weather conditions encountered during UAV rescue missions.

- The Weather Suppression module effectively mitigates the adverse effects caused by weather and illumination, resulting in detection results that closely resemble those achieved under normal weather conditions. By separating the noise map and illumination map from the backbone network, we enhance the quality of image features, thus improving the robustness of our model.
- The Multi-level Attention module dynamically adjusts the weights using the attention mechanism after the Feature Pyramid Network (FPN), facilitating effective cross-level feature aggregation. This module improves the model's ability to focus on relevant features and adaptively fuse them, enhancing the overall detection capability.
- Our model excels in human action detection in UAV images. Through comprehensive comparison experiments and ablation studies, we demonstrate the superior performance of our model, highlighting the significant contributions made by the Weather Suppression and Multi-level Attention modules.

2 Related Work

2.1 Object Detection

In the past decade, object detection has made significant progress, with the mainstream detectors being divided into two-stage and single-stage approaches. Single-stage detectors are generally faster but slightly less accurate compared to two-stage detectors. R-CNN [13] is a milestone in introducing deep learning methods to the field of object detection, and it belongs to the two-stage method. The subsequent advancements include Fast R-CNN [12], which integrates the classification head and regression head into the network, and Faster R-CNN [37], which proposes the Region Proposal Network (RPN) to generate region proposals, forming an end-to-end object detection framework. Many of the later two-stage methods are built upon the improvements made by Faster R-CNN. For example, Mask R-CNN [15] reduces the accuracy loss of ROI pooling by introducing ROI-align. The pioneering work on single-stage detectors is YOLO [35], which has led to the development of numerous detectors in the YOLO series, including the widely used YOLOv3 [36] and the latest YOLOx [10]. Additionally, there are other mainstream basic methods, including SSD [25] and RetinaNet [21]. However, after YOLOv3, the integration of Anchor Box, Focal Loss, and Feature Pyramid Network (FPN) gradually became more prevalent. Due to the speed and increased accuracy offered by single-stage detectors, UAV image detection has mostly focused on improving single-stage detectors [19].

In the context of the detection task, the model neck plays a critical role as it serves as the connection between the backbone and the detection head. The backbone is responsible for extracting feature maps, often using the classification field's outputs directly. On the other hand, the detection head typically consists of both classification and regression modules. Therefore, the model neck, which aggregates feature map information, is crucial for effective object detection. Initially, methods like SSD [25] utilized the multi-level output of the backbone for detection but lacked explicit feature aggregation, leading to the evolution of the model neck. Subsequently, the Feature Pyramid Network (FPN) emerged as the mainstream approach for model necks, and it has been adopted by many models such as YOLOv3 [36], RetinaNet [21], and Faster R-CNN [37]. FPN utilizes a top-down aggregation process to enhance feature maps. Later, PANet [6] introduced bottom-up aggregation to further improve performance. Building upon these advancements, subsequent methods introduced additional techniques to refine feature fusion in the model neck. For example, ASFF [24] employed an attention mechanism to determine fusion weights, while NAS-FPN [11] and BiFPN [43] focused on identifying important blocks for repeated fusion. In a similar vein, the study mentioned in [20] leveraged average pooling and deconvolution to enhance FPN performance, particularly for detecting small objects in UAV images. These developments highlight the continuous pursuit of refining the model neck to improve object detection performance.

2.2 UAV Image Detection

The flight height of UAVs can vary from a dozen to hundreds of meters, and the cameras mounted on gimbals can capture images at different angles. As a result, UAV images contain objects with large scale variations and often include side and top views, making them challenging for conventional object detection methods. Fortunately, UAV images typically have fewer overlapping objects due to the shooting angle, reducing the need to focus extensively on handling overlapping objects [19]. Earlier research in UAV image detection often involved adapting general object detection methods to UAV settings [40,41]. Popular UAV datasets, such as VisDrone [53], UAVDT [7], and UAV123 [29], provide labeled data for various object classes, enabling common tracking and detection tasks. To address the specific challenges of UAV image detection, researchers have made improvements to single-stage detectors based on methods like SSD and YOLO. For example, [4,38] enhance the SSD method, while [35,44] modify the YOLO method to make single-stage detectors more suitable for UAV image detection. The UAV-Gesture dataset [32] contains a variety of drone command signals, but it focuses on close-range interactions between drones and people, with low-altitude flights and minimal background changes, resembling typical surveillance videos rather than distinct UAV image scenarios. On the other hand, the Okutama-Action dataset [1] consists of drone videos that specifically capture human actions, encompassing various heights, angles, and backgrounds. [27] utilizes the Okutama-Action dataset and their own dataset to apply UAV image human action detection to real-world rescue operations, demonstrating the practical implications of their research.

2.3 Weather Impact

When operating outdoors, especially during search and rescue missions, UAVs often encounter challenging environmental conditions such as unfavorable illumination (e.g., strong light or nighttime) and adverse weather (e.g., cloudy or foggy conditions). These factors can have a significant impact on the visibility and appearance of objects captured in UAV images, leading to less accurate detection results [46]. Strong light and nighttime conditions can cause objects to lose texture details, undergo color distortions, and blend into the background. Unfavorable weather conditions such as clouds and fog can further obscure parts of objects, making them appear incomplete and reducing overall visibility. These factors pose challenges to accurate detection in UAV images. While illumination variations and weather conditions also affect image quality in general object detection, they become critical factors in UAV image detection, particularly in response to special events where UAVs play a crucial role in search and rescue operations [18]. Consequently, suppressing the adverse effects of unfavorable illumination and weather can improve the overall image quality and, in turn, enhance the performance of object detection algorithms in UAV images [17,49].

3 Method

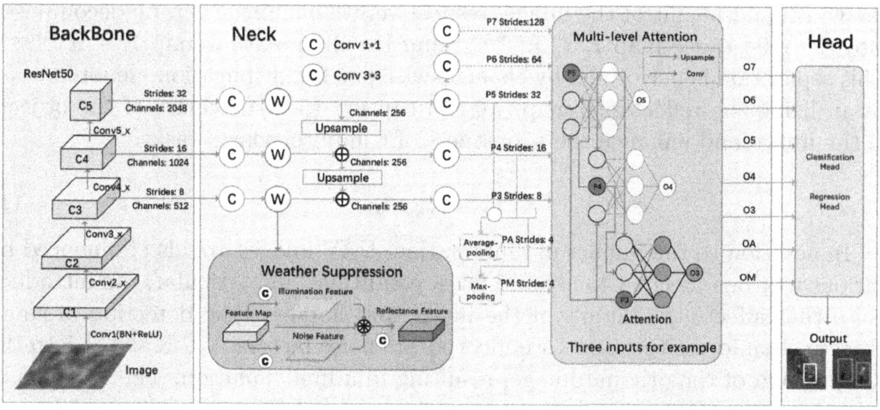

Fig. 2. Overview of our model. The model backbone is ResNet50 with multi-level output, the neck contains a Weather Suppression module and a Multi-level Attention module, and the head includes the Focal Loss.

3.1 Overview

Our model is designed to perform human action detection on UAV images. It takes a single image as input and provides the action category and bounding box for each person as output. The overview of our model is illustrated in Fig. 2. To strike a balance between performance and speed, we employ a medium-sized ResNet50 [16] as the backbone network. Following the approach of YOLO [35], SSD [25], and RetinaNet series, we utilize multi-level output. To optimize computational efficiency, we output C3-C5 features into the model neck. Within the model neck, we incorporate two key modules: the Weather Suppression module and the Multi-level Attention module. The Weather Suppression module aims to minimize the influence of illumination and weather conditions on the image. On the other hand, the Multi-level Attention module intelligently assigns weights for multi-level feature synthesis, promoting effective feature aggregation. For the model head, we adopt a RetinaNet-like classification head and regression head. To optimize the learning process, we utilize the Focal Loss [21], which reinforces the learning of challenging samples. This is particularly important for human action detection, as the differences between various actions can be subtle.

3.2 Weather Suppression

According to the Retinex theory [17,49], an image can be decomposed into a reflectance map and an illumination map. The reflectance map represents the inherent properties of the objects in the image and remains relatively constant, while the illumination map captures the variations in light intensity. In our model, we consider the original image as $I_o \in \mathbb{R}^{w \times h \times 3}$, where w and h represent the width and height of the image, respectively. This image can be decomposed into the reflectance map $I_r \in \mathbb{R}^{w \times h \times 3}$ and the illumination map $I_i \in \mathbb{R}^{w \times h \times 1}$. This separation is performed by element-wise pixel multiplication, denoted as $\odot$. By utilizing the reflectance map, we can capture the true colors of the objects in the image and enhance the robustness of computer vision tasks.

$$I_o = I_r \odot I_i \tag{1}$$

In addition to the impact of illumination, UAV images are also influenced by various weather effects. Cloudy and foggy conditions, in particular, can introduce noise that affects the details of the image and hampers the detection of small objects. To address this issue, we incorporate noise factors $N \in \mathbb{R}^{w \times h \times 1}$ into the composition of the original image, resulting in a final equation. The addition of noise factors to the original image formulation is a key step in our model, as it helps to enhance the robustness and accuracy of object detection in challenging weather conditions.

$$I_o = I_r \odot I_i + N \tag{2}$$

Thus, we can obtain the reflection map from the original image using a simple operation. However, it is important to note that to avoid encountering division by zero in the illumination map, we define $\tilde{I_i} = \frac{1}{I_i}$.

$$I_r = (I_o - N) \odot \tilde{I_i} \tag{3}$$

Unlike the approach taken in Darklighter [49], where the original images are processed separately, we adopt a integrated strategy to handle the adverse effects of illumination and weather. Separately processing the original images would require a extra network for iterative optimization, which cannot be easily integrated into an end-to-end framework with subsequent tasks. In our model, we aim to address these challenges of illumination and weather simultaneously within the detection task. To achieve this, we feed the images directly into the backbone network for feature extraction, similar to regular detection tasks. In this way, the feature maps obtained after the backbone contain not only the features of the reflection map but also the features of the illumination map and noise. Therefore, we perform the weather suppression process after the backbone. As depicted in Fig. 2, for each level of feature maps that require further processing, we employ the Weather Suppression module. This module is responsible for separating the features of the illumination map and noise, and then calculating the features of the reflection map for subsequent processing. It is important to note that the Weather Suppression module of each level does not share parameters.

3.3 Multi-level Attention

In our own efforts, we have developed a multi-level attention mechanism. Our model neck incorporates the attention module after the Weather Suppression and FPN modules. This attention module learns the importance of each level and assigns fusion weights more rationally. Unlike top-down, bottom-up, or search-based approaches, the weights determined by the attention module are based on network learning, making them more reasonable and effective. By combining top-down and bottom-up feature flows, our model gains a better understanding of the contextual relationships within an image. The top-down pathway allows the model to incorporate high-level semantic information, while the bottom-up pathway captures fine-grained details. This contextual awareness enables the model to make more informed decisions about object detection. The attention mechanism we've introduced facilitates the adaptive fusion of features from different levels. This is a crucial advantage, as it allows the model to dynamically adjust the contribution of each feature level based on the specific requirements of the detection task. This adaptability ensures that the model allocates more resources to relevant levels, leading to improved performance.

Given our specific task of human action detection in UAV images, which involves relatively small objects and considerations for network size, we choose to perform fusion at the third to fifth levels. Human actions in UAV images can vary significantly in scale due to differences in altitude, angles, and distances. The combination of top-down and bottom-up feature propagation helps our model handle these scale variations effectively. The multi-level attention mechanism contributes to the model's ability to generalize well across different scenarios. By adaptively selecting and fusing features, the model can learn to extract relevant information regardless of changes in lighting, weather conditions, or object sizes. This enhanced generalization is particularly valuable for real-world applications where UAVs may encounter diverse environments.

$$O3 = Attn(P3, Upsample(P4), Upsample(P5)) \tag{4}$$

$$O4 = Attn(Conv(P3), P4, Upsample(P5)) \tag{5}$$

$$O5 = Attn(Conv(P3), Conv(P4), P5) \tag{6}$$

Certainly, our model includes deeper networks for feature fusion from level 3 to level 7, as well as Max-pooling and Average-pooling operations. These components will be discussed and evaluated in our ablation experiments to assess their impact on the model's performance.

4 Experiment

4.1 Dataset

We conducted our experiments using the Okutama-action dataset [1]. The dataset was captured by UAVs at a resolution of 4k and a frame rate of 30 FPS. The data was collected at various flight heights ranging from 10 to 45 m, with the camera angle set at either 45°C or 90°C. The dataset consists of 22 scenes, each involving up to 9 actors.

For our baseline human action detection, we used a subset of the dataset that contains 77,365 frames of images with a resolution of 1280×720. Among these frames, 60,039 were used for training and 17,326 for testing. The dataset includes 12 common actions, namely Handshaking, Hugging, Reading, Drinking, Pushing/Pulling, Carrying, Calling, Running, Walking, Lying, Sitting, and Standing. Additionally, the dataset also includes instances of humans with undefined actions, which we refer to as "Others", following the approach in [27]. Hence, in total, we have 13 action categories. However, the presence of multiple simultaneous actions in the dataset complicates the classification task. To address this, we reclassify the action labels into atomic categories: Running, Walking, Lying, Sitting, and Standing. This categorization is similar to the six categories proposed in [27], with the addition of Waving in their case. We exclude Waving as a separate category since it can still be performed alongside other actions. As a result, we eliminate all action labels except for a single label, "Person", which represents human detection. Finally, we conducted experiments on the Okutama-action dataset using three different label divisions: the original 13 categories, the reclassified 5 categories, and the single category (Person) only. These divisions are depicted in Fig. 3.

4.2 Pre-processing

The images in the Okutama-action dataset are frames extracted from videos. To enhance efficiency and reduce redundancy, we adopt the common practice of selecting one frame every ten frames for training purposes. Most networks used in our experiments resize the input images to a range of 300 to 600 pixels. Compared to other datasets like MS COCO [22] and Pascal VOC [9], the Okutama-action dataset has higher resolution images with smaller objects. Resizing the images using a standard method would result in severe compression of object pixels, leading to incorrect detection. To overcome this issue, we employ a segmentation approach as a pre-processing step. We divide a 1280×720 pixel image into six

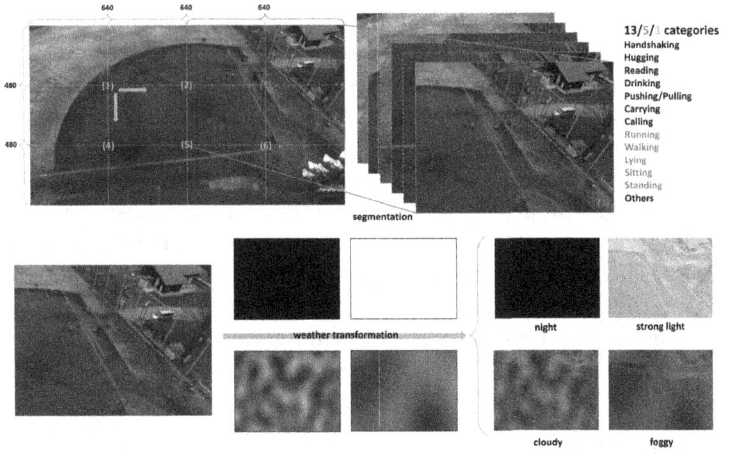

Fig. 3. Okutama-action dataset and data pre-processing.

segments of size 640×480, with a 50% overlap in both horizontal and vertical directions. This results in a 3×2 grid segmentation, as illustrated in Fig. 3. This segmentation strategy offers several advantages. First, it serves as a data augmentation technique by providing a larger training dataset, which helps the deep network in learning discriminative features. Second, segmentation preserves the details of the objects without compression, taking advantage of the high-resolution nature of the images and facilitating the detection of small objects. Lastly, the 50% overlap ensures that objects are not split across different sub-images, reducing the risk of missing detections due to cutoff boundaries.

The Okutama-action dataset covers scenes with varying weather conditions (sunny and cloudy) and illumination conditions (morning and afternoon). However, these conditions may not include extreme disturbances encountered in challenging environments like search and rescue scenarios. To address this limitation and simulate harsh conditions, we employ a method inspired by [5] for image transformation. Specifically, we perform pixel-level transformations to synthesize four different severe conditions: extreme illumination (strong light and night) and unfavorable weather (cloudy and foggy). The image transformation follows an image blending algorithm [42], and the process is described in Eq. 7. This transformation allows us to augment the dataset with images that capture the challenges posed by extreme lighting and adverse weather conditions, enabling the model to learn robustness and improve its performance in real-world scenarios.

$$\Phi = \alpha I + (1 - \alpha)N + P \tag{7}$$

The image transformation process involves several components: the original image I, the noise map N for simulating different illumination and weather conditions, the perturbation factor P for brightness correction, the weight ratio

α, and the resulting transformed image Φ. The noise maps N are categorized into four types. For illumination transformations, we use all-0 or all-255 matrices to represent strong light and night conditions. For weather transformations, we generate simulated clouds or fog using Perlin noise [33], as shown in Fig. 3. The perturbation factor P is set as an integer multiple of the all-1 matrix, which is used to correct the overall brightness of the image. The weight ratio α is set to 0.3, highlighting the impact of harsh weather during the transformation process. During training and testing, the experiments are divided into two groups: the first group does not undergo any weather transformation, while the second group undergoes weather transformation. In the second group, each image is randomly transformed using one type of noise, providing variability and diversity in the dataset.

4.3 Experiments

The evaluation metrics used for assessing the performance of the human action detector in our experiments are based on the mean Average Precision (mAP), which is also commonly used in the MS COCO dataset evaluation [22]. To calculate the mAP, we use the Intersection over Union (IoU) metric, which measures the overlap between the predicted bounding box and the ground truth box. Specifically, we choose an IoU threshold of 0.5, which is a common practice in evaluating UAV image detectors [1,27,47,48].

Our standard model employs features from P3 to P5, with a weather suppression module for adverse weather conditions. Models for our experiment comparison are as follows. The data pre-processing part is consistent for all reproduction models.

- Okutama-action [1]: These results are obtained from the original Okutama-action paper. The method used is similar to SSD, and the experiments are conducted using both 12 categories and 1 category. The results for the 12 categories can be roughly compared with our 13 categories.
- Drone-surveillance [27]: This paper focuses on using UAV images in the rescue field and utilizes both the Okutama-action dataset and their own dataset. For the Okutama-action dataset, they conduct experiments using both 6 categories and 1 category. The results for the 6 categories can be roughly compared with our 5 categories.
- DIF R-CNN [47]: This method is a pedestrian detection approach based on Faster R-CNN with context information enhancement. The authors initially conducted experiments using 1 category of the Okutama-action dataset, but the experiment setup was not consistent with the standard split of the training and testing sets. In a subsequent paper [48], they updated the experiment and provided revised results using the same method, which we cite for comparison.
- TDFA [48]: This method is a two-stream video-based detection approach. The best results on the Okutama-action dataset require a total of 9 frames of input images. Since the video-based method cannot be directly compared with other methods, we consider its experiments with single-image input separately.

- SSD [25]: We reproduce the standard SSD method on the Okutama-action dataset, using 512-pixel input resolution, which has shown improved results.
- RetinaNet [21]: We reproduce the RetinaNet model, using ResNet as the backbone.
- YOLOv3 [36]: We reproduce the popular YOLOv3 method, utilizing DarkNet53 as the backbone and 608-pixel input resolution.
- YOLOx [10]: We reproduce the state-of-the-art YOLOx model, focusing on the S-mode variant with memory consumption comparable to the other models.

Table 1. Comparison results of human action detection experiments on Okutama-action dataset. * indicates a rough comparison.

mAP@0.5	1 category	5 categories	13 categories
Okutama-action [1]	72.3%	–	18.8%*
Drone-surveillance [27]	–	35.0%*	–
DIF R-CNN [47]	84.1%	–	–
TDFA [48]	78.7%	–	–
SSD [25]	80.3%	39.5%	20.5%
RetinaNet [21]	85.5%	40.6%	20.7%
YOLOv3 [36]	78.3%	35.8%	17.8%
YOLOx [10]	84.3%	36.4%	19.4%
Ours w/o Weather Suppression	**85.7%**	**43.8%**	**23.9%**

Table 2. Comparison results of human action detection experiments on Okutama-action dataset with weather transformation.

mAP@0.5	1 category	5 categories	13 categories
SSD [25]	80.0%	39.4%	20.3%
RetinaNet [21]	85.4%	41.4%	22.8%
YOLOv3 [36]	77.1%	33.6%	19.7%
YOLOx [10]	82.3%	35.4%	22.0%
Ours w/o Weather Suppression	85.4%	41.0%	23.2%
Ours w/ Weather Suppression	**85.6%**	**45.2%**	**24.7%**

In Table 1, we can see that our model, which includes the Multi-level Attention module, achieves state-of-the-art results in all three divisions (1 category, 5 categories, and 13 categories). The second-best results are obtained by RetinaNet. Our model outperforms the second-place model by 0.2% in the 1 category division, and by 3.2% in both the 5 and 13 categories divisions.

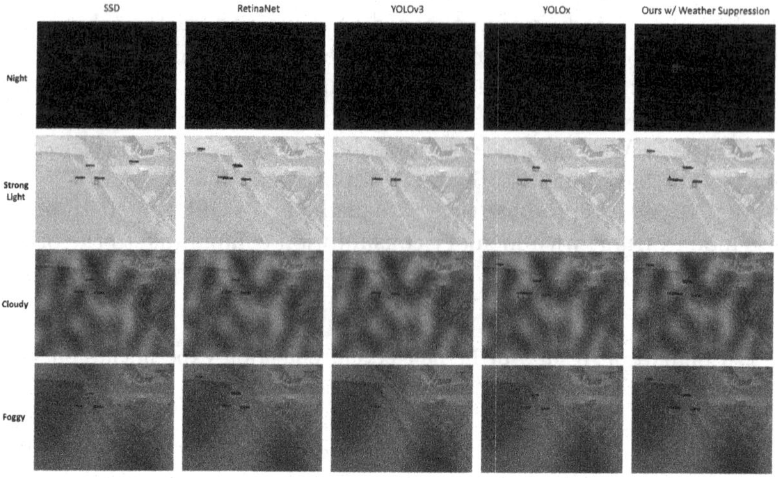

Fig. 4. Comparison of model detection results under harsh conditions (More comparison results in https://youtu.be/Os9RCcDWgz4.)

In Table 2, we present the results of the models under harsh illumination and weather conditions, including both with and without the Weather Suppression module. In the 1 category division, the detection results of each model are slightly reduced, indicating that harsh conditions have a small impact on the detection results. Our Weather Suppression module only reduces the detection results by 0.1% compared to normal conditions, and the model without the Weather Suppression module still outperforms the other models.

In the 5 categories division, the detection results of each model decrease significantly, and our model without the Weather Suppression module experiences a decrease of 2.8%. However, with the addition of the Weather Suppression module, our model achieves 45.2% results, surpassing all other models and even exceeding the results under normal conditions.

It is worth noting that in the 13 categories division, some models perform better under unfavorable conditions. This is due to the label definition of the Okutama-action dataset, where the 13 categories include the simultaneous execution of multiple actions. The labels are chosen for the least obvious action categories in order to maintain the highest classification diversity. For example, if a person is sitting and reading, the label assigned is "reading", resulting in a wrong detection result for "sitting". This label definition leads to some misunderstandings in the detection results because the network detects the most obvious categories.

Overall, the atomic classification of 5 categories is more persuasive. Our model performs competitively in all divisions and shows promising results, especially with the inclusion of the Weather Suppression module, which helps mitigate the impact of harsh illumination and weather conditions on the detection performance.

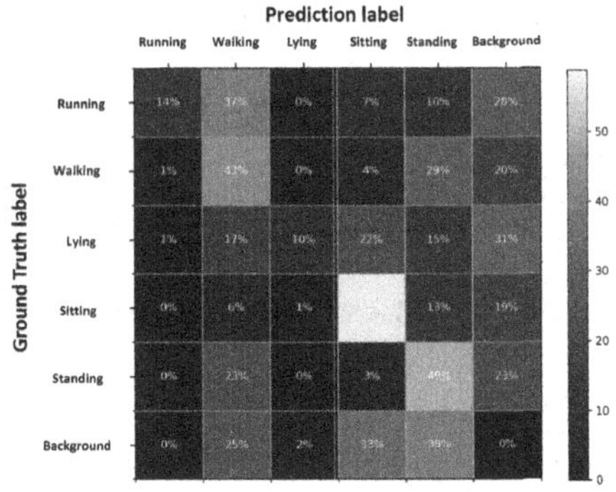

Fig. 5. Confusion matrix for 5 categories.

Table 3. Model Comparison.

	FPS	GFLOPs	Params(M)	Memory(G)
DIF R-CNN [47]	4.55	–	–	–
TDFA [48]	12.50	–	–	–
SSD [25]	17.24	102.80	24.39	11.99
RetinaNet [21]	12.50	61.32	36.10	1.40
YOLOv3 [36]	18.52	58.15	61.52	6.96
YOLOx [10]	16.13	9.99	8.94	6.24
Ours w/o Weather Suppression	11.77	61.59	31.40	2.56
Ours w/ Weather Suppression	11.76	72.74	36.71	2.75

The 5 categories atomic division is considered to have practical significance, and it is visually demonstrated in Fig. 4. Notably, only our model achieves correct detections in the night condition, indicating the contribution of the Weather Suppression module. Our model demonstrates advantages under various other harsh conditions as well. The confusion matrix in Fig. 5 reveals that Walking, Running, and Standing remain the most challenging categories to classify due to their similar features. This aligns with our perception that these actions are harder to distinguish based on a single image. Lying has distinct features but is more difficult to detect, while Sitting exhibits the best distinguishing features.

Table 3 presents a comparison of several model values. Our model achieves a more balanced situation across various measures while still achieving state-of-the-art results. In terms of complexity, YOLOx has the best performance but requires higher memory usage. YOLOv3 and SSD also demand high memory usage. RetinaNet is the closest in complexity to our model, while DIF R-CNN is slower due to its two-stage detection. The balanced nature of our model makes it well-suited for deployment on UAV platforms with limited hardware resources.

Table 4. Ablation experiments.

mAP@0.5	1 category	5 categories	13 categories	settings
1	83.6%	41.6%	18.4%	P3-P7
2	83.9%	39.5%	19.3%	P3-P5&PA
3	82.6%	38.2%	20.2%	P3-P5&PM
4	85.7%	43.8%	23.9%	P3-P5

In the ablation experiments presented in Table 4, we explore different feature fusion strategies in the attention module. We first investigate the inclusion of higher-level features such as P6 and P7 but find that they do not yield better results. This can be attributed to the fact that objects in UAV images are smaller in size, and the higher-level features may overlook these small objects. Next, we incorporate Average-pooling and Max-pooling in the attention module, but again, the results do not improve significantly. While pooling features have been shown to enhance model performance in some methods, in our model, attention plays a crucial role in rationalizing feature fusion weights. The inclusion of pooling features can disrupt this fusion process. Based on these findings, our final adopted model utilizes P3-P5 as feature inputs in the attention module, which yields the best results.

5 Conclusions

In summary, our proposed Multi-level Attention network with Weather Suppression addresses the challenges of all-weather action detection in UAV rescue scenarios. Through rigorous experiments, we demonstrate the effectiveness of our approach in mitigating the effects of illumination and weather conditions, improving model performance, and achieving state-of-the-art results. The balanced performance of our model across various metrics makes it a promising solution for deployment on UAV platforms in search and rescue applications.

References

1. Barekatain, M., Martí, M., et al: Okutama-action: an aerial view video dataset for concurrent human action detection. In: CVPR Workshops, pp. 2153–2160 (2017)
2. Bozcan, I., Kayacan, E.: AU-AIR: a multi-modal unmanned aerial vehicle dataset for low altitude traffic surveillance. In: ICRA, pp. 8504–8510 (2020)
3. Bozcan, I., Kayacan, E.: Context-dependent anomaly detection for low altitude traffic surveillance. CoRR abs/2104.06781 (2021)
4. Budiharto, W., Gunawan, A.A.S., et al.: Fast object detection for quadcopter drone using deep learning. In: ICCCS, pp. 192–195 (2018)
5. Cai, Y., Du, D., et al.: Guided attention network for object detection and counting on drones. In: MM, pp. 709–717 (2020)

6. Cai, Z., Vasconcelos, N.: Cascade R-CNN: delving into high quality object detection. In: CVPR, pp. 6154–6162 (2018)
7. Du, D., Qi, Y., Yu, H., Yang, Y., Duan, K., Li, G., Zhang, W., Huang, Q., Tian, Q.: The unmanned aerial vehicle benchmark: object detection and tracking. In: Ferrari, V., Hebert, M., Sminchisescu, C., Weiss, Y. (eds.) ECCV 2018. LNCS, vol. 11214, pp. 375–391. Springer, Cham (2018). https://doi.org/10.1007/978-3-030-01249-6_23
8. Erdelj, M., Natalizio, E.: UAV-assisted disaster management: applications and open issues. In: ICNC, pp. 1–5 (2016)
9. Everingham, M., Gool, L.V., et al.: The pascal visual object classes (VOC) challenge. Int. J. Comput. Vis. **88**, 303–338 (2010)
10. Ge, Z., Liu, S., et al.: YOLOX: exceeding YOLO series in 2021. CoRR abs/2107.08430 (2021)
11. Ghiasi, G., Lin, T., Le, Q.V.: NAS-FPN: learning scalable feature pyramid architecture for object detection. In: CVPR, pp. 7036–7045 (2019)
12. Girshick, R.B.: Fast R-CNN. In: ICCV, pp. 1440–1448 (2015)
13. Girshick, R.B., Donahue, J., et al.: Rich feature hierarchies for accurate object detection and semantic segmentation. In: CVPR, pp. 580–587 (2014)
14. Goodfellow, I.J., Pouget-Abadie, J., et al.: Generative adversarial networks. Commun. ACM. **63**, 139–144 (2020)
15. He, K., Gkioxari, G., et al.: Mask R-CNN. In: ICCV, pp. 2980–2988 (2017)
16. He, K., Zhang, X., Ren, S., Sun, J.: Deep residual learning for image recognition. In: CVPR, pp. 770–778 (2016)
17. Land, E.H.: The retinex theory of color vision. Sci. Am. **237**, 108–129 (1977)
18. Li, T., Liu, J., et al.: UAV-human: a large benchmark for human behavior understanding with unmanned aerial vehicles. In: CVPR, pp. 16266–16275 (2021)
19. Li, Z., Liu, X., et al.: A lightweight multi-scale aggregated model for detecting aerial images captured by UAVs. J. Vis. Commun. Image Represent. **77**, 103058 (2021)
20. Liang, X., Zhang, J., et al.: Small object detection in unmanned aerial vehicle images using feature fusion and scaling-based single shot detector with spatial context analysis. IEEE Trans. Circuits Syst. Video Technol. **30**, 1758–1770 (2020)
21. Lin, T., Goyal, P., et al.: Focal loss for dense object detection. In: ICCV, pp. 2999–3007 (2017)
22. Lin, T.-Y., et al.: Microsoft COCO: common objects in context. In: Fleet, D., Pajdla, T., Schiele, B., Tuytelaars, T. (eds.) ECCV 2014. LNCS, vol. 8693, pp. 740–755. Springer, Cham (2014). https://doi.org/10.1007/978-3-319-10602-1_48
23. Liu, C., Szirányi, T.: Real-time human detection and gesture recognition for on-board UAV rescue. Sensors. **21**, 2180 (2021)
24. Liu, S., Huang, D., Wang, Y.: Learning spatial fusion for single-shot object detection. CoRR abs/1911.09516 (2019)
25. Liu, W., et al.: SSD: single shot multibox detector. In: Leibe, B., Matas, J., Sebe, N., Welling, M. (eds.) ECCV 2016. LNCS, vol. 9905, pp. 21–37. Springer, Cham (2016). https://doi.org/10.1007/978-3-319-46448-0_2
26. Mabrouk, A.B., Zagrouba, E.: Abnormal behavior recognition for intelligent video surveillance systems: a review. Expert Syst. Appl. **91**, 480–491 (2018)
27. Mishra, B., Garg, D., et al.: Drone-surveillance for search and rescue in natural disaster. Comput. Commun. **156**, 1–10 (2020)
28. Moranduzzo, T., Melgani, F.: Detecting cars in UAV images with a catalog-based approach. IEEE Trans. Geosci. Remote. Sens. **52**, 6356–6367 (2014)

29. Du, D., et al.: The unmanned aerial vehicle benchmark: object detection and tracking. In: Ferrari, V., Hebert, M., Sminchisescu, C., Weiss, Y. (eds.) ECCV 2018. LNCS, vol. 11214, pp. 375–391. Springer, Cham (2018). https://doi.org/10.1007/978-3-030-01249-6_23
30. Papaioannidis, C., Mademlis, I., Pitas, I.: Autonomous UAV safety by visual human crowd detection using multi-task deep neural networks. In: ICRA, pp. 11074–11080 (2021)
31. Perera, A.G., Law, Y.W., Chahl, J.: Drone-action: an outdoor recorded drone video dataset for action recognition. Drones. **3**, 82 (2019)
32. Perera, A.G., Law, Y.W., Chahl, J.: UAV-GESTURE: a dataset for UAV control and gesture recognition. In: Leal-Taixé, L., Roth, S. (eds.) ECCV 2018. LNCS, vol. 11130, pp. 117–128. Springer, Cham (2019). https://doi.org/10.1007/978-3-030-11012-3_9
33. Perlin, K.: Improving noise. In: ACM Transactions of Graph, pp. 681–682, July 2002
34. Radovic, M., Adarkwa, O., Wang, Q.: Object recognition in aerial images using convolutional neural networks. J. Imaging. **3**, 21 (2017)
35. Redmon, J., Divvala, S.K., et al.: You only look once: unified, real-time object detection. In: CVPR, pp. 779–788 (2016)
36. Redmon, J., Farhadi, A.: Yolov3: an incremental improvement. CoRR abs/1804.02767 (2018)
37. Ren, S., He, K., et al.: Faster R-CNN: towards real-time object detection with region proposal networks. In: NIPS, pp. 91–99 (2015)
38. Rohan, A., Rabah, M., Kim, S.H.: Convolutional neural network-based real-time object detection and tracking for parrot AR drone 2, pp. 69575–69584 . IEEE Access (2019)
39. Semsch, E., Jakob, M., et al.: Autonomous UAV surveillance in complex urban environments. In: IAT, pp. 82–85 (2009)
40. Sevo, I., Avramovic, A.: Convolutional neural network based automatic object detection on aerial images. IEEE Geosci. Remote. Sens. Lett. **13**, 740–744 (2016)
41. Sommer, L.W., Schuchert, T., Beyerer, J.: Fast deep vehicle detection in aerial images. In: WACV, pp. 311–319 (2017)
42. Szeliski, R.: Computer Vision: Algorithms and Applications, 1st edn. Springer, Heidelberg (2010). https://doi.org/10.1007/978-3-030-34372-9
43. Tan, M., Pang, R., Le, Q.V.: Efficientdet: scalable and efficient object detection. In: CVPR, pp. 10778–10787 (2020)
44. Tijtgat, N., Volckaert, B., Turck, F.D.: Real-time hazard symbol detection and localization using UAV imagery. In: VTC, pp. 1–5 (2017)
45. Wen, X., Shao, L., et al.: Efficient feature selection and classification for vehicle detection. IEEE Trans. Circuits Syst. Video Technol. **25**, 508–517 (2015)
46. Wu, Z., Suresh, K., et al.: Delving into robust object detection from unmanned aerial vehicles: a deep nuisance disentanglement approach. In: ICCV, pp. 1201–1210 (2019)
47. Xie, H., Chen, Y., Shin, H.: Context-aware pedestrian detection especially for small-sized instances with deconvolution integrated faster RCNN (DIF R-CNN). Appl. Intell. **49**, 1200–1211 (2019)
48. Xie, H., Shin, H.: Two-stream small-scale pedestrian detection network with feature aggregation for drone-view videos. Multidimens. Syst. Signal Process. **32**, 897–913 (2021)
49. Ye, J., Fu, C., et al.: Darklighter: light up the darkness for UAV tracking. In: IROS, pp. 3079–3085 (2021)

50. Yu, W., Yang, T., Chen, C.: Towards resolving the challenge of long-tail distribution in UAV images for object detection. In: WACV, pp. 3257–3266 (2021)
51. Zhang, C., Ge, S., et al.: Accurate UAV tracking with distance-injected overlap maximization. In: MM, pp. 565–573 (2020)
52. Zhang, X., Izquierdo, E., Chandramouli, K.: Dense and small object detection in UAV vision based on cascade network. In: ICCV, pp. 118–126 (2019)
53. Zhu, P., Wen, L., et al: Vision meets drones: a challenge. CoRR abs/1804.07437 (2018)

Learning Dense UV Completion for 3D Human Mesh Recovery

Qingping Sun[1], Yanjun Wang[2], Zhenni Wang[1], and Chi-Sing Leung[1(✉)]

[1] Department of Electrical Engineering, City University of Hong Kong, Hong Kong, China
qingping.sun@my.cityu.edu.hk, eeleungc@cityu.edu.hk

[2] Department of Electronic Engineering, Shanghai Jiao Tong University, Shanghai, China
c.w.a.6@sjtu.edu.cn

Abstract. Human mesh reconstruction from a single image is a challenging task due to the occlusion caused by self, objects, or other humans. Existing methods either fail to separate human features accurately or lack proper supervision for feature completion. In this paper, we propose Dense Inpainting Human Mesh Recovery (DIMR), a two-stage method that leverages dense correspondence maps to handle occlusion. Our method utilizes a dense correspondence map to separate visible human features and completes human features on a structured UV space with an attention-based feature completion module. We also design a feature inpainting training procedure that guides the network to learn from unoccluded features. We evaluate our method on several datasets and demonstrate its superior performance under heavily occluded scenarios compared to other methods. Extensive experiments show that our method obviously outperforms prior methods on heavily occluded images and achieves comparable results on the standard benchmarks. Moreover, our method is comparable with previous methods on no heavily occluded images.

Keywords: Occlusion · IUV · Inpainting · Skinned Multi-Person Linear Model (SMPL)

1 Introduction

Recovering human mesh only from a single image is meaningful for numerous applications like AR/VR, human-related interaction, and digital try-on systems. However, it's a challenging task due to the complicated articulations of the human body and the ambiguity between the 3D model and the 2D projection. Recently, numerous works have been proposed to resolve these issues, and they

Q. Sun and Y. Wang—Equal contributions.

B. Luo et al. (Eds.): ICONIP 2023, CCIS 1963, pp. 558–569, 2024.
https://doi.org/10.1007/978-981-99-8138-0_44

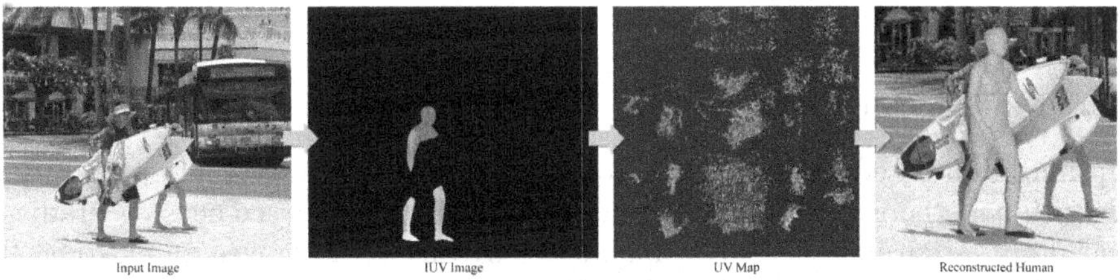

Fig. 1. Illustration of our insight.

can generally be categorized into optimization-based methods and learning-based methods. The optimization-based methods [1–3] utilize conventional optimization algorithms to estimate the parametric human body model. The learning-based methods directly utilize deep learning algorithms to reconstruct the 3D human model, which can be further divided into the model-based methods [4–9] and the model-free methods [10–12]. The model-based methods estimate the parametric human body model, which is similar to the optimization-based methods, while the model-free methods directly estimate the 3D representation of the human body, such as mesh, point cloud, and signed distance function. Although impressive progress has been achieved, most of these algorithms focus on non-occluded cases or minor occlusion cases. However, occlusion is inevitable due to the interaction between the human with the human and the human with the object. If an occlusion occurs, the performance of the models may be degraded apparently.

To resolve these issues, we propose Dense Inpaiting Human Mesh Recovery (DIMR) to solve different occlusions. As shown in Fig. 1, our method consists of two key steps: 1) estimate the dense UV map of the obscured body and mask out the occlusion regions; 2) perform the UV-based feature wrapping and completion for future mesh reconstruction. To take full advantage of dense correspondence maps to separate human body features, we utilize a dedicated network for dense correspondence map regression. Then, a feature-wrapping process is designed to accurately separate human features and project them to a structured UV map. With these wrapped features, we further propose an attention-based mechanism to complete the occluded features according to the visible parts. After that, we derive part-wise human features, which are used to regress the parameters of a Skinned Multi-Person Linear Model (SMPL). To enhance the occlusion handling ability, we design a feature inpainting training procedure that guides the network for feature completion with the unoccluded features. To validate our method, we conduct extensive experiments on the 3DOH [13], 3DPW [14], 3DPW-OC [14] and 3DPW-PC [14] datasets. The results indicate that our model yields a higher accuracy under heavily occluded scenarios compared to other occlusion handling methods and maintains a comparable accuracy on general human datasets.

2 Related Work

2.1 Mainstream Human Mesh Recovery Methods

Optimization-based methods [1–3] refer to using the conventional optimization algorithm methods to fit a parametric model to the 2D observed cues via gradient backpropagation in an iterative manner. SMPLify [2] leverages an existing 2D keypoints estimation algorithm to estimate the 2D keypoints and then utilizes an optimization approach to fit SMPL to the detected 3D keypoints. With the development of deep learning techniques, learning-based methods have gained significant traction, which can generally be further divided into two categories: model-based methods [4–9] and model-free methods [10–12]. HMR [4] is a model-based method that leverages CNN to extract features from the 2D image and then estimate the pose and shape parameters of SMPL. GraphCMR [11] is a model-free method that first leverages CNN to extract the global features of the input image and then attaches these features to the mesh vertices of the initialization pose (T-pose) and regresses the 3D location of each vertex.

2.2 Occlusion-Aware Human Mesh Recovery

Most recent methods are proposed for non-occlusion or minor occlusion cases. Some algorithms [7–9, 12, 13, 15, 16] are elaborately devised for the occlusion scenarios. ROMP [8] is devised for multi-human reconstruction. To resolve the multi-human overlap problem, ROMP proposes a collision-aware representation to construct a repulsion field of body centers, which is used to push apart the close body center. BEV [9] is an extension of ROMP that proposes a Bird's-Eye-View representation to reason about depth explicitly. METRO [12] is similar to PARE, but it uses the Transformer modules to model the relationship between the vertices and joints. To tackle the human-object occlusion, OOH [13] regards it as an image inpainting. It first estimates a UV map to represent the object-occluded human body and then uses the UV map to guide the feature inpainting.

3 Our Method

3.1 UV Based Feature Warpping

The initial UV maps generated by DensePose [17] are part-specific, where the density of part UV maps is unbalanced. For example, the UV maps for the hands are denser than those for the limbs, shown in the Fig. 2. Therefore, it's challenging for our algorithm to make inter-part feature completion directly from the initial UV map. To resolve this issue, we map the part-specific UV maps into a single holistic map, considering the position and proportion of different parts. After this arrangement, the processed UV map could preserve more neighboring body-part relationships. Based on the elaborately designed UV map rearrangement, we can map the estimated image features F_{img} into the balance UV space F_{UV} that can be formulated as:

$$\begin{aligned} u &= U(x, y), \\ v &= V(x, y), \\ F_{UV}(u, v) &= F_{img}(x, y), \end{aligned} \tag{1}$$

where x and y are under the image coordinate, u and v are under the UV coordinate, and U and V denote the UV mapping function that converts the $x - y$ coordinates into the $u - v$ coordinates.

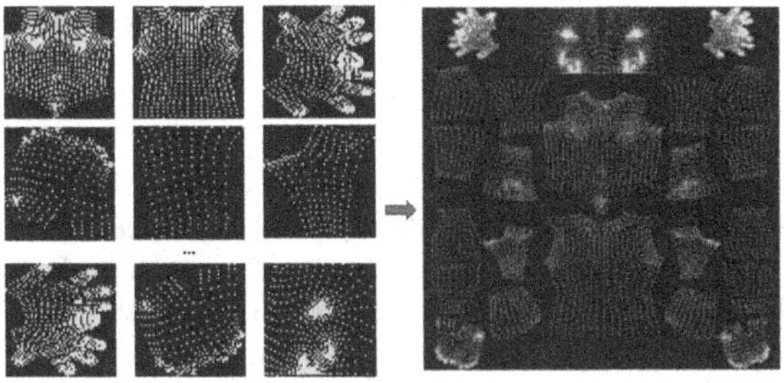

Fig. 2. Illustration of UV map rearrangement. The original DensePose [17] UV map (left) is part-based, with density varying widely on different parts of the body. We rearrange the map together according to their actual body proportions and relative position.

3.2 Wrapped Feature Completion

Given the balanced UV space, the wrapped feature completion module aims to utilize the visible features to reason the occluded features. The module first uses convolution blocks to downsample the wrapped feature map and combine the neighboring features. After that, we propose an attention module to consider features across the feature map. This method applies two 1×1 convolution layers to the downscaled feature map F_{uv} to derive body part attention weight $\mathbf{A}$ and attention value $\mathbf{V}$. Then, it weighted sums the attention value through element-wise matrix multiplication formulated as:

$$\varphi_i = \sum_{h,w} \sigma(\mathbf{A}_i) \circ \mathbf{V}, \tag{2}$$

where i denotes i-th body-part, $\circ$ denotes element-wise multiplication, and σ is softmax function that normalizes the attention map. As a consequence of feature wrapping, the part features of different instances are expected to be situated in the same location. Consequently, the attention map for a given part should emphasize a fixed position as the primary reference for the part feature φ_i. Additionally, since various instances may exhibit diverse occlusions, the attention map should also have several other focus areas to obtain information from other parts. It is noteworthy that we train this model without explicit supervision, as the location of the part features is expected to be similar.

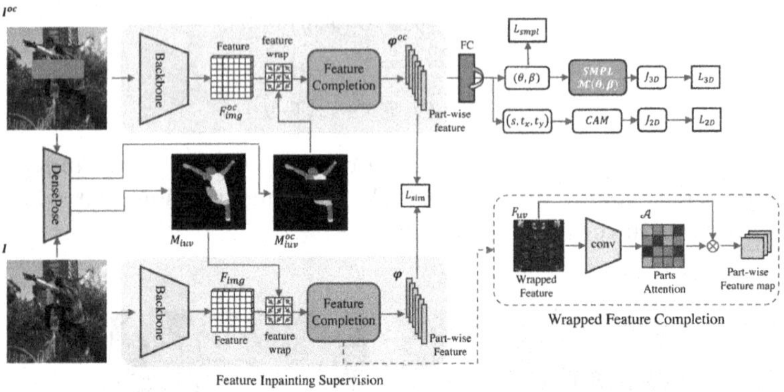

Fig. 3. Pipeline overview. Our model comprises a UV feature wrap module, a feature completion model, and an FC layer for SMPL parameter regression. Our network takes the image I and the predicted IUV image M_{iuv} as input. The IUV image wraps the image feature to UV space. The attention-based feature completion module completes the occluded feature based on the visible ones. The inpainting training takes the original and occluded image as input and uses the original feature as the supervision for the occluded feature.

3.3 UV Based Feature Inpainting

As shown in Fig. 3, this training method takes two inputs, the original image I and the occluded image I^{occ} augmented by synthetic occlusion. We also add occlusion as a binary mask on the predicted IUV image, denoted as M_{iuv} and M_{iuv}^{occ}. The network processes both pairs of inputs. The unoccluded part feature vector φ serves as the supervision to regulate the occluded vector φ^{oc}. The objective is to predict similar part features even under occlusion. We model this as a similarity loss, which is formulated as:

$$L_{sim} = \frac{1}{p}\sum_{i=0}^{p-1} \frac{\boldsymbol{\varphi}_i^{oc}\boldsymbol{\varphi}_i}{\|\boldsymbol{\varphi}_i^{oc}\|_2\|\boldsymbol{\varphi}_i\|_2}. \tag{3}$$

This training strategy introduces the ground truth features to the network, helping the model to inpaint the missing features.

3.4 Loss Functions

We supervise our training with SMPL parameters loss L_{smpl}, 3D keypoints loss L_{3D}, 2D keypoints L_{2D} and similarity loss L_{sim} that can be formulated as:

$$\begin{aligned} L_{smpl} &= \|(\boldsymbol{\theta},\boldsymbol{\beta}) - (\hat{\boldsymbol{\theta}},\hat{\boldsymbol{\beta}})\|_2, \\ L_{3D} &= \left\| J_{3D} - \hat{J}_{3D} \right\|_2, \\ L_{2D} &= \left\| J_{2D} - \hat{J}_{2D} \right\|_2. \end{aligned} \tag{4}$$

The full loss for our method is defined as:

$$L = L_{smpl} + \lambda_{3D} L_{3D} + \lambda_{2D} L_{2D} + \lambda_{sim} L_{sim}, \quad (5)$$

where λ_{2D}, λ_{3D} and λ_{sim} are regularization parameters.

3.5 Implementation Details

We utilize HRNet [18] as the image encoder, which is pre-trained on COCO poses estimation [19]. The input images are resized to 224 × 224. To train the DIMR, we use Adam [20] optimizer with a learning rate of $2.5e^{-4}$, and the batch size is set to 128. Loss weights are set to $\lambda_{3D} = 5$, $\lambda_{2D} = 5$, and $\lambda_{sim} = 1$ to ensure that each weighted loss item is of the same magnitude. For the training details, firstly, we train our model without parallel inpainting for 50K iterations with random scale, rotation, flipping, and channel noise augmentation. After that, we finetune the model in parallel training for 15K iterations with synthetic occlusion.

Table 1. Performance evaluation on occlusion datasets. For a fair comparison, we train our models without 3DPW datasets when evaluated on both 3DPW subsets, and we train our models without the 3DOH train set when evaluating on a 3DOH test set. The Top-1 values are bolded, and the Top-2 values are underlined. * indicates the IUV map is generated by DensePose [17] and † indicates the IUV map is generated from the mesh. All metrics in *mm*.

Methods	3DPW-PC			3DPW-OC			3DOH	
	PA-MPJPE↓	MPJPE↓	PVE↓	PA-MPJPE↓	MPJPE↓	PVE↓	PA-MPJPE↓	MPJPE↓
HMR-EFT [24]	–	–	–	60.9	94.9	111.3	66.2	101.9
SPIN [5]	82.6	129.6	157.6	60.8	95.6	121.6	68.3	104.3
PyMAF [6]	81.3	126.7	154.3	–	–	–	–	–
ROMP [8]	79.7	119.7	152.8	65.9	–	–	–	–
PARE [7]	–	–	–	56.6	90.5	107.9	<u>57.1</u>	<u>88.6</u>
OCHMR [15]	77.1	114.2	150.6	–	–	–	–	–
DIMR*	<u>71.8</u>	<u>113.9</u>	<u>149.0</u>	<u>56.3</u>	<u>89.2</u>	<u>102.5</u>	58.8	91.7
DIMR†	**54.6**	**88.9**	**104.3**	**50.6**	**75.2**	**94.9**	**50.9**	**79.3**

4 Experiments

Similar to the previous methods [4,7,8], our model is trained on datasets Human3.6M [21], COCO [19], and MuCo [22]. To evaluate our algorithm, following ROMP [8] and PARE [7], we divide 3DPW [14] into two different subsets, including 3DPW-PC for person-person occlusion, 3DPW-OC for object occlusion, to evaluate the capability of our model on handling occlusion. We also evaluate our method on 3DOH, which is an object occlusion dataset. To assess the general performance of our model, we also evaluate it on the 3DPW full set. The mean per-joint position error (MPJPE), Procrustes-aligned MPJPE (PA-MPJPE), and the per-vertex error (PVE) are reported on these datasets to evaluate our method. For qualitative assessment, we employ OCHuman [23], a person occlusion dataset (Table 2).

Table 2. Quantitative comparison with general methods. We compare the performance of our method with the general methods [5,6,10,12,24] and the occlusion-aware methods [7,8,15,16] on 3DPW dataset. Three evaluation metrics, PA-MPJPE, MPJPE, and PVE, are reported. The Top-1 values are bolded, and the Top-2 values are underlined. * indicates the IUV map is generated by DensePose [17], and † indicates the IUV map is generated from the mesh. All metrics in *mm*.

Methods	PA-MPJPE↓	MPJPE↓	PVE↓
Methods for General			
DecoMR [10]	61.7	–	–
SPIN [5]	59.2	96.9	116.4
PyMAF [6]	58.9	92.8	110.1
HMR-EFT [24]	52.2	85.1	98.7
METRO [12]	<u>47.9</u>	<u>77.1</u>	**88.2**
Methods for Occlusion			
OCHMR [15]	58.3	89.7	107.1
Pose2UV [16]	57.1	–	–
ROMP [8]	53.3	85.5	103.1
PARE [7]	50.9	82	<u>97.9</u>
DIMR*	52.7	86	102.5
DIMR†	**47.8**	**73.0**	99.5

4.1 Comparison to the State-of-the-Art

Occlusion Evaluation. Table 1 shows the comparison of our models with the other SOTA methods on the occlusion datasets. On the 3DPW-PC and 3DPW-OC datasets, our baseline model, DIMR*, already achieves state-of-the-art performance. On the 3DPW-PC dataset, our model exhibits better performance than OCHMR [15], a heatmap-based method designed for inter-person occlusion, with a higher PA-MPJPE score. This suggests that our feature extraction method is more accurate compared to the center heatmap-based method, and in terms of accurate pose prediction, our dense map-based method performs better. Compared to another dense correspondence method like PyMAF [6], our model is clearly better, as a dedicated network for map prediction is more robust under human occlusion. On the 3DOH dataset, our baseline model, DIMR*, demonstrates comparable performance with the state-of-the-art method PARE [7]. Note that DIMR† achieves a substantial increase in accuracy, as evidenced by a 50% lower MPJPE on the 3DPW-OCC dataset compared to PARE. On the PA-PMJPE of 3DPW-PC, our model increases performance by 20%. This dramatic increase in performance on person-occlusion subsets demonstrates that with an accurate dense correspondence map, our network can accurately separate visible human features from occluded people.

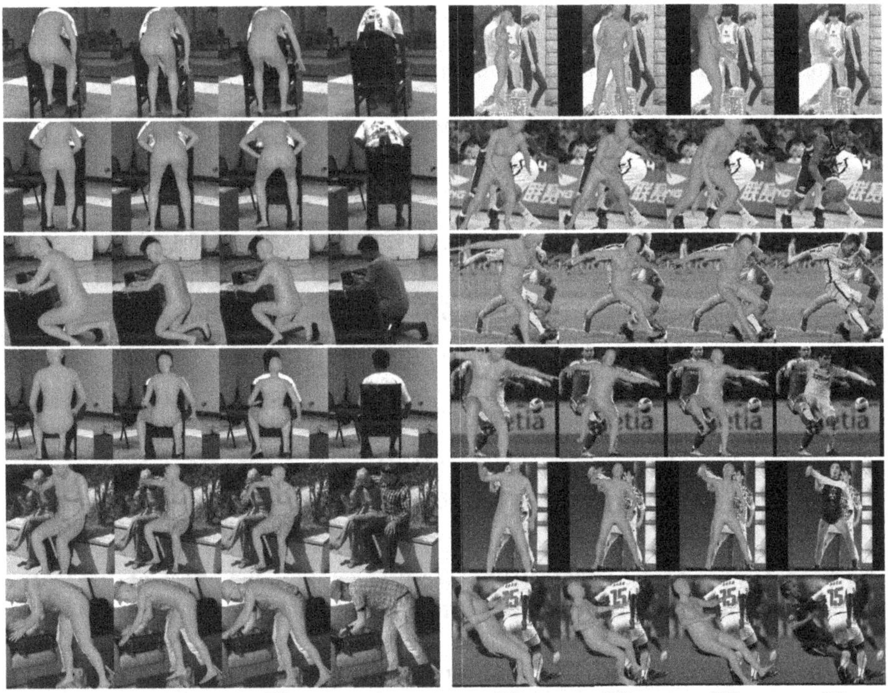

Fig. 4. Qualitative results on the 3DOH (left column 1–4), 3DPW (left column 5–6), and OCHuman (right columns) datasets. Compared to the other methods [7,25], DIMR demonstrates higher accuracy in detecting limbs under various types of occlusion. In the instances where human subjects heavily overlap, our model can effectively predict the correct limbs with the assistance of a dense correspondence map.

General Comparison. We also compare our method with other state-of-the-art methods on the 3DPW, which is a general dataset. As shown in Table 4, our results indicate that DIMR achieves comparable performance when compared to state-of-the-art methods. Furthermore, we observe that our model outperforms the occlusion-specific methods [8,15,16]. We attribute this success to the ability of our model to incorporate both global and local information through our designed architectures.

Qualitative Comparison. Figure 4 shows the qualitative results compared with CLIFF [25] and PARE [7]. We demonstrate our result on the OCHuman [23], 3DPW [14] and 3DOH [13] datasets. DIMR predicts accurate human limbs under object occlusions. Under hard cases like heavy human occlusion, our model is able to predict the correct human with the corresponding limbs because of the accurate feature separation based on the dense map.

Table 3. Ablation study on different input feature choices. We compare three settings that remove one of our design elements: feature wrapping, UV structure, and neighboring UV mapping. All metrics in *mm*.

Settings	3DPW		
	MPJPE↓	PA-MPJPE↓	PVE↓
w/o UV feature	103.4	59.2	129.5
w/o UV structure	80.8	51.4	106.1
w/o neighboring UV	75.0	49.6	103.1
DIMR	**73.0**	**47.8**	**99.5**

4.2 Ablation Study

We conduct our ablation study on the 3DPW and 3DPW-OC datasets. In the ablation study, we mainly utilize the ground truth IUV setting for comparison.

Effect of UV Feature Wrapping. We first explore the effect of our feature wrapping technique by modifying the input to our feature completion module. The IUV image provides the network with strong prior knowledge, including dense human information. The wrapping process provides the network with a piece of structured feature information. To assess the individual value of these features, we design three alternative methods that remove one of these features. As shown in Table 3, in the first row, we substitute the wrapped feature F_{UV} with the wrapped input image I_{UV}, by wrapping I with M_{IUV}. This setting mainly removes the features of the UV map, and in the second row, we remove the inverse-wrapped feature and instead concatenate the M_{IUV} with the image feature map F_{img}. We perform feature completion on the original feature map. This setting removes the structured UV map representation but preserves the image feature and dense correspondence, and in the last row, we replace the UV map with a randomly organized UV map, which contains less structured neighboring information. The input feature of these three settings is illustrated in Fig. 5. Comparing to the original DIMR, setting (a) leads to the most performance decrease, proving that the model requires local information for accurate regression. Setting (b) shows limited performance even if it utilizes the ground truth IUV image. Setting (c) brings a considerable performance increase compared to (b), indicating the effectiveness of UV wrapping. However, the neighboring UV brings more feature fusion.

Effect of Wrapped Feature Completion. To analyze the design of our feature completion module. We compare our proposed design with two other settings in Table 4. In the first setting, we remove the attention layer and only use the down-sampled F^d_{UV} for the regression, thereby eliminating long-range feature connection. We further replace the convolution down-sample layer with an average-pooling layer, which removes the feature extraction within the body

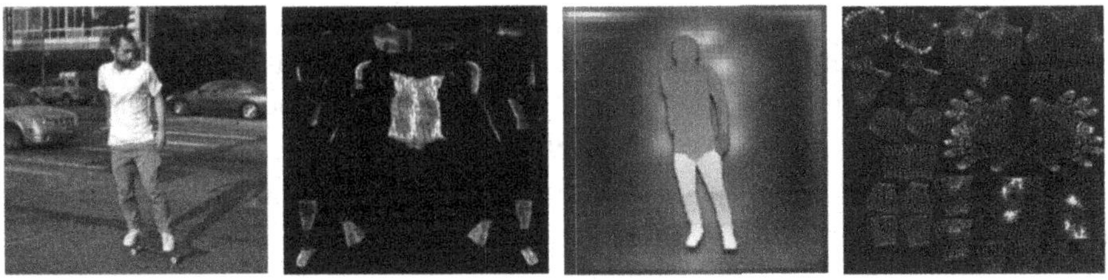

Fig. 5. The illustration of our modified inputs for the ablation study. (a) replaces the wrapped feature F_{UV} with the wrapped input image I_{UV}; (b) removes the inverse-wrapped feature and instead concatenates the M_{IUV} with the image feature map F_{img}; (c) replaces the UV map with a randomly organized UV map.

Table 4. Ablation study on the input feature. We compare our original setting with the model without attention and the model without attention and convolution.

Settings	3DPW		
	MPJPE↓	PA-MPJPE↓	PVE↓
DIMR w/o atten	79.6	51.1	107.6
DIMR w/o atten. conv	87.3	53.3	110.8
DIMR	**73.0**	**47.8**	**99.5**

Table 5. Ablation study on inpainting training. We compare the models performance with synthetic occlusion and inpainint training on 3DPW, 3DPW-OC.

Dataset	Settings	PA-MPJPE↓	MPJPE↓
3DPW-OC	synthetic + inpaint	**53.4**	**82.3**
	synthetic	53.4	84.2
3DPW	synthetic + inpaint	50.4	**79.7**
	synthetic	**49.0**	81.7

part. Table 5 shows our results of different settings. The setting combining the convolution layer and attention layer achieves the best results, suggesting that both the local and long-range feature fusion are critical for optimal performance.

Effect of Wrapped Feature Inpainting. In this experiment, we compare the model trained with inpainting training and synthetic occlusion to the models with only synthetic occlusion. As shown in Table 5, the model utilizing inpainting has better accuracy, proving that intermediate supervision is crucial for synthetic occlusion augmentation.

5 Conclusion

In this paper, we propose a novel human mesh recovery method named DIMR, which is robust to object and human occlusions. DIMR combines two fundamental ideas: separating the human feature from the occlusion and completing features from invisible cues. We leverage the accurate IUV map to isolate the part features of the target individual from those of other humans. The attention-based feature completion module is then utilized to effectively recover the occluded features using the cues from other visible parts. In addition, we adopt an inpainting training strategy to enhance performance. Quantitative and qualitative results prove that our model achieves state-of-the-art performance among occlusion-handling methods and has comparable results on general datasets. Future works can focus on designing occlusion-robust IUV prediction modules for more accurate human mesh recovery under occlusion.

References

1. Lassner, C., Romero, J., Kiefel, M., Bogo, F., Black, M.J., Gehler, P.V.: Unite the people: closing the loop between 3d and 2d human representations. In: Proceedings of the IEEE Conference on Computer Vision and Pattern Recognition (2017)
2. Bogo, F., Kanazawa, A., Lassner, C., Gehler, P., Romero, J., Black, M.J.: Keep It SMPL: automatic estimation of 3D human pose and shape from a single image. In: Leibe, B., Matas, J., Sebe, N., Welling, M. (eds.) ECCV 2016. LNCS, vol. 9909, pp. 561–578. Springer, Cham (2016). https://doi.org/10.1007/978-3-319-46454-1_34
3. Fang, Q., Shuai, Q., Dong, J., Bao, H., Zhou, X.: Reconstructing 3d human pose by watching humans in the mirror. In: Proceedings of the IEEE/CVF Conference on Computer Vision and Pattern Recognition, pp. 12814–12823 (2021)
4. Kanazawa, A., Black, M.J., Jacobs, D.W., Malik, J.: End-to-end recovery of human shape and pose. In: Proceedings of the IEEE Conference on Computer Vision and Pattern Recognition, pp. 7122–7131 (2018)
5. Kolotouros, N., Pavlakos, G., Black, M.J., Daniilidis, K.: Learning to reconstruct 3d human pose and shape via model-fitting in the loop. In: International Conference on Computer Vision, pp. 2252–2261 (2019)
6. Zhang, H., et al.: PYMAF: 3d human pose and shape regression with pyramidal mesh alignment feedback loop. In: International Conference on Computer Vision (2021)
7. Kocabas, M., Huang, C.H.P., Hilliges, O., Black, M.J. :Pare: part attention regressor for 3d human body estimation. In: International Conference on Computer Vision, pp. 11127–11137 (2021)
8. Sun, Y., Bao, Q., Liu, W., Fu, Y., Black, M.J., Mei, T.: Monocular, one-stage, regression of multiple 3d people. In: Proceedings of the IEEE/CVF International Conference on Computer Vision (ICCV), pp. 11179–11188 (2021)
9. Sun, Y., Liu, W., Bao, Q., Fu, Y., Mei, T., Black, M.J.: Putting people in their place: Monocular regression of 3d people in depth. In Proceedings of the IEEE/CVF Conference on Computer Vision and Pattern Recognition, pp. 13243–13252 (2022)
10. Zeng, W., Ouyang, W., Luo, P., Liu, W., Wang, X.: 3d human mesh regression with dense correspondence. In: Proceedings of the IEEE/CVF Conference on Computer Vision and Pattern Recognition, pp. 7054–7063 (2020)
11. Kolotouros, N., Pavlakos, G., Daniilidis, K.: Convolutional mesh regression for single-image human shape reconstruction. In: Proceedings of the IEEE/CVF Conference on Computer Vision and Pattern Recognition, pp. 4501–4510 (2019)
12. Lin, K., Wang, L., Liu, Z.: End-to-end human pose and mesh reconstruction with transformers. In: Proceedings of the IEEE/CVF Conference on Computer Vision and Pattern Recognition, pp. 1954–1963 (2021)
13. Zhang, T., Huang, B., Wang, Y.: Object-occluded human shape and pose estimation from a single color image. In: Proceedings of the IEEE/CVF Conference on Computer Vision and Pattern Recognition, pp. 7376–7385 (2020)

14. von Marcard, T., Henschel, R., Black, M., Rosenhahn, B., Pons-Moll, G.: Recovering accurate 3D human pose in the wild using IMUs and a moving camera. In: European Conference on Computer Vision (2018)
15. Khirodkar, R., Tripathi, S., Kitani, K.: Occluded human mesh recovery. In: Proceedings of the IEEE/CVF Conference on Computer Vision and Pattern Recognition (2022)
16. Huang, B., Zhang, T., Wang, Y.: Pose2uv: single-shot multiperson mesh recovery with deep UV prior. IEEE Trans. Image Process. **31**, 4679–4692 (2022)
17. Kokkinos, I., Guler, R.A., Neverova, N.: Densepose: dense human pose estimation in the wild (2018)
18. Cheng, B., Xiao, B., Wang, J., Shi, H., Huang, T.S., Zhang, L.: Higherhrnet: scale-aware representation learning for bottom-up human pose estimation. In: Proceedings of the IEEE/CVF Conference on Computer Vision and Pattern Recognition, pp. 5386–5395 (2020)
19. Lin, T.-Y., et al.: Microsoft COCO: common objects in context. In: Fleet, D., Pajdla, T., Schiele, B., Tuytelaars, T. (eds.) ECCV 2014. LNCS, vol. 8693, pp. 740–755. Springer, Cham (2014). https://doi.org/10.1007/978-3-319-10602-1_48
20. Kingma, D.P., Ba, J.: Adam: a method for stochastic optimization. In: Proceedings of International Conference Learning Representation (2015)
21. Ionescu, C., Papava, D., Olaru, V., Sminchisescu, C.: Human3.6m: large scale datasets and predictive methods for 3d human sensing in natural environments. IEEE Trans. Pattern Anal. Mach. Intell. **36**(7), 1325–1339 (2013)
22. Mehta, D., et al.: Single-shot multi-person 3d pose estimation from monocular RGB. In 2018 International Conference on 3D Vision (3DV), pp. 120–130. IEEE (2018)
23. Zhang, S.H., et al.: Pose2seg: detection free human instance segmentation. In: Proceedings of the IEEE/CVF Conference on Computer Vision and Pattern Recognition, pp. 889–898 (2019)
24. Joo, H., Neverova, N., Vedaldi, A.: Exemplar fine-tuning for 3d human model fitting towards in-the-wild 3d human pose estimation. In: International Conference on 3D Vision, pp. 42–52. IEEE (2021)
25. Li, Z., Liu, J., Zhang, Z., Xu, S., Yan, Y.: Cliff: carrying location information in full frames into human pose and shape estimation. https://info.arxiv.org/help/cs/index.html (2022)

Correction to: Correlation-Distance Graph Learning for Treatment Response Prediction from rs-fMRI

Francis Xiatian Zhang, Sisi Zheng, Hubert P. H. Shum, Haozheng Zhang, Nan Song, Mingkang Song, and Hongxiao Jia

Correction to:
Chapter 24 in: B. Luo et al. (Eds.): *Neural Information Processing*, CCIS 1963,
https://doi.org/10.1007/978-981-99-8138-0_24

In an older version of this paper, the name of the first author was incorrect. It has now been updated.

The updated version of this chapter can be found at
https://doi.org/10.1007/978-981-99-8138-0_24

B. Luo et al. (Eds.): ICONIP 2023, CCIS 1963, p. C1, 2024.
https://doi.org/10.1007/978-981-99-8138-0_45

Author Index

A
Alexandre, Luís A. 428
Asadi, Houshyar 313

B
Bulanda, Daniel 229

C
Cai, Fuhan 245
Calhoun, Vince D. 518
Cao, Jian 75
Cao, Xinyan 75
Chalup, Stephan 190
Che, Jinming 75
Chen, Di 506
Chen, Feiyu 528
Chen, Jiejie 177
Chen, Xiong 494
Chen, Yuming 124

D
Dai, Jinyue 528
Dai, Liting 38
Deng, Zhiyi 528
Di, Haoda 48
Ding, Jinhong 287
Duch, Włodzisław 428
Dydek-Dyduch, Ewa 26

F
Fan, Yi 506
Fan, Zhuoyao 48
Fang, Xiangzhong 245
Feng, Yuan 287
Fu, Guohong 389
Fu, Zhihui 455
Furman, Łukasz 428

G
Gomolka, Zbigniew 26
Gong, Chen 389
Gong, Xiao-Feng 518

H
Hajnowski, Marcin 428
Han, Baoai 494
Han, Kai 62
Han, Yue 518
Han, Zerui 287
Hao, Shuohui 62
Hao, Ying-Guang 518
Hasan, Md Mahbub 350
Hasan, Md Rakibul 350
He, Hongcai 528
Hoang, Thuong 313
Horzyk, Adrian 229
Hossain, Md Zakir 350
Hou, Zeng-Guang 326
Huang, He 258
Huang, Jiaqiang 100
Huang, Junjian 100

J
Jia, Hongxiao 298
Jia, Xiyuan 136
Jiang, Biao 506
Jiang, Ping 177
Jiang, Zhuohang 48
Jin, Zhong-Wei 402

K
Kang, Xiaoyang 338
Khan, Mehshan Ahmed 313
Kong, Dexin 389
Kosno, Jakub 229
Kroon, Steve 190
Kuang, Li-Dan 518
Kuderov, Petr 270

B. Luo et al. (Eds.): ICONIP 2023, CCIS 1963, pp. 571–573, 2024.
https://doi.org/10.1007/978-981-99-8138-0

L
Leung, Chi-Sing 558
Li, Binghao 540
Li, Binghua 287
Li, Changjin 75
Li, Jin 218
Li, Ning 483
Li, Sunan 415
Li, Weiping 470
Li, Wei-Xing 518
Li, Xinhong 494
Li, Yifan 287
Li, Yuanzhang 362
Li, Yun 218
Li, Zhao 362
Lian, Hailun 415
Lim, Chee Peng 313
Lin, Jie 494
Lin, Jinlong 75
Lin, Qiu-Hua 518
Lin, Zhishu 376
Lipiński, Piotr 112
Liu, Chang 15
Liu, Duo 245
Liu, Ge 245
Liu, Jia-Wen 402
Liu, Kang 38
Liu, Mengqiu 86
Liu, Quan 3
Liu, Shiyu 205
Liu, Yao 540
Liu, Yi 62
Liu, Zhaoyi 48
Liu, Zhe 62
Long, Yufan 48
Lu, Bao-Liang 402
Lu, Cheng 415
Luo, Wendian 48
Luo, Yangjie 338
Luo, Yiqian 483
Lv, Qiyun 136
Lv, Shaogao 205

M
Ma, Ming 287
Miao, Gongxun 86
Mo, Tong 470
Mohaghegh, Mahsa 443
Muller, Matthew 190

N
Nafisi, Vahid Reza 443
Nahavandi, Saeid 313
Ni, Hao 136
Nojima, Ryo 162

P
Pan, Yudong 483
Panov, Aleksandr I. 270
Peng, Qinmu 376

Q
Qian, Pengwei 218
Qiang, Jipeng 218
Qiao, Xu 362

R
Ratajczak, Ewa 428
Ren, Wei 75
Ren, Yanzhen 494
Roozbehi, Ali 443

S
Sałabun, Wojciech 150
Sammut, Claude 540
Shao, Jie 528
Sheng, Victor S. 62
Shi, Xin 389
Shum, Hubert P. H. 298
Słupiński, Mikołaj 112
Song, Mingkang 298
Song, Nan 298
Song, Yuqing 62
Starzyk, Janusz A. 229
Su, Jianqiang 326
Sun, Qingping 558

T
Tan, Yuan 362
Tian, Jun 455
Tołpa, Krzysztof 428
Tu, Weiping 494

V
Volovikova, Zoya 270

W
Wang, Haizhou 48
Wang, Jiaxi 470

Wang, Jiaxing 326
Wang, Junkongshuai 338
Wang, Lihua 162
Wang, Lu 338
Wang, Weiqun 326
Wang, Wentao 455
Wang, Xin 86, 136
Wang, Yanjun 558
Wang, Yihan 326
Wang, Yong 415
Wang, Yuelin 258
Wang, Yuetian 376
Wang, Zhenni 558
Wang, Zhuo 3
Wei, Linsen 205
Więckowski, Jakub 150
Wu, Guohua 86, 136
Wu, Tianrui 177

X

Xia, Likun 287
Xiao, Li 494
Xiong, Lianjin 483
Xu, Chao 38
Xu, Jiamiao 376
Xu, Tianxiang 15
Xu, Yehan 287
Xu, Zenglin 205

Y

Yan, He 75
Yan, Yongxu 455
Yang, Chuanwu 376
Yang, Jinyue 100
Yang, Xiuping 494
Yang, Yu-Bin 506
Yang, Yuhong 494
Yao, Lina 540
Ye, Shuo 376
Yi, Weiyan 494
You, Xinge 376
Yu, Nan 389
Yu, Xiao 362
Yuan, Lifeng 86
Yuan, Peipei 376
Yuan, Yun 389
Yuan, Yun-Hao 218
Yun, Wenhao 86, 136

Z

Zeslawska, Ewa 26
Zhang, Haozheng 298
Zhang, Jianlin 15
Zhang, Kun 15
Zhang, Li 362
Zhang, Lihua 338
Zhang, Francis Xiatian 298
Zhang, Xiongzhen 3
Zhang, Yangsong 483
Zhang, Zhen 86, 136
Zhang, Zhongqiang 245
Zhang, Zhuzhu 177
Zhao, Haiming 177
Zhao, Yan 415
Zhao, Yongjian 75
Zheng, Sisi 298
Zheng, Wei-Long 402
Zhong, Yao 100
Zhou, Yifan 350
Zhu, Yan 62
Zhu, Yi 218
Zhu, Yuanheng 124
Zia, Ali 350
Zong, Yuan 415

Printed in the United States
by Baker & Taylor Publisher Services